T0074326

Lecture Notes in Electrical Engineering

Volume 203

For further volumes:
http://www.springer.com/series/7818

Sang-Soo Yeo · Yi Pan
Yang Sun Lee · Hang Bae Chang
Editors

Computer Science
and its Applications

CSA 2012

 Springer

Editors

Sang-Soo Yeo
Department of Electrial Engineering
  and Computer Science
Mokwon University
Daejeon
Republic of South Korea

Yi Pan
Department of Computer Science
Georgia State University
Atlanta, GA
USA

Yang Sun Lee
Division of Computer Engineering
Mokwon University
Daejeon
Republic of South Korea

Hang Bae Chang
Division of Business Administration
Sangmyung University
Seoul
Republic of South Korea

ISSN 1876-1100   ISSN 1876-1119   (electronic)
ISBN 978-94-007-5698-4   ISBN 978-94-007-5699-1   (eBook)
DOI 10.1007/978-94-007-5699-1
Springer Dordrecht Heidelberg New York London

Library of Congress Control Number: 2012949371

Printed on acid-free paper

Springer is part of Springer Science+Business Media (www.springer.com)

Orgainzation Committee

Welcome Message from the General Chairs

CSA 2012

Welcome to the 4th FTRA International Conference on Computer Science and its Applications (CSA 2012), going to be held in the Jeju Island, Korea.

CSA 2012 is the next event in a series of highly successful International Conference on Computer Science and its Applications, previously held as CSA 2008 (Australia, October, 2008), CSA 2009 (Jeju, December, 2009), and CSA 2011(Jeju, December, 2011).

The CSA 2012 will be the most comprehensive conference focused on the various aspects of advances in computer science and its applications. The CSA 2012 will provide an opportunity for academic and industry professionals to discuss the latest issues and progress in the area of CSA.

We would like to thank all authors of this conference for their paper contributions and presentations; and we would like to sincerely appreciate the following prestigious invited speakers who kindly accepted our invitations, and helped to meet the objectives of the conference:

- **Dr.Yi Pan**

Georgia State University, USA

We also sincerely thank all our chairs and committees, and these are listed in the following pages. Without their hard work, the success of CSA 2012 would not have been possible. Finally, we would like to thank the SGSC 2012, NT3CA 2012, WMS 2012, WPS 2012 workshop chairs, and special session organizers of SS-STS, SS-CASDAI, SS-SPIVN, for their great contributions.

With best regards,
Looking forward to seeing you at CSA 2012
    Sang-Soo Yeo, Mokwon University, Korea (Leading Chair)
    Odej Kao, Technische Universitat Berlin, Germany
    Yi Pan, Georgia State University, USA
    CSA 2012 General Chairs

Welcome Message from the Program Chairs

CSA 2012

On behalf of the CSA 2012 organizing committee, it is our pleasure to welcome you to the 4th International Conference on Computer Science and its Applications (CSA 2012).

The success of a conference is mainly determined by the quality of its technical program. This year's program will live up to high expectations due to the careful selection by the Program Committee. They have spent long hours in putting together an excellent program and deserve a big applause.

The conference received 163 submissions, and all were reviewed by the Program Committee. In the review process, we assigned at least three Program Committee members to each paper. After careful deliberation and peer reviews, we selected 52 papers for presentation and inclusion of the conference proceedings, whose acceptance rate is around 32 %.

There are many people who contributed to the success of CSA 2012. We would like to thank the many authors from around the world for submitting their papers. We are deeply grateful to the Program Committee for their hard work and enthusiasm that each paper received a thorough and fair review. Finally, we would like to thank all the participants for their contribution to the conference.

Sincerely yours,

Yang Sun Lee, Mokwon University, Korea
Gregorio Martinez, University of Murcia, Spain
Hong Ji, BUPT, China
Jinhua Guo, University of Michigan-Dearborn, USA
Robert C. H. Hsu, Chung Hua University, Taiwan
CSA 2012 Program Chairs

Conference Organization

CSA 2012

Organizing Committee

Steering Chairs

James J. (Jong Hyuk) Park, Seoul National University of Science and Technology, Korea
Han-Chieh Chao, National Ilan University, Taiwan
Mohammad S. Obaidat, Monmouth University, USA

General Chairs

Sang-Soo Yeo, Mokwon University, Korea
Odej Kao, Technische Universitat Berlin, Germany
Yi Pan, Georgia State University, USA

General Vice-Chair

Changhoon Lee, SeoulTech, Korea

Program Chairs

Yang Sun Lee, Mokwon University, Korea
Gregorio Martinez, University of Murcia, Spain
Hong Ji, BUPT, China
Jinhua Guo, University of Michigan - Dearborn, USA
Robert C. H. Hsu, Chung Hua University, Taiwan

Workshop Chair

Hangbae Chang, Sangmyung University, Korea
Weili Han, Fudan University, China
Waltenegus Dargie, Technical University of Dresden, Germany
WenZhan Song, Georgia State University, USA

International Advisory Board Committee

Hamid R. Arabnia, The University of Georgia, USA
Doo-soon Park, SoonChunHyang University, Korea
Hsiao-Hwa Chen, Sun Yat-Sen University, Taiwan
Philip S. Yu, University of Illinois at Chicago, USA
Yi Pan, Georgia State University, USA

Salim Hariri, University of Arizona, USA
Leonard Barolli, Fukuoka Institute of Technology, Japan
Jiankun Hu, RMIT University, Australia
Shu-Ching Chen, Florida International University, USA
Lei Li, Hose University, Japan
Victor Leung, University of British Columbia, Canada

Publicity Chairs

Amiya Nayak, University of Ottawa, Canada
Weiwei Fang, Beijing Jiaotong University, China
Sung-Bae Cho, Yonsei University, Korea
Min-Woo Cheon, Dongshin University, Korea

Local Arrangement Chair

Sang Oh Park, KISTI, Korea
Taeshik Shon, Ajou University, Korea
Namje Park, Jeju National University, Korea
SoonSeok Kim, Halla Universty, Korea

Program Committee

Hoon Choi, Chungnam National University, Korea
Evi Syukur, University of New South Wales, Australia
Prakash Veeraraghavan, La Trobe University, Australia
M. Dominguez-Morales, University of Seville, Spain
Min-Woo Cheon, Dongshin University, Korea
Gregorio Martinez, University of Murcia (UMU), Spain
Fei Yan, Wuhan University, China
Chan Yeob Yeun, Khalifa University of Science, Technology & Research
(KUSTAR), UAE
Neal N. Xiong, Georgia State University, USA
Yasuhiko Morimoto, Hiroshima University, Japan
Dieter Gollmann, TUHH, Germany
Jordi Forne, Technical University of Catalonia, Spain
Jose Onieva, University of Malaga, Spain
Jan de Meer, Brandenburg Technical University(BTU), Germany
HeeSeok Kim, Korea University, Korea
Zubair Baig, King Fahd University of Petroleum and Minerals, Saudi Arabia
El-Sayed El-Alfy, King Fahd University of Petroleum and Minerals, Saudi Arabia
Mohammed Houssaini Sqalli, King Fahd University of Petroleum and Minerals,
Saudi Arabia
Talal Mousa Alkharobi, King Fahd University of Petroleum and Minerals, Saudi
Arabia
Chih-Lin Hu, National Central University, Taiwan
Atul Sajjanhar, Deakin University, Australia
Tanveer Zia, Charles Sturt University, Australia
Claudio Ardagna, University of Milan, Italy
Chuan-Ming Liu, National Taipei University of Technology, Taiwan
Yao-Nan Lien, National Chengchi University, Taiwan
Honggang Wang, University of Massachusetts, USA
Yuh-Shyan Chen, National Taipei University, Taiwan
Rongxing Lu, University of Waterloo, Canada
Chang Wu Yu, Chung Hua University, Taiwan ROC
Chih-Shun Hsu, Shih Hsin University, Taiwan ROC
Qingyuan Bai, Fuzhou University, China
Anirban Mondal, Indraprastha Institute of Information Technology, Delhi India
Ramamohanarao Kotagiri, The University of Melbourne, Australia
Debajyoti Mukhopadhyay, Balaji Institute of Telecom & Management Pune, India
Biswajit Panja, University of Michigan, USA
Ashkan Sami, Shiraz University, Iran
Mei-Ling Shyu, University of Miami, USA
Choochart Haruechaiyasak, NECTEC, Thailand
Masao Ohira, Nara Institute of Science and Technology, Japan

Alton Chua Yeow Kuan, Nanyang Technological University, Singapore
Theng Yin Leng, Nanyang Technological University, Singapore
Margaret Tan, Nanyang Technological University, Singapore
Richi Nayak, Queensland University of Technology, Australia

Welcome Message from the Workshop Chairs

NT3CA 2012

It is our pleasure to welcome you to The 2012 FTRA International Workshop on New Technology Convergence, Cloud, Culture, and Art (NT3CA 2012) that is being held in Jeju, Korea, November 22–25, 2012.

The 2012 FTRA International Workshop on International Workshop on New Technology Convergence, Cloud, Culture, and Art (NT3CA 2012), co-sponsored by FTRA will be held in Jeju, Korea, November 22–25, 2012. The NT3CA 2012 is a workshop for scientists, engineers, researchers, and practitioners throughout the world to present the latest research, new trends of IT such as cloud computing or social network, culture technology, ideas and applications in all areas of new technology convergence, cloud, culture, and art. The topic can be many related services based on cloud computing environment such as security, social network, communication technique, evaluation, technical tuning, virtualization, framework, software technique, and so on. This workshop will be a good chance to discuss the opinions among the experts. The NT3CA 2012 is co-sponsored by FTRA. In addition, the conference is supported by KITCS.

We would like to send our sincere appreciation to all participating members who contributed directly to NT3CA 2012. We would like to thank all Program Committee members for their excellent job in reviewing the submissions. We also want to thank the members of the organizing committee, all the authors and participants for their contributions to make NT3CA 2012 a grand success.

Sincerely yours,
Jin-Mook Kim, Sunmoon University, Korea
NT3CA 2012 General Chair

NT3CA 2012 Workshop Organization

Organizing Committee

General Chairs

Sang-Soo Yeo, Mokwon University, Korea
Jin-Mook Kim, Sunmoon University, Korea

Program Chairs

Juyeon Jo, University of Nevada, USA
Hwa-Young Jeong, Kyung Hee Unversity, Korea

Program Committee

Jeong-Woo Byun, Kyung Hee university, Korea
Gi-Hwon Kwon, Kyong Gi University, Korea
Jae-Soo Yoo, Chung Buk National University, Korea
Hwang-Rae Kim, Kong Ju National University, Korea
Koo-Rack Park, Kon Ju National University, Korea
Young-Ae Jeong, Sunmoon University, Korea
Jong-hee Kim, Sunmoon University, Korea
Sung-Sik Hong, Hyejeon Colleage, Korea
Woon-Ho Choi, Seoul National University, Korea
Kim Dong-Kun, Kyung Hee University, Korea
Jeong-Won An, Osan University, Korea
Hyun-Jung Yoon, Yewon Art University, Korea
Hye-Jung Jung, Pyeong Taek University, Korea
Chang-Yong Lee, Kong Ju National University, Korea
Sunny Kim, Texas Tech University, USA

Welcome Message from the Workshop Chairs

SGSC 2012

On behalf of the program chairs and technical program committees, it is our pleasure to welcome you to the Workshop on Smart Grid Security and Communications (SGSC-2012). This workshop is in conjunction with the 4th FTRA International Conference on Computer Science and its Applications (CSA 2012), to be held in Jeju, Korea, November 22–25, 2012. This workshop aims at bringing together researchers from academic and industry laboratories for a face-to-face meeting to discuss common research interests. This activity is a catalyst for discussing relevant research questions and engagement in the growing area of smart grid and related to the major area of cyber security and communication. SGSC 2012 is not an end in itself but a means by which research communities achieve their new funding streams and novel collaborations, and establish a long-lasting research network. We hope to continue last year's success and have another great event this year. We would also like to thank the program committee of SGSC 2011 and the external reviewers for their constant support. It is our great honor to invite you to attend this workshop. Enjoy the program and your stay in Jeju!

Sincerely yours,
Bo-Chao Cheng, National Chung-Cheng University, Taiwan
SGSC 2012 General Chair

SGSC 2012 Workshop Organization

Organizing Committee

General Chairs

Garry W. Chang (FIEEE), National Chung-Cheng University, Taiwan
James J. (Jong Hyuk) Park, SeoulTech, Korea

Organizing Chair

Bo-Chao Cheng, National Chung-Cheng University, Taiwan

Program Chairs

Steven Low, Caltech, USA (FIEEE)
Paolo Tenti, University of Padova, Italy (FIEEE)
Boming Zhang, Tsinghua University, China (FIEEE)
Huy Kang Kim, Korea University, Korea

Publicity Chairs

Ping-Hai Hsu, ITRI, Taiwan
Ezendu Ariwa, London Metropolitan University, UK

Program Committee

Huan Chen, National Chung-Hisng Univerity, Taiwan
Kwang Hyuk Im, PaiChai University, Korea
Abdelmajid Khelil, TU Darmstadt, Germany
Chang Oh Kim, NCSOFT, Korea
Ibrahim Korpeoglu, Bilkent University, Turkey
Jung-Shian Li, National Cheng-Kung University, Taiwan
Kwang-Hui Lee, Changwon National University, Korea
Yong-hoon Lim, KEPRI, Korea
Jiunn-Liang Lin, ITRI, Taiwan
Young Hoon Moon, Korea Telecom, Korea
Jiyoung Woo, Korea University, Korea
James C.N. Yang, National Dong Hwa University, Taiwan
Chien-Chung Shen, University of Delaware, USA

Contents

Part III   Signal Processing for Image, Video, and Networking

Part IV   Science, Technology and Society

Part I
Computer Science and its Applications

Utilizing TPM Functionalities on Remote Server

Norazah Abd Aziz and Putri Shahnim Khalid

Abstract Trusted Platform Module (TPM) has become an essential functionality in the information security world today. However, there are legacy computers that do not have TPM onboard and would still want to use the TPM functionalities without having to replace the hardware. Also, TPMs are not available for virtual machines hence there is a need to provide integrity of the virtual machine platforms. This paper introduces a framework to provide a remote server with TPM capabilities for the legacy computer and also virtual machines to be able to utilize TPM functionalities. In this framework, there is also a need to provide fault tolerance mechanism to ensure reliability of the server and also scalability feature is incorporated to cater for growing number of users. The main component of the framework is the 'vTPM Manager' module which resides in the remote TPM server. This vTPM Manager handles the creation and deletion of virtual TPMs, providing fault tolerance mechanism and also scalability feature for the whole system. By using this framework, users who do not have a TPM residing in their device would be able to remotely access the TPM server to utilize the TPM functionalities with the assurance of a fault tolerance mechanism and the number of users is unlimited since it is scalable.

Keywords Fault-tolerance · Scalability · TPM instances · Migration · vTPM Manager · Virtual TPM

N. Abd Aziz (✉) · P. S. Khalid
MIMOS Berhad, Technology Park Malaysia, 57000 Kuala Lumpur, Malaysia
e-mail: azahaa@mimos.my

P. S. Khalid
e-mail: shahnim.khalid@mimos.m

S.-S. Yeo et al. (eds.), *Computer Science and its Applications*,
Lecture Notes in Electrical Engineering 203, DOI: 10.1007/978-94-007-5699-1_1,
© Springer Science+Business Media Dordrecht 2012

1 Introduction

Trusted Computing (TC) is a technology developed and promoted by non-profit industry consortium. The technology aims to enhance the security of hardware and software building blocks. The consortium known as Trusted Computing Group (TCG) [1] has come up with specifications on Trusted Platform Module (TPM) which has potentials to be used for security and trust related services like remote attestation and key management. In order to utilize the Trusted Computing functionalities, new PCs and laptops are equipped with a TPM [2] on the motherboard by many hardware manufacturers.

However, legacy computers and older motherboards do not have TPMs onboard. This poses a problem for users wanting to utilize the Trusted Computing functionalities without having to replace all the equipments. Furthermore, most virtual machine environments are not equipped with a TPM. Since virtual machines are used widely in cloud computing environment, it is necessary to apply TPM functionalities to provide integrity of the virtual machine platforms.

In this paper, we introduce a framework to provide a remote server with TPM capabilities in the form of hardware TPM and/or software based TPM. Software based TPM in this context is referred as virtual TPM (vTPM) in this paper. Users can connect to the remote server and use the trusted computing capabilities. We will discuss about fault tolerance mechanisms to ensure users can connect to the server at all times through virtual TPM (vTPM) instances. In addition, we also discuss on the scalability of the servers in order to cater for high number of users and their associated vTPM instances at one time.

This paper is organized as follows. Section 1 starts with this brief introduction and followed by current related work in Sect. 2. Section 3 of the paper explains about the fault tolerance mechanism and scalability of the server. The basic framework of the attempt implementation containing the process flow of system requirement is presented in Sect. 4. The paper continues to describe the concept of remote server with TPM capabilities implementation handled by a vTPM Manager module. Finally Sect. 5 describes the current implementation. The paper ends with a conclusion.

2 Related Work

A system which enables the trusted computing for an unlimited number of virtual machines was proposed by [3]. Their approach is to virtualize the TPM, so the TPM functionalities are available to operating systems and applications running in virtual machines. We adopted their approach which provides added functions to create and destroy virtual TPM instances as well as to maintain the migration of a virtual TPM instance with its respective virtual machine. The difference is they implement multiplexing of request from clients to their associated vTPM instances but our approach only interacts with the clients during initialization process.

The paper in [4] is extended from [3] by adding built-in attestation mechanism. They have introduced a ticket-based remote attestation scheme. Compared to our framework, their vTPM instances management resides in Virtual Machine (VM). But, similar with our approach, the software TPM is also always protected by the hardware TPM. During vTPM spawns, its PCR values are initialized with values from the underlying hardware TPM.

The security and reliability issues in client virtualization were also discussed in [5]. Their proposed solutions leverage on Intel vPro and TPM in order to overcome the issues and using trusted VM container through remote attestation protocol verification. Our approach is not limited to the Intel technology and platform and hence is more feasible.

In [6], the paper describes their approach to secure cloud-based system using trusted computing. Their design mainly focuses on the virtual DRTM (Dynamic Root of Trust for Measurement). By virtualizing the DRTM, they control the locality by modifying the Xen vTPM Manager. Locality is based on the memory addressing which corresponds to different levels in a system, for example security kernel at Locality 0 while application at Locality 3 and so on. Our approach differs in that we are not modifying the way the guest OS access the data based on these locality using a certain algorithm that has to be embedded in the hypervisor (such as Xen).

3 Fault Tolerance, Scalability and Attestation

3.1 Fault Tolerance

Fault-tolerance is the property of a software or hardware that enables a system to operate continuously in the event of failure of (or one or more faults within) some of its components [7]. In other word, it is designed to recover from failure immediately with no loss of service. There are a few levels of fault tolerance based on the ability to continue operation in the event of a power failure in time basis. The levels are defined by whether the fault tolerance feature is provided by software, embedded in hardware, or by both combinations.

A fault-tolerance mechanism consists of three types: replication, redundancy and diversity. Fault-tolerance replication is requesting or directing tasks in parallel from multiple identical instances of the same system or subsystem based on the best output [7]. Similarly, fault-tolerance redundancy is also providing multiple identical instances but switching to one of the remaining instances in case of a failure [8]. In other word, it uses multiple nodes that are ready to provide service in order to recover from service failure of a single node. In contrast to replication and redundancy type, diversity provides multiple different implementations of the same specification.

In this paper we focus on fault tolerance using hardware which is provided to ensure the TPM server is available at all times. As mention earlier, trust and security is the main concern for the framework, hence replication of the vTPM instances is required to ensure users can connect to the server at all times even when there is a failure.

3.2 Scalability

According to [9], scalability is desirable in technology as well as business settings because both benefit significantly from the ability to easily increase volume without impacting the contribution margin. Scalability is the ability of a hardware or system to adapt to increasing demand due to the growing amount of context volume or size in order to meet user capacity. Our framework is designed to be scalable in the sense that the system can be upgraded easily and transparently to the users without shutting down the system. Hence, further investment to the system for adding new processors, devices and storage to the system has no additional cost.

Scalability can be measured in various dimensions [9], but this paper focus on load scalability. It means that the system easily expands and organizes its resource to sustain heavier or lighter number of inputs for modification and deletion activities. In our approach, the network scalability addresses the issue of retaining performance levels while adding additional servers to a network. Additional servers are typically added to a network when additional processing power is required.

3.3 Attestation

One of the most important uses of TPM is to enable a computing platform to attest its integrity to another entity. The attestation protocol involves measuring various 'properties' of the platform and storing the values in the TPM. When a remote entity asks for assurance of the integrity of the platform, the measurements are verified and sent over to the other entity. Our framework is designed to implement this feature in virtualization environment which is used to assure users of the integrity of the spawned vTPM. In our framework attestation protocol is run by enhanced virtualization API named as TMCI and a vTPM to verify the integrity of the associated VM. When a request for a VM is received, the TMCI will first ask the vTPM to attest the integrity of the VM. The VM will be created and given to the user if the attestation is successful. Otherwise another VM has to be created.

4 Framework

There arise desires for a service which provides TPM capabilities in the form of hardware and/or virtual TPM that would enable users who do not have TPM hardware in their device to access TPM functionalities as and when they require it. Therefore, our framework introduces a server which users can access remotely to use the TC functionalities. The server which is called TPM server is embedded with TPM hardware or a software-based TPM (vTPM) or both to cater for the users' needs. The concern is how to manage the multiple users and the multiple instances of vTPM. In some applications such as cloud computing, there can be millions of users connecting to the server at any one time. Hence, the TPM server must be scalable. The TPM server will also have to be available at all times since all the keys and state files are saved in the server and any disruption to the server may lead to loss of information on the user side. These functionalities are handled by a vTPM Manager (vMgr) on the TPM server which will be discussed further in the next section.

Figure 1 illustrates components of our framework which contains a fault-tolerance mechanism in order to maintain the availability of TPM server. The framework consists of users which are running on different types of hardware, but not limited to desktops, virtual machines, laptops and mobile devices. Users who need to access the TPM server connects through the network. Other than vMgr to handle all the resources related to vTPM, there is also a migration controller component which manages the migration of the TPM server to another physical location. However, in this paper we will not discuss further on migration controller component. Since our approach for virtualization environment is specific for cloud computing, we have enhanced the virtualization API. The enhanced virtualization API is called TMCI and is placed in the framework as shown in Fig. 1.

TMCI is the virtualization API which processes the URI and then communicates bidirectional with vMgr and public API in Libvirt command. Public API will pass the request to the driver API and look into the corresponding Virtual Machine (VM) driver in Libvirt. In our approach, QEMU-KVM is used for the VM creation, spawn or destroy activities through Libvirt. Before creating and destroying the VM, TMCI will verify the integrity of the vTPM instances with vMgr. If the verification fails vMgr will send an error message to TMCI and TMCI will cut off the communication with Libvirt. Since the communication to TMCI is disconnected, Libvirt cannot create or destroy the VM.

The fault-tolerance component supported by TPM Server comprises of TPM Primary Server and TPM Backup Server. The TPM Primary Server is the main server and backup is provided by duplicating the server in the TPM Primary Backup Server. For scalability purposes, TPM Secondary Server is connected to the TPM Server Primary and its backup is provided in the TPM Secondary Backup Server. The scalability function does not limit the number of servers; hence there

Fig. 1 Basic implementation framework

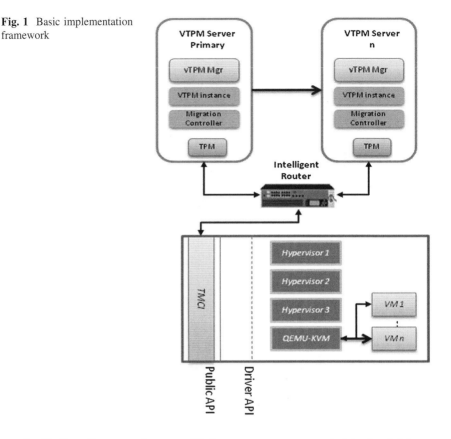

can be Tertiary Server and so on. All the servers are connected to the database where all the states are saved periodically.

4.1 vTPM Manager

vTPM Manager (vMgr) is the main feature in our framework. It is tasked to handle the creation or spawning of vTPM instances for individual users. Each user will be assigned a dedicated vTPM instance, which is linked to the TPM hardware. vMgr will assign dedicated port to the vTPM for the user to connect. vMgr is also responsible to update the storage or database with all the vTPM information and keys, which are also called the 'state files'. vMgr is the point of contact for the users to create, reactivate, suspend, resume, terminate and destroy the vTPM instances. The main process of vMgr is defined in Fig. 2.

The creation of a vTPM instance starts when the vMgr receives a request from user. vMgr is a service module which keeps running as a daemon to receive any request from the users. The request can be for creation, resume, suspend or destroy

Fig. 2 Creation and
management of vTPM
instance flow diagram

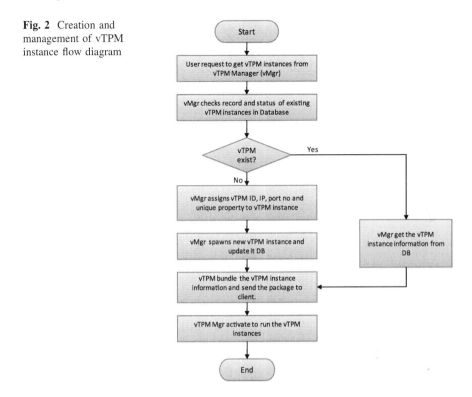

the vTPM instance. Each new vTPM will be assigned parameters which include vTPM ID, port number, IP and unique property value by vMgr. Then, vMgr will spawn the new vTPM instance. The parameter will be stored in the database and also sent to the user.

In addition to the creation of vTPM instances and managing its status, vMgr also handle scalability issue. When a TPM server has run out of system resource, another TPM server is needed to answer the users' request to spawn a vTPM. Here, the main vMgr will manage and decide which TPM server should accept the request. For this, the main vMgr must first register itself to the system. Before the process of ensuring scalability in spawning that vTPM can happen, the main vMgr must register itself to the system. Once registration has been completed, as a request arrives, the main vMgr would either spawn the vTPM instance or pass the request to a Secondary vMgr if it has exhausted its resources.

The main vMgr handles all registration of the Secondary vMgr which has just joined to the system. Once the main TPM server has reached its maximum spawned vTPM, the vMgr that handles vTPM instances in that server interacts with the registered Secondary vMgr located at a different TPM server and request that vMgr to spawn a new vTPM instance using parameters given during the interaction process given by the main vMgr. A centralized database is used to store all data of the Secondary vMgr. The data consist of the IP and port number, as well as an ID of the Secondary vMgr.

Therefore, when the main vMgr wants the Secondary vMgr to spawn a new vTPM instance, the main vMgr will retrieve the represented data of the Secondary vMgr from the database, and uses that data to send request to the Secondary vMgr to spawn a vTPM instance. Once the Secondary vMgr retrieved the request, it will then spawn a new vTPM based on the parameters given by the main vMgr and updates the centralized data indicating that the spawned vTPM ID is being spawned from that server and managed by the new vMgr.

Another function of vMgr is to handle the fault tolerance mechanism. This is separated into two processes: replication and fault tolerance itself. As mentioned earlier, the process starts with user sending request for a vTPM to the vTPM Manager Primary (vMP) to spawn a vTPM instance. Next, the vMP will generate IP and port number for the vTPM and record the information into the database. After the updating, vMP will spawn a vTPM instance. Then, the vTPM creates the state file and updates the vMP. The vMP will check updates of the state file periodically. If there is an update, vMP will send updated vTPM state file to vTPM Manager Primary Backup (vMPB). The vMPB will retrieve the virtual machine ID or user ID from the vTPM state file. The vMPB will get virtual machine data or user data from database. The vMPB will spawn vTPM accordingly and end the process.

The process above will ensure that the backup server will always have updated state files of all the vTPM instances. In the event of failure of the main TPM server, an intelligent device routes the request to the backup server. Hence the users can still access their vTPM instances even though main TPM server is down. When this happens, the backup TPM server will take over the functions of the main TPM server. Once the faulty machine is has recovered, the vMgr will ensure that the record is synchronized again between the main TPM server and its backup server.

4.2 Prototype Implementation

Referring to the above framework, installation of all components must be established in order to proceed with the development. The current phase of implementation setup is focusing on providing the environment development for TPM server and its components. Some of the components are based on [10, 11] implementation such as OpenSSL engine, TrouSerS TCG Software Stack, TPM Tools, the TPM Device Driver and the modified TPM Emulator. The main components are vMgr daemon and database such as using mySQL. The vMgr daemon is configured to be able to accept requests from any virtualization API for future implementation of cloud computing.

In our prototype, we use Libvirt [12] as the virtual machine control engine. Libvirt is a library which provides the functions to provision, create, modify, monitor, control, migrate and terminating an operating system running on a virtual machine. Due to the use of Uniform Resource Identifier (URI), Libvirt is able to support different types of hypervisors such as QEMU-KVM and Xen. In the prototype, Libvirt is used directly by TMCI.

We have conducted the system test on the components after it is integrated into our proposed framework to ensure the system runs well. The test involved up to 1000 instances of vTPM and works as expected. Each component after the integration in our framework can still run its original and enhanced functions.

5 Conclusion

In this paper, we have provided a framework for users to access TPM functionalities on a remote server. Typical scenario would be when the users of the system are using legacy computers or devices which does not have a built-in TPM but would still want to benefit from the Trusted Computing technology. By providing this framework however, the system needs to cater for the growing number of users and also the recovery from failure. Hence we have provided ways for the server to overcome this using the scalability feature and also the fault tolerance mechanism. With this approach, users can access the servers at any time with the ease and reliability that comes with the whole framework.

References

1. Trusted Computing Group: http://trustedcomputinggroup.org
2. TPM Main: Part 1 design principles. 1.2 revision 85 edition, (2005)
3. Berger, S., Caceres, R., Goldman, K.A., Perez, R., Sailer, R., Doorn, L.v.: vTPM: virtualizing the trusted platform module. In: 15th USENIX security symposium (2006)
4. Stumpf, F., Benz, M., Hermanowski. M., and Eckert, C.: Approach to a trustworthy system architecture using virtualization. ATC 2007, LNCS 4610, pp. 191–202, Springer (2007)
5. Wang, W., Zhang, Y., Lin, B., Wu, X.Y., Miao, K.: Secured and reliable VM migration in personal cloud, 2nd international conference on computer engineering and technology (ICCET), IEEE (2010)
6. Dai, W., Jin, H., Zou, D., Xu, S., Zhen, W. and Shi, L.; TEE: A virtual DRTM based execution environment for secure cloud-end computing. Proceeding CCS'10 proceedings of the 17th conference on computer and communications security, ISBN: 978-1-4503-0244-9, ACM (2010)
7. Shilin, Z., Mei, G.: Distributed multimedia content processing based on web service. Proceeding of international forum on computer science-technology and applications, ISBN: 978-0-7695-3930-0, IEEE (2009)
8. Morel, G., Pétin, J.F., Johnson, T.L.: Reliability, maintainability, and safety. Springer handbook of automation (2009)
9. Clarke, J., Dede, C.: Robust designs for scalability. AECT research symposium, June 22–25, Bloomington, Indiana (2006)

10. Norazah, A.A., Lucyantie, M.: Identity credential issuance with trusted computing, 2nd international conference on computing and informatics, ICOCI'09 (2009)
11. Lucyantie, M., Norazah, A.A., Habibah, H., Mohd Anuar, M.I.: Attestation with trusted configuration machine. Proceeding of international conference on computer applications and industrial electronics ICCAIE, ISBN: 9781457720574, IEEE (2011)
12. The virtualization API, http://libvirt.org/

Security and QoS relationships in Mobile Platforms

Ana Nieto and Javier Lopez

Abstract Mobile platforms are becoming a fundamental part of the user's daily life. The human–device relationship converts the devices into a repository of personal data that may be stolen or modified by malicious users. Moreover, wireless capabilities open the door to several malicious devices, and mobility represents an added difficulty in the detection of malicious behavior and in the prevention of the same. Furthermore, smartphones are subject to quality of service (QoS) restrictions, due to users' needs for multimedia applications and, in general, the need to be always-on. However, Security and QoS requirements are largely confronted and the mobility and heterogeneous paradigm on the Future Internet makes its coexistence even more difficult, posing new challenges to overcome. We analyze the principal challenges related with Security and QoS tradeoffs in mobile platforms. As a result of our analysis we provide parametric relationships between security and QoS parameters focusing on mobile platforms.

Keywords Security · QoS · Tradeoffs · Mobile platforms

1 Introduction

Security risks in mobile platforms are increasingly a customer concern. In particular, the theft of personal data is a widely discussed issue. As a consequence, some mobile platforms have begun to develop specific solutions to avoid the theft of

A. Nieto (✉) · J. Lopez
Computer Science Department, University of Malaga, Málaga, Spain
e-mail: nieto@lcc.uma.es

J. Lopez
e-mail: jlm@lcc.uma.es

S.-S. Yeo et al. (eds.), *Computer Science and its Applications*,
Lecture Notes in Electrical Engineering 203, DOI: 10.1007/978-94-007-5699-1_2,
© Springer Science+Business Media Dordrecht 2012

private data from mobile terminals. Indeed, threats in mobile platforms open up a new market for anti-virus providers, whose products have been adapted to protect mobile platforms (e.g. McAfee Mobile Security). These new services are of particular interest in corporate environments, where personal devices can inadvertently introduce malware into the system. In addition, from a commercial point of view, new emerging technologies (e.g. NFC) open the door to new ways to trick the user [6, 11].

Furthermore, the widespread use of multimedia applications does necessitate the presence of mechanisms to ensure the quality of service (QoS), and more generally the quality of experience (QoE). These applications have the added difficulty of being deployed in resource-constrained devices, so more requirements have to be taken into account apart from those concerned with improving the multimedia capabilities. Therefore it is not only the network parameters that must to be controlled. In mobile platforms the QoS mechanisms have to integrate network parameters (e.g. bandwidth) and device parameters (e.g. battery power). Furthermore, as user's satisfaction is closely related with the success of the platform, it is fundamental to add QoE measurements to the quality study. However, security mechanisms can damage the user's opinion [13]. In fact, security mechanisms tend to consume network and local resources and can easily affect the normal performance of devices.

The main objective behind the approach presented here is to provide an analysis about security and QoS tradeoffs in mobile platforms where the aforementioned concepts will be widely exposed. Indeed, although such concepts are separately analyzed in several papers, in the end they have to coexist in the same environment. Thus, understanding these relationships previous to running it on the architecture is fundamental to savings in cost and ensuring better security and performance. As a result, we provide a parametric relationship scheme between security and QoS, which also considers the effect that such dependencies have on the user perception.

The remainder of the paper is organized as follows: Sect. 2 provides an analysis on the current challenges related with Security and QoS. In Sect. 3, an analysis of Security and QoS tradeoffs in mobile platforms is shown. Finally, Sect. 4 discusses related work and Sect. 5 concludes the paper.

2 Security and QoS Challenges

Figure 1 shows a simplified diagram with the main Security and QoS challenges identified based on the current literature. We consider three sections: commercial purposes, development and communication.

Commercial Purposes: Identifies popular characteristics developed with the aim of improving functionality of personal devices to be more attractive for the user. For example, in the QoS context, the use of multiple radio antennas allows better transmission services at higher speeds and also offers the possibility for

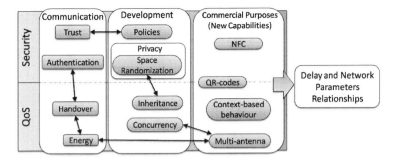

Fig. 1 Security and QoS challenges

early detection of collisions, among other things [12], but increase the complexity of the system [2, 21]. Regarding security, NFC allows secure e-payment, and can be built in the device as it is anti-tampering, but it can also affect the system's overhead by increasing the transaction time and therefore the response time [5]. To the contrary, QR-codes can be used (like tokens) in authentication mechanisms to reduce the response time [18]. However, both NFC and QR-codes open the door to new threats [19] and ways to trick the user [5, 10]. We also note that context-based services are also of growing interest [22], because they allow a more realistic behavior based on knowledge.

Development: Takes into account the security risks due to a wrong implementation of security requirements [9, 20]. Related with it, [14] highlights the importance of establishing different users' permission levels to prevent that, once the attacker finds a bug in the system, it can take absolute control. Also it is necessary to pay attention to problems caused by incompatibility of functions. For example, privacy mechanisms based on space randomization (e.g. ASLR) can be unusable when inheritance-based mechanisms (e.g. Zygote) are used to allow two processes to share the same memory space to reduce the overhead [17]. Note that privacy is becoming a major issue, because user participation in mobile platforms requires it. But, privacy is being continuously threatened. For example, in [8], the authors show that it is possible to recover information from mobile platforms even though it has been deleted.

Communication: Considers the requirements and mechanisms to protect network communication at a low cost. From a QoS point of view, solutions to provide end-to-end QoS guarantees should consider local QoS requirements, such as memory, or energy consumption. The last one is critical, not only to improve the QoE, but also because without it the device is useless. Traditional QoS requirements and energy consumption tradeoffs are studied in [1, 3, 21]. In particular, [1] provides a solution based on predicting the behavior of the system, which is not always a feasible option. From a security point of view, trust mechanisms are fundamental to ensure the survival of a communication platform based on the interaction between entities through the Internet. However, it implies in several cases the use of certificates or complex authentication schemes that are not supported on mobile phones.

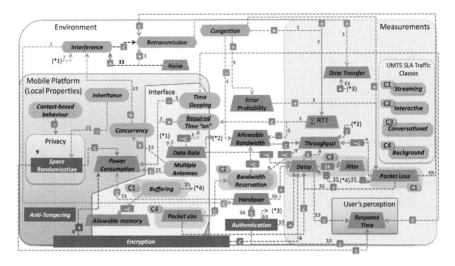

Fig. 2 Security and QoS parametric relationships in mobile platform

3 Parametric Relationships on Mobile Platforms

This section analyses the dependencies between Security and QoS parameters illustrated in Fig. 2. Below, the mathematical definition of each relation is described.

3.1 Mathematical Definition

In our previous work [16] we defined a set of dependency relationships between parameters (1, 2, 5, 6). However, we need to add new equations to the current formulation to express the specific dependencies on mobile platforms. Below, we work with a formulation based on basic expressions (1–4) in order to clarify the dependencies diagram (Fig. 2).

Basic expressions		Complex expressions (based on 1–4)	
$D^+ :: aD^+b \Rightarrow (\Delta a \to \Delta b)$	(1)	$D^c :: (\Delta a \to \Delta b) \wedge (\nabla a \to \nabla b) \equiv aD^+b \wedge aD^{\neg+}b$	(5)
$D^- :: aD^-b \Rightarrow (\Delta a \to \nabla b)$	(2)	$D^t :: aD^c b \wedge bD^c a$	(6)
$D^{\neg+} :: aD^{\neg+}b \Rightarrow (\nabla a \to \nabla b)$	(3)	$D^{\neg c} :: (\Delta a \to \nabla b) \wedge (\nabla a \to \Delta b) \equiv aD^-b \wedge aD^{\neg-}b$	(7)
$D^{\neg-} :: aD^{\neg-}b \Rightarrow (\nabla a \to \Delta b)$	(4)	$D^{i+} :: (\Delta a \to \Delta b) \wedge (\nabla a \to \Delta b) \equiv aD^+b \wedge aD^{\neg-}b$	(8)
		$D^{i-} :: (\Delta a \to \nabla b) \wedge (\nabla a \to \nabla b) \equiv aD^-b \wedge aD^{\neg+}b$	(9)

In order to get a basic set of equations, we add to the formulation in [16] the Eqs. (3 and 4), corresponding to inverse positive and negative respectively. D^{-+} (3) means that the decrement of the first parameter causes the decrement of the second parameter, while in D^{--} (4) the decrement of the first parameter causes the increment of the second parameter. Moreover, complex equations are obtainable from basic equations by adding to the formulation in [16] the Eqs. 7, 8 and 9, corresponding to inverse complete, independent positive and independent negative respectively. D^{-c} (7) means that both parameters are related negatively (D^-) and inverse negatively (D^{--}). The independent relationships (8 and 9) have been added to reflect the dependencies in which regardless the change of value in the first parameter the result is always the increasing (D^{i+}) or decreasing (D^{i-}) of the second parameter. This happens, for example, with the relationship between parameters Delay and Jitter, as we shall see.

3.2 Dependency Relationships Diagram

Fig. 2 shows the dependency relationships diagram that we will now explain. Each dependency is marked with the dependency symbol corresponding to the dependency relationship (+, −, ¬ +, ¬ −, c, t, ¬ c, i + , i−) and the reference to the article where it appears. However, some of them are based on known for-mulations for the calculation of some parameters, specifically 10, 11 and 12. These last dependencies have been highlighted with the symbol *. There are also some dependencies that are explained in the text, and that appear without being refer-enced in any paper previously mentioned here. Moreover, the diagram also inte-grates the SLA traffic classes named in [15], which are: Interactive, Background, Streaming and Conversational.

$$Delay \ = \ \#bits/DataRate \tag{10}$$

$$Jitter \ = \ |DelayT_0 - \ DelayT_1| \tag{11}$$

$$Throughput(per \ user) \ = \ DataRate/\#Users \tag{12}$$

As we can see, delay, throughput and power consumption are highly influenced by the rest of parameters and characteristics. On the one hand, delay severely affects network performance. As we can see in Fig. 2, both streaming and con-versational traffic are affected whether delay increases or not. Note that buffering can help to minimize the delay if the data can be pre-processed while it is in the buffer, and also helps to decrease packet loss when an adequate buffer size is defined. However, the buffering technique demands memory in order to work. We also observe that the handover increases the delay, as do the authentication mechanisms. Moreover, although increasing data rate can decrease the delay (13), it is important to note that it also may cause interferences because high speeds introduce noise. In addition, long packet size can increase the delay because it introduces more data to be sent in the same packet (14). So, depending on the intermediary communication mechanisms, it can require a greater amount of time to be processed (e.g. decode/coding data). In addition, when the receptor fails, the entire packet has to be sent again.

$$(\Delta DataRate \rightarrow \nabla Delay) \wedge (\nabla DataRate \rightarrow \Delta Delay) \equiv DataRate D^{-c} Delay \quad (13)$$

$$(\Delta \#bits \rightarrow \Delta Delay) \wedge (\nabla \#bits \rightarrow \nabla Delay) \equiv PacketSize D^{c} Delay \qquad (14)$$

On the other hand, although increasing the packet size means the delay increases, when the packet size is very small it may cause throughput degradation because each packet requires that a header is sent, increasing the volume of data to be sent. Therefore, header content is not considered as useful data for communication at service level, and thus the throughput decreases. However, if the packet size is too big then the throughput can be damaged too. For example, if the packet size is static and the data to be sent is less than the packet size, then the packet has to be completed with garbage data to achieve the total size, and such data cannot be counted as useful data. The problem is greater if bandwidth reservation mechanisms are static. In such cases, the bandwidth that has been previously reserved is unusable for other devices. As a consequence, the greater the packet size the higher the bandwidth reservation, decreasing the network's resources.

Note that, when the delay increases then the throughput decreases because the channel is probably saturated. Contrarily, when the delay decreases the throughput can be increased because there are more available resources for data transmission and fewer errors are likely to occur. If the throughput is poor, the service is not receiving sufficient data to work properly. This can damage the user's perception of the service, which is also affected when the response time increases. Packet loss affects both; throughput and delay. When packet loss is high (e.g. due to congestion or the high error probability), the delay increases and, contrarily, the throughput decreases (in the case that the re-send data is not considered for throughput calculation). Indeed, if the packet loss increases, then the number of retransmissions also increases, thus increasing the data transfer and also the power consumption.

Regarding the power consumption, it is strongly decreased by the time that the antennas are active (required time-on). Thus, although local security mechanisms increase the computational requirements, the power consumption is increased mainly due to those operations related with data transmission. Note that, by decreasing the required time-on, the power consumption can also decrease if the network interface is disabled in such situations. Indeed, if the data rate increases, then the required time-on decreases, but it is possible that noise appears when speed increases. LTE, one 4G technology, requires ICIC techniques to avoid interferences, precisely due to high speeds. However, it does not mean that, because of this, LTE terminals consume less energy. On the contrary, LTE technology is able to use multiple antennas, improving performance in communications, but also requiring more energy for transmission. Besides, multiple antennas also increase the complexity of the terminal, where concurrent operations can coexist increasing the interference probability.

Finally, in general, security mechanisms increase the response time. It is particularly true when additional messages to establish a secure communication channel are required. The rising amount of data to be sent inevitably affects power consumption, but also causes delay, which increases as a consequence of the growing traffic. Cryptographic techniques also affect power consumption, but sent data requires even more energy than that. Therefore, when authentication mechanisms have to be performed during the handover, the overall performance of the network can be severely damaged. Indeed, the handover process involves several operations to be effective (message interchange between entities), and, therefore, it also increases the power consumption by itself. Lastly, privacy mechanisms based on space randomization (e.g. ASLR) provide local security for user's data. However, context-based services require the storing of the user's preferences in the mobile platform or sending it to an external server in order to work. In both cases the user's privacy is affected.

4 Related Work

There are some papers related with the study of Security and QoS tradeoffs, although they have been developed within a specific scenario. Therefore they do not provide a general view of the current state of the art in mobile platforms. For example, [13] analyzes how the authentication mechanisms affects delay, and how it affects the user's perception. Moreover, the end-to-end secure protocol proposed in [7] for Java ME-based mobile data collection also considers the balance between flexibility, efficiency, usability and security. In said work, the effect that different encryption algorithms (e.g. AES, RSA) have on performance has been studied. Moreover, [4] propose SECR3T, a secure communication system over 3G networks that considers QoS restrictions. For example, the paper shows the effect that encryption protocols have on delay (minimum and maximum), and how it affects different types of data (audio and video). Power consumption is also considered, and the authors conclude that it greatly depends on the implementation of the protocol used (e.g. TLS, ECDH). Moreover, as we have seen, in [18] the advantages of using QR-codes for authentication in Cloud computing environments in order to increase the performance are highlighted, and in [5] the impact of NFC technology on delay and computation time is shown.

5 Conclusions

In this paper we have seen how several studies have focused on both security and QoS concepts, although with different aims, deploying mechanisms or defining models to solve specific problems. We have carried out an analysis of such mechanisms and also detected further challenges to be addressed. Moreover, we have grouped this knowledge in a comprehensive dependency relationship map, in order to enable the future development of tools to support developers in the development of secure and efficient services for mobile platforms. The final diagram shows that delay, throughput and power consumption are highly influenced by the rest of the parameters and characteristics. Note that, in this paper we provide several examples where security and QoS tradeoffs are present. However, in such papers the tradeoffs are very specific, mainly focusing on a particular problem or scenario and are not considered together. This lack of abstraction makes the complete understanding of security and QoS tradeoffs in mobile networks more difficult. The novelty of this paper is therefore, precisely, in providing such a global vision where, in addition, parametric relationships are provided within an understandable logic.

Acknowledgments This work has been partially supported by the Spanish Ministry of Economy and Competitiveness through the projects SPRINT (TIN2009-09237) and IOT-SEC (ACI2009-0949), being the first one also co-funded by FEDER. Additionally, it has been funded by Junta de Andalucia through the project PISCIS (TIC-6334). The first author has been funded by the Spanish FPI Research Programme.

References

1. Anastasi, G., Conti, M., Gregori, E., Passarella, E.: Balancing energy saving and QoS in the mobile internet: an application-independent approach. In: Proceedings of the 36th Annual Hawaii International Conference on System Sciences, 10 pp. IEEE (2003)
2. Aziz, D., Sigle, R.: Improvement of lte handover performance through interference coordination. In: IEEE 69th Conference on Vehicular Technology, pp. 1–5. IEEE (2009)
3. Bellasi, P., Bosisio, S., Carnevali, M., Fornaciari, W., Siorpaes, D. Constrained power management: application to a multimedia mobile platform. In: Design, Automation Test in Europe Conference Exhibition (DATE), March 2010, pp. 989–992 (2010)
4. Castiglione, A., Cattaneo, G., Maio, G.D., Petagna, F.: Secr3t: Secure end-to-end communication over 3G telecommunication networks. In: Fifth International Conference on Innovative Mobile and Internet Services in Ubiquitous Computing (IMIS), 2011, pp. 520–526, 30 June 2011–2 July 2011
5. Damme G.V., Wouters, K.: Practical experiences with NFC security on mobile phones. Katholieke Universiteit Leiden (2009)
6. Dai Z., Dino A.: Apple ios 4 security evaluation. Trail of Bits LLC, 2011
7. Gejibo, S., Mancini, F., Mughal, K., Valvik, R., Klungsøyr, J.: Challenges in implementing an end-to-end secure protocol for Java ME-based mobile data collection in low-budget settings. In: Engineering Secure Software and Systems, pp. 38–45 (2012)
8. Glisson, W, Storer, T, Mayall, G, Moug, I., Grispos, G: Electronic retention: what does your mobile phone reveal about you? Int. J. Inf. Secur. **10**, 337–349 (2011). 10.1007/s10207-011-0144-3
9. Grace, M., Zhou, Y., Wang, Z., Jiang, X.: Systematic detection of capability leaks in stock android smartphones. In: NDSS 2012
10. Kieseberg, P, Leithner, M, Mulazzani, M, Munroe, L, Schrittwieser, S, Sinha, M, Weippl, E.: Qr code security. In: Proceedings of the 8th International Conference on Advances in Mobile Computing and Multimedia, MoMM'10, pp. 430–435. ACM, New York, NY, USA (2010)

11. Kiminki, S., Saari, V., Hirvisalo, V., Ryynanen, J., Parssinen, A., Immonen, A., Zetterman, T.: Design and performance trade-offs in parallelized RF SDR architecture. In: IEEE Sixth International ICST Conference on Cognitive Radio Oriented Wireless Networks and Communications (CROWNCOM), 2011, pp. 156–160 (2011)
12. Kumar, S.P.S., Anand, S.V.: A novel scalable software platform on android for efficient QoS on android mobile terminals based on multiple radio access technologies. In: Wireless Telecommunications Symposium (WTS), 2011, pp. 1–6, April 2011
13. Lorentzen, C., Fiedler, M., Johnson, H., Shaikh, J., Jorstad, I. On user perception of web login—a study on qoe in the context of security. In: 2010 Australasian Telecommunication Networks and Applications Conference (ATNAC), pp 84–89, 31 Oct 2010–3 Nov 2010
14. Miller, C., Honoroff, J., Mason, J.: Security evaluation of apples iphone. Independent Security Evaluators, 19 July 2007
15. Mohan, S., Agarwal, N.: A convergent framework for QoS-driven social media content delivery over mobile networks. In: IEEE 2011 2nd International Conference on Wireless Communication, Vehicular Technology, Information Theory and Aerospace & Electronic Systems Technology (Wireless VITAE), pp. 1–7 (2011)
16. Nieto, A., Lopez, J.: Security and QoS tradeoffs: Towards a Fi perspective. In: 2012 26th International Conference on Advanced Information Networking and Applications Workshops (WAINA), pp. 745–750, March 2012
17. Oberheide, J.: A look at ASLR in android ice cream sandwich 4.0. The Duo Bulletin, 2012
18. Oh, D.-S., Kim, B.-H., Lee, J.-K. A study on authentication system using QR code for mobile cloud computing environment. In: Oh, D.-S., Kim, B.-H., Lee, J.-K. (eds.) Future Information Technology, vol. 184 of Communications in Computer and Information Science, pp. 500–507. Springer, Heidelberg (2011)
19. Roland, M., Langer, J., Scharinger, J.: Security vulnerabilities of the NDEF signature record type. In: 2011 3rd International Workshop on Near Field Communication (NFC), pp. 65–70, Feb 2011
20. Seriot, N.: iphone privacy. Black Hat DC, p. 30, 2010
21. Uddin, M.M., Haseeb, S., Ahmed, M., Pathan, A.-S.K.: Comprehensive QoS analysis of mipl based mobile IPv6 using single vs. dual interfaces. In: 2011, International Conference on Electrical, Control and Computer Engineering (INECCE), pp. 388–393, June 2011
22. Wac, K., van Halteren, A., Konstantas, D.: QoS-predictions service: Infrastructural support for proactive QoS- and context-aware mobile services (position paper). In: On the Move to Meaningful Internet Systems, OTM 2006 Workshops, pp. 1924–1933. Springer, Heidelberg

A Hybrid Natural Computing Approach for the VRP Problem Based on PSO, GA and Quantum Computation

Kehan Zeng, Gang Peng, Zhaoquan Cai, Zhen Huang
and Xiong Yang

Abstract In this paper, a novel hybrid natural computing approach, called PGQ, combining Particle Swarm Optimization (PSO), Genetic Algorithm (GA) and quantum computation, is introduced to solve the Vehicle Routing Problem (VRP). We propose a quantum approach, called QUP, to update the particles in PSO. And, we add GA operators to improve population quality. The simulation results show that the PGQ algorithm is very effective, and is better than simple PSO and GA, as well as PSO and GA mixed algorithm.

Keywords VRP · PSO · GA · Quantum computation

1 Introduction

The Vehicle Routing Problem (VRP), as known as a hot topic since it was discovered, has many significant applications in economics, logistics and industry engineering. In general, a solution to VRP is a set of tours for several vehicles

K. Zeng (✉) · G. Peng · Z. Cai · Z. Huang · X. Yang
Department of Computer Science, Huizhou University, Huizhou 516007, People's Republic of China
e-mail: zengkehan@hzu.edu.cn

G. Peng
e-mail: peng@hzu.edu.cn

Z. Cai
e-mail: cai@hzu.edu.cn

Z. Huang
e-mail: hz9714105@hzu.edu.cn

X. Yang
e-mail: xyang.2010@hzu.edu.cn

S.-S. Yeo et al. (eds.), *Computer Science and its Applications*,
Lecture Notes in Electrical Engineering 203, DOI: 10.1007/978-94-007-5699-1_3,
© Springer Science+Business Media Dordrecht 2012

from a depot to customers and return to the depot without exceeding the capacity constraints of each vehicle, at the minimum cost.

Among the approaches solving the VRP, natural computing based algorithms show standout effectiveness and efficiency, and gradually predominates the trend of solutions for the VRP. Some state-of-the-art algorithms are listed as follows.

Yu Bin et al. proposed an improved ant colony optimization for the VRP, which is better than some other meta-heuristic approaches [1]. Humberto César Brandão de Oliveira and Germano Crispim Vasconcelos proposed a hybrid search method associating non-monotonic Simulated Annealing to Hill-Climbing and Random Restart. They compared their algorithm with the best results published in the literature for the 56 Solomon instances, and it was shown how statistical methods can be used to boost the performance of the method [2]. Gong et al. proposed an discrete PSO approach for the VRP with Time Windows (VRPTW), and the simulation results and comparisons illustrated the effectiveness and efficiency of the algorithm by Solomon's benchmarks testing [3]. Yucenur and Nihan proposed a geometric shape-based genetic clustering algorithm for the multi-depot vehicle routing problem [4] and which is much more effective and timesaving than the nearest neighbor algorithm. Lau et al. proposed a fuzzy logic guided genetic algorithms (FLGA) to solve the VRP with Multiple Depots [5]. The role of fuzzy logic is to dynamically adjust the crossover rate and mutation rate. They compared the algorithm with some benchmarks and which showed that FLGA outperforms other search methods. Christos D. Tarantilis et al. proposed a template-based tuba search algorithm for consistent VRP and got good experiment results [6].

In recent years, the idea of integration is widely applied in algorithm design. Accordingly, hybrid natural computing approaches increasingly catch much focus for solving the VRP.

In this paper, a novel hybrid approach integrating PSO with GA and quantum computing is introduced. We name it PGQ. And, a quantum approach to update the particles is proposed, which is called QUP. The simulation of PGQ on the benchmark dataset vrpnc1 shows it is very effective and efficient.

2 The PGQ Algorithm

The PGQ algorithm consists of the main body PSO, in which the particles are updated by a quantum approach; and includes the genetic operators, selection, crossover and mutation. A particle is a string of customer's numbers, composed by all tours. In the rest presentation, g-best and p-best are the global best particle and the individually historical best particle, respectively of PSO.

2.1 Particle Initialization

A particle is initialized by the following process:

Step 1: a vehicle departs from the depot;
Step 2: repeat selecting an unvisited customer randomly and delivering goods until a customer's quantity requirement of goods exceeds the current quantity of goods carried by the vehicle;
Step 3: the vehicle return to the depot and a tour is generated;
Step 4: repeat Steps 1–3 until all customers are visited once.

2.2 Quantum Approach Updating Particles

The randomness of quantum movement is imposed on particle updating process. The procedure is described as follows:

Step 1: generate two random number u and f between 0 and 1.
Step 2: $g = 0.5 * \frac{dcn-cin}{dcn} + 0.5$, where g is called random coefficient; dcn is the initialized total generation number; cin is the number of current generation;
Step 3: $p = f * ptc(cin) + (1 - f)*g\text{-}best$, where $ptc(cin)$ is the cinth particle; p is the combination of from the 0th percentile to the fth percentile of $ptc(cin)$ and from the fth percentile to the 100th percentile of g-best; if some customer reoccurs, delete the repetitive one and add in unvisited customers, assuring that all customers are visited once.
Step 4: if $u \geq 0.5$ then

$$ptc(cin) = ptc(cin) + g * \ln\frac{1}{u} * p \tag{1}$$

else then

$$ptc(cin) = p + g * \ln\frac{1}{u} * ptc(cin) \tag{2}$$

The two formulae determine different changes of particles. Formula (1) means select $\left(g * \ln\frac{1}{u}\right)\%$ part of p randomly to add into $ptc(cin)$ and delete the repetitive customers in $ptc(cin)$ and add in unvisited customers, assuring that all customers are visited once. Formula (2) has the same way operation, while exchanging positions of $ptc(cin)$ and p.

This quantum approach to update particles is called QUP.

2.3 Mutation Operator

The mutation is to select two tours randomly in a particle and exchange two customers from each tour randomly. The mutation probability $mp = low + cin * \frac{(up-low)}{dcn}$, where $low = 0.1$ and $up = 0.9$.

2.4 Crossover Operator

The Crossover is imposed on each particle with the g-best, and the crossover probability is 1. The procedure of crossover is described as follows:

Step 1: two crossover points are generated randomly;
Step 2: compare the customer series between crossover points, between the particle and the g-best;
Step 3: if the two series have some same customers, those customers are taken out from the particle;
Step 4: link the customer series of the g-best into the particle after the last customer.

2.5 The Process of PGQ

The process of PGQ is summarized as follows:

Step 1: 1000 particles are generated randomly as the initial population;
Step 2: compute the fitness of each particle, which is defined as the reciprocal of a solution, the sum of lengths of all tours in a particle;
Step 3: find the historically best record of each particle (p-best) and the global best particle (g-best);
Step 4: select 50 particles by using roulette-wheel selection;
Step 5: perform the Quantum approach updating particles (QUP);
Step 6: update the p-best and the g-best;
Step 7: perform crossover and mutation;
Step 8: generate 950 new particles randomly;
Step 9: repeat Steps 2–8 until the generation number meets dcn.

Table 1 Comparisons among GA, PSO, PG and PGQ on vrpnc1

Population scale (generation number)	Algorithm	Average (error %)	Best (error %)
300	GA	1190.65 (126.96)	1022.89 (94.98)
(500)	PSO	1183.71 (125.64)	1036.63 (97.60)
	PG	601.57 (14.67)	563.66 (7.44)
	PGQ	569.06 (8.47)	535.46 (2.07)
400	GA	1090.87 (107.94)	980.65 (86.93)
(500)	PSO	1080.14 (105.90)	978.32 (86.49)
	PG	597.86 (13.96)	547.36 (4.34)
	PGQ	557.43 (6.25)	531.16 (1.25)
500	GA	970.75 (85.04)	880.64 (67.87)
(800)	PSO	973.54 (85.58)	891.42 (69.92)
	PG	587.54 (12.00)	542.44 (3.40)
	PGQ	536.98 (2.35)	524.61 (0)

3 Simulation

We implement the PGQ algorithm by Matlab and compare it with simple PSO and simple GA, as well as PSO and GA mixed algorithm (PG), on the benchmark dataset vrpnc1. Table 1 shows that whatever population scale and generation number are, PGQ outperform others; and PGQ can reach the optimum when population scale is 500 and generation number is 800. And, along with increase of the population scale and generation number, the average solution and the best solution of PGQ gradually approach the optimum.

The analysis of the above observations is summarized as: the increase of the population scale and generation number can explore more solution space and exploit deeper in a local area, thus more accurate results can be obtained; and, contrasting to simple GA and simple PSO, PG incorporates the advantages of GA and PSO, performing better; and, PGQ integrates good attributes of GA, PSO and quantum computation and outperforms them; and, the randomness of quantum movement can increase the randomness of particles changing and avoid PSO to be stranded in a local optimum.

Therefore, PGQ perform well for the VRP and can reach the global optimum when the parameters such as population scale, generation number, are set properly.

4 Conclusions

This paper addresses the VRP using a hybrid natural computing approach integrating PSO, GA and quantum computation. We name it PGQ. The randomness of quantum movement is applied to update the particles, based on which, we propose QUP. The simulation results show PGQ is very effective and is better than other heuristic algorithms.

References

1. Yu, B., Yang, Z.Z., Yao, B.: An improved ant colony optimization for vehicle routing problem. Eur. J. Oper. Res. **196**, 171–176 (2009)
2. Humberto, C.B.O., Germano, C.V.: A hybrid search method for the vehicle routing problem with time windows. Ann. Oper. Res. **180**, 125–144 (2010)
3. Gong, Y.J., Zhang, J., Liu, O., Huang, R.Z., Chung, H.S.H, Shi, Y.S.: Optimizing the vehicle routing problem with time windows: a discrete particle swarm optimization approach. IEEE Trans. Syst. Man Cybern. Part C Appl. Rev. **42**, 254–267 (2012)
4. Yucenur, G.N., Nihan, C.D.: A new geometric shap-based genetic clustering algorithm for the multi-depot vehicle routing problem. Expert Syst. Appl. **38**, 11859–11865 (2011)
5. Lau, H.C.W., Chan, T.M., Tsui, W.T., Pang, W.K.: Application of genetic algorithms to solve the multidepot vehicle routing problem. IEEE Trans. Autom. Sci. Eng. **7**, 383–392 (2010)
6. Christos, D.T., Stavropoulou, F., Panagiotis, P.R.: A template-based Tabu Search algorithm for the consistent vehicle routing problem. Expert Syst. Appl. **39**, 4233–4239 (2012)

Finger Triggered Virtual Musical Instruments

Chee Kyun Ng, Jiang Gi Fam and Nor Kamariah Noordin

Abstract With the current human movement tracking technology it is possible to build a real-time virtual musical instrument with a gestural interface which is similar to a real musical instrument. In this paper, a simple finger triggering based controller for virtual musical instruments is presented. The virtual musical instruments that can be operated by using this controller are piano and drum. This system consists of three main components; finger data glove system, musical notes recognition system, and data transceiver system. Finger triggering devices are mounted to each finger in a data glove. This finger data glove has the capability to get data from triggered devices and the attached microcontroller in the data glove system is used to receive the data before transmitted to the musical notes recognition system through wireless or wired transmission. The musical notes recognition system matches the received triggering signal data with the predefined musical notes data using matching algorithm. When the data is matched, the musical notes data will be playing the associated musical notes. The developed system has the ability to switch the virtual musical instruments between piano and drum. Nevertheless, this developed finger triggered virtual musical instruments has the real-time capability to analyze the sounds or musical notes that have yet encountered in the studio.

Keywords Musical instrument · Musical notes · Recognition system · Data glove

C. K. Ng (✉) · J. G. Fam · N. K. Noordin
Department of Computer and Communication Systems Engineering,
University Putra Malaysia (UPM), 43400 Serdang, Selangor, Malaysia
e-mail: mpnck@eng.upm.edu.my

J. G. Fam
e-mail: ggizsy@gmail.com

N. K. Noordin
e-mail: nknordin@eng.upm.edu.my

S.-S. Yeo et al. (eds.), *Computer Science and its Applications*,
Lecture Notes in Electrical Engineering 203, DOI: 10.1007/978-94-007-5699-1_4,
© Springer Science+Business Media Dordrecht 2012

1 Introduction

Music is an art and the harmonious voice of creation which the medium of music is sound. The objects and artifacts are always used by humans being to make music. There are many ways to generate sound such as using reeds, bells, strings and etc. Human being always uses these ways to make musical instruments. After the electricity has been discovered, human being started to use electricity to generate sound. The electricity medium enabled human being to build many type of devices for electrical instruments. In the first half of the twentieth century there are a lot of new musical instruments sprung up and became more complicated. Most of electrical instruments are based on a keyboard, which has been proven to be a flexible interface. As it can be found for centuries there are many types of musical instruments such as organ, harpsichord, piano and etc. [1].

Most of modern music is using electronic musical instrument to compose the rhythm. The research on music has become active and aggressive after the development of new electronic musical instruments, controller and synthesizer. When the software based synthesizers are emerged and the personal computers have become more affordable and faster, musicians started to implement their own digital musical instruments by using controllers other than the keyboard itself. They started to develop the controller system with the help of gestural control technology which is a set of techniques use the extraction of control signals from the performer without physical contact with the musical instruments. Researchers are more alert in giving musicians a great deal of freedom to play the musical instruments. Musician blame that there are limited control and gesture set on musical instruments such as allowing musicians to express, play and composes music creatively during their singing and dancing. A survey shows that there is a research direction toward creating musical data glove that can interface with the virtual musical instruments. However, electronic musical instruments up to date do not provide the features of gestures into data glove adaptively but only increase the range of sounds which is an extension of the sounds that are produced by the original musical instrument [2].

In this paper, the development of a data glove for playing musical instruments using a simple finger triggering system is proposed. The approach taken is to ensure that the data glove is capable of providing freedom for musician to play music and perform dance movement simultaneously. Besides that, the developed data glove system also has the capability of playing different types of musical instrument. This system consists of three main components; finger data glove system, musical notes recognition system, and data transceiver system. In the finger data glove system, the data from the finger triggering are collected from the designed hand glove which based on either conductive sleeve or push button switch. These data are diagnosed and translated by a Programmable Integrated Circuit (PIC) microcontroller attached to the hand glove, and generating signals which are in digital data format before transmitted to the musical notes recognition system through either wireless or wired transmission. The developed musical notes recognition system is capable of diagnosing the received data for matching with the predefined musical notes data by using matching algorithm. When the data is matched, the associated musical notes will be executed before played at the sound system.

2 The Conceptual Design of Finger Triggering Based Virtual Musical Instruments

In this section, the conceptual design of a finger triggering based virtual musical instruments is described. The overall system is comprised of three main interfaces; finger data glove, musical notes recognition and data transceiver systems. In the finger data glove system, a pair of musical gloves includes finger sleeves is used to achieve finger triggering from the user's hands. The musical notes recognition system is a software based application programming interface (API) that has been configured to match the received triggering inputs with the corresponding predefined musical notes data stored in the system's library. The transceiver unit system is used in unidirectional communication to send the triggering signals from finger data glove system to musical note recognition system in either wired or wireless.

Initially, a microcontroller in the finger data glove circuit provides triggering signals that received from the finger sleeves to the transceiver unit. The triggered signals then enable user-interaction with the computer based recognition API through transceiver unit. In the musical note recognition API, the received triggering signal from its transceiver unit will invoke the API to display the simulation action and play the corresponding melody note when it successfully matches the predefined musical note. The overall system description can be shown in Fig. 1.

2.1 Finger Data Glove System

The finger data glove system consists of a simple data glove with electronic components, microcontroller circuit and transceiver unit. This gadget is mountable to user hand as shown in Fig. 2. The data glove has the triggering elements for each finger that selectively stimulated when power supplies are provided. These triggering elements can be flex sensors, pushbutton switches or conductive sleeves that are installed on tip of each finger and hand. Each of these finger triggering elements is connected to microcontroller separately. The received triggering signal is then being processed by using microcontroller before transmitted out through a transceiver unit. To compare the efficiency of the system, three different finger triggering elements are used. These finger triggering elements are flex sensors, pushbutton switches and conductive sleeves. Each triggering element uses its own triggering method to synthesize the stimulated signal for playing a basic octave of musical notes. Actually, only a minimum of eight fingers are needed to trigger the octave of musical notes which is listed in Table 1.

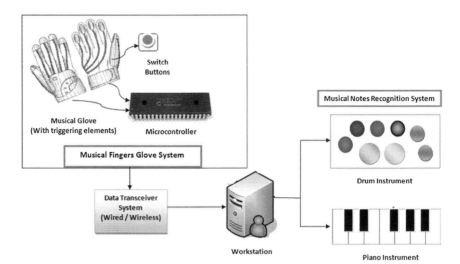

Fig. 1 System overview

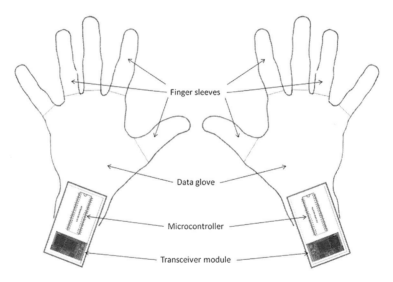

Fig. 2 Finger data glove layout

2.2 Data Transceiver System

The data transceiver system is used to perform unidirectional communication by transmitting triggering signals from finger data glove system to musical notes recognition system either in wired or wireless as shown in Fig. 3. To compare the efficiency of the system, two different transceiver modules are used. Wired communication uses USB RS232 converter to transfer the triggering data to

Finger	Note	Finger	Note
Table 1 A complete set of musical notes triggered in a pair of gloves (a) left hand and (b) right hand			
a. *Left hand*		b. *Right hand*	
1	Do	1	So
2	Re	2	La
3	Mi	3	Ti
4	Fa	4	Do

musical notes recognition system. The USB RS232 converter is installed to finger data glove system and connected to microcontroller as shown in Fig. 3a. In contrast, as for wireless communication, the Bluetooth module is used to transfer the triggering data to musical notes recognition system. The Bluetooth module is installed to finger data glove system and connected to microcontroller as shown in Fig. 3b.

2.3 Musical Notes Recognition System

The musical notes recognition system is a system that analyzes the received triggering data from finger data glove system. The musical notes recognition system gets the triggering data from finger data glove system by using data transceiver system. The system has an API that will perform data matching based on the received triggering data with the predefined musical notes database. The matching algorithm is an algorithm that matches the finger triggering input obtained from the finger data glove system to the predefined musical notes in order to play the musical instruments that have been stored in the musical notes recognition system.

At the start of the matching algorithm, the finger triggering data is received by finger data glove system and then sent to musical notes recognition system via data transceiver system in the converted letter form. Matching algorithm will match the received letters with the letters that have been predefined in the application program. An appropriate musical note will be played after a matched is found. The corresponding musical notes of each letter for piano and drum are shown in Table 2.

3 Results and Discussions

The schematic diagram for the whole finger data glove system is shown in Fig. 4. The whole finger data glove system is controlled by a programmable PIC microcontroller, PIC16F877A. Its inputs are connected to PIC pins of RAs and RBs via J7 connector. The wired or wireless transceiver module is connected to

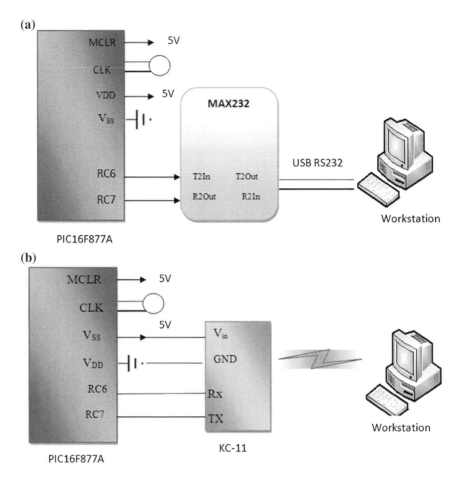

Fig. 3 Data transceiver systems, **a** USB RS232 and **b** Bluetooth KC11

Table 2 The corresponding musical notes of each letters for (a) piano and (b) drum

Letters	Notes	Name of piano notes	Letters	Notes	Name of drum sets
a. *Piano*			b. *Drum*		
A	60	Middle C	A	55	Snare drum
B	62	Middle D	B	56	Floor tom
C	64	Middle E	C	39	Bass drum
D	65	Middle F	D	42	Hi-hat
E	67	Middle G	E	46	Ride cymbal
F	69	Middle A	F	87	Rack tom
G	71	Middle B	G	53	Crash
H	72	Middle C1	H	59	Cross stick

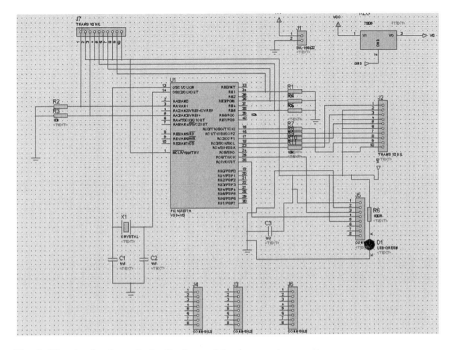

Fig. 4 The circuit schematic for the finger data glove system

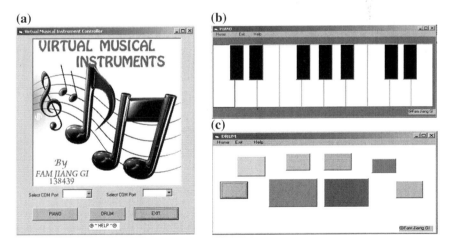

Fig. 5 The APIs of musical notes recognition system, **a** welcome, **b** piano and **c** drum

PIC pins of RC6 and RC7. The extra connector of PIC RCs pins are used via J2 for the switching among the developed various virtual musical instruments which are piano and drum in this paper. The whole finger data glove system is supplied with 9 V battery. Since the operation voltage range for the PIC16F877A is from 2 to 5.5 V, a voltage regulator LM7805 IC is used to convert the 9 to 5 V.

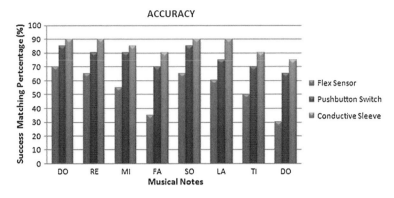

Fig. 6 The accuracy performance for three different types of musical glove

This 5 V supplied input is connected to PIC pin V_{DD} via J1 socket. The PIC is also connected with a 20 MHz crystal oscillator in series with two capacitors to ground. It is used to create a 20 MHz oscillation for supporting the serial communication.

The musical notes recognition system is a software based API that has been configured to match the received triggering inputs with the corresponding predefined musical notes stored in the system's library. There are total three main APIs for the musical notes recognition system. The APIs are including welcome API, piano API and drum API which are shown in Fig. 5.

The accuracy performances are obtained from the experiments that have been tested with three triggering elements such as flex sensors, pushbutton switches and conductive sleeves. It is evaluated by accumulating the success percentage of the notes can hit by wearing these three different types of musical glove. The system is tested 20 times on three users, in which each user is required to perform Do, Re, Mi, Fa, So, La, Ti and Do notes for virtual piano instrument. Figure 6 shows the performance evaluation on accuracy in terms of percentage for three different types of fingers triggering elements.

4 Conclusions

In this paper, a simple finger triggered glove-based virtual musical instruments that is capable of play music has been presented. The presented virtual musical instruments give musicians a great deal of freedom to play the music instruments and performing dance movement simultaneously. The recognition is based on the triggering finger elements and transmits data to the musical notes recognition system through transceiver system. Another advantage of the presented musical system is being able to work without using computer keyboard. It requires the finger data glove to trigger the virtual musical instruments.

References

1. Bongers, B.: Physical interfaces in the electronic arts. interaction theory and interfacing techniques for real-time performance, In: Wanderley, M.M., BattierTrends, M. (eds.) Gestural Control of Music. Cambridge University Press, New York (2000)
2. Mulder, A.: Virtual Musical Instruments: Accessing the Sound Synthesis Universe as a Performer. In: Proceedings of the First Brazilian Symposium on Computer Music (1994)

A Low Complexity Multi-Layered Space Frequency Coding Detection Algorithm for MIMO-OFDM

Jin Hui Chong, Chee Kyun Ng, Nor Kamariah Noordin and Borhanuddin Mohd. Ali

Abstract In this paper, a low complexity multi-layered space frequency OFDM (MLSF-OFDM) coding scheme is presented with the proposed two detection algorithms, fast QR decomposition detection algorithm or denoted as FAST-QR and enhanced FAST-QR (E-FAST-QR). Both algorithms not only reduce the implementation complexity of QR decomposition but also show a good performance in terms of bit error rate (BER). Hence, the proposed detection algorithms can be used to maintain guaranteed quality of service (QoS) in MIMO-OFDM system.

Keywords MIMO-OFDM · MLSF-OFDM · MLSTBC-OFDM · QR · SIC-ZF

1 Introduction

Conventional single-input single-output (SISO) system, which is a wireless communication system with a single antenna at the transceiver, is vulnerable to multipath fading effect. Multipath is the arrival of the multiple copies of transmitted

J. H. Chong (✉) · C. K. Ng · N. K. Noordin · B. Mohd. Ali
Faculty of Engineering, Department of Computer and Communication Systems Engineering,
University Putra Malaysia, UPM Serdang 43400 Selangor, Malaysia
e-mail: chongjinhui@yahoo.com

C. K. Ng
e-mail: mpnck@eng.upm.edu.my

N. K. Noordin
e-mail: nknordin@eng.upm.edu.my

B. Mohd. Ali
e-mail: borhan@eng.upm.edu.my

S.-S. Yeo et al. (eds.), *Computer Science and its Applications*,
Lecture Notes in Electrical Engineering 203, DOI: 10.1007/978-94-007-5699-1_5,
© Springer Science+Business Media Dordrecht 2012

signal at receiver through different angles, time delay or differing frequency (Doppler) shifts due to the scattering of electromagnetic waves. The idea of multiple antenna array has evolved into multiple-input multiple-output (MIMO) system, which provides transmit and receive diversities. It increases robustness to the effect of multipath fading in wireless channels, besides yielding higher capacity, spectral efficiency and better bit error rate (BER) performance over conventional SISO systems. Orthogonal frequency division multiplexing (OFDM) has been widely developed due to its immunity against delay spread. In addition, OFDM is suitable for frequency-selective MIMO systems to form MIMO-OFDM since the introduction of orthogonal subcarriers makes system design and implementation simple [1, 2].

Recently, a combination of space–time coded OFDM with Vertical Bell Laboratories Layered Space–Time (V-BLAST) OFDM known as multi-layered space–time block code OFDM (MLSTBC-OFDM) has been introduced in [3] to achieve spatial multiplexing and diversity gains simultaneously. This enables an increase in system capacity, satisfactory link quality in terms of bit error rate (BER) and combats error propagation by increasing the diversities of the V-BLAST OFDM layer. Moreover, the number of received antennas can be reduced compared to traditional V-BLAST by exploiting the nature of STBC orthogonality. It is also shown in [4] that space-frequency coded OFDM (SF-OFDM) transmitter diversity performs better than space–time coded OFDM (ST-OFDM) system in MIMO-OFDM channel with higher Doppler frequency.

In this paper an implementation of a multi-layered space-frequency OFDM (MLSF-OFDM) that integrates SF-OFDM with V-BLAST OFDM in a layered architecture is presented. The MLSF-OFDM system is modeled over Monte-Carlo time-variant channel model with different maximum Doppler frequency. A fast QR decomposition detection algorithm, denoted as FAST-QR, and the enhanced FAST-QR (E-FAST-QR) algorithm are proposed for MLSF-OFDM system. It is shown that the computational complexity of proposed FAST-QR algorithm is approximately 48 % lower than the conventional QR decomposition detection algorithm. Besides, the result shows that the BER performance of proposed FAST-QR algorithm degrades marginally compared to zero forcing with successive interference cancellation (SIC-ZF) detection algorithm.

2 MLSF-OFDM System Model

The Monte-Carlo time-variant channel is assumed to be constant over OFDM symbol time slot for MLSF-OFDM system. It is also assumed that the guard interval length is larger than the channel maximum delay spread. The delay spread function of Monte-Carlo time-variant channel can be characterized as a linear superposition of L elementary paths, each characterized by propagation delay τ_l, Doppler shift v_l and complex amplitude a_l in l-th channel path as

Fig. 1 MLSF-OFDM
transmitter model

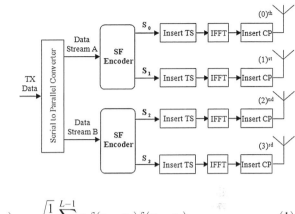

$$h_L(v, \tau) = \sqrt{\frac{1}{L}} \sum_{l=0}^{L-1} a_l \delta(v - v_l) \delta(\tau - \tau_l) \qquad (1)$$

where $\delta(\bullet)$ denotes the Dirac delta function and L is the time domain channel response length.

2.1 Transmitter Model

A block diagram of a MLSF-OFDM transmitter model with four transmit antennas is shown in Fig. 1. A single main data stream is de-multiplexed into two sub-data streams. The sub-data stream A with J data symbols is coded into two vectors, S_0 and S_1 by the SF encoder block and sub-data stream B with J data symbols is coded into two vectors, S_2 and S_3 by the SF encoder block. As the SF layer spans two adjacent sub-carriers in each OFDM time slot, the MLSF-OFDM transmission is denoted as S_{MLSF} for one OFDM symbol period. After that, all sub-data streams from each layer are sent to its respective OFDM transmitter, which carries J sub-carriers. Frequency domain sub-data streams are summed into an OFDM symbol after passing through inverse fast Fourier transformation (IFFT) blocks. A cyclic prefix (CP) is added in the guard interval.

Special training sequence (TS), which was proposed in [5], is used in the MLSF-OFDM transmitter. With the special TS, the problem of singular matrix in the least square (LS) channel estimation implementation can be avoided. The special training sequence at m-th transmit antenna, denoted as $X_m[K]$, with zeros in every odd sub-carrier index is given as

$$X_m[K] = \begin{cases} A \cdot e^{j2\pi((K/2) + m(J/2M))^2/J} & K \in \gamma \\ 0 & K \in \delta \end{cases} \qquad (2)$$

where $K = 0, ..., J-1$ and $m = 0, ..., 3$. γ and δ are the set of the even and odd sub-carrier indices, respectively. A is amplitude of pilot symbol and M is the number of transmit antennas. It can be seen that the training sequence of the

Fig. 2 MLSF-OFDM
receiver model

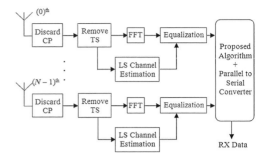
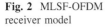

subsequent antenna is only a shifted version of the training sequence of the first antenna to the left-hand side. Moreover, the desirable peak-to-average power ratio of one is still retained.

2.2 Receiver Model

A block diagram of a MLSF-OFDM receiver model with N ($N \geq 2$) receive antennas is shown in Fig. 2. During the reception, each antenna receives the signal transmitted from four transmit antennas. First, the CP of each received signal is removed. After passing through the fast Fourier transform (FFT) blocks, the sub-carriers are separated. Then, the received symbol matrix is sent to the decoder where the proposed algorithm is performed. After the FFT is applied, the demodulated special TS of the j-th sub-carrier of the OFDM symbol in the n-th receive antenna, denoted as $Y_n[j]$, is given as

$$Y_n[j] = \sum_{m=0}^{3} H_{n,m}[j]X_m[j] + W_n[j] \qquad (3)$$

where $j = 0, \ldots, J-1$ and $H_{n,m}[j]$ is the j-th channel coefficient in the frequency domain between the m-th transmit and the n-th receive antenna. $X_m[j]$ is the j-th transmitted training sequence from m-th transmit antenna. $W_n[j]$ is the j-th additive white Gaussian noise (AWGN) of n-th receive antenna with unit variance, σ^2 and zero mean. $j = 0, \ldots, J-1$ is the OFDM sub-carriers index, the subscripts $m = 0$, 1, 2, 3 and $n = 0, \ldots, N-1$ are transmit antenna and the receive antenna indices respectively.

3 Fast QR Decomposition Detection Algorithm

The proposed fast QR decomposition detection algorithm, denoted as FAST-QR, for MLSF-OFDM with N receiver is discussed in detail here. The algorithm used in this scheme can be described as follows. The QR decomposition is applied to the j-th channel matrix $\mathbf{H}_j$ to start the MLSF-OFDM detection algorithm. QR decomposition is performed with $\mathbf{H}_j = \mathbf{Q}_j\mathbf{R}_j$, where $\mathbf{Q}_j$ is a unit norm orthogonal columns $2\,N \times 4$ matrix and $\mathbf{R}_j$ is a 4×4 upper triangular square matrix. As $\mathbf{H}_j$ is orthogonal, $\mathbf{Q}_j$ and $\mathbf{R}_j$ can be simplified to

$$
\mathbf{H}_j =
\begin{bmatrix}
Q_{0,0} & Q_{N,0}^* & Q_{0,2} & Q_{N,2}^* \\
\vdots & \vdots & \vdots & \vdots \\
Q_{N-1,0} & Q_{2N-1,0}^* & Q_{N-1,2} & Q_{2N-1,2}^* \\
Q_{N,0} & -Q_{0,0}^* & Q_{N,2} & -Q_{0,2}^* \\
\vdots & \vdots & \vdots & \vdots \\
Q_{2N-1,0} & -Q_{N-1,0}^* & Q_{2N-1,2} & -Q_{N-1,2}^*
\end{bmatrix}
\begin{bmatrix}
R_{0,0} & 0 & R_{0,2} & R_{0,3} \\
0 & R_{0,0} & -R_{0,3}^* & R_{0,2}^* \\
0 & 0 & R_{2,2} & 0 \\
0 & 0 & 0 & R_{2,2}
\end{bmatrix}
$$

$$(4)$$

The QR decomposition can be calculated with the modified Gram-Schmidt (MGS) method. Note that $R_{0,1} = R_{2,3} = 0$, so the propagation error effect of $S_1(j)$ and $S_3(j)$ is reduced. Besides, $R_{1,1} = R_{0,0}$, $R_{1,2} = -R_{0,3}^*$, $R_{1,3} = -R_{0,2}^*$ and $R_{3,3} = R_{2,2}$. The $R_{0,0}$ and $R_{2,2}$ are non-complex number. Therefore, it is only necessary to calculate two columns of $\mathbf{Q}_j$ and two rows of $\mathbf{R}_j$ (irrespective of the rows of $\mathbf{H}_j$). Thus, the complexity of QR decomposition computation is reduced. The j-th transmitted symbols are decoded as

$$
\hat{s}_0(j) = Q\left(\frac{y_0(j) - \hat{s}_2(j)R_{0,2} - \hat{s}_3(j)R_{0,3}}{R_{0,0}}\right)
$$

$$
\hat{s}_1(j) = Q\left(\frac{y_1(j) - \hat{s}_2(j)(-R_{0,3}^*) - \hat{s}_3(j)R_{0,2}^*}{R_{0,0}}\right)
$$

$$
\hat{s}_2(j) = Q\left(\frac{y_2(j)}{R_{2,2}}\right)
$$

$$(5)$$

$$
\hat{s}_3(j) = Q\left(\frac{y_3(j)}{R_{2,2}}\right)
$$

where $Q(\cdot)$ denotes the quantization operation appropriate to the constellation in use.

On the other hand, the Enhanced Fast-QR or E-FAST-QR detection algorithm enhances the performance of the FAST-QR algorithm by reducing the BER of SF layer B. In this algorithm, a new modified received $N \times 2$ matrix, denoted as $\mathbf{Z}_j$, is created by subtracting the data symbols of SF layer A from $\mathbf{Y}_j$, which results in the following matrix

$$
\begin{bmatrix} Z_0(j)\,Z_0(j+1) \\ \vdots \\ Z_{N-1}(j)\,Z_{N-1}(j+1) \end{bmatrix} = \begin{bmatrix} Y_0(j)\,Y_0(j+1) \\ \vdots \\ Y_{N-1}(j)\,Y_{N-1}(j+1) \end{bmatrix}
$$
$$
- \begin{bmatrix} H_{0,0}(j)\,H_{0,1}(j) \\ \vdots \\ H_{N-1,0}(j)\,H_{N-1,1}(j) \end{bmatrix} \begin{bmatrix} s_0(j) - s_1^*(j) \\ s_1(j)\,s_0^*(j) \end{bmatrix} \qquad (6)
$$

The data symbols of SF layer B are decoded from $\mathbf{Z}$ and the detection scheme is given by

$$
\vec{s}_2(j) = Q\left(\sum_{k=0}^{N-1} \left(H_{k,2}^*(j)Z_k(j) + H_{k,3}(j)Z_k^*(j+1) \right) \right)
$$
$$
\vec{s}_3(j) = Q\left(\sum_{k=0}^{N-1} \left(H_{k,3}^*(j)Z_k(j) - H_{k,2}(j)Z_k^*(j+1) \right) \right)
\qquad (7)
$$

4 Performance Evaluations

The simulation is performed based on the following design parameters as shown in Table 1. The performance evaluations are simulated using 4×4 MIMO system over Monte-Carlo time-variant channel with different maximum Doppler frequency. The performance of each layer of MLSF-OFDM is compared to MLSTBC-OFDM with FAST-QR. The channel impulse response (CIR) of L-path propagation environment used in our simulation is derived from the COST 207 rural area channel model [6].

Figure 3 shows the BER performance comparison for SF layer A with various schemes in environment of different maximum Doppler frequencies. With maximum Doppler frequency of 50 Hz (slow fading), it can be observed that the BER performance of SF layer A with E-FAST-QR is same as FAST-QR since E-FAST-QR does not enhance the performance of SF layer A. Meanwhile, the BER performance of SF layer A with FAST-QR and E-FAST-QR slightly outperforms SIC-ZF with highest complexity for signal-to-noise-ratio (SNR) less than 10.5 dB. This is because the ZF equalizer faces problems at lower SNR due to noise amplification [7]. Moreover, it can be seen that the BER of SF layer A with all considered schemes increases in environment with maximum Doppler frequency of 500 Hz (fast fading).

Figure 4 illustrates the BER performance comparison for SF layer B with various schemes in environment of different maximum Doppler frequencies. With

Table 1 Simulation parameters

Design parameters	Value
Carrier frequency, f_c	2.4 GHz
FFT size, J	512
Spectrum width of the system	10 MHz
Sampling duration	100 ns
Cyclic prefix	¼
Useful OFDM symbol time	51.2 μs
Guard time assuming 25 %	12.8 μs
OFDM symbol duration	64.0 μs
Modulation scheme	QPSK
Time domain channel response length, L	6 samples
Maximum Doppler frequency, v_D	50, 500 Hz
Special training sequence	Nguyen [Nguyen04]
Channel estimation	Least-square
Pilot distance in time domain	4

Fig. 3 BER performance comparison for layer A with various schemes

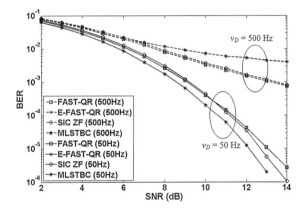

maximum Doppler frequency of 50 Hz (11.25 km/h), it can be observed that the BER performance of SF layer B with E-FAST-QR outperforms FAST-QR as E-FAST-QR enhances the performance of SF layer B. Meanwhile, the BER performance of SF layer B with E-FAST-QR slightly outperforms SIC-ZF with highest complexity for SNR less than 10.5 dB [8]. This is because the ZF equalizer faces problems at lower SNR du to noise amplification [7]. For SNR more than 10.5 dB, SIC-ZF performs better than E-FAST-QR. In contrast, the BER performance of SF layer B with E-FAST-QR slightly outperforms SIC-ZF for all SNR in environment with maximum Doppler frequency of 500 Hz (112.5 km/h) due to noise amplification of ZF equalizer.

Figure 5 plots the comparison of the complexity of FAST-QR and QR decomposition with the modified Gram-Schmidt (MGS) for the increasing number of receive antennas. It can be observed that the complexity of FAST-QR reduces approximately 48 % compared to the QR decomposition with MGS. For

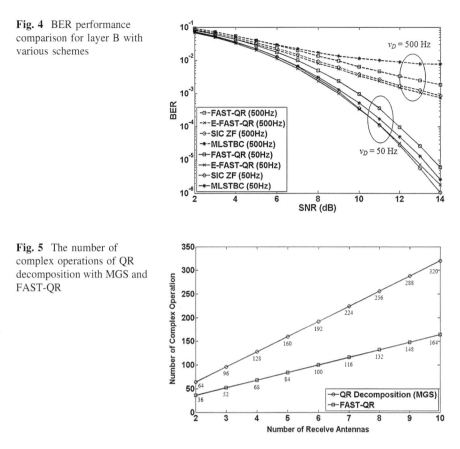

Fig. 4 BER performance comparison for layer B with various schemes

Fig. 5 The number of complex operations of QR decomposition with MGS and FAST-QR

FAST-QR, it is only necessary to calculate two columns of Q_j and two rows of R_j with irrespective of the number of receive antennas.

5 Conclusion

In this paper, a MLSF-OFDM scheme, which performs better than MLSTBC-OFDM scheme in MIMO-OFDM channel with higher Doppler frequency, has been introduced. Besides, two detection algorithms, denoted as FAST-QR and E-FAST-QR have been proposed. The implementation complexity of QR decomposition has been reduced by using the proposed FAST-QR detection algorithms. Both algorithms also show a good performance in terms of BER compared to SIC-ZF. Hence, the proposed detection algorithms can be used for MIMO-OFDM system to maintain guaranteed QoS and reduce complexity for real time services.

References

1. Telatar, E.: Capacity of multi-antenna Gaussian Channels. Eur. Trans. Telecommun. **10**, 585–595 (1999)
2. Foschini, G.J., Gans, M.J.: On limits of wireless communications in a fading environment when using multiple antennas. Wireless Pers. Commun. **6**, 311–335 (1998)
3. Lei, M., Harada, H.: Performance comparison of multi-layer STBC OFDM and V-BLAST OFDM, In: 64th IEEE Vehicular Technology Conference (VTC-2006 Fall), pp. 1–5 (2006)
4. Lee, K.F., Williams, D.B.: A space-frequency transmitter diversity technique for OFDM systems. IEEE GLOBECOM **3**, 1473–1477 (2000)
5. Nguyen, V.D., Patzold, M.: Least square channel estimation using special training sequences for MIMO-OFDM systems in the presence of inter-symbol interference, In: Nordic Radio Symposium (NRS), Oulu, Finland (2004)
6. COST 207: Digital Land Mobile Radio Communications—Final report, Office for Official Publications of the European Communities, Luxembourg (1989)
7. Barry, J.R., Messerschmitt, D.G., Lee, E.A.: Digital Communication, Kluwer, Boston (2004)
8. Yeo, S.Y., Baek, M.S., Song, H.K.: Improved and trustworthy detection scheme with low complexity in VBLAST system. In: Xiao, B., et al. (eds.) ATC 2007, LNCS, vol. 4610, pp. 269–275. Springer, Heidelberg (2007)

Dynamic Transmit Antenna Shuffling Scheme for MIMO Wireless Communication Systems

Jin Hui Chong, Chee Kyun Ng, Nor Kamariah Noordin
and Borhanuddin Mohd. Ali

Abstract In this paper, two novel dynamic transmit antenna shuffling schemes, namely 'Optimal' and 'Max STBC', are presented in order to reduce the interference in V-BLAST/STBC scheme. These antenna shuffling schemes with the channel state information (CSI) from the receiver, which significantly improves performance of the MIMO system in Rayleigh flat-fading channels by selecting the appropriate antenna shuffling pattern, have been compared with a few other related schemes in terms of BER and system capacity. The 'Optimal' antenna shuffling scheme improves the BER performance significantly with a gain of 2 dB at BER of 10^{-3} compared to all other schemes, while the 'Max STBC' antenna shuffling scheme enhances the V-BLAST/STBC system capacity by 4 %.

Keywords MIMO · V-BLAST · STBC · QR · ZF · CSI · Antenna shuffling scheme

J. H. Chong (✉) · C. K. Ng · N. K. Noordin · B. Mohd. Ali
Faculty of Engineering, Department of Computer and Communication Systems Engineering,
University Putra Malaysia, UPM Serdang 43400, Selangor, Malaysia
e-mail: chongjinhui@yahoo.com

C. K. Ng
e-mail: mpnck@eng.upm.edu.my

N. K. Noordin
e-mail: nknordin@eng.upm.edu.my

B. Mohd. Ali
e-mail: borhan@eng.upm.edu.my

S.-S. Yeo et al. (eds.), *Computer Science and its Applications*,
Lecture Notes in Electrical Engineering 203, DOI: 10.1007/978-94-007-5699-1_6,
© Springer Science+Business Media Dordrecht 2012

1 Introduction

The V-BLAST/STBC scheme, which was introduced in [1, 2], is a combination of the Alamouti's Space-Time Block Code (STBC) and Vertical Bell Laboratories Layered Space-Time (V-BLAST) schemes. It increases Multiple-Input Multiple-Output (MIMO) system capacity and maintains reliable bit error rate (BER) performance by achieving spatial multiplexing and diversity gains simultaneously. However, the V-BLAST layer in V-BLAST/STBC scheme is more susceptible to multipath fading and noise. The performance of V-BLAST layer significantly influences the overall V-BLAST/STBC scheme performance as the data rate of V-BLAST layer is higher than STBC layer. In [3], a hybrid MIMO transmission antenna shuffling scheme was proposed, which enhances the V-BLAST/STBC scheme performance by allocating the V-BLAST layer with the strongest sub-channels power. However, allocating V-BLAST layer with the strongest sub-channels power will increase the interference to STBC layer. The spatially-multiplexed V-BLAST and STBC layers in the V-BLAST/STBC scheme assume each other as interferer. The interference from V-BLAST layer is considered as noise to STBC layer, which is allocated with weakest power link sub-channels. Besides, interference between the transmitted symbols in V-BLAST layer will also impact the overall system performance.

In order to reduce the interference V-BLAST/STBC scheme, a dynamic transmit antenna shuffling scheme, which takes the effect of interference from all other sub-data streams into account, is considered. A dynamic transmit antenna shuffling scheme with little feedback information from the receiver has been introduced by [4–6]. This antenna shuffling scheme significantly improves performance of the MIMO system in Rayleigh flat-fading channels by selecting the appropriate antenna shuffling pattern. In this paper, two novel techniques of dynamic antenna shuffling scheme namely 'Max STBC' and 'Optimal', which is applicable to V-BLAST/STBC scheme, are presented. Smart channel allocations to appropriate transmit antennas for each encoded STBC or V-BLAST layer data stream is made based on the channel state information (CSI) from the receiver. The proposed antenna shuffling schemes, will give higher MIMO system capacity and lower data loss. The 'Optimal' antenna shuffling scheme with low complexity feedback requirements improves the BER performance significantly with a gain of 2 dB at BER of 10^{-3} compared to the V-BLAST/STBC transceiver scheme without the antenna shuffling capability. Besides, the 'Max STBC' antenna shuffling scheme increases the system capacity of V-BLAST/STBC transceiver scheme by 4 %. Therefore, the proposed dynamic transmit antenna shuffling schemes improve the MIMO transmit diversity and system capacity with little feedback information.

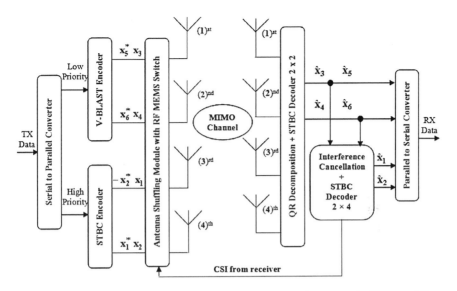

Fig. 1 V-BLAST/STBC scheme with proposed dynamic transmit antenna shuffling capability

2 Proposed Dynamic Transmit Antenna Shuffling Schemes

In this section, two novel techniques of dynamic antenna shuffling schemes, STBC layer protection based dynamic transmit antenna shuffling scheme ('Max STBC') and optimal dynamic transmit antenna shuffling scheme ('Optimal') are presented. The block diagram of the dynamic transmit antenna shuffling capable V-BLAST/STBC transceiver model with four transmit and four receive antennas is shown in Fig. 1. The received matrix of V-BLAST/STBC 4×4 scheme without dynamic transmit antenna shuffling scheme over two consecutive symbol periods can be expressed as

$$y = Hx + z \tag{1}$$

where y is the 4×2 received signal matrix and $\mathbf{x}$ is the V-BLAST/STBC 4×2 transmission matrix. $\mathbf{H}$ is the 4×4 Rayleigh flat-fading channel gain matrix, which is assumed to be constant across two consecutive symbol transmission periods. The entries of $\mathbf{H}$ are circularly symmetric, independent and identically distributed (i.i.d.) Gaussian random variables with zero-mean and unit variance. $\mathbf{Z}$ is the additive white Gaussian noise (AWGN) 4×2 complex matrix with unit variance, σ^2 and zero mean. After receiving the CSI from receiver, the transmitter will select an appropriate antenna shuffling pattern according to the algorithm of dynamic transmit antenna shuffling schemes. The low complexity decoding algorithm is used in V-BLAST/STBC scheme.

2.1 STBC Layer Protection Based Dynamic Transmit Antenna Shuffling Scheme ('Max STBC')

In the 'Max STBC' dynamic transmit antenna shuffling scheme, the receiver estimates all the CSI and obtains the power from each sub-channel. After that, the order of the sub-channels power is fed back to the transmitter. It is not important that the magnitude of sub-channels power varies, as long as the order of the sub-channels power does not change, so the receiver does not need to feed back the transmitter in every time slot.

The transmitter allocates channels with stronger transmit links power to the STBC layer based on the power order of transmit links. Then, extra gain is provided to the data symbols of STBC layer. In other word, the 'Max STBC' antenna shuffling scheme significantly reduces BER of high priority data compared to V-BLAST/STBC scheme without dynamic transmit antenna shuffling scheme.

2.2 Optimal Dynamic Transmit Antenna Shuffling Scheme ('Optimal')

The 'Optimal' dynamic transmit antenna shuffling scheme takes the effect of interference from all other sub-data streams into considerations. This is an important factor for the overall performance enhancement since interference caused by other sub-data streams is considered as noise. The detected data symbol, which experiences less interference from other data symbols, achieves a higher post-detection signal-to-noise-ratio (SNR) [7]. Post-detection SNR is the SNR at the output of the signal processing circuitry since it is measured after the detection circuitry.

As there is no redundant information in V-BLAST layer, V-BLAST layer is more susceptible to multi-path fading and noise. Therefore, in the 'Optimal' antenna shuffling scheme, the transmitter allocates the sub-channels to V-BLAST layer, which carries more data than STBC layer, in such a way that the post-detection SNR value of detected data symbols of V-BLAST layer is maximized to achieve lower BER. This is done to minimize the interference from STBC layer as higher post-detection SNR reduces probability of decoding error in V-BLAST layer.

The receiver estimates all the CSI and obtains the post-detection SNR value from each detected data symbol. After that, the order of the post-detection SNR value is fed back to the transmitter. It is not important that the channel coefficient varies, as long as the order of the post-detection SNR value does not change. Therefore, the receiver does not need to send feedback information to transmitter in every time slot.

3 Performance of Proposed Dynamic Transmit Antenna Shuffling Schemes

Simulations have been performed for 4×4 MIMO system to evaluate the performance of the proposed dynamic transmit antenna shuffling schemes in V-BLAST/STBC scheme in terms of BER by using 4-ary quadrature amplitude modulation (4-QAM) constellation. The performance of the proposed antenna shuffling schemes is compared with the scheme proposed in [3], low complexity QR and zero forcing (ZF) schemes over Rayleigh flat-fading channel, where both low complexity QR and ZF are noted as no antenna shuffling capacity schemes [8].

Figure 2 shows the BER performance comparison for V-BLAST layer with various dynamic transmit antenna shuffling schemes. It can be observed that the performance of V-BLAST layer with low complexity QR without antenna shuffling capability scheme outperforms ZF scheme by more than 2 dB at BER of 10^{-3}. This is due to the interference cancellation activity after decoding the candidate of STBC layer by the low complexity QR algorithm, while there is no interference cancellation for ZF algorithm. However, the performance of V-BLAST layer with the 'Max STBC' antenna shuffling scheme is the worst and it degrades the system gain by about 1 dB at BER of 10^{-3} compared to the low complexity QR scheme. This is due to the allocation of stronger link power sub-channels for the data symbols of STBC layer only, and consequently less protection is provided to the data symbols of V-BLAST layer.

On the other hand, it can also be seen that the BER performance of V-BLAST layer with the 'Optimal' dynamic transmit antenna shuffling scheme is the best among the antenna shuffling schemes, with a gain of about 1 dB at BER of 10^{-4} compared to Freitas's scheme as proposed in [3]. This happened since interference is not considered in Freitas's scheme [3], whereas in 'Optimal' antenna shuffling scheme, the post-detection SNR of V-BLAST layer in detector is maximized to reduce the probability of decoding error in V-BLAST layer. Consequently, the performance of V-BLAST layer with 'Optimal' antenna shuffling scheme is enhanced.

Figure 3 depicts the comparison of BER performance for STBC layer with various dynamic transmit antenna shuffling schemes. It can be observed that the performance of STBC layer with the low complexity QR scheme without antenna shuffling capability outperforms ZF scheme with a gain of about 1 dB at BER of 10^{-4}. This is because the low complexity QR scheme performs an interference cancellation by subtracting the data symbols of V-BLAST layer from the original received matrix, while there is no interference cancellation for ZF algorithm.

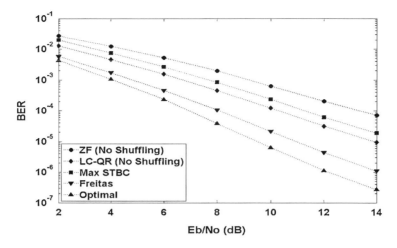

Fig. 2 Comparison of BER performance for V-BLAST layer with different dynamic transmit antenna shuffling scheme

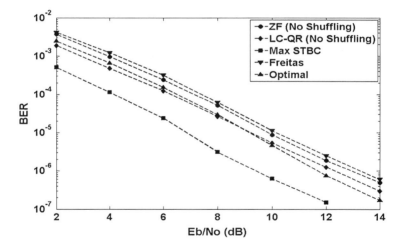

Fig. 3 Comparison of BER performance for STBC layer with different dynamic transmit antenna shuffling scheme

On the other hand, the performance of STBC layer with 'Max STBC' dynamic transmit antenna shuffling scheme is the best among the antenna shuffling schemes. It achieves more than 3 dB gains compared to Freitas's scheme at BER of 10^{-4}. This is due to the provided extra protection to the data symbols of STBC layer by allocating stronger sub-channels to STBC layer.

Figure 4 illustrates the comparison of average BER performance for various dynamic transmit antenna shuffling schemes. It can be seen that average BER performance of 'Optimal' antenna shuffling scheme is the best, and it achieves performance improvement with a gain of more than 0.5 dB at BER of 10^{-3}

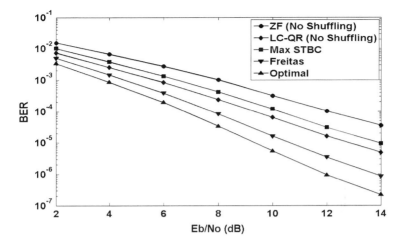

Fig. 4 Comparison of average BER performance for different dynamic transmit antenna shuffling scheme

compared to Freitas's scheme. However, the complexity of the 'Optimal' antenna shuffling scheme is much higher than the other schemes, so there is a trade-off between BER performance and complexity.

In contrast, the average BER performance of 'Max STBC' dynamic transmit antenna shuffling scheme is the lowest, and it degrades approximately 1 dB compared to low complexity QR scheme without antenna shuffling capability at BER of 10^{-3}. This is because of the degradation of BER performance of V-BLAST layer, which carries more data symbols than STBC layer over two consecutive symbol period. Hence, it leads to the decrease of overall system error performance of 'Max STBC' antenna shuffling scheme.

For the 10 and 1 % outage capacity comparison for theoretical maximum limit in MIMO system, various schemes with and without dynamic transmit antenna shuffling capability are shown in Fig. 5. It can be observed that the capacity of 'Max STBC' antenna shuffling scheme is higher than low complexity QR without the antenna shuffling scheme in 1 % outage capacity although the overall BER is higher as shown in Fig. 4.

In contrast, the 'Optimal' antenna shuffling scheme with the lowest BER suffers from the lowest system capacity among the dynamic transmit antenna shuffling schemes. The system capacity of 'Optimal' antenna shuffling scheme degrades marginally compared to low complexity QR without the antenna shuffling capability. Hence, it is a trade-off between BER performance and system capacity for 'Optimal' and 'Max STBC' antenna shuffling schemes.

Figure 6 illustrates the capacity of cumulative distribution function (CDF) in MIMO system for various schemes with and without dynamic transmit antenna shuffling capability at SNR of 20 dB. The results show that the capacity of 'Max STBC' antenna shuffling scheme is higher than low complexity QR without the antenna shuffling scheme in outage probability less than 10 %. Hence, the 'Max

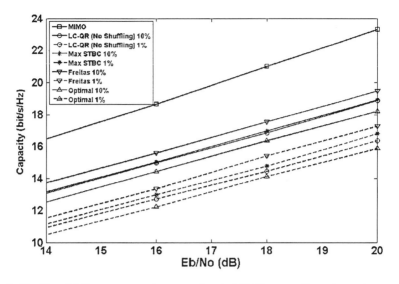

Fig. 5 The 10 and 1 % outage capacity comparison in MIMO system for various schemes with and without dynamic transmit antenna shuffling capability

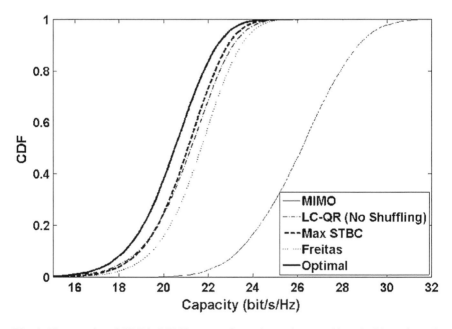

Fig. 6 The capacity of CDF in MIMO system for various schemes with and without dynamic transmit antenna shuffling capability at SNR of 20 dB

STBC' antenna shuffling scheme increases the system capacity of V-BLAST/ STBC transceiver scheme by 4 %. In contrast, the capacity of 'Optimal' antenna shuffling scheme is the lowest among the considered schemes.

4 Conclusion

In this paper, two novel dynamic transmit antenna shuffling schemes have been introduced for V-BLAST/STBC scheme. These antenna shuffling schemes, namely 'Optimal' and 'Max STBC', have been compared with a few other related schemes in terms of BER and system capacity. The 'Optimal' antenna shuffling scheme shows the best performance in terms of BER compared to all other schemes, while the 'Max STBC' antenna shuffling scheme enhances the V-BLAST/STBC system capacity. Therefore the 'Optimal' antenna shuffling scheme can be used for systems with QoS requirements when the system capacity or complexity is not the main issue. In contrast, the 'Max STBC' antenna shuffling scheme can be used for systems that need high capacity to accommodate the growing demand for real high bandwidth network with less stringent QoS requirement.

References

1. Mao, T., Motani, M.: STBC-VBLAST for MIMO wireless communication systems, IEEE. Int. Conf. Commun. (ICC 2005) **4**, 2266–2270 (2005)
2. Longoria-Gandara, O., Sanchez-Hernandez, A., Cortez, J., Bazdresch, M., Parra-Michel, R.: Linear Dispersion codes generation from hybrid STBC-VBLAST architectures. In: 4th International Conference Electrical and Electronics Engineering (ICEEE 2007), pp 142–145 (2007)
3. Freitas, W.C., Cavalcanti, F.R.P., Lopes, R.R.: Hybrid MIMO Transceiver scheme with antenna allocation and partial CSI at transmitter side. In: IEEE 17th International Symposium on Personal, Indoor and Mobile Radio Communications (PIMRC 2006), pp. 1–5 (2006)
4. Shim, S., Kim, K., Lee, C.: An efficient antenna shuffling scheme for a DSTTD system. IEEE Commun. Lett. **9**, 124–126 (2005)
5. Zhou, L., Shimizu, M.: A novel condition number-based antenna shuffling scheme for D-STTD OFDM system, In: IEEE 69th Vehicular Technology Conference (VTC Spring 2009), pp. 1–5 (2009)
6. Lee, H., Powers, E.J.: Low-complexity mutual information-based antenna grouping scheme for a D-STTD system. In: Proceeding of IEEE GLOBECOM 2006 (2006)
7. Bellamy, J.C.: Digital Telephony. Wiley, New York (2000)
8. Sandeep, G., Ravi-Teja, C., Kalyana-Krishnan, G., Reddy, V.U.: Low complexity decoders for combined space time block coding and V-BLAST. In: Wireless Communications and Networking Conference (WCNC 2007), pp. 582–587 (2007)

A Low Complexity V-BLAST/STBC Detection Algorithm for MIMO System

Jin Hui Chong, Chee Kyun Ng, Nor Kamariah Noordin and Borhanuddin Mohd. Ali

Abstract In this paper, a new low complexity detection algorithm for V-BLAST/STBC scheme based on QR decomposition, denoted as low complexity QR (LC-QR), is presented. The performance of the proposed LC-QR decomposition detection algorithm in V-BLAST/STBC transceiver scheme is investigated with other MIMO systems, such as ZF, MMSE and QR decomposition schemes. It is shown that the BER performance in V-BLAST/STBC scheme is better than V-BLAST scheme while its system capacity is higher than orthogonal STBC scheme when the LC-QR is exploited.

Keywords MIMO · V-BLAST · STBC · MMSE · ZF · QR

1 Introduction

The V-BLAST/STBC scheme, which was introduced in [1] is a combination of the Alamouti's Space–Time Block Code (STBC) and Vertical Bell Laboratories Layered Space–Time (V-BLAST) schemes. A number of research efforts on

J. H. Chong (✉) · C. K. Ng · N. K. Noordin · B. Mohd. Ali
Faculty of Engineering, Department of Computer and Communication Systems Engineering,
University Putra Malaysia, UPM Serdang 43400 Selangor, Malaysia
e-mail: chongjinhui@yahoo.com

C. K. Ng
e-mail: mpnck@eng.upm.edu.my

N. K. Noordin
e-mail: nknoordin@eng.upm.edu.my

B. Mohd. Ali
e-mail: borhan@eng.upm.edu.my

S.-S. Yeo et al. (eds.), *Computer Science and its Applications*,
Lecture Notes in Electrical Engineering 203, DOI: 10.1007/978-94-007-5699-1_7,
© Springer Science+Business Media Dordrecht 2012

V-BLAST/STBC scheme have been carried for Multiple-Input Multiple-Output (MIMO) antenna system with the goal of maximizing the system capacity and reducing its bit error rate (BER). The V-BLAST/STBC scheme improves the performance of MIMO by combining spatial multiplexing and diversity technique together [2]. However, the spatially-multiplexed V-BLAST and STBC layers in the V-BLAST/STBC scheme assume each other as interferer. Therefore, the transmitted symbols must be decoded with well-known detection techniques such as zero-forcing (ZF), minimum mean-squared error (MMSE) and QR decomposition which are employed in V-BLAST scheme [3]. The QR decomposition of $A \times B$ channel matrix $\mathbf{H}$ is a factorization $\mathbf{H} = \mathbf{QR}$, where $\mathbf{Q}$ is $A \times B$ unitary matrix and $\mathbf{R}$ is $B \times B$ upper triangular matrix. The computational implementation of QR decomposition is less than ZF and MMSE [4], thus the complexity of V-BLAST/STBC scheme can be reduced using QR decomposition.

In this paper, a new detection algorithm based on QR decomposition, denoted as low complexity QR (LC-QR), is presented. The computational complexity (total number of arithmetic operations) of the proposed LC-QR algorithm will be significantly lower than the conventional ZF, MMSE and QR decomposition detection algorithms. The performance of V-BLAST/STBC transceiver scheme with proposed LC-QR algorithm is compared with other MIMO systems, such as V-BLAST and orthogonal space–time block codes. The BER performance of V-BLAST/STBC scheme is better than V-BLAST scheme while the system capacity of V-BLAST/STBC scheme is higher than STBC scheme when the LC-QR is exploited.

2 Proposed Low Complexity V-BLAST/STBC Detection Algorithm with QR Decomposition

The block diagram of a V-BLAST/STBC transceiver model with M ($M \geq 3$) transmit and N ($N \geq M - 1$) receive antennas is shown in Fig. 1. A single main data stream is de-multiplexed into two sub-data streams according to the data priority. The high priority data is assigned to STBC layer for extra gain while low priority data is sent to V-BLAST layer with higher capacity. The proposed LC-QR detection algorithm with M transmit and N receive antennas is discussed in detail in this section. Instead of decoding the transmitted symbols with the equivalent channel matrix H with dimension $2N \times 2(M - 1)$, the original channel matrix $\mathbf{H}$ with dimension $N \times M$ is utilized. If $N < M - 1$, the LC-QR algorithm is invalid and not applicable. The flow chart of LC-QR algorithm is shown in Fig. 2.

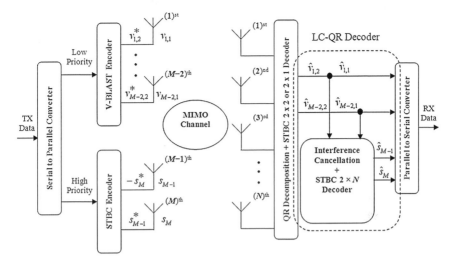

Fig. 1 V-BLAST/STBC scheme consisting of a transmitter and a QR decomposition receiver with proposed algorithm

Fig. 2 Flow chart of the LC-QR algorithm

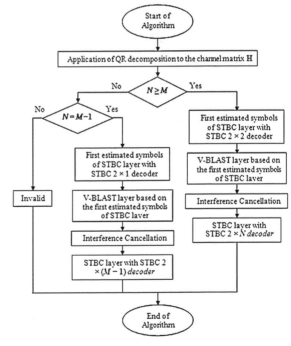

3 Comparison of the Complexity of LC-QR with ZF, MMSE and QR Decomposition Algorithms

The complexity of ZF, MMSE, QR decomposition with H and LC-QR with $\mathbf{H}$ is analyzed here. It is assumed that the channel matrix $\mathbf{H}$ with dimension $N \times M$ is used to analyze the complexity of the LC-QR. The equivalent channel matrix H with dimension $2N \times 2(M - 1)$ is used to analyze the complexity of ZF, MMSE and QR decomposition. It is observed that there are zeros in channel matrix H, therefore the multiplication and addition with zero are not taken into account in ZF and MMSE complexity calculation. The comparison of number of complex arithmetic operation and reduction of complexity for ZF, MMSE, QR decomposition with H and LC-QR detection algorithm with $\mathbf{H}$ is shown in Tables 1 and 2 respectively. From Table 2, it can be observed that the proposed LC-QR detection algorithm significantly reduces the arithmetic complexity compared to ZF, MMSE and QR decomposition algorithm.

4 Performance of Proposed Low Complexity QR Decomposition

Simulations have been performed for MIMO system using MATLAB to evaluate the performance of the V-BLAST/STBC scheme for ZF, MMSE and QR decomposition with H and LC-QR with $\mathbf{H}$ in terms of BER. The 4-ary quadrature amplitude modulation (4-QAM) constellation is used in these simulations with Rayleigh flat-fading channel for Figs. 3 and 5 as well as Naftali channel [5] for the rest of the figures. The entries of Rayleigh flat-fading channel matrix are circularly symmetric, i.i.d. Gaussian random variables with zero-mean and unit variance.

Figure 3 shows the BER performance comparison of ZF V-BLAST 4×4, O-STBC 4×4 with rate 3/4 and V-BLAST/STBC 4×4 with LC-QR in Rayleigh flat-fading channel environment, which is constant across four consecutive symbol transmission periods. It could be seen that BER performance of O-STBC 4×4 with symbol rate 3/4 is the best among the considered schemes as it is a pure spatial diversity scheme with full diversity gain. Moreover, O-STBC does not suffer from inter-symbol interference (ISI) as the transmitted symbols are orthogonal to one another. In contrast, ZF V-BLAST 4×4 with symbol rate four is the worst among the schemes because it is a pure spatial multiplexing scheme which suffers from poor diversity gain. Besides, interference between transmitted symbols in ZF V-BLAST scheme greatly reduces the BER performance. It can be seen that the V-BLAST/STBC 4×4 with symbol rate three shows a compromise of BER performance with respect to pure spatial multiplexing or diversity scheme.

Figure 4 depicts the 10 and 1 % outage capacity for basic MIMO (theoretical limit), ZF V-BLAST, O-STBC and V-BLAST/STBC with LC-QR in 4×4 system. The spectral efficiency of ZF V-BLAST changes a lot with different outage

Table 1 Comparison of number of complex arithmetic operation for ZF, MMSE, QR decomposition and LC-QR detection algorithm

Complex arithmetic operation	ZF with H	MMSE with H	QR decomposition with H	LC-QR with $\mathbf{H}$
Addition	$4(4\,N-1) \times (M-1)^2 - 2\,(M-1)$	$4(4\,N-1) \times (M-1)^2$	$2(2\,N-1) \times (M-1)$ $+ (M-2) \times (2\,M-3)$	$N(3\,M-2) + (M-3)$ $\times (M-1)$
Subtraction	–	–	$2\,M-3$	$2(M-2) + 2\,N + 1$
Multiplication	$4\,N(M-1) \times (4\,M-3)$	$4\,N(M-1)$ $\times (4\,M-3)$	$(M-1) \times (4\,N + 2\,M-3)$	$M^2 + 3MN-M + 4$
Division	–	–	$2(M-1)$	$2(M-2)$
Householder reflection	–	–	$8\,N(M-1)^2 - (8/3)(M-1)^3$	$NM^2 - M^3/3$
Gaussian elimination	$8(M-1)^3$	$8(M-1)^3$	–	–
Total	$8(M-1)^3 + 4(8\,N-1)$ $(M-1)^2 + 2(2\,N-1)$ $(M-1)$	$8(M-1)^3$ $8(M-1)^3 + 4$ $(8\,N-1)$ $(M-1)^2 + 4\,N$ $(M-1)$	$8\,N(M-1)^2 - (8/3)(M-1)^3$ $+ (2\,M-3)^2 + 2$ $(M-1)(4\,N + 1)-1$	$NM^2 - M^3/3 + (M + 2)$ $(2\,N-1) + 2(M-2)$ $(M + 2) + 4\,N$ $(M-1) + 10$

Table 2 Reduction of complexity for ZF, MMSE and QR decomposition compared to LC-QR detection algorithm

Complex arithmetic operation	Reduction of complexity compared to ZF and MMSE with H (%)				Reduction of complexity compared to QR decomposition with **H** (%)			
	$M = 4$		$M = 5$		$M = 4$		$M = 5$	
	$N = 4$	$N = 5$	$N = 5$	$N = 6$	$N = 4$	$N = 5$	$N = 5$	$N = 6$
Addition	93.2	92.2	94.0	94.1	17.3	17.2	21.5	21.1
Subtraction	–	–	–	–	–	–	–	–
Multiplication	89.7	90.3	92.7	93.0	–	–	8.3	8.1
Division	–	–	–	–	33.3	33.3	25.0	25.0
Householder reflection	–	–	–	–	80.6	79.9	82.1	81.8
Gaussian elimination	–	–	–	–	–	–	–	–
Total	87.9	87.7	90.9	90.7	51.5	53.0	59.3	60.5

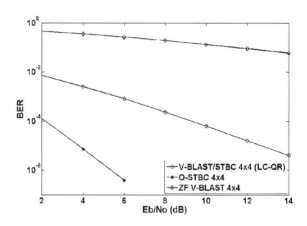

Fig. 3 BER performance of various 4 × 4 MIMO schemes

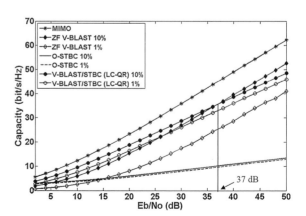

Fig. 4 The 10 and 1 % outage capacity comparison for various 4 × 4 MIMO schemes

Fig. 5 BER performance of various algorithms in V-BLAST/STBC scheme with Rayleigh flat-fading channel model

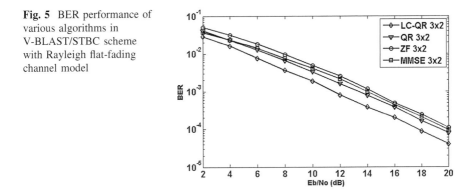

probability. For instance, ZF V-BLAST requires 8 dB to maintain the capacity of 15 bps/Hz when it proceeds from 10 to 1 % outage probability. This is caused by lack of diversity of ZF V-BLAST. In contrast, the V-BLAST/STBC with LC-QR just requires 3 dB to maintain at the capacity of 15 bps/Hz. Last but not least, the O-STBC is the most stable one, as the curve of 10 % outage capacity is very close to the curve of 1 % outage capacity. Besides, it can be observed that the capacity of ZF V-BLAST with 10 % outage probability is the highest among the considered schemes for SNR higher than 37 dB.

It can be concluded that Figs. 3 and 4 present the tradeoffs among ZF V-BLAST, O-STBC and V-BLAST/STBC with LC-QR in 4×4 system. The O-STBC achieves the best BER performance but the system capacity is the lowest among the schemes. The system capacity of ZF V-BLAST with 10 % outage capacity is the highest for SNR above 37 dB but the BER performance is the worst among the schemes. In contrast, the system capacity of V-BLAST/STBC with LC-QR is close to MIMO and better than ZF V-BLAST for SNR below 37 dB. Moreover, the BER performance of V-BLAST/STBC with LC-QR is significantly better than ZF V-BLAST as V-BLAST/STBC achieves spatial multiplexing and diversity gain simultaneously.

Figure 5 shows the BER performance of various algorithms in V-BLAST/STBC scheme with Rayleigh flat-fading channel model, which is constant across two consecutive symbol transmission periods. The LC-QR 3×2 with **H** outperforms ZF 3×2 and MMSE 3×2 with H by more than 2 dB gain at BER of 10^{-3} [6]. At the same time, the LC-QR 3×2 with **H** outperforms QR 3×2 with H by approximately 1.5 dB gain at BER of 10^{-3}. This is because the estimated candidate of STBC layer, which is more robust than V-BLAST layer, is decoded first. After decoding the estimated candidate of STBC layer, interference cancellation activity is performed to produce a new modified received signal matrix with less interference. For ZF and MMSE algorithm, there is no interference cancellation activity. It is clear that V-BLAST layer performance dominates over final decision of STBC layer. With increasing SNR, the probability of error in decoding the V-BLAST layer is reduced, the probability of correct decoding is increased at the STBC layer. As the V-BLAST layer transmits four data symbols while STBC layer

Fig. 6 BER performance of various algorithms in V-BLAST/STBC scheme under different maximum delay spread environment

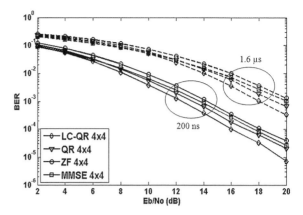

transmits two data symbols over two consecutive symbol transmission periods, thus a better BER performance of V-BLAST layer with LC-QR leads to overall V-BLAST/STBC system performance improvement.

Figure 6 illustrates the BER performance comparison of various algorithms in V-BLAST/STBC scheme with Naftali channel model under different maximum delay spread environment. The LC-QR outperforms ZF and MMSE by approximately 2 dB gain at BER of 10^{-3} for both indoor (200 ns) and outdoor (1.6 μs) environment. At the same time, the LC-QR 4 × 4 outperforms QR 4 × 4 by approximately 1 dB gain at BER of 10^{-3} for both indoor and outdoor environment. In order to maintain the BER of 10^{-3}, the LC-QR with maximum delay spread = 1.6 μs requires 6 dB gain compared to indoor environment with maximum delay spread = 200 ns.

5 Conclusions

In this paper, it is illustrated that V-BLAST/STBC scheme, which achieves spatial multiplexing and diversity gain simultaneously, increases system capacity to accommodate the ever growing demand for real time system with tolerably lower QoS. It is also shown that the system capacity of V-BLAST/STBC scheme is close to MIMO and better than ZF V-BLAST for SNR below 37 dB. Moreover, a new V-BLAST/STBC detection algorithm that utilizes LC-QR decomposition has been introduced. The LC-QR significantly reduces the arithmetic operation complexity and remains a satisfactory BER performance compared to ZF, MMSE, QR decomposition, SIC ZF, SIC MMSE and sorted QR decomposition.

References

1. Mao, T., Motani, M.: STBC-VBLAST for MIMO wireless communication systems. In: IEEE International Conference on Communications (ICC 2005), vol. 4, pp. 2266–2270 (2005)
2. Longoria-Gandara, O., Sanchez-Hernandez, A., Cortez, J., Bazdresch, M., Parra-Michel, R.: Linear dispersion codes generation from hybrid STBC-VBLAST architectures. In: 4th International Conference Electrical and Electronics Engineering (ICEEE 2007), pp 142–145 (2007)
3. Sandeep, G., Ravi-Teja, C., Kalyana-Krishnan, G., Reddy, V.U.: Low Complexity Decoders for Combined Space Time Block Coding and V-BLAST, Wireless Communications and Networking Conference (WCNC 2007), pp. 582–587 (2007)
4. Wai, W.K., Tsui, C.Y., Cheng, R.S.: A low complexity architecture of the V-BLAST system. In: Proceeding of IEEE Wireless Communications and Networking Conference, vol. 1, pp. 310–314 (2000)
5. Noordin, N.K., Ali, B.M., Jamuar, S.S., Ismail, M.: Space-time and space-frequency OFDM with convolutional precoding over fading channels. In: IEEE Region 10 Conference (TENCON 2008), pp. 1–6 (2008)
6. Liu, L., Wang, Y.Z.: Spatially selective STBC-VBLAST in MIMO communication system. In: International Conference on Communications, Circuits and Systems (ICCCAS 2008), pp. 195–199 (2008)

A Grid-Based Cloaking Scheme
for Continuous Location-Based Services
in Distributed Systems

Hyeong-Il Kim and Jae-Woo Chang

Abstract Recently, many people are using location-based services (LBSs). However, since users continuously request LBS queries to LBS server by sending their exact location data, their privacy information is in danger. Therefore, cloaking schemes have been proposed to protect the user's location privacy. However, because the existing techniques generate a cloaking area in a centralized manner, an anonymizer suffers from performance degradation and security problem. Therefore, a cloaking scheme based on distributed system is required for the safe and comfortable use of LBSs by mobile users. In this paper, we propose a grid-based cloaking scheme for continuous location-based services in distributed systems. The proposed scheme stores information and performs operations in a distributed manner to create a cloaking area. Finally, we show from a performance analysis that our cloaking scheme shows better performance than the existing centralized cloaking schemes.

Keywords Cloaking scheme · Privacy protection · Continuous LBS

1 Introduction

Recently, with the development of wireless communication technologies and the wide spread of smart phones, many applications based on users' location information are in the spotlight. Such applications providing additional services by

H.-I. Kim · J.-W. Chang (✉)
Department of Computer Engineering, Chonbuk National University,
Chonju, Chonbuk, South Korea
e-mail: jwchang@chonbuk.ac.kr

H.-I. Kim
e-mail: melipion@chonbuk.ac.kr

S.-S. Yeo et al. (eds.), *Computer Science and its Applications*,
Lecture Notes in Electrical Engineering 203, DOI: 10.1007/978-94-007-5699-1_8,
© Springer Science+Business Media Dordrecht 2012

using the users' location information are called LBS (Location-based Service). When a user wants to enjoy LBS such as finding nearest POI (Point of Interest), the user sends a request to LBS server with his/her location information which is received from a mobile device equipped with GPS. However, in this case, the user's private information is in danger as the exact location of the user is sent to LBS server. If an adversary or malicious LBS provider gets this information, they can know the user's private information like lifestyle, disease and religion. Therefore, a method for users' privacy protection is required for the safe and comfortable use of LBS by mobile users.

Several cloaking schemes [1, 2] have been proposed to protect the user's location privacy. These schemes generate and send a cloaking area satisfying the K-anonymity property to LBS server instead of sending the user's exact location. K-anonymity property is simply defined as a mechanism when a cloaking area includes not only the user who requests the LBS query but also the $k - 1$ other users around him/her. By using the cloaking area, the location exposure probability of a user can be reduced by $1/k$. This is because an adversary cannot distinguish the query issuer among other users inside the cloaking area. But the general cloaking scheme cannot support a user who requests LBS requests continuously. The reason is that only the query issuer is inside the cloaking area in every service time while other users can be changed in each service time. In this case, the adversary may find out the query issuer by comparing the user list of each service time inside the cloaking area.

There are two main schemes considering this problem. Advanced K-Anonymity Area (Advanced KAA) was proposed in [3] whereas Grid-based Continuous Cloaking (GCC) was presented in [4]. To support the continuous LBS, both schemes create a cloaking area which can protect the user's privacy by calculating the anonymity level of a generated cloaking area by using Entropy [5]. However both schemes suffer from following problems as they make use of the centralized system. First, because all communications with mobile users and computations are performed by an anonymizer, there occurs performance degradation due to the bottleneck on the anonymizer side. Second, because all location information of users is stored in the anonymizer, there can be severe security threats when the anonymizer is attacked by an adversary.

To solve this problem, we propose a grid-based cloaking scheme for continuous location-based services in distributed systems. The proposed scheme stores information and performs operations in distributed manner to create a cloaking area. In addition, the proposed scheme communicates hierarchically when the query issuer aggregates the information to prevent the query issuer's bottleneck.

The rest of the paper is organized as follows. Section 2 highlights the related works. The proposed cloaking scheme is described in Sect. 3. Extensive experimental evaluation is presented in Sect. 4. Finally, Sect. 5 concludes the paper.

2 Related Work

There exist some schemes that create a cloaking area considering K-anonymity to protect the query issuer's privacy in centralized manner. In Gruteser et al. [1], proposed a cloaking scheme based on Quad-tree for the first time. The technique proposed in [2] by Mokbel et al. creates a cloaking area by using the grid-based pyramid data structure. However, cloaking schemes in centralized system suffer from performance degradation due to the bottleneck and security problems of an anonymizer.

To solve this problem, there exist some techniques in which the query issuer creates a cloaking area by communicating with other mobile users without an anonymizer. Chow et al. proposed the cloaking scheme in [6] in which a query issuer finds k − 1 other mobile users and creates a cloaking area. The scheme proposed in [7] by Ghinita et al. converts a user's location into 1-dimensional coordinates by using Hilbert curve. Then, it makes up the Chord, the distributed hash table, and creates a cloaking area. However these cloaking schemes have a shortcoming that they cannot guarantee a user's privacy protection when the user requests queries continuously. The reason is that only the query issuer is located inside the cloaking area in every service time while other users can be changed each service time.

To tackle this problem, Xu et al. proposed Advanced KAA [3] which calculates the anonymity level by using Entropy [5]. Given a query issuer N, a cloaking area A for N, a set of users inside the A, $S(A) = \{N_1, N_2, ..., N_m\}$, and $p_i (1 \leq i \leq m)$, the probability of N_i being N, the entropy of cloaking area A is calculated by using (1).

$$H(A) = - \sum_{i=1}^{m} p_i \log p_i \qquad (1)$$

To calculate each user's probability p_i, it first generates the transition matrix M, where each element is the number of samples for each participant of time T_{i-1} being each user of T_i. Then the P_i at T_i is calculated by the product of the P_{i-1} and M. The anonymity level of the cloaking area is computed by using (2).

$$D(A) = 2^{H(A)} \qquad (2)$$

The cloaking area creation algorithm of Advanced KAA is as follow. When a user requests a query, an anonymizer finds the minimum circle which includes k−1 nearby other users of the query issuer. After that, the anonymizer doubles the found circle and sets it as C_{bound}. Next, it finds the minimum circle satisfying the requested K-anonymity, C_{min}, inside the C_{bound} and sets C_{min} as a cloaking area A_0. To generate a cloaking area A_i (i > 0), the anonymizer gets the updated location of the users included in A_{i-1} and sets the C_{bound} containing those users at T_i. Then, it finds the minimum circle satisfying the requested K-anonymity inside the C_{bound} and sets it as a cloaking area A_i. However, Advanced KAA requires much time to create a cloaking area because it considers all circles that can be composed inside the C_{bound} to find the minimum circle.

GCC proposed in [4] by Lee et al. solved the shortcomings of Advanced KAA. GCC creates the cloaking area based on the Gird-based cell expansion to reduce the cloaking area creation time. We briefly explain the cloaking area creation algorithm of GCC as follows. When a user requests a query, an anonymizer expands a rectangular bounding area, C_{bound}, towards 4 directions until it covers at least $k-1$ other nearby users of the query issuer. Next, it sets the cloaking area A_0 satisfying the requested K-anonymity by reducing the C_{bound} on each side. To generate a cloaking area A_i ($i > 0$), the anonymizer gets the updated location of the users included in A_{i-1} and it sets the C_{bound} which contains those users at T_i. Then the anonymizer finds the cloaking area A_i satisfying the requested K-anonymity by expanding an area from the query issuer. However, because both [3] and [4] operate on centralized system with an anonymizer, they suffer from performance degradation and security problems of an anonymizer.

3 Grid-Based Cloaking Scheme for Continuous Queries

In this section, we first present motivations of our research then describe the system architecture. After that, we explain our proposed method DGCC (Distributed Grid-based Continuous Cloaking).

3.1 Motivations

Existing cloaking schemes supporting continuous LBS operate in a centralized system. The centralized system has some drawbacks. First, an anonymizer suffers from performance degradation due to the bottleneck. This is because the anonymizer performs every process that includes communicating with mobile users, creating cloaking area and pruning candidates sent from the LBS server. Second, the anonymizer can be exposed to the security problems. In centralized system, anonymizer stores all location information of mobile users to create a cloaking area. Hence an attacker or malicious LBS provider could invade the mobile users' privacy by abusing the location information stored in the anonymizer.

Therefore, we propose a distributed grid-based cloaking scheme in which a query issuer creates a cloaking area to support continuous LBS. We design a cloaking area creation scheme considering three requirements. First, as the capability of a mobile storage is limited, it distributes the information needed for creating cloaking area to participants. Second, operations such as generating transition matrix and calculating users' probabilities are performed not only by the query issuer but also by participants included in the cloaking area creation step as the capability of a mobile computation is limited. In the end, it chooses aggregate nodes considering the composed clusters to prevent bottleneck of a query issuer. By sending messages through aggregate node of the mobile users, communications concentrated on the query issuer can be distributed.

3.2 System Architecture

In our work, we first convert the mobile users' location into Hilbert value by using Hilbert curve. Then, we construct clusters based on the virtual ring topology by using the cluster formation method in [7]. In our system, the procedure of processing a query is as follow. First, a query issuer who is registered in the certification server requests information of nearby mobile users by contacting a representative node of corresponding cluster in Chord protocol. Second, the representative node retrieves nearby mobile users and returns the information of them to the query issuer. Third, above process is repetitively performed until the number of mobile user's information exceeds the required K-anonymity that the query issuer requests. Once enough user information is acquired, the query issuer creates a cloaking area and sends a query to LBS server. Fourth, LBS server processes the query over transmitted cloaking area and returns the candidates to the query issuer. In the end, the query issuer prunes the candidates and gets the actual result.

3.3 Distributed Grid-Based Cloaking Area Creation Algorithm

In this section, we present our DGCC (Distributed Grid-based Continuous Cloaking) which is an enhanced version of GCC to support the continuous LBS in distributed manner. DGCC is divided into two parts. One is for service request time T_0, another is for service running time T_i. Step 1–step 5 explains the cloaking area creation process after a query issuer request LBS service.

Step 1: Cloaking area creation for T_0

At T_0, the query issuer first expands a rectangular bounding area, C_{bound}, towards 4 directions (North, East, West, South) until it covers at least $k - 1$ other nearby users of the query issuer. Next, it sets a cloaking area A_0 satisfying the requested K-anonymity by reducing the C_{bound} on each side. After that, during the service running times, T_i, a cloaking area is created through step 2–5 in distributed manner while preserving the user's privacy. The detailed explanation for each step is as follows.

Step 2: Temporal cloaking area creation

The mobile users included in the cloaking area at T_0, namely participants, move along their route as time goes by and send their updated location at T_i to the query issuer. Based on the received locations, the query issuer sets temporal minimum

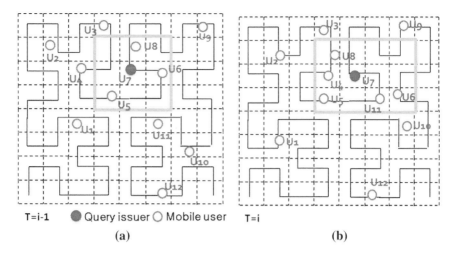

T=i-1   ● Query issuer ○ Mobile user   T=i

(a)   (b)

Fig. 1 An example of C_{bound} creation (**a**) Mobile users at T_{i-1} (**b**) Mobile users at T_i

bounding rectangle including updated location of all participants as C_{bound}. The query issuer calculates Hilbert values of grid-cells inside C_{bound} and sends each Hilbert value to the representative node for the cell in order to find both participants and newly included users in the C_{bound}. If the query issuer receives a list of users inside C_{bound}, it stores the information of users inside C_{bound}.

Figure 1 gives an example of creating C_{bound}. Figure 1a shows the cloaking area at T_{i-1} when the query issuer U_7 requests K-anonymity $= 4$. Figure 1b shows C_{bound} at T_i which includes all participants of T_{i-1}. The query issuer receives a list of users inside C_{bound} through communication with the representative node. Finally, the query issuer stores the information of mobile users, U_4, U_5, U_6, U_7, U_8 and U_{11}.

Step 3: Aggregate node selection for each cluster

The query issuer sends α, where α means the number of samples to generate a transition matrix, and aggregate table to participants of T_{i-1}. This is for calculating the probability of the users in distributed manner. Aggregate table stores the information of participants, aggregate node and the number of users to aggregate. Aggregate node is a node that aggregates the probabilities of participants belonging to the same cluster. Aggregate node is randomly selected for each cluster when the query issuer sends α and aggregate table to participants.

Users' probabilities can be aggregated as follow. At first, the query issuer classifies participants and randomly selects aggregate nodes for each cluster. Then, the query issuer notifies aggregate nodes of the number of users to aggregate. Also, the query issuer informs other participants of their aggregate node. After that each participant calculates users' partial probabilities (this will be covered in step 4) and sends it to the aggregate node. In the end, each aggregate node aggregates the partial probabilities sent from participants and sends it to the query issuer.

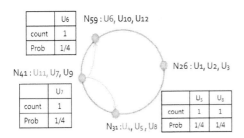

Fig. 2 An example of Chord structure at T_i

For example, Fig. 2 shows the Chord structure that reflects mobile users of Fig. 1b. There are 4 clusters and every participant keeps their previous probability at T_{i-1}. The query issuer U_7 transmits aggregate table to participants so that communication is performed hierarchically.

Step 4: Calculating users' probabilities in C_{bound}

To compute the entropy of a cloaking area being generated, this step first calculates the probabilities of participants at T_{i-1} being each user at T_i. For this purpose, transition matrix M in which each row is made of α samples is needed. In M, each row means the probability that a participant being each user in C_{bound}. To calculate the probabilities, we devised the method that each participant takes charge of each probability in a distributed manner. For this, the participant uses α and aggregate table. Using such information, the participant generates partial transition matrix M' and calculates the partial probabilities by using (3). Partial probability is [1*n] matrix where n means the number of users included in C_{bound} at T_i. In the (3), before $_{Uprob}$ means the probability of the participant at T_{i-1} and M'_i means partial transition matrix at T_i.

$$\text{Prob}_i = \text{before}_{Uprob} * M'_i / \alpha \tag{3}$$

Next, participants in each cluster transmit the calculated partial probabilities to corresponding aggregate node. Then, aggregate nodes merge partial probabilities and send it to the query issuer. The query issuer aggregates all partial probabilities received from every aggregate node and finally get the probabilities of users included in C_{bound}.

Step 5: Cloaking area creation for T_i

Once the probabilities of users included in C_{bound} are calculated, the query issuer expands the cloaking area from the cell where the query issuer exists. At this time, by considering the nearest user first, we can keep the cloaking area small. For every expansion, the query issuer computes the entropy of the cloaking area by using (1) and calculates the anonymity level k' by using (2). When the $k' \geq$ K-anonymity,

Table 1 Environment setting and variables

Environment setting		Variables	
Item	Capacity	Parameter	Value
CPU	Intel Core2 Quad CPU Q6600 2.40 GHz	K-anonymity	5–20
Memory	2 Gb	Total service time	5
OS	Windows XP professional	# of mobile users	5,000
Compiler	Microsoft Visual Studio 6.0	Grid cell size	512×512

this area is set to be the cloaking area of the query issuer. Next, the query issuer sends each user's probability to corresponding user for later service time.

4 Performance Evaluation

We implement our scheme under the environment setting as shown in Table 1. Moving object data sets are generated by using the Network-based Generator [9] on the real road map of Oldenburg, Germany, a city about 15×5 km^2. We set the total size of the map as 1 so as to more easily understand the proportion of a generated cloaking area. The variables used in performance evaluation are also shown in Table 1.

We present a set of experiments that demonstrates the performance of our DGCC (Distributed Grid-based Continuous Cloaking). As the DGCC is the first work supporting continuous LBS in distributed system, we compare our scheme against KAA and GCC studied in centralized system. We evaluate the performance in terms of the memory size needed and the number of processed messages for creating a cloaking area, size of cloaking area, and cloaking time varying K-anonymity.

Figure 3 shows the average size of a cloaking area varying K-anonymity. All schemes create bigger cloaking area when K-anonymity increases. The reason of that KAA creates smallest cloaking area is it considers all possible candidates in C_{bound} to create smallest cloaking area. Although KAA creates smallest cloaking area, DGCC creates smaller cloaking area than that of GCC. This is because some users' locations are mapped into one point when DGCC converts users' location into Hilbert values to use the encrypted location. Hence, DGCC can search more users when it expands cells and creates smaller cloaking area than GCC.

Figure 4 shows the cloaking time for varying K-anonymity. Cloaking time includes processing time and communication time. All schemes need much cloaking time as K-anonymity increases. In terms of the processing time, KAA shows the worst performance as it considers all possible candidates to create the smallest cloaking area. On the other hand, DGCC and GCC show better performance than that of KAA because they rapidly create a cloaking area by using grid-based cell expansion. Meanwhile, DGCC needs less time than GCC because the query issuer processes the query with participants in distributed manner. On the

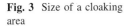

Fig. 3 Size of a cloaking area

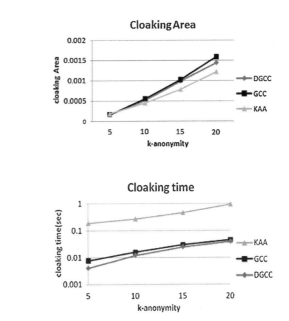

Fig. 4 Cloaking time

other hand, in terms of communication time, DGCC spends much communication time than KAA and GCC because it needs additional information transmission to calculate participants' probabilities. However, in terms of cloaking time, DGCC shows better performance than the existing works as the processing time is of greater proportion than that of communication time.

5 Conclusion

With the development of wireless communication technologies and popularity of smart phones, many LBS applications are in the spotlight. Existing mechanisms for the safe and comfortable use of LBS by mobile users have been studied only in centralized manner. In this paper, we propose a grid-based cloaking scheme for continuous location-based services in distributed systems. The proposed scheme stores information and performs operations in distributed manner to create a cloaking area. In addition, the proposed scheme uses aggregate nodes to prevent problems such as query issuers' bottleneck which may occur in distributed system. Through performance analysis, we show that our cloaking scheme efficiently creates a cloaking area while guaranteeing K-anonymity property than the existing centralized schemes.

As a future work, we need to study on a cloaking scheme that generates a cloaking area as a set of road segments to protect the information of road segments that a query issuer moves along.

Acknowledgments This research was supported by Basic Science Research Program through the National Research Foundation of Korea (NRF) funded by the Ministry of Education, Science and Technology (2010-0023800).

References

1. Gruteser, M., Grunwald, D.: Anonymous usage of location-based services through spatial and temporal cloaking. In: Proceedings of the International Conference on Mobile Systems, Applications and Services, pp. 31–42 (2003)
2. Mokbel, M., Chow, C., Aref, W.: The new casper: query processing for location services without compromising privacy. In Proceedings of the International Conference on Very Large Data Bases, pp. 763–774 (2006)
3. Xu, T., Cai, Y.: Location anonymity in continuous location-based services. ACMGIS, pp. 221–238 (2007)
4. Lee, A., Kim, H., Chang, J.: Grid-based cloaking area creation scheme supporting continuous location-based services. J. Korea Spatial Inf. Soc. **11**(3), 19–30 (2009)
5. Serjantov, A., Danezis, G.: Towards an information, theoretic metric for anonymity. Privacy Enhancing Technologies workshop (PET 2002), Vol. 2482 of LNCS, pp. 41–53 (2002)
6. Chow, C.Y., Mokbel, M.F., Liu, X.: Peer-to-Peer spatial cloaking algorithm for anonymous location-based services. In: Proceedings of the ACM-GIS, pp. 171–178 (2006)
7. Ghinita, G., Kalnis, P., Skiadopoulos, S.: MobiHide: a mobilea Peer-to-Peer system for anonymous location-based queries. In: Proceedings of SSTD, Vol. 4605, pp. 221–238 (2007)
8. Stoica, I., Morris, R., Karger, D., Kaashoek, M., Balakrishnan, H.: Chord: a scalable Peer-to-Peer lookup service for internet application. In: Proceedings of IEEE/ACM TON, Vol. 11 No.1, pp. 17–32 (2003)
9. Brinkhoff, T.: A framework for generating network-based moving objects. GeoInformatica **6**(2), 153–180 (2002)

Occluded and Low Resolution Face Detection with Hierarchical Deformable Model

Xiong Yang, Gang Peng, Zhaoquan Cai and Kehan Zeng

Abstract This paper presents a hierarchical deformable model for robust human face detection, especially with occlusions and under low resolution. By parsing, we mean inferring the parse tree (a configuration of the proposed hierarchical model) for each face instance. In modeling, a three-layer hierarchical model is built consisting of six nodes. For each node, an active basis model is trained, and their spatial relations such as relative locations and scales are modeled using Gaussian distributions. In computing, we run the learned active basis models on testing images to obtain bottom-up hypotheses, followed by explicitly testing the compatible relations among those hypotheses to do verification and construct the parse tree in a top-down manner. In experiments, we test our approach on CMU+MIT face test set with improved performance obtained.

Keywords Face detection · Image parsing · Hierarchical representation · Bottom-up/top-down computing · Active basis

X. Yang (✉) · G. Peng · Z. Cai · K. Zeng
Department of Computer Science, Huizhou University, Huizhou 516007, China
e-mail: xyang.2010@Springer.com1007129119@qq.com

G. Peng
e-mail: peng@Springer.com

Z. Cai
e-mail: cai@Springer.com

K. Zeng
e-mail: zengkehan@Springer.com

S.-S. Yeo et al. (eds.), *Computer Science and its Applications*,
Lecture Notes in Electrical Engineering 203, DOI: 10.1007/978-94-007-5699-1_9,
© Springer Science+Business Media Dordrecht 2012

1 Introduction

Face detection plays an important role in the field of object detection/recognition. Although a great deal of research [1–3] has already been performed to advance the field and improve the capability and robustness of face detection, there are still many years from closing the performance gap between the human visual system and engineered object detection systems. Face detection is a challenging vision task when appearing with occlusions and under low resolution. This paper presents a hierarchical deformable model accounting for these variations and an algorithm combining bottom-up hypotheses and top-down verification to construct parse trees [4] of faces in testing images. Here, the key observation is that occlusion and low resolution make the image data very different from the regulars. Figure 1 is some examples, in which (a) shows complete faces in clear resolution (b) is the incomplete faces due to occlusions, pose variations, etc. (c) is the faces lacking detectable features in low resolution. The face images in Fig. 1b, c make face detection a challenging task.

In the literature, face detection methods can be classified into two categories: (1) Image patch based methods [1, 2] for frontal faces as in Fig. 1a. They are sensitive to local variations such as occlusions. (2) Component based approaches [1, 3] for faces with some occlusions. The performance depends on the defined components and the binding strategies. Both of them cannot handle faces under low resolution in Fig. 1c.

As shown in Fig. 2, this paper builds a hierarchical deformable model for human faces. Consider the face node in the model, we can intuitively identify three detection ways (from itself, parent and children nodes) and thus integrate them to handle occlusion and low resolution. In the model each node such as header-shoulder, eyes, etc. is represented by active basis templates [5], see Fig. 3. In computing, we scan the testing images by learned active basis detectors to obtain bottom-up hypotheses, followed by explicitly testing the compatible relations among them to do verification in a top-down manner. Compared with AdaBoost [2], which is one of the most robust and popular approaches, the results on CMU+MIT face test set justify the robustness of our methodology.

2 Graph Model for Face in Hierarchy

In general, hierarchical representation is a type of Bayesian generative modeling [4], which can be represented by a graph $G = <V, E>$, V consists of L layer nodes in a hierarchical order, that is, $V = V^1 \cup V^2 \cup \cdots \cup V^L$, and each node in V^i can decompose into a set of nodes in V^{i-1}, $i \in [2, L]$ (the circles in Fig. 2) or terminate directly into a set of instances (the rectangles in Fig. 2), E includes two types of relations, $E = E^{\leftrightarrow} \cup E^{\updownarrow}$, $E^{\updownarrow}$ represents decompositions between nodes at two successive layers (the solid lines between circles), and $E^{\leftrightarrow}$ represents spatial

Fig. 1 Face examples in natural images. **a** Normal faces; **b** Incomplete faces; **c** Low resolution faces

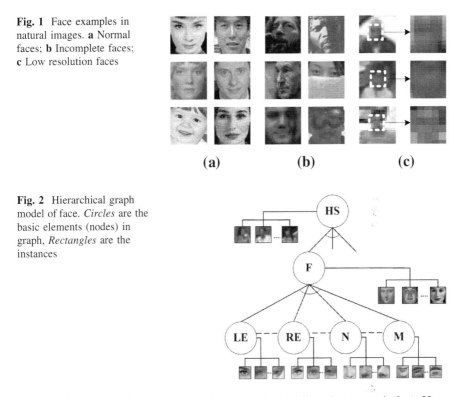

(a) (b) (c)

Fig. 2 Hierarchical graph model of face. *Circles* are the basic elements (nodes) in graph, *Rectangles* are the instances

relations between nodes at the same layer (the dashed lines between circles). Here each node $v \in V$ is modeled by a deformable template in the form of active basis (see Fig. 3), and their spatial relations including relative locations and scales are captured by Gaussian model.

3 Active Basis Model

The active basis model [5] is a deformable model which uses a small number of Gabor wavelet elements as visual primitives for modeling object category at selected locations and orientations. Let Λ be the domain of the image patch M and $\{B_{x,y,s,o}\}$ the dictionary of Gabor wavelet elements. The (x, y, s, o) are densely sampled: $(x, y) \in \Lambda$, s is a fixed size (about 1/10 of the length of Λ) and $o \in \{i\pi/N, i = 0, \cdots, N - 1\}$ (e.g. $N = 15$). These Gabor wavelet elements form an over-complete dictionary for modeling M. Then we obtain the sparse coding scheme, $M = \sum_{i=1}^{n} c_i B_i + U$, where n is the number of selected bases, $B_i \in \{B_{x,y,s,o}\}$ are Gabor wavelet elements corresponding to sketchable part of image M, c_i are coefficients, and U represents the unexplained residual image. Each element B_i is allowed to slightly perturb its size, location and orientation before they are linearly combined to generate the observed image, and thus active basis template is flexible and robust for intra-class structural variations.

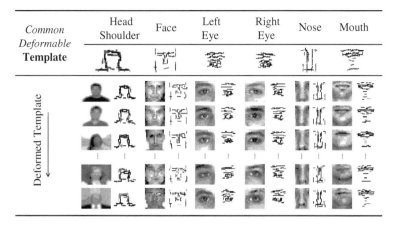

Fig. 3 Active basis models. The *first row* displays the common deformable templates. The *second row* is some images together with the corresponding deformed templates. *Downward arrow* means the descending order of matching scores

4 Problem Formulation

Given an input image M and the hierarchical model $G = <V, E>$ as shown in Fig. 2, the goal is to find all instances of faces and align the other five nodes in G, if they appear. The output is a parse tree [4], pt, according to the hierarchical model in Fig. 2. The solution space for the parse tree pt consists of all the valid configurations Ω of the six nodes in natural images. Note that the definition of detection here actually goes beyond the traditional one in which only a bounding box is outputted [1]. The Bayesian formulation is

$$pt^* = \arg\max_{pt \in \Omega} p(pt|M; G) \tag{1}$$

Intuitively, as shown in Fig. 2, we can detect faces in three kinds of ways: (i) Computing based on the whole face image patches directly. This can handle faces in Fig. 1a; (ii) Based on binding some detected face components such as two eyes. This can handle faces in Fig. 1b; (iii) In terms of predicting from detected head-shoulders. This can handle faces in Fig. 1c. The nature of hierarchical modeling and the integration of the three ways would lead a robust face detection framework.

5 Two Computing Processes in Hierarchical Model

The inference algorithms in hierarchical representation can be roughly classified into two types of computing processes of bottom-up (based on low-level features) and top-down (based on high-level cues), which are complex and interactive

Fig. 4 ROC curves
comparing our approach
(*solid curve*) with AdaBoost
(*dotted curve*)

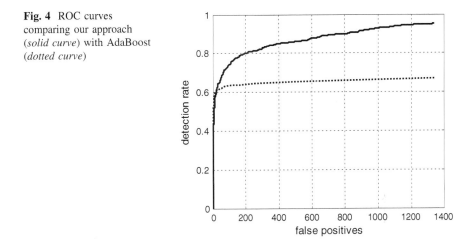

operations in human perception shown by cognition and neuroscience experiments
[6]. We integrate the two processes to parse faces in natural images. (i) Bottom-up
detection process uses active basis detectors exhaustively scan testing images at
multiple scales, the score of a detected hypothesis A^t (indexed by t) is the logarithm of posterior probability ratio given a small image patch M_Λ,

$$s_A^t = \log\frac{p(A^t|M_\Lambda)}{p(\bar{A}^t|M_\Lambda)} \tag{2}$$

$\bar{A}^t$ means competitive hypothesis. The process generates an excessive number of
hypotheses which may overlap or conflict with each other. (ii) Top-down prediction process validates the bottom-up hypotheses A_i^t, A_j^t according to the prior
Gaussian model of spatial relations $e_{i,j} \in E$ between them by following the
Bayesian posterior probability,

$$s_{pt}^t = \log\frac{p(pt^t|A_i^t, A_j^t, e_{i,j})}{p(\overline{pt}^t|A_i^t, A_j^t, e_{i,j})} = \log\frac{p(A_i^t, A_j^t, e_{i,j}|pt^t)p(pt^t)}{p(A_i^t, A_j^t, e_{i,j}|\overline{pt}^t)p(\overline{pt}^t)} \tag{3}$$

$\overline{pt}^t$ is the competitive hypothesis. The pt^t is accepted if s_{pt}^t is larger than a specific
threshold.

6 Experimental Procedure and Results

First, prepare training data, including the image patches of face, head-shoulder and
face components and the spatial relations among them [7]. Second, learn active
basis templates from the prepared image patches (Fig. 3). Third, learn the prior
Gaussian model of spatial relations of face. At last we run the learned active basis

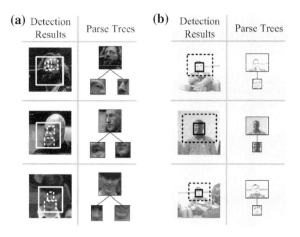

Fig. 5 False negative examples of AdaBoost are detected with parse trees outputted by our approach successfully. The hypotheses in dashed rectangles predict face regions in solid rectangles. (**a**) Incomplete faces (**b**) Faces in low resolution

detectors to obtain bottom-up hypotheses, followed by testing the compatible relations among the hypotheses to do verification in a top-down manner.

Testing data is the CMU + MIT frontal face set including 734 faces with occlusions, multi-resolution and pose variations. 2898 non-face patches are randomly cropped as negative samples from the same image set. Fig. 5 shows several difficult examples which are detected with parse trees outputted by our approach successfully, though they are missed by AdaBoost [2], Fig. 4 is the ROC curves of AdaBoost and our approach, when the number of false alarm is larger than 20, our approach improves the detection rate greatly.

The presented approach not only has the merits of component based method, which is more robust to occlusions and pose variations than the global holistic method, but also improves the performance of pure bottom-up process and outputs the configurations of the faces even either in low resolution or with occluded components. Since occlusions and low resolution are the common problems in detection and recognition tasks, our approach can be generalized to the objects in other categories.

Acknowledgments This work was supported by National Natural Science Foundation of China (No. 61170193), Scientific Research Starting Foundation for Ph.D. in Huizhou University (No. C510.0210).

References

1. Degtyarev, N., Seredin, O.: Comparative testing of face detection algorithms. Image Signal Process. **6134**, 200–209 (2010)
2. Viola, P., Jones, M.J.: Robust real-time face detection. Int. J. Comput. Vis. **57**(2), 137–154 (2004)
3. Heisele, B., Serre, T., Poggio, T.: A component-based framework for face detection and identification. Int. J. Comput. Vis. **74**(2), 167–181 (2007)
4. Zhu, S.C., Mumford, D.: A stochastic grammar of images. Found. Trends. Comput. Graph. Vis. **2**(4), 259–362 (2006)
5. Wu, Y.N., Si, Z.Z., Gong, H.F., Zhu, S.C.: Learning active basis model for object detection and recognition. Int. J. Comput. Vis. **90**(2), 198–235 (2010)
6. Thorpe, S., Fize, D., Marlet, C.: Speed of processing in the human visual system. Nature **381**, 520–522 (1996)
7. Yao, B., Yang, X., Zhu, S.C.: Introduction to a large-scale general purpose ground truth database: methodology, annotation tool and benchmarks. In: 6th International Conference on EMMCVPR, pp. 169–183. Springer, Berlin (2007)

Negotiated Economic Grid Brokering for Quality of Service

Richard Kavanagh and Karim Djemame

Abstract We demonstrate a Grid broker's job submission system and its selection process for finding the provider that is most likely to be able to complete work on time and on budget. We compare several traditional site selection mechanisms with an economic and Quality of Service (QoS) oriented approach. We show how a greater profit and QoS can be achieved if jobs are accepted by the most appropriate provider. We particularly focus upon the benefits of a negotiation process for QoS that enables our selection process to occur.

Keywords Negotiation · Grid brokering · Quality of service · Job admission control · Provider selection

1 Introduction

Grids enable the execution of large and complex programs in a distributed fashion. It is however, common that resources are provisioned in a best effort approach only, with no guarantees placed upon service quality. It has also been known for some time that guaranteed provision of reliable, transparent and quality of service (QoS) oriented resources is the next important step for Grid systems [1, 2].

In real world commercial and time-critical scientific settings guarantees that computation is going to be completed on time are required. It is therefore important to establish at submission time the requirements of the users in terms of

R. Kavanagh (✉) · K. Djemame
School of Computing, University of Leeds, Leeds LS2 9JT, UK
e-mail: screk@leeds.ac.uk

K. Djemame
e-mail: k.djemame@leeds.ac.uk

S.-S. Yeo et al. (eds.), *Computer Science and its Applications*,
Lecture Notes in Electrical Engineering 203, DOI: 10.1007/978-94-007-5699-1_10,
© Springer Science+Business Media Dordrecht 2012

completion time and cost/priority of the work. In establishing and handling this Grids can be moved away from the best-effort service which limits their importance, as users' reluctance to pay or contribute resources for late returning of results is mitigated [3].

We present two motivational scenarios that illustrate this need for time guarantees.

The first is a commercial scenario such as animation where frames maybe computed overnight before the animation team arrive, partial completion of the work delays or stops the team from starting the next day's work [4]. The second scenario is in an academic environment where it is common before conferences for Grids to become overloaded [5]. It therefore makes sense to prioritise jobs based upon when the results are required. In order that prioritisation is provided correctly an economic approach is used to ensure users truthfully indicate their priorities [6, 7].

This paper's main contribution is that we report upon our study of QoS provision due to enhanced job admission control, within our newly implemented Grid brokering system. We demonstrate the improvement in QoS by submitting jobs for estimates in our negotiation based system and then selecting the best provider for computation.

The remaining structure of the paper is as follows. In Sect. 2 we describe the pricing model and illustrate how broker profit relates to QoS provision. In Sect. 3 we discuss the provider selection policies under test. We then in Sect. 4 discuss the experimental setup and report upon the results in Sect. 5. In Sect. 6 we discuss the related works and in Sect. 7 we conclude our work and discuss future work.

2 Pricing Model and Negotiation

In this paper we introduce a WS-Agreement (Negotiation) [8] based job submission and brokering system that is part of the ISQoS (Intelligent Scheduling for Quality of Service) broker [9]. In the first stage the broker acquires the job submission templates from each provider. It then fills the templates with the user's preferences. These preferences include:

1. Budget—The user's maximum price they are willing to pay.
2. A due date and deadline—A preferred time and the last point in where the job is still of use.
3. Task description/s—Job Submission Description Language (JSDL) document/s describing the work to be performed.
4. File size and execution requirement—Estimates for each task within a job.

The task descriptions focus upon describing Bag of Task based applications, which are the predominate form of workload upon Grids [5]. We hence use the word job to describe the bag of tasks as a whole. These workloads are formed by

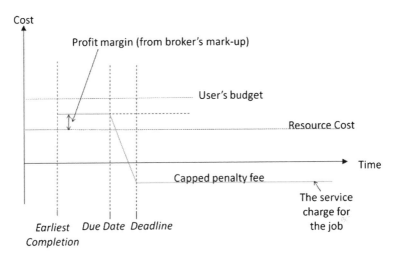

Fig. 1 The pricing model

sets of tasks that execute independently of one another, without communication and are hence considered to be "embarrassingly parallel" [10].

The broker then requests offers from providers in the tender market [11]. Each provider calculates a schedule that is suitable for the completion of the work and submits its offer back to the provider. This offer includes the estimated completion time for the job, the overall cost and completion time estimates for each individual task.

The broker then applies a mark-up (see Fig. 1) performs an assessment and submits the best offer to the user for acceptance. In cases where work is impossible to complete (see Fig. 2) the broker can send recommendations based upon the existing offers. In this case indicating the increase in time and/or budget that is required in order to complete the work on the Grid under its current state/load. This multiple rounds of negotiation is however out of the scope of this paper and during experimentation we simply reject the job as changing input values to simulate the user's preferences would be highly subjective. It should be noted however providers will not accept work that will go past the deadline so the offer collection phase aids the finding of a suitable provider for the work to be completed upon.

The service charge to the user drops with time after the due-date to a fixed value at the deadline (see Fig. 1). We chose zero for this cap because it locks the breakeven point to a specific place between the due date and deadline [9]. The service charge is useful as the broker has to pay the providers for the resources used unless a fault occurs or the provider does not perform the required amount of work before the deadline. It also generates a buffer in both economic and temporal terms around the ideal zone for offers, by generating a maximum resource cost before the broker starts using its own markup and a point in time where the job breaks even (see Fig. 2).

Fig. 2 Offer evaluation

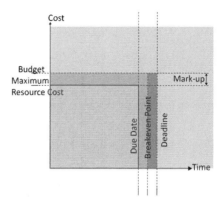

3 Provider Selection Policies

The broker in order to make a profit by generating the appropriate level of QoS must decide which jobs are practical to compute within the allotted time and which provider should compute the job. This brings about various selection strategies for the work to be computed. We introduce several strategies and list them in three categories, namely classical, flooding and selective.

The first classic strategies relate to current mechanisms for submitting to the Grid. They do not require any data from the offers, hence represent a situation with direct submission without negotiation. This can be achieved either randomly or by submitting based upon the current load of the provider.

Randomly: In order to submit randomly offers are first asked for and then an offer is chosen randomly. We chose this way to keep the pattern of submission as similar as possible to the others in this experiment. The framework does however allow for direct submission thus ignoring the negotiation phase.

Current Load: In this scenario we hook into the Ganglia [12] information provider. We use an average of the cpu_user value across all workers for a given provider. This closely as possible represents if a CPU is busy or not as per the UK's NGS [13] load monitor tool.[1] The user CPU usage is taken so as to ignore as much as possible minor non-Grid system activities taking place upon the worker nodes.

The second set of strategies floods the Grid and tries to optimize greedily upon either time or profit, these represent naïve optimization strategies.

The Earliest First and *Highest Profit*: These mechanisms sort the offers (by either profit or completion time) and select the topmost offer. This strategy makes no account for the broker's profit and so long as the budget and the deadline constraints are met then the job is accepted.

---

[1] http://www.ngs.ac.uk/load-monitor
    http://nationalgridservice.blogspot.co.uk/2011/03/loaded.html

The last set of strategies named selective aim to filter out the worst offers and ensure only jobs likely to make the broker sufficient profit are accepted. These mechanisms are *Highest Profit (Profitable Only)*, *Hybrid Offer Filter*, *Load Based Selection (Profitable Only)*, *Random (Profitable Only)* and a *Near Going Rate* mechanism.

Highest Profit (Profitable Only): This extends the highest profit approach and checks to see if the broker will make a profit before accepting.

A *Near Going Rate* mechanism and *Hybrid Offer Filter* [9]: They have been configured to initially sort by profit and select only profitable jobs. The difference from other profit driven strategies is derived from how they perform selection from this sorted list.

Near Going Rate: This establishes from a history of the last n records the current rate at which profit is accumulated. It then establishes a minimum value below this that is acceptable. If the new offer is above this threshold then it is accepted.

Hybrid Offer Filter: If the constraints are fully met then the job is automatically accepted. If the offer is constrained by either time or budget then a going rate assessment is performed. The main aim of this variation is to ensure if the arrival rate slows then unconstrained (fully profitable) job are always accepted. This is particularly advantageous if different mark-ups/priorities are in use and other factors such as differing network transport cost compared to the cost of computation.

Random (Profitable Only) and Load Based Selection (Profitable Only): These extend the classical methods by allowing them to submit to the site chosen by their ranking mechanism and then checking to see if the broker will make a profit.

4 Experimental Setup

We perform experimentation to discover the best selection strategy for selecting between Grid providers, with the aim of enhancing QoS provision. We focus this experimentation upon high load scenarios where correct selection is most required. The high load ensures far more jobs are available than can be computed on time, hence to ensure time constraints are met, which is directly linked to the broker's profit in the pricing model then some jobs must be rejected.

The configuration of the experiments performed is described in this section.

We sent *100 jobs* with *8 tasks* each into a Grid with *2 providers*. Each provider had *4 virtual machines*, of which one also acted as a head node. Jobs were submitted with a 30 s gap between submissions, from a separate broker virtual machine instance. This being shorter than the time it takes to compute a job meant the Grid fills and resources become scarce as per a time sensitive, high utilization scenario presented earlier. Each provider is configured to use the round robin scheduling algorithm.

The virtual machines ran Ubuntu 11.10 (64 bit) server, with full virtualization and ran upon 4 physical hosts. The virtual environment was constructed using

OpenNebula [14] 2.0 and Xen 4.0.1 [15]. Each head node had 1 GB of RAM allocated and worker nodes 768 MB. Each processor ran at a speed of 2.4 GHz.

The ISQoS Grid uses WS-Agreement for Java v1.0 for the Broker and Provider agreement process and Ganglia 3.2.0 was used as the information provider.

Jobs were setup to be none data intensive and the *stage in/out size* was 1 megabyte. This mitigates issues with considering the network configuration of the virtual cluster on the cloud testbed. The *compute size* of a job was set to 3,000. This value derives from a reference processor of 3,000 MHz multiplied by an expected duration of 1 min. This means upon the resources available, tasks are expected to last approximately 1 min and that if a job was allocated to a single machine it would take 8 min to complete.

Each job's *due date* was set to the submission time +8 and its *deadline* was set to the submission time +12, with the knowledge that the Grid would soon be overtaxed.

Each job was given a *budget* of 20,000 which was chosen to be sufficiently high so as not to act as a selection pressure. A fixed *mark-up* for the broker of 20 % was chosen, which means the broker breaks even 16.67 % of the way between the due date and deadline [9], so the provider must complete work before this point to remain in profit. A static resource price was chosen that bills time for both the use of network and resource time equally at 1 unit per second.

We performed 6 runs of each trace that is used in the experiment 95 % confidence intervals are marked on the graphs. The first 9 accepted jobs of the traces have been ignored to counteract effects of starting with an unloaded Grid.

5 Results

In this section we look at several key metrics aiming at service quality and suitability for the broker, namely the job acceptance, slack, start delay and the overall profit.

The broker's profit directly relates to meeting the QoS requirements, in Fig. 3 we observe a distinction between mechanisms where profit checking is permissible or not. Highest profit (profitable only) tends to go past the due date making it less suitable. Adaptations of classical submission strategies do well, but tend to have a wide variance in slack and job acceptance (Fig. 4) as compared to the Hybrid Offer filter. This is also reflected in the overall profit (Fig. 5), with the Hybrid and Going Rate approaches winning out, some 31.6 % above their nearest rivals. The Hybrid approach however works much better than the going rate in lower arrival rate situations [9]. The load based and random selection mechanisms appear to be very similar. It is suspected that the load based selection mechanism does not accurately reflect the queue length/amount of work to be performed when nearing full capacity (as per the experiment), hence acts more as a means of random number generation.

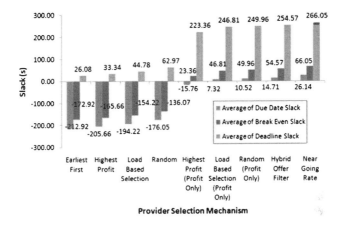

Fig. 3 Average slack

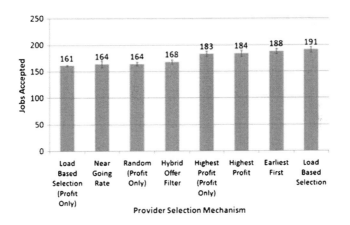

Fig. 4 Job acceptance

The start delay (Fig. 6), is used here as a metric for understanding the pressures upon resources on the Grid. Selection based strategies fair best, while random and the highest profitable job strategies perform worst with notable variance. The deviation from the ordering as compared to how many jobs accepted should also be noted as it gives some notion of the differing quality of site selection.

6 Related Work

The brokering mechanism we present revolves around its pricing mechanism so we focus our discussion there. Related models rarely capture the user's real requirements, as we have done. Early models focus purely upon slowdown such as

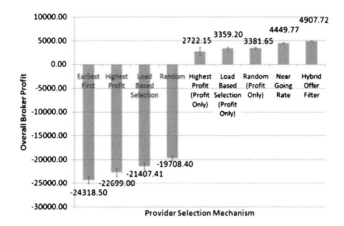

Fig. 5 Overall broker profit

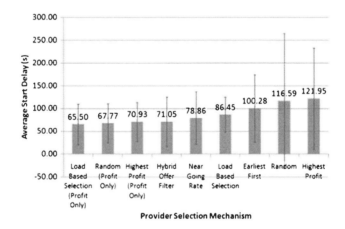

Fig. 6 Average start delay

First Reward and Risk Reward [16] and First Price [17], thus are very system centric. Another pitfall we've avoided is that penalty bounds are also not always set, such as in [16, 17] and LibraSLA [18]. Pricing mechanisms however, should have properties such as budget balance and individual rationality among others [19]. First Profit, First Opportunity and First Opportunity Rate [20] like our work uses the same scheduling algorithm to schedule as they do for admission control. However, our broker's mark-up, gives it rational for participation in the market while also generating a marked difference in providing a boundary of acceptable QoS. The Aggregate Utility [4] model has a lot of flexibility in specifying user requirements at the expense of complexity for the end user. Resource Aware Policy Administrator (RAPA) [21], focuses upon divisible load and caps the maximum deadline in order to limit the maximum penalty paid. Nimrod/G [22] is a early work with a limited pricing mechanism and no SLAs.

7 Conclusions and Future Work

We have shown how classical job submission strategies do not fare well in a QoS oriented approach even when providers do not accept jobs past their deadline requirement. Filtering upon profit that is directly linked to QoS vastly improves the situation. The correct use of the pricing model for job selection so that it reflects future scheduled work also significant enhances QoS provision. Our future work will include dynamic pricing to reflect the current Grid workload better, performing tests upon a bigger Grid infrastructure and investigating deployment in the Cloud.

References

1. Liu, C., Baskiyar, S.: A general distributed scalable grid scheduler for independent tasks. J. Parallel Distrib. Comput. **69**(3), 307–314 (2009)
2. Battre, D., et al.: Planning-based scheduling for SLA-awareness and grid integration. In: PlanSIG 2007 the 26th Workshop of the UK Planning and Scheduling Special Interest Group. 2007. Prague, Czech Republic
3. Kokkinos, P., Varvarigos, E.A.: A framework for providing hard delay guarantees and user fairness in Grid computing. Future Gen Comput. Syst. **25**(6), 674–686 (2009)
4. AuYoung, A., et al.: Service contracts and aggregate utility functions, in 15th IEEE International Symposium on High Performance Distributed Computing (HPDC-15). 2005, IEEE: New York, pp. 119–131
5. Iosup, A., Epema, D.: Grid computing workloads. Internet Comput. IEEE **15**(2), 19–26 (2011)
6. Buyya, R., Abramson, D., Venugopal, S.: The grid economy. Proc. IEEE **93**(3), 698–714 (2005)
7. Lai, K.: Markets are dead, long live markets. SIGecom Exch. **5**(4), 1–10 (2005)
8. Open Grid Forum, WS-Agreement Negotiation Version 1.0. 2011. p. 63
9. Kavanagh, R., Djemame, K.: A grid broker pricing mechanism for temporal and budget guarantees. In: 8th European Performance Engineering Workshop (EPEW'2011). 2011. Borrowdale, The Lake District, UK: Springer
10. Lee, Y.C., Zomaya, A.Y.: Practical scheduling of bag-of-tasks applications on grids with dynamic resilience. IEEE Trans. Comput. 56(6):815–825 (2007)
11. Buyya, R., et al.: Economic models for resource management and scheduling in Grid computing. Concurr. Comput.: Pract. Experience **14**(13–15), 1507–1542 (2002)
12. Ganglia Project.: Ganglia Monitoring System. 2012; Available from: http://ganglia.sourceforge.net/
13. NGS.: National Grid Service. 2009; Available from: http://www.ngs.ac.uk
14. OpenNebula Project.: OpenNebula Homepage. 2012; Available from: http://opennebula.org/
15. Citrix Systems.: Home of the Xen hypervisor. 2012; Available from: http://www.xen.org/
16. Irwin, D.E., Grit, L.E., Chase, J.S.: Balancing risk and reward in a market-based task service. In: Proceedings of 13th IEEE International Symposium on High performance Distributed Computing, 2004
17. Chun, B.N., Culler, D.E.: User-centric performance analysis of market-based cluster batch schedulers. In: 2nd IEEE/ACM International Symposium on Cluster Computing and the Grid, 2002
18. Chee Shin, Y, Buyya, R.: Service level agreement based allocation of cluster resources: handling penalty to enhance utility. In: Cluster Computing, 2005

19. Schnizler, B., et al.: Trading grid services—a multi-attribute combinatorial approach. Eur. J. Oper. Res. **187**(3), 943–961 (2008)
20. Popovici, F.I., Wilkes, J.: Profitable services in an uncertain world. in Supercomputing, 2005. In: Proceedings of the ACM/IEEE SC 2005 Conference. 2005
21. Han, Y., Youn, C.-H.: A new grid resource management mechanism with resource-aware policy administrator for SLA-constrained applications. Futur. Gener. Comput. Syst. **25**(7), 768–778 (2009)
22. Abramson, D., Giddy, J., Kotler, L.: High performance parametric modeling with Nimrod/G: Killer application for the global Grid? In: International Parallel and Distributed Processing Symposium (IPDPS). Cancun, Mexico, 2000

Differential Fault Analysis on HAS-160 Compression Function

Jinkeon Kang, Kitae Jeong, Jaechul Sung and Seokhie Hong

Abstract In FDTC 2011, Hemme et al. proposed differential fault analysis on SHA-1 compression function. Based on word-oriented fault models, this attack can recover the chaining value and the input message block of SHA-1 compression function with 1,002 random word fault injections. In this paper, we show that their attack can be applied to HAS-160 compression function. As a result, our attack can extract the chaining value and the input message block of it with about 1,000 random word fault injections. This is the first known cryptanalytic result on HAS-160 by using side channel attacks.

Keywords HAS-160 · Differential fault analysis · Cryptanalysis

1 Introduction

HAS-160 is a cryptographic hash function which is designed for use with the Korean KCDSA digital signature algorithm [1]. HAS-160 has the Merkle-Damgård structure that processes 512-bit input message blocks and produces a 160-bit hash

J. Kang · K. Jeong · S. Hong (✉)
Center for Information Security Technologies, Korea University, Seoul, South Korea
e-mail: shhong@korea.ac.kr

J. Kang
e-mail: jinkeon.kang@gmail.com

K. Jeong
e-mail: kite.jeong@gmail.com

J. Sung
Department of Mathematics, University of Seoul, Seoul, South Korea
e-mail: jcsung@uos.ac.kr

S.-S. Yeo et al. (eds.), *Computer Science and its Applications*,
Lecture Notes in Electrical Engineering 203, DOI: 10.1007/978-94-007-5699-1_11,
© Springer Science+Business Media Dordrecht 2012

value. HAS-160 is derived from SHA-1, with some changes intended to increase its security. Until now, there were many cryptanalytic results on HAS-160. For example, Yun et al. found a collision in the first 45 steps of HAS-160 with complexity 2^{12} [2]. This was improved until 53 steps by Cho et al. in [3]. Mendel et al. presented the first actual colliding message pair for 53 steps and showed that the attack can be applied to 59 steps [4]. In Sasaki and Aoki [5], proposed preimage attacks on 52 steps. Hong et al. presented improved preimage attack for 68 steps in [6]. To our knowledge, there is no cryptanalytic result on HAS-160 by using side channel attacks.

Differential fault analysis (DFA), one of the side channel attacks, was first proposed by Biham and Shamir on DES in 1997 [7]. This attack exploits faults within the computation of a cryptographic algorithm to reveal the secret information. So far, DFAs on many block ciphers such as AES [8], ARIA [9], SEED [10] and SHACAL-1 [11] have been proposed.

On the other hand, Hemme et al. introduced DFA on SHA-1 compression function [12]. Based on [11], they assume that a random 32-bit fault is injected to the input register of the target step. To recover the chaining value and the input message block of SHA-1 compression function, this attack requires 1,002 random word fault injections.

In this paper, we show that the attack proposed in [12] can be applied to HAS-160 compression function. The structure of HAS-160 is similar to that of SHA-1 except some modifications of message scheduling and amounts of bit rotations. Thus, we apply the attack proposed in [12] to HAS-160 compression function. As a result, our attack can extract the chaining value and the input message block of it with about 1,000 random word fault injections. This is the first known cryptanalytic result on HAS-160 by using side channel attacks.

This paper is organized as follows. In Sect. 2, we give brief description of HAS-160 compression function and the attack procedure of [12]. Our attack on HAS-160 compression function is described in Sect. 3. Finally, we give our conclusion in Sect. 4.

2 Preliminaries

2.1 Description of HAS-160 Compression Function

HAS-160 is an iterated cryptographic hash function that processes 512-bit input message blocks and produces a 160-bit hash value. The design of HAS-160 is based on design principles of SHA-1. The overall structure is similar to SHA-1, with some modifications such as simpler message scheduling and variable amounts of bit rotations. Since our attack considers only the compression function of HAS-160, we focus on it in this section. For the description of HAS-160, see [1].

Table 1 Message ordering of HAS-160

Step	0	1	2	3	4	5	6	7	8	9	10	11	12	13	14	15	16	17	18	19
0–19	18	0	1	2	3	19	4	5	6	7	16	8	9	10	11	17	12	13	14	15
20–39	18	3	6	9	12	19	15	2	5	8	16	11	14	1	4	17	7	10	13	0
40–59	18	12	5	14	7	19	0	9	2	11	16	4	13	6	15	17	8	1	10	3
60–79	18	7	2	13	8	19	3	14	9	4	16	15	10	5	0	17	11	6	1	12

512-bit message block M is divided into sixteen 32-bit message words $[0], X[1], \ldots, X[15]$. The additional 4 message words are generated from the 16 input message word as follows.

$$X[16] = X[l(j+01)] \oplus X[l(j+02)] \oplus X[l(j+03)] \oplus X[l(j+04)],$$
$$X[17] = X[l(j+06)] \oplus X[l(j+07)] \oplus X[l(j+08)] \oplus X[l(j+09)],$$
$$X[18] = X[l(j+11)] \oplus X[l(j+12)] \oplus X[l(j+13)] \oplus X[l(j+14)],$$
$$X[19] = X[l(j+16)] \oplus X[l(j+17)] \oplus X[l(j+18)] \oplus X[l(j+19)]$$

The message expansion of HAS-160 is a permutation of 20 expanded message words in each round. Within each round, there is a separate message ordering where each of the 20 values are processed. The message orders are specified in Table 1.

As shown in Fig. 1, the step function of HAS-160 compression function starts from a fixed initial value IV of five 32-bit register A_0, B_0, C_0, D_0, E_0 and updates them in total 80 steps. The step function of step i is described as follows $(i = 0, \ldots, 79)$. Here, K_i is a round constant and message ordering $l(i)$ is described in Table 1.

$$A_i = A_{i-1}^{\lll s_1[i]} + f_i(B_{i-1}, C_{i-1}, D_{i-1}) + E_{i-1} + X[l(i)] + K_i;$$
$$B_i = A_{i-1}; C_i = B_{i-1}^{\lll s_2[i]}; D_i = C_{i-1}; E_i = D_{i-1};$$

HAS-160 uses eighty boolean functions $f_0, f_1, \ldots, f_{79}$. Each function f_i, where $0 \le i \le 79$, operates on three 32-bit words (x, y, z) and produces a 32-bit word as output. The function $f_i(x, y, z)$ is defined as follows.

$$f_i(x, y, z) = (x \wedge y) \vee (\neg x \wedge z)(0 \le i \le 19: \text{round } 1),$$
$$f_i(x, y, z) = x \oplus y \oplus z(20 \le i \le 39, 60 \le i \le 79: \text{round } 2, 4),$$
$$f_i(x, y, z) = y \oplus (x \wedge \neg z)(40 \le i \le 59: \text{round } 3).$$

The amount of bit rotation s_1 is different in each step of a round.

$$s_1[i \bmod 20] = \{5, 11, 7, 15, 6, 13, 8, 14, 7, 12, 9, 11, 8, 15, 6, 12, 9, 14, 5, 13\}$$

The rotation value s_2 is different in each round as follows.

$$s_2[i] = 10(10 \le i \le 19: \text{round } 1), \quad 38; s_2[i] = 17(20 \le i \le 39: \text{round } 2),$$
$$s_2[i] = 25(40 \le i \le 59: \text{round } 3), \quad 38; s_2[i] = 30(60 \le i \le 79: \text{round } 4)$$

Fig. 1 Step function of
HAS-160

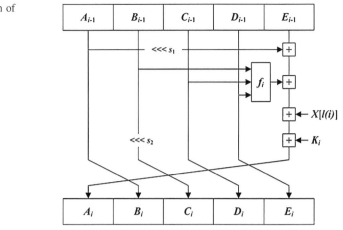

2.2 DFA on SHA-1 Compression Function

In this subsection, we present DFA on SHA-1 compression function proposed in
[12]. This attack can recover the 160-bit chaining value and the 512-bit input
message block by injecting 1,002 random word faults. This attack is divided into two
sub-steps. The first sub-step is to compute the input value $(A_{79}, B_{79}, C_{79}, D_{79}, E_{79})$ of
step 79. For computing this value, the following propositions are used.

Proposition 1 [12] Let $0 < t < 32$ and $0 \leq X, Y, \Phi, \Psi < 2^{32}$ be integers, such that
the following system of equations holds:

$$Y - X = \Phi \text{ and } (Y \lll t) - (X \lll t) = \Psi, \tag{1}$$

where "$-$" denotes subtraction mod 2^{32} and "$\lll t$" denotes left rotation of a 32-
bit number by t. Further let $0 \leq X_0, \Phi_0, \Psi_0 < 2^{32-t}$ and $0 \leq X_1, \Phi_1, \Psi_1 < 2^t$ be
integers, such that $X = X_1 \cdot 2^{32-t} + X_0$, $\Phi = \Phi_1 \cdot 2^{32-t} + \Phi_0$, $\Psi = \Psi_0 \cdot 2^t + \Psi_1$.
Then the following relations hold:

(1) (a) $(\Psi_1 - \Phi_1) \bmod 2^t \in \{0, 1\}$,
   (b) $X_0 < 2^{32-t} - \Phi_0$ if $(\Psi_1 - \Phi_1) \bmod 2^t = 0$.
   (c) $X_0 < 2^{32-t} - \Phi_0$ if $(\Psi_1 - \Phi_1) \bmod 2^t = 0$.
(2) (a) $(\Psi_0 - \Phi_0) \bmod 2^{32-t} \in \{0, 1\}$,
   (b) $X_1 \geq 2^t - \Psi_1$ if $(\Phi_0 - \Psi_0) \bmod 2^{32-t} = 1$,
   (c) $X_1 < 2^t - \Psi_1$ if $(\Phi_0 - \Psi_0) \bmod 2^{32-t} = 0$.

By using Proposition 1, we can find out the value of X. In detail, depending on
the (known) values of Φ and Ψ, we can compute the partial information on X. If
we consider the more fault injections, we get the more information on X.

Proposition 2 [11] Let x be an unknown 32-bit value and (δ, Δ) be known 32-bit values. We assume that the following equation holds.

$$(x \oplus \delta) - x = \Delta. \tag{2}$$

Then the ith bit of x, x_i, has the following properties.

$$x_i = \begin{cases} 0 \text{ or } 1 & \text{if } i = 31 \\ 0 \text{ or } 1 & \text{if } \delta_i = 0(0 \leq i \leq 30) \\ \delta_{i+1} \oplus \Delta_{i+1} & \text{if } \delta_i = 1(0 \leq i \leq 30) \end{cases}.$$

By using Proposition 2, we can determine some bits of the unknown value x. Analogous to Proposition 1, if we consider the more fault injections, we get the more information on x.

Applying Proposition 1 and 2 to SHA-1 compression function, we can compute candidates of $(A_{79}, B_{79}, C_{79}, D_{79}, E_{79})$. As simulation results, we get about 2^{28} candidates of $(A_{79}, B_{79}, C_{79}, D_{79}, E_{79})$ with 867 fault injections.

The goal of the second sub-step is to recover the input message block M. This sub-step is based on the DFA attack on SHACAL-1 proposed in [11]. Note that the structure of SHACAL-1 is based on SHA-1 compression function. Particularly, the position of round key in SHACAL-1 is equal to that of message word in SHA-1 compression function. Thus, the attack procedure of recovering message word in [12] is similar to that of recovering round key in [11]. In this sub-step, Proposition 2 is only used. Applying Proposition 2 to SHA-1 compression function, they can compute the 512-bit input message block. From simulation results, it requires 135 fault injections.

Hence, this attack requires total $1,002(= 867 + 135)$ random word fault injections in order to recover the chaining value and the input message block of SHA-1 compression function.

3 DFA on HAS-160 Compression Function

Now we are ready to introduce DFA on HAS-160 compression function. Our attack is based on the attack proposed in [12]. The structure of HAS-160 is identical to that of SHA-1 except some modifications of message scheduling and amounts of bit rotations. This minor difference does not affect applying the attack proposed in [12].

Similar to the attack proposed in [12], our attack is divided into two sub-steps. The first sub-step is to compute the input value $(A_{79}, B_{79}, C_{79}, D_{79}, E_{79})$ of step 79. In the second sub-step, we recover the input message block M.

Our attack procedure of the first sub-step is as follows.

1. With the unknown initial chaining value $(A_0 \parallel B_0 \parallel C_0 \parallel D_0 \parallel E_0)$ and the input message block M, obtain a right compression value $Y = Y_A \parallel Y_B \parallel Y_C \parallel Y_D \parallel Y_E = (A_{80} + A_0) \parallel (B_{80} + B_0) \parallel (C_{80} + C_0) \parallel (D_{80} + D_0) \parallel (E_{80} + E_0)$.

2. Inject random word faults to the intermediate value A_{79} (see Fig. 2a) and calculate faulty compression values $Y^* = Y_A^* \parallel Y_B^* \parallel Y_C \parallel Y_D \parallel Y_E = (A_{80}^* + A_0) \parallel (B_{80}^* + B_0) \parallel (C_{80} + C_0) \parallel (D_{80} + D_0) \parallel (E_{80} + E_0)$.

3. Construct the following equation.

$$A_{79}^* - A_{79} = Y_B^* - Y_B, \left(A_{79}^* \lll 13\right) - (A_{79} \lll 13) = Y_A^* - Y_A.$$

   From Proposition 1, compute the partial information on A_{79} and get candidates of B_0 by using an equation $B_0 = Y_B - B_{80} = Y_B - A_{79}$.

4. Repeat the step 2 and 3 with the difference that we inject random word faults to the intermediate value A_{78} instead of A_{79}. Construct the following equation.

$$\left(A_{78}^* \lll 30\right) - (A_{78} \lll 30) = Y_C^* - Y_C,$$
$$\left(A_{78}^* \lll 5\right) - (A_{78} \lll 5) = Y_B^* - Y_B.$$

   From Proposition 1, compute the partial information on $A_{78} \lll 30$ and get candidates of C_0 by using an equation $C_0 = Y_C - C_{80} = Y_C - (A_{78} \lll 30)$.

5. For each faulty result of step 4, by considering $\left(A_{80}^* - A_{80}\right)$, set up the Eq. (2) with $\quad x = B_{79} \oplus C_{79} \oplus D_{79}, \quad \delta = B_{79}^* \oplus B_{79} = ((Y_C^* - C_0) \lll 2) \oplus ((Y_C - C_0) \lll 2)$ and $\Delta = \left(A_{80}^* - A_{80}\right) - \left(\left(A_{79}^* \lll 13\right) - (A_{79} \lll 13)\right) = \left(Y_A^* - Y_A\right) - (((Y_B^* - B_0) \lll 13) - ((Y_B - B_0) \lll 13)))$. From Proposition 2, get candidates of $BCD_{79} := B_{79} \oplus C_{79} \oplus D_{79}$.

6. Repeat the step 2, 3 and 5 with the difference that we inject random word faults to the intermediate value A_{77}. Construct the following equation.

$$\left(A_{77}^* \lll 30\right) - (A_{77} \lll 30) = Y_D^* - Y_D, \left(A_{77}^* \lll 14\right) - (A_{77} \lll 14)$$
$$= ((Y_C^* - C_0) \lll 2)) - ((Y_D - D_0) \lll 2).$$

   From Proposition 1, compute the partial information on $A_{77} \lll 30$ and get candidates of D_0 by using an equation $D_0 = Y_D - D_{80} = Y_D - (A_{77} \lll 30)$. Similar to the step 5, by considering $\left(A_{79}^* - A_{79}\right)$, set up the Eq. (2) with $x = B_{78} \oplus C_{78} \oplus D_{78}, \quad \delta = B_{78}^* \oplus B_{78} = ((Y_D^* - D_0) \lll 2) \oplus ((Y_D - D_0) \lll 2)$ and $\quad \Delta = \left(A_{79}^* - A_{79}\right) - \left(\left(A_{78}^* \lll 5\right) - (A_{78} \lll 5)\right) = \left(Y_B^* - Y_B\right) - (((Y_C^* - C_0) \lll 7 - ((Y_C - C_0)) \lll 7))$. From Proposition 2, get candidates of $BCD_{78} := B_{78} \oplus C_{78} \oplus D_{78}$.

7. Calculate the value E_0 and E_{79} with candidates of B_0, C_0, D_0, BCD_{78} and BCD_{79} obtained in the previous steps.

$$E_0 = Y_E - A_{80} = Y_E - D_{79} = Y_E - (BCD_{79} \oplus ((Y_C - C_0) \lll 2) \oplus (Y_D - D_0)))$$

$$E_{79} = D_{78} = BCD_{78} \oplus B_{78} \oplus C_{78} = BCD_{78} \oplus (((Y_D - D_0) \lll 2) \oplus (Y_E - E_0)))$$

From the first sub-step, we can obtain candidates of $(B_0, C_0, D_0, E_0, E_{79})$. For each candidate, we can compute $(A_{79}, B_{79}, C_{79}, D_{79}, E_{79})$ by using the following equation.

$$A_{79} \| B_{79} \| C_{79} \| D_{79} \| E_{79} = Y_B - B_0 \| (Y_C - C_0) \lll 2 \| Y_D - D_0 \| Y_E - E_0 \| E_{79}$$

Similarly, we can compute a faulty value by using the following equations.

$$A_{79}^* \| B_{79}^* \| C_{79}^* \| D_{79}^* \| E_{79}^* = Y_B^* - B_0 \| (Y_C^* - C_0) \lll 2 \| Y_D^* - D_0 \| Y_E^* - E_0 \| E_{79}^*$$

Here, $E_{79}^* = E_{79} + Y_A^* - Y_A - ((Y_B - B_0) \lll 13) - ((Y_B^* - B_0) \lll 13) + (((Y_C - C_0) \lll 2) \oplus (Y_D - D_0) \oplus (Y_E - E_0)) - (((Y_C^* - C_0) \lll 13) \oplus (Y_D^* - D_0) \oplus (Y_E^* - E_0))$. Now we have a number of candidates for the input values of step 79.

The goal of the second sub-step is to recover the input message block $M = (X[1], X[2], \ldots, X[16])$ by using candidates for the input values of step 79 obtained in the first sub-step. Our attack procedure of the second sub-step is as follows.

1. Inject random word faults to B_{76} (see Fig. 2b) and then by considering $(A_{78}^* - A_{78})$, set up the Eq. (2) with $x = B_{77} \oplus C_{77} \oplus D_{77} = (D_{79} \lll 2) \oplus E_{79} \oplus E_{78}$, $\delta = C_{77}^* \oplus C_{77} = C_{79}^* \oplus C_{79}$ and $\Delta = (A_{78}^* - A_{78}) - ((A_{77}^* \lll 14) - (A_{77} \lll 14)) = (B_{79}^* - B_{79}) - ((C_{79}^* \lll 16) - (C_{79} \lll 16))$. From Proposition 2, get two candidates of E_{78}.

2. By considering $(A_{77}^* - A_{77})$, set up the Eq. (2) with $x = B_{76} \oplus C_{76} \oplus D_{76} = (E_{79} \lll 2) \oplus E_{78} \oplus E_{77}$, $\delta = B_{76}^* \oplus B_{76} = (E_{79}^* \lll 2) \oplus (E_{79} \lll 2)$ and $\Delta = A_{77}^* - A_{77} = (C_{79}^* \lll 2) - (C_{79} \lll 2)$. With the result in step 1, one can obtain 2^2 candidates for (E_{77}, E_{78}).

3. Message words $X[1]$ and $X[6]$ are used in step 77 and 78, respectively. They are expressed by the following equations.

$$X[1]38; = A_{79} - (A_{78} \lll 5) - (B_{78} \oplus C_{78} \oplus D_{78}) - E_{78} - K_{79}$$
$$38; = A_{79} - (B_{79} \lll 5) - ((C_{79} \lll 2) \oplus D_{79} \oplus E_{79}) - E_{78} - K_{79},$$
$$X[6]38; = A_{78} - (A_{77} \lll 14) - (B_{77} \oplus C_{77} \oplus D_{77}) - E_{77} - K_{78}$$
$$38; = B_{79} - (C_{79} \lll 16) - ((D_{79} \lll 2) \oplus E_{79} \oplus E_{78}) - E_{77} - K_{78}.$$

With 2^2 candidates for (E_{77}, E_{78}), one can deduce 2^2 candidates for message words $(X[1], X[6])$.

4. Repeat above procedure by injecting faults to the intermediate values $B_{74}, B_{72}, \ldots, B_{60}$. Then, we can get 2^{18} candidates for input message words used in step 61–78.

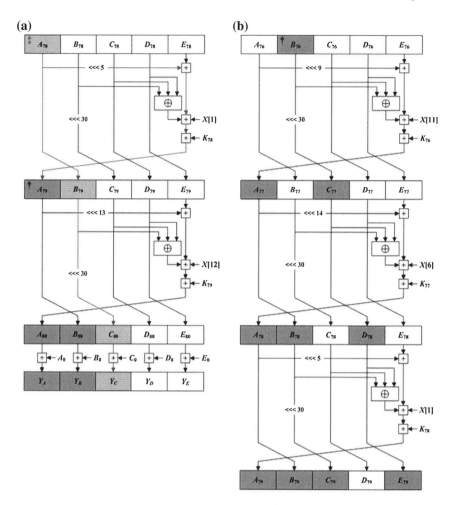

Fig. 2 DFA on HAS-160 compression function (**a** first sub-step, **b** second sub-step)

According to the message expansion of HAS-160, the 512-bit input message block can be directly calculated from message words used in step 61–78. After about 2^{18} operations, only one right input message block is remained. We can calculate the chaining value by inverting the HAS-160 compression function with obtained message words. If the resulting values are not equal to the values obtained in the first sub-steps, discard the candidates and repeat the second sub-step with the another candidates for $(A_{79}, B_{79}, C_{79}, D_{79}, E_{79})$.

We simulated our attack on a PC using Visual studio 2010 on Intel Core^{TM} i7 CPU 950 3.07 GHz with 4.00 GB RAM. As simulation results, our attack requires about 1,000 random word fault injections on average with a probability of 0.80.

4 Conclusion

In this paper, we proposed DFA on HAS-160 compression function. Our attack is based on DFA on SHA-1 compression function. As a result, we require about 1,000 random word fault injections to recover the chaining value and the input message block of HAS-160 compression function. This result is the first known cryptanalytic result by using side channel attacks.

Acknowledgments This research was supported by the Ministry of Knowledge Economy (MKE), Korea, under the Information Technology Research Center (ITRC) support program (NIPA-2012-H0301-12-3007) supervised by the National IT Industry Promotion Agency (NIPA).

References

1. Telecommunications Technology Association, Hash Function Standard Part 2: Hash Function Algorithm Standard, HAS-160 (2000)
2. Yun, A., Sung, S., Park, S., Chang, D., Hong, S., Cho, H.: Finding collision on 45-step HAS-160. In: Won, D., Kim, S. (eds.) ICISC 2005. LNCS, vol. 3935, pp. 146–155. Springer, Heidelberg (2006)
3. Cho, H., Park, S., Sung, S., Yun, A.: Collision search attack for 53-step HAS-160. In: Rhee, M., Lee, B. (eds.) ICISC 2006. LNCS, vol. 4296, pp. 286–295. Springer, Heidelberg (2006)
4. Mendel, F., Rijmen, V.: Colliding message pair for 53-step HAS-160. In: Nam, K., Rhee, G. (eds.) ICISC 2007. LNCS, vol. 4817, pp. 324–334. Springer, Heidelberg (2007)
5. Sasaki, Y., Aoki, K.: A Preimage Attack for 52-Step HAS-160. In: Lee, P., Cheon, J. (eds.) ICISC 2008. LNCS, vol. 5461, pp. 302–317. Springer, Heidelberg (2009)
6. Hong, D., Koo, B., Sasaki, Y.: Improved Preimage Attack for 68-Step HAS-160. In: Lee, D., Hong, S. (eds.) ICISC 2009. LNCS, vol. 5984, pp. 332–348. Springer, Heidelberg (2010)
7. Biham, E., Shamir, A.: Differential fault analysis of secret key cryptosystems. In: Burton, K. (ed.) Crypto 1997. LNCS, vol. 1294, pp. 513–525. Springer, Berlin (1997)
8. Tunstall, M., Mukhopadhyay, D., Ali, S.: Differential fault analysis of the advanced encryption standard using a single fault. In: Ardagna, C, Zhou, J. (eds.) WISTP 2011. LNCS, vol. 6633, pp. 224–233. Springer, Heidelberg (2011)
9. Li, W., Gu, D., Li. J.: Differential fault analysis on the ARIA algorithm. Inf. Sci. **10**(178), 3727–3737 (2008)
10. Jeong, K., Lee, Y., Sung, J., Hong, S.: Differential fault analysis on block cipher SEED. Math. Comput. Model. **55**, 26–34 (2012)
11. Li, R., Li. C., Gong, C.: Differential fault analysis on SHACAL-1. In: Breveglieri, L., Koren, I., Naccache, D., Oswald, E., Seifert, J. (eds.) FDTC 2009, pp. 120–126. IEEE Computer Society, Los Alamitos (2009)
12. Hemme, L., Hoffman, L.: Differential Fault Analysis on the SHA1 Compression Function. In: Breveglieri, L., Guilley, S., Koren, I., Naccache, D., Takahashi, J. (eds.) FDTC 2011, pp. 54–62. IEEE Computer Society, Los Alamitos (2011)

A New Energy-Efficient Routing Protocol Based on Clusters for Wireless Sensor Networks

Min Yoon and Jae-Woo Chang

Abstract Recently, Wireless Sensor Network (WSN) has been broadly studied in ubiquitous computing environments. In WSN, it is important to reduce communication overhead by using an energy-efficient routing protocol because the resources of the sensor node are limited. Although there exist some cluster-based routing protocols, they have some problems. First, the random selection of a cluster head incurs a node concentration problem. Secondly, they have a low reliability for data communication due to the less consideration of node communication range. To solve these problems, we, in this paper, propose a new energy-efficient cluster-based routing protocol. To resolve the node concentration problem, we design a new cluster head selection algorithm based on node connectivity and devise cluster maintenance algorithms. Moreover, to guarantee data communication reliability, we use message success rate, which is one of popular measures for data communication reliability, in order to select a routing path. Through our performance analysis, we show that our protocol outperforms existing schemes in terms of communication reliability and energy efficiency.

Keywords Sensor network · Routing protocol · Cluster based routing protocol · Selection of cluster head · Message success

M. Yoon · J.-W. Chang (✉)
Department of Computer Engineering, Chonbuk National University, Jeonju,
Republic of Korea
e-mail: jwchang@jbnu.ac.kr

M. Yoon
e-mail: myoon@jbnu.ac.kr

S.-S. Yeo et al. (eds.), *Computer Science and its Applications*,
Lecture Notes in Electrical Engineering 203, DOI: 10.1007/978-94-007-5699-1_12,
© Springer Science+Business Media Dordrecht 2012

1 Introduction

Wireless Sensor Network (WSN) has been broadly studied in ubiquitous computing environment. The WSN can be applied to various types of applications, such as environment management and military monitoring [1–3]. However, the sensor nodes which form WSN have resource constraints such as limited power, slow processor and less memory. So, it is essential to improve the energy efficiency to enhance the quality of application service. Energy consumption in WSN is of three types; aggregation, communication, and data processing. Among them, the amount of energy consumption for communication is the greatest; hence it is important to reduce communication overhead by using an energy-efficient routing protocol. There have been much research in the field of cluster based routing protocol [4–8]. The protocols choose a portion of nodes in the whole network as the cluster heads, and construct the clusters based on the cluster heads. However, the existing cluster-based routing protocols have some problems. First, the random selection of cluster head incurs a node concentration problem. Secondly, they have a low reliability for data communication due to the less consideration of node communication range. Actually, the existing methods do not consider the real obstacle, such as wall, making unstable network. Finally, data communication overhead is greatly increased because of sending all sensor node information to the sink node for constructing clusters.

To resolve these problems, we, in this paper, propose a new energy efficient routing protocol based on clusters. First, to solve the node concentration problem, we design a new cluster head selection algorithm based on node connectivity and devise cluster splitting and merging algorithms to construct the optimal clusters from initial clusters. Secondly, to guarantee data communication reliability, we use message success rate, which is one of the popular measures for data communication reliability, in order to select a routing path. Finally, to reduce data communication overhead, we consider only the information of neighboring nodes at both cluster construction and cluster head selection phases.

The rest of the paper is organized as follows. In Sect. 2, we present related work. In Sect. 3, we propose our scheme for providing energy efficiency in WSNs. In Sect. 4, we provide performance analysis of our scheme by comparing it with the existing schemes in terms of power consumption and communication reliability. In Sect. 5, we conclude this research with some future directions.

2 Related Work

There exists much research in the field of efficient energy consumption in WSN. Among the existing methods, cluster based routing methods is more popular in order to enhance the network lifetime. The cluster head deals with the data from cluster members and aggregates the data for sending them to the sink node.

This method significantly reduces the communication overhead to enhance the life time of sensor nodes. The most related work to our research are Low-Energy Adaptive Clustering Hierarchy (LEACH) and LEACH enhancements which we describe briefly as below.

LEACH [6] is one of the most popular hierarchical routing protocols for wireless sensor networks. The idea is to form the clusters of the sensor nodes based on received signal strength indicator (RSSI) and use local cluster heads as routers to the sink. The LEACH protocol includes two stages; node clustering and information transmission. In the clustering stage, a fraction P of all sensor nodes are chosen to serve as cluster heads. In this method, each node produces a random number between 0 and 1. Then the produced number is compared to a predetermined threshold to determine whether that node is a cluster head or not. If the random number be smaller than T, it is selected as cluster head. The other nodes must be allocated to a cluster head that is closest to them. In the information transmission stage, the cluster heads aggregate the data received from their cluster members and send the aggregated data to the base station by single hop communication. LEACH outperforms traditional clustering algorithms by using adaptive clusters and rotating cluster head, which can distribute energy consumption among all the sensor nodes. In addition, LEACH is able to perform local computation in each cluster to reduce the amount of data that must be transmitted to the sink node. This achieves a large reduction in the energy dissipation since computation is much cheaper than communication. However, LEACH assumes every node can directly reach a base station. However, one hop transmission directly to a base station can be a high-power operation and is especially inefficient when considering the amount of redundancy typically found in sensor networks. Moreover, the randomness in LEACH does not provide any optimal guarantees and may encounter rounds of communication.

Kang et al. [7] presented Low-Energy Adaptive Clustering Hierarchy-Centralized (LEACH-C) as a new cluster-based routing protocol, in order to distribute cluster heads evenly over the network and reduce energy dissipation. During the initial stage, each node has to send information about its current location and energy level to the base station. Therefore, the sensor nodes with remaining energy below the average node energy are restricted from becoming a cluster head. The base station runs an optimization algorithm to select cluster heads and divide the network into a set of clusters. Thus, LEACH-C requires the position of each node at the beginning of each round. However, the number of nodes in each cluster is not guaranteed to be a given value while forming clusters. Also, an expensive global positioning system (GPS) is required for sending initial information.

Xaingning et al. [8] described Multihop-LEACH protocol which improves communication mode from single hop to multi-hop between cluster head and sink. This method selects cluster heads based on the LEACH protocol. However, when transmitting data from the cluster head to the sink, the cluster head forwards the information by multi-hop communication among cluster heads.

3 Cluster-Based Routing Protocol Using Message Success Rate

Existing clustering based routing methods cause the unbalanced network problem because the cluster head is arbitrarily selected and the sensor nodes are concentrated in the partial cluster. In addition, LEACH and Multihop-LEACH do not consider the communication range of cluster head while generating route. Moreover, LEACH-C manages the information of all sensor nodes in the sink node, therefore the message transmission overhead increases.

To resolve these problems, we, in this section, propose a new cluster-based routing protocol using message success rate. First, to solve the node concentration problem, we design a new cluster head selection algorithm based on node connectivity and devise cluster split and merge algorithms to construct the optimal clusters from temporal clusters. Secondly, to guarantee data communication reliability, we use message success rate for selecting a routing path. Finally, to reduce data communication overhead, we consider only the information of neighboring nodes at both cluster construction and cluster head selection phases. Our cluster based routing algorithm consists of four phases; network information generation, cluster head selection, cluster formation, and cluster management. Each of the four phases of our algorithm, we describe as follows.

3.1 Network Information Generation

In this phase, each sensor nodes generate the network information by using the message success rate. For this, each sensor node searches for its own adjacent node. And it transmits the message to the adjacent node and measures the message success rate (MSR). By using the message success rate, each sensor nodes set the routing multi-hop path to sink node.

Definition 1 Message Success Rate (MSR) can be determined as given below. Here let *sPacekt* be number of sending packets to neighbor, *rPacket* be number of receiving packets in neighbor, and t be the total time of communicate between neighboring nodes.

$$AvgSuccRate = \frac{\sum rPacket(t)}{\sum sPacket(t)}$$

Using MSR, sensor nodes set the parent–child node and the routing path to the sink node through the parent nodes. This resembles Maximum Spanning Tree (MST) technique [9]. And each sensor node produces its own network information through the routing path. The network information of a sensor node is generated by communicating it with adjacent nodes. Table 1 shows network information in sensor nodes.

Table 1 Network information

Information	Description
NodeID	Sensor node ID
HopCount	Number of hops between sink node
NeighborID	Neighbor nodes' id within communication bound
NeighborCount	Number of neighbor nodes
NeighborMegSuccRate	Message success rate (MSR) with neighbor nodes
InitialChildnodeCount	Initial child nodes' count
InitialChildnodeID	Initial child nodes' ID
InitiaParentnodeID	Initial parent node ID

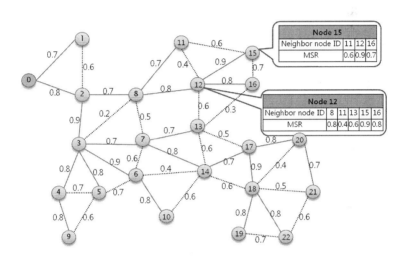

Fig. 1 Network information for constructing routing path

Figure 1 shows an example of the routing path construction with the network information. In Fig. 1, every sensor nodes measure the message success rate with the adjacent nodes. In this figure, the numeric value between sensor nodes is the message success rate between two nodes. The dotted line is communication path and solid line is the routing path. They are chosen based on the message success rate. For example, the node 15 has the node 11, 12, 16 as the adjacent nodes. And the node 12 is set to the parent node because its message success rate is the highest among them. Meanwhile, since the node 8 has the highest message success rate with the node 12, it is selected as the parent node of the node 12.

3.2 Cluster Head Selection

This phase selects cluster head node by considering the routing path and network information as mentioned above. Since the cluster head represents the cluster, it receives data from the nodes in the same cluster, aggregates the data, and sends them to the sink. According to the [6], the optimal rate of the number of cluster heads is 5 % over the total sensor node, so we assume that the optimal rate of cluster head over cluster members is 5 % in our scheme. The optimal rate of cluster head means the minimum number of head nodes in a cluster for reducing amount of energy consumption efficiently. The optimal rate of cluster head will be discussed in detail in Sect. 4. Because all nodes included in a cluster transmit data through the cluster head, it is essential to choose the optimal cluster head for reducing the amount of the energy consumption. The considerations for choosing the cluster head are as follows. First, the connectivity between a cluster head and neighboring nodes has to be high, meaning that the density of the cluster head node has to be high. Secondly, the message success rate has to be guaranteed to be higher than a given threshold value. Finally, a cluster head has to be positioned in the center of the network so that the number of hops may be uniform.

Definition 2 Weight of cluster head can be determined as below. Here let *NeighborCount* be number of neighbors, *NeighborMSR* be message success rate between neighboring nodes, and *HopCount* (number of hops between sink node) be its own number of hops. And let α, β, and γ be the threshold value for the cluster head selection ($\alpha + \beta + \gamma = 1$).

$$Weight = \alpha * NeighborCount + \beta * \sum (NeighborMSR) - \gamma * HopCount$$

3.3 Cluster Formation

In this phase, we construct initial clusters and use splitting and merging algorithms to generate the optimal cluster. The cluster generation phase consists of 3 steps; initial cluster construction, cluster splitting, and cluster merging. First, the initial cluster construction step constructs the clusters with the optimal rate of cluster head. The leaf nodes which do not have a child node traverse an upper node until satisfying the optimal cluster rate. This means that the lower nodes send the count of node to parent node and the parent node determine whether the count is large than the optimal cluster rate. If the optimal cluster rate is satisfied, a cluster is constructed. Secondly, the cluster splitting step splits into two clusters some initial clusters which do not satisfy the splitting rate δ. In particular, if the cluster has many nodes, the nodes can run out of energy sooner than others. So, the cluster having many nodes has to be split. When splitting a cluster, the parent node in the cluster chooses some part of cluster for splitting by considering the child nodes' information. In this step, the parent node chooses a part of cluster members whose lower nodes are more than those for the other parts. The chosen nodes are constructed as a new cluster. So, the clusters with the even number of nodes are constructed. Finally, the cluster merging step merges clusters which do not satisfy the merging threshold ε. To merge adjacent clusters, it considers the number of an adjacent cluster's members.

Fig. 2 Initial cluster
construction

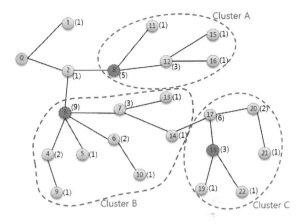

Figure 2 shows an example of construction of initial cluster when the optimal cluster rate is given 5. In this figure, the leaf nodes without a child node, i.e., nodes 1, 5, 9, 10, 11, 13, 15, 16, 19, 21, and 22, traverse their upper nodes until satisfying the optimal cluster rate 5. Since the node 17 satisfies the optimal cluster rate 5, it constructs a *cluster C* with its child nodes, i.e., node 17, 18, 19, 20, 21, 22, and sends the information on the cluster construction to its parent node. Then, the node 14 which receives the information from the node 17 sets the count of cluster member 1. In the same way, the network has *A, B, C* clusters with the respective cluster head node 3, 8, 18. In this figure, the *number* next to a node indicates the sum of the number of its child nodes and its own node for the cluster construction. Figure 3 shows an example of cluster splitting step when the threshold of splitting cluster is given 9. As shown in this figure, the *cluster B* doesn't satisfy the splitting threshold 9, so it is split into two clusters by considering the number of child nodes from the parent node in a cluster, In this figure, the node 7 and its child nodes are constructed as a new cluster (*cluster B-2*). In *cluster B-2*, the node 7 is selected as a cluster head by considering the weights in the cluster members. Figure 4 shows an example of cluster merging step when the merging threshold is given 3. In this figure, the number of member*s in cluster B-2* is lower than the threshold, the merging step is performed. In this figure, the *cluster B-2 is merged with the cluster A* among the adjacent *cluster of B-2, i.e., A, B-1, and C*. After two clusters are merged, a new *cluster A* chooses a cluster head by comparing node 7 with node 8. Finally, because node 1 and 2 do not satisfy the optimal cluster rate, they are clustered as *cluster D* with sink node.

3.4 Cluster Management

In this phase, we provide the periodic head replacement and the reconfiguration of the whole cluster for fault recovery. The node chosen as a cluster head node consumes a lot of energy than other nodes. Therefore, it is necessary to replace the

Fig. 3 Cluster splitting

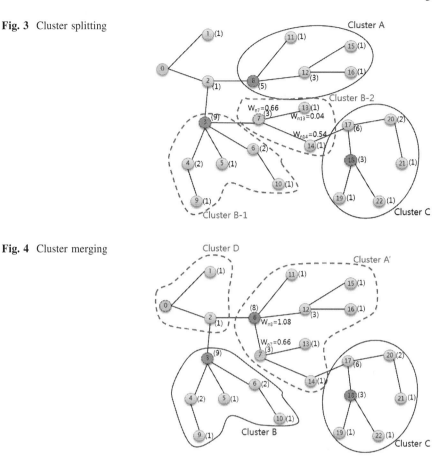

Fig. 4 Cluster merging

cluster head node periodically in order to extend the total network lifetime. The existing methods, like LEACH and multihop-LEACH, incur a node concentration problem and much energy consumption because of replacing cluster head at random. In this phase, we replace the cluster head node periodically by considering the connectivity between neighboring nodes and the number of hops, so that we can achieve good energy consumption. If the replaced head node has less energy than a given threshold, the whole cluster can be reconstructed by considering the remaining energy of nodes.

4 Performance Analysis

We compare our cluster-based routing scheme with the existing routing methods, such as LEACH and multihop-LEACH, under various setting. For this, we measure energy consumption, reliability of communication, standard deviation of

(a)   **(b)**   **(c)**

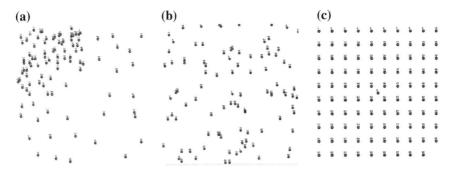

Fig. 5 Three data distribution. **a** skewed, **b** random, **c** uniform

cluster members. For this experiment, we also determine the optimal threshold for the weight used in selecting cluster head and the optimal number of cluster members in a cluster.

4.1 Experiment Setup

We develop our scheme in Network embedded system C (NesC) and ran the experiments on TOSSIM [10] simulator and TinyOS [11] embedded operating system with an Intel Xeon 3.0 GHz with 2 GB RAM. We set the maximum communication range with 25 m and the maximum of communication range of MicaZ [12] sensor node. We consider the scenario where every sensor node generates a message and send it to the sink node periodically. The energy consumption model used in the experiment is the free space model like the Eqs. (1) and (2). The Eq. (1) is for the amount of energy used during sending a message, and the Eq. (2) is for energy required during receiving a message. In this equation, $E_{elec}(k)$ means the number of k bits message using the transmit electronics, and $E_{amp}(k, d)$ means the number of k bits message using transmit amplifier. In transmit amplifier, it considers distance between sensor nodes.

$$\begin{aligned} E_T(k, d) &= E_{T\_elec}(k) + E_{T\_amp}(k, d) \\ &= E_{elec} * k + E_{amp} * k * d^2 \end{aligned} \tag{1}$$

$$\begin{aligned} E_R(k) &= E_{R\_elec(k)} \\ &= E_{elec} * k \end{aligned} \tag{2}$$

In our experiment, we set the three types of data distribution, i.e., uniform, Gaussian and skewed distribution, where each distribution has 100 nodes within 100 m * 100 m network area as shown in Fig. 5. Since LEACH-C is based on the GPS, we do not include it in our experiment.

Fig. 6 Deviation of weight
for choosing cluster head

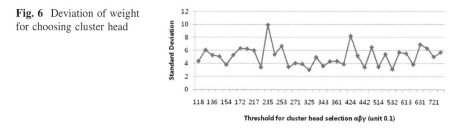

4.2 Cluster Weight

We provide the optimal weight for choosing cluster heads discussed in Sect. 3 and
the optimal rate of cluster head to construct cluster efficiently. Figure 6 shows the
standard deviation of number of cluster members by considering the optimal
weight, such that α is number of neighbor, β is message success rate with neighbor,
and γ is number of hops. Since the cluster head can be replaced according to the
weight, the weight affects the construction of cluster. As shown in the figure, we
provide 36 combination weights according to α, β, and γ ($\alpha + \beta + \gamma = 1$). For
example, 118 in X axis means that α is 0.1, β is 0.1, and γ is 0.8. When $\alpha = 0.3$,
$\beta = 0.2$, and $\gamma = 0.5$, the deviation gets a minimum value. So, the combination of
$\alpha = 0.3$, $\beta = 0.2$, and $\gamma = 0.5$ is determined as the optimal weight for choosing a
cluster head.

Figure 7 provide the deviation of cluster head rate. The optimal rate is calculated by Eq. (3). In (3), *numberofNodes* means total number of sensor nodes and
rateofClusterHeader means a rate of cluster header in whole network. Therefore,
optimalRate should be the optimal rate that all clusters have sensor node uniformly. We evaluate the deviation of number of cluster members when the rate of
cluster head is ranged from 2 to 10 %. It is shown that the lowest deviation is 4.2
when the rate of cluster head is 17 %. This means clusters are constructed well
when the rate is 17 %.

$$optimalRate = \frac{numberofNodes}{rateofClusterHeader} \tag{3}$$

4.3 Cluster Efficiency

To construct clusters in a uniform manner achieves good energy consumption. To
evaluate the good construction of clusters, we measure the standard deviation of
cluster distribution for our scheme and the existing methods. As shown in Fig. 8,
for all distribution cases, our scheme achieves much lower deviation on cluster
than the existing methods. In the uniform distribution, our method achieves 5.3
deviations whereas LEACH and multihop-LEACH are 8.5 and 9.1, respectively. In
terms of cluster efficiency, our method outperforms the existing methods because

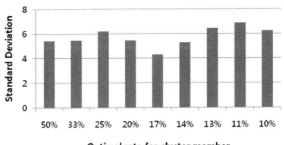

Fig. 7 Deviation of cluster head rate

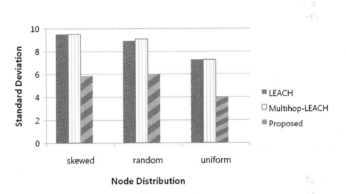

Fig. 8 Deviation of cluster distribution

our method performs the cluster splitting and merge algorithm for the optimal construction of clusters.

4.4 Energy Efficiency

The main goal of our method is to reduce communication cost for good energy consumption. So, we measure the average communication distance and the energy consumption of sensor nodes. The communication energy forms 80 % of the total energy consumption of a sensor node and the energy is consumed proportionally to the communication distance. If the communication distance is short, the communication energy can be saved. Figure 9 shows the energy consumption of sensor node. As shown in this figure, our scheme consumes lower energy than the existing methods. This is because our scheme has shorter communication distance according to the good selection of cluster heads.

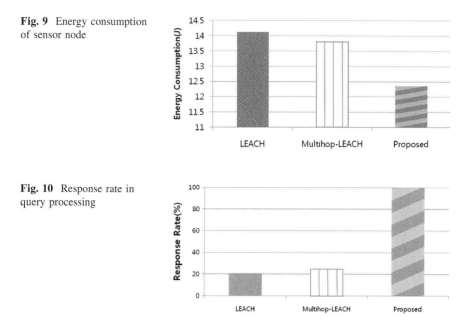

Fig. 9 Energy consumption of sensor node

Fig. 10 Response rate in query processing

4.5 Response Reliability

To verify the reliability of a routing protocol, we measure a response rate for query processing. For this, when a query is given to each sensor node, we measure the rate of messages returned to the sink node In Fig. 10, our scheme returns 99 % query result whereas LEACH and multihop-LEACH return 20 and 24 % results, respectively. The reason is that our scheme constructs a route such that an actual message is reachable by using message success rate. In addition, our scheme uses a recovery algorithm to reconstruct a new path when unexpected situations occur.

5 Conclusion

In this paper, we proposed a new energy-efficient routing scheme based-on cluster. In our scheme, we design a new cluster head selection algorithm based on node connectivity and devise cluster split and merge algorithms to conduct the optimal clusters from initial clusters. We use message success rate which is one of the popular measures for data communication reliability, to guarantee data communication reliability for network. To reduce data communication overhead, we use only information of neighbor nodes at both cluster construction and cluster head selection. Through our performance analysis, we show that our scheme constructs clusters in a more uniform way and provides better energy efficiency than the existing methods. In addition, our scheme provides 4–5 times more reliable rate than LEACH and multihop-LEACH. Hence our scheme can prolong the lifetime of resources-constraint WSNs than LEACH and multihop-LEACH.

As our future work, we will apply our scheme to real wireless sensor network platform in order to verify the reliability of our protocol.

Acknowledgments This research was supported by Basic Science Research program through the National Research Foundation of Korea (NRF) funded by the Ministry of Education, Science and Technology (grant number 2010-0023800).

References

1. Karl, H., Willig, A.: A short survey of wireless sensor networks, TKN-03-018 (2003)
2. Romer, K.: Programming paradigms and middleware for sensor networks. GI/ITG Workshop on Sensor Networks, Karlsruhe, pp. 49–54 (2004)
3. Survey: Wireless Sensor Networking out of the Lab, into Production. http://www.millennial.net/newsandevents/pressreleases/050824.asp, February 2006
4. Akkaya, K., Younis, M.: A survey on routing protocols for wireless sensor networks. Ad Hoc Netw. **3**(5), 352–349 (2005)
5. Abbasi, A.A., Younis, M.: A survey on clustering algorithms for wireless sensor networks. Comput. Commun. **30**(14–15), 2826–2841 (2007)
6. Woo, A., Culler, D.E.: A transmission control scheme for media access in sensor networks. In: Proceedings of the ACM/IEEE International Conference on Mobile Computing and Networking, ACM Press, New York, pp. 221–235 (2001)
7. Heinzelman, W.B., Chandrakasan, A.P., Balakrishnan, H.: An application-specific protocol architecture for wireless microsensor networks. In: Proceedings of the IEEE International Conference on Transactions on Wireless Communications, IEEE, vol. 1, no. 4, pp. 660–670 (2002)
8. Xiangning, F., Yulin, S.: Improvement on LEACH protocol of wireless sensor network. In: International Conference on Sensor Technologies and Applications, IEEE, pp. 260–264 (2007)
9. Eppstein, D.: Spanning Trees and Spanners. Handbook of Computational Geometry, Elsevier, Amsterdam, pp. 425–461 (1999)
10. Hill, J., Szewczyk, R., Woo, A., Hollar, S., Culler, D., Pister, K.: System Architecture Directions for Networked Sensors, vol. 35, no. 11, pp. 93–104. ACM, New York (2000)
11. Levis, P., Lee, N.: TOSSIM: A Simulator for TinyOS Networks, September, 2003
12. http://www.xbow.com/Products/Product_pdf_files/Wireless_pdf/MICAz_Datashee

MIDI-to-Singing Online Karaoke for English M-Learning

Hung-Che Shen and Chung-Nan Lee

Abstract This research reports a proposal for learning English by using MIDI-to-Singing Karaoke in a mobile environment. Such an application is about an ongoing work of a "MIDI-to-Singing" system which is capable of generating synthesized singing based on a given MIDI melody. Considering that the mobile-phone has had a boom in the recent years, we have developed an online Karaoke that makes lifelong learning something natural and effective for English m-learning. In this demonstration, MIDI music, synthesized singing voice, lyrics and background photos are well aligned to compose a Karaoke video. Based on a HTML5 GUI design, we describe existing implementations for putting subtitles and captions alongside the HTML5 <video> tag inside Web pages and a proposal for standardizing such Karaoke application. We believe that using the MIDI-to-Singing online Karaoke as an educational tool will contribute to the success of English learning students.

Keywords MIDI-to-Singing · HTML5 · Mobile devices · m-Learning

H.-C. Shen (✉)
Computer Science Department, I-Shou University, Kaohsiung, Taiwan, Republic of China
e-mail: shungch@isu.edu.tw

C.-N. Lee
Computer Science Department, National Sun Yat-Sen University, Kaohsiung, Taiwan, Republic of China
e-mail: cnlee@mail.cse.nsysu.edu.tw

S.-S. Yeo et al. (eds.), *Computer Science and its Applications*,
Lecture Notes in Electrical Engineering 203, DOI: 10.1007/978-94-007-5699-1_13,
© Springer Science+Business Media Dordrecht 2012

1 Introduction

Karaoke originated in Japan. The word comes from 'kara', empty and 'oke', orchestra. In this form of singing, the music tracks for specific songs are provided by the media, be it CD, computer or tapes, the vocals or the voice is provided by the individual playing Karaoke. Most Karaoke systems require the singer to use a microphone, however, with online Karaoke, one can sing along without the microphone. Karaoke has been accepted as an aiding tool for English learning [1]. According to Gipe [2], "Music can provide opportunities for students who experience difficulty with reading and writing to think and learn through another language that is repetitive, melodious, and emotional." Based on what Murphey [3] referred to as the "song-stuck-in-my-head phenomenon". Songs, with their repeated lyrics and rhythms, have been examined by researchers as possible tools for enhancing learning/memory of vocabulary development and other language competencies such as writing/reading skills, grammatical structures and pronunciation.

Most schools have data-projectors that can be hooked up to the computers with internet access in the classroom, with monitor screen projected on the screen. The words can be displayed on the screen as students sing along. This provides new possibilities, opportunities and challenges for the educational landscape. One of them can be Karaoke. The use of karaoke in the classroom can make the entire learning process more enjoyable [4]. In the last few years we've been starting to accept the computer as an artificial singer that performs MIDI-to-Singing as an English teaching aid. Some audio and video samples of English learning in vocabulary, grammar and pronunciation are provided at the MIDI-to-Singing online Karaoke website.

http://140.127.182.30/CS/cs.php?id=3 [5]

Based on a developed MIDI-to-Singing songs collection, our task in this paper is made somewhat easier by providing online English Karaoke for m-learning. M-Learning is often defined as learning that takes place with the help of portable electronic tools [6]. It is claimed by Attewell [7] that m-learning can help learners to learn independently, improve their literacy, remain more focused, increase motivation, encourage collaboration, raise self-confidence, etc. Although many educationalists in the field see great potential for the use of mobile devices in m-learning, there are currently very few successful implementations to consider as the best practice. We propose that using online Karaoke as a major tool for English m-learning, since m-Learning could be thought to be a form of 'informal' learning and this learning activities can take place when the users are on the move. In our experimental study, it is found out that the potential of using mobile phones in learning English language through Karaoke is promising. Usually, reading the text book in classroom will helps students familiarize themselves with the adapted Karaoke songs. Once they become familiar with the lyrics or text, they can sing along using Karaoke with mobile phones easily.

Three core areas of mobile learning are (1) Authoring and Publishing, (2) Content development, (3) Delivery and Tracking. Therefore, this work is organized as follows. In Sect. 2, we present authoring and publishing of MIDI-to-Singing Karaoke songs. After that, the design of content development is described in Sect. 3. Section 4 is about the delivery and tracking of Karaoke content. Finally, experimental results of m-learning are given in Sect. 5. Section 6 concludes this paper.

2 Authoring and Publishing

This section covers the MIDI-to-Singing system overview. This system allows user to create synthesized singing voices with a synchronized a MIDI score. In order to publish Karaoke on the website, the system can export files such as song.mp3 and song.lrc.

2.1 MIDI-to-Singing System Overview

The overview of a "MIDI-to-Singing" system is illustrated in Fig. 1. The MIDI-to-Singing system is implemented in the C++ programming language and a MIDI sequencer called "MaxSeq" [8]. MaxSeq is a windows sequencer written in C++ using MFC. Current SVS products are well suited to studio situations but would not be appropriate for live performance, where speed of lyrical entry and musical score alignment are the primary considerations. Therefore, we integrate both MaxSeq and Flinger [9] to extend their capabilities.

A MIDI-to-Singing system is the integration of a MIDI sequencer with singing voice synthesis. The final audio output is the mix of MIDI format music with a wave file of singing voice as shown in Fig. 2.

In order to show Karaoke lyrics on the website, a LRC file format is adopted as the MIDI-to-Singing output. Simple LRC format was introduced by Kuo (Djohan) Shiang-shiang's Lyrics Displayer [10]. It was one of the first programs, if not the first, that attempted to simulate Karaoke performance. It usually displays a whole line of lyrics, but it is possible to display a word at a time, such as one would see in modern Karaoke machines, by creating a time tag for each word rather than each line. The Line Time Tags are in the format [mm:ss.xx] where mm is minutes, ss is seconds and xx is hundredths of a second. The following is a normal example.

[mm:ss.xx] lyrics line 1
[mm:ss.xx] lyrics line 2
...
[mm:ss.xx] last lyrics line

In GUI considerations, we design a lead sheet notation for the composed MIDI music. A lead sheet is a musical composition represented by a chord progression, lyrics and a melody line, usually on a single musical staff. In order to change the lyrical content at performance time, we provide "cut and paste" of lyrics between

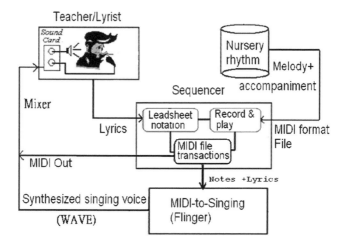

Fig. 1 MIDI-to-Singing system overview

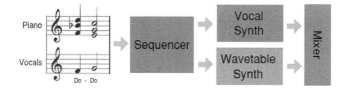

Fig. 2 The mixer of MIDI music and a wave files

an editor and a text box. Figure 3 shows the interface for inputting lyrics for a MIDI melody. A large amount of nursery rhyme MIDI songs with lyrics that we use are downloaded from the Internet website [11]. Lyrics are given in an ordinary text file, with manually inserted white spaces for separating syllables and the individual notes. To obtain different voices, the user can select a singer from the "singer menu" for singing.

3 Content Development

In the following, a couple of online Karaoke GUI design issues are explained briefly.

3.1 GUI Design

The GUI design of online MIDI-to-Singing online Karaoke will contain the following features:

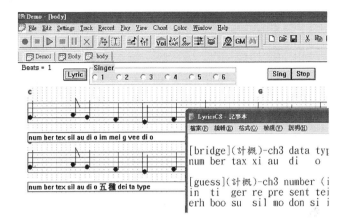

Fig. 3 The MIDI-to-Singing interface for inputting lyrics

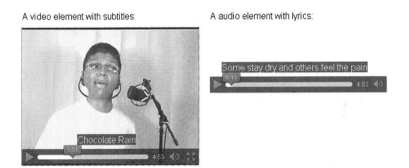

Fig. 4 Video and audio accessibility approaches with the new HTML5 video tag

- User can select a song from a list of songs which are systematically categorized.
- The user can create and manage songs of his/her own playlist/list of favorite songs.
- When a particular song is selected, the lyrics of the selected song can be seen on the screen.
- The lyrics of the song are highlighted a word at a time in synchronization with the music.
- Download a set of lessons to help students learn the basics of English language online or in their mobile devices (offline).
- Pronunciation exercises where students can record their singing pronunciation, and if possible, compare it with the correct one.
- Questions and answers that validate the understanding of reviewed videos.

Figure 4 is the demo that was created by Silvia Pfeiffer [12] for Mozilla web browser to explore video and audio accessibility approaches with the new HTML5 video tag.

```
<video class="v" src="chocolate_rain/chocolate_rain.ogv"
       poster="chocolate_rain/chocolate_rain.png" controls aria-label="Chocolate Rain video"
       autobuffer title="Chocolate Rain video">
  <itext id="lyrics" lang="en" type="text/lrc" display="yes"
         src="chocolate_rain/chocolaterain.lrc" category="LRC"></itext>
</video>

<audio class="a" src="chocolate_rain/chocolate_rain.ogg"
       controls aria-label="Chocolate Rain song"
       autobuffer title="Chocolate Rain song">
  <itext id="lyrics" lang="en" type="text/lrc" display="yes"
         src="chocolate_rain/chocolaterain.lrc" category="LRC"></itext>
</audio>
```

Fig. 5 HTML5 audio and video player

With some customized HTML5 [13] and JavaScript, the online Karaoke interface can be customized as we want. Figure 5 shows the main elements of the above Web page. It contains subtitles for the video and lyrics for the audio, both using a file written in lrc format the subtitles and lyrics are displayed as overlays for the video and audio elements into div elements that have been pre-defined for them.

4 Delivery and Tracking

We used a pre-constructed blog in order to suggest ideas regarding the use of mobile phones in the English learning process and to inquire about this use. The teacher informed the students that they could ask questions in the blog, add remarks, add feedback, comment, document events and actions and write about their feelings regarding the experiment. They also asked the students to write about their expectations and the activities that they would like to be engaged with.

The teachers interviewed each participant for 20 min about her/his English learning using mobile phones. The interviews were semi structured to analyze and characterize the students' perceptions of this learning.

In order to reach this aim the authors have sought answers to the following questions:

• What are the opinions of students about the mobile phone-based learning system?
• Are there significant differences between the pretest and posttest results?
• What are the students' suggestions for the development of the system?
• How often should the words be sent?
• During which hours of the day should the words be sent?

5 Experimental Results

This analysis was based on the students' comments in the blog. Then, we used open and axial coding to categorize the students' perceptions of the English learning qualities and the characteristics of the learning environment when the learning took place in the mobile phone environment. This coding was done based on the students' comments in the blog and in the interviews. Our students perceive the environment of English learning using mobile phones as having different characteristics than the traditional one. These characteristics enrich their learning and make it more enjoyable. Some students identified herself with the experiment because it makes learning English easier, simpler, and collaborative. It should be noted that we consider the novelty of the experiment to be the main reason for the student's astonishment, because many students mentioned this novelty in the blog. We can see that the novelty of the experiment is a main factor for the students' participation and involvement. Moreover, the students expected the experiment to be enjoyable and a fun one, and this expectation was realized because of the interesting learning qualities and the mobile phone features. These fun feelings motivated the students making them like the experiment and identify with it, and consequently believe in its success. The students were also interested in where they learn the mathematical material because: they enjoy learning outdoors, outdoor learning answers children's natural curiosity and enthusiasm. Learning with fun encouraged them to learn.

6 Conclusion

This is part of ongoing research and development MIDI-to-Singing system. It has being used as a teaching aid in university classroom. To have successful permanent learning in anytime, anywhere and any device, the MIDI-to-Singing online Karaoke website could be particularly useful in increasing vocabulary and enhancing fluency in reading. As English learners immerse themselves in music and interact with text, a new dimension of their learning behavior starts to emerge. During Karaoke process, learners are engaged in both aspects of reading and singing that promote fluency. Additionally, the reading skills acquired using a Karaoke build confidence among users who are then able to transfer the skills to academic reading. English learners who dislike reading aloud in front of class members and others, find use of Karaoke less threatening as they sing aloud in a small group or with the MIDI-to-Singing virtual singer or with music in the background as opposed to total silence.

HTML5 is a powerful tool for listening to audio via the Internet, as powerful as anything the "cloud" services have to offer and much more versatile. Coding audio into HTML5 pages is fairly straightforward. Finally, we feel low-resolution scrubbing is particularly suited for mobile devices, as it reduces both bandwidth and CPU load. Our implementation should work with minimal modification on HTML5 supported mobile devices, such as the iPad, and it would be interesting to evaluate such a Karaoke audio/video implementation. Many online-video navigation tasks are demonstrated that low-resolution real-time scrubbing can significantly improve Karaoke performance, and provided a simple HTML5 compatible implementation. Given today's prevalence of online streaming video sites, we feel online Karaoke m-learning is important and timely contribution.

References

1. Gupta, A.: Karaoke: a tool for promoting reading. Read Matrix **6**(2), 80–89 (2006)
2. Gipe, J.P.: Multiple Paths to Literacy: Classroom Techniques for Struggling Readers. Prentice Hall, Merrill (2002)
3. Murphey, T.: The song stuck in my head phenomenon: a melodic din in the LAD? System **18**(1), 53–64 (1990)
4. Shen, H.C., Lee, C.N.: Playing MIDI-to-singing songs in computer science class to reinforce concepts. In: 6th International WOCMAT and New Media Conference, Chungli, Taiwan (2010)
5. MIDI-to-Singing English Karaoke website. http://140.127.182.30/CS/cs.php?id=3 (2012)
6. Quinn, C. mLearning: Mobile, Wireless, in Your Pocket Learning. Linezine, from http://www.linezine.com/2.1/features/cqmmwiyp.htm (2007)
7. Attewell, J.: Mobile Technologies and Learning: A Technology Update and m-Learning Project Summary. Technology Enhanced Learning Research Centre, Learning and Skills Development Agency. Learning and Skills Development Agency, London (2005)
8. Messick, P.: Maximum MIDI, pp. 105–204. Manning Publication, Greenwich (1998)
9. Flinger www. http://cslu.cse.ogi.edu/tts/flinger/ (2006)
10. Information about LRC. http://www.mobile-mir.com/en/HowToLRC.php (2012)
11. Information about nursery rhyme. http://www.kididles.com/ (2012)
12. Pfeiffer, S. Video demo, http://www.annodex.net/~silvia/itext/chocolate_rain.html (2012)
13. HTML5 in General: http://www.w3.org/TR/html5 (2012)

A New Cloaking Method Based on Weighted Adjacency Graph for Preserving User Location Privacy in LBS

Miyoung Jang and Jae-Woo Chang

Abstract The propagation of position identifying devices, such as Global Positioning System (GPS), becomes increasingly a privacy threat in location-based services (LBSs). However, in order to enjoy such services, the user must precisely disclose his/her exact location to the LBS. So, it is a key challenge to efficiently preserve user's privacy while accessing LBS. For this, the existing method employs a framework that not only hides the actual user location but also reduces bandwidth consumption. However, it suffers from privacy attack for a long term observation of user behaviour. Therefore, we aim to provide the solutions which can preserve user privacy by utilizing k-anonymity mechanism. In this paper, we propose a weighted adjacency graph based k-anonymous cloaking technique that can provide protection to user and also reduce bandwidth usages. Our cloaking approach efficiently supports k-nearest neighbor queries without revealing private information of the query initiator. We demonstrate via experimental results that our algorithm yields much better performance than the existing one.

Keywords Location privacy · Location-based services (LBS) · Privacy preservation · Location cloaking · k-Anonymity · Voronoi diagram

M. Jang · J.-W. Chang (✉)
Department of Computer Engineering, Chonbuk National University,
Jeonju, South Korea
e-mail: jwchang@jbnu.ac.kra

M. Jang
e-mail: brilliant@jbnu.ac.kra

S.-S. Yeo et al. (eds.), *Computer Science and its Applications*,
Lecture Notes in Electrical Engineering 203, DOI: 10.1007/978-94-007-5699-1_14,
© Springer Science+Business Media Dordrecht 2012

1 Introduction

Location-based services allow users to connect with others based on their current locations. In most cases, people use their positioning devices (e.g., Mobile phone, PDA, Navigators) to find out his/her location like restaurants, bars and stores they visit. However, frequent and continuous accesses to the services expose users to privacy risk. A location-based service provider might be able to collect private and delicate information about a user's choices and habits from the user's location. For example, for a k-nearest neighbor query (KNN), the user (u) requests to LBS about the nearest dental clinic from his/her location and based on u's location LBS server returns nearest dental clinic name to u. For this LBS server can infer user health conditions that might be harmful for user's privacy. Due to an increasing awareness of privacy risks, users might desist from accessing LBSs, which would prevent the proliferation of these services [1].

Current research aligns on developing techniques to elaborate on k-anonymity that preserve user privacy during the access of LBSs. Sweeny proposed k-anonymity [2] approach which tries to blur a user's location among k-1 users. Most of the cloaking techniques enclose more non-result objects with original result due to the achievement of a user's privacy. However, larger result set size preserves more privacy, but consumes more network bandwidth and device battery as well. Furthermore, the crucial matter is how to minimize the number of non-result objects during cloaking period.

The location cloaking approach called 2-Phase Asynchronous Search (2PASS) has been proposed that minimize the bandwidth usages as well as preserve user privacy. Basically, 2PASS follows client–server model. It is based on a notion of Voronoi cells and each cell contains one object that is the nearest neighbor of any point in its cell. The user can fix the cloaked area to the Voronoi cell of the nearest neighbor object, if he/she knows the Voronoi cells in advance. 2PASS is able to save the bandwidth usages and processes a KNN query in two steps. In the first step, the user requests the Voronoi cell information corresponding to the query. In the second step, it selects objects to request. Although 2PASS optimizes the bandwidth usages, it suffers from privacy attack.

In this paper, we propose a weighted adjacency graph based k-anonymous cloaking technique that can reduce bandwidth usages and provide protection to user. We follow user-cloaking-server model where the trusted third party (location cloaker, LC) performs location cloaking for the user. Our algorithm computes a KNN query in three phases. In the first phase, the user requests the location-cloaker (LC) and LC requests the WAG information corresponding to the query. LC selects objects to request in the second phase and returns the actual result to the user in the third phase. We also add k-anonymity property for enhancing users' privacy while accessing LBSs. Our contributions can be summarized as follows:

- We propose a weighted adjacency graph based k-anonymous cloaking technique to reduce bandwidth usages of requested services.

- We use location cloaker between user and LBS server to handle cloaking operation.
- We also consider k-anonymity property for enhancing user's privacy.

The rest of this paper is organized as follows. In Sect. 2, we discuss related cloaking methods. We discuss our detailed system architecture and propose a weighted adjacency graph based k-anonymous cloaking technique in Sect. 3. Section 4 is devoted to experimental results. Finally, we conclude our work with future direction in Sect. 5.

2 Related Work

Recently, considerable research interest has focused on preventing identity inference in location-based services. The main concern is to allow the mobile user to request services without compromising his/her privacy. We classify these techniques into three main groups: pseudonym, dummy and cloaking.

2.1 Pseudonym-Based Techniques

Pseudonym combines the location and user identity called Mix zone [3]. For this, the server only receives the location without the user identity. However, such a technique is limited to those location-based services that do not require the user's identity.

2.2 Dummy-Based Techniques

In dummy based technique [4], it generates false user locations called dummies and combine them with the actual user location into the request. When a user generates a query of asking near-by restaurants, this approach sends the user's location with dummies' locations, and then the LBS provider returns lists of restaurants, which are close from each of the locations in the query (the user's and dummies' locations). The user can refer to the list corresponding to his/her actual location, i.e., filter out the other lists. However, the server may infer the actual user location from dummies after monitoring long-term movement patterns of the user.

2.3 Cloaking-Based Techniques

In cloaking based technique [5–9], it generates an area (circular or rectangular) that encloses an actual query issuer with other users based on his/her privacy requirement (e.g., k-anonymity, granularity metric etc.). By this, the user can hide his/her location from adversary or LBS server.

To the best of our knowledge, there exists only one research on result-aware location cloaking approach (called 2PASS), proposed by Hu and Xu [9]; based on client–server model the algorithm minimizes the number of requested objects directly. Basically, 2PASS [9] is based on the notion of Voronoi cells and each cell contains one object that is the nearest neighbor of any point in its cell. To access the Voronoi cell information, they develop a weighted adjacency graph (WAG). WAG is a weighted undirected graph that stores the Voronoi diagram and Delaunay triangulation. The specialty of this graph is to notify that the WAG vertices are weighted based on Voronoi cell area size. User can compute out the objects to request from the server by using WAG-tree index. To reduce computational overhead, 2PASS also proposes a method to partition the entire WAG into WAG snippets of reasonable size so that the user receives only the snippets surrounding the query location.

3 K-Anonymous Cloaking Technique Based on Weighted Adjacency Graph

3.1 Motivation

Most of the related work [5–9] has been proposed for different privacy metrics (e.g., k-anonymity, l-diversity etc.); their objective is always to minimize the size of the cloaked region, and thus minimizing the communication bandwidth between the user and the service provider. However, previous work did not consider user's location privacy and the reduction of bandwidth usage at the same time. To the best of our knowledge, there exists only one research using Voronoi diagram based WAG-tree index for saving bandwidth usage as well as considering user location privacy, which was proposed by Hu and Xu [9]. However, it supports client–server model, where the user directly send a query to the LBS server and LBS finds out the nearest objects (e.g., hospitals, restaurants, etc.) for the user. As a result, LBS server might infer the actual user position and acts as an adversary.

To solve the problems of existing work, we first propose a new method which is based on WAG indexing and it follows trusted third party (cloaker) model. By this, a cloaker can handle cloaking procedure rather than user. Our method computes a KNN query in three phases. In the first phase, the user requests the location cloaker (LC) and LC requests the WAG information corresponding to the query. LC selects objects to request in the second phase and returns the actual result to the user in the third phase. In this paper, we improve a WAG-tree index by considering k-anonymity property due to enhancing user privacy while accessing LBSs. In this way, our method is able to provide users' location privacy as well as to save bandwidth usages.

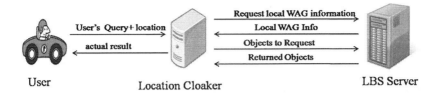

Fig. 1 System architecture

3.2 System Architecture

We first describe our system architecture that is based on client-cloaker-server architecture. Our approach to address the range nearest neighbor queries is based on the weighted adjacency graph (WAG) that encloses Voronoi diagram.

Using WAG we propose k-anonymous cloaking framework that adopt trusted third party model. This choice is made by the following reasons. First, the third party location cloaker (LC) acts as a mediator between user and the server, and performs location cloaking. Second, by LS, we ensure that a third party will honor the user privacy requirements. In this paper, we focus on granularity metric like 2PASS [9], so location cloaking generates a minimized cloaked region that encloses the user's genuine location and whose area is no less than a user specified threshold τ. Figure 1 shows the system architecture of our method. In the first phase, a user sends a query to LC and LC requests LBS for iWAG (improved WAG) information, where the weight of a vertex is based on the area of the corresponding Voronoi cell and the number of users on that cell. In the second phase, LC selects objects from the iWAG (e.g. three restaurants Pizza Hut, MacDonald and KFC) and requests them for their complete contents (e.g., customer reviews and reservation status etc.). In third phase, LC sends the exact answer to the user. In our system architecture, LC is responsible for cloaking procedure instead of user. Thus, the LBS server may not be used to infer a query issuer user.

To access the Voronoi cell information, each vertex is also assigned a non-negative weight. The difference is that the WAG vertices are weighted based on the size of Voronoi cell area and the number of users on that cell. So, we call this improved WAG (iWAG).

3.3 K-WAG Algorithm

In this section, we propose weighted adjacency graph based k-anonymous cloaking technique (k-WAG) to solve the problems of the 2PASS [9]. In k-WAG, user sends a query with privacy requirement to location cloaker (LC) and LC requests the objects based on the Voronoi cell information to satisfy the privacy requirement on the cloaked region. To minimize the object number while still meeting the privacy threshold τ and k-anonymity requirement, the criteria of object selection are a combination of the following: (i) the sum of the areas of Voronoi cells from the selected objects must exceed τ and the number of users $\geq$ k on that cell; (ii) the genuine nearest neighbor o* must be selected; and (iii) these Voronoi cells must be connected, i.e., no cell is isolated from the rest of the cells.

To achieve this, we describe the iWAG generation algorithm and iWAG-tree construction algorithms. First, the iWAG generation algorithm make use of the weight (w) for each object which is combination of Vw and Va. The Vw is calculated based on Voronoi cell area size and Va means number of user (Un) on that Voronoi cell. We set the priority for Voronoi cell area size and the number of user in that cell. For example, if we consider the total priority, $p = (\alpha + \beta) = 1$, then the preference of the number of user (β) is get priority than the preference of Voronoi cell area size (α). Therefore, the following equation holds true

$$W = (Vw \times \alpha) + (U\_n/(total\ U\_n) \times \beta) \qquad (1)$$

Figure 2a shows Voronoi diagram with eight users. We calculate objects weight based on Eq. (1). For example, if we consider object a, then weight of aw = 0.206. By this, we get all of objects' weight as shown in Fig. 2b. Next, we describe three algorithms. Firstly, Algorithm 1 describes the iWAG generation procedure. It calculates each Voronoi cell area size in line 1 and then computes how many users exist on that cell in line 2. Based on (line 1 and 2), it calculates vertex weight by using Eq. (1) in line 3. It sets all vertex weight in line 4 and ends the algorithm. Actually, we consider the total vertex weight is 1. Our objective is to find out the valid weight connected component based on iWAG. For this, we follow approximate minimum valid weight connected component (MVWCC) algorithm [9].

---

iWAG Generation Algorithm
Input: V_a : Voronoi cell
U_n : # of user for each Voronoi cell
Output: V_w : vertex weight
1: compute area size for each Voronoi cell (V_a)
2: calculate number of user on that cell (U_n)
3: give vertex weight (V_w)
4: set WAG (sum of the vertex weight)   //end of algirhtm

Second, the iWAG-tree is constructed by top-down approach. At first, it checks two conditions: (i) area range and (ii) object's (Point of Interest) number. Then it builds snippets if satisfying given conditions. Otherwise it partitions the whole space into four parts. The algorithm maintains objects whose Voronoi cells in the whole space overlap this sub-space; it is essentially the k-range nearest neighbor

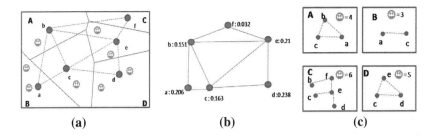

Fig. 2 Voronoi diagram and iWAG. **a** Voronoi diagram, **b** improved weighted adjacency graph (iWAG), **c** iWAG snippets

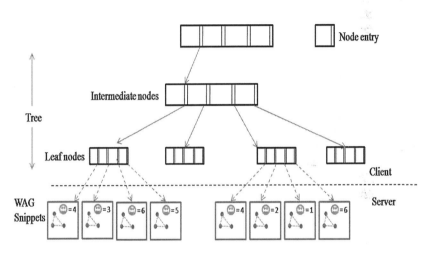

Fig. 3 iWAG tree

(kRNN) of this space. The algorithm recursively computes range nearest neighbor of a child node until satisfying the certain criterion. Figure 3 shows the iWAG-tree and snippet pointed by it.

Finally, our k-WAG query processing algorithm works as follows. The whole iWAG tree is sent to LC during the system initialization time. Based on the kNN, the LC traverses the iWAG tree and finds out the snippet that contains the query point. And the LC matches the privacy requirements (k-anonymity, the area size (τ) etc.) that are sent by the user. If the area of this snippet is still smaller than the user specified requirements, the user will locate the lowest-level child node of this snippet whose area-exceed privacy requirements. And the user requests all snippets rooted at this node, called host snippets. The LC then adds the received host snippets into a single iWAG and calculates the minimum valid connected components by using MVWCC algorithm [9]. In this process, LC does cloaking procedure instead of user and LC does not provide any location information or privacy requirements of user to the server.

Table 1 Parameter settings

Parameters	Range
Total user	239,668
Query number	70,000
Granularity threshold (τ)	0.000001, 0.00001, 0.0001, 0.001
Maximum area of WAG snippet	0.001
K-Anonymity	2, 4, 6, 8, 10
Average number of user in each cell	2

Fig. 4 Query processing time versus τ

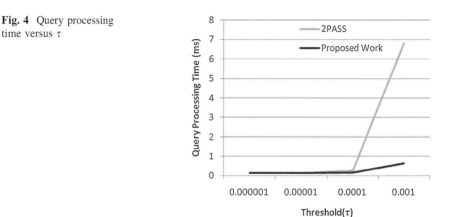

4 Performance Analysis

4.1 Experimental Setup

We run the experiment by using Visual Studio 2010 on Window XP operating system with Intel® Xeon® CPU 2.00 GHz and 2 GB memory. Our system consists of three main components; the mobile user, the location cloaker and the LBS server. For the mobile user, we assume that each Voronoi cell encapsulates at least 2 users. We use the real data set of Northern East America (NE) that contains 119,898 point of interest (POIs). For easy presentation, the coordinates of these objects are normalized to a unit square.

We compare our work with the exiting approach 2PASS [9] in terms of response time and bandwidth size. The parameter settings are summarized in Table 1.

4.2 Performance Evaluation

Figure 4 demonstrates the query processing time with different τ value. Since a bigger threshold value usually contains more underlying network area, it takes longer to process. As shown in Fig. 4, the increase of former metric is quite

Fig. 5 Average number of results versus τ

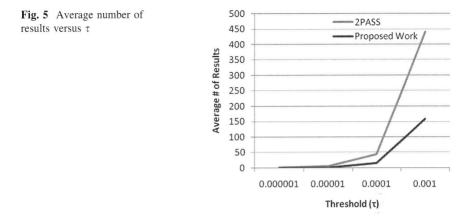

moderate until $\tau \leq 0.0001$ On the other hand, the latter metric linearly increases as τ grows. The response time of 2PASS is larger than that of our scheme, which is mainly due to the more number of objects in our scheme. As a consequence, 2PASS also consumes more bandwidth than the proposed one.

We calculate the bandwidth size based on average number of objects returned by varying with different thresholds. Figure 5 depicts the result set size with different τ. We observe that a bigger τ value generates a larger candidate result set.

5 Conclusion

We identify the limitations of privacy-ware data access in location-based services. We propose a weighted adjacency graph based k-anonymous cloaking technique that allows k-anonymity property for providing the location privacy of all users in the network. Our technique follows third party based approach where the location cloaker handles cloaking process instead of user. We also minimize the bandwidth consumption by using iWAG-tree index from which the location cloaker can compute out the objects to request from the server. Through our experimental performance evaluations, we have shown that our cloaking method is much more efficient in terms of both response time and bandwidth consumption than the 2PASS.

In future work, we plan to extend our work beyond a larger geographical area. We also plan to conduct the extensive performance evaluations by our experiments with various data sets and study the behavior of our work when the user issues a series of requests within a short period.

Acknowledgments This research was supported by Basic Science Research program through the National Research Foundation of Korea (NRF) funded by the Ministry of Education, Science and Technology (grant number 2010-0023800).

References

1. Muntz, W.R., Barclay, T., Dozier, J., Faloutsos, C., Maceachren, A., Martin, J., Pancake, C., Satyanarayanan, M.: IT Roadmap to a Geospatial Future. The National Academics Press, Washington (2003)
2. Sweeney, L.: k-Anonimity: a model for protecting privacy. Int. J. Uncertain. Fuzziness Knowl. Based Syst. **10**(5), 557–570 (2002)
3. Beresford, A., Stajano, F.: Location privacy in pervasive computing. IEEE Pervasive Comput. **2**(1):46–55 (2003)
4. Kido, H., Yanagisawa, Y., Satoh, T.: An anonymous communication techniques using dummies for location based services. In: Proceeding 2nd ICPS, pp. 88–97 (2005)
5. Mokbel, M.F., Chow, C.Y., Aref, W.G.: The new casper: query processing for location services without compromising privacy. In: Proceedings of the International Conference Very Large Database (VLDB), pp. 763–774 (2006)
6. Kalnis, P., Ghinita, G., Mouratidis, K., Papadias, D.: Preventing location-based identity inference in anonymous spatial queries. IEEE Trans. Knowl. Data Eng. **19**(12), 1719–1733 (2007)
7. Bamba, B., Liu, L., Pesti, P., Wang, T.: Supporting anonymous location queries in mobile environments with privacy grid. In: Proceeding of the International Conference World Wide Web, pp 237–246, April 2008
8. Mokbel, M.F.: Towards privacy-aware location based database servers. In: Proceeding of the 22nd IEEE International Conference on Data Engineering (ICDE) Workshop, Atlanta, Georgia, USA (2006)
9. Hu, H., Xu, J.: 2PASS: Bandwidth-optimized location cloaking for anonymous location-based services. IEEE Trans. Parallel Distrib. Syst. **21**(10), 1458–1472 (2010)

Simulation Videos for Understanding Occlusion Effects on Kernel Based Object Tracking

Beng Yong Lee, Lee Hung Liew, Wai Shiang Cheah and Yin Chai Wang

Abstract Occlusion handling is one of the most studied problems for object tracking in computer vision. Many previous works claimed that occlusion can be handled effectively using Kalman filter, Particle filter and Mean Shift tracking methods. However, these methods were only tested on specific task videos. In order to explore the actual potential of these methods, this paper introduced 64 simulation video sequences to experiment the effectiveness of each tracking methods on various occlusion scenarios. Tracking performances are evaluated based on Sequence Frame Detection Accuracy (SFDA). The results showed that Mean shift tracker would fail completely when full occlusion occurred. Kalman filter tracker achieved highest SFDA score of 0.85 when tracking object with uniform trajectory and no occlusion. Results also demonstrated that Particle filter tracker fails to detect object with non-uniform trajectory. The effect of occlusion on each tracker is analyzed with Frame Detection Accuracy (FDA) graph.

Keywords Computer vision · Object tracking · Occlusion handling

B. Y. Lee (✉) · L. H. Liew
Universiti Teknologi MARA (UiTM), Kota Samarahan, Sarawak, Malaysia
e-mail: bengyong@gmail.com

L. H. Liew
e-mail: lhliew@sarawak.uitm.edu.my

W. S. Cheah · Y. C. Wang
Universiti Malaysia Sarawak (UNIMAS), Kota Samarahan, Sarawak, Malaysia
e-mail: wscheah@fit.unimas.my

Y. C. Wang
e-mail: ycwang@fit.unimas.my

S.-S. Yeo et al. (eds.), *Computer Science and its Applications*,
Lecture Notes in Electrical Engineering 203, DOI: 10.1007/978-94-007-5699-1_15,
© Springer Science+Business Media Dordrecht 2012

1 Introduction

Object tracking is a process of continuously identifying the location of tracked objects in a video from frame to frame since the moment of first appearance of the object until it exit the video. However, during the time of the object presence in the video, the tracked object can be partially or fully occluded by other object in a video. Occlusions happen due to three reasons. Firstly, an object can be occluded when it is blocked by the background structure such as the building pillar or furniture in a room [1]. Second, occlusion can happen when other moving foreground objects overlap the tracked object [2, 3]. Lastly, occlusion happens when track features are blocked from camera view when the tracked objects turn away from camera. This is known as self-occlusion [3].

Many tracking methods have been proposed in handling occlusion using selected video samples. The selected video samples are usually obtained from actual recording by the authors or from benchmark dataset such as PETS [4] and ETISEO [5]. These video dataset provide a good impression on the performance of the proposed tracking method in real world. However, the complex scenario in the video such as shadow, illumination changes and moving background could obscure the evaluation of the actual performance of the tracking methods.

Therefore, we propose to run the experiments on simulation videos sequences. According to Taylor et al. [6], simulated video data is ideal to provide a good indication of which algorithms work well in a given scenario. In addition, simulated video can provides accurate ground truth for performance evaluation. In simulation videos, an ideal environment without any noise and distraction can be created. The results contained under such environment could reflect the actual performance of the tested tracking methods. It will also be easier to analyze the effect of occlusion in simulation videos since the environment and interaction between objects can be controlled.

2 Kernel Based Object Tracking

Tracking methods can be classified into three categories based on the features used in tracking, namely the point tracking, silhouette tracking and kernel based tracking (Yilmaz et al. [7]). Among the three tracking methods, kernel based tracking is widely used because it could provide high accuracy tracking and it has lower computational cost than silhouette based tracking. Many of the current works on kernel based tracking focused on Kalman Filter, Particle Filter and Mean Shift tracking (Comaniciu et al. [8]).

Comaniciu et al. [9] and Yilmaz [10] have developed trackers based on Mean Shift to derive target object candidate based on appearance model similarity. The results from their research show that the Mean Shift tracker (MS) is robust to partial occlusion, background clutter, target scale variations and rotations in depth.

However, both authors did not discussed on the Mean Shift tracker performance under full occlusion scenario.

Mirabi and Javadi [11] and Wang et al. [12] experimented Kalman Filter tracker (KF) on real-world video sequences. Their proposed system is proven to track object successfully under difficult situations such as noise, shadow and illumination changes. However their experiments used limited video data and the tracking result is not discussed quantitatively.

Particle filters provide robust tracking of moving objects in a cluttered environment especially in tracking moving object that move in non-linear and non-Gaussian trajectory [13]. Cho et al. [13] has used Particle Filter tracker (PF) with images' grey level model to track the moving object while Liang et al. [14] combined Particle Filter tracker with color and shape model to track objects in video sequences.

Based on the reviews of these previous works [9–14], the accuracy of the tracking experiments was not discussed nor compare statistically with other tracking methods. Besides, authors used real-world video for testing cannot extensively discuss the limitation of their proposed method because the result could be affected by too many factors such as background noise, object's shadow, object's shape transformation and illumination changes.

Therefore, to reduce complexity in video samples, we use simulated video sequences to control the video scene. With controlled scene setup, the results obtained from the experiments can reflect the actual ability of trackers in handling occlusion.

3 Video Data

All simulation videos in this paper were created using OpenGL and Visual C++. Each video sequences has 60 frames with resolution of 784×562 pixels. Two set of simulation video sequences with moving object are generated. The first set simulates a moving ball with uniform trajectories in eight directions (Fig. 1) while second set with non-uniform trajectories (Fig. 2).

Each set of the video sequences has four occlusion scenarios, i.e. no occlusion, partial occlusion, full occlusion and long occlusion. Occlusions in video sequences are created by placing a red cube with various sizes at the middle of the video frame. To simulate partial occlusion, we placed a small cube with half the size of the moving ball in the video. To simulate full occlusion, we used a cube with equal size of the moving ball, while to simulate long occlusion, the size of the red cube used is double the size of the moving ball in the video.

A total of 64 video sequences were created for this paper. Only simple sphere is used to represent a moving object in each video. The rationale of using only sphere is to avoid difficulty in tracking object that could be caused by severe appearance change when an object moves around in the videos. With simple object shape, the tracking result can reflect the actual tracking ability in various occlusion scenarios.

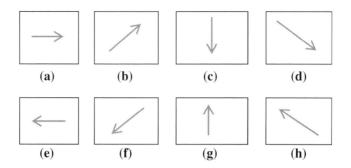

Fig. 1 Uniform ball trajectories: **a** from left to right, **b** from bottom left to top right, **c** from top to bottom, **d** from top left to bottom right, **e** from right to left, **f** from top right to bottom left, **g** bottom to top, **h** from bottom right to top left

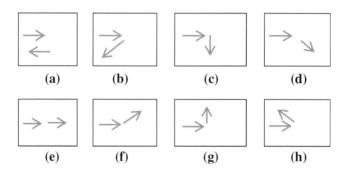

Fig. 2 Non-uniform ball trajectories; all video begin with a ball moving from left to the center of the video frame, followed by: **a** back to left, **b** move to bottom left, **c** move to bottom, **d** move to bottom right, **e** move to right, **f** move to top right, **g** move to top, **h** move to top left

4 Tracking Performance Measurement

In this paper, we use Frame Detection Accuracy (SFDA) in Eq. (1) and Frame Detection Accuracy (FDA) in Eq. (2) proposed by Kasturi et al. [1] to evaluate the tracking effectiveness of each tracking method. SFDA and FDS are highly cited protocol for performance evaluation of object detection and tracking in video sequences [15, 16].

To calculate the result for both mentioned measurements, ground truth is generated using object detection algorithm based on background subtraction [17].

$$SFDA = \frac{\sum_{t=1}^{t=Nframce} FDA(t)}{\sum_{t=1}^{t=Nframce} \exists (N_G^{(t)} or N_D^{(t)})} \tag{1}$$

$$FDA(t) = \frac{Overlap\_Ratio}{\frac{N_G^{(t)} + N_D^{(t)}}{2}} \tag{2}$$

$$Overlap\_Ratio = \sum_{i=1}^{N_{mapped}^{(t)}} \frac{\left|G_i^{(t)} \cap D_i^{(t)}\right|}{\left|G_i^{(t)} \cup D_i^{(t)}\right|} \qquad (3)$$

- G_i denotes the ith ground-truth object at the sequence level;
- $G_i^{(t)}$ denotes ith ground-truth object in frame t;
- D_i denotes the ith detected object at the sequence level
- $D_i^{(t)}$ denotes the ith detected object in frame t;
- $N_G^{(t)}$ and $N_D^{(t)}$ denote the number of ground-truth objects and the number of detected objects in frame t, respectively;
- $N_G^{(t)}$ and $N_D^{(t)}$ denote the number of ground-truth objects and the number of detected objects in frame t, respectively;
- $N\ framce$ is the number of frames in the sequence;
- $N_{mapped}^{(t)}$ refers to sequence level detected object and ground truth pairs;
- $N_{mapped}^{(t)}$ refers to frame t mapped ground truth and detected object pairs.

5 Experimental Results

The tracker algorithms in MATLAB script are modified and customized based on available sources to suit the experiments. Kalman filter (KF) tracker used in the experiments is modified from Kashanipour [17], the PF tracker used is based on Paris [18] and the Mean Shift (MS) tracker is based on Bernhard [19].

Tracking results of various tracker used for the experiments are shown in Tables 1 and 2 for set 1 videos (videos with object moving in uniform trajectory) and set 2 videos (video with object moving in non-uniform trajectory) accordingly.

The first 3 row on the top in Table 1 show the SFDA for the three tracking methods used in this experiment while tracking moving object in eight different video sequences (Seq. a–Seq. h) while no occlusion presence in the video. Each video sequences shows a moving object with unique trajectory as shown in Fig. 1. Table 2 show the SFDA for tracking moving object with non-uniform trajectory as shown in Fig. 2.

Generally, without occlusion in video sequences, KF has the best tracking result, followed by MS and PF has the poorest result. Based on data in row 2, Table 2, PF perform poorly to track moving object in the video sequences when the object moved in non-uniform trajectory. A close examination found that PF will lose track of a tracked object when the object suddenly changed its trajectory after a long period of consistent trajectory. Stretched consistent trajectory caused the distribution area of the particle become contracted and cover only a small area in the video frame. Therefore, when the trajectory of the object changed suddenly, the PF fail to track the moving object as shown in Fig. 3.

Table 1 Tracking result (SFDA) for simulation videos with moving object in uniform trajectory (set 1)

Occlusion	Tracker	Uniform trajectory							
		Seq a	Seq b	Seq c	Seq d	Seq e	Seq f	Seq g	Seq h
None	KF	0.84	0.85	0.85	0.85	0.84	0.85	0.85	0.85
	PF	0.59	0.06	0.37	0.05	0.16	0.09	0.61	0.15
	MS	0.63	0.42	0.41	0.41	0.63	0.46	0.44	0.49
Partial	KF	0.79	0.81	0.79	0.81	0.79	0.81	0.78	0.81
	PF	0.17	0.02	0.12	0.00	0.10	0.02	0.06	0.03
	MS	0.56	0.36	0.34	0.36	0.59	0.40	0.34	0.41
Full	KF	0.59	0.61	0.50	0.67	0.64	0.60	0.45	0.57
	PF	0.39	0.13	0.38	0.06	0.30	0.06	0.23	0.09
	MS	0.26	0.21	0.19	0.21	0.26	0.23	0.17	0.22
Long	KF	0.17	0.18	0.11	0.18	0.17	0.19	0.12	0.19
	PF	0.11	0.07	0.11	0.09	0.13	0.07	0.11	0.11
	MS	0.17	0.14	0.11	0.14	0.16	0.15	0.09	0.15

Table 2 Tracking result (SFDA) for simulation videos with moving object in non- uniform trajectory (set 2)

Occlusion	Tracker	Non Uniform trajectory							
		Seq a	Seq b	Seq c	Seq d	Seq e	Seq f	Seq g	Seq h
None	KF	0.83	0.83	0.84	0.85	0.84	0.84	0.84	0.83
	PF	0.03	0.03	0.02	0.01	0.02	0.02	0.02	0.01
	MS	0.63	0.55	0.51	0.52	0.61	0.55	0.55	0.55
Partial	KF	0.80	0.80	0.78	0.80	0.79	0.81	0.80	0.81
	PF	0.04	0.03	0.02	0.01	0.01	0.02	0.01	0.01
	MS	0.59	0.52	0.48	0.50	0.56	0.51	0.51	0.52
Full	KF	0.64	0.58	0.51	0.55	0.59	0.60	0.59	0.65
	PF	0.03	0.04	0.02	0.01	0.01	0.02	0.01	0.02
	MS	0.53	0.46	0.27	0.28	0.26	0.28	0.27	0.47
Long	KF	0.17	0.18	0.17	0.18	0.17	0.18	0.17	0.18
	PF	0.03	0.03	0.01	0.01	0.01	0.01	0.01	0.02
	MS	0.33	0.27	0.18	0.18	0.17	0.18	0.18	0.25

6 Effect of Occlusion

In Set 1 videos, all trackers produced lower SFDA when occlusion occurred. Partial occlusion has lowest impact on all trackers, followed by full occlusion and long occlusion has the most impact on all trackers.

In most cases, MS has better performance than PF in tracking object in Set 1 videos. However, when full occlusion occurred, PF could recover from occlusion and continued to track the moving object after the occlusion. On the other hand, MS could not recover under the same scenario. This is confirmed by the tracking result in Table 2 for *Seq. a, Seq c, Seq. e* and *Seq. g*. The FDA of three trackers for video *Seq. a* in Set 1 is shown in Fig. 4.

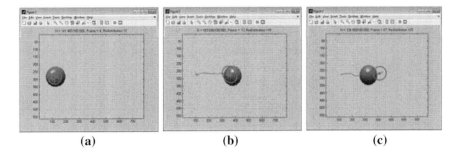

| (a) | (b) | (c) |

Fig. 3 Particle distribution: **a** particles cover a large area at the initial state; **b** when object trajectory remains consistent between frames, the particle area shrunk; **c** particle area become so small and fail to detect the moving object change direction

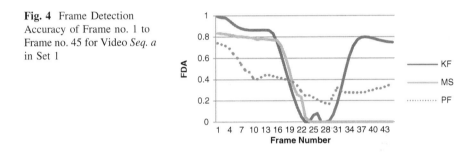

Fig. 4 Frame Detection Accuracy of Frame no. 1 to Frame no. 45 for Video *Seq. a* in Set 1

Tracking result for Set 2 videos, the video sequences generated to simulate moving object with un-uniform trajectory are shown in Table 2. With no occlusion in the video sequences, tracking result from KF and MS does not show much difference in both Set 1 and Set 2 videos. This implies that the trajectory of the moving object does not affect the effectiveness of KF and MS trackers.

In Table 2, the SFDA of MS in video *Seq. a* and *Seq. b* with full occlusion is 0.53 and 0.46 accordingly. These results are higher when compared to the result in Table 1, when full occlusion occurs in Set 1 videos. MS has better tracking result in Seq. a and Seq. b with full occlusion in Set 2 video because the location of the moving object reappear after full occlusion is near to the location where the object being occluded. Therefore, this implies that the location where a tracked object reappears after occlusion could affect MS tracking result.

7 Conclusion

In this paper, 64 simulated video were design and generated to test the tracking capability of three popular trackers observed from review of previous works. Experiments are conducted using the Kalman filter, Particle filter and Mean Shift tracker.

Results obtained from the experiments imply that occlusion does affect the effectiveness of trackers. Partial occlusion has the least effect on tracking and long occlusion has the most effect by reducing the effectiveness of all trackers. Other than the size of occlusions, the location where a tracked object reappears after occlusion could also affect the tracking result in object tracking.

References

1. Kasturi, R., Goldgof, D., Soundararajan, P., Manohar, V., Garofolo, J., Bowers, R., Boonstra, M., Korzhova, V., Zhang, J.: Framework for performance evaluation of face, text, and vehicle detection and tracking in video: data, metrics, and protocol. IEEE Trans. Pattern Anal. Mach. Intell. **31**(2), 319–336 (2009)
2. Di Caterina, G., Soraghan, J. J.: An improved mean shift tracker with fast failure recovery strategy after complete occlusion. In: IEEE International Conference on Advanced Video and Signal-Based Surveillance, pp. 130–135 (2011)
3. Gao, J.: Self-occlusion immune video tracking of objects in cluttered environment. In: IEEE International Conference on Advanced Video and Signal-Based Surveillance, pp. 79–84 (2003)
4. PETS2009: Benchmark data. http://www.cvg.rdg.ac.uk/PETS2009/a.html
5. ETISEO: Evaluation For Video Understanding. http://www-sop.inria.fr/orion/ETISEO/
6. Taylor, G.R., Chosak, A.J., Brewe, P.C.: OVVV: using virtual worlds to design and evaluate surveillance systems. In: Proceedings of the IEEE Computer Society Conference on Computer Vision and Pattern Recognition, pp. 1–8 (2007)
7. Yilmaz, A., Javed, O., Shah, M.: Object tracking: a survey. In: ACM Computing Surveys, vol. 38, no. 4 (2006)
8. Comaniciu, D., Ramesh, V.: Kernel-based object tracking. IEEE Trans. Pattern Anal. Mach. Intell. **25**(5), 564–577 (2003)
9. Comaniciu, D., Ramesh, V., Meer P.: Real-time tracking of non-rigid objects using mean shift. In: Proceedings of IEEE Conference on Computer Vision and Pattern Recognition, pp. 142–149 (2000)
10. Yilmaz, A.: Object tracking by asymmetric Kernel mean shift with automatic scale and orientation selection. In: Proceedings of IEEE Conference on Computer Vision and Pattern Recognition, CVPR '07, pp. 1–6 (2007)
11. Mirabi, M., Javadi, S.: People tracking in outdoor environment using Kalman filter. In: Proceedings of Third International Conference on Intelligent Systems, Modelling and Simulation (ISMS), pp. 303–307 (2012)
12. Wang, H., Huo, L., Zhang, J.: Target tracking algorithm based on dynamic template and Kalman filter. In: Communication Software and Networks (ICCSN), 2011 IEEE 3rd, pp. 330–333 (2011)
13. Cho, J.U., Jin, S.H., Pham, X.D., Jeon, J.W.: Object tracking circuit using particle filter with multiple features. In: SICE-ICASE, 2006. International Joint Conference, pp. 1431–1436 (2006)
14. Liang, N., Guo, L., Wang Y.: An improved object tracking method based on particle filter. In: Consumer Electronics, Communications and Networks (CECNet), 2012 2nd International Conference on, pp. 3107–3110 (2012)
15. Li, Y., Huang, C., Nevatia, R.: Learning to associate: hybrid boosted multi-target tracker for crowded scene. In: IEEE Conference on Computer Vision and Pattern Recognition, pp. 2953–2960 (2009)
16. Manohar, V., Soundararajan, P., Dmitry Goldgof, H.R., Kasturi, R., Garofolo, J.: Performance evaluation of object detection and tracking in video. In: Narayanan P.J. et al. (eds.) ACCV 2006. LNCS, vol. 3852, pp. 151–161. Springer, Heidelberg (2006)

17. Kashanipour, A.: 2D Target Tracking Using Kalman Filter, MATLAB Central. http://www.mathworks.com/matlabcentral/fileexchange/14243-2d-target-tracking-using-kalman-filter
18. Paris, S.: Particle filter color tracker. In: MATLAB Central. http://www.mathworks.com/matlabcentral/fileexchange/17960-particle-filter-color-tracker
19. Bernhard, S.: Mean-Shift Video Tracking, MATLAB Central. http://www.mathworks.com/matlabcentral/fileexchange/355, March 2012

Originator Recognition (or) Path Recovery Mechanism for Load-Based Routing Protocol

Gee Keng Ee, Chee Kyun Ng, Fazirulhisyam Hashim,
Nor Kamariah Noordin and Borhanuddin Mohd. Ali

Abstract 6LoWPAN has become a new technology to provide the internet connectivity to the traditional WSN. The introduction of 6LoWPAN adaptation layer enables the smooth delivery of packet from network layer to MAC layer. In this paper, the LOAD-based routing protocol with the proposed originator recognition (OR) path recovery mechanism is introduced. The proposed OR path recovery mechanism modifies the IETF conceptual LOAD protocol message by inserting an identity key in the generated RERR message. The identity key is then used by the originator of a failed data packet to initialize path recovery during the link failure. The developed OR-LOAD has examined under the 6LoWPAN environment in Qualnet simulator. Its performance is evaluated and compared to AODV in terms of packet delivery ratio and throughput.

Keywords WSN · 6LoWPAN · AODV · LOAD · Path recovery mechanism

G. K. Ee (✉) · C. K. Ng · F. Hashim · N. K. Noordin · B. Mohd. Ali
Faculty of Engineering, Department of Computer and Communication Systems Engineering,
University Putra Malaysia (UPM), 43400 Serdang, Selangor, Malaysia
e-mail: darrow_keng@yahoo.com

C. K. Ng
e-mail: mpnck@eng.upm.edu.my

F. Hashim
e-mail: fazirul@eng.upm.edu.my

N. K. Noordin
e-mail: nknordin@eng.upm.edu.my

B. Mohd. Ali
e-mail: borhan@eng.upm.edu.my

S.-S. Yeo et al. (eds.), *Computer Science and its Applications*,
Lecture Notes in Electrical Engineering 203, DOI: 10.1007/978-94-007-5699-1_16,
© Springer Science+Business Media Dordrecht 2012

1 Introduction

The IPv6 over low power wireless personal area network (6LoWPAN) provides a wireless sensor network (WSN) node with internet protocol (IP) communication capabilities by putting an adaptation layer above the IEEE 802.15.4 link layer to enable header compression, packet fragmentation and reassembly, and layer-two forwarding [1]. While 6LoWPAN adaptation layer focusing in enabling the packet delivery from the network layer to the MAC layer, in order to provide the delivered packet sent from a source node to a destination node, 6LoWPAN routing becomes a research issue. There are two 6LoWPAN routing schemes, mesh-under and route-over, and their protocol layer stacks are shown in Fig. 1. The mesh-under approach performs its routing at adaptation layer, and hence no IP routing is performed within LoWPAN where it is directly based on the IEEE 802.15.4 MAC addresses (16-bit or 64-bit logical address). In contrast, the route-over approach performs its routing at network layer, and IP address is used to perform IP routing with each node served as an IP router [2].

Currently, there are two IETF working groups (WGs) work for the 6LoWPAN routing protocol, Routing Over Low-power and Lossy networks (ROLL) and Mobile Ad hoc Network (MANET). ROLL introduces the IPv6 Routing Protocol for Low-power and Lossy networks (RPL) for route-over routing protocol in 6LoWPAN. The RPL uses the directed acyclic graphs (DAGs) to perform IPv6 forwarding mechanism between routers, hosts and edge routers [3]. In contrast, MANET has proposed to simplify the existing IETF routing protocols for 6LoWPAN. Due to the lower overhead of reactive routing protocol, the existing IETF routing protocol such as Ad-hoc On-Demand Distance Vector (AODV) [4] routing is considered as a strong candidate for 6LoWPAN. However, researches are still in progress to study the suitability of AODV routing protocol for route-over routing in 6LoWPAN. As there has yet any mandate in the 6LoWPAN mesh-under routing protocol, rather than only focus on route-over routing protocol, MANET modifies the mechanism of AODV, and uses link-layer addressing (64 bits or 16 bits) for 6LoWPAN mesh-under routing. The modified version of AODV is named as 6LoWPAN Ad-Hoc On-Demand Distance Vector Routing (LOAD) protocol [5] and is still under progress.

Hence, LOAD routing protocol is proposed by MANET to route the delivered packet from a source node to a destination node via the link-layer addresses. Figure 2 shows the exchange of LOAD protocol message mechanism. When a source node needs to send a data to a destination, it first uses route request (RREQ) message to find a path to the particular destination. The destination node will respond to the source node with route reply (RREP) message that contains the route cost information. The data packet is then sent multi-hopping through a route that has the less weak links (link whose LQI is worse than a certain threshold value) and less hop counts from the source node to the destination node. For each

(a)

Application Layer
Transport Layer (TCP/UDP)
Network Layer (IPv6)
Adaptation Layer – **Routing**
802.15.4 MAC Layer
802.15.4 PHY Layer

(b)

Application Layer
Transport Layer (TCP/UDP)
Network Layer (IPv6) - **Routing**
Adaptation Layer
802.15.4 MAC Layer
802.15.4 PHY Layer

Fig. 1 Routing schemes in 6LoWPAN **a** mesh-under and **b** route-over

Fig. 2 LOAD protocol message exchange mechanism

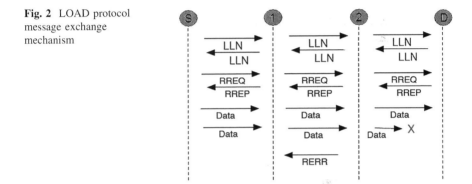

successful data transmission, a MAC layer acknowledgement, termed as Link Layer Notification (LLN), is used for the confirmation. The missing LLN message of an irreparable broken link that caused by interference signal will require the upstream node, node 2 in this case, to send a route error (RERR) message to the source node of the failed data packet. However, the conceptual LOAD does not define the method to notify the data source node with the received RERR during the link break. Hence, there is no any path recovery mechanism in the conceptual LOAD when the link break occurs.

This paper presents a new LOAD-based routing protocol that characterized with the proposed Originator Recognition (OR) path recovery mechanism. Unlike the conceptual LOAD that uses link-layer addresses for mesh-under routing, the proposed OR-LOAD routing protocol uses IP addressing for route-over routing in order to provide a global routing within or outside the 6LoWPAN. By referring to the routing information kept in OR-LOAD routing table, the OR-LOAD protocol routes a data packet from a source node to a destination node through multiple hops. When a link break occurred during the data transmission, the proposed OR path recovery mechanism is initialized. It notifies the originator of a failed data packet to reinitialize a new route discovery in order to look for an alternative path to the currently unreachable destination node.

Final Destination	Next Hop Address	Valid	Out Interface	Life Time	Locally Repairable	Weak Link	Hop count	Last Hop Count

Fig. 3 OR-LOAD routing table

(a)

RREQ ID	Originator Address	Previous Hop Address	Forward Route Cost	Reverse Route Cost	Valid Time

(b)

Destination Address	TTL	Times

Fig. 4 OR-LOAD route request **a** sent and **b** seen tables

2 Data Structures of the Proposed OR-LOAD Routing Protocol

By default, Qualnet simulator does not provide 6LoWPAN protocol stack as a platform for the developed OR-LOAD routing protocol. Thus, the 6LoWPAN adaptation layer with its main mechanisms, fragmentation and reassembly function, have been developed as presented in [6]. The developed adaptation layer will be then integrated with other existing protocol layers in Qualnet to set up a complete 6LoWPAN protocol stack for every sensor node. In order to fully implement the conceptual LOAD with the proposed OR path recovery mechanism, some modifications have been done in the conceptual LOAD data structures and routing algorithm. Three data tables have been defined in the proposed OR-LOAD routing protocol such that OR-LOAD routing table, OR-LOAD route request sent table and OR-LOAD route request seen table as shown in Figs. 3 and 4. All these data tables are maintained in participating nodes. The OR-LOAD routing table is used as reference for a router to make a routing decision during the data transmission while the OR-LOAD route request sent and seen tables are used to handle the RREQ and RREP protocol messages during a route discovery process.

Unlike the conceptual LOAD protocol that uses link-layer addresses for routing, the proposed OR-LOAD protocol uses IP addresses to replace the link-layer address field in the conceptual LOAD protocol messages such as RREQ and RREP. As both RREQ and RREP have same header fields, Fig. 5 shows both protocol messages that used in OR-LOAD protocol during a route discovery process. When there is an irreparable broken link during the data transmission, the proposed OR path recovery mechanism enables the source node (originator) to be notified by RERR and thus reinitiates a new route discovery process for an

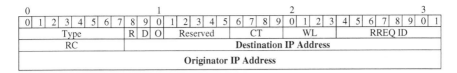

Fig. 5 OR-LOAD RREQ or RREP message format

Fig. 6 OR-LOAD RERR message format

alternative path to the currently unreachable destination. Figure 6 shows the modified RERR message that used for OR path recovery mechanism. There is a new header field, originator IP address field, added in the RRER message as an identity key for the source of failed data packet. The operation of OR path recovery mechanism is described in next section.

3 Mechanism of the Proposed OR-LOAD Routing Protocol

There are three operation states involved in the proposed OR path recovery mechanism; memorizing state, encapsulation state and recognition state. Figure 7 shows the summary of the proposed OR path recovery mechanism in LOAD.

During the memorizing state, the intermediate node will memorize the originator address header field information from a buffered data packet. This header field information is then stored into a new temporary buffer. When a link between two nodes is broken where the local repair is fail, the generated RERR message will trigger the encapsulation state. In the encapsulation state, the memorized originator address header field in the temporary buffer is encapsulated into the generated RERR message. The added originator address header field acts as an identity key for the receiving node of RERR message.

During the RERR message forwarding towards the originator, recognition state will be initialized. In this state, each previous hop that receives the RERR message from the next hop towards a destination will use the added originator address header to identify the destination of received RERR message, where the destination of RERR message is the originator of the failed data packet. In the case that the receiving node of RERR message is the originator of the failed data packet, the originator will reinitiate a new route discovery process for a new alternative path to the currently unreachable destination, if there is a data packet still needed to be transmitted to the destination. In contrast, if the receiving node of RERR message

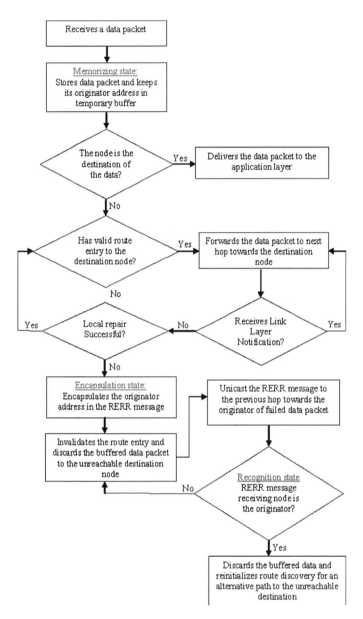

Fig. 7 Summary of OR path recovery mechanism

is not the originator of the failed data packet, the previously buffered data packet for the unreachable destination as indicated in the RERR header field will be dropped. Then, the received RERR message will be forwarded to the previous hop unicastly toward the originator of the failed data packet.

Table 1 List of simulation parameters

Parameter	Value
Radio type	802.15.4 radio
MAC protocol	802.15.4
Network protocol	IPv4
IP fragmentation unit	1280 bytes
Routing protocol	AODV or OR-LOAD
Number of nodes	5, 10, 15, 20, 25, 30, 35, 40, 45, 50 nodes
Scenario dimension	$500 \times 500m^2$
Simulation time	30 min
Packet size	1280 bytes
Node placement model	Random
Application protocol	CBR

4 Results and Discussion

The proposed OR-LOAD scheme is simulated by Qualnet simulator. Table 1 shows the list of simulation parameters. With the increasing number of nodes that placed with the specific dimension, more interference signals will cause the link breaks occur frequently. As using IPv6 over IEEE 802.15.4 standard involves the kernel modification of Qualnet, which is disallowed due to company privacy, the Maximum Transmission Unit (MTU) of IPv6 packet is emulated by 1280 bytes IPv4 packet.

The developed OR path recovery mechanism is evaluated under the increasing interference environment with node deployment scalability up to 50 nodes. As the IETF conceptual LOAD has yet mandated, the performance of OR-LOAD will be compared with AODV that uses the precursor list for its path recovery mechanism, in order to verify the lightweight of OR-LOAD in 6LoWPAN as shown in Fig. 8. Figure 8a shows OR-LOAD has higher packet delivery ratio compared to AODV under 6LoWPAN environment. Thus, OR-LOAD achieves packet delivery ratio about 6 % higher than AODV at 10 nodes. With the increasing number of nodes, OR-LOAD achieves packet delivery ratio about 100 % better than AODV. Moreover, OR-LOAD performs an average of 14.2 % higher packet delivery ratio than AODV. This is because the link breaks happen more frequently when the number of nodes is increased, where OR-LOAD shows its significance in path recovery compared to AODV under 6LoWPAN environment. In the throughput comparison as illustrated in Fig. 8b, OR-LOAD performs better where it has an average of 28.3 % higher throughput than AODV. This is due to less packet loss in OR-LOAD. Nevertheless, both Fig. 8a, b shows that OR-LOAD and AODV reach the saturated level when 50 nodes are used. It can be concluded that the AODV and OR-LOAD support the 6LoWPAN network scalability up to 50 nodes.

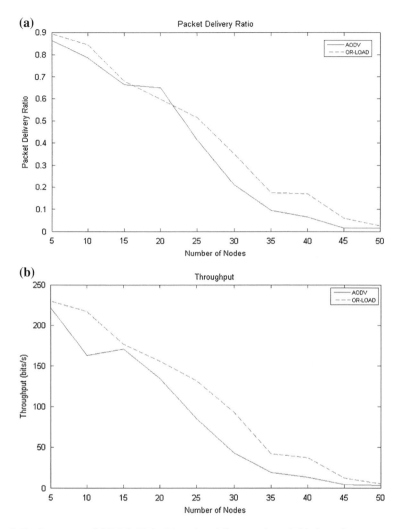

Fig. 8 Performances of OR-LOAD in (**a**) packet delivery ratio and (**b**) throughput

5 Conclusions

This paper presents the path recovery mechanism for the broken links in 6LoWPAN using the proposed OR-LOAD routing protocol. By using the 6LoWPAN protocol stack, the IETF conceptual LOAD has been modified from the aspect of its data structures. Unlike the conceptual LOAD, the newly proposed OR-LOAD routing protocol which uses route-over approach is targeting to route the data packet globally. It modifies the RERR message format by adding a new originator address header field in it. During link break of an irreparable link, the originator that notified by the RERR message will be identified by the added address field header.

The originator after notifying the link failure will reinitialize a new route discovery process to find an alternative path to the currently unreachable destination. Simulation results show that the proposed OR-LOAD outperforms than AODV with average of 14.2 and 28.3 % in terms of packet delivery ratio and throughput respectively, thus OR-LOAD has a higher reliability than AODV.

References

1. Montenegro, G., Kushalnagar, N., Hui, J., Culler, D.: Transmission of IPv6 Packets over IEEE 802.15.4 Networks. RFC4944 (2007)
2. Chowdhury, A.H., Ikram, M., Cha, H.S.: Route-over vs Mesh-under Routing in 6LoWPAN. In: International Wireless Communications and Mobile Computing Conference (IWCMC), pp. 1208–1212. (2009)
3. Oliveira, L.M.L., et al.: Routing and mobility approaches in IPv6 over LoWPAN mesh networks. Int. J. Commun. Syst. **24**, 1445–1466 (2011)
4. Perkins, C., Belding-Royer, E., Das, S.: Ad hoc On-Demand Distance Vector (AODV) Routing. RFC 3561 (2003)
5. Kim, K., Daniel Park, S., Montenegro, G., Yoo, S., Kushalnagar, N.: 6LoWPAN Ad Hoc On-Demand Distance Vector Routing (LOAD). draft-daniel-6lowpan-load-adhoc-routing-03 (2007)
6. Chan, C.W., Ee, G.K., Ng, C.K., Hashim, F., Noordin, N.K.: Development of 6LoWPAN adaptation layer with fragmentation and reassembly mechanisms by using Qualnet Simulator, In: Abd Manaf, A., et al. (eds.): ICIEIS 2011, CCIS, vol. 254, pp. 199–212. Springer, Heidelberg (2011)

Analysis of Correlation Peak Position Modulation

Jihah Nah and Jongweon Kim

Abstract This paper extends the peak position modulation (PPM) technique introduced by Nah and Kim (Image Watermarking for Identification Forgery Prevention) and provides a detailed performance analysis. The proposed correlation peak position modulation (CPPM) significantly improves the tradeoff between fidelity and robustness in comparison with traditional watermarking approaches such as spread spectrum modulation and position based watermarking. CPPM ensures robustness and sufficient capacity which is modulated by user information. Theoretical analyses and experimental results were provided to compare with the performance of the spread spectrum technique and position based watermarking. These analyses and experiments confirm the proposed algorithm is a competitive method for image watermarking.

Keywords Peak position modulation · Image watermarking · Spread spectrum · Collusion attack

1 Introduction

As the manipulation of digital content becomes easier, copyright infringement becomes a serious problem. For this reason, watermarking techniques have been proposed for copyright protection and authentication of digital media [1–4].

J. Nah
Digital Copyright Protection Research Institute,
Sangmyung University, Seoul, Korea
e-mail: jihah.nah@gmail.com

J. Kim (✉)
Department of Copyright Protection, Sangmyung University, Seoul, Korea
e-mail: jwkim@smu.ac.kr

S.-S. Yeo et al. (eds.), *Computer Science and its Applications*,
Lecture Notes in Electrical Engineering 203, DOI: 10.1007/978-94-007-5699-1_17,
© Springer Science+Business Media Dordrecht 2012

The advantage of watermarking is that it can hide content-user information, and can trace a user who has illegally copied and distributed content.

Spread spectrum modulation is widely used in robust watermarking designs. Various spread spectrum-based watermarking techniques have been proposed, using discrete cosine transform (DCT) [5], Fourier–Mellin [6], wavelets [7], etc. Spread spectrum-based watermarking schemes achieve better imperceptibility and resilience, but have a major weakness in terms of their non-blind detection. Other shortcomings are their low capacity and high decoding overhead. Moreover, the detection reliability cannot be improved by any practical watermarking scheme at a high embedding rate. In a collusion attack, the embedded watermarks may not be detected due to the number of watermarked images from different attackers.

Maity, Kundu, and Das proposed M-ary modulation for robust spread-spectrum watermarking with improved capacity [8–10]. They embedded watermark information in the wavelet coefficients of the specific subbands. M-ary modulation for M-band DWT (MbDWT) is used for improving the payload capacity.

Borges and Mayer introduced the position-based watermarking (PBW) technique [11]. PBW represents a message according to the arbitrary locations of embedded small watermark blocks in the original image. PBW exhibits robustness to collusion attacks.

Theoretical analyses and experimental results for a correlation peak position modulation (CPPM) were provided to compare with the performance of the spread spectrum technique and position based watermarking. These analyses and experiments confirm the proposed algorithm is a competitive method for image watermarking. Using only one key, CPPM provides sufficient capacity and is robust against noise. Experimental results show that CPPM is suitable for watermark embedding.

2 Proposed CPPM Method

CPPM can provide more robustness than traditional spread spectrum techniques, because it has a longer sequence length for the same image, and different users are apt to have different image correlation peak positions. This is due to different users having different pieces of information, each of which is loaded at a different position in the image [3, 4].

The quality of digital content, and the detection performance of embedded information, can be ensured with the use of long sequences as watermarks by Eqs. (1) and (2). In other words, CPPM achieves superior robustness to other schemes for the same amount of information because the length of the watermark sequence can be maximized. For example, when 16 bits of information are embedded in a 256×256 image, CPPM has a longer sequence, with a watermark block size of 256×256, than spread spectrum watermarking methods, which have a block size of 64×64.

CPPM is introduced to ensure appropriate payloads and robustness. The existence of a spread spectrum sequence can generally be identified using correlation. CPPM uses the position of a correlation peak as embedded information. The basic equation is given as:

$$\rho_{I_w w}(u, v) = \iint I_w(x, y) w(x + u, y + v) dx dy \tag{1}$$

where $I_w(x,y)$ is a marked image, and w is a watermark generated by a pseudo-random number (PRN) sequence. The variables u and v represent the amount of shift in the 2D coordinate system.

The peak position, representing the embedded information, is then given by:

$$p = u + v \times width = B_n 2^n + B_{n-1} 2^{n-1} + \cdots + B_1 2^1 + B_0 \tag{2}$$

where B_n represents the nth bit.

The message is modulated as the peak position using Eqs. (1) and (2). The information capacity per watermark block can be maximized because the position of the correlation peak is shifted according to the embedded information.

3 Error Probability and Capacity of CPPM

The watermarked signal I_w is obtained by additive embedding, $I_w = I + w$. The received image is $I_R = I + w + n$, where some additive channel noise, n. It is assumed that I, w and n are zero mean and statistically mutual independent sequences.

The normalized correlation detection statistics is represented by Eq. (3), considering additive noise, n, in the channel.

$$\rho = \frac{Corr(I_R, w)}{Corr(w, w)} = \frac{Corr(I + w + n, w)}{\sigma_w^2} \tag{3}$$

It is assumed that both I and n are derived from a Gaussian distribution. In other words, $I_i \approx N(0, \sigma_I^2)$ and $n_i \approx N(0, \sigma_n^2)$. The random variable ρ will represent a normal distribution for large N according to the Central Limit Theorem.

$$\rho \approx N(\mu_\rho, \sigma_\rho^2), \ \mu_\rho = E\{\rho\}, \ \sigma_\rho^2 = \frac{\sigma_I^2 + \sigma_n^2}{N\sigma_w^2} \tag{4}$$

In the proposed method, the several information bits are represented by only a watermark sequence like a bit information of spread spectrum method so, we can consider the proposed system is a spread spectrum method for embedding a bit information. Therefore, the error probability of detecting a bit with value 0 is given by Eq. (6) when a bit with value 1 is transmitted.

$$p_e = \Pr\{\rho\} \tag{5}$$

$$p_e = \frac{1}{2} erfc\left(\sqrt{\frac{\sigma_w^2 N}{2(\sigma_I^2 + \sigma_n^2)}}\right), \tag{6}$$

where $erfc(x) = \frac{2}{\sqrt{\pi}} \int_x^\infty e^{-t^2} dt$ is the complementary error function.

Similarly, the error probability is obtained equally when a bit 0 is transmitted, a bit 1. At this time, it is considered that the message bits are derived from an even probability density function with zero mean. The proposed system is the same as 1 bit information transmission in spread spectrum, even though the entire message has several bits information. That is the entire message is only 1 bit. So, the probability of the error message is given by $p_{CPPM} = 1 - (1 - p_e)^N$ where N is the detection trials for correlation peak position. At this time, the entire message is represented by 1 bit and the effects of channel coding are disregarded.

In CPPM, N detections have to be considered for the total error probability. Missing one correlation peak position causes the error message. Therefore, the probability of the error message is shown in Eq. (7). At this point, locations of the image are provided for embedding the message in CPPM.

$$p_{CPPM} = 1 - \left(1 - \frac{1}{2} erfc\left(\sqrt{\frac{\sigma_w^2 N}{2(\sigma_x^2 + \sigma_n^2)}}\right)\right)^N \tag{7}$$

On the other hand, the image size N and the number K of watermark blocks affect the capacity of the CPPM simply. Those elements are used for embedding the message. Without superposition of the watermark block over another block, the resulting capacity is given by Eq. (8).

$$Cap = K \times \left(\log_2\left(\frac{N}{K}\right) + 1\right) \tag{8}$$

4 Performance Analysis

4.1 Error Probability and Capacity Analysis

Bit error probabilities are derived for spread spectrum (SS, [11]), Position Based Watermarking (PBW, [11]) and CPPM. The spatial blind insertion is only analyzed for SS, PBW and CPPM.

The error probability has the same value when a bit 0 is transmitted and the message b is −1, considering that the message bits are derived from an even probability density function with zero average. The probability of receiving an

Fig. 1 Error probabilities for
the SS, PBW and CPPM

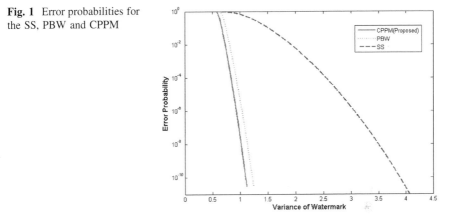

error message is obtained by $p_{ss} = 1 - (1 - p_e)^L$ when the entire message is L bits
and the effects of channel coding are disregarded.

$$p_{ss} = 1 - \left(1 - \frac{1}{2} erfc \left(\sqrt{\frac{\sigma_w^2 N}{2L(\sigma_x^2 + \sigma_n^2)}} \right) \right)^L \tag{9}$$

where N means dimension of watermarked images.

When the spread spectrum method is used, the probability for incorrect
detection of at least one bit is shown in Eq. (9).

In PBW, N detections can be taken into account for the total error probability.
One location error results in the error message. Therefore, the probability of the
error message is shown in Eq. (10). At this point, locations of the image are
provided for embedding the message considering all location in PBW.

$$p_{PBW} = 1 - \left(1 - \frac{1}{2} erfc \left(\sqrt{\frac{\sigma_{w_p}^2 S}{2(\sigma_x^2 + \sigma_n^2)}} \right) \right)^N \tag{10}$$

where S means dimension of watermark blocks.

The error probability curves for three algorithms is provided in Fig. 1 by
plotting Eqs. (7), (9) and (10) with respect to σ_w^2 and by using the parameters
$N = 512 \times 512$, $S = 32 \times 32$, $L = 28$, $K = 2$, $\sigma_x^2 + \sigma_n^2 = 3000$.

The performance of the spread spectrum technique is indicated by the bold
dashed line in Fig. 1. The performance of PBW is represented by bold dotted lines,
when $\sigma_{w_p}^2 = \sigma_w^2$. The CPPM performance is indicated by bold solid lines, for
$\sigma_{w_c}^2 = \sigma_w^2$.

In case of the spread spectrum, PBW and CPPM, error probabilities are $p_{ss} =$
0.2936, $p_{PBW} = 4.9353 \times 10^{-11}$ and $p_{CPPM} = 2.1818 \times 10^{-11}$ respectively. Their
condition is 16 bits capacity, the 256×256 dimension of watermarked image and

Fig. 2 Capacity as function
of K with $N = 512^2$ and
$S = 32^2$

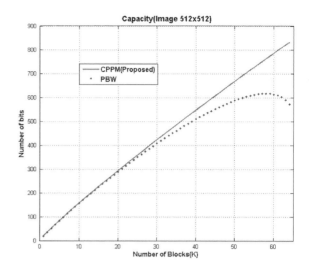

$\sigma_w^2 = \sigma_{w_p}^2 = \sigma_{w_c}^2 = 3$. In case of minimum error probability of PBW (K = 1, 2 bits), it is value is the same as the error probability of CPPM (16 bits).

The capacity of the PBW is affected by the image size. Also, the size S of the watermark block and the number K affect the capacity. The interference caused by the embedded blocks of K can be eliminated avoiding superposition between the pixels residing in several watermark blocks. However, the capacity upper bound of proposed algorithm is also reduced. The resulting capacity is given by Eq. (11)

$$Cap = \log_2[Comb(A_p, K)] \tag{11}$$

where $Comb(A_p, K) = A_p!/[(A_p - K)!K!]$ and $A_p = N - (K - 1)\left[\left(2S^{\frac{1}{2}} - 1\right)^2 - 1\right]$

is the number of exploitable positions which K watermark blocks are embedded into. At this time, superposition is not allowed.

The capacity of PBW and the proposed algorithm is shown in Fig. 2 using Eqs. (8) and (11), for $N = 512^2$ and $S = 32^2$.

The proposed method shows slightly higher capacity than PBW when the number of blocks is less than 30, and if the number is getting greater than 30, the capacity of the CPPM is getting better. Especially, the capacity of PBW is saturated around K = 57 and then the capacity is rapidly decreased. However, the capacity of the CPPM is theoretically increased according the number of blocks is increasing.

4.2 Experiments

Table 1 shows fidelity comparison about spread spectrum, PBW and CPPM. For a fair comparison, dimension of watermarked images is set to $N = 256^2$. PSNR

Table 1 Fidelity comparison (PSNR : dB)

Methods	Image						
	Capacity (bits)	Lena	Barbara	Peppers	F-16	Sailboat	Average
SS	28	29.52	30.81	31.41	33.87	31.74	32.14
PBW	28	37.95	39.24	39.89	42.30	40.15	40.33
Proposed	44	40.18	40.21	40.16	38.81	40.71	40.01

Table 2 Fidelity and BER comparison after compression in the DWT domain

Performance		No. of message			
		M-ary [8]		CPPM_S3 (proposed)	
		2	4	2	4
QF35	PSNR (dB)	35.501	34.354	37.29	35.80
	BER (%)	2.45	3.45	0	0
QF30	PSNR (dB)	–	–	39.21	35.38
	BER (%)	–	–	18.18	18.18

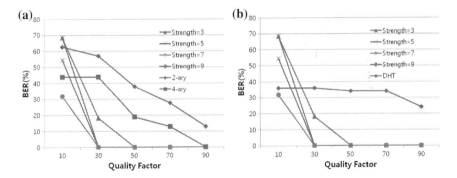

Fig. 3 Performance of JPEG compressions. **a** BER of CPPM and M-ary, **b** BER of CPPM and DHT

values of CPPM are similar to PBW and better than spread spectrum. At the same time, it has more capacity than others.

Table 2 compares the performance of the proposed method with another algorithm developed by Maity et al. [8, 10] in the discrete wavelet transform (DWT) domain, where BER indicates the bit error rate at quality factor 35 for a JPEG 2000 compression operation. Maity et al. embedded watermark information in wavelet coefficients of the specific subbands, and used M-ary modulation to improve the payload capacity. The proposed CPPM shows better fidelity and error performance at strength level 3 (CPPM_S3) than the M-ary method. However, lowering the quality factor to 30 produces a rapid increase in BER.

To test the method's robustness against a compression attack, watermarks were embedded in a cover image of size 256 × 256. Figure 3 shows the results of

Table 3 BER comparison when the watermark is embedded with Gaussian

Error performance	Quality factor		
	QF30	QF35	QF50
BER (%)	50.0	27.27	20.46

compression at various quality factors. The error performance of our method is better than both M-ary and the Discrete Hadamard Transform (DHT) [10]. However, if a PRN is generated with a Gaussian distribution instead of $[1, -1]$, then the error performance of CPPM against JPEG compression is worse.

Table 3 shows the BERs are 27.27 and 20.46 % when the watermarked images are compressed with QF35 and QF50, respectively. Even the watermark with a Gaussian distribution is embedded more strongly before the JPEG compression, but the watermarked image is more susceptible to errors after compression.

5 Conclusions

In this paper, we provide theoretical analysis and extensions for the CPPM technique. Evaluations of robustness performance to AWGN are presented. We demonstrated that CPPM can provide more robustness than traditional spread spectrum-based watermarking, due to its longer sequence length for the same sized image or video. Another advantage of the proposed method is its error tolerance within some rotation range, which is a result of using a $2^l \times 2^m$ unit sample space instead of 1×1 pixels.

To evaluate the performance of the proposed method, watermark information was embedded in the spatial domain and wavelet-transformed domain using CPPM. Improvements in capacity and robustness performance under compression were presented. Performance experiments were conducted by comparing the BER of the spread spectrum, PBW, M-ary, and CPPM methods considering watermark block size, embedding strength, and transparency.

The results of these experiments showed that CPPM has more sufficient capacity than the spread spectrum, PBW, and M-ary methods. In the case of the same-sized image, a watermark block size of 256×256 can represents 16 bits, four 128×128 sized watermark blocks can represent 56 bits, and $2^k \times 2^k$ times $2^{n-k} \times 2^{n-k}$ sized watermark blocks can represent $2^{2k} \times 2(n-k)$ bits. Another advantage is that there are only 1–2 bit errors, as CPPM causes decimal errors in the range 1–2. Therefore, error correction is very simple. Experimental results show that the proposed CPPM is a competitive watermarking algorithm for images and video signals.

The proposed algorithm can be applied to detect piracy or to trace an illegal distributor.

Acknowledgments This research project was supported by the Ministry of Culture, Sports and Tourism (MCST) and Korea Copyright Commission in 2011.

References

1. Kim, J., Kim, N., Lee, D., Park, S., Lee, S.: Watermarking two dimensional data object identifier for authenticated distribution of digital multimedia contents. Signal Process. Image Commun. **25**, 559–576 (2010)
2. Li, D., Kim, J.: Secure image forensic marking algorithm using 2D barcode and off-axis hologram in DWT-DFRNT domain. Appl. Math. Inf. Sci. **6**(2S), 513–520 (2012)
3. Nah, J., Kim, J., Kim, J.: Video forensic marking algorithm using peak position modulation. J. Appl. Math. Inf. Sci. **6**(5S) (2012) (to be published)
4. Nah, J., Kim, J., Kim, J.: Image watermarking for identification forgery prevention. J. Korea Contents Assoc. **11**(12), 552–559 (2011)
5. Cox, I.J., Kilian, J., Leighton, T., Shamoon, T.: Secure Spread spectrum watermarking for multimedia. IEEE Trans. Image Process. **6**(12), 1673–1687 (1997)
6. Ruanaidh, J.J.K.O., Pun, T.: Rotation, scale and translation invariant spread spectrum digital image watermarking. Signal Process. **66**(III), 303–317 (1998)
7. Grobois, R., Ebrahimi, T.: Watermarking in JPEG 2000 domain. In: Proceeding of the IEEE Workshop on Multimedia Signal Processing, 2001, pp. 3–5
8. Maity, S.P., Nandi, P.K., Das, T.S.: Robustness improvement in spread spectrum watermarking using M-ary modulation. In: Proceedings of 11th National Conference on Communication NCC 2005, pp. 569–573
9. Maity, S.P., Kundu, M.K., Mandal, M.K.: Capacity improvement in spread spectrum watermarking using biorthogonal wavelet. In: Proceedings of 12th National Conference on Communication NCC 2006, pp. 114–118
10. Maity, S.P., Kundu, M.K.: Perceptually adaptive spread transform image watermarking scheme using Hadamard transform. J. Inf. Sci. **181**(3), 450–465 (2011)
11. Borges, P.V.K., Mayer, J.: Informed position based watermarking. In: IEEE Proceedings of the 17th Brazilian Symposium on Computer Graphics and Image Processing, 2004

Simulation Study on Distribution of Control Points for Aerial Images Rectification

Lee Hung Liew, Beng Yong Lee, Yin Chai Wang and Wai Shiang Cheah

Abstract A raw uncalibrated aerial image acquired from a non-metric digital camera carried by an aircraft normally has lens and perspective distortions. However, geometric distortions are not occurred individually but accumulated irregularly in aerial images. Ground control points are important features used and geometric transformation is the essential process in non-parametric approach for aerial image rectification. The efficiency of the rectification would be affected if the control points are allocated in the image without considering the proper distribution. A simulation study is conducted using grid image and aerial images to examine the effect of different distribution patterns of control points. It demonstrates that lower order global transformation has limitation in rectifying images with complex distortions and images with different distribution patterns of control points have different deformation rates.

Keywords Aerial image rectification · Geometric distortion · Ground control point · Geometric transformation

L. H. Liew (✉) · B. Y. Lee
Universiti Teknologi MARA (UiTM), Kota Samarahan, Sarawak, Malaysia
e-mail: lhliew@sarawak.uitm.edu.my

B. Y. Lee
e-mail: bylee@sarawak.uitm.edu.my

Y. C. Wang · W. Cheah
Universiti Malaysia Sarawak (UNIMAS), Kota Samarahan, Sarawak, Malaysia
e-mail: ycwang@fit.unimas.my

W. Cheah
e-mail: wscheah@fit.unimas.my

S.-S. Yeo et al. (eds.), *Computer Science and its Applications*,
Lecture Notes in Electrical Engineering 203, DOI: 10.1007/978-94-007-5699-1_18,
© Springer Science+Business Media Dordrecht 2012

1 Introduction

Satellite and aircraft-based system are the two major platforms that commonly used to collect remote sensing images [1]. Aerial images provide visual records which are closer to the area of interest and are widely used for various purposes such as generating digital map, monitoring crop growth and ecology. Aerial images are generally captured with a camera mounted vertically which is attached to an aircraft with flying height between 200 and 15,000 m [2]. Although spatial information retrieved from aerial images could assist in human activities, such images are exposed to geometric distortions [3]. Therefore, a raw uncalibrated aerial image could not be used directly without rectification.

Image rectification is a pre-processing step to correct the geometric errors of a distorted image [4]. A geometrically rectified image then can be used to extract distance, polygon area and direction information with certain accuracy [5]. There are two main approaches used for image rectification: parametric and non-parametric approaches. This paper focuses on non-parametric approach which is usually implemented in geographic information system (GIS) and remote sensing software [5]. In non-parametric approach, ground control points (GCPs) are used as reference points and an appropriate mathematical function is applied to generate geometric transformation approximation. The accuracy of geometric rectification depends on the quality, quantity and distribution of control points [6]. However, current studies on image rectification are mostly focus on transformation models.

This paper attempts to study the effect of control points' distribution for aerial images rectification. The efficiency of the rectification would be affected if the control points are allocated in the image without considering the proper distribution. Besides that, it is believed that distortions occur differently in an aerial image. The deformation rate of a distorted image is studied based on the distribution of the control points through the simulation study.

This paper is organized as follow. In Sect. 2, the problems of aerial images are discussed. In Sect. 3, the image rectification using non-parametric approach is presented. The simulation study on distribution of control points using grid image and aerial images is explained. In Sect. 5, the experimental results are analyzed to compare the rectification rate and deformation rate of the different distribution patterns. In Sect. 6, conclusion of this paper is summarized.

2 General Problems of Aerial Images

Aerial images like other remotely sensed images face the problems of geometric distortion. The occurrence of geometric distortions are due to different sources such as lens, earth curvature, topographic relief and inconsistencies in the attitude of the aircraft which involve roll, yaw and pitch [7]. There are lots of studies have been carried out to research on image distortions. For instance, Jensen [8] discussed separately the effect of various geometric distortions; Guennadi [9] studied the effect of different sources of image geometric distortions and evaluate sensitivity of the corresponding correction models based on Ikonos satellite images; Yang et al. [10] proposed real-time rectification based on the analysis of the geometric distortions and Zhao et al. [11] investigated image distortions in multiformat aerial digital camera.

The two distortions that usually detected in aerial images are barrel distortion and perspective distortion. Barrel distortion is caused by the wide angle lens used in the non-metric digital camera. Such distortion modifies straight lines to be bended and points to be moved from their correct position [12]. Perspective distortion is due to unsteadily moves of the aircraft which caused by atmospheric perturbations [8]. It is claimed that geometric distortions are more seriously found in aerial images than satellite images [5] due to the stability of the two platforms and their height relation to the distance from the Earth [8].

Generally, geometric distortions are not happened individually in a raw uncalibrated aerial image but accumulated in the image. Furthermore, distortions occur differently and irregularly in the whole image [5]. Such accumulated and irregular distortions have increased the complexity of deformation in the images.

3 Images Rectification Using Non-parametric Approach

The general procedure flow of non-parametric aerial images rectification includes ground control points detection and selection, corresponding ground control points matching, geometric transformation determination, rectification evaluation with residual error estimation and image resampling and interpolation. Such approach not considers the imaging mechanisms that cause the distortions [13].

Ground control points (GCPs) are points on the surface of the earth where both image coordinates and map coordinates can be identified [8]. The rectification of an aerial image could be performed by generating a mapping transformation through the matching of control points from the reference image and the raw aerial image.

Geometric transformation is a vector function that maps the pixel in a reference image to a new position in a target image [14]. It converts the distorted image coordinates into the desired coordinates based on the reference image. Four commonly employed geometric transformations which could be found in ArcGIS software: first, second, third order polynomial transformations and adjust transformation are tested in the simulation study. First order polynomial transformation is a combination of translation, scaling, rotation and shearing transformation [15]. Second and third order polynomial transformations are usually applied to handle nonlinear distortions. Adjust transformation optimizes both global and local accuracy with a polynomial transformation and followed by a triangulated irregular network interpolation to adjust the control points locally to have a better match for the target control points [16].

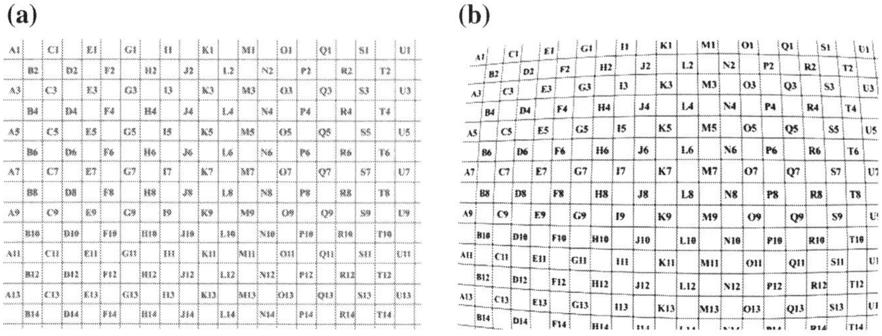

(a)

(b)

Fig. 1 Grid images used in the simulation study: **a** reference image, **b** distorted image with accumulation of the barrel, vertical and horizontal perspective effect

The distribution of control points and the geometric transformation play important roles in the images rectification. The proper distribution of GCPs and appropriate geometric transformation could be studied through the consideration of the expected distortions. Hence, this paper presents a simulation study which firstly explores the effect of control points' distribution patterns with commonly used geometric transformations in rectifying distorted grid image of barrel and perspective. The simulation study is then proceeds with aerial images by referring to the first experiment as initial guide.

4 Simulation Study on Distribution of Control Points

The simulation study consists of two experiments: rectification of grid image and rectification of aerial images. The experiments are carried out with a set of 30 corresponding control points which are manually selected and matched between the reference image and the distorted image. The distributions of control points are based on uniform, border, corner and centre patterns. Uniform distribution denotes that control points are located regularly at the whole image; border distribution means that control points are placed at image boundary; corner distribution indicates that control points are positioned at the four ends of the image while control points with centre distribution are assembled at the middle of the image.

Grid image is used in the simulation study because barrel and perspective effects as the expected distortions in aerial images could be easily detected in grid image. The output of this experiment is used as initial guide for aerial images rectification. The reference image and the distorted image used are illustrated in Fig. 1. The reference image is created with 21 × 14 grid lines. The distorted image is built using the same image for reference but with barrel, vertical perspective and horizontal perspective effects combined. Both images are fixed in size of 4272 × 2848 pixels, which are equal in size with the aerial images.

Table 1 Total RMSE (in pixels) for rectification of grid image

Image	Distribution pattern	Geometric transformation			
		1st order	2nd order	3rd order	Adjust
Grid image	Uniform	54.56734	33.26257	2.64072	0.98069
	Border	56.11092	24.09833	2.18036	1.10062
	Corner	56.85835	16.06877	3.10684	1.47094
	Centre	4.76054	2.16590	1.64263	0.51380

Second experiment involves four aerial images: aerial image 1, aerial image 2, aerial image 3 and aerial image 4. The aerial images of a flat area are selected for the experiment. These aerial images are captured using a non-metric digital camera which is carried by a helicopter with flying height of 1200 m approximately. A QuickBird image with a spatial resolution of 0.5 m is used as the reference image.

5 Experimental Results and Analysis

The performance of rectification with different control points' distribution is measured using the total root mean square error. The total RMS error is measured between the control points from the reference image and the rectified one as denoted in Eq. 1. Besides the rectification result, another test on distance analysis of control points before rectification is carried out. The distance analysis is evaluated using Euclidean distance between the control points of reference image and distorted one. The purpose of the distance analysis is to validate the significant of rectification results by considering the actual differential of the control points. The distance analysis also could indicate the deformation rate in an image according to the distribution patterns.

$$RMSE = \sqrt{\frac{1}{N}\left(\sum_i \|(x,y)_i - (x',y')_i\|^2\right)} \tag{1}$$

Tables 1, 2, 3 and 4 show the rectification results of grid image, aerial image 1, aerial image 2, aerial image 3 and aerial image 4. The first order polynomial transformation gives the highest total RMSE in all the distribution patterns with different distortions. However, the total RMSE in all the distribution patterns tested is reduced when second and third polynomial transformations are used. The lowest total RMSE is achieved when using adjust transformation in all the distribution patterns.

Among the four types of distribution patterns, it is noticed that centre distribution gives the lowest total RMSE and uniform distribution demonstrate the second lowest total RMSE in each geometric transformation. In order to validate the significant of rectification results, the actual differential between the control

Table 2 Total RMSE (in meters) for rectification of aerial image 1

Image	Distribution pattern	Geometric transformation			
		1st order	2nd order	3rd order	Adjust
Aerial image 1	Uniform	3.46245	1.54836	1.40481	0.18410
	Border	4.7857	1.41403	1.35385	0.48387
	Corner	4.36289	1.29788	1.09743	0.37872
	Centre	0.90795	0.85130	0.74821	0.17352

Table 3 Total RMSE (in meters) for rectification of aerial image 2

Image	Distribution pattern	Geometric transformation			
		1st order	2nd order	3rd order	Adjust
Aerial image 2	Uniform	3.38343	2.22872	1.79943	0.62751
	Border	4.16849	2.49074	2.00422	0.74626
	Corner	3.66203	2.19075	2.07959	1.14972
	Centre	1.31739	1.21663	1.16919	0.23642

Table 4 Total RMSE (in meters) for rectification of aerial image 3

Image	Distribution pattern	Geometric transformation			
		1st order	2nd order	3rd order	Adjust
Aerial image 3	Uniform	9.52202	2.92183	2.57022	0.71273
	Border	13.41082	4.2343	3.90368	0.77795
	Corner	11.93433	4.78258	4.24444	0.75125
	Centre	2.98112	2.61788	2.35714	0.56889

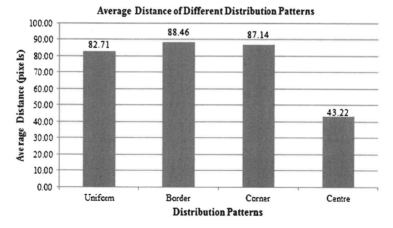

Fig. 2 Distance analysis of control points with different distribution patterns for grid image

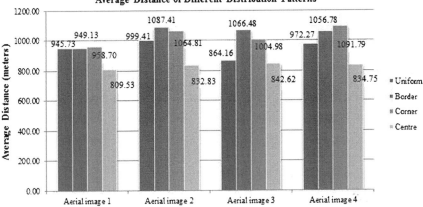

Fig. 3 Distance analysis of control points with different distribution patterns for aerial images

Table 5 Total RMSE (in meters) for rectification of aerial image 4

Image	Distribution pattern	Geometric transformation			
		1st order	2nd order	3rd order	Adjust
Aerial image 4	Uniform	2.95554	1.55296	1.34436	0.12666
	Border	4.27473	1.5641	1.33336	0.14999
	Corner	4.83573	2.14395	1.85796	0.39109
	Centre	0.57735	0.53580	0.47600	0.06259

points of reference image and distorted images are analyzed. This distance analysis could reflect the deformation rate before rectification and the results are presented in Figs. 2 and 3 for grid image and aerial images respectively.

From Fig. 2, it is found out that the average distance of control points under centre distribution pattern is the lowest for grid image with barrel and perspective distortions. This means that control points distributed at centre of the grid image are basically much less deformed than control points that are placed uniformly or at border and corner. Consequently, the total RMSE of the rectification for centre distribution is obviously low. On the other hand, the average distance of control points distributed at the border and corner of the grid image are higher than the average distance of control points with uniform distribution. This reflects that both border and corner of the image contain higher deformation rate. Uniform distribution with control points positioned consistently at the entire image causes the deformation rate has been equally averaged. The same outcome is also achieved when using aerial images as shown in Fig. 3. It is demonstrated that image centre has lowest deformation rate and control points used in aerial image rectification with uniform distribution could provide unbiased result (Table 5).

6 Conclusion

In non-parametric approach of aerial image rectification, ground control points are important features. The effect of control points' distribution is studied through a simulation study with rectification of grid image and aerial image. The deformation rate at each pattern of distribution also has been considered through the distance analysis of the control points to verify the rectification achieved. It is experimentally demonstrated that the extremely low of total RMSE for centre distribution is influenced by its lowest deformation rate at the image centre. Higher deformation rate is found at the border and corner of the image. Uniform distribution provides better distribution of control points for rectification by the consideration of averaging overall deformation rates at the entire image.

This study would able to assist in identifying a proper distribution of control points for aerial images rectification through the concern of expected distortions and commonly used geometric transformations. Using single geometric transformation for the whole image rectification might not appropriate since distortions occur irregularly in the image. Image centre could be used as the image region which is less distorted. In future, further investigation on the effect of other factors such as different density of control points should be carried out.

References

1. Xiang, H., Tian, L.: Method for automatic georeferencing aerial remote sensing (RS) images from an unmanned aerial vehicle (UAV) platform. Biosyst. Eng. **108**(2), 104–113 (2011)
2. Ronald, E.J.: Guide to GIS and image processing, Clark Labs, Clark University. http://gis.vsb.cz/vojtek/content/gitfast_p/files/source/IDRISI.pdf (2006)
3. Eltohamy, F., Hamza, E.H.: Effect of ground control points location and distribution on geometric correction accuracy of remote sensing satellite images. In: 13th International Conference on Aerospace Sciences & Aviation Technology (ASAT-13) (2009)
4. Xiangyang, S., Conggui, L., Yizhen, S.: Comparison and analysis research on geometric correction of remote sensing images. In: International Conference on Image Analysis and Signal Processing (IASP) (2010)
5. Wang, C., Zhang, Y., Wu, Y., Gu, Y.: Highly accurate geometric correction for seriously oblique aero remote sensing image based on the piecewise polynomial model. J Comput Inf Syst **7**(2), 342–349 (2011)
6. Wu, D., Liu, Y., Guo, L., Cheng,F., Li, G.: Spatial projection rectification for densifying ground control points. In: 8th International Symposium on Spatial Accuracy Assessment in Natural Resources and Environmental Sciences, pp. 51–58 (2008)
7. Richards, J.A., Jia, X.: Remote Sensing Digital Image Analysis: An Introduction, 4th edn. Springer, Heidelberg (2006)
8. Jensen, J.R.: Geometric Correction. Introductory Digital Image Processing: A Remote Sensing Perspective. Prentice Hall, Upper Saddle River (2005)
9. Guennadi, A.G.: Geometric accuracy of Ikonos: zoom in. IEEE Trans. Geosci. Remote Sens. **42**(1), 209–214 (2004)

10. Yang, X., Meng, F., Hu, L., Li, J.: Real-time image geometric rectification for scene matching based on aircraft attitude. In: International Symposium on Computer Science and Computational Technology (ISCSCT), vol. 1, pp. 805–808 (2008)
11. Zhao, L., Cheng, X., Wu, Z.: Study on image distortion caused by camera incline in multi-format aerial digital camera. In: 1st International Conference on Information Science and Engineering (ICISE), pp. 1452–1455 (2009)
12. Linder, W.: Digital Photogrammetry: A Practical Course, 2nd edn. Springer, Heidelberg (2006)
13. Pai, D.T.: Auto rectification for robotic helicopter aerial imaging. Thesis Master of Science in Computer Science, San Diego State University (2010)
14. Siddiqi, A.M., Saleem, M., Masud, A.: A local transformation function for images. In: International conference on Electrical Engineering (ICEE '07), pp. 1–5 (2007)
15. Zhang, Y.: Image processing using spatial transform. In: International Conference on Image Analysis and Signal Processing (IASP), pp. 282–285 (2009)
16. ArcGIS Desktop Help 9.2. Georeferencing a raster dataset. http://webhelp.esri.com/arcgisdesktop/9.2/index.cfm?id=2710&pid=2701&topicname=Georeferencing_a_raster_dataset (2011)

A Secure Image Watermarking Using Visual Cryptography

Xun Jin and JongWeon Kim

Abstract In this paper we proposed a secure image watermarking algorithm for digital image using the visual cryptography. The proposed algorithm operates in discrete cosine transform (DCT) and discrete fractional random transform (DFRNT) domain which is characterized by fractional order α and random seed β. In the watermark embedding process of the proposed algorithm some blocks of a cover object is selected randomly, and they are transformed to DFRNT domain. The watermark is generated by the visual cryptography and embedded into the DCT coefficient of the DFRNTed block. The watermark extraction process is the reverse process of embedding process. The proposed watermarking algorithm has high security by the visual cryptography and the randomness of discrete fractional random transform. The experimental results show that the proposed watermarking algorithm is imperceptible and moreover is robust against JPEG compression, common image processing distortions.

Keywords Image watermarking · Visual cryptography · Discrete fractional random transform · Discrete cosine transform · Random block selection

1 Introduction

As the manipulation of digital content becomes easier, copyright infringement becomes a serious problem. It is important to detect copyright violation and control unauthorized access to digital content. Digital watermarking, which

X. Jin · J. Kim (✉)
Deptartment of Copyright Protection, Sangmyung University, Seoul, Korea
e-mail: jwkim@smu.ac.kr

X. Jin
e-mail: jinxun0110@gmail.com

S.-S. Yeo et al. (eds.), *Computer Science and its Applications*,
Lecture Notes in Electrical Engineering 203, DOI: 10.1007/978-94-007-5699-1_19,
© Springer Science+Business Media Dordrecht 2012

embeds copyright information into digital content, is widely used for authentication and copyright protection of digital content [1–4].

Digital watermarking is a process in which digital content such as image, video, audio, and even text is protected by embedding information such as a hidden copyright message into the content. Such a watermark should be imperceptible to others while being perceptible to the copyright holder who possesses the proper private information key [5].

In the case of image content, digital watermark signals are commonly embedded in the spectral or frequency domain. In particular, several researchers have reported that embedding a watermark in the spectral domain is more robust. Various spectral domain approaches have been used to transform a host image to spectral domains such as the discrete cosine transform (DCT) domain, discrete wavelet transform (DWT) domain, discrete Fourier transform (DFT) domain, and discrete fractional Fourier transform (DFRFT) domain. The watermark is then embedded into the transformed image using certain algorithms. Finally, the watermarked image is transformed back to the spatial domain.

Recently, a watermarking algorithm based on discrete fractional random transform (DFRNT) was reported [6], this algorithm exploits the inherent randomness of the transform itself. Generally, intrinsic randomness improves the robustness of the watermarking against attacks. However, this algorithm requires a copy of the original image to extract the embedded watermark. Such a non-blind watermarking algorithm is unsuitable for industrial applications. As we have seen, most research teams have proposed the secure watermarking algorithms how to embed the watermark information securely not how to generate secure watermarks.

In this paper, we propose a novel watermark generation scheme using the visual cryptography. The proposed algorithm does not require a copy of the original image to extract the watermark, and it provides strong security through the use of a random kernel matrix. We have demonstrated that the proposed blind watermarking scheme is highly secure on account of the inherent randomness of DFRNT and that it is robust under several attacks.

2 Basic Theories

In this section, we are going to review the basic theories about the visual cryptography and discrete fractional random transform briefly.

2.1 Visual Cryptography

The visual cryptography is introduced by Naor and Shamir [7], which a kind of cryptography that encode image I_O into image I_1 and image I_2. Only with I_1 or I_2, human visual system cannot recognize the information encoded into the two

Fig. 1 The scheme of visual cryptography

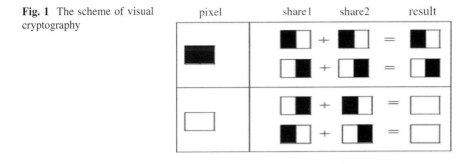

images. When I_1 and I_2 are overlapped together, the encoded information can be recognized directly by the human visual system. The I_O, I_1 and I_2 are composed of black and white pixels as illustrated in Fig. 1.

Each pixel in the original image is represented by at least one subpixel in each of the n transparencies or shares generated. Each share is comprised of collections of m black and white subpixels where each collection represents a particular original pixel [7–9].

A digital image copyright protection scheme based on visual cryptography is proposed by Hwang [9]. Zaghloul et al. [10] also a visual cryptography watermarking algorithm which is improved from Hwang's method. The algorithm ensures the security because of the visual cryptography. However it has some problems that the algorithm did not embed information into the cover object and it has to know the verification information for each object, which means non-blind watermarking scheme.

2.2 Discrete Factional Random Transform

It has been reported that a DFRNT can essentially be derived from the discrete fractional Fourier transform (DFRFT) [11]. The randomness is generated by using a random matrix. The overall process is quite similar to that by which the transform matrix of the DFRFT is obtained. The DFRNT can be defined by a diagonal symmetric random matrix. The matrix Q is generated by an $N \times N$ real random matrix P that satisfies the following relation

$$Q = \left(P + P^T\right)/2 \tag{1}$$

where $Q_{lk} = Q_{kl}$. We define the kernel matrix of the DFRNT R^α in such a way that Q commutes with R^α, i.e.,

$$R^\alpha Q = Q R^\alpha \tag{2}$$

In Eq. (2), these two matrices have the same eigenvectors. By the characteristic of a symmetric matrix, the eigenvectors of the matrix $\{V_{Rj}\}$ ($j = 1, 2, \cdots, N$) are real-orthonormal to each other. The eigenvector matrix $\{V_R\}$ is obtained by combining these column vectors:

$$V_R = [V_{R1}, V_{R2}, \cdots, V_{RN}] \tag{3}$$

The coefficient matrix that corresponds to the eigenvalues of the DFRNT is defined as

$$D_\alpha^R = diag\left[1, \exp\left(-j\frac{2\pi\alpha}{M}\right), \cdots, \exp\left(-j\frac{2(N-1)\pi\alpha}{M}\right)\right] \tag{4}$$

where M and α indicate the periodicity and the fractional order of the DFRNT, respectively. Then, the kernel transform matrix of the DFRNT can be expressed as

$$R^\alpha = V_R D_\alpha^R V_R^T \tag{5}$$

and the DFRNT of a two-dimensional image X is expressed as

$$X_R = R^\alpha X (R^\alpha)^T \tag{6}$$

The DFRNT is linear, unitary, index additive, and energy conserved. However, its kernel transform matrix is random, and this affords high security in information security applications such as digital watermarking.

3 Proposed Watermarking Scheme

The proposed watermarking scheme uses the visual cryptography for making a watermark and two transformations for embedding the watermark. The watermark information is expressed by a Quick Response (QR) code and the visual cryptography divides the code to visual crypto watermark by random numbers. All embedding and extraction processes are simply carried out in the DFRNT domain. For the same random number β, a change in the value of fractional order α produces an entirely different transformed image and can make the watermark undetectable. The fractional order α of the DFRNT and the random seed β used to generate the random kernel matrix are used as secret keys. The visual crypto watermark is embedded invisibly into the DCT coefficients of DFRNT pairs of blocks which are randomly selected from the host image.

3.1 Visual Crypto-Watermark Generation

The proposed algorithm generates the watermark using the visual cryptography. The visual crypto-watermark is generated by random numbers (RN) and QR code which expresses watermark information. The visual crypto VF generation process is as follows (Fig. 2).

QR code   Random number   Verification information

Fig. 2 Visual cryptography of QR code

Step 1:   Generate RN composed of value 0 and 1 with seed β_3.
Step 2:   Generate VF with RN and QR code. If $RN_i = 0$ and QR_i is equal to 0, or $RN_i = 1$ and QR_i is equal to 1, then assign the VF_i to be 1. If $RN_i = 1$ and QR_i is equal to 0, or $RN_i = 0$ and QR_i is equal to 1, then assign the VF_i to be 0.
Step 3:   Embed VF into the DCT coefficient of the DFRNTed block.

3.2 Watermark Embedding Process

The proposed algorithm embeds the visual crypto watermark into DCT coefficients of DFRNTed blocks. The block is randomly selected from the original image and the blocks are not overlapped to each other. The embedding process is as follows;
Step 1:   Generate random numbers with seed β_1 to select blocks randomly.
Step 2:   The selected blocks are transformed to DFRNT domain with the fractional order α and the random seed β_2.
Step 3:   The transformed blocks are transformed to DCT domain.
Step 4:   Generate random numbers RN with seed β_3.
Step 5:   Generate verification information VF with QR code and RN.
Step 6:   Quantize the DCT coefficient.
Step 7:   Embed VF into DCT coefficient.

Figure 3 shows the embedding process, as illustrated in the above steps.

3.3 Watermark Extraction Process

The extraction process is inverse process of the embedding process and as follows;
Step 1:   Generate random numbers with seed β_1 to select blocks randomly.
Step 2:   The selected blocks are transformed to DFRNT domain with the fractional order α and the random seed β_2.
Step 3:   The transformed blocks are transformed to DCT domain.
Step 4:   Quantize the DCT coefficient to extract VF.

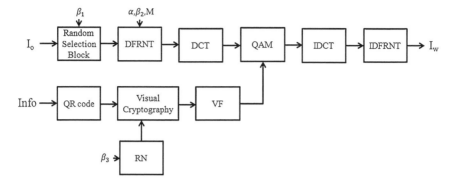

Fig. 3 Watermark embedding process

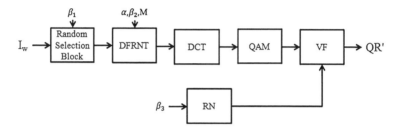

Fig. 4 Watermark extracting process

Fig. 5 Original image (*left*) and watermarked image (*right*)

Step 5:   Generate random numbers RN with seed β_3.
Step 6:   Extract watermark from VF and RN.

Figure 4 depicts the extraction process, as illustrated in the above steps.

Fig. 6 Original watermark (*left*) and extracted watermark (*right*)

Table 1 BERs according to attacks (Lena)

Attack	BER	PSNR
JPEG (70 %)	0	36.92
JPEG (40 %)	1.6	35.13
JPEG (30 %)	4.32	34.32
JPEG (20 %)	12.48	33.11
Gaussian (0.001)	2.88	29.78
Salt and pepper (0.01)	1.12	25.31
Cropping (10 %)	0.32	16.78
Rotation (45)	4.16	15.06

Fig. 7 Bit error rate of JPEG compression

Fig. 8 Bit error rate of other noises and attacks

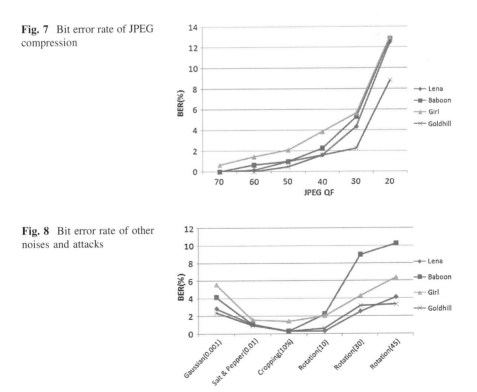

4 Performance Evaluations

The proposed watermarking scheme used for measuring the performance of the cover image "Lena", "Baboon", "Girl", and "Goldhill" images with 512×512 size. And the watermark is the 25×25 QR code. Figure 5 shows the original Lena image and the watermarked Lena image, and Fig. 6 shows the extracted QR watermark after JPEG compression with quality factor (QF) = 60, and the bit error rate is 0.16 % that means only one pixel error (a pixel in the circle).

The Table 1 shows the results after several attacks for Lena image. The algorithm recovers the QR code within 5 % after the several attacks except the JPEG with QF = 20. Actually, QF = 20 means the image is almost distorted. In this evaluation, we used the QR code as a watermark. QR code has the error correction capability by itself. The 25×25 QR code can contain 272 bits when ECC level is L (about 7 %) and 224, 176, 128 bits when the levels are M (=15 %), Q (=25 %), and H (=30 %), respectively. For example, if the algorithm recovers the QR code within 15 % bit error rate (BER), the 224 bits watermark information can be extracted.

Figure 7 shows the BERs against the QFs of JPEG compression. BERs for those images are getting less when the QF is getting higher and this means the proposed watermarking scheme is robust to JPEG compression.

Figure 8 depicts the BERs after several attacks such as Gaussian noise, Salt and pepper noise, cropping, and rotation with angles 10, 30 and 45. The results show the proposed algorithm is robust to noise and geometrical attacks.

As a result, the proposed algorithm is robust to frequency attacks because of the use of DCT and DFRNT, and to geometrical attacks because of the embedding of a random selected block.

5 Conclusions

In this paper, we proposed a digital watermarking scheme using DFRNT, DCT and Visual Cryptography. The proposed algorithm selects blocks randomly and transformed to DFRNT domain and DCT domain, then verification information of watermark image is embedded into DCT coefficient. The fractional order α and random seed β are used as the secret keys required to access the watermarked image in the proposed algorithm. The experimental results indicate that the proposed algorithm is robust against frequency and geometrical attacks.

Improvements in capacity and robustness performance under compression and noise attacks were presented. Performance experiments were conducted by comparing the bit error rate value and variety attacks. The results show the proposed method is robust against JPEG compression but fragile against Gaussian noise.

Acknowledgments This research project was supported by Ministry of Culture, Sports and Tourism (MCST) and from Korea Copyright Commission in 2011.

References

1. Kim, J., Kim, N., Lee, D., Park, S., Lee, S.: Watermarking two dimensional data object identifier for authenticated distribution of digital multimedia contents. Signal Process. Image Commun. **25**, 559–576 (2010)
2. Li, D., Kim, J.: Secure image forensic marking algorithm using 2D barcode and off-axis hologram in DWT-DFRNT domain. Appl. Math. Inf. Sci. **6**(2S), 513s–520s (2012)
3. Cox, I.J., Kilian, J., Leighton, T., Shamoon, T.: Secure spread spectrum watermarking for multimedia. IEEE Trans. Image Process. **6**(12), 1673–1687 (1997)
4. Nah, J., Kim, J., Kim, J.: Video forensic marking algorithm using peak position modulation. J. Appl. Math. Inf. Sci. **6**(6S), (2012) (to be published)
5. Cox, I., Kilian, J., Shammon, T.: Secure spread spectrum watermarking for images, audio and video. In: Proceedings of ICIP-96, Lausanne, 1996, pp. 243–246
6. Guo, J., Liu, Z., Liu, S.: Watermarking based on discrete fractional random transform. Opt. Commun. **272**(2), 344–348 (2007)
7. Naor, N., Shamir, A.: Visual Cryptography, Advances in Cryptology: Eurocrypt'94, pp. 1–129. Springer, Berlin (1995)
8. Hawkes, L., Yasinsac, A., Cline, C.: An Application of Visual Cryptography to Financial Documents. Technical Report TR001001, Florida State University (2000)
9. Hwang, R.: A digital image copyright protection scheme based on visual cryptography. Tambang J. Sci. Eng. **3**(2), 97–106 (2000)
10. Zaghloul, R.I., Al-Rawashdeh, E.F.: HSV image watermarking scheme based on visual cryptography. World Acad. Sci. Eng. Technol. **44**, 482–485 (2008)
11. Liu, Z., Zhao, H., Liu, S.: A discrete fractional random transform. Opt. Commun. **255**(4–6), 357–365 (2005)

Bit Error Rate Analysis of Partial Packets in Ad hoc Wireless Networks

Jia Lu, XinBiao Gan, Gang Han, Baoliang Li and Wenhua Dou

Abstract An Ad hoc wireless network environment is constructed and the number of errors in partial packets, bit error time series, is collected to study bit error rate of wireless burst-error channels. It is verified that the bit error time series is chaotic by computing maximal lyapunov exponent. The self-similarity of the bit error time series is also validated using the rescaled range (R/S) method, the aggregated variance time (V/T) method, absolute moment method and periodic diagram method respectively. Our research results show that we can utilize the complex non-linearity system theory such as chaos and self-similarity to study the number of errors in a partial packet. It is very important for evaluating bit error rate of wireless burst-error channels and prediction of the bit error ratio of a partial packet in Ad hoc wireless network.

Keywords Partial packets · Maximal lyapunov exponential · Chaos · Self-similar

J. Lu (✉) · X. Gan · G. Han · B. Li · W. Dou
School of Computer, National University of Defense Technology,
Changsha 410073 Hunan, People's Republic of China
e-mail: lujia661@126.com

X. Gan
e-mail: XinBiaoGan@nudt.edu.cn

G. Han
e-mail: GangHan@nudt.edu.cn

B. Li
e-mail: BaoliangLi@nudt.edu.cn

W. Dou
e-mail: WHdou@nudt.edu.cn

S.-S. Yeo et al. (eds.), *Computer Science and its Applications*,
Lecture Notes in Electrical Engineering 203, DOI: 10.1007/978-94-007-5699-1_20,
© Springer Science+Business Media Dordrecht 2012

1 Introduction

Self-similarity is an important concept and it has been applied to data communications traffic analysis widely. In mathematics, a self-similar object is exactly or approximately similar to a part of itself. Many objects in the real world or man-made environment are statistically self-similar [1].

It is observed that the pattern of data traffic is well modeled by self-similar processes in a wide variety of real-world networking situations. The total traffic on network is self-similar [2–4]. In addition, the network threat time series shows self-similarity [5]. However, the self-similarity of bit error rate in Ad hoc wireless network has not been studied.

The aim of this paper is to explore the self-similarity of bit error rate in Ad hoc wireless network. We set up an Ad hoc wireless network environment and collect partial packets. The number of errors in partial packets is used to form bit error time series. We compute maximal lyapunov exponent of the bit error time series and find chaos in it. Then, the self-similarity of the bit error time series is validated using the rescaled range (R/S) method, the aggregated variance time (V/T) method, absolute moment method and periodic diagram method respectively. Our research results show there is chaos in the bit error time series.

A further analysis indicates that the self-similarity of bit error rate in Ad hoc wireless network may be caused by the self-similarity of data communications traffic.

In the following section, we describe our measurement setup and the time series collection briefly. Section 3 tests chaotic for the bit error time series. The self-similarity of the bit error time series is verified in Sect. 4; Conclusions are provided in Sect. 5.

2 Experiment Setup and Time Series Collection

We set up an Ad hoc wireless network environment which is consisted of 7 nodes and one of them is monitoring node. For each node, we employ the *TP link-wn650 g* wireless network card and the network card driver is *MadWifi-0.9.4*.

In order to compute number of errors in partial packets, we modify *ieee80211 monitor.c* and *ieee80211 proto.c* contained in *MadWifi-0.9.4* appropriately and set the wireless network card of monitoring node in monitor mode. The monitoring node output all partial packets into a file *errorTrace*. Then, we write a program built with *VC++6.0* to read these packers from *errorTrace* and combine them and its corresponding correct packets by the bitwise "$\oplus$" to compute number of errors. In order to simplify calculation process, we use data packets with all data bytes set to 0×00. Because of the scrambling procedure, it is not expected that the contents of data packets have any significant impact on the experiment results [6]. The length of data packets is 164 bytes.

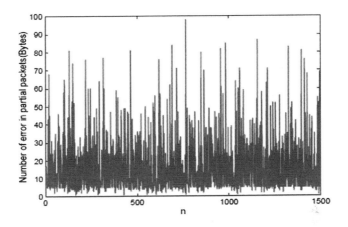

Fig. 1 The bit error time series

The number of errors in partial packets is used to form bit error time series. In this experiment, we receive 6737 partial packets and all of them are used to form bit error time series. Due to space limit, we only plot 1500 elements in Fig. 1 and all 6737 partial packets are used in latter calculation.

3 Chaotic Tests of the Bit Error Time Series

Chaos means a state of disorder and it has been used to model error processes in wireless bursts channels. The distribution of the lengths for both runs and bursts presents a heavy-tail behavior and chaotic maps have been used to model error processes in wireless bursts channels [7]. However, the heavy-tail behavior was observed only based on complementary cumulative distribution function (CCDF) and the author just concerned the lengths of both runs and bursts. In this paper the chaotic of the bit error time series is verified by computing maximal lyapunov exponent.

One method that has been widely used to verify chaos is the maximum lyapunov exponent, which provides a qualitative and quantitative characterization of the amount of chaos in a system [8]. A system with one or more positive lyapunov exponents is defined to be chaotic. We employ Wolf's algorithm to compute the maximum lyapunov exponent in this paper. In Fig. 2, we plot the maximum lyapunov exponent against embedding dimensions for various delay time. From this figure, it is very clear that the bit error time series has chaos because its maximum lyapunov exponent is greater than 0.

The bit error time series exhibits chaos. Since self-similarity is closely linked to chaos theory and the geometry of chaotic systems over time reveals a self-similar structure [9], we will discuss the self-similarity of the bit error time series in the next section.

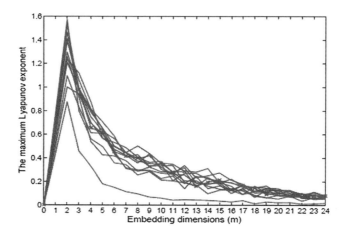

Fig. 2 The maximum lyapunov exponent against embedding dimensions for various delay time

4 Self-Similarity of the Bit Error Time Series

Self-similarity is notions pioneered by Benoit B. Mandelbrot [10]. It describes the phenomenon where a certain property of an object is preserved with respect to scaling in space and/or time [11].

A simple, direct parameter, characterizing the degree of self-similar, is the *hurst* parameter. If $0:5 < H < 1$, then the process is self-similar and large values indicate stronger self-similarity [8]. A number of methods have been proposed to estimate *hurst* parameter [12, 13]. The rescaled range (R/S) method, aggregated variance time (V/T) method, absolute moment method and periodic diagram method [12] are employed to estimate *hurst* parameter of the bit error time series.

The R/S method is based on empirical observations and it is a statistical measure of the variability of a time series [12]. Its purpose is to provide an assessment of how the apparent variability of a series changes with the length of the time-period being considered. Using this method, H is roughly estimated through the slope of the linear line in a log–log plot, depicting the R/S statistics over the number of points of the aggregated series. In Fig. 3, we plot the R/S statistics of bit error time series against the number of points of the aggregated series on a log–log plot. From Fig. 3, it is very clear that the bit error time series has self-similarity since its trace has slope much greater than 1/2. The *hurst* parameter estimated with the R/S method is 0.6214.

The aggregated variance time method is based on the slowly decaying variance property [12], which indicates long-range dependency. The slope of the straight line in a log–log plot, depicting the sample variance over the block size of each aggregation, is used for estimating H. H is given by $H = 1 - \beta/2$. In Fig. 4, we plot the variance of bit error time series against aggregate Levels on a log–log plot. It is very clear that the bit error time series has self-similarity since its trace has slope

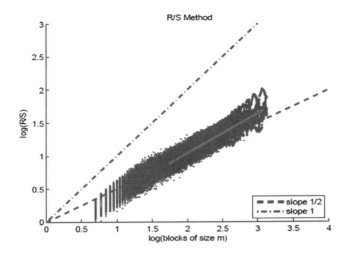

Fig. 3 Estimating the *hurst* parameter of a given sequence with RS method

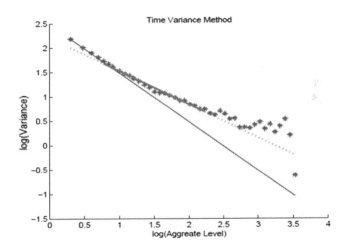

Fig. 4 Estimating the *hurst* parameter of a given sequence with aggregate variance method

much greater than −1 in Fig. 4. The *hurst* parameter estimated with the aggregated variance time method is 0.6581.

The absolute moment method is related to the variance method computed for the first moment [12]. The slope of the straight line in a log–log plot, depicting the first moment of the aggregated block over the block size, provides an estimator for H, by H = 1 + α. In Fig. 5, we plot the absolute moment of bit error time series against aggregate levels on a log–log plot. From Fig. 5, its very clear that the bit error time series has self-similarity since its trace has slope much greater than −1/ 2. The *hurst* parameter estimated with the absolute moment method is 0.6443.

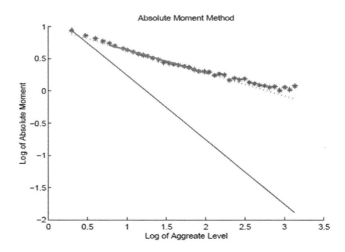

Fig. 5 Estimating the *hurst* parameter of a given sequence with absolute moment method

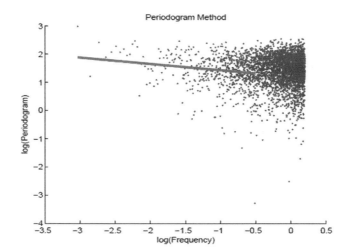

Fig. 6 Estimating the *hurst* parameter of a given sequence with periodogram method

The periodogram method is based on the power-spectrum singularity a 0-property [12]. The slope of the straight line, approximating the logarithm of the spectral density over the frequency as the frequency approaches 0, yields H. In Fig. 6, we plot the periodogram of bit error time series against Frequency on a log–log plot. The *hurst* parameter estimated with the periodogram method is 0.6119.

The *hurst* parameters obtained from the different methods are 0.6214, 0.6581, 0.6443 and 0.6119 respectively. All values are very similar with *hurst* values of about 0.6 and higher. This suggests the presence of self-similar behavior in the bit error time series. Since self-similar time-series can be forecasted, the bit error time series in wireless burst-error channels is forecastable.

It is noticeable that the *hurst* parameter of the bit error time series is smaller than the *hurst* parameter of the total traffic on Ad hoc wireless network [3]. Generally speaking, the more traffic on Ad hoc wireless network, the greater probability bit errors occurrence in wireless burst-error channels. Therefore, we put forward the hypothesis that self-similarity of traffic brings about the self-similarity of the bit error rate in Ad hoc wireless network. In addition, since there is radio noise in the radio communication procedure, the bit error rate of wireless burst-error channels in Ad hoc wireless network shows weaker self-similarity than the total traffic on Ad hoc wireless network.

5 Conclusions

Lots of works have been done for the self-similarity of Ad hoc wireless network traffic. We study the bit error rate of wireless burst-error channels in Ad hoc wireless network, based on the *errorTrace* collected in Ad hoc wireless network environment. The bit error time series is defined for the first time. The chaotic of the bit error time series is verified by computing maximal lyapunov exponent. Using methods of estimating *hurst* parameters such as the rescaled range (R/S) method, the aggregated variance time (V/T) method, absolute moment method and periodic diagram method respectively, self-similar behavior in the bit error time series is validated respectively. Given the stronger self-similarity of the total traffic on Ad hoc wireless network, the reason behind self-similarity of bit error rate in Ad hoc wireless network is discussed.

References

1. Stallings, W.: High-Speed Networks: TCP/IP and ATM Design Principles. Prentice-Hall, Upper Saddle River (1998)
2. Willinger, W., Taqqu, M., Sherman, R., Wilson, D.: Self-similarity through high-variability: statistical analysis of Ethernet LAN traffic at the source level. In: Proceedings of ACM SIGCOMM'95, pp. 100–113. Cambridge (1995)
3. Liang, Q.: Ad Hoc wireless network traffic self-similarity and forecasting. IEEE Comm. Lett. **6**, 297–299 (2002)
4. Kalden, R. Ekstrm, H.: Searching for mobile mice and elephants in GPRS networks. Mobile Comput. Commun. Rev. **8**(4), 37–46 (2004)
5. Xuan, L., Lu, X., Yu, R., Zhao, X.: Selfsimilarity analysis of network threat time series. J. Commun. (2002)
6. Han, B., Ji, L., Lee, S., Bhattacharjee, B., Miller, R.R.: All bits are not equal—A study of IEEE 802.11 communication bit errors. In: Infocom, pp. 1602–1610 (2009)
7. Kopke, A., Willig, A.,Karl, H.: Chaotic Maps as Parsimonious Bit Error Models of Wireless Channels. In: Proceedings of ACM Infocom'03, pp. 513–523 (2003)
8. Yang, C., Wu, Q.: (2010) A robust method on estimation of Lyapunov exponents from a noisy time series, Nonlinear Dyn. 279–292

9. Marks-Tarlow, T.: The fractal geometry of human nature. In: Robertson, R., Combs, A. (eds.) Chaos Theory in Psychology and the Life Sciences. Lawrence Erlbaum, Mahwah. pp. 275–283 (2009)
10. Mandelbrot, B.B.: The Fractal Geometry of Nature. W.H.Freeman, New York (1982)
11. Crovella, M.E., Bestavros A.: (1996) Self-similarity in World Wide Web Traffic: Evidence and Possible Causes. In: Proceedings of ACM SIGMETRICS'96. 5(6)
12. Kalden, R., Ibrahim, S.: Searching for Self-Similarity in GPRS. The 5th annual Passive and Active Measurement Workshop, April 2004
13. Jeong, H.-D.J, Lee, J.-S.R., McNickle, D., Pawlikowski, K.: Comparision of Various Estimators in Simulated FGN. Simul. Model. Pract. Theory **15**, 1173–1191 (2007)

An Interaction System Architecture and Design between Smart Computing and Cloud Computing

Tae-Gyu Lee, Seong-Hoon Lee and Gi-Soo Chung

Abstract This paper discusses both cloud computing and smart computing and then debates structural and functional issues about how the two computing systems interact for offering user-friendly information solutions. Smart Table is proposed to configure cloud resources. Smart control algorithm is proposed to choose the best resources. The information interaction between cloud computing and smart computing services is realized for efficient construction of information systems and for user's convenience of information services.

Keywords Cloud computing · Smart computing · Cloud architecture · Sensitive information architecture

1 Introduction

This paper analyzes mutually exclusive features and interdependent complementary capabilities based on the intrinsic features of each of cloud computing and smart computing. Based on this analysis, it discusses the structural and functional approach for how to interact with each other in order to present a convenient and efficient information solution for the user by establishing the information interaction between the two systems. More efficient system configuration and convenient user services

T.-G. Lee (✉) · G.-S. Chung
Korea Institute of Industrial Technology (KITECH), Ansan, Korea
e-mail: tigerlee88@empal.com

S.-H. Lee
BaekSeok University, Cheonan, Korea
e-mail: shlee@bu.ac.kr

S.-S. Yeo et al. (eds.), *Computer Science and its Applications*,
Lecture Notes in Electrical Engineering 203, DOI: 10.1007/978-94-007-5699-1_21,
© Springer Science+Business Media Dordrecht 2012

are realized by creating a synergy by combining benefits of both systems through the information interaction between cloud systems and smart services.

The characteristics of cloud computing are represented as virtualization, immediacy, scalability, etc. These features minimize the cost of information systems based on cloud economics and make a possible to build infinitely scalable mission-critical projects [1]. Nevertheless, the restrictions of an integrated management through the virtualization of hardware resources distributed on the Internet, remains as the ongoing challenge topic to maximize the real-time immediacy in information building systems. Cloud systems are interested in the service development on the large-scale projects or team-based projects, so they are still to be desired for the appropriate service infrastructure and interfaces for personal, in fact [2, 3].

Smart Computing is to provide the intelligent and optimal information services to meet different situation logic of all existing computing capabilities and user information services. Smart computing presents the ultimate goal of providing the accurate information services to users not just to provide information quickly and to manage resources effectively.

Definition 1 Smart computing is a personalized user-oriented computing by using intelligence and lightweight of computing resources such as hardware, software, and user service.

It presents the optional alternatives of the system resources to assign for the user.

It provides an auto-selection technique that is optimized to user environment based on the optional resources.

An important feature of smart computing services is to provide appropriate intelligent information services tailored to the user's current situations through the specific information devices for users' personal [4]. In order to support such a smart service, the agile and easy computing system is needed for the client user's perspective, and the complex and intelligent computing system is required in order to provide an optimal alternative for each situation for the service center or server's perspective [5]. In other words, it implies that the information systems should be equipped with the proper transparency and the accuracy of information service scenarios in terms of users. Furthermore, real computing resources are hidden for users in terms of smart service center, but those are faced with the challenge of flexibility to deal with according to the user's information service scenarios that can accurately identify and analyze the situation of the user terminal. However, current smart computing systems are focused on the features and interfaces of user terminal. Therefore, there are no enough how to configure and support the background support systems of them [2–4].

This study finds the main features and functional issues from the conventional cloud and smart information systems. It also presents a system configuration method to realize an effective information interactions and an efficient information system implementation through an appropriate structural combination of cloud and smart computing systems.

Fig. 1 Overview of smart
computing over cloud
computing

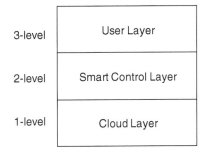

We will describe this paper in the following order. Section 2 analyzes the structural and functional features of the current cloud and smart computing systems, and describes the major issues. Section 3 proposes Smart Table and Smart control algorithm, which configure smart computing. Section 4 will define and analyze the performance and system complexity functions for providing resources in cloud computing and smart-cloud computing. Finally, Sect. 5 concludes this paper and describes future works.

2 Interaction System Architecture of Smart and Cloud Computing

This section describes how to interact between the smart computing and cloud computing issues. Figure 1 shows three-tier layered architecture that organizes smart user and smart control logic over cloud computing resource.

The user layer supports the smart interface based on user scenarios. The smart interface notifies to user for the results that is automatically selected by the interface agent, or presents an optional optimum alternative to user for manual selection. Smart control layer is a smart resource selection process to build the best resource scenario on the cloud resource layer of the lower layer for supporting user scenarios of the upper layer. Smart logic and Smart Table are key components. The cloud layer is consisted of SaaS, PaaS, IaaS, etc. of the existing cloud service systems. Depending on smart user requirements, it operates the cloud resource service optionally integrated with SaaS, PaaS, and IaaS.

Figure 2 shows the horizontal structure for information interaction of smart computing and cloud computing. Smart computing components are classified into smart hardware, smart software, and smart users. Those are mapped with IaaS, PaaS, and SaaS as a component of cloud computing, respectively. Depending on the user's information service needs, it supports smart IaaS, smart PaaS, and smart SaaS as cloud services combined with smart logic to provide the system resources that the user requires.

Smart computing supports a smart thin client, a foreground configuration near to a smart user, and a smart mobile interface. Cloud computing center supports of

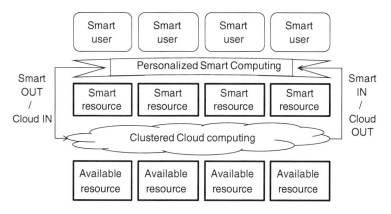

Fig. 2 Interaction architecture of cloud and smart computing

the thick server center, a smart user's background configuration, and the resource virtual interface based on flexibility and immediacy.

Cloud computing receives input information (smart request; cloud IN) such as the smart user's mobility location, the smart device's resource situation, and smart scenarios from smart computing because it effectively supports high-performance computing and resource efficiency. In addition, it sends the virtualization interface (virtual API) and parameters (smart response; cloud OUT) including the optimized resource information to support the smart computing performance and resource efficiency.

Because smart computing supports intelligent and accurate information services to individual users, it receives the input information (cloud request; smart IN) such as the selectable available resources and the priority parameters on the mobile path of smart user from cloud computing. In addition, it sends the smart interfaces (smart API) and parameters (cloud response; smart OUT), which indicate allocation and performance values of the virtual resource of cloud computing based on a smart scenario forecasting.

2.1 Features Supported by Cloud Computing

Cloud computing performs a central role as the following features in information interaction as the background role of smart computing:

• Hiding and abstraction are supported.

Cloud computing conceals the physical status of resources allocated to a user through resource virtualization and provides a view of logical resource allocation.

- Dynamic scalability and flexibility are supported

Depending on the customer's business needs, it can quickly build computing environments and can be extended in automation-based. Users can use the computing resources at any time (immediacy) as well as the amount they want (flexibility) from the cloud computing. Cloud computing resources can be extended continuously through the expansion of the available cloud virtual resources associated to Internet without getting to the computing resources limitations of an individual or a company.

- It builds a billing system as On-demand way.

Smart-cloud service pricing method is by billing as much as the used. For example, as the electricity and gas of house is used by billing as much as the used, all of the services of IT will be the used concept of borrowing as much as the need without having to purchase in the future.

- It realizes the convenience based on the automated setup.

In cloud computing era, most computing will accelerate automation. For example, if the existing operators buy a computer or server, they manually installed OS, platform, DB, various development tools, etc. However, in cloud computing environments, at the same time that a user selects the required server, the computing environment including the OS that the user needs is automatically installed. In addition, professional engineers who are responsible for all the hard works as server management, firewall settings, load-balancing settings, App Install, Configuration settings, etc. can be easily a replacement as the unprofessional user using the Web UI. In conclusion, most of the IT-related tasks that manually performed by existing users are automated and they can be set up easily through the Web UI.

In the operating process of an existing server hosting, the steps to create a server usually have been taken over at least a few days up to several weeks so that he can use the server after a user ordered the request application server. However, they can be completed within tens of minutes to minutes through the cloud setup automation, and the ease of operation can be achieved by just manipulating the touch of a button through automation when a operator manages the physical resources like a server.

2.2 Features Supported by Smart Computing

Smart computing plays a role in information interaction for the foreground role of the cloud computing as the following features:

- It achieves the information accuracy based on intelligence process (accuracy).

Smart Computing optionally presents the alternatives of resource allocation and management that are optimized for the situation logic of smart user's individual.

Table 1 Smart table for resource context

SEQ	RCODE	RC	Distance (sec)	Size	Cost	Provider	Address
1	GPO:LPO:RNO	Memory	0.001	10,000 KB	1,000₩	LG	Seoul
2	GP1:LP1:RN1	CPU	0.02	500MHZ/MIPS	1$	LOTTEE	Tokyo
3	GP2:LP2:RN2	NET	0.03	10 Mbps	1$	SKT	Beijing
4	GP3:LP3:RN3	HDD	0.1	2 GB	0.5$	KT	LA

Thus, it pursues the accuracy of its information services by mutual dependency with performance optimization and resource efficiency of cloud computing.

- It builds transparency and ease of use (usability) of a smart interface.

Smart computing supports an important user notification and order interface for hidden virtualization resource management of cloud computing to build easy guide interface of Smart user.

- It builds the smart processes based on predictability and Intelligence (scenario of accumulated Know-how).

It supports a resource scheduling method of the cloud systems of helping the resource allocation and reservation by defining a set of events based on smart scenarios to enhance the ability of information service of smart devices [6].

- It supports smart mobility.

Smart computing systems, by the intelligent mobile service/platform agents, strengthen a mobile information service by understanding the status of virtual resources of the cloud and by operating optimally the resource allocation of smart computing.

3 Smart Logic and Smart Control Algorithm

Smart Table (ST) is maintained for each Smart user independent with MapReduce of the Cloud [7]. Smart Table constitutes resource allocation table based on resource allocation scenarios to support smart user's smart work scenarios. Smart Table, as shown in Table 1, configures the extraction table by the unit of smart user from MapReduce.

The column heads of Smart Table as Table 1 are as follows: RCODE is the identified unique code for cloud resource; Resource Class (RC) is a category such as memory, CPU, network device (Router), HDD, etc.; Distance (sec) indicates the time length consuming from Smart user to target resource; Size indicates the size of each resource by unit; Cost is the cost for using of resources; Provider is a company providing the resources; Address is the location of the resource. The Measured variables will applied to the resource request parameters in the following Smart Table creation function:

$$f_s = Creat\_ST(Cloud\_ID, Cloud\_class, Select\_rule,$$
$$Auto\_ \mod e, Resouce\_REQ\_Param)$$

Cloud's MapReduce is pursuing the management efficiency of the large data sets of overall system, while Smart Table is pursuing the management efficiency of a personal set of the smart user's information scenarios.

If the dataset of ST is growing, the searching data structure is needed to reduce the complexity of the resource selection process. Selection rule can optionally operate among weights policy, FIFO (or sequence), Worst-fit, Best-fit, etc. To ensure the integrity of ST's resource data, the synchronization between ST and Cloud resource information causes problems.

In Fig. 3, the smart control algorithm is the process of resource delivery, which receives the smart resource requests from Smart user (Smart IN) and returns the available resources for Smart user. In initialization, ST stores the extracted resource information from the set of cloud resources based on the scenario forecasts or user's preferences. Smart logic specifies the selection rule to select the available resources. First, it searches for the target resource in the set of available resources within ST. Depending on whether the automatic option is selected, the target resources is selected automatically or is selected manually by a Smart user's intervention. If the resource selection is determined, it checks the availability of resources in the resource Cloud, and passes the results of available resource parameter to smart user.

4 Functions and Comparisons

It compares Smart Model (SM) and Non-Smart-Model (NSM). Traditional cloud models manage resources through a single virtual model. The proposed smart control scheme is looking resource alternatives for available cloud resources in Smart Table as a optional virtualization model, and accesses directly to the cloud resources and performs operations such as data input–output and program execution.

The existing MapReduce iteratively and incrementally assigns requirement resources based on linear search technique. Thus, it has the resource setup complexity $O(n^2)$. However, the proposed ST is an alternative configuring the status information of available resources to independent with MapReduce to optimize the cost of resource discovery. Thus, it has the resource setup complexity $O(n)$.

Resource efficiency is the ratio of used time of the target resource k compared to that of overall computing time as Eq. 1.

$$E_k = \frac{UsedTime}{UsedTime + UnusedTime} \tag{1}$$

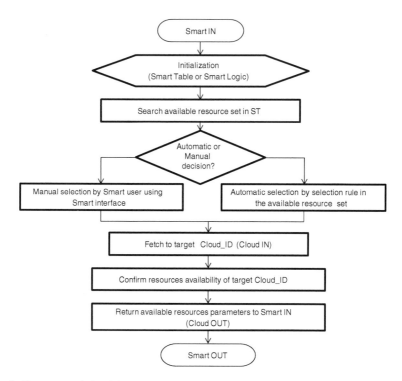

Fig. 3 Smart control algorithm

Average resource efficiency of the entire system is shown in the following Eq. 2.

$$E_{avg} = \frac{1}{m} \bullet \sum_{k=1}^{m} E_k \tag{2}$$

Ratio of cost reduction is as follows.

$$C_f = \frac{C_{SM}}{C_{NSM}} \tag{3}$$

If $C_f = 1$, the resource allocation cost of SM is equal to that of NSM. If $C_f < 1$, the resource allocation cost of SM is less than that of NSM. If $C_f > 1$, the resource allocation cost of SM is more than that of NSM.

5 Conclusions and Future works

In order to create synergies of cloud computing and smart computing system, this study presents an alternative that realizes an efficient information interaction system through the structural integration of cloud computing and smart computing systems. It also performs an optimized resource selection process of cloud computing resources based on a Smart Table and smart logic according to the parameters including user's resource demand passed.

The information interaction architecture can achieve a more efficient and balanced configuration of total information system by providing smart computing as a personalized systematic alternative and by providing a cloud computing as an aggregation systematic alternative. As shown in Sect. 4, the proposed system reduces the resource navigation complexity of cloud system and maximizes economy and performance of smart users. Topological configuration for the interaction provides the foreground services of smart computing and gives the background services of cloud computing.

References

1. Wikipedia Cloud computing. http://en.wikipedia.org/wiki/Cloud_computing (2012). Accessed March 2012
2. Borthakur, D.: The Hadoop Distributed File System: Architecture and Design. The Apache Software Foundation, (2007)
3. Mather, T., Kumaraswamy, S., Latif, S.: Cloud Security and Privacy, O'Reilly Media Inc, (2009)
4. Wikipedia Smart device. http://en.wikipedia.org/wiki/Smart_device (2011). Accessed December 2011
5. Dutta, K., Guin, R.B., Banerjee, S., Chakrabarti, S., Biswas, U.: A smart job scheduling system for cloud computing service providers and users: modeling and simulation. In: 1st International Conference on Recent Advances in Information Technology (RAIT-2012), March 2012
6. Chang, Y.-S., Shih, P.-C.: A resource-awareness information extraction architecture on mobile grid environment. J. Netw. Comp. Appl., Elsevier Ltd 682–695 (2010)
7. Apache Hadoop. http://hadoop.apache.org/

A Virtual Machine Scheduling Algorithm for Resource Cooperation in a Private Cloud

Ruay-Shiung Chang, Yao-Chung Chang and Ren-Cheng Ye

Abstract In recent years, virtualization has been widely applied in cloud computing because of its ability to increase resource utilization. With the scale of cloud computing architecture becoming larger, efficient resource allocation has also become more important. Existing scheduling algorithms for virtual machines cannot use new information to decide upon allocation of the appropriate physical machines because current scheduling algorithms lack the ability to be updated with up-to-the-minute information about each physical machine when making allocations. This situation means a physical machine can be assigned too many virtual machines, thereby causing overloading situations. Therefore, a more efficient and flexible architecture to allocate resources is needed. In this study, we present a cloud architecture and Layered Calculation Virtual Machine Allocation (LCVMA), to perform exceptionally well in terms of achieving above goals. With this architecture and algorithm, we can identify the physical machines with low workloads, and service providers can allow users to use resources more efficiently. The threshold in our mechanism presents possibilities for reducing overload situations. Resource utilization and allocation can therefore become more efficient and economical.

R.-S. Chang · R.-C. Ye
Department of Computer Science and Information Engineering,
National Dong Hwa University, Hualien, Taiwan, ROC
e-mail: rschang@mail.ndhu.edu.tw

R.-C. Ye
e-mail: m9821044@ems.ndhu.edu.tw

Y.-C. Chang (✉)
Department of Computer Science and Information Engineering,
National Taitung University, Taitung, Taiwan, ROC
e-mail: ycc@nttu.edu.tw

S.-S. Yeo et al. (eds.), *Computer Science and its Applications*,
Lecture Notes in Electrical Engineering 203, DOI: 10.1007/978-94-007-5699-1_22,
© Springer Science+Business Media Dordrecht 2012

Keywords Virtualization · Cloud computing · Private cloud · Virtual machine scheduling algorithm

1 Introduction

According to NIST definition, cloud computing can be associated with four types of deployments [1]. Virtualization is a significantly important technique, which providers can make use of to set up multiple virtual machines in a given physical machine. By so doing, resource utilization can be increased. To achieve the goal of energy conservation, virtualization is essential. With the rapid growth in the development of cloud computing, the consumption of energy also increases [2, 3]. Thus, new issues like the creation of "Green Clouds" and the reduction of power consumption have also arisen. As a result, many researchers have concentrated on these two subjects [4–8].

To increase resource utilization efficiency, we propose both architecture and a scheduling algorithm to address the aforementioned problem. The proposed algorithm, called Layered Calculation Virtual Machine Allocation, performs exceptionally well in terms of achieving this goal. In fact, it not only can decrease scheduling time, but also increase resource utilization. With this architecture and algorithm, we can identify the physical machines with low workloads, and service providers can allow users to use resources more efficiently. As a result resources can be used flexibly and economically.

The rest of this study is organized as follows: Sect. 1 provides background information about virtualization and scheduling. Section 2 discusses related work for other scheduling methods. The proposed architecture and algorithm, Layered Calculation Virtual Machine Allocation, are introduced in Sect. 3. The implementation and subsequent results are set out and analyzed in Sect. 4. Finally, conclusions and future work suggestions comprise Sect. 5.

2 Related Works

In any operating system, scheduling all jobs which are to be executed on computers is critical [9–13]. Whenever a job is begun, demands are placed on the existing resources, which are always limited. Therefore, while another job would like to use the resources, it will have to wait for the previous job to be completed; it stands to reason that lots of jobs will be waiting. As a result job scheduling becomes the most significant issue. The following section introduces some existing job scheduling methods.

(1) First-Come First Serve (FCFS)

FCFS is an intuitional job scheduling method, but it is not very efficient. If there is a job which requires resources for a long time, other jobs will have to wait until it is completed. Although FCFS does not result in starvation, the overall performance is not especially good. Moreover, if there is a job involving a small calculation assigned to a higher performance resource, a job requiring a large calculation is precluded from using this resource, resulting in the performance of the whole system slowing down. Our proposed algorithm searches for the physical node with more free space to increase efficiency in the overall system. The situation of insufficient resources will thereby be eliminated.

(2) Work queue with Replication [14]

WQR is a job scheduling method based on the work queue algorithm. It will put more jobs in a faster processor, with the jobs being assigned in random order and sent to the resources. WQR replicates the jobs to available resources, and the amount of replication is decided by the user. When the replication of a job is completed, the job scheduler will kill other replications. The disadvantage is that if more jobs are put in the same processor, the processor may become overloaded due to the fact that the overall job workload is difficult to predict. Thus, we define an upper bound threshold for the workload in our proposed model to ensure that such overloading situations will not happen.

(3) Fastest Processor to Largest Task First (FPLTF) [15]

FPLTF is a basic concept which may require the setting of two parameters, such as CPU speed and job workload. The largest job is assigned to the fastest processor; however, if there are a lot of jobs with heavy workloads, the overall performance will be poor. Dynamic Fastest Processor to Largest Task First (DFPLTF) was introduced to solve this problem; it assigns the highest priority to the largest job, and it will predict processor speed information and job workload. Although DFPLTF's performance is not bad, it may suffer some problems resulting from the definition of job size. In cloud computing, the workload of virtual machines is hard to analyze; therefore, the upper bound threshold has been defined and a ranking system has been used to classify the physical machines to avoid this situation.

(4) Balanced Ant Colony Optimization (BACO) [16]

BACO is based on the ACO algorithm; it defines a pheromone indicator, using it as the pheromone of a resource. The pheromone indicator is calculated by the sum of the transmission time and execution time of a given job assigned to resources. The pheromone indicator is defined as:

$$PI_{ij} = \left[\frac{M_j}{bandwidth_i} + \frac{M_j}{CPU\_speed_i * (1 - load_i)} \right]^{-1}$$

$$\rightarrow PI_{ij} = \left[M_j * \left(\frac{CPU\_speed_i * (1 - load_i) + bandwidth_i}{CPU\_speed_i * (1 - load_i) * bandwidth_i} \right) \right]^{-1}$$

where PI_{ij} is the pheromone indicator for job j assigned to the resource i; M_j is the size of a job; and CPU_speed_i, $load_i$ and $bandwidth_i$ are the status of resource i. The larger the PI_{ij} value, the more appropriate it is for resource i to execute job j. Although BACO's performance is not bad, it is difficult to estimate a job's execution time. This situation could result in incorrect allocations.

(5) Resource Centric Score Algorithm for Virtual Machine Placement [17]

This algorithm is expected to find a physical node which still has remaining free space for a virtual machine. In addition to the resource centric score, it makes use of two functions: the *Fit* function and *OverFit* function. The OverFit function computes the physical node which is the most suitable for the virtual machine, and will be calculated by the following formula:

$$\text{overfit(host, Vm)} = \frac{|\text{VM.goals\_set}\backslash\text{host.goals\_set}|}{|\text{host.goals}_{set}|}$$

Lastly, the whole score is shown as:

hn_score = fit(host,VM) * Score(host) * overfit(host, VM)

the algorithm can only assign one virtual machine at a time. Thus, if an application needs to set up many virtual machines, the algorithm will not be efficient.

In response to this problem, our model defines the rank and threshold, and dispatches the virtual machines in accordance with these parameters. This effectively assigns many virtual machines into physical nodes, and uses the threshold to avoid the problem of overloading.

3 Proposed Mechanism

A typical cloud computing architecture system, as shown in Fig. 1, consists of the cloud header, the virtual machine allocator, the physical resource information monitor, and sub-networks. The cloud header manages cloud computing and is responsible for receiving user requirements. The virtual machine allocator is responsible for seeking useful resources from the private cloud to initiate VMs and allocating applications onto these VMs. The physical resource information monitor is used to monitor and provide the status of physical machines in every sub-network.

This mechanism has an issue: because an application may use several virtual machines to execute its job, there are similar virtual machines, related to the application's requirements during the allocation. To solve this issue, we first make a judgment on the configuration of a threshold. When the SNS is calculated, we will estimate the sub-network to define a ranking for physical machines with PRS, as we need to identify which physical machines have lower workloads. When the lower workload physical machines are identified, we define their rank and set up the virtual machines according to this rank. When the physical machines with thresholds that are too high arrive, we reassign the remaining virtual machines

Fig. 1 The proposed cloud architecture

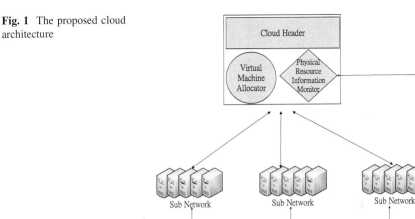

with our algorithm. Therefore, through this mechanism, we are able to dramatically reduce the time required for given assignments and also the frequency of allocation.

Our algorithm is described below: To search for an appropriate resource, a virtual machine can be associated with three parameters; CPU, Bandwidth and Memory. We attempt to calculate a sub-network score which consists of APCS, APBS and APMS. The initial value is calculated as follows:

$$SNS_i = \alpha APCS_i + \beta APBS_i + \gamma APMS_i \tag{1}$$

where the Sub-Network Score (SNS_i) is the score of sub-network i. The sub-network possessing a larger value of SNS_i has a higher priority to be chosen because its overall average performance for sub-network i is better than that of other sub-networks.

The Average Physical resource CPU Score ($APCS_i$) is the average available CPU speed of the physical resource in sub-network i. The Average Physical resource Bandwidth Score ($APBS_i$) is the average available Bandwidth of the physical resource in sub-network i. The Average Physical resource Memory Score ($APMS_i$) is the average available memory of the physical resource in sub-network i, and α, β, γ are the weights of each parameter which can be defined by the user, so long as they satisfy $\alpha + \beta + \gamma = 1$.

$APCS_i$ is computed from every PCS in sub-network i, where PCS_k is the score of the available CPU speed (denoted CPU_Speed_k) of each physical machine k in the sub- network i:

$$PCS_k = \frac{CPU\_Speed_k(1 - C\_load_k)}{N_{vm}} \cdot a \tag{2}$$

where C_load_k is the CPU_k utilized rate, N_{vm} is the number of virtual machines in physical machine k, and a is the coefficient compared to the whole cloud system. We define PCS_k as the available CPU of a resource.

APCS$_i$ is the average PCS$_k$ of sub-network i, and APCS$_i$ is defined as:

$$APCS_i = \frac{\sum_{k=1}^{M-1} PCS_k}{M-1} \tag{3}$$

where M is the number of physical machines in the sub network.

Similarly, APBS$_i$ is computed from every PBS in the sub-network i, and PBS$_k$ is the score of the available Bandwidth$_k$ of each physical machine k in the sub-network i.

$$PBS_k = \frac{Bandwidth_k(1 - B\_load_k)}{N_{vm}} \, b \tag{4}$$

where B_load$_k$ is the Bandwidth$_k$ utilized rate, N_{vm} is the number of virtual machines in physical machine k, and b is the coefficient compared to the whole cloud system. We define PBS$_k$ as the available Bandwidth of a resource.

APBS$_i$ is the average PBS$_k$ of the sub-network i, and APBS$_i$ is defined as:

$$APBS_i = \frac{\sum_{k=1}^{M-1} PBS_k}{M-1} \tag{5}$$

Lastly, APMS$_i$ is computed from every PMS in the sub network i, and PMS$_k$ is the score of available Memory$_k$ of each physical machine k in the sub-network i:

$$PMS_k = \frac{Memory_k(1 - M\_load_k)}{N_{vm}} \, c \tag{6}$$

where M_load$_k$ is the Memory$_k$ utilized rate, N_{vm} is the number of virtual machines in physical machine k, and c is the coefficient compared to the whole cloud system. We define PMS$_k$ as the available Memory of a resource.

APMS$_i$ is the average PMS$_k$ of the sub network i, and APMS$_i$ is defined as:

$$APMS_i = \frac{\sum_{k=1}^{M-1} PMS_k}{M-1} \tag{7}$$

According to the algorithm, PCS, PBS and PMS will form a physical resource score called PRS. After all of the parameters are calculated, they are associated with α, β, γ to form SNS. The virtual machine allocator compares all SNS to choose an appropriate resource.

"Local update" and "global update" are used in this algorithm. After the virtual machine is added to a resource in a sub-network, the condition of the resource will be changed and the local update will be applied to update all three parameters of the sub-network, which are PCS, PBS and PMS. Similarly, after the virtual machine has completed its work and exited the foreign sub-network, the global update will access all resources in the whole cloud system and update the PCS, PBS and PMS of all sub-networks.

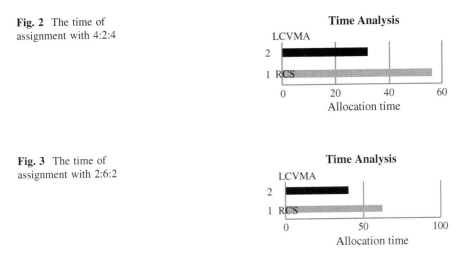

Fig. 2 The time of assignment with 4:2:4

Fig. 3 The time of assignment with 2:6:2

4 Experimental Results and Analysis

In the case of LCVMA algorithm, there was no load in terms of CPU, Bandwidth and a Memory level over 90 % because our upper bound threshold was defined as 80 %. Due to this threshold, we expected to be able to control every physical node for Quality of Service purposes. We wanted every physical node to work efficiently, so defining a threshold was necessary. If the physical nodes workloads were too high, this would have reduced the performance. However, it appears that some physical nodes had higher workloads than others, however this situation was predictable. Some physical nodes' workloads may have been lower than 80 %, but the whole sub-network SNS was lower than that obtained using the other algorithm. Therefore, through the LCVMA's calculations and assignment of virtual machines, workloads were increased. Still, we considered that if our private cloud had many sub-networks, this situation would be significantly reduced. Moreover, through the intervention of local and global updates, we were able to control every physical node and could avoid assigning too many virtual machines to any node.

In addition to comparing LCVMA with RCS, we observed other values of the parameters, a, b, c using 4:3:3 as well as 2:6:2 in our proposed algorithm, LCVMA. Having reset the values of the two sets of parameters, we compared the results with the 4:2:4 situation. Under the 4:3:3 and 2:6:2 settings, although there were no overloading conditions, the loads could not be balanced because we defined a strict threshold with CPU, Bandwidth and Memory. We had to satisfy the three thresholds simultaneously, and then allocate the virtual machines into the physical machines. Therefore, under these alternate settings, the load balance we achieved with the 4:2:4 ratio in our proposed algorithm was hard to achieve.

The last comparison considered the time required for assignment and we again compared the ratios of 4:2:4, 2:6:2 and 4:3:3, as shown in Figs. 2, 3 and 4. We expected that our algorithm would be five times lower than RCS, but we noted that

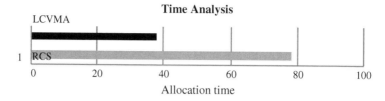

Fig. 4 The time of assignment with 4:3:3

the actual situation did not conform to our prediction because some virtual machines were recalculated by LCVMA; if the whole private cloud system was full and an appropriate physical node could not be found, the application waited for the global update information and the subsequent reallocation.

5 Conclusion and Future Work

We proposed system architecture and a scheduling algorithm to quickly find physical machines with free resources, which can then be efficiently used to allocate virtual machines. The algorithm in our proposal not only increases the utilization of physical machines, but also prevents overloading situations. For service providers and cloud managers, our system architecture and algorithms enable them to let users use resources more efficiently and economically. Compared to the other algorithm called Resource Centric Score, our proposed Layered Calculation Virtual Machine Allocation (LCVMA) is more effective in terms of both workload control and assignment time. In the future, we expect to improve our algorithm in terms of load balance, and use virtual local area network (VLAN) in our architecture. By doing so, all virtual machines from the same application will be able to communicate easily with each other. Because every virtual machine from the same application may not be assigned into the same physical nodes or the same private networks, the use of VLAN will assist them when interconnecting with each other.

Acknowledgments This work was supported in part by Taiwan National Science Council (Grant 100-2221-E-259-011 and NSC100-2221-E-143 -003).

References

1. Cloud computing, Wikipedia. http://en.wikipedia.org/wiki/Cloud_computing
2. Virtualization, Wikipedia. http://en.wikipedia.org/wiki/Virtualization
3. Maheswaran, M., Ali, S., Siegel, H.J., Hensgen, D., Freund, R.: Dynamic matching and scheduling of a class of independent tasks onto heterogeneous computing system. J Parallel Distrib. Comput. **59**, 107–131 (1999)

4. Wang, X., Zhang, B., Chen, H., Jin, X., Luo, Y., Li, X., Wang Z: Detecting and analyzing VM-exits. In: Computer and Information Technology (CIT), 2010 IEEE 10th International Conference, pp. 2273–2277 (2010)
5. Jang, J.-W., Jeon, M., Kim, H.-S., Jo, H., Kim, J.-S., Maeng, S.: Energy reduction in consolidated servers through memory-aware virtual machine scheduling. IEEE Trans. Comput. **60**, 552–564 (2011)
6. Feller, E., Rilling, L., Oorin, C., Lottiaux, R., Leprince, D.: Snooze: a scalable, fault-tolerant and distributed consolidation manager for large-scale clusters. In: Green Computing and Communications (GreenCom), 2010 IEEE/ACM International Conference and International Conference on Cyber, Physical and Social Computing (CPSCom), pp. 125–132 (2010)
7. Lin, B., Dinda, P.A., Lu, D.: User-driven scheduling of interactice virtual machines. In: Proceedings of the Fifth IEEE/ACM International Workshop on Grid Computing, pp. 380–387 (2004)
8. Andrew, J.Y., von Laszewski, G., Wang, L., Sonia, L-A., Carithers, W.: Efficient resource management for cloud computing environments. In: IEEE, International Conference on Green Computing (2010)
9. Hu, J., Gu, J., Sun, G., Zhao, T.: A scheduling strategy on load balancing of virtual machine resources in cloud computing environment. In: International Symposium on Parallel Architectures, Algorithms and Programming (PAAP), pp. 89–96 (2010)
10. Xu, Z., Hou, X., Sun, J.: Ant algorithm-based task scheduling in grid computing. In: Electrical and Computer Engineering, 2003. IEEE CCECE 2003. Canadian Conference, vol. 2, pp. 1107–1110 (2003)
11. Sodan, A.: Adaptive scheduling for QoS virtula machines under different resource availability—first experience. Workshop on Job Scheduling Strategies for Parallel Processing, Canada (2009)
12. Ongaro, D., Cox, A.L., Rixner, S.: Scheduling I/O virtual machine monitors. In: ACM SIGPLAN/SIGOPS International Conference on Virtual Execution Environments (2008)
13. Kim, H., Lim, H., Jeong, J., Jo, H., Lee, J.: Task-aware virtual machine scheduling for I/O performance. In: ACM SIGPLAN/SIGOPS International Conference on Virtual Execution Environment, pp. 101–110 (2009)
14. Paranhos, D., Cirne, W., Brasileiro, F.: Trading cycles for information: using replication to schedule bag-to-tasks application on computational grids. In: International Conference on Parallel and Distributed Computing (Euro-Par). Lecture Notes in Computer Science, vol. 2790, pp. 169–180 (2003)
15. Saha, D., Menasce, D., Porto S. et al.: Static and dynamic processor scheduling disciplines in heterogeneous parallel architectures. J Parallel Distrib Comput **28**(1), 1–18 (1995)
16. Chang, R-S., Chang, J-S., Lin, P-S.: Balanced job assignment based on ant algorithm for computing grids. In: Asia-Pacific Service Computing Conference, pp. 291–295, 11–14 December 2007
17. Jonathan, R-C.: A trust aware distributed and collaborative scheduler for virtual machine in cloud. LIFO, ENSI de Bourges, RR. September (2011)

Heterogeneous Core Network Architecture for Next-Generation Mobile Communication Networks

Yao-Chung Chang

Abstract With the increasing value-added services of information and mobile networks, the problem of insufficient bandwidth is posing the greatest challenge. To resolve this issue, the LTE (Long Term Evolution) of the 3G Partnership Project (3GPP) and the IEEE 802.16 Worldwide Interoperability for Microwave Access (WiMAX) are using IP-based next-generation network architectures to actively develop the 4G mobile communications network. The main purpose of this study was to examine the development and applications of the key technologies for the next generation heterogeneous mobile communications network architecture. On the basis of the Estinet7.0 simulation platform, this study simulated the implementation of interworking LTE and WiMAX scenarios. The core network architectures of LTE and WiMAX and their components' functions were first introduced, and then the interoperability scenarios of 4G network technology co-existence and integration were analyzed. Next, this study explored the handover mechanism and other factors of consideration. Based on the signal sign, both in concept and in practice, a rule-based communications service architecture for the handover authentication mechanism was proposed, which could provide a heterogeneous network integration service architecture for the seamless transition of LTE and WIMAX networks without damaging the respective domains of the networks. The proposed architecture is expected to extend the interoperability between LTE and WiMAX systems in a simpler and faster way in the future.

Keywords LTE · EPC network · WiMAX · Heterogeneous mobile core network

Y.-C. Chang (✉)
Department of Computer Science and Information Engineering, National Taitung
Unviersity, Taitung, Taiwan, R.O.C
e-mail: ycc@nttu.edu.tw

S.-S. Yeo et al. (eds.), *Computer Science and its Applications*,
Lecture Notes in Electrical Engineering 203, DOI: 10.1007/978-94-007-5699-1_23,
© Springer Science+Business Media Dordrecht 2012

1 Introduction

According to the relevant literature, the 3GPP-proposed UMB LTE-related technology cannot be peered with the original 2G or 3G network wireless access technology. IEEE has also developed the 802 wireless network-related standards and technology. Hence, how to develop a co-existence mechanism for heterogeneous mobile network core systems is a considerably important issue for the future development of 4G networks.

LTE is the wireless communications technology for 3GPP [1]. Derived from the GSM/UMTS mobile wireless communications technology, the core network architecture of LTE is an evolved packet core network (EPC). Different from the previous technologies, LTE uses the more powerful wireless transmission throughput provided by OFDMA, which has stronger multi-route resistance capabilities. Coupled with MIMO antenna technology, LTE can considerably improve wireless transmission efficiency. The LTE system architecture is composed of an EPC core network and an evolved universal terrestrial radio access network (E-UTRAN) [2, 3]. The LTE system components include: an E-UTRAN NodeB (eNodeB), a serving gateway (S-GW), a packet data network gateway (PDN-GW), a mobility management entity (MME), a policy and charging rules function (PCRF), and a home subscriber server (HSS). The programs for the control plane and the user plane are located between the MME and the S-GW. The MME/S-GW connection conducts security-related authentication, authorization and billing at the mobile user end, establishes, amends and dismantles the EPS Bearer and controls the roaming function of mobile phone users, including the handover with the eNodeB of the base station. When it is in idle mode, it implements functions such as tracking user positions and user calls.

WiMAX is short for Worldwide Interoperability for Microwave Access. It is a new wireless broadband technology with the characteristics of a long transmission distance and a large transmission amount. It has been mainly proposed and maintained by the WiMAX Forum [4, 5]. WiMAX and LTE both use OFDMA technology. The difference is that the IP-based technology of WiMAX has improved performance, a longer transmission distance (up to 48 km) and a higher transmission speed (134 Mbps). The mobile WiMAX system architecture is composed of an access services network gateway ASN-GW) connected with a number of connectivity service network (CSN) components and a base station (BS). The ASN-GW has different functions based on the different profile specifications of mobile WiMAX. The most commonly used is Profile-C, which is divided into the ASN-GW and the BS, with handover control at the BS. The ASN-GW connects the access network and core network components through the platform program using control signals (signaling) and subscriber packet/traffic. Figure 1 illustrates the LTE and WiMAX mobile communications core network architectures.

The remainder of this paper is organized as follows: Sect. 2 outlines the interoperability of LTE and WiMAX core network. Then, the coexistence

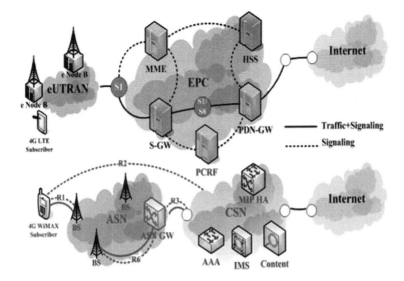

Fig. 1 Core network architecture

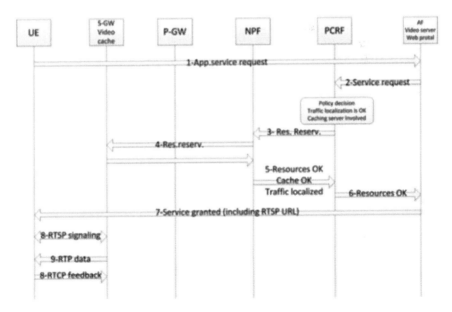

Fig. 2 Network policy function

scenarios of 4G network technology are analyzed in Sect. 3. Section 4 proposes the coexistence and integration considerations for heterogeneous mobile core network. Finally, conclusions of this paper are drawn in Sect. 5 (Fig. 2).

2 Interoperability for Heterogeneous Mobile Core Network

There are numerous wireless access technologies that realize effective interoperability between LTE and WiMAX [6]. The major needs and considerations for interoperability are mainly as follows:

1. Mobility support (LTE/WiMAX Handover), and the reduction of service notification programs to subscribers during handover.
2. The establishment of protocols for cooperation and roaming between LTE and WIMAX systems. Manufacturers on both sides should provide subscribers with identical advantages, such as the same manufacturer in an interoperable network.
3. The processing of roaming subscribers' accounts and billing.
4. Keeping subscriber identities the same in an LTE/WiMAX co-existing environment.
5. Creating a subscriber database that is independent from the networks and that can be safely shared in an interoperable network. Such a subscriber database could be the HSS from 3GPP or the AAA server from IETF.

The combination of different technologies should provide more efficient service and mode options and provide more effective handover mechanisms. However, the defined network nodes and mechanisms should support the signal exchange between heterogeneous backbone networks. For balanced considerations, the link and integration methods can be roughly divided into the following types: Open Link, Loose Link, Close Link and Very Close Link. Each link approach has different advantages and disadvantages; therefore, the link type can also affect the performance of the integrated network. It is worthy of exploration whether networks can be integrated by a either a single link method or different linking methods to improve the performance of an integrated network.

3 Coexistence Issues of Heterogeneous Mobile Core Network

3.1 Network Policy

Ouellette et al. [6] proposed a new method, which establishes a network policy function (NPF) in the LTE network. NSP defines policies relating to subscribers and services. With the subscriber group and the authenticated subscriber group, it can successfully provide the S-GW service quality guarantee based on the following viewpoints:

1. After separating services from the access network, an integration service is necessary.
2. Separate the management of the subscriber/network, and simplify the roaming steps.

3. The commercial role is separated into independent PCC/QoS network domains to simplify the design of complex sharing programs.
4. The local mobile anchor point access network is in the roaming program, allowing the user equipment to move to another AEG based on network conditions.

This method can prevent the transmission failure of large-volume multi-media files over long distances. Similar to the architecture of a content distribution network (CDN), if the multi-media network can support LTE and WIMAX at the same time, the mobility will be improved.

3.2 Handover for LTE and WiMAX

When a mobile device or an LTE UE wants to access data via a wireless network, a connection must be obtained from the neighboring eNodeB. First, the UE conducts the downlink synchronization of the time sequence and frequency with the base station and accepts information from the base station broadcasting system. In the respect of uploading, the propagation delay of wireless waves in the air will result in the round-trip delay of the UE and BS signals. Hence, before allowing uploading access to the UE, the BS must first estimate the time sequence differences for the uplink synchronization. When the UE and BS are in the process of data communication, the movement of the user will result in handover to another BS. In this case, it has to send out an RR to the new BS in order to complete the synchronization of the uplinked data. The handover process as stipulated in 3GPP, from the switch-off communication of the mobile device with the original service BS to the uplink synchronization with the new BS, is illustrated below:

Keshava Murthy [7] made a number of suggestions. First, subscriber identification programs and UE data management can be done in one of the specific LTE and WiMAX networks. Next, in the feedback of information, both networks can use the IPSec security program to protect the user plane and the control plane. In the respect of identity authentication, despite different algorithms and programs, the same protocols can be used. Authentication, authorization and billing can be applied in the network components of ASN-GW/MME and AAA/HSS. The access gateway, packet classification, priority sequence, speed adjustment, scheduling and distinguishing service in the IP layer's QoS mechanism must provide end-to-end QoS support that is realized in both networks. Both networks can support an identical framework that uses the QoS of well-defined network nodes and control application function settings to filter packet routing data. Such a framework can be used in both LTE and WiMAX networks. It can be learnt from the above that LTE and WiMAX are highly similar, and that the integration of LTE and WIMAX is possible. Although how to integrate these two systems has not been described in detail, it can be concluded that the integration of LTE and WIMAX networks will be a big step toward the future development of communication networks (Fig. 3).

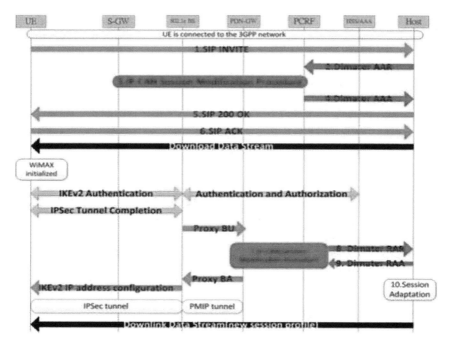

Fig. 3 LTE–WiMAX handover process

Regarding the handover methods, [8] proposed a handover mechanism. First, the UE should be loaded with the SIP-related service applications of 802.16(e) MS. The subscriber end can be connected with the WiMAX network in order to adjust the billing and accounting protocol management schemes of the core networks at both ends through PCRF. It can also communicate with the HSS/AAA server to connect with the host end.

Keshava Murthy [7] completed the LTE-WIMAX handover smoothly. During the handover, the LTE and WIMAX authentication is saved to reduce unnecessary resource wastage. Regarding large networks, whether the handover from WIMAX to LTE can be realized is not clearly illustrated. However, this concept shows that the handover from LTE to WIMAX is feasible. After modification, it can be applied in the standards. The handover from mobile WiMAX to 3GPP is shown in Fig. 4.

4 Design and Integration of Heterogeneous Mobile Core Network

4.1 Coexistence of LTE EPC and WiMAX [9–13]

The practical functions of the LTE EPC and WiMAX mobile communications core network architectures are analyzed as below:

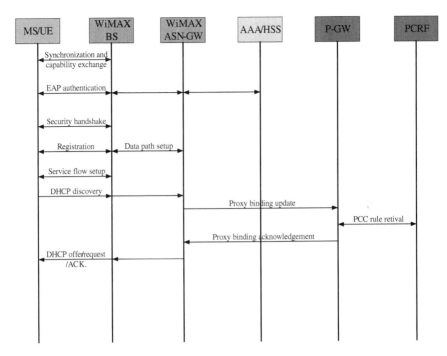

Fig. 4 WiMAX to 3GPP handover process

- ASN and serving gateway: the interface between the core network and the BS; responsible for the management of inter-base BS handover.
- Home agent (HA) and PDN gateway: responsible for connecting with the Internet and providing fixed IP address-related information requested for the mobility of mobile subscribers.
- Integrated service controller (ISC): responsible for providing AAA (authentication, authorization, and accounting) functions for LTE networks and WiMAX networks.
- Integrated home subscriber server (IHSS): responsible for recording the repository profiles of mobile subscribers in both LTE networks and WiMAX networks, and responsible for providing functions to manage mobile subscriber IDs and quality of service.
- Integrated policy controller (IPC): responsible for providing a real time resource management and distribution mechanism for the core network, applications and mobile subscribers.
- Subscriber data broker (SDR): responsible for recording the repository profile of LTE and WiMAX mobile subscribers and coordinating with IHSS, ISC and IPC to provide mobile subscribers with heterogeneous network roaming services. (Figs. 5, and 6).

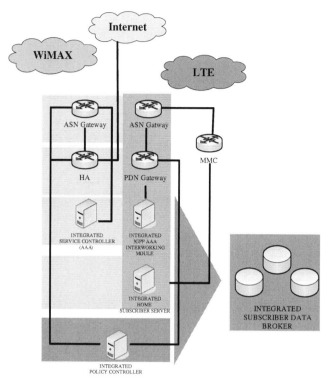

Fig. 5 Integration for LTE EPC and WiMAX

4.2 Simulation of LTE EPC and WiMAX Core Network

Using the EstiNet-7.0 [14] platform for simulation, this study first discussed the
LTE and WiMAX co-existing architecture and relevant protocols. Data transmission followed the TCP/IP protocol with a shared host and router. The LTE
system consisted of a packet data network gateway, a serving gateway, an eNodeB, and a UE. The WiMAX system consisted of an 802.16 base station, and an
802.16 mobility station. The transmission routes are respectively as follows:

- LTE transmission route: Host →RouterB →EPC Gateway →LTE Link → e-
  NodeB → UE
- WiMAX transmission route: Host → RouterA → WiMAX BS → WiMAX
  Link → MS (Fig. 7).

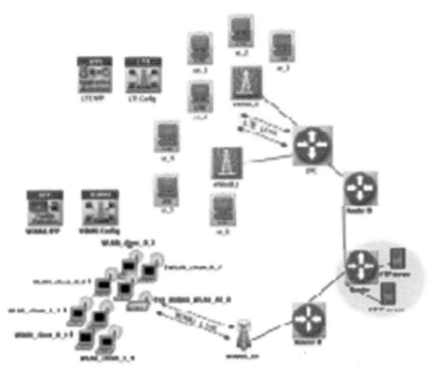

Fig. 6 Simulation network architecture

4.3 Handover Issues of LTE and WiMAX

NPF is as defined in [7]. Although multiple NPFs may result in the delay of service processes, NPF can relieve the burden of PCRF without changing the function of PCRF. This will maintain the service and quality for LTE network users. The administrator can use a single NPF for management and modification without managing large amounts of PCRF data or worrying about unintended modification of the PCRF contents. Hence, in addition to being used as the relay station during the handover process, NPF can also expand to many additional service locations (e.g. video caches). Figure 8 illustrates the modified handover process by the following steps:

Step 3. The PCRF sends out an RR (reserve request) to the NPF. The RR contains numerous AVP multi-media components. The modification events, such as information regarding the notification of IP-CAN type conversions, are registered in the UE network.

Step 5. After the successful storage of resources, the authentication and reserve answer are sent to the PCRF for confirmation of the resource storage.

Steps 8, 9. The request and answer are sent from the NPF.

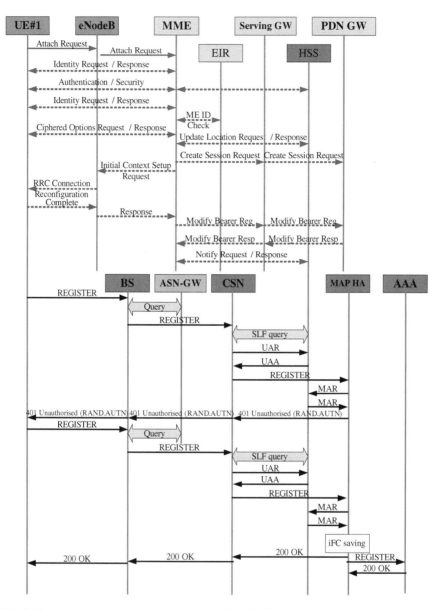

Fig. 7 Roaming communication process for mobile subscriber

The relevant simulation of the LTE handover service suggested that the eNodeB can manage the wireless resources and connect with the UE to transmit packets. The UE will scan the signal-to-noise ratios of neighboring base stations every two seconds. If the SNR of a new BS is higher, it will request the original BS for a handover. In the simulation of the handover between LTE and WIMAX, this study modified the operating methods proposed in [6]. In addition, [7] expressly

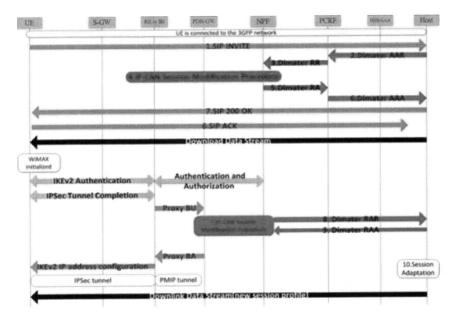

Fig. 8 Modified handover process

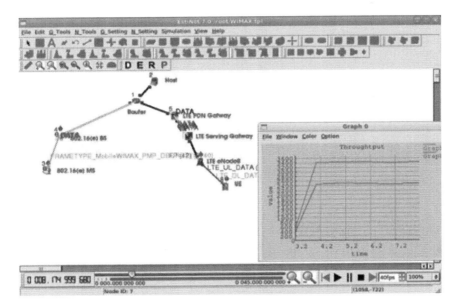

Fig. 9 Throughput of LTE and WiMAX

stated, "after separation of the services from the access network, it is necessary to integrate all services". This study proposed a new NPF for the connection of S-GW and PCRF to examine the development and applications of the key technologies for the next generation heterogeneous mobile communications network architecture. The proposed architecture is expected to extend the interoperability between LTE and WiMAX systems in a simpler and faster way in the future.

Figure 9 compares the throughput of LTE and WiMAX, which was around 2,400 in both cases. Since the QoS-related transmission information was added in WiMAX, it was higher than LTE by about 1,000 kbps in throughput.

5 Conclusion

This paper studies the core network architecture evolved from LTE and IEEE 802.16. Furthermore, the integration and coexistence of 4G access technologies are invested and discussed in this paper. Moreover, a novel handoff procedure for heterogeneous LTE and WiMAX core network is presented and analyzed. This paper uses Estinet simulator to demo the integration and handoff latency between LTE and WiMAX core network. In comparison to that of other generations of wireless communication system for homogeneous environments, this paper performs the testing and performance analysis for mobile applications running on Next-Generation Mobile Communication Core Networks.

Acknowledgments The author would like to thank Dr. Yu-Shan Lin and Mr. Jhih Jie Yang for this work, especially National Science Council of Taiwan under Contract No. NSC 100-2221-E-143-003 for financially supporting this research.

References

1. Hayes, S.: Evolution of the 3GPP system. In: 3GPP Workshop on LTE, June 1 (2010)
2. Classon, B.: Overview of UMTS air-interface evolution. In: Proceedings of IEEE Vehicular Technology Conference, September (2006)
3. Ghosh, A., Ratasuk, R., Mondal, B., Mangalvedhe N., Thomas, T.: LTE-advanced next-generation wireless broadband technology. IEEE Wirel. Commun. **17**(3), 10–22 (2010).
4. WiMAX Introduction. http://zh.wikipedia.org/wiki/WiMAX
5. WiMAX Forum. http://www.wimaxforum.org/home/
6. Ouellette, S., Marchand, L., Pierre, S.: A potential evolution of the policy and charging control/QoS architecture for the 3GPP IETF-based evolved packet core. IEEE Commun. Mag. 49(5), 231–239 (2011)
7. Keshava Murthy, K.S.: NextGen wireless access gateway analysis of combining WiMAX and LTE gateway functions. In: Proceeding of 2nd International Conference on Internet Multimedia Services Architecture and Applications(IMSAA 2008), pp. 1–6, December 10–12 (2008)

8. Ulvan, A., Bestak, R., Ulvan, M.: Handover scenario and procedure in LTE-based femtocell networks. In: Third Joint IFIP Wireless and Mobile Networking Conference (WMNC), pp.1–6, October 13–15 (2010)
9. Corici, M., Magedanz, T., Vingarzan, D., Weik, P.: Enabling ambient aware service delivery in heterogeneous wireless environments. In: IEEE Global Telecommunications Conference (GLOBECOM 2010), pp. 1–6 (2010)
10. Motorola. Coexistence studies of contiguous aggregation deployment scenarios for LTE-A. 3GPP doc. R4-060913, TSG RAN WG4, mtg. #51, San Francisco, CA, May 2009. ftp://ftp.3gpp.org (2009)
11. Taaghol, P., Salkintzis, A., Iyer, J.: Seamless integration of mobile WiMAX in 3GPP networks. IEEE Commun. Mag. **46**(10), 74–85 (2008)
12. Zein, N.: Coexistence of 4G access technologies. In: Proceeding of IET Conference on Next Generation Networks. December 1 (2009)
13. Yahiya, T.A., Chaouchi, H.: On the integration of LTE and mobile WiMAX networks. In: Proceedings of 19th International Conference on Computer Communications and Networks (ICCCN), pp. 1–5 (2010)
14. EstiNet Network Simulator and Emulator. http://www.estinet.com/

A Novel Learning Algorithm Based on a Multi-Agent Structure for Solving Multi-Mode Resource-Constrained Project Scheduling Problem

Omid Mirzaei and Mohammad-R. Akbarzadeh-T.

Abstract The resource-constrained project scheduling problem (RCPSP) includes activities which have to be scheduled due to precedence and resource restrictions such that an objective is satisfied. There are several variants of this problem currently, and also different objectives are considered with regards to the specific applications. This paper tries to introduce a new multi-agent learning algorithm (MALA) for solving the multi-mode resource-constrained project scheduling problem (MMRCPSP), in which the activities of the project can be performed in multiple execution modes. This work aims to minimize the total project duration which is referred to its makespan. The experimental results show that our method is a new one for this specific problem and can outperform other algorithms in different areas.

Keywords Multi-agent systems · Machine learning · Multi-mode resource-constrained project scheduling problem · MMRCPSP

O. Mirzaei (✉)
Department of Computer Engineering, Mashhad Branch,
Islamic Azad University, Mashhad, Iran
e-mail: O.Mirzaei@mshdiau.ac.ir

M.-R. Akbarzadeh-T.
Center of Excellence on Soft Computing and Intelligent Information Processing, Ferdowsi
University of Mashhad, Mashhad, Iran
e-mail: akbarzadeh@ieee.org

S.-S. Yeo et al. (eds.), *Computer Science and its Applications*,
Lecture Notes in Electrical Engineering 203, DOI: 10.1007/978-94-007-5699-1_24,
© Springer Science+Business Media Dordrecht 2012

1 Introduction

Project scheduling has become a popular subject in recent years both in science and practice. It has drawn increasing attentions in many real life applications and industries such as project management and crew scheduling, fleet management and also machine assignment, construction engineering, automobile industry, software development and the last but not least, make-to-order firms in which the capacities have been reduced in order to cope with lean management concepts. Furthermore, the resource-constrained project scheduling problem (RCPSP) is proven to be an NP-hard optimization problem [1].

Recently, in the literature of project scheduling, the RCPSP problem has been considered as a standard problem in the field. Within the classical type of this problem, the activities of the project have to be scheduled in such a way that an objective is satisfied. The most common objective in classical mode is makespan minimization. Thus, one has to consider not only technological precedence constraints but also the limitations of the renewable resources required to accomplish the activities. The precedence relations between the activities are demonstrated using a graph representation which is called Activity-On-Node (AON) Diagram. There exist several extensions on this single problem. The classical type of this problem can be extended to multi-mode resource-constrained project scheduling problem (MMRCPSP) in which each task can be done in many different execution modes [2]. Each mode stands for another way of mixing different levels of resource requirements with an affiliated duration.

According to the categorization scheme suggested by Slowinski [3], renewable, nonrenewable, and doubly constrained resources [4] are the three classifications of resources necessary for the execution of a project. Renewable re-sources (such as hour, day, week and month) are available on a period-by-period basis while nonrenewable resources (such as money, energy and raw material) are limited on a total project basis. Doubly constrained are those resources which are limited on both total project basis and per-period basis.

A broad variety of methods have been proposed for the multi-mode resource-constrained project scheduling problem so far. These algorithms are able to be classified into three main groups: exact algorithms [4], heuristic algorithms, and agent-based algorithms [5]. The heuristic approaches themselves can be divided into two strategies: classical meta-heuristic (for instance genetic algorithms [6], tabu search [2], simulated annealing [2, 7], ant colony [8] and bee colony [9]), and nonstandard meta-heuristic (for instance local search-oriented solutions [10] and population-based algorithms [11]).

The schedule generation scheme which is used by different algorithms can be performed either in serial [12] or parallel [13]. Each method makes use of one of these schemes to construct schedules and obtain the overall project makespan. With regards to empirical experiments which have been done so far, the parallel schedule generation scheme do not always lead to optimal solutions [5]. Hence, we have chosen the serial schedule generation scheme in this work.

This study concerns an agent based solution for the multi-mode resource-constrained project scheduling problem. The activities of the project are considered as agents who will make a multi-agent system considering the AON network. Each agent has two devices for making its decisions: (1) learning automaton and (2) heuristic-based stochastic local dispatcher. LA shows significant theoretical convergence properties in both single and multi automata environments. Considering this matter, a motivation for using them is that they are excellent tools for multi-agent reinforcement learning solutions [14]. The local dispatcher is considered in each agent to add a degree of randomness in selecting the order of activities and also the execution mode of each activity. A global dispatcher is also considered the same as [5] to avoid the algorithm from getting stuck into local optima.

In comparison with our main Ref. [5], there are two main differences. Firstly, the schedules are constructed locally here which increases the flexibility of the solution and allows the project to be dynamic. Secondly, local dispatchers have been added to each agent as their second tool for making their corresponding decisions. These dispatchers are heuristic-based stochastic, which will lead to a degree of randomness in the agents' decisions. Randomness has been incorporated for the cases in which the agents cannot make a decision through their learning devices (rationality) and the heuristic value leads to faster convergence of the algorithm.

The organization of this paper runs as follow: Section 2 provides the description of the problem and also its model. Section 3 contains our proposed multi-agent based learning algorithm for the multi-mode resource constrained project scheduling problem. Section 4 includes some experiments and comparative outcomes. Lastly, Sect. 5 states some conclusions and discusses future work.

2 Problem Description and Model

The standard MMRCPSP can be formulated as follows. We consider a project which consists of J activities (jobs) labeled $j = 1, 2, \ldots, J$. The processing time (or duration) of an activity j is denoted as p_j. When an activity begins, it may not be interrupted and its mode may not be changed. As mentioned before, there are precedence constraints between some of the activities due to technological requirements. These precedence constraints are given by sets of immediate predecessors P_j, demonstrating that if all of activity j's predecessors are not completed, it may not begin before. These can be all represented using an activity-on-node network (diagram) which is assumed to be acyclic. There are two more activities which are called source and sink which display the start and end of the project. These activities are "dummy" with duration of zero and no resource needs. Figure 1 demonstrates a sample AON network in the MMRCPSP problem.

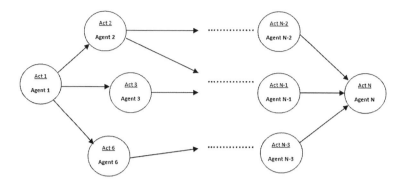

Fig. 1 A sample AON network for MMRCPSP problem

Each activity needs a definite number of resources to be done with the exclusion of dummy source activity and also the sink activities. R represents the set of renewable resources. For each renewable resource $r \in R$, the per period availability is invariable and given by K_r^p. N indicates a set of nonrenewable resources. The overall availability of each nonrenewable resource $n \in N$ for the whole project is shown by K_r^v [4].

Any of these activities can be done in a set of different modes of execution. A combination of various resources and/or levels of resource requirements with a specific duration are referred to as a mode [4]. Activity j may be carried out in M_j modes marked as $m = 1, 2, \ldots, M_j$ and its duration which is done in mode m is given by d_{jm}. Moreover, whenever activity j is carried out in mode m, it uses K_{jmr}^p units of renewable resource r each time it is in process, where we presume w. log $K_{jmr}^p \leq K_r^p$ for each renewable resource $r \in R$ [5]. Otherwise, activity j could not be carried out in mode m. Furthermore, it uses K_{jmr}^v units of nonrenewable resources $n \in N$.

There may be different objectives for this kind of problem. These include objectives based on renewable and nonrenewable resources and also robustness based objectives. Here, the objective of our work is to lessen the makespan of the project. Our work is based on this assumption that the parameters are nonnegative and integer-valued.

3 Multi-Agent Learning Algorithm

The first decision making unit which is used by agents is learning automaton. This unit is an adaptive one, which is located in an accidental environment and learns the best possible action based on past actions and environmental feedback. Properly, it can be illustrated by a quadruple $\{A, f, d, U\}$, in which A stands for a set of actions which are possibly taken, the random reinforcement signal given by the environment is shown by f, d represents the probability distribution over all actions and the learning scheme which is used to update d is demonstrated by U. At each instant k, $k = 1, 2, 3, \ldots$, the automaton selects an action considering its action probability vector d(k) [5].

$$d(k) = [d_1(k), d_2(k), \ldots, d_r(k)], \quad \sum d_i(k) = 1 \tag{1}$$

The environment receives the selected action as input and its response (feedback) to these actions serves as input to the automaton.

Many automaton update schemes with different properties have been studied and planned up to now. Among these, linear reward-penalty, linear reward-inaction and linear reward-ε-penalty are some important instances of linear update schemes. The purpose of all these schemes is fundamentally to boost the opportunity to choose an action when it brings about a success and decrease it when it results in a failure. The general algorithm is given by below equations:

If a_m is the action taken at time t:

$$d_m(t+1) = d_m(t) + \alpha_{reward} f(t)(1 - d_m(t)) - \alpha_{penalty}(1 - f(t))d_m(t) \tag{2}$$

If $a_j \neq a_m$:

$$d_j(t+1) = d_j(t) - \alpha_{reward} f(t)d_j(t) + \alpha_{penalty}(1 - f(t))[(q-1)^{-1} - d_j(t)] \tag{3}$$

The parameters that illustrate the reward and penalty are α_{reward} and $\alpha_{penalty}$. If the algorithm indicates $\alpha_{reward} = \alpha_{penalty}$, it is a sign of linear reward-penalty (L_{R-P}); a linear reward-inaction (L_{R-I}) is involved when $\alpha_{penalty} = 0$, and $\alpha_{penalty}$ is called linear reward-ε-penalty (L_{R-P}) when it is small compared to α_{reward}. $f(t) \in [0, 1]$ is the reward given by the environment as feedback for the action taken at instant t, and the number of actions is shown by q.

The ε-optimality property of (L_{R-I}) method in all stationary environments has made us apply it for learning the activity order and the best execution modes of activities [5]. LRO is the learning rate (reward parameter) which is used for learning the activity order and LRM is the one that is used for learning the execution mode of each activity.

In the proposed algorithm, a local frame is developed and enlarged along the execution of the project. This frame is used for making local schedules of the activities which have been added to it before. This procedure is repeated for some iterations (itrs) to complete the learning of agents. If an agent is observed for the first time and it is added to scheduled activities list, all of its successors are added to the frame if their predecessors presented in the frame before. To present an initial outline of our proposed algorithm, we have demonstrated it in Fig. 2.

According to Fig. 3, each agent makes use of two learning automata to choose its own execution mode and also the order of visiting its successor activities. The algorithm is willing to choose this order with regards to the agent's decision. Agents make this decision through consulting their learning devices (i.e. learning automata). These learning automata pick out an alternative based on their action

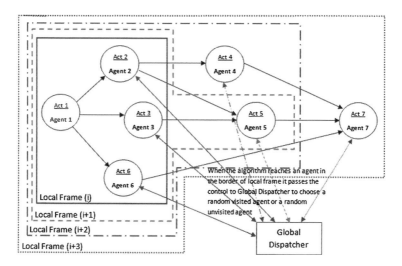

Fig. 2 The proposed multi-agent algorithm with its local frames

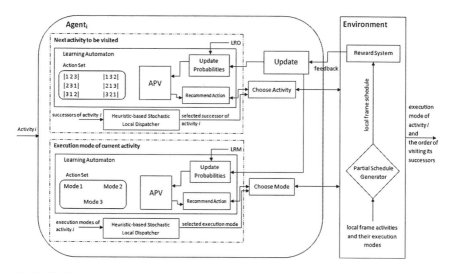

Fig. 3 The inner structure of an agent

probability vector. The length of this vector depends on operational choices. For instance, if the agent should decide between two choices for visiting its successor activities, the length of this vector will be two. The reward system makes use of the information from partially made schedules, and it will update the action probability vector of learning automata in the corresponding agent consistent with the reinforcement (reward) rules in Eqs. (2) and (3). If the makespan of the constructed schedule at instant t was:

- Better : $f(t) = 1$
- Equal :

$$f(t) = f_{eq}(f_{eq} \in [0, 1])) \tag{4}$$

- Worse : $f(t) = 0$

The speed of learning can be easily changed by modifying both f_{eq} and the learning rates LRO and LRM. A higher value for f_{eq} can make the learning faster, particularly for this kind of problems where efforts to develop novel schedules only rarely bring about improvements in quality [5]. The settings of the two learning rates are reliant on each other. A suitable arrangement of these values will be vital for achieving a good general performance.

Hence, the agents use a degree of rationality in making their decisions through learning automata and then, they get closer to optimal decisions by receiving feedbacks from the environment. Another decision making unit has been inserted in each agent named "heuristic-based stochastic local dispatcher" to add a degree of randomness to decisions made by agents. If the decisions made by rationality are wrong to a certain probability, the randomness prevents the agent from getting stuck into local optima. So, if this unit takes the control to some probability (γ), it will make the decisions according to a heuristic value. Moreover, there is also a global dispatcher in our proposed algorithm which is used when we reach an agent in the border of local frame and its successors are not presented within it. In this case, the control is given to this global dispatcher. It has a specific probability (δ) of selecting a random suitable agent from the list of previously observed agents, or else it selects a random suitable unvisited agent. This probability can be either static or dynamic through the process. The inner structure of each agent is illustrated in Fig. 3.

Once again, it should be mentioned that when an agent wants to make a decision through using its local dispatcher for choosing the execution mode of the current activity or the next activity which wants to be scheduled, it uses a heuristic value. This value is inversely proportional to the activities duration and their resource requirements. It means that when the local dispatcher wants to choose the next activity for scheduling, it makes this decision with regards to activities duration and their resource requirements, i.e. the activities which have the shortest duration and the lowest resource requirements in their best execution modes have a bigger probability to be selected. The aforementioned heuristic value can be well understood referring to the below equation:

$$h_{i,j} = \frac{1}{D_{i,j} \times RR_{i,j}} \tag{5}$$
$$i = 1, 2, \ldots, Successors, \ j = 1, 2, \ldots, Execution \ Modes \ of \ Activity \ i$$

Table 1 Average computation times

Algorithm	SGS	c15 10_1	c21 31_2	j10 61_7	j12 31_10	j18 30_9	j20 24_5	j30 49_7	m1 57_1	m5 21_4	n3 54_10	r4 39_3
CPSO [16]	Serial	7.57	16.38	6.15	8.44	17.94	16.34	31.09	4.52	13.33	18.24	17.92
HGA [17]	Both	10.23	18.86	9.05	10.43	19.49	18.46	34.07	7.32	16.31	21.03	20.12
MARLA [5]	Serial	1.29	2.09	0.80	1.21	2.22	2.43	5.15	1.65	1.79	3.01	1.99
MALA [proposed]	Serial	1.59	3.01	0.86	1.26	3.10	4.11	16.56	2.38	2.60	4.15	3.56

Where D and RR refer to the activities duration and their resource requirements respectively.

4 Experimental Results

Throughout this section, the performance of the multi-agent learning algorithm will be evaluated for the multi-mode resource-constrained project scheduling problem (MMRCPSP). Our algorithm has been implemented in MATLAB 2009 on a system with an Intel(R) Core(TM)2 Duo processor, 4 GB RAM and a 64-bit windows 7 operating system. The famous sets of instances produced by the project generator ProGen for the MMRCPSP have been used to test the proficiency of the proposed algorithm [15] and the infeasible instances have been expelled from the experiments.

We have chosen two heuristic algorithms and one multi-agent algorithm to show the performance of our proposed method. The first algorithm is a kind of nonstandard meta-heuristic solution and is based on particle swarm which had been led to good results according to its authors' claims [16]. The second one is based on genetic algorithm and is a classical meta-heuristic approach [17]. Finally, the multi-agent algorithm is another agent-based solution which was claimed to be very effective for this kind of problem [5]. The following empirically obtained parameters have been used to get the whole results: LRO = 0.4, LRM = 0.4, $f_{eq} = 0.01$, $\gamma = 0$, $\delta = 0.5$ and itrs = 3.

The initial criterion which has been used for comparing the efficiency of the algorithms from the view point of computational burden is the average computation times. These values have been measured accurately for 10 times of execution on numerous datasets and are gathered in Table 1. The instances of datasets have been specified below each one.

It is clear from Table 1 that our suggested algorithm has much lower computational times in all of the datasets than two heuristic algorithms. Moreover, we have separately applied all of the algorithms on j10 dataset to see the distribution of computation times on it. In Table 2, we present the results of this simulation. It

Table 2 Distribution of the computation times (%)—Dataset = j10

Algorithm	SGS	[0.1,0.3)	[0.3,0.5)	[0.5,0.7)	[0.7,1)	[1,3)	[3,5)	[5,7)	[7,9)	[9, 12]
CPSO [16]	Serial	0	0	0	0	0	52.38	23.80	19.04	4.78
HGA [17]	Both	0	0	0	0	0	0	28.57	66.66	4.77
MARLA [5]	Serial	0	0	4.76	95.24	0	0	0	0	0
MALA [proposed]	Serial	0	0	28.57	71.43	0	0	0	0	0

is totally clear from the table that the multi-agent algorithm has much lower computation times in contrast with heuristic ones. If we take it into consideration more precisely, we can find out that our proposed algorithm has a higher distribution over [0.5, 0.7) and a lower distribution over [0.7, 1). This means that our proposed method suggests solutions with shorter computation times in most of the cases when we compare it with the other heuristic algorithms.

Finally, to compare the performance of our algorithm with other algorithms we have chosen five datasets from the PSPLIB. Then, we have calculated the optimal solutions found by each of the algorithms, the average deviation from the optimal solutions and also the maximum deviation in all of the experiments. The outcomes are all gathered in Table 3. Considering this table, it can be well understood that the population-based algorithm which is based on particle swarm outperforms the others up to the point that the number of activities do not exceed from twelve in j12 dataset.

Putting all the outcomes together, it is clear that the proposed algorithm has higher computational time but better efficiency in comparison with the other competitive multi-agent solution [5] for the static multi-mode resource-constrained project scheduling problem. The utmost performance of the suggested algorithm can be obtained in a real life problem where the AON network is not static and also some other assumptions have been regarded.

5 Conclusions and Future Work

Various applications of project scheduling can be found in different economic environments such as development projects, construction engineering, software development, and also make-to-order companies. Consequently, in this atmosphere which is extremely competitive, if one can develop efficient algorithms which are able to deal with various execution modes for activities, they'll play an important role in decision making process. Specifically, those which have considered real life conditions in their solutions. Furthermore, other optimization areas

Table 3 Comparison of proficiency in different algorithms

Algorithm	SGS	j10			j12			j16			j18			j20		
		Opt. (%)	Avg. Dev. (%)	Max. Dev. (%)	Opt. (%)	Avg. Dev. (%)	Max. Dev. (%)	Opt. (%)	Avg. Dev. (%)	Max. Dev. (%)	Opt. (%)	Avg. Dev. (%)	Max. Dev. (%)	Opt. (%)	Avg. Dev. (%)	Max. Dev. (%)
CPSO [16]	Serial	99.25	0.03	0.05	98.47	0.09	0.12	85.91	0.44	0.47	79.89	0.89	0.92	74.19	1.10	1.13
HGA [17]	Both	98.51	0.06	0.09	96.53	0.17	0.19	90.00	0.41	0.44	84.96	0.63	0.65	80.32	0.87	0.91
MARLA [5]	Serial	98.70	0.05	0.06	98.17	0.10	0.13	92.18	0.24	0.26	86.23	0.21	0.23	81.59	0.85	0.88
MALA [proposed]	Serial	98.73	0.05	0.06	98.25	0.09	0.11	92.31	0.22	0.24	86.42	0.18	0.20	81.71	0.80	0.83

such as bin packing and knapsack problem can make use of the solutions proposed for this problem [18].

Indeed, the MMRCPSP problem is a really challenging problem and in this study a novel multi-agent learning algorithm has been developed to solve this problem. First of all, we assign an agent to each activity in the AON network. Then we construct a local frame and develop it step by step as the project goes on. Then, the agents make their decisions based on their learning automata (rationality) or their local dispatcher (heuristic-based randomness). The decisions are the next activity to be visited and also the execution mode of the current activity. The partial schedules constructed from the local frame are used to update the action probability vectors of learning automata in all the agents of the frame.

To evaluate our new method, we have applied it on numerous datasets from the PSPLIB which are the most popular datasets for this problem. The experimental outcomes demonstrate that our algorithm works better than the other approaches in the performance of solutions for the MMRCPSP problem. Moreover, it consumes less computational time than the other two heuristic methods we have considered for our simulations.

References

1. Blazewicz, J., Lenstra, J.K., Kan, A.H.G.R.: Scheduling subject to resource constraints: classification and complexity. Discrete Appl. Math. **5**, 11–24 (1983)
2. Mika M, Waligora G, Weeglarz J.: Simulated annealing and tabu search for multi-mode resource-constrained project scheduling with positive discounted cash flows and different payment models. Eur. J. Oper. Res. **164**(3 SPEC), 639–668 (2005)
3. Slowinski, R., Soniewicki, B., Weglarz, J.: DSS for multi objective project scheduling. Eur. J. Oper. Res. **79**, 220–229 (1994)
4. Hartmann, S., Drexl, A.: Project scheduling with multiple modes. A comparison of exact algorithms. Networks **32**(4), 283–297 (1998)
5. Wauters, T., Verbeeck, K.: Vanden Berghe G, De Causmaecher P.: Learning agents for the multi-mode project scheduling problem. J Oper Res Soc **62**, 281–290 (2011)
6. Hartmann, S.: Project scheduling with multiple modes: a genetic algorithm. Ann. Oper. Res. **102**(1–4), 111–135 (2001)
7. Bouleimen, K., Lecocq, H.: A new efficient simulated annealing algorithm for the resource-constrained project scheduling problem and its multiple mode version. Eur. J. Oper. Res. **149**(2), 268–281 (2003)
8. Merkle, D., Middendorf, M., Schmeck, H.: Ant colony optimization for resource-constrained project scheduling. IEEE Trans. Evol. Comput. **6**, 333–346 (2002)
9. Ziarati, K., Akbari, R., Zeighami, V.: On the performance of bee algorithms for resource-constrained project scheduling problem. Appl Soft Comput J **11**(4), 3720–3733 (2011)
10. Valls, V., Quintanilla, M. S., Ballestin, F.: Resource-constrained project scheduling: a critical activity reordering heuristic. Eur. J. Oper. Res. Forthcoming (2004)
11. Debels D, Reyck B. De, Leus R, Vanhoucke M.: A hybrid scatter search/Electromagnetism meta–heuristic for project scheduling. Eur. J. Oper. Res. To appear (2004)
12. Kelley Jr, J.E.: The critical-path method: resources planning and scheduling. In: Industrial Scheduling. Prentice-Hall, New Jersey, pp. 347–365 (1963)

13. Bedworth, D., Bailey, J.: Integrated production control systems management, analysis design. Wiley, New York (1982)
14. Verbeeck, K., Nowe, A., Vrancx, P., Peeters, M.: Reinforcement learning theory and applications. Multi-automata learning, Chapter 9. I-Tech Education and Publishing: Vienna, pp 167–185 (2008)
15. Project Scheduling Problem Library. http://129.187.106.231/psplib/main.html
16. Jarboui, B., Damak, N., Siarry, P., Rebai, A.: A combinatorial particle swarm optimization for solving multi-mode resource-constrained project scheduling problems. Appl. Math. Comput. **195**(1), 299–308 (2008)
17. Lova, A., Tormos, P., Cervantes, M., Barber, F.: An efficient hybrid genetic algorithm for scheduling projects with resource constraints and multiple execution modes. Int. J. Prod. Econ. **117**(2), 302–316 (2009)
18. Hartmann, S.: Packing problems and project scheduling models: an integrating perspective. J Oper Res Soc **51**, 1083–1092 (2000)
19. Jedrzejowicz P, Ratajczak-Ropel E.: Solving the RCPSP/max problem by the team of agents. In: Agent and Multi-Agent Systems: Technologies and Applications, Lecture Notes in Computer Science, Volume 5559/2009, pp. 734–743 (2009)

Security Analysis of the Keyschedule of ARIA-128

HyungChul Kang, Yuseop Lee, Kitae Jeong, Jaechul Sung
and Seokhie Hong

Abstract A Korean standard block cipher ARIA is a 128-bit block cipher supporting 128-, 192- and 256-bit secret keys. It is well known that an attacker should obtain at least four consecutive round keys or particular two round keys in order to recover the secret key of ARIA-128. In this paper, we propose the method to recover the secret key by using only the last round key. The proposed method is based on a guess-and-determine attack, which requires the computational complexity of $O(2^{68})$. To our knowledge, this result did not introduced so far.

Keywords Block cipher · ARIA-128 · Keyschedule · Guess-and-determine attack

1 Introduction

ARIA is a 128-bit block cipher that supports 128-, 192-, 256-bit secret keys [1]. In 2004, the Korean Agency for Technology and Standards selected this algorithm as a standard cryptographic technique (KS × 1213). It has an involutional SPN.

H. Kang (✉) · Y. Lee · K. Jeong · S. Hong
Center for Information Security Technologies (CIST), Korea University, Seoul, Korea
e-mail: kanghc@korea.ac.kr

Y. Lee
e-mail: yusubi@korea.ac.kr

K. Jeong
e-mail: kite.jeong@gmail.com

S. Hong
e-mail: shhong@korea.ac.kr

J. Sung
Department of Mathematics, University of Seoul, Seoul, Korea
e-mail: jcsung@uos.ac.kr

S.-S. Yeo et al. (eds.), *Computer Science and its Applications*,
Lecture Notes in Electrical Engineering 203, DOI: 10.1007/978-94-007-5699-1_25,
© Springer Science+Business Media Dordrecht 2012

The number of rounds is 12, 14 and 16, depending on the size of secret key. According to the size of secret key, we call this algorithm ARIA-128/192/256, respectively.

So far, there were several cryptanalytic results on ARIA-128. They are based on boomerang attacks [2], impossible differential cryptanalysis [3–5], linear cryptanalysis [6], and integral attack [7]. Note that all of them recovered only the round keys of reduced rounds of ARIA. Moreover, they did not consider the procedure to recover the secret key.

It is well known that, in order to recover the secret key of ARIA-128, an attacker should obtain at least four consecutive round keys or particular two round keys. For example, Wei et al. proposed differential fault analysis on the full ARIA-128 [8]. They recovered the last four round keys by using 45 fault injections on average. Then, the secret key could be computed by using the recovered round keys. On the other hand, Biryukov et al. claimed that an attacker should obtain particular two round keys in order to recover the secret key of ARIA-128 [9]. In detail, when he obtain two round keys from the same words W_0, W_1 (the description of (W_0, W_1) is introduced in Sect. 2), it is possible to extract the secret key of ARIA-128. To our knowledge, there is no result considering only one round key.

In this paper, we study the keyschedule of ARIA-128 and show that it is possible to recover the secret key of ARIA-128 by using only the last round key. First, we assume that an attacker can obtain the last round key of ARIA-128. Then, he constructs 128 equations for the secret key and last round key. Based on a guess-and-determine attack, he can recover the secret key of ARIA-128 with the computational complexity of $O(2^{68})$. This is the first known cryptanalytic result considering only one round key.

The remainder of this paper is organized as follows. First, we briefly present the keyschedule of ARIA-128 in Sect. 2. In Sect. 3, we introduce the method recovering the secret key of ARIA-128 by using the last round key. Finally, we give our conclusion in Sect. 4.

2 Description of the Keyschedule of ARIA-128

A 128-bit block cipher ARIA-128 supports the 128-bit secret key and has an involution SPN structure consisting of 12 rounds. Since our attack considers only the keyschedule of ARIA-128, we focus on the description of it in this section. For the detail description of ARIA-128, see [1].

The keyschedule of ARIA-128 consists of two parts: the initialization part and the round key generation part. In the initialization part, the secret key is expanded. First, the 128-bit secret key SK is loaded to a 256-bit as follows.(KL, KR)

$$(KL, KR) = (SK, 0).$$

Then, as shown in Fig. 1, four 128-bit values (W_0, W_1, W_2, W_3) are computed as follows. Here, CK_i is a 128-bit constant $(1 \leq i \leq 3)$.

$$W_0 = KL,$$
$$W_1 = F_o(W_0, CK_1) \oplus KR,$$
$$W_2 = F_e(W_1, CK_2) \oplus W_0,$$
$$W_3 = F_o(W_2, CK_3) \oplus W_1$$

Figure 2 present F_o and F_e. They consist of a constant addition, a substitution layer (SL) and a diffusion layer (DL). In a substitution layer (SL), two types of 8×8 S-boxes S_1, S_2 and their inverses S_1^{-1}, S_2^{-1} are used. The diffusion layer (DL) employs an 16×16 involution binary matrix as follows.

$$DL: \begin{pmatrix} y_0 \\ y_1 \\ y_2 \\ y_3 \\ y_4 \\ y_5 \\ y_6 \\ y_7 \\ y_8 \\ y_9 \\ y_{10} \\ y_{11} \\ y_{12} \\ y_{13} \\ y_{14} \\ y_{15} \end{pmatrix} = \begin{pmatrix} 0&0&0&1&1&0&1&0&1&1&0&0&0&1&1&0 \\ 0&0&1&0&0&1&0&1&1&1&0&0&1&0&0&1 \\ 0&1&0&0&1&0&1&0&0&0&1&1&1&0&0&1 \\ 1&0&0&0&0&1&0&1&0&0&1&1&0&1&1&0 \\ 1&0&1&0&0&1&0&0&1&0&0&1&0&0&1&1 \\ 0&1&0&1&1&0&0&0&0&1&1&0&0&0&1&1 \\ 1&0&1&0&0&0&0&1&0&1&1&0&1&1&0&0 \\ 0&1&0&1&0&0&1&0&1&0&0&1&1&1&0&0 \\ 1&1&0&0&1&0&0&1&0&0&1&0&0&1&0&1 \\ 1&1&0&0&0&1&1&0&0&0&0&1&1&0&1&0 \\ 0&0&1&1&0&1&1&0&1&0&0&0&0&1&0&1 \\ 0&0&1&1&1&0&0&1&0&1&0&0&1&0&1&0 \\ 0&1&1&0&0&0&1&1&0&1&0&1&1&0&0&0 \\ 1&0&0&1&0&0&1&1&1&0&1&0&0&1&0&0 \\ 1&0&0&1&1&1&0&0&0&1&0&1&0&0&1&0 \\ 0&1&1&0&1&1&0&0&1&0&1&0&0&0&0&1 \end{pmatrix} \begin{pmatrix} x_0 \\ x_1 \\ x_2 \\ x_3 \\ x_4 \\ x_5 \\ x_6 \\ x_7 \\ x_8 \\ x_9 \\ x_{10} \\ x_{11} \\ x_{12} \\ x_{13} \\ x_{14} \\ x_{15} \end{pmatrix}$$

In the round key generation part, total thirteen round keys ek_i of the encryption process are generated as follows ($1 \le i \le 13$).

$$ek_1 = (W_0) \oplus (W_1^{\ggg 19}), \quad ek_2 = (W_1) \oplus (W_2^{\ggg 19})$$
$$ek_3 = (W_2) \oplus (W_3^{\ggg 19}), \quad ek_4 = (W_0^{\ggg 19}) \oplus (W_3)$$
$$ek_5 = (W_0) \oplus (W_1^{\ggg 31}), \quad ek_6 = (W_1) \oplus (W_2^{\ggg 31})$$
$$ek_7 = (W_2) \oplus (W_3^{\ggg 31}), \quad ek_8 = (W_0^{\ggg 31}) \oplus (W_3) \tag{1}$$
$$ek_9 = (W_0) \oplus (W_1^{\lll 61}), \quad ek_{10} = (W_1) \oplus (W_2^{\lll 61})$$
$$ek_{11} = (W_2) \oplus (W_3^{\lll 61}), \quad ek_{12} = (W_0^{\lll 61}) \oplus (W_3)$$
$$ek_{13} = (W_0) \oplus (W_1^{\lll 31})$$

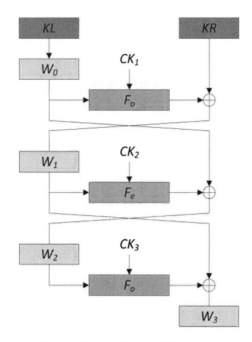

Fig. 1 Initialization part of the keyschedule of ARIA-128

(a)   **(b)**

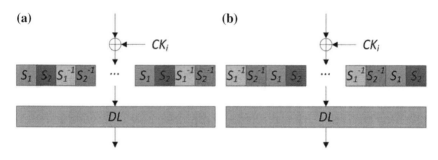

Fig. 2 (a) F_o and (b) F_e

3 Recovery of the Secret Key by Using the Last Round Key

In this section, we propose a method to recover the secret key of ARIA-128 by using the last round key. Throughout this paper, we use the following notations.

w_k, ck_k, c_k the k-th bit of W_0, CK_1, C, respectively $(k = 0, \cdots, 127)$

$w_{i \sim j}, ck_{i \sim j}, c_{i \sim j}$ the i-th bit to jth of W_0, CK_1, C, respectively

$S_{m,l}(), S_{m,l}^{-1}()$ the l-th bit of output of S-boxes S_m and S_m^{-1}, respectively $(m = 1, 2, l = 0, 1, \cdots, 7)$

$\underline{x_k}, \underline{x_{i \sim j}}$ the guessed or determined bits of $x_k, x_{i \sim j}$, respectively

3.1 Construction of 128 Equations

Note that, from Eq. (1), ek_1, ek_5, ek_9 and ek_{13} are computed by using W_0 and W_1. Here, KR is equal to zero in the initialization part of the keyschedule. Applying this property, we can construct equations for W_0. That is, these equations consist of only the secret key. In this section, we focus on only the case of ek_{13}. The reason is that most known cryptanalysis on block ciphers generally recovers the last round key. Thus, the assumption that we obtain the exact value of ek_{13} is reasonable. Hence, we show the procedure to recover the secret key of ARIA-128 under the assumption that we get the exact value of ek_{13}. Under other assumptions that we obtain the exact value of ek_1, ek_5, ek_9, the procedures are explained in a similar fashion.

First, ek_{13} and W_1 are computed as follows (see Fig. 1 and 2).

$$ek_{13} = (W_0) \oplus (W_1^{\lll 31}),$$
$$W_1 = F_o(W_0, CK_1) \oplus 0 \Rightarrow W_1 = DL(SL(W_0 \oplus CK_1)).$$

By using the above equations, we get the following equation (see Fig. 3).

$$ek_{13} = (W_0) \oplus \left(DL(SL(W_0 \oplus CK_1))^{\lll 31} \right)$$

$$\Rightarrow (ek_{13})^{\ggg 31} = (W_0)^{\ggg 31} \oplus DL(SL(W_0 \oplus CK_1)).$$

Since DL has involutional $(DL = DL^{-1})$ and linear properties, we obtain the following equation.

$$DL\left((ek_{13})^{>>>31} \right) = DL\left((W_0)^{>>>31} \right) \oplus SL(W_0 \oplus CK_1).$$

Here, $DL(ek_{13}^{\ggg 31})$ is a known value from our assumption. Thus, we denote this value C. That is, the above equation is modified to Eq. (2).

$$C = DL\left((W_0)^{\ggg 31} \right) \oplus SL(W_0 \oplus CK_1) \tag{2}$$

Since C is an 128-bit value, we can construct total 128 equations from Eq. (2). For example, an equation for c_0 is constructed as follows.

$$S_{1,0}(w_{0\sim7} \oplus ck_{0\sim7}) \oplus w_1 \oplus w_{17} \oplus w_{33} \oplus w_{41} \oplus w_{73} \oplus w_{81} \oplus w_{121} = c_0$$

3.2 Recovery of W_0

Now we are ready to recover the secret key 128-bit SK of ARIA-128. Recall that SK is equal to W_0. Thus, we need to only recover W_0. This step is based on a guess-and-determine attack. In detail, to obtain the whole W_0, we guess a 74-bit partial value of W_0. This procedure is summarized in Table 1.

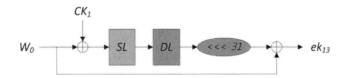

Fig. 3 Computation of ek_{13}

Table 1 Our attack procedure to recover W_0

Step	Guessed bits (Length of bits)	Determined bits (Length of bits)	Probability of filtering	Number of Candidates	Computational Complexity
(1)	$w_{9\sim15}, w_{17\sim23}, w_{33\sim39}, w_{73\sim79},$ $w_{89\sim95}, w_{113\sim119}, w_{121\sim127},$ $w_{16} \oplus w_{24} \oplus w_{40} \oplus w_{80}$ $\oplus w_{96} \oplus w_{120} \oplus w_0$ (50 bits)	$w_{80\sim87}$ (8 bits)	–	2^{50}	2^{50}
(2)	$w_{49\sim55}, w_{105\sim111}, w_{120}$ (15 bits)	$w_{1\sim7}$ (7 bits)	–	2^{65}	2^{65}
(3)	w_0 (1 bit)	$w_{41\sim47}$ (7 bits)	–	2^{66}	2^{66}
(4)	–	w_{40} (1 bit)	2^{-6}	2^{60}	2^{66}
(5)	$w_{57\sim63}, w_{16}$ (8 bits)	$w_{65\sim71}$ (7 bits)	–	2^{68}	2^{68}
(6)	–	w_{56} (1 bit)	2^{-6}	2^{62}	2^{68}
(7)	w_{32} (1 bit)	$w_{97\sim103}$ (7 bits)	–	2^{63}	2^{63}
(8)	–	w_{72} (1 bit)	2^{-6}	2^{57}	2^{63}
	–	w_{112} (1 bit)	2^{-6}	2^{51}	2^{57}
(9)	w_8 (1 bit)	$w_{25\sim31}$ (7 bits)	–	2^{52}	2^{52}
(10)	–	w_{24} (1 bit)	2^{-6}	2^{46}	2^{52}
	–	w_{48} (1 bit)	2^{-6}	2^{40}	2^{46}
	–	w_{64} (1 bit)	2^{-6}	2^{34}	2^{40}
	–	w_{88} (1 bit)	2^{-6}	2^{28}	2^{34}
	–	w_{96} (1 bit)	2^{-6}	2^{22}	2^{28}
	–	w_{104} (1 bit)	2^{-6}	2^{16}	2^{22}
(11)	–	–	2^{-16}	1	2^{16}

First, we consider the following equations for $c_{80 \sim 87}$.

$$S_{1,0}^{-1}(w_{80 \sim 87} \oplus ck_{80 \sim 87}) \oplus w_9 \oplus w_{17} \oplus w_{33} \oplus w_{73} \oplus w_{89} \oplus w_{113} \oplus w_{121} = c_{80},$$

$$S_{1,1}^{-1}(w_{80 \sim 87} \oplus ck_{80 \sim 87}) \oplus w_{10} \oplus w_{18} \oplus w_{34} \oplus w_{74} \oplus w_{90} \oplus w_{114} \oplus w_{122} = c_{81},$$

$$S_{1,2}^{-1}(w_{80 \sim 87} \oplus ck_{80 \sim 87}) \oplus w_{11} \oplus w_{19} \oplus w_{35} \oplus w_{75} \oplus w_{91} \oplus w_{115} \oplus w_{123} = c_{82},$$

$$S_{1,3}^{-1}(w_{80 \sim 87} \oplus ck_{80 \sim 87}) \oplus w_{12} \oplus w_{20} \oplus w_{36} \oplus w_{76} \oplus w_{92} \oplus w_{116} \oplus w_{124} = c_{83},$$

$$S_{1,4}^{-1}(w_{80 \sim 87} \oplus ck_{80 \sim 87}) \oplus w_{13} \oplus w_{21} \oplus w_{37} \oplus w_{77} \oplus w_{93} \oplus w_{117} \oplus w_{125} = c_{84},$$

$$S_{1,5}^{-1}(w_{80 \sim 87} \oplus ck_{80 \sim 87}) \oplus w_{14} \oplus w_{22} \oplus w_{38} \oplus w_{78} \oplus w_{94} \oplus w_{118} \oplus w_{126} = c_{85},$$

$$S_{1,6}^{-1}(w_{80 \sim 87} \oplus ck_{80 \sim 87}) \oplus w_{15} \oplus w_{23} \oplus w_{39} \oplus w_{79} \oplus w_{95} \oplus w_{119} \oplus w_{127} = c_{86},$$

$$S_{1,7}^{-1}(w_{80 \sim 87} \oplus ck_{80 \sim 87}) \oplus w_{16} \oplus w_{24} \oplus w_{40} \oplus w_{80} \oplus w_{96} \oplus w_{120} \oplus w_0 = c_{87}.$$

For these equations, we guess a 50-bit value $(w_{9 \sim 15}, w_{17 \sim 23}, w_{33 \sim 39}, w_{73 \sim 79}, w_{89 \sim 95}, w_{113 \sim 119}, w_{121 \sim 127}, w_{16} \oplus w_{24} \oplus w_{40} \oplus w_{80} \oplus w_{96} \oplus w_{120} \oplus w_0)$. Then, we can determine an 8-bit value $w_{80 \sim 87}$. Thus, we obtain 2^{50} candidates of a 57-bit value $(w_{9 \sim 15}, w_{17 \sim 23}, w_{33 \sim 39}, w_{73 \sim 79}, w_{80 \sim 87}, w_{89 \sim 95}, w_{113 \sim 119}, w_{121 \sim 127})$. Note that we use the guessed value $w_{16} \oplus w_{24} \oplus w_{40} \oplus w_{80} \oplus w_{96} \oplus w_{120} \oplus w_0$ in Step (11). The computational complexity of this step is $O(2^{50})$ (see Step (1) in Table 1).

In Step (2), we consider the following seven equations for $c_{120 \sim 126}$.

$$S_{2,0}^{-1}((w_{120}, \underline{w_{121 \sim 127}}) \oplus ck_{120 \sim 127}) \oplus w_1 \oplus \underline{w_9} \oplus \underline{w_{33}} \oplus w_{49} \oplus \underline{w_{89}} \oplus w_{105} \oplus \underline{w_{113}} = c_{120},$$

$$S_{2,1}^{-1}((w_{120}, \underline{w_{121 \sim 127}}) \oplus ck_{120 \sim 127}) \oplus w_2 \oplus \underline{w_{10}} \oplus \underline{w_{34}} \oplus w_{50} \oplus \underline{w_{90}} \oplus w_{106} \oplus \underline{w_{114}} = c_{121},$$

$$S_{2,2}^{-1}((w_{120}, \underline{w_{121 \sim 127}}) \oplus ck_{120 \sim 127}) \oplus w_3 \oplus \underline{w_{11}} \oplus \underline{w_{35}} \oplus w_{51} \oplus \underline{w_{91}} \oplus w_{107} \oplus \underline{w_{115}} = c_{122},$$

$$S_{2,3}^{-1}((w_{120}, \underline{w_{121 \sim 127}}) \oplus ck_{120 \sim 127}) \oplus w_4 \oplus \underline{w_{12}} \oplus \underline{w_{36}} \oplus w_{52} \oplus \underline{w_{92}} \oplus w_{108} \oplus \underline{w_{116}} = c_{123},$$

$$S_{2,4}^{-1}((w_{120}, \underline{w_{121 \sim 127}}) \oplus ck_{120 \sim 127}) \oplus w_5 \oplus \underline{w_{13}} \oplus \underline{w_{37}} \oplus w_{53} \oplus \underline{w_{93}} \oplus w_{109} \oplus \underline{w_{117}} = c_{124},$$

$$S_{2,5}^{-1}((w_{120}, \underline{w_{121 \sim 127}}) \oplus ck_{120 \sim 127}) \oplus w_6 \oplus \underline{w_{14}} \oplus \underline{w_{38}} \oplus w_{54} \oplus \underline{w_{94}} \oplus w_{110} \oplus \underline{w_{118}} = c_{125},$$

$$S_{2,6}^{-1}((w_{120}, \underline{w_{121 \sim 127}}) \oplus ck_{120 \sim 127}) \oplus w_7 \oplus \underline{w_{15}} \oplus \underline{w_{39}} \oplus w_{55} \oplus \underline{w_{95}} \oplus w_{111} \oplus \underline{w_{119}} = c_{126}.$$

In these equations, underlined symbols mean the guessed or determined values in the previous steps. Recall that we have 2^{50} candidates of the 57-bit value in Step (1). Thus, we guess only a 15-bit value $(w_{49 \sim 55}, w_{105 \sim 111}, w_{120})$ for each candidate. Then, we can determine a 7-bit value $w_{1 \sim 7}$. As shown in Table 1, we get 2^{65} candidates of an 79-bit value $(w_{1 \sim 7}, w_{9 \sim 15}, w_{17 \sim 23}, w_{33 \sim 39}, w_{49 \sim 55}, w_{73 \sim 87}, w_{89 \sim 95}, w_{105 \sim 111}, w_{113 \sim 127})$. The computational complexity of this step is $O(2^{65})$.

Similarly to Step (2), Step (3) consider the following 7 equation for $c_{0 \sim 6}$.

$$S_{1,0}\big((w_0, \underline{w_{1\sim7}}) \oplus ck_{0\sim7}\big) \oplus \underline{w_1} \oplus \underline{w_{17}} \oplus \underline{w_{33}} \oplus w_{41} \oplus \underline{w_{73}} \oplus \underline{w_{81}} \oplus \underline{w_{121}} = c_0,$$

$$S_{1,1}\big((w_0, \underline{w_{1\sim7}}) \oplus ck_{0\sim7}\big) \oplus \underline{w_2} \oplus \underline{w_{18}} \oplus \underline{w_{34}} \oplus w_{42} \oplus \underline{w_{74}} \oplus \underline{w_{82}} \oplus \underline{w_{122}} = c_1,$$

$$S_{1,2}\big((w_0, \underline{w_{1\sim7}}) \oplus ck_{0\sim7}\big) \oplus \underline{w_3} \oplus \underline{w_{19}} \oplus \underline{w_{35}} \oplus w_{43} \oplus \underline{w_{75}} \oplus \underline{w_{83}} \oplus \underline{w_{123}} = c_2,$$

$$S_{1,3}\big((w_0, \underline{w_{1\sim7}}) \oplus ck_{0\sim7}\big) \oplus \underline{w_4} \oplus \underline{w_{20}} \oplus \underline{w_{36}} \oplus w_{44} \oplus \underline{w_{76}} \oplus \underline{w_{84}} \oplus \underline{w_{124}} = c_3,$$

$$S_{1,4}\big((w_0, \underline{w_{1\sim7}}) \oplus ck_{0\sim7}\big) \oplus \underline{w_5} \oplus \underline{w_{21}} \oplus \underline{w_{37}} \oplus w_{45} \oplus \underline{w_{77}} \oplus \underline{w_{85}} \oplus \underline{w_{125}} = c_4,$$

$$S_{1,5}\big((w_0, \underline{w_{1\sim7}}) \oplus ck_{0\sim7}\big) \oplus \underline{w_6} \oplus \underline{w_{22}} \oplus \underline{w_{38}} \oplus w_{46} \oplus \underline{w_{78}} \oplus \underline{w_{86}} \oplus \underline{w_{126}} = c_5,$$

$$S_{1,6}\big((w_0, \underline{w_{1\sim7}}) \oplus ck_{0\sim7}\big) \oplus \underline{w_7} \oplus \underline{w_{23}} \oplus \underline{w_{39}} \oplus w_{47} \oplus \underline{w_{127}} \oplus \underline{w_{87}} \oplus \underline{w_{127}} = c_6.$$

We can determine a 7-bit value $w_{41\sim47}$ by guessing an 1-bit value w_0. Thus we obtain 2^{66} candidates of an 87-bit value $(w_{0\sim7}, w_{9\sim15}, w_{17\sim23}, w_{33\sim39}, w_{41\sim47}, w_{49\sim55}, w_{73\sim87}, w_{89\sim95}, w_{105\sim111}, w_{113\sim127})$. The computational complexity of this step is $O(2^{66})$.

In Step (4), we reduce the number of candidates by using the following 8 equations for $c_{40\sim46}$.

$$S_{2,0}\big((w_{40}, \underline{w_{41\sim47}}) \oplus ck_{40\sim47}\big) \oplus \underline{w_1} \oplus \underline{w_{41}} \oplus \underline{w_{49}} \oplus \underline{w_{81}} \oplus \underline{w_{89}} \oplus \underline{w_{105}} \oplus \underline{w_{121}} = c_{40},$$

$$S_{2,1}\big((w_{40}, \underline{w_{41\sim47}}) \oplus ck_{40\sim47}\big) \oplus \underline{w_2} \oplus \underline{w_{42}} \oplus \underline{w_{50}} \oplus \underline{w_{82}} \oplus \underline{w_{90}} \oplus \underline{w_{106}} \oplus \underline{w_{122}} = c_{41},$$

$$S_{2,2}\big((w_{40}, \underline{w_{41\sim47}}) \oplus ck_{40\sim47}\big) \oplus \underline{w_3} \oplus \underline{w_{43}} \oplus \underline{w_{51}} \oplus \underline{w_{83}} \oplus \underline{w_{91}} \oplus \underline{w_{107}} \oplus \underline{w_{123}} = c_{42},$$

$$S_{2,3}\big((w_{40}, \underline{w_{41\sim47}}) \oplus ck_{40\sim47}\big) \oplus \underline{w_4} \oplus \underline{w_{44}} \oplus \underline{w_{52}} \oplus \underline{w_{84}} \oplus \underline{w_{92}} \oplus \underline{w_{108}} \oplus \underline{w_{124}} = c_{43},$$

$$S_{2,4}\big((w_{40}, \underline{w_{41\sim47}}) \oplus ck_{40\sim47}\big) \oplus \underline{w_5} \oplus \underline{w_{45}} \oplus \underline{w_{53}} \oplus \underline{w_{85}} \oplus \underline{w_{93}} \oplus \underline{w_{109}} \oplus \underline{w_{125}} = c_{44},$$

$$S_{2,5}\big((w_{40}, \underline{w_{41\sim47}}) \oplus ck_{40\sim47}\big) \oplus \underline{w_6} \oplus \underline{w_{46}} \oplus \underline{w_{54}} \oplus \underline{w_{86}} \oplus \underline{w_{94}} \oplus \underline{w_{110}} \oplus \underline{w_{126}} = c_{45},$$

$$S_{2,6}\big((w_{40}, \underline{w_{41\sim47}}) \oplus ck_{40\sim47}\big) \oplus \underline{w_7} \oplus \underline{w_{47}} \oplus \underline{w_{55}} \oplus \underline{w_{87}} \oplus \underline{w_{95}} \oplus \underline{w_{111}} \oplus \underline{w_{127}} = c_{46}.$$

Guessing w_{40}, we can check 2^{66} candidates from Step (3) by using these equations. Since the filtering probability of this step is 2^{-7}, this step results in 2^{60} candidates of 88-bit value $(w_{0\sim7}, w_{9\sim15}, w_{17\sim23}, w_{33\sim47}, w_{49\sim55}, w_{73\sim87}, w_{89\sim95}, w_{105\sim111}, w_{113\sim127})$. In this case, the computational complexity is $O(2^{67})$. However, we can easily determine w_{40} by using 7-bit partial input/output values of S-box S_2. Thus, we do not need to guess w_{40}. Hence, the computational complexity is $O(2^{66})$ (see Step (4) in Table 1).

The procedures of Step (5)–(10) are explained in a similar fashion. After Step (10), we get 2^{16} candidates of the whole W_0. See Table 1 for the detailed result of each step.

Finally, we consider 16 equations for $(c_7, c_{15}, c_{23}, c_{31}, c_{39}, c_{47}, c_{55}, c_{63}, c_{71}, c_{79}, c_{87}, c_{95}, c_{103}, c_{111}, c_{119}, c_{127})$. Note that these equations are not used in the previous steps. Thus, we can check the survived candidates of W_0 with these equations. Since the filtering probability is $2^{-16}(=(2^{-1})^{16})$, we expect that the right W_0 only pass this test. Hence, our attack algorithm can recover the 128-bit secret key of ARIA-128.

The computational complexity of this attack depends on heavily Step (5) and (6). In Step (5) and (6), the computational complexity is $O(2^{68})$, respectively. Thus the total computational complexity of our attack is $O(2^{68})$.

4 Conclusion

In this paper, we introduced a method to recover the 128-bit secret key of ARIA-128 by using only the last round key. In detail, when the attacker knows the last round key of ARIA-128, it is possible to compute the 128-bit secret key with the computational complexity of $O(2^{68})$.

This is the first known cryptanalytic result considering only one round key of ARIA-128. Although the computational complexity of our attack is impractical, we believe that our method can be applied to known cryptanalysis on ARIA-128 where only the last round key is recovered.

Acknowledgments "This research was supported by the MKE(The Ministry of Knowledge Economy), Korea, under the ITRC(Information Technology Research Center) support program supervised by the NIPA(National IT Industry Promotion Agency)" (NIPA-2012-C1090-1101-0004).

References

1. NSRI.: Specification of ARIA (ARIA v. 1.0). http://210.104.33.10/ARIA/doc/ARIA-specification-e.pdf. 2005
2. Fleischmann, E., Forler, C., Gorski, M., Lucks, S.: New boomerang attacks on ARIA. In: Indocrypt'10, LNCS 6498, pp. 163–175, Springer, Berlin 2010
3. Wu, W., Zhang, W., Feng, D.: Impossible differential cryptanalysis of reduced-round ARIA and Camellia. J. Comput. Sci. Technol. **22**(3), 449–456 (2007)
4. Li, R., Sun, B., Zhang, P., Li, C.: New impossible differential of ARIA. Cryptology ePrint Archive, Report 2008/227, 2008
5. Li, S., Song, C.: Improved impossible differential cryptanalysis of ARIA. In: ISA'08, pp. 129–132, IEEE Computer Society, 2008
6. Liu, Z., Gu, D., Liu, Y., Li, J., Li, W.: Linear cryptanalysis of ARIA block cipher. In: ICISC'11, LNCS 7043, pp. 242–254, Springer, 2011
7. Li, P., Sun, B., Li, C.: Integral cryptanalysis of ARIA. In: Inscrypt'09, LNCS 6151, pp. 1–14, Springer, 2010
8. Wei, L., Dawu, G., Juanru, L.: Differential fault analysis on the ARIA algorithm. Inform. Sci. **178**(19), 3738–3737 (2008)
9. Biryukov, A., Cannière, C., Lano, J., Ors, S., Preneel, B.: Security and performance analysis of ARIA. http://www.cosic.esat.kuleuven.be/publications/article-500.ps. 2004

Estimating Number of Columns in Mixing Matrix for Under-Determined ICA Using Observed Signal Clustering and Exponential Filtering

Charuwan Saengpratch and Chidchanok Lursinsap

Abstract Under-determined Independent Component Analysis arises in a variety of signal processing applications, including speech processing. In this paper, we proposed a new approach focusing on the estimation of the number of columns and their values of the mixing matrix. The method is based on the observation that the observed vectors must be clustered along the direction of the column vectors of the mixing matrix. A new clustering measure and cluster direction finding are introduced. The propose algorithms are tested with real speech signals and compared with both AICA method and Information Index Removal, Perturbed Mean Shift Algorithm. Our result gives the correct number of columns with higher accuracy under the performance measure of algebraic matrix distance index.

Keywords Independent component analysis · Under-determined ICA · Blind source separation · Linear transformation · Sparse representation

1 Introduction

Knowledge of Independent Component Analysis (ICA) has a great importance for potential speech signal processing, brain signal processing, feature extraction, and acoustic processing. Defined ICA is very closely related to the method called Blind Source Separation (BSS). ICA problem arises in many application domains, such as speech separation, array antenna processing, multi-sensor biomedical records and financial data analysis [1]. Blind Source Separation was using a two stage sparse representation approach and presented an algorithm for estimating the mixing

C. Saengpratch (✉) · C. Lursinsap
Faculty of Science, Chulalongkorn University, Bangkok, Thailand
e-mail: c.saengpratch@hotmail.com

S.-S. Yeo et al. (eds.), *Computer Science and its Applications*,
Lecture Notes in Electrical Engineering 203, DOI: 10.1007/978-94-007-5699-1_26,
© Springer Science+Business Media Dordrecht 2012

matrix [2]. The over-complete were introduced that the source separation problem consist of estimating the original sources, the observed signals, then were introduced the over-complete source separation case where there are the mixture less than source signals [3]. A novel BSS algorithm for de-mixing under-determined presented anechoic mixtures [4].

In this paper, we proposed a new method to find the number of columns in the mixing matrix and estimated the value of each element. The remainder of this paper has the following structure. Section 2 describes the problem with constraints. Section 3 focuses on the main idea and the detail of the proposed algorithms. The experimental result is presented in Sect. 4. Section 5 concludes the paper.

2 Problem Statement

Under-determined ICA is considered in this paper. Let $\mathbf{S} = [\mathbf{s_1}, s_2, \ldots, s_N]^T \in R^{m \times T}$ and $\mathbf{X} = [x_1, x_2, \ldots, x_M]^T \in \mathbf{R}^{n \times M}$ be the source signals of n dimensions and observed signals of m dimensions, respectively. Each observed signal x_i is computed by using a mixing matrix $\mathbf{A} = [a_1, a_2, \ldots, a_m] \in \mathbf{R}^{n \times m}$ for n < m as follows.

$$X = AS \tag{1}$$

The source signals are based on the sparseness assumption [5]. The problem of how to recover the source signals after estimating the mixing matrix will not be considered here. However, we adopted the recovering method from [6].

3 Main Concept and Algorithms

Our method to find the columns of mixing matrix A is based on this simple observation. Let $X = [x_{i,1}, x_{i,2}, \ldots, x_{i,3}]^T$ be an observed signal i for $1 <= i <= n$ in a p-dimensional space. The mixing matrix A maps each incoming signal s_i to a new location considered as an observed signal x_i in lower dimensions. An observed signal can be viewed as a vector. These observed vectors are clustered along each new independent basis represented by each column vector of mixing matrix A. Therefore, to compute each column vector of A, a new similarity measure for all observed vectors clustered at any column must be introduced. Our method consists of the following main steps.

1. Computing the direction of each observed vector x_i in terms of radian degrees.
2. Grouping all observed vectors based on their directions computed from Step 1.
3. Finding the direction of each group and use it as a column vector of the mixing matrix A.

In this paper, we consider only the case where the observed vectors are in two dimensions. Hence, $X_i = [x_{i,1}, x_{i,2}]^T$. The detail of the first two steps is given in the following Algorithm.

3.1 Grouping Observed Vectors Algorithm

Let W be an empty set.
For each x_i, $1 \leq i \leq n-1$ do

$$\theta_i = \frac{180}{\pi} x \left(\arctan \left(\frac{x_{i,2}}{x_{i,1}} \right) \right)$$

$$\theta_{i+1} = \frac{180}{\pi} x \left(\arctan \left(\frac{x_{i+1,2}}{x_{i+1,1}} \right) \right)$$

If $\left| \dfrac{\theta_i}{\theta_{i+1}} \right| \leq \left| \theta_i - \theta_{i+1} \right|$ then

$W = W \cup \{x_i, x_{i+1}\}$.

End
Let $k = 1$.
Let group $= 0$.
While $W \neq \varnothing$ do

Let $\mathbf{x}_f^{(W)}$ be the first element in W.

Let θ_{old} be the radian degree of $\mathbf{x}_f^{(W)}$.

Let B_k be a new empty set.

$B_k = B_k \cup \left\{ \mathbf{x}_f^{(W)} \right\}$.

$W = W - \left\{ \mathbf{x}_f^{(W)} \right\}$.

For each $x_j \in W$ do

Let θ_{new} be the radian degree of x_j.

If $\left| \dfrac{\theta_{new}}{\theta_{old}} \right| \geq \left| \theta_{new} - \theta_{old} \right|$ then

$B_k = B_k \cup \left\{ x_j \right\}$.

$W = W - \left\{ x_j \right\}$.

EndIf
EndFor
$k = k + 1$.
group $=$ group $+ 1$.
EndWhile

Variable group in the algorithm is for counting the total number of groups. After grouping the observed vectors according to their radian degrees, the actual direction of each group will be derived and used as the column vector in mixing matrix A. Not every group generated by Grouping Observed Vectors Algorithm is feasible enough to derive a column vector in mixing matrix A. Only the group with high density of clustering will be selected. The detail of how to find the actual column of mixing matrix A is the following.

3.2 Finding Actual Column Algorithm

Let Q be an empty set.
For $1 \leq i \leq group$ do

$\quad t = |B_i|$.

$\quad q = 1 - e^{-\frac{1}{t}}$.

$q' = 1 - q$.

$\quad$ If $q \geq q'$ then

$\quad Q = Q \cup \{B_i\}$.

EndFor
For each element $B_j \in Q$ do

$\quad$ Compute eigenvector of B_j .

EndFor

Steps 4 and 5 are used to measure the density of clustering of each group. The eigenvectors extracted from steps 10 in Finding Actual Column Algorithm above are used as column vectors in mixing matrix A. Figure 1 shows an example as the result of the above algorithms. Figure 1a is the observed vectors. After being grouped by Grouping Observed Vectors Algorithm, Fig. 1b is obtained. The actual column vectors of the mixing matrix are shown in Fig. 1c.

4 The Experimental Results

The algorithms are tested with a real speech as shown in Fig. 2a. There are three sources of signals and two observed signals. The source signals are mixed by using the following mixing matrix.

$$\mathbf{A} = \begin{bmatrix} 0.7071 & -0.4472 & -0.9487 \\ 0.7071 & 0.8944 & 0.3162 \end{bmatrix} \tag{2}$$

The wave form of each observed signal is shown in Fig. 2b. The distribution of both observed vectors are plotted as shown in Fig. 2c. After applying Grouping Observed Vectors Algorithm to all observed vectors, three groups are obtained as shown in Fig. 3a. The direction of each eigenvector in each group is computed and illustrated in Fig. 3b. Our result was compared with the results produced by AICA method [7] and the Information Index Removal and Perturbed Mean Shift Algorithm [8]. To measure the accuracy of the estimated mixing matrix, Algebraic

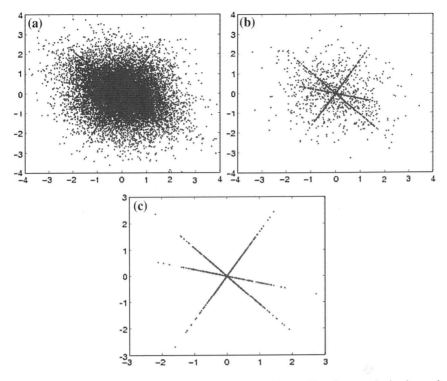

Fig. 1 An example of the result from the proposed algorithm. **a** The given synthetic observed vectors. **b** The result after grouping. **c** The computed eigenvectors which are used as column vectors in the mixing matrix

Matrix-Distance Index (AMDI) [7] was used. The following matrix is the result from our algorithms. Table 1 summarizes the accuracy of each method.

$$\hat{\mathbf{A}} = \begin{bmatrix} -0.9487 & 0.7071 & -0.4472 \\ 0.3162 & 0.7071 & 0.8944 \end{bmatrix} \qquad (3)$$

5 Conclusions

This paper presents a new approach for de-mixing mixtures under-determined Independent Component Analysis. This technique is based on the observation that the observed vectors usually clustered along the direction of column vectors in the mixing matrix. Our algorithms can find the correct number of column vectors in the mixing matrix with the lowest AMDI value. However, a further modification must be studied to scope with higher dimensional signals.

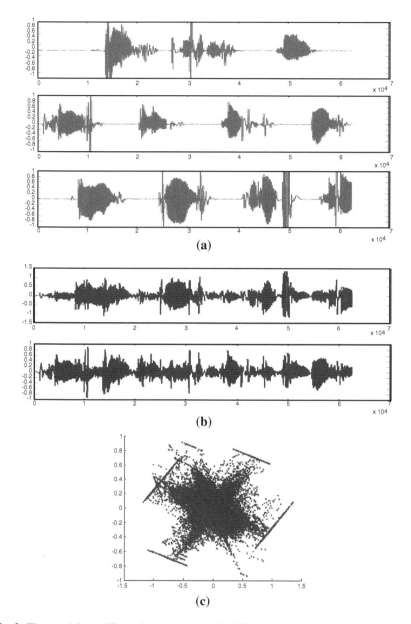

Fig. 2 The tested data. **a** Three given source signals. **b** Two observed signals. **c** The distribution plot of observed vectors

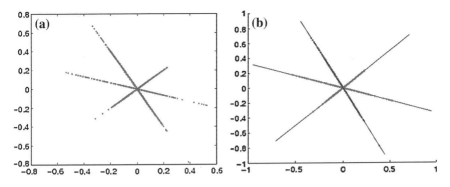

Fig. 3 The result from our algorithm. **a** Groups of observed vectors in the experiment obtained after applying **Grouping Observed Vectors Algorithm**. **b** The direction of each eigenvector in each group

Table 1 The comparison of the estimated mixing matrix by different methods

Comparative methods	Data 2D ($n = 3$)
Our proposed method	2.1950e-05
The perturbed mean shift	2.6000e-05
The AICA method	<0.001

Acknowledgments The Commission on Higher Education, Thailand for supporting by grant fund under the program Strategic Scholarships for Frontier Research.

References

1. Chinnasarn, K., Lursinsap, C., Palade, V.: Blind separation of mixed kurtosis signed signals using partial observations and low complexity activation functions. Int. J. Comput. Intell. Appl. **4**, 207–223 (2004)
2. Yuanqing, L., Amari, S., Cichocki, A., Daniel, W.C.H., Shengli, X.: Under-determined blind source separation based on sparse representation. IEEE Trans. Signal Process. **54**, 423–437 (2006)
3. Mitianoudis, N., Stathaki, T.: Over-complete source separation using laplacian mixture models. IEEE Signal Process. Lett. **12**, 277–280 (2005)
4. Rayan, S., Yilmaz, O., Martin, J.M., Rafeef, A.: Blind separation of anechoic under-determined speech mixtures using multiple sensors. In: IEEE International Symposium on Signal Processing and Information Technology (2006)
5. Paul, D.O., Barak, A.P.: The LOST algorithm: finding lines and separating speech mixtures. EURASIP J. Adv. Signal Process. **2008**,784296 Hindawi Publishing Corporation (2008)
6. Steven, V.V., Ignacio, S.: A spectral clustering approach to under-determine postnonlinear blind source separation of sparse sources. IEEE Trans. Neural Netw. **17**, 811–814 (2006)

7. Khurram, W., Fathi, M.S.: Algebraic independent component analysis: an approach for separation of over-complete speech mixtures. In: IEEE-INNS Joint International Conference on Neural Networks (2003)
8. Panyangam, B., Chinnasarn, K., Lursinsap, C.: Estimating columns of under-determined mixing matrix by information index removal and perturbed mean shift algorithm. In: IEEE Third International Conference on Natural Computation (2007)

Realization of Coordinative Control Between Multi Readers and Multi RF-SIM Cards Under Mobile RF-SIM Mode

Songsen Yu, Yun Peng and Xiaopeng Huang

Abstract RF-SIM card is a highly integrated system which packets a 2.45 GHZ radio frequency interface chip and a high-security payment chip to an ordinary SIM card. In order to solve the identifications and accesses of multi readers to multi RF-SIM cards. We make full use of the ability of 2.45 GHZ frequency which supports multi channels to divide RF-SIM channel into three types of channel: reservation channel, coordination channel and working channel. The RF-SIM reader should reverse a coordination channel before inquiring RF-SIM cards. After the reservation then it inquires the RF-SIM cards within its communication area. RF-SIM card takes turns to select a coordination channel to response card randomly. After the successful handshaking then RF-SIM card turns to a working channel to transmit datum. This method can effectively solve the identifications and accesses of multi readers to multi RF-SIM cards. The approach of its realization is proposed.

Keywords RF-SIM card · Multi-reader · Anti-collision coordination

S. Yu (✉)
Department of Computer, South China Normal University, Foshan 528225, China
e-mail: SongsenYu@Springer.com

Y. Peng
College of Computer Information Engineering, Jiangxi Normal University,
Nanchang 330072, China
e-mail: YunPeng@Springer.com

X. Huang
Eastcompeace Smart Card Co. Ltd, ZhuHai, 519060, China
e-mail: XiaopengHuangLNCS@Springer.com

S.-S. Yeo et al. (eds.), *Computer Science and its Applications*,
Lecture Notes in Electrical Engineering 203, DOI: 10.1007/978-94-007-5699-1_27,
© Springer Science+Business Media Dordrecht 2012

1 Introduction

The fusion of RFID non-contact technology and the contact technology of SIM card is the core issue of the mobile payment technology development. The mobile payment services develop on the basis of the fusion of contact and non-contact technologies. Through the combination of both of them, not only the applications of RFID technology in near field payment, entrance guard, attendance report, bus services are realized, but also satisfied the services of traditional mobile phones. The most important is that it generates a blue sea service. Specifically speaking, through the menu of mobile phone you can inquire the balance of smart card, transaction records etc. [1–3]. The functions of recharging account and the real-time interactions with traditional bank card, metro card and enterprise card are realized by load credit over the air.

2 Structure of RF-SIM System

2.1 Constitution of the System

RF-SIM looks the same as ordinary telecom card (see as Fig. 1) whose interface complies with ISO7168 standards. RF-SIM card is a highly integrated card with different encapsulation methods highly integrates a 2.45 GHZ radio frequency interface chip, a high secure payment chip and an ordinary SIM chip (the chip of user identification module). It completely retains the functions of ordinary SIM card and also provides secure radio frequency channels and payment services of financial level. Its structure is shown in Fig. 2.

In the implementation of RF-SIM card, the security chip controls the security of non-contact channel and expands the applications of value-added services such as CMS2AC etc. SIM chip realizes both the traditional mobile corresponding applications and the expanded applications. 2.45 GHZ radio frequency chip acts as the radio frequency corresponding channel. The approach of the realization of payment applications of RF-SIM card and attributions of the chip can be decided according to the specific means provided by the manufacturer. The I/O manager of the card supervises the interfaces such as SCD, SCR, SCC etc. These interfaces can be successively used in the interactions with the mobile phones, radio frequency reader, and the application extensible chip. The manufacturer needs to analyze and process the APDU of the interface.

2.2 Structure of Software

RF-SIM card adopts the hierarchical method in the design of the structure of software, as in Table 1. The structure consists of a low level hardware interface driver (HIS), a chip operation system (COS) and a top application layer [4–6].

Fig. 1 Appearance of
RF-SIM

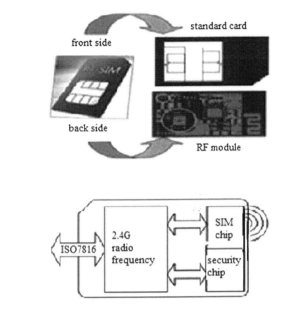

Fig. 2 Structure of RF-SIM

HIS are the interfaces of hardware drivers, such as ISO/IEC7816 reading and writing driver, RF 2.45 GHZ transceiver driver, and drivers for reading, writing, and erasing operations on the internal memory RAM, EEPROM/FLASH ROM. Chip Operation System lies between HIS and application layer. It offers transmission managements, assignments of demands, security managements and file managements to every application. The application layer is based on COS, applications are completely separated from each other.

Applying hierarchical design to the structuring of software enables, on one hand, the codes of different layers perform independently and achieve high cohesion and low coupling as far as possible, on the other hand, it is convenient for collaborative developments. The application layer developers just need to concentrate on application related parts. COS developers concentrate on providing necessary supports in data transmission, security management, file management and HIS developers concentrate on the development of hardware drivers, and providing supports and services that COS needs.

2.3 Information Stream of RF-SIM

The information stream of RF-SIM includes [8, 9]:

(1) GSM service flow
   The specific procedures of the functions of data authentication in fundamental GSM network such as making telephone calls, sending messages are as follows: (a) mobile phone transmitting the instructions of SIM to security chip, (b) the

Table 1 Software structure of RF-SIM

Application 1	Application 2	. . .	Application n
COS (*Chip Operation System*)			
HSI (*Hardware Software Interface*)			

RF2.4G	ISO7816	EEPROM	...	RAM	FLASH ROM

security chip distinguishes the instruction type after receiving the instructions, if the instructions are for SIM card, then they are transmitted to SIM module without any process. (c) SIM module conducts relative processing after receiving the instructions of SIM, then transmits it to security chip, eventually reverses it back to mobile phone.

(2) Payment services flow of mobile phone

RF channel is used to realize the functions of electronic purse such as all-in-one card, and micro-payment etc. specific process is as follows: (a) RF reader transmits the instructions of electronic purse to RF-SIM module through 2.45 GHZ frequency. (b) The security chip processes the instructions received from RF module and reverses to RF module. (c) RF module transmits them back to RF reader.

(3) Value-added services flow

Applications of value-added services mainly base on STK (subscriber identity module tool kit) select different operations by STK menu. Specific procedure is as follows: (a) mobile phone transmits the instructions of STK menu to security chip. (b) The security chip module processes the instructions from the mobile phone then reverses them back.

RF-SIM card adopts a highly secure payment master chip and forms a highly secure payment master chip centered integral structure, which makes sure that all inflow and outflow datum are audited and confirmed to ensure the security of datum in the chip.

3 Coordination Principles

GFSK modulation is adopted in the RF part of RF-SIM card system, modulation frequency is added to carrier frequency therefore the frequency of symbol "1" is fc + fs and that of symbol "0" is fc−fs, where fc represents the carrier frequency and fs represents the signal frequency. The modulation mode is shown in Fig. 3 [7].

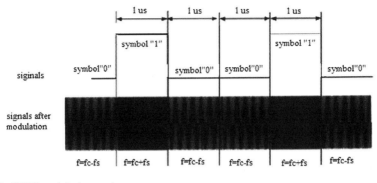

Fig. 3 GFSK modulation mode

In the existing standards of RF interface of RFID-SIM, RFID-SIM card uses 2.45 GHZ ISM frequency which is divided into two types: working channel and beacon channel. Working channel supports eight frequencies, beacon channel supports one frequency. Working channel is mainly used for services of communication and beacon channel is used for auxiliary inquiry. Therefore the inquiry of RF-SIM reader to RF-SIM card can be conducted in only one beacon channel. Collisions are inevitable when there are multiple RF-SIM readers within the working area, which makes it incompetent for the identification and access of multi readers to multi RF-SIM cards.

Considering 2.45 GHZ frequency has a available bandwidth of 83.5 MHZ, it can support 167 beacon channels if the bandwidth of each beacon channel is 500 KHZ and there will be enough beacon channels for frequency hopping when disturbances occur. To make full use of the capability of supporting multi channels of 2.45 GHZ frequency, we divide RF-SIM channel into three types of beacon channels: reservation channel, coordination channel and working channel.

Reservation channel: mainly used for RF-SIM reader multicasting the message of the coordination channel it will occupy to the adjacent ones.

Coordination channel: mainly used for the polling of RF-SIM reader to RF-SIM card.

Working channel: mainly used for communicating service.

The processes under this model are as follows:

(1) RF-SIM reader: RF-SIM reader monitors the reserved channel by carrier sensing, it notifies the adjacent readers of the coordination channel it will occupy when the reserved channel is free and continue detecting the reserved channel for a while (about a message round-trip time: the summation of the time for the message reaching the adjacent readers and that for the adjacent readers transmitting the collision information of coordination channel). RF-SIM reader will occupy the related coordination channels for the polling for RF-SIM cards when there is no collision of coordination channels. However if collision occurs or the coordination channel is busy, it will receive the collision information

Fig. 4 RF-SIM reader

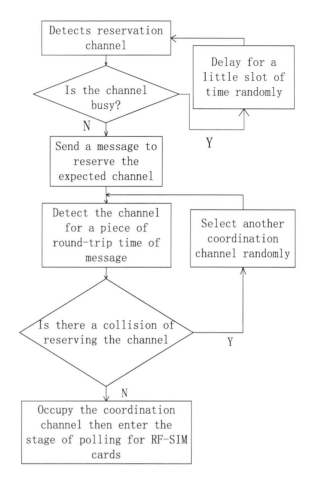

from the adjacent readers, then delays for a random piece of time until there is no collision and the coordination channel is not busy.

RF-SIM reader will recognizes the RF-SIM cards in the coordination channel after reserving the coordination channel successfully.

(2) RF-SIM card: RF-SIM card detects a coordination channel randomly, if it receives a packet of polling message from RF-SIM reader then it will reply a responsive message of successful receipt according to the rule of random ALOHA and then enter a stage of mutual identify authentication and data procession stage, otherwise it will detects another coordination channel randomly.

The schematic diagram of how it works is shown in Figs. 4 and 5

Fig. 5 RF-SIM card

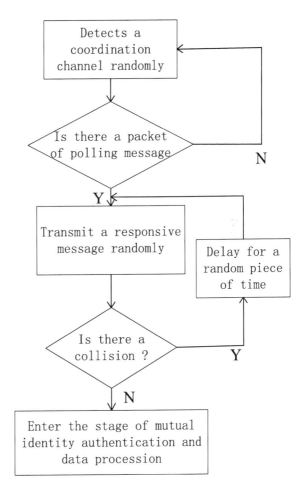

3 Formats of Instructions

To simplify the description of the realization of the coordinative control methods between multi-reader and multi-RF-SIM under the mode of moving RF-SIM, the relative instructions are defined as follows:

BROAD (R-ID, X_band): Broadcast Instruction. It is used for RF-SIM reader broadcasting the message of coordination channel which it wants to use to the adjacent readers.

BACK (X_band): Response Instruction to channel collision. RF-SIM reader receives the broadcast instructions from neighbors and checks out whether there is a collision of using the coordination channel, if so, it will response this instruction.

INQUIRY(R_ID, X_band): Card Inquiry Instruction. The inquire instruction from RF-SIM reader to RF-SIM card through the coordination channel X_band.

CACK (C_ID, W_band, X_band): the response instruction. RF-SIM card detects a coordination channel X_band randomly, if receives an inquiry instruction from a RF-SIM reader, it then replies a response instruction CACK. Where W-band is the working coordination channel it expected.

RC-REQ (W_band, C_ID): the instruction of RF-SIM reader requiring for a link with RF-SIM card. The require instruction for a link with destination card which is sent by RF-SIM reader through the working channel W-band.

CR-RSP (W_band, R_ID): the instruction of RF-SIM card responses to the connection request of the reader. The response sent by RF-SIM card to the destination reader through working channel W-band.

APDATA (R/C_ID, len, payload, checksum): the instruction of data transmission. It is used for mutual transmission of data between RF-SIM reader and card after the connection is established.

The definitions of each parameter are as follows: R-ID is the number of RF-SIM reader; C-ID is the RF-SIM card number; X_band represents the coordination channel; W-band represents working channel; R/C-ID represents the number of RF-SIM reader or card; len represents the length of data in bytes; payload represents the data transmitted; checksum is the summation of checksums of R/C-ID, len and payload fields.

4 Realization of this Protocol

In sight of the formats of instructions, the specific realization of the protocol of RF-SIM reader part and card part is as follows:

```
RF-SIM_Reader ()
{set an initial value for X_band;
Repeat

  BROAD (R_ID, X_band);
  Wait a back-and-forth slot-time...
  If Receive RACK (X_band)
  Change the value of X_band

Until no Receive RACK (X_band);
Repeat

  INQUIRY(R_ID,X_band);

Until Receive CACK(C_ID,W_band,X_band);

  RC_REQ(W_band,C_ID);
  If Receive CR_RSP(W_band,R_ID)

APDATA(R/C_ID,len,payload,checksum)

}
RF-SIM_Card()
```

```
{set an initial value for X_band;
Repeat

  CACK(C_ID,W_band,X_band);
  Wait a back_and_forth slottime...
  If no Receive RC_REQ(W_band,C_ID)

  Change the value of X_band;

Until Receive RC_REQ(W_band,C_ID);
CR_RSP(W_band,R_ID0;
APDATA(R/C_ID,len,payload,checksum)
}
```

5 Conclusion

SIM card is an application-specific integrated circuit card for identity authentication. It is an important carrier of functional expansion. RF-SIM card of 2.45 GHZ frequency is an ingenious combination of 2.45 GHZ radio frequency technology and SIM card. It not only completely retains the service functions of ordinary SIM card, but also provides secure radio-frequency channels and payment services of financial level. The available bandwidths of 2.45 GHZ frequency are 83.5 MHZ. It has a strong anti-interfere ability for there are many channels for frequency hopping when interference emerges. This paper proposes a control method of multi readers enquire multi cards by taking the advantages of its ability of supporting multi channels. This method can effectively solve the identification and access operation of multi-reader to multi-card and help to promote the development of mobile phone payment services. It makes mobile phone not only a communication tool but also an important information processing reader by combining finance and media.

Acknowledgments This work is supported in part by the development of GuangDong strategic emerging industries of special funds in 2011(2011168012), Ministry of industry and information In Internet of Things special funds in 2011, National Natural Science Foundation of China (61172156, 61174123, 61102034), Guangzhou Application Foundation key project (11C42090780).

References

1. Miao, Z.: Design and implementation of smart card web server and application framework for multi-application. Degree of Master Thesis, Beijing University of Posts and Telecommunications, pp. 50–63 (2010)
2. Huang, Y.: Research and implementation of COS Based on ACI. Degree of Master Thesis, Guangxi Normal University, pp. 75–86 (2010)

3. Wang, C.: A COS design for smart card with two communication interface. Degree of Master Thesis, XiaMeng University, pp. 37–53 (2008)
4. Zhang, W., Xie, Y.: The analysis of existing condition of mobile RFID smart card and researches on its development tendency. Guangdong Telecommun. Technol. **10** (2010)
5. Wang, C.: Bi-interface multi-application oriented design of operation system of SIM card. PhD Thesis of Xiamen University (2008)
6. Sheng, Q., Li, L., Tang, H.: 2.4 GHZ RF-SIM card applied in mobile phone payment. Telecommun. Inf. **1** (2010)
7. The Interface Specification of China Mobile RFID-SIM Card Payment System, version number: 0.7.0 (2009)
8. Ni, W., Liu, Y., Liu, W., Yang, Y.: Design and implementation of RFID based multilevel anti-counterfeiting system. Comput. Eng. Des. **30**(15) (2009)
9. Wang, Y., Yang, J., Zhan, Y., Wan, P.: Collision avoidance MAC protocol for RFID reader networks. J. Univ. Electron. Sci. Technol. China **40**(3) (2011)

Effects of Smart Home Dataset Characteristics on Classifiers Performance for Human Activity Recognition

Iram Fatima, Muhammad Fahim, Young-Koo Lee
and Sungyoung Lee

Abstract Over the last few years, activity recognition in the smart home has become an active research area due to the wide range of human centric-applications. A list of machine learning algorithms is available for activity classification. Datasets collected in smart homes poses unique challenges to these methods for classification because of their high dimensionality, multi-class activities and various deployed sensors. In fact the nature of dataset plays considerable role in recognizing the activities accurately for a particular classifier. In this paper, we evaluated the effects of smart home datasets characteristics on state-of-the-art activity recognition techniques. We applied probabilistic and statistical methods such as the Artificial Neural network, Hidden Markov Model, Conditional Random Field, and Support Vector Machines. The four real world datasets are selected from three most recent and publically available smart home projects. Our experimental results show that how the performance of activity classifiers are influenced by the dataset characteristics. The outcome of our study will be helpful for upcoming researchers to develop a better understanding about the smart home datasets characteristics in combination with classifier's performance.

I. Fatima (✉) · M. Fahim · Y.-K. Lee · S. Lee
Department of Computer Engineering, Kyung Hee University, Yongin, Korea
e-mail: iram.fatima@oslab.khu.ac.kr

M. Fahim
e-mail: fahim@oslab.khu.ac.kr

Y.-K. Lee
e-mail: yklee@oslab.khu.ac.kr

S. Lee
e-mail: sylee@oslab.khu.ac.kr

S.-S. Yeo et al. (eds.), *Computer Science and its Applications*,
Lecture Notes in Electrical Engineering 203, DOI: 10.1007/978-94-007-5699-1_28,
© Springer Science+Business Media Dordrecht 2012

1 Introduction

A smart home is an intelligent agent that perceives state of resident and the physical environments using sensors. Recent advancements in the field of machine learning and data mining have enabled activity recognition research using smart homes sensing data to play a direct role in improving the general quality of health care. It is one of the best solutions to provide a level of independence and comfort in the homes of elderly people rather than requiring them to reside at health care centers [1]. The advancement of sensor technology has proven itself to be robust, cost-effective, easy to install and less intrusive for inhabitants. This fact is supported by a large number of applications developed using activity recognition to provide solutions to a number of real-world problems such as remote health monitoring, life style analysis, interaction monitoring, and behavior mining [2, 3].

A diverse set of machine learning and data mining algorithms have been previously used to identify the performed activities from the smart home datasets. The quest to optimize the performance of classifiers has a long and varied history. The diverse characterized data of smart homes require intelligent machine learning and data mining algorithms for automated analysis in order to make logical inferences from the stored raw data that may results in activity classification [4]. With the passage of time researchers found refinements that result in more accurate classification on comparable datasets. We study the relationship between the distribution of data, on the one hand, and classifier performance, on other. It is shown that predictable factors such as the available amount of training data, the spatial variability of data samples, deployed sensors in smart homes and the total activity occurrences in the dataset influence the performance of classifiers to a significant degree.

To select an appropriate classifier for certain type of data, there is a need to understand the behavior of classifiers on different data characteristics. Despite the great work and diversity in the existing classification methods, no significant work is done so far to assist a researcher in selecting a suitable classification technique for a particular nature of smart home dataset. The process of selecting an appropriate classifier is still a trial and error process that clearly depends on the relationship between the classifier and the data. The focus of this study is to facilitate the researchers in order to understand the effects of dataset characteristics on different classifiers for activity recognition. A particular dataset cannot be classified with same accuracy from all classifiers. Some vital dataset characteristics are dataset duration, performed activities, deployed sensors, activated sensors for a particular activity, total occurrences of single activity, and closely correlated activities. We compared state-of-the-art classifiers such as Artificial Neural Network (ANN) [5], Hidden Markov Model (HMM) [2], Conditional Random Field (CRF) [3], and Support Vector Machines (SVM) [6]. These four selected schemes are applied on four datasets selected from three most significant smart home projects such as CASAS [7], ISL [8] and House_n [9] smart homes. The main subject of this paper is to provide a systematic and unbiased evaluation of the existing activity classification schemes to resolve the uncertainties associated with the choice of classifier and the nature of smart home dataset. The results show that neither of the classifier is best for all datasets, the classification accuracy of each classifier depends on the underline data characteristics. We also illustrate that dataset characteristics highly affect the classifiers' individual class level assignments along with their overall performances.

The rest of the paper is organized as follows. We describe related work in Sect. 2. In Sect. 3, we discussed the smart home datasets with their important characteristics for activity recognition. In Sect. 4, we introduced four classifiers with their preferred settings for our experiments. The analyzed results of the CASAS, ISL, and House_n smart home datasets are presented in Sect. 5. Finally, conclusion and future works are given in Sect. 6.

2 Related Work

Several studies have been conducted to determine effective and accurate activity classification methods for smart home datasets. In [10] authors study the impact of semi-Markov models classification accuracy using datasets from ISL smart homes. They consider availability of labeled data, the importance of training time and speedy inference for experimental purpose. In their analysis they showed that CRF outperforms other semi-Markov models. The authors in [6] apply SVM to identify daily living activities on their health smart home dataset. They selected a set of features from dataset according to their domain of interest before using multi-SVM for effective activity classification as compare to other classifiers. The work in [11] applied the ANN for to cluster analysis of human activities of daily living within their own developed smart home environment. Specifically their approach is GSOM-based data mining to cluster analysis of human activities effectively.

The authors in [12] proposed a data mining framework to recognize activities based on raw data collected from CASAS smart home. The framework synthesizes the sensor information and extracts the useful features as many as possible. They compared several machine learning algorithms on the selected features to compare the performance of activity recognition. They discussed the performance of the machine learning algorithm on the basis of their selected feature based on information gain and mRMR. The authors in [13] employ data mining techniques to look at the problem of sensor selection for activity recognition in smart homes along with classifier selection. They examine the issue of selecting and placing sensors in a CASAS smart home in order to maximize activity recognition accuracy. In [2] authors used ISL smart home dataset to show the potential of generative and discriminative models for recognizing activities. They presented that CRFs are more sensitive to overfitting on a dominant class than HMM.

The commonly observed methodologies in literature for smart home datasets are with only limited number of algorithms from the machine learning repository

Table 1 The CASAS datasets Twor2009, and Tulum2009. The 'Num' column shows activity count, the 'Time' column shows activity time in minutes, and the 'Sensor' column shows activity sensor events

Twor2009				Tulum2009			
Activities	Num	Time	Sensor	Activities	Num	Time	Sensor
Idle	–	8240.93	73043	Idle	–	102986.4	203408
Bed toilet transition	39	94.55	2241	Wash dishes	71	1204.84	24869
Meal preparation	118	6207.32	41730	Watch TV	528	4955.43	52222
R1 sleeping in bed	35	18428.36	29503	Enter home	73	119.42	604
R2 sleeping in bed	35	18572.11	29604	Leave home	75	101.58	1854
Cleaning	2	49.75	1540	Cook breakfast	80	1440.31	33435
R1 work	59	5902.51	45675	Cook lunch	71	972.74	24527
R2 work	44	2530.03	17955	Group meeting	11	1847.04	31084
R1 bed to toilet	34	337.06	2298	R1 eat breakfast	66	932.87	20077
R1 personal hygiene	45	663.32	5818	R1 snack	491	4461.85	81183
R2 personal hygiene	39	1029.47	8237	R2 eat breakfast	47	497.06	13649
Study	9	922.71	8133	–	–	–	–
Wash_Bathtub	1	33.09	219	–	–	–	–
Watch TV	31	3228.77	17879	–	–	–	–

and select the one which gives relatively better results for their particular domain. No existing work has intensions to analyze the classifiers to show the effects of data characteristics. Our study will help the researchers in choosing an appropriate classifier based on a particular type of dataset.

3 Smart Home Datasets

Smart home datasets are generally associated with high-dimensional features and multiple classes. To comprehensively evaluate the performance of various classification schemes on smart home datasets, we analyzed four datasets from three smart home projects. We selected *Tulum2009* and *Twor2009* from CASAS smart home project. The dataset duration is 83 and 46 days respectively and deployed sensors are 20 and 71 respectively. From ISL and House_n smart homes, *House A* and *Subject 1* datasets are evaluated. The duration of these datasets is 24 and 16 days respectively and deployed sensors are 14 and 28 respectively. The detail analysis and our calculated data dimensions of datasets are shown in Tables 1 and 2.

Analyzed data characteristics show that each dataset is different from others in respective total time duration, deployed sensors, activity count, activity time, and activity sensor events. All these data attributes effect internal processing of classifiers based on their design intensions.

Table 2 The ISL & House_n datasets House A, and Subject 1

HouseA				Subject1			
Activities	Num	Time	Sensor	Activities	Num	Time	Sensor
Idle	–	5817.23	23	Idle	–	20930.6	731
Leaving	33	19664.27	84	Toileting	82	161.5	323
Toileting	114	155.38	402	Washing dishes	7	42.23	67
Showering	23	136.38	59	Preparing breakfast	14	182.42	147
Brush teeth	10	9.78	22	Preparing lunch	17	524.37	497
Sleeping/Go to bed	24	7914.37	183	Preparing dinner	8	136.72	122
Prepare dinner	9	306.47	128	Preparing a snack	13	58.43	66
Snack	12	24.33	50	Preparing a beverage	15	55.47	77
Prepare breakfast	20	39.42	122	Dressing	24	88.7	121
Eating	1	22.56	0	Bathing	18	343.93	224
Drink	20	12.23	63	Grooming	37	216.98	302
Load washing machine	3	4.01	7	Cleaning	8	149.67	145
Load dishwasher	5	31.85	15	Doing laundry	19	146.12	172
Unload dishwasher	4	15.23	27	Going out to work	12	2.87	25
Store Groceries	1	1.183	3	–	–	–	–
Unload washing machine	4	3.27	9	–	–	–	–
Receive guest	3	424.93	65	–	–	–	–

The 'Num' column shows activity count, the 'Time' column shows activity time in minutes, and the 'Sensor' column shows activity sensor events

4 Classifiers for Activity Recognition

In this section, we briefly introduce the applied classifier for the domain of activity recognition with preferred settings for our experiments. The detail of each classifier is given as:

ANN: It is an information processing network of artificial neurons connected with each other through weighted links. In activity recognition, multilayer neural network with back propagation learning algorithm is utilized to recognize the human activities [5]. The structure of the network, number of hidden layers, and number of neuron in each layer effects the learning of different activities. The activation of the neurons in the network depends on the activation function. We train multi-layer neural network through back propagation learning method and weights are updated by the following equation:

$$\Delta w_{ki} = -c \left[-2 \sum_j \left\{ \left(y_{j(desired)} - y_{j(actual)} \right) f'\left(act_j\right) w_{ij} \right\} f'\left(act\right)_i x_k \right] \quad (1)$$

Where Δw is the weights adjustment of the network links. In our network, we used one hidden layer, twenty neurons, tangent sigmod function as an activation function as given below:

$$\varphi(v) = \tanh\left(\frac{v}{2}\right) = \frac{1 - \exp(-v)}{1 + \exp(-v)} \tag{2}$$

Learning of the network is limited to maximum 1000 epochs. The multi-layer neural network can be seen as an intuitive representation of a multi layer activity recognition system. The number of correctly classified activities depends on the number of training instances during the learning phase.

HMM: It is a generative probabilistic graph model that is based on the Markov chains process [2]. Model is based on the number of states and their transition weight parameters. Parameters are learned thorough observation and following parameters are required to train the model:

$$\lambda = \{A, B, \pi\} \tag{3}$$

Where λ is graphical model for activity recognition, A is a transition probability matrix, B represents the output symbol probability matrix, and π is the initial state probability [2]. We used Baum-Welch algorithm to determine the states and transition probabilities during training of HMM. The *ith* classification of an activity is given as:

$$\lambda_i = \{A_i, B_i, \pi_i\}, \quad i = 1, \ldots, N \tag{4}$$

CRF: It is a discriminative probabilistic graph model for labeling the sequences. The structure of the CRF is similar to HMM but learning mechanism is different due to absence of the hidden states [2]. In CRF model, the conditional probabilities of activity labels with respect to sensor observations are calculated as follows:

$$p(y_{1:T}|x_{1:T}) = \frac{1}{Z(x_{1:T}, w)} exp\left\{\sum_{j=1}^{N_f} w_j F_j(x_{1:T}, Y_{1:T})\right\} \tag{5}$$

In Eq. 5, Z denotes normalized factor and $F_j(x_{1:T}, Y_{1:T})$ is a feature function. To make the inference in the model, we compute the most likely activity sequence as follows:

$$y_{1:T}^* = \text{argmax}_{y'_{1:T}} p\left(y'_{1:T}|x_{1:T}, w\right) \tag{6}$$

SVM: SVM is statistical learning method to classify the data through determination of a set of support vectors and minimization of the average error [6]. It can provide a good generalization performance due to rich theoretical bases and transferring the problem to a high dimensional feature space. For a given training set of sensors value and activity pairs, the binary linear classification problem require the following maximum optimization model using the Lagrangrian multiplier techniques and Kernel functions as:

$$\text{Maximize(w.r.t.}\alpha)\sum_{i=1}^{n}\alpha_i - \frac{1}{2}\sum_{i=0}^{n}\sum_{j=1}^{n}\alpha_i y_i \alpha_j K(x_i, x_j) \tag{7}$$

$$\text{Subject to}: \sum_{i=1}^{n}\alpha_i y_i = 0, 0 \leq \alpha_i \leq C \tag{8}$$

Where K is the kernel function that satisfies $K(x_i, x_j) = \Phi^T(x_i)\Phi(x_j)$. In our case, we used radial basis function (RBF) for recognizing the activities.

$$K(x_i, x_j) = \exp\left(\frac{-\parallel x_i - x_j \parallel^2}{(2\sigma^2)}\right) \tag{9}$$

Activity recognition is multi-class problem so we adopt "one-versus-one" method to classify the different activities. Classification of the final activity class is based on the voting mechanism and maximum vote of a class determined the activity label.

5 Results and Evaluation

In this section, we show the effect of data dimensions through demonstrating how performances of activity recognition techniques are influenced by the dataset characteristics. We split the dataset using the 'leave one day out' approach; therefore, the sensor readings of 1 day are used for testing and the remaining days for training. Figs. 1, 2, 3 and 4 show the experimental results for the *Tulum2009*, *Twor2009*, *House A* and *Subject 1* datasets characteristics respectively. In each dataset, for each activity, accuracies of ANN, HMM, CRF and SVM are illustrated in the following paragraphs.

In our experiments ANN shows high diversity in its performance. It performs better on the set of activities whose training instances are high in the dataset while its performance is insignificant for the recognition of those activities whose training examples are few in the dataset. Overall performance of ANN on Tulum2009 is 81.09 % and it correctly classified "R1 Snack" activity; however, it could not recognize the "Group Meeting" activity. The training instances for these activities are 491 and 11 respectively that affects the ANN classification process. In case of *Twor2009*, training instances for "Meal Preparation" are 118 and ANN outperforms all other classifier on the identification of this activity. While on the same dataset, it could not classify "Study" activity due to less training instances. For *House A*, the overall performance of ANN is low 41.11 % as compared to other datasets. However, ANN is the only classifier that 100 % classify the "Toileting" activity as its training instances (i.e., 114 samples) are very high as compare to other activities. In case of Subject 1 too, training instances of "Toileting" activity are more (i.e., 82 samples) and ANN performance is better than other classifiers for the

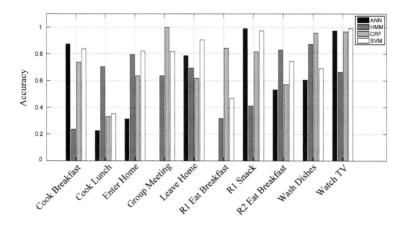

Fig. 1 *Tulum2009* activity based accuracy of classifiers

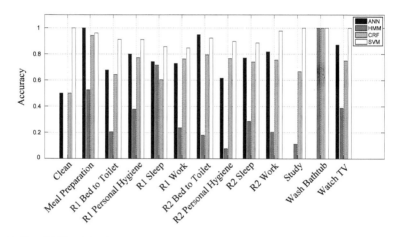

Fig. 2 *Twor2009* activity based accuracy of classifiers

recognition of this activity. Although the overall accuracy of ANN varies from dataset to dataset but the better performance of ANN is consistently depend on the high number of training instance in the dataset for a particular activity.

For HMM, the number of deployed sensors effect the activity class distribution by observing their variation during the performed activities. For example, HMM performs well 57.83 % in case of *House A* due to small number of deployed sensors (i.e., 14 sensors). It correctly classified "Take Shower", "Unload Dishwasher" and "Store Groceries". In case of *Tulum2009*, it outperforms other classifier for "Cook Lunch" and "R2 Eat Breakfast" activities with overall accuracy 56.84 %, the number of deployed sensors are 20 in this case. For *Subject 1* and *Twor2009* accuracy of HMM is not significant, the number of deployed sensors in these smart homes are 28 and 71. The large number of miss classified

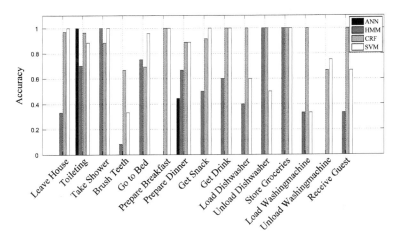

Fig. 3 *House A* activity based accuracy of classifiers

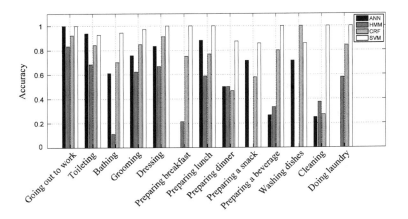

Fig. 4 *Subject 1* activity based accuracy of classifiers

activities are the result of HMM distributions modeling as they are observed in the dataset.

The performance of Conditional Random Fields (CRF) is affected by a set of data characteristics. Its performance does not depend only on single data dimension like senor count or activity occurrence however; its internal processing is based on conditioning of a set of data attributes. CRF outperforms all classifiers in case of *Tulum2009* for "Group Meeting", "R1 Eat Breakfast" and "Wash Dishes". However, other classifiers are better for "Cook Lunch", "Enter Home" and "Leave Home", "R1 Snack" and "R2 Eat Breakfast". In case of *House A*, for "Brush teeth", "Load washingmachine" and "Receive Guest" CRF is superior. For *Subject 1* its performance is high only for "Washing dishes" however for "Cook Lunch", "Enter Home" and "Leave Home", "R1 Snack" and "R2 Eat

Table 3 Overall classifiers accuracy

Classifier Dataset	ANN	HMM	CRF	SVM
Tulum2009	0.8109	0.5684	0.8374	0.8889
Twor2009	0.7983	0.3421	0.7780	0.9307
HosueA	0.4111	0.5783	0.9230	0.8919
Subject 1	0.6836	0.5200	0.7745	0.9563

Breakfast" other classifiers performed better. In case of *Twor2009*, CRF shows low performance for all activities except "Wash Bathtub".

SVM efficiently identified activities in case of *Subject 1* and *Twor2009*, it outperforms all classifier in these datasets except for "Washing Dishes" and "R2 Bed to Toilet" respectively. The performance of SVM is high if performed activities in the dataset are highly discriminative however it is hard for SVM to differentiate between activities that are closely correlated in various data dimensions. Due to this reason in *House A* and *Tulum2009*, for some activities other classifiers are better than SVM as discussed in above paragraphs. For example, in *Tulum2009* it confused "R1 Eat Breakfast" with "R1 Snack" activity both activities are very similar to each other, the second most confused activity is "Cook Lunch". SVM performance is affected if the performed activities are closely interrelated in respect to data dimensions.

The overall comparison results of different classifiers are presented in Table 3. It specifies the overall accuracy associated with each of the dataset over the four learning techniques. The above results and statistics clearly show that dataset characteristics highly affect the classifiers' individual class level assignments and thus their overall performances.

6 Conclusion and Future Work

In this paper, we have presented the influence of smart home dataset characteristics on the performance of the classifiers. We conclude that the nature of a given dataset plays an important role on the classification accuracy of algorithms; therefore, it is imperative to choose an appropriate algorithm for a particular dataset. We have identified some general characteristics of a dataset that can be useful in selecting the most suitable algorithm as per the nature of underlying dataset. To assess the performance of the four machine learning methods, we applied each classifier on four different datasets from three smart home projects. We analyzed the results of the experiments and provided the explanation of those results for each classifier. It facilitates the researchers to understand the variability in classifiers performances influenced by dataset characteristics.

In future, we would like to devise a framework that can recommend the most suitable classifier for the candidate dataset by analyzing the patterns in the dataset.

Acknowledgments This research was supported by the Ministry of knowledge Economy (MKE), Korea, under the Information Technology Research Center (ITRC) support program supervised by the National IT Industry Promotion Agency (NIPA)" [NIPA-2009-(C1090-0902-0002)].

References

1. Mihail, P., Elena, F.: Linking clinical events in elderly to in-home monitoring sensor data: a brief review and a pilot study on predicting pulse pressure. J. Comput. Sci. Eng. **2**, 180–199 (2008)
2. van Kasteren, T.L.M., Englebienne, G., Krose, B.J.A.: An activity monitoring system for elderly care using generative and discriminative models. J. Pers. Ubiquitous Comput. **14**, 489–498 (2010)
3. Krose, B.J.A., Kasteren, T.L.M., Gibson, C.H.S., Dool, T.: CARE: Context Awareness in Residences for Elderly. In: The Proceeding of 6th International Conference of the International Society for Gerontechnology, pp. 101–105 (2008)
4. Munguia Tapia, E., Intille, S. S., Larson, K.: Activity recognition in the home setting using simple and ubiquitous sensors. Lecture Notes in Computer Science, pp. 158–175 (2004)
5. Helal, S., Kim, E., Hossain, S.: Scalable approaches to activity recognition research. In: Proceedings of the Workshop of How to do Good Activity Recognition Research? Experimental Methodologies, Evaluation Metrics, and Reproducibility Issues (2010)
6. Fleury, A., Vacher, M., Noury, N.: SVM-based multimodal classification of activities of daily living in health smart homes: sensors, algorithms, and first experimental results. IEEE Trans. Inf. Technol. Biomed. **14**, 274–283 (2010)
7. Cook. D.: CASAS smart home project [Online]. http://www.ailab.wsu.edu/casas/ (2012). Accessed 10 June 2012
8. Kasteren, T., Noulas, A., Englebienne, G., Ben, K.: Accurate activity recognition in a home setting. In: Proceeding of International Conference Ubiquitous Computing, pp 21–24 (2008)
9. Larson, K.: House_n [Online]. http://architecture.mit.edu/house_n/ (2012). Accessed 10 June 2012
10. van Kasteren, T.L.M., Englebienne, G., Krose, B.J.A.: Activity recognition using semi-markov models on real world smart home datasets. J. Ambient Intell. Smart Environ. **2**, 311–325 (2010)
11. Zheng, H., Wang, H., Black, N.: Human activity detection in smart home environment with self-adaptive neural networks. In: Proceedings of the IEEE International Conference on Networking, Sensing and Control, pp. 1505–1510 (2008)
12. Chen, C., Das, B., Cook, D.J.: A data mining framework for activity recognition in smart environments. In: Proceeding of the International Conference on Intelligent Environments, pp. 80–83 (2010)
13. Cook, D., Holder, L.: Sensor selection to support practical use of health-monitoring smart environments. J. Data Min. Knowl. Discov. (2011)

Activity Recognition Based on SVM Kernel Fusion in Smart Home

Muhammad Fahim, Iram Fatima, Sungyoung Lee and Young-Koo Lee

Abstract Smart home is regarded as an independent healthy living for elderly person and it demands active research in activity recognition. This paper proposes kernel fusion method, using Support Vector Machine (SVM) in order to improve the accuracy of performed activities. Although, SVM is a powerful statistical technique, but still suffer from the expected level of accuracy due to complex feature space. Designing a new kernel function is difficult task, while common available kernel functions are not adequate for the activity recognition domain to achieve high accuracy. We introduce a method, to train the different SVMs independently over the standard kernel functions and fuse the individual results on the decision level to increase the confidence of each activity class. The proposed approach has been evaluated on ten different kinds of activities from CASAS smart home (Tulum 2009) real world dataset. We compare our SVM kernel fusion approach with the standard kernel functions and get overall accuracy of 91.41 %.

Keywords Activity recognition · Smart home · SVM · Kernel fusion

M. Fahim (✉) · I. Fatima · S. Lee · Y.-K. Lee
Ubiquitos Computing Laboratory, Department of Comptuer Engineering,
Kyung Hee University, Suwon, Korea
e-mail: fahim@oslab.khu.ac.kr

I. Fatima
e-mail: iram.fatima@oslab.khu.ac.kr

S. Lee
e-mail: sylee@oslab.khu.ac.kr

Y.-K. Lee
e-mail: yklee@oslab.khu.ac.kr

S.-S. Yeo et al. (eds.), *Computer Science and its Applications*,
Lecture Notes in Electrical Engineering 203, DOI: 10.1007/978-94-007-5699-1_29,
© Springer Science+Business Media Dordrecht 2012

1 Introduction

Activity recognition is active area of research since the inclusion of smart home concept for providing ubiquitous lifecare services. It can provide a valuable health monitoring services for ageing society to improve their quality of life. In the recent years, several smart homes have been developed such as CASAS and MavHome [1] at Washington State University, Aware Home [2] at Georgia Tech University, Adaptive House [3] at University of Colorado, House_n [4] at Massachusetts Institute of Technology (MIT), and House A [5] at Intelligent Systems Laboratory. The nomenclature of the activity recognition has two broad categories, visual sensing and ubiquitous sensing technologies. In first category, camera based techniques are used to monitor the behavior of inhabitants. These are not practical due to privacy reason, day/night vision problem and jumble environment. In second category, ubiquitous sensors are embedded on the different objects to recognize the daily life activities. They are cost-effective, easy to install and less intrusive to the privacy of inhabitants. Activity recognition is a big challenge by using ubiquitous sensors due to complex and highly diverse life styles.

Several machine learning and statistical approaches have been proposed to recognize the human activities and achieves acceptable accuracy with particular attentions [6, 7]. This paper investigates the theoretically rich statistical method SVM due to its quality of generalization and ease of training as compared to traditional method artificial neural network. We adopt decision fusion mechanism to increase the accuracy of recognition rate. A kernel fusion method is introduced to transform the activity recognition problem into higher feature spaces. The output of each individual is combined for the final consensus about the activity class label. Our approach is able to recognize activities more efficiently in a reasonable amount of time using fast training method Sequential Minimal Optimization (SMO) instead of Quadratic Programming (QP). For empirical evaluation, we performed the experiments on the CASAS smart home (Tulum 2009) real world activity dataset. Results show that our approach yields a significant improvement in the accuracy as compared to the single kernel function.

We structure our paper as follow: Section 2 provides information about some of the existing approaches for human activity recognition. Section 3 presents our proposed approach for activity recognition in smart homes. In Sect. 4, we illustrate the experimental results followed by comparison and discussion. And finally the conclusion and future work are drawn in Sect. 5.

2 Related Work

There have been a number of machine learning methods for recognizing activities, such as Markov models [8], dynamic Bayesian network [9] and frequent activity patterns methods [10]. Rashidi et al. [10] track the regular activities to monitor functional health and to detect changes in an individual's patterns and lifestyle. They described activity mining and tracking approach based on Markov models and validates their algorithms on data collected in physical smart environments. Kasteren et al. [9] described the use of probabilistic model dynamic Bayesian network using less parameters to give better results. They showed how the use of an observation history of sensors increases the accuracy in case of the static model. Furthermore, the use of the observation history allows their model to capture more correlations in the sensor pattern. The authors in [11] proposed unsupervised approach to find frequent periodic activity patterns and resident implicit and explicit feedback can automatically update their model to reflect the changes. Their model did not take any assumptions into account about the activity structure, but leaves it completely to their algorithms and discover the resident's activity patterns in smart home environment.

Fusion techniques play an important role to achieve high accuracy as compared to single classifier and successfully produced more accurate results in different application domains such as image processing [12], gene functional classification [13]. In the context of sensory data, Xin at el. [14] address the fusion process of contextual information derived from the sensor data. They analyzed the Dempster-Shafer theory and merged with weighted sum to recognize the activities of daily living for inhabitants in smart homes. Their framework is capable to handle the sensors uncertainty and demonstrated the conjunction of sensors to support the reliable decision making. They demonstrate the concept of their work with the help of tutorial type exercise. They showed that the number of sensors and reliability of each sensor has a significant impact on the overall results in term of decision making.

In the previous works, it has been shown that many machine learning algorithms are proposed to recognize the daily life activities in the Smart homes. Our objective is to utilized theoretical rich model (i.e., SVM) and combines fusion technique to provide more robust and accurate activity recognition approach. We believe that, nature of the problem can be solved more precisely by fusion techniques along with SVM method.

3 The Proposed Approach

In our proposed approach, four individual SVM kernels are designed to use the preprocessed annotated smart home dataset in parallel. The output of each SVM kernel is a decision about the activity class label. In order to increase the confidence of final predicted class, decision fusion technique is applied to recognize the human daily life activities with more accuracy. The architecture of proposed approach is shown in Fig. 1 and details as follow.

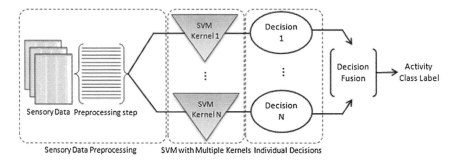

Fig. 1 The proposed architecture

3.1 Sensory Data Preprocessing

Ubiquitous sensors in the smart home continuously sense the environment and generate signals according to the subject interactions. Log files are maintained with attributes *start time, end time, sensor id* and *sensor value*. In order to recognize the performed activities, recorded dataset is preprocessed into the form of $\{(x_1, y_1), \ldots, (x_n, y_n)\}$. The x_i.values are the vectors of form $(x_i, 1, x_i, 2, \ldots, x_i, m)$ whose components are embedded sensors such as stove-sensor, refrigerator-sensor, and door-sensor. The values of "y" are drawn from a discrete set of classes $\{1, \ldots, K\}$ as a "Meal preparation", "Cleaning", "Laundry" and so on.

3.2 SVM with Multiple Kernels

Given training set of sensors value and activity pairs [i.e., (x_i, y_i)], the SVM for the binary linear classification problem require the following optimization model including the error-tolerant margin.

$$Minimize \frac{1}{2} w^T w + C \sum_{i=1}^{n} \xi_i \tag{1}$$

$$Subject\ to : y_i\left(w^T x_i + b\right) \geq 1 - \xi_i,\ and\ \xi_i \geq 0 \tag{2}$$

Where "w" is a weight vector and b is bias. C is the error penalty and ξ_i are slack variables, measuring the degree of misclassification of the sample x_i. The maximum margin is obtained by minimizing the first term of objective function, while the minimum total error of all training examples is assured by minimizing the second term. Activity recognition is multi-class problem so we adopt "one-versus-one" approach for this purpose and trained the SVM through SMO for efficient performance [15]. The above optimization model can be simplified by using the Lagrangian multiplier techniques and Kernel functions:

$$Maximize(w.r.t\ \alpha) \sum_{i=1}^{n} \alpha_i - \frac{1}{2} \sum_{i=0}^{n} \sum_{j=1}^{n} \alpha_i y_i \alpha_j y_j K(x_i, x_j) \tag{3}$$

$$Subject\ to: \sum_{i=1}^{n} \alpha_i y_i = 0, 0 \le \alpha_i \le C \tag{4}$$

where K is the kernel function that satisfies $K(x_i, x_j) = \Phi^T(x_i)\Phi(x_j)$. It is a function that transforms the input data to a high-dimensional space where the separation could be linear [16]. SVM can provide a good generalization performance, but oftently far from the expected level of accuracy, due to their approximation algorithms and high complexity of data [17]. In order to achieve high accuracy in the classification, we trained following multiple kernels [15] and fused the individual results.

$$K(x_i, x_j) = x_i^T x_j \tag{5}$$

$$K(x_i, x_j) = (x_i^T x_j + 1)^p \tag{6}$$

$$K(x_i, x_j) = \exp\left(\frac{-\parallel x_i - x_j \parallel^2}{(2\sigma^2)}\right) \tag{7}$$

$$K(x_i, x_j) = \tanh(kx_i^T x_j - \delta) \tag{8}$$

Equations 4, 6, 7, and 8 shows the linear, polynomial, Gaussian (RBF) and multi-layer perceptron (MLP) kernel functions respectively. An "ideal" kernel function assigns a higher similarity score to any pair of objects that belong to the same class. In this problem domain, kernel function behaved differently on the performed activities due to complex situation that may arise because of so many ways of doing an individual activity. The level of accuracy for performed activities is dependent on the different kernel function and discussed more in Sect. 4.

3.3 Decision Fusion

The output of the SVM kernels are the class labels and it has very clear inference about the activity. Each kernel sources are independent of each other and contain the class label about the certain activity. Before assigning the final class label, we get the confidence with the help of max rule as below:

$$Activity\ Label = argmax\left(\sum_{i=1}^{4} SVM(K_i)\right) \tag{9}$$

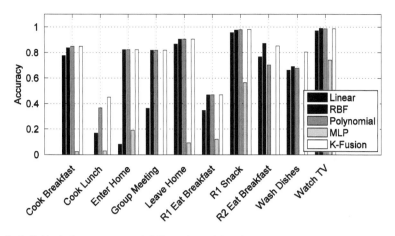

Fig. 2 Individual class accuracy of different kernel functions

4 Evaluation and Results

We performed experiments over the CASAS smart home dataset (*Tulum 2009*) collected in a Washington State University family apartment with full-time residents [4]. Twenty different kinds of temperature and motion sensors are deployed at various locations. Two volunteer was performing the common house hold activities during the 83 days stay. In that time, volunteer's annotated activities are preparing breakfast, lunch, snack, wash dishes, watch television, group meeting, and leave home. During our experiments, we recognized activities of both volunteers according to uniquely assigned labels in the annotated file. The approach has been implemented in MATLAB 7.6. The configuration of the computer is Intel Pentium(R) Dual-Core 2.5 GHz with 3 GB of memory and Microsoft Window 7. We used 70 % of dataset for training and 30 % data for testing as an evaluation criterion for recognizing the daily living activities. Figure 2 shows the average accuracy of kernel fusion in comparison to recognition rate as compared to the individual kernel functions.

In Fig. 2, different kernel functions show the performance variations for the recognition of same activities. The overall performance of "MLP" is low as compared to other individual kernel functions. "RBF" usually perform well in most of the cases but "kernel fusion" outperforms all individual kernel methods. In the proposed method, fusing the individual decisions, strengthen the confidence to assign final activity class label. Although, "RBF" is better than "kernel fusion" in case of "*R2 Eat Breakfast*" activity but its overall accuracy, 90.41 %, is significantly better than "RBF" and other individual kernel functions as shown in Table 1.

Table 1 illustrates the accuracies achieved by different kernel functions along with our proposed kernel fusion method. We analyze from our experiments nature of the different activities differ in different situations and may learn properly over

Table 1 Overall kernel functions accuracy

Kernels	Accuracy (%)
Linear kernel	81.03
RBF kernel	88.89
Polynomial kernel	88.49
MLP kernel	46.33
Kernel fusion	90.41

the different kernel functions. A significant improvement is achieved by fusing the individual kernel outputs.

5 Conclusion and Future Work

Although a lot of work has been done to recognize daily life activities but few methods investigated fusion techniques that can help to improve the accuracy. In this paper, we investigate the fusion of kernel methods to overcome the learning effects of different kernel function for individual activities. We proposed a method to train independent kernel functions with SVM and the decision of each individual's is combined at decision level by max rule. Our study found that it increases the overall accuracy of recognized activities and also performs well in case of those activities where single kernel methods suffer from accuracy problems. To investigate further, we intend to define our own activity domain specific kernel function to refine the accuracy rate.

Acknowledgments This research was supported by the The Ministry of Knowledge Economy (MKE), Korea under the Information Technology Research Center (ITRC) support program supervised by the National Industry Promotion Agency (NIPA) [NIPA-2010-(C1090-1021-0003)].

References

1. Cook, D.: CASAS smart home project [Online]. http://www.ailab.wsu.edu/casas/ (2012). Accessed 10 June 2012
2. Jones, B.: Aware home [Online]. http://awarehome.imtc.gatech.edu/ (2012). Accessed 10 June 2012
3. Mozer, M.C.: The adaptive house [Online]. http://www.cs.colorado.edu/~mozer/index.php?dir=/Research/Projects/Adaptive (2012). Accessed 10 June 2012
4. Larson, K.: House_n [Online]. http://architecture.mit.edu/house_n/ (2012). Accessed 10 June 2012
5. Kasteren, T., Noulas, A., Englebienne, G., Ben, K.: Accurate activity recognition in a home setting. In: UbiComp, pp. 21–24 (2008)
6. Rashidi, P., Cook, D.J.: Activity recognition based on home to home transfer learning. In: 24 AAAI Conference on Artificial Intelligence, Georgia, USA (2010)

7. Xin, H., Chris, N., Maurice, M., Sally, M., Bryan, S., Steven, D.: Evidential fusion of sensor data for activity recognition in smart homes. J. Elsevier Pervasive Mob. Comput. **5**, 236–252 (2009)

8. van Kasteren, T.L.M., Englebienne, G., Kröse, B.J.A.: An activity monitoring system for elderly care using generative and discriminative models. J. Pers. Ubiquitous Comput. **14**, 489–498 (2010)

9. van Kasteren, T.L.M., Krösem, B.J.A.: Bayesian activity recognition in residence for elders. In: International Conference of Intelligent Environments, pp. 209–212 (2007)

10. Rashidi, P., Cook, D.J.: Mining and monitoring patterns of daily routines for assisted living in real world settings. In: The Proceedings of ACM International Health Informatics Symposium, pp. 336–345 (2010)

11. Rashidi, P., Cook, D.J., Schmitter, E.M.: Discovering activities to recognize and track in a smart environment. IEEE Trans. Knowl. Data Eng. **23**, 527–539 (2011)

12. Wang, Q., Shen, Y.: The effects of fusion structures on image fusion performances. In: Proceedings of the 21st IEEE Instrumentation and Measurement Technology Conference, pp. 468–471 (2004)

13. Chen, B., Sun, F., Hu, J.: Local linear multi-SVM method for gene function classification. In: World Congress on Nature and Biologically Inspired Computing, pp. 183–188 (2010)

14. Xin, H., Chris, N., Maurice, M., Sally, M., Bryan, S., Steven, D.: Evidential fusion of sensor data for activity recognition in smart homes. J. Elsevier Pervasive Mob. Comput. **5**, 236–252 (2009)

15. Scholkopf, B.: Advances in kernel methods: support vector learning, ISSBN: 9780585128290. MIT Press, Cambridge (1999)

16. Fleury, A., Noury, N., Vacher, M.: Supervised classification of activities of daily living in health smart homes using SVM. In: Proceeding of 31st International Conference of the IEEE Engineering in Medicine and Biology Society, pp. 2–6 (2009)

17. Li, M., Yang, J., Hao, D., Jia, S.: ECoG recognition of motor imagery based on SVM ensemble. In: Proceedings of the IEEE International Conference on Robotics and Biomimetics, pp. 1967–1972 (2009)

Secure Anonymous Conditional Purchase Order Payment Mechanism

Wei-Chen Wu and Horng-Twu Liaw

Abstract In this paper, an Anonymous Conditional Purchase Order (ACPO) payment system is proposed which is an electronic payment system suitable for the real world. It not only keeps the high efficiency of previous buyer-driven systems but also strengthens the anonymity between customers and banks. The customers can advertise and put ACPO on the web according to their requirements, and then the merchants can evaluate which ACPO is appropriate for trade in any particular web site. Besides, timestamp will solve the problem of non-dispute and make our scheme more flexible. For example, customers can modify their ACPO to suit the merchant if they are not getting any response from the merchant with whom they want to trade.

Keywords Electronic payment system · Anonymity · Buyer-driven

1 Introduction

The definition of the electronic payment system is for it to provide a secure environment which satisfies the requirements of an electronic payment system, including privacy, authentication, non-dispute, integrity, auditability, efficiency,

W.-C. Wu (✉)
Computer Center, Hsin Sheng College of Medical Care and Management, Taoyuan, Taiwan, Republic of China
e-mail: wwu@hsc.edu.tw

H.-T. Liaw
Department of Information Management, Shih Hsin University, Taipei, Taiwan, Republic of China
e-mail: htliaw@cc.shu.edu.tw

S.-S. Yeo et al. (eds.), *Computer Science and its Applications*,
Lecture Notes in Electrical Engineering 203, DOI: 10.1007/978-94-007-5699-1_30,
© Springer Science+Business Media Dordrecht 2012

availability, and reliability. Now, more and more technology industry, financial institutions, academic research centers, and governments are devoted to research of the electronic payment system. Up to the present, there are several famous electronic payment systems such as SET, CAFE Ticks, PayWord, iKP, Millicent, and Mondex [1–3, 4]. The advantages, disadvantages, suitable conditions, roles of the transaction, and requirements of each system are respectively different. Recently, Kelsey et al. [5] and Foo et al. [6] proposed a Conditional Purchase Orders (CPO) payment system that created a buyer-driven system of commerce. Buyers can post their requirements, much like classified advertisements, and sellers can evaluate these requirements and decide whether or not to fulfill them. Moreover, the system provides a mechanism to bind the buyer to a seller once the seller meets those requirements: a buyer cannot renege on a deal once a buyer has accepted unless the buyer fails to meet the stated requirements. Relative to the traditional payment, it is not only a buyer-driven system but handled that what goods and prices by merchant.

In 1983, Chaum [7] introduced the first blind signature scheme to ensure that the user's private information would not be revealed when he was proceeding with casting or purchasing over the Internet that allows the spender to spend electronic cash anonymously. The system provides some levels of anonymity against a collaboration of shops and banks. Such signatures require that a signer to sign a message without knowing its contents. Electronic cash systems, which use the blind signature scheme, require many public keys corresponding to the face-value of coin. To decrease the number of public keys, the partially blind signature scheme [8] were developed.

The problems of electronic payment systems are similar to those of ordinary payment systems in the world, which include counterfeit money, money laundry, and pocketing monies. Furthermore, new types of problems occur with electronic payment systems, including double spending, malicious tracing of customers, and doing customers an injustice. Malicious persons can successfully speculate in dishonest transactions because that merchant and the customer aren't trading face to face; hence, how to design a perfect electronic payment system, which avoids dispute after transaction and ensures the rights and responsibilities of both sides, becomes more and more important. The rest of this paper is organized as follows: in Sect. 2, we discuss previous work on this subject. Section 3 proposes an Anonymous Conditional Purchase Order (ACPO) payment system based on the secure blind signature scheme. Finally, concluding remarks are given in Sect. 4.

2 Previous Work

The concept of the electronic payment system based on Conditional Purchase Orders (CPO) was first introduced by Kelsey et al. [5]. A CPO is posted by the customer and placed on a public FTP server or WWW server. Then potential products' suppliers (merchants) can browse these CPOs on the Internet and bind the CPO with which he wants to deal. Binding conditional purchase orders this way can make product suppliers and customers transform transaction messages into a signed commitment efficiently.

Foo et al. [6] modify the scheme proposed by Kelsey et al. [5] and introduce the concept of passive entities into the electronic payment system. In general, the payment system would consist of five phases as follows: the registration phase, withdrawal phase, negotiation phase, payment phase and the goods delivery phase. Passive entities transmit the messages without executing any encrypted and decrypted processing during the passive phase of the protocol. For this reason, the electronic payment system with the concept of the passive entity will be more efficient. The following is the definition of passive entities:

- Passive entities don't execute any encryption or decryption during the passive phase of the protocol.
- Passive entities must transmit or receive messages with other relative entities. In other words, passive entities must communicate with other participants during the protocol phase.
- Before the payment phase, the relative entities must agree with their commitments, which they made previously.

Our scheme proposes an efficient fair blind signature scheme which makes it possible for a government or a judge to recover the link between a signature and the instance of the signing protocol which produces that signature when the unlink ability property is abused. Hence, the preparation phase of our scheme is the same as that of mentioned that is based on the theory of quadratic residues [9–12]. Under a modulus n, an integer x is a quadratic residue (QR) in Z_n^* if and only if there exists an integer y in Z_n^* such that $y^2 \equiv x \pmod{n}$ where Z_n^* is the set of all positive integers which are less than and relatively prime to n. Given x and n, it is computationally infeasible to derive y if n contains large prime factors and the factorization of n is unknown [10].

3 Anonymous Conditional Purchase Order Scheme

We propose an Anonymous Conditional Purchase Order (ACPO) payment system which assumes that in the system environment of our proposed scheme exists a secure public key signature scheme and a public key cryptosystem and works perfectly based on the secure blind signature scheme [8]. The proposed scheme consists of four participants, four assumptions, and six phases. All participants must register with the certificate issuer, and acquire the certificate of form X.509 V3, both of which are issued by the publisher of the certificate. In addition, the bank will keep records of all transaction commitments that the participants produce in each phase.

The underlying assumptions of our proposed scheme are: (1) All customers and merchants have opened a bank account; (2) There exists a secure public key signature scheme and a public key cryptosystem such as RSA algorithm [13, 14]; (3) There exists a secure anonymous channel; (4) There exists a secure blind signature scheme [8]; (5) There exists a secure hash function [14]. In addition, the used notations are described as follows:

Acc_C	The bank account of the customer.
$E_{Key}(Message)$	Message, encrypted by a public key.
$D_{Key}(Message)$	Message, decrypted by a private key.
ID_{Mi}	The ID number of ith merchant.
$S_{Key}(Message)$	Message, signed by a private key.
$V_{Key}(Message)$	Message, verified by a public key.
PK_C, RK_C	The customer's public key and private key.
PK_{CA}, RK_{CA}	The CA's public key and private key.
PK_B, RK_B	The bank's public key and private key.
PK_M, RK_M	The merchant's public key and private key.
$Cert_{CA}$	The CA's certificate.
$Cert_C$	The customer's certificate.
$Cert_M$	The merchant's certificate.
$Cert_B$	The bank's certificate.
$TimeS_i$	The ith timestamp is generated.

3.1 Preparation Phase

The aim of this phase is to create arguments for our proposed scheme. Firstly, the banks and the certification authority (CA) choose suitable arguments. Then all merchants that take part in our scheme must request a certificate from the CA. Finally, customers apply for the general certificate and the anonymous certificate from the CA. The banks and the CA need to create arguments and the procedures are described as follows:

(1) The bank publishes a one-way hash function H and its public key $n (= p_1 p_2)$, where p_1 and p_2 are large primes and $p_1 \equiv p_2 \equiv 3(mod\ 4)$.
(2) Similarly, the CA also publishes a random string $\bar{\omega}$, its public key $n'(= p_3 p_4)$ and $Cert_{CA}$ where p_3 and p_4 are large primes and $p_3 \equiv p_4 \equiv 3(mod4)$ and $n' = p_3 p_4 > n$.
(3) The bank prepares information for requesting a certificate *OtherRequestInfo*. Furthermore, it concatenates two primes with *OtherRequestInfo* and transmits a message which has been encrypted with the public key PK_{CA}.

$$PK_{CA}(p_1, p_2, OtherRequestInfo)$$

The encrypted message is sent to the certificate issuer. The CA confirms the message and responds sending certificate $Cert_B$ to bank.

(4) All merchants taking part in our scheme must apply for certificate $Cert_M$ from the CA.

The customers must apply for the general certificate and the anonymous certificate from the CA and the procedures are described as follows:

(1) The customers download the CA's certificate from the web site.
(2) The customers select a random integer r, prepare information for requesting certificate *OtherRequestInfo*, and transmit a message which has been encrypted with public key PK_{CA} and encoded with ASN.1 to the CA.
(3) After the CA receives the request message for the certificate, it calculates identification of the customer ID and anonymous identification of customer ID' with the following formulas.

$$ID = H(Acc_C||\varpi)2mod\ n\prime$$

$$ID\prime = H(ID||r)2mod\ n$$

Then the CA delivers the general certificate and the anonymous certificate to the customer.

$$(Cert_C(ID, r, Other\ Info), Cert'_C(ID', Other\ Info))$$

All of the above certificates are conformed to the X.509 V3 standard. The data *OtherInfo* contains the following fields: Certificate Version, Certificate Serial Number, Issuer Name, Subject Name, Subject Public Information, Extensions, and Signature. The anonymous certificate $Cert'_C$ doesn't include the privacy of the personal data, which is different from the general certificate. The general certificate $Cert_C$ is used for the payment phase in which the merchant can use the information of the general certificate to deal with the customer's order.

After registering with the CA, the customer will get the anonymous certificate and the general certificate. The information in the anonymous certificate is only used in the withdrawal phase and in the payment phase. The banks cannot obtain the user's personal information from the anonymous certificate.

3.2 Withdrawal Phase

To withdraw e-cash coins from the bank, the customer produces and sends a request message to the bank. The customer blinds the request message with blinding factor R_C which is randomly chosen by the customer. After the bank signs the blind coins $(C,R_B,R_C,info)$, the customer withdraws e-cash coins from the bank and uses them in the payment phase.

(1) Customer → Bank: The customer prepares the related parameters to calculate the blinding message $\propto$. The customer randomly chooses two integers R_C and M in Z_n^*. The message *info* includes the customer account number Acc_C, the unit of money w and the withdrawal date *Date*

$$\alpha = H(M)\, R_C^{H(info)}\, mod\, n$$

(2) Bank → Customer: The bank produces the private key x by the following formula.

$x = 1/F(H(info))\ mod\ LCM(p\text{-}1,q\text{-}1)$

F is the blinding function which used for building message.

Then the bank calculates the blinding signature s_b with the private key x. $s_b = \alpha^x\ mod\ n$ and sends the blinding signature s_b to the customer.

(3) Customer → Bank: After the customer receives the blinding signature s_b from the bank, the customer unblinds a signature s by the following formulas.

$$s = s_b/R_C mod\, n$$

The customer verifies the signature s and then sends s and *info* to the bank.

(4) Bank → Customer: After the bank verifies the signature s, the e-cash coins will be produced by the following formula.

$$C = H\big(info, R_B, H(M)\big)^{XC} mod\, n$$

The message R_B is an index for the relation of e-coins in the database. Then, the bank stores R_B and C into its database, and deducts w dollars from the account Acc_C. Then, the bank transmits R_B and C to the customer. After the above steps, the customer will obtain the valid coins $(C,R_B,R_C,info)$.

3.3 Order Posting Phase

In this phase, the customer will fill out the CPO and send the related message to the bank via an anonymous channel. Then the bank verifies the signature of the transaction message and produces a commitment to the customer order.

(1) Customer → Bank: Firstly, the customer will produce the related information (*OI*) of the CPO, including identification order information (*OID*), the information of an order for goods (*OINFO*), and the expiration date of the order (*EXPR*). The information *OI* provides product information and product conditions which the merchant can look up on the web site. The customer signs information *OI* and anonymous certificate (*Cert'_C*) with his public key and transmits an encrypted message to the bank through the anonymous channel.

$$E_{PKB} = \left(OI, Cert'_C, S_{RKC}(OI, Cert'_C)\right)$$

(2) Bank → Customer: After the bank receives message, the bank decrypts the message and verifies whether or not the customer's anonymous certificate is valid. If the certificate is valid, the bank will produce an order commitment to the customer and post the commitment on the web site, which is exclusively for the use of registered participants.

3.4 Order Binding Phase

All merchants taking part in our scheme must pay a service charge to the bank. Then they can use the binding function that the bank provides, including elements such as connection to the exclusive binding web site, browsing the CPOs, binding the CPOs, and looking up the CPOs on the web site. Once the merchant finds a CPO appropriate for trade, he must download the commitment which the customer produces from the order posting phase.

(1) Merchant↔Web: All merchants taking part in our scheme can browse the CPOs on an exclusive binding web site and look up the conditions in the CPO. If the merchant has the capability of providing a suitable product, he must download the commitment, described as follows:

$$\left(OI, Cert'_C, TimeS_1, S_{RKC}(OI, Cert'_C), S_{RKB}\left(S_{RKC}(OI, Cert'_C), TimeS_1\right)\right)$$

TimeS_1 is timestamp which is provided by the bank and ensures that the merchant doesn't download the commitment repeatedly.

(2) Merchant → Bank: The merchant has to verify that the commitments of the purchase order haven't been altered by a third party. This can be done in the following manner:

$$V_{PKB}\left(S_{RKB}\left(S_{RKC}(OI, Cert'_C), TimeS_1\right)\right)$$

Then the merchant produces the following message and sends it to the bank.

$$E_{PKB}\left(OI, Cert_M, S_{RKM}\left(S_{RKB}\left(S_{RKC}\left(OI, Cert'_C\right), TimeS_1\right), Cert_M, TimeS_2\right)\right)$$

According to OID in OI, the bank looks up $S_{RKB}(S_{RKC}(OI, Cert'_C), TimeS_1)$ in the database and verifies whether the conditions that the customer requested in the order posting phase are the same. If the merchant is suitable for the customer, the bank will suspend browsing the CPO.

(3) Bank → Merchant: After the bank verifies the above formula, the bank will check $S_{RKM}(S_{RKB}(S_{RKC}(OI, Cert'_C), TimeS_1), Cert_M, TimeS_2)$. If yes, the formula will become the commitment of the merchant and send the successful flag to the merchant and insert the commitment of the merchant into the CPO. Then the customer can inquire about his CPO during a valid period.

Our proposed scheme assumes that there exists a public key signature scheme and a public key cryptosystem. In addition, nobody can forge and steal the private key from the merchant. Under these circumstances, the merchant cannot disclaim any message that is signed with his private key.

3.5 Payment Phase

The aim of this phase is for the customer to pay the bill with e-cash, which is withdrawn from the bank in the withdrawal phase. The customer will pay the merchant for the desired product, which was chosen in the order binding phase.

(1) Customer → Merchant: After the customer receives the merchant's commitment, the customer will send the merchant's commitment $S_{RKB}(S_{RKC}(OI,-Cert'_C), TimeS_1)$, the e-cash $H(info, R_B, H(M))$, C, and the certificate $Cert_C$ to the merchant.

(2) Merchant → Customer: The merchant will check whether the received commitment is the same as the commitment which the merchant had downloaded in the order binding phase. If the two commitments are the same, the merchant will use the following formula to check whether the e-cash is valid. $C^{ec} = H(info, R_B, H(M)) mod\ n$, finally, the merchant will deliver the product to the customer according to the personal information given in the certificate.

(3) Merchan → Bank: If the merchant wants to deposit e-cash into the bank, the merchant must send the e-cash to the bank. Then the bank will verify the validity and correctness of the money. After finishing the verification, the bank will transfer the total amount of the e-cash to the merchant's account.

(4) Merchant → Web: The merchant puts the information product, which encrypted by a public key Key which exist in $Cert'_C$, on an exclusive binding web site.

(5) Customer → Web: The Customer downloads the information product on an exclusive binding web site and decrypts it use his private key.

Table 1 A comparison of buyer-driven systems

Requirement	Foo et al. [6]	Kelsey et al. [5]	Our scheme
Anonymity (Bank)	No	No	Yes
Anonymity (Merchant)	Yes	No	Yes
Confidentiality	Yes	Yes	Yes
Authentication	Bank + Sign	Bank + Sign	CA + PKI
Non-Dispute	Yes	Yes	Enhanced
Integrity	Yes	Yes	Yes
Auditability	Yes	Yes	Yes
Efficiency	Yes	Yes	Enhanced
Availability	No	No	Yes

The customers transmit the CPO to the bank through the anonymous channel, hence the bank cannot trace where the customer comes from or learn the customer's information.

4 Concluding Remarks

In this paper, we propose an Anonymous Conditional Purchase Order (ACPO) payment system, which is an electronic payment system suitable for the real world. Table 1 shows the comparison of our scheme with those proposed by [5, 6]. In future, we will address ourselves to designing a payment system, which enables multi-merchants to deal with one CPO. It will solve the problems of red tape, with enterprise and government bids, and of the high transaction costs to buyers and sellers.

References

1. Hauser, R., Steiner, M., Waidner, M.: "Micro-payments based on ikp. In: Proceedings of the 14th Worldwide Congress on Computer and Communications Security Protection, pp. 67–82 (1996)
2. Pedersen, T.: Electronic payments of small amounts. In: Proceedings of the Cambridge Workshop on Security Protocols, Lecture Notes in Computer Science 1189. Springer-Verlag, Berlin, pp. 59–68 (1997)
3. Rivest, R.L., Shamir, A.: PayWord and MicroMint: two simple micropayment schemes. MIT, Cambridge (1996)
4. LeVeque, W.J.: Fundamentals of number theory. Addison-Wesley, Reading (1997)
5. Kelsey, J., Schneier, B.: Conditional purchase orders. In: Proceedings of the 4th ACM conference on computer and communications security (CCS '97). ACM, New York, pp. 117–124 (1997)
6. Foo, E., Boyd, C.: Passive entities: a strategy for electronic payment design. In: ACISP 2000. Lecture Notes in Computer Science 1841. Springer-Verlag, Berlin, pp. 134–148 (2000)

7. Chaum, D.: Blind signatures for untraceable payments. Advances in Cryptology—CRYPTO'82. Lecture Notes in Computer Science. Springer-Verlag, Berlin, pp. 199–203 (1983)
8. Abe, M., Fujisaki, E.: How to date blind signatures. Advances in Cryptology—Proceedings of ASIACRYPT'96. Lecture Notes in Computer Science 1163. Springer-Verlag, Berlin, pp. 244–251 (1996)
9. Menezes, A.J., Oorschot, P.C., Vanstone, S.: Handbook of applied cryptography. CRC Press, Boca Raton (1996)
10. Peralta, R.C.: A simple and fast probabilistic algorithm for computing square roots modulo a prime number. IEEE Trans. Inf. Theory 32, 846–847 (1986)
11. Rabin, M.O.: Digitalized signatures and public-key functions as intractable as factorization. Technical Report, MIT/LCS/TR212, MIT Lab., Computer Science (1979)
12. Diffie, W., Hellman, M.: New directions in cryptography. IEEE Trans. Inf. Theory 22(6), 644–654 (1976)
13. Rivest, R.L., Shamir, A., Adleman, L.: A method for obtaining digital signatures and public-key cryptosystems. Commun. ACM 21(2), 120–126 (1978)
14. Rivest, R.L.: The MD5 message-digest algorithm. RFC 1321, IETF (1992)

Trustworthiness Inference of Multi-tenant Component Services in Service Compositions

Hisain Elshaafi and Dmitri Botvich

Abstract This paper presents a novel approach to the inference of trustworthiness of individual components shared between multiple composite services in distributed services environments. In such environments, multiple component services are orchestrated from distributed providers to create new value-added services. A component service can be shared by several distributed compositions. A composite service is offered to its consumers who rate its reliability and satisfaction after each transaction. However, since composite services are provided as an integrated service it is not possible to attribute failures or causes of dissatisfaction to individual components in isolation. A collaborative detection mechanism can provide a solution to the evaluation of component trustworthiness based on consumer reporting of composite service execution results.

Keywords Trustworthiness · Reliability · Reputation · Composite service

1 Introduction

Service Oriented Computing is increasingly popular, with increased attention from industry. A key concept is that services can be composed into business processes to create new higher level services. The paradigm enables flexible use of resources for optimising operations within and across organisations.

H. Elshaafi (✉) · D. Botvich
Telecommunications Software and Systems Group, Waterford Institute of Technology, Waterford, Ireland
e-mail: helshaafi@tssg.org

D. Botvich
e-mail: dbotvich@tssg.org

S.-S. Yeo et al. (eds.), *Computer Science and its Applications*,
Lecture Notes in Electrical Engineering 203, DOI: 10.1007/978-94-007-5699-1_31,
© Springer Science+Business Media Dordrecht 2012

In the distributed services environments, services are offered for consumers either directly or as part of *composite services* (*CSs*). Therefore, a component service may simultaneously exist and can be invoked in multiple CSs. Meanwhile, a CS may contain several component services. This two-directional relationship although may cause complexity in the management of those services, it can also provide benefits as well. One such benefit to the operation of CSs is the correlation between the reliabilities of CSs sharing one or more components in order to infer the reliability of their components. Consequently, the CSs optimise their trustworthiness through the awareness of component reliabilities during selection and afterwards via runtime adaptation by replacing unsatisfactory components.

The trustworthiness of a service relies on multiple properties of that service. For example, authentication is a necessary step in establishing trust as it assures the consumer that the service provider is who it claims it is. However, an authenticated service may not behave in the way it is required or expected in terms of reliability, reputation, etc. We define trust and trustworthiness as follows. *Trust* is a relationship between two or more entities that indicates the contextual expectations from an entity towards another in relation to reliance in accomplishing a certain action at a certain quality. *Trustworthiness* of an entity is the level of trust that the trusting entity or its agent has in that entity. These definitions are in line with existing approaches considering trust as multidimensional concept, such as in [1–3].

Reliability is one of the important properties in trustworthy services. The reliability of CSs depends on how reliable is each component service. We define reliability of a service as the percentage of its successful completion of executions. Reputation is the data available about a service from consumer satisfaction ratings that can be used together with other types of data in determining the service's trustworthiness. Component services in a composition may vary in their importance to the CS as a whole. For example, in a travel service a user may not appreciate all component services to the same extent such as car rental, health insurance, flight booking, etc. Therefore, a CS is a unit composed of unequal subunits in terms of their contribution to the trustworthiness of the CS. In this paper we provide an approach to inferring the reliabilities and consumer satisfaction (reputation) of individual component services based on the reliability and satisfaction of composite services sharing component services to varying degrees and in multiple CS process structures.

The paper is organised as follows. Related work is described in Sect. 2. A motivating scenario is discussed in Sect. 3. Section 4 describes the requirements of inference operations of trustworthiness properties for components from those of CSs. The aggregation of component trustworthiness properties is discussed in Sect. 5. Section 6 describes the proposed framework and the procedure of the trustworthiness inference. Conclusions and future work are discussed in Sect. 7.

2 Related Work

To our knowledge there is no previous work on inferring the reliability of component services based on the perceived reliability of their CSs. Techniques of aggregation of Quality of Service (QoS) including reliability in some BPEL and OWL-S supported workflow constructs were described by Hwang et al. [4]. Cardoso et al. [5] present a predictive QoS model to compute the reliability and other QoS attributes for workflows automatically based on atomic task QoS. Hwang et al. [6] propose a solution to the dynamic Web service selection problem in a failure-prone environment. It aims to determine a subset of Web services to be invoked at runtime that can successfully orchestrate a CS based on aggregated reliabilities. Grassi and Patella [7] present reliability prediction mechanisms that aim to satisfy the requirements for decentralization and autonomy. They propose mechanisms to recursively aggregate the reliability of a CS based on those of its components.

Nepal et al. [8] present an approach to the propagation of reputation values from composite to component services that aims to fairly distribute those values. However, it does not consider the distributed collaboration proposed in our work. Additionally, the reputation of a CS as presented is considered the average of that of its component services. This can cause inaccuracies since the resulting reputation value can provide a camouflage to low reputation components and cause frequent bad consumer ratings for a seemingly high reputation CSs.

The terms *trust* and *trust models* are used in Web service standards (e.g. WS-Trust [9]) but they are limited to the context of verifying the identity of a service [2]. However, establishing a service's identity does not necessarily mean that the service is trustworthy. For instance, an authenticated service could be temporarily unreliable or unavailable. Takabi et al. [3] discuss the barriers and possible solutions to providing trustworthy services in cloud environments. They describe the need by multiple service providers to collaborate and compose value-added enterprise services. They propose that a trust framework should be developed to allow for efficiently capturing parameters for establishing trust and to manage evolving trust requirements.

Our previous work [10] discusses trustworthiness in CSs. It describes a trustworthiness monitoring and prediction software module that receives data on service properties including QoS, consumer ratings and security events. The data, which indicates violations or adherence to contracts, is used to predict service trustworthiness. In [11] we describe the aggregation techniques for several trustworthiness properties of composite services that are assembled from more fine-grained component services.

3 Motivating Scenario of Multi-tenant Component Services

Figure 1 shows an illustrative example using BPMN notations [12] for two e-commerce product purchasing CSs. The CSs are offered to consumers by orchestrating distributed component services. The CSs contain eleven tasks of the workflow that can be fulfilled by component services. The service labels are generally self-explanatory. For example, 'Product Specification' is a service provided by one component vendor for multiple abstract categories of products depending on the CS. CS1 differs from CS2 in that it supports currency conversion and debit payment and allows choice between direct buying or bidding on prices. In CS2, consumers can choose to buy by credit payment service requiring other component services as well for applying for a loan and checking consumer credit history and loan rate. Alternatively, the consumer can purchase the product through a debit payment. The component services are invoked in business processes with multiple workflow constructs as discussed in Sect. 5 and indicated in the figure including Sequence, Synchronized Parallel and Exclusive Choice. The structure of the workflows affects the operation of the CS including the prediction of its reliability and the inference of component reliabilities.

The CS interacts with the consumers in a black-box fashion and the component providers are invisible to the consumers. The services are implemented using a BPMN based process with a Web front-end for the consumers that collects their data and submits it to the BPMN execution engine for processing orders. An example of such engine is Activiti [13]; a BPMN 2 process engine for Java. In our scenario, the engine provides interfaces to receive reliability and satisfaction rating by consumers.

4 Requirements of Trustworthiness Update for Components

An approach to determine the trustworthiness of component services based on their CSs has to strive to meet certain requirements in order to be usable in the services environments. These requirements include the following:

- *Fairness.* The varying contribution of components to the trustworthiness of the CSs presents a challenge to fairly distribute the consumer ratings to individual components. The trustworthiness update mechanisms must depend on a variety of information to identify a low performing service. Such information may include the importance of each component in a CS, the fluctuation of consumer ratings and multiple sources of ratings for the component e.g. multitenant component service.
- *Protection from Threats.* The approach needs to be robust and aware of current threats that exist as a result of vulnerabilities in the trustworthiness prediction mechanisms especially when monitoring of the behaviour of individual components is not possible. Examples of possible threats include; *Free-riding* where

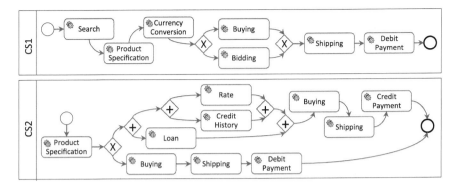

Fig. 1 Example e-commerce scenario composite services

a component service unjustifiably benefits from the high trustworthiness ratings of a CS, and *Camouflaging* in which low trustworthiness of a component is hidden through insertion into a trustworthy CS. Research on threats to trust and reputation systems also covers some other relevant threats such as in [14, 15].

- *Reusability.* The distribution of the trustworthiness updates to components results in reliable component trustworthiness levels that can be used again to compute trustworthiness of existing or new composite services.

5 Composition Constructs and Aggregation of Trustworthiness

5.1 Composition Constructs

Component services are executed in a business process. The process is viewed externally as a Web service. A CS consists of one or more path constructs. Each construct contains one or more service activities. A component service is selected for each activity. The following are common constructs (illustrated in Table 1):

- *Sequence.* Services are invoked one after another.
- *Synchronised Parallel (AND split/AND join).* Two or more services are invoked in parallel and their outcome is synchronised. All services must be executed successfully for the next task (service) to be executed.
- *Loop or Iteration.* A service is invoked in a loop until a condition is met. We assume that the number of iterations or its average is known.
- *Exclusive Choice (XOR Split/XOR join).* A service is invoked instead of others if a condition is met. We assume that the likelihood of each alternative service to be invoked is known.

Table 1 Aggregation of reliability and reputation per process construct

Construct		Reliability	Reputation
Sequence		$\prod_{i=1}^{n} r_i$	$\prod_{i=1}^{n} p_i^{\omega_i}$
Synchronized parallel		$\prod_{i=1}^{n} r_i$	$\prod_{i=1}^{n} p_i^{\omega_i}$
Loop		r_i^n	$p_i^{n \cdot w_i}$
Exclusive choice		$\sum_{i=1}^{n} \rho_i \cdot r_i$	$\left(\sum_{i=1}^{n} \rho_i \cdot p_i \right)^{\omega_i}$
Unsynchronized parallel		$1 - \prod_{i=1}^{n}(1 - r_i)$	$\prod_{i=1}^{n} p_i^{\omega_i}$
OMulti-choice sync. merge		$\sum_{j \subset S} \left(\rho_j \cdot \prod_{i \in j} r_i \right)$	$\sum_{j \subset S} \left(\rho_j \cdot \prod_{i \in j} p_i^{\omega_i} \right)$
Unordered sequence		$\prod_{i=1}^{n} r_i$	$\prod_{i=1}^{n} p_i^{\omega_i}$

n = no. of construct components; ρ_i = probability of execution of component v_i; ρ_j = probability of execution of subset j of construct components

- *Unsynchronised Parallel* (*AND split/OR join*). Two or more services are executed in parallel but no synchronisation of the outcome of their execution
- *Multi-choice with Synchronised Merge* (*OR split/AND join*). Multiple services may be executed in parallel. Subsequent services can be executed when all executions are completed
- *Unordered Sequence*. Multiple services are executed sequentially but arbitrarily

We use θ to denote a service construct in a composition. In BPMN [12], AND join/split gateway is signified with '+', OR with 'O' and XOR with '×'. An empty gateway '◇' means it waits for one incoming branch before triggering the outgoing flow. More complex patterns are supported by modelling languages and products to varying degrees. The types of workflow constructs are investigated in other works such as Workflow Patterns Initiative [16].

5.2 Reliability

1. *Sequence, Synchronized Parallel and Unordered Sequence.* A failure of a component means failure of subsequent dependent components. This is unlike some other types of constructs (e.g. Unsynchronized Parallel) where subsequent

components may be partially independent of the failure of the construct components and can be executed if a minimum set of components succeeds. Therefore, we calculate the reliability of these constructs as a product of that of its components.

$$r_\theta = \prod_{i=1}^{n} r_i \tag{1}$$

2. *Loop.* The reliability of a Loop containing n iterations of a service v_i is the same as a Sequence construct of n copies of v_i i.e. $r_\theta = r_i^n$.
3. *Exclusive Choice.* The construct reliability is the sum of that of the exclusive components multiplied by their probabilities of execution in the CS.
4. *Unsynchronized Parallel.* Since an Unsynchronized Parallel construct only fails if all constituent services fail, its reliability is calculated as follows:

$$r_\theta = 1 - \prod_{i=1}^{n} (1 - r_i) \tag{2}$$

5. *Multi-choice with Synchronized Merge.* In each subset of components that may be executed in parallel, all its components must be executed successfully. Therefore, like in the case of reputation of this construct we sum the probabilities of each subset multiplied by the reliability of that subset. In a construct θ with a set S of components and two or more probable subsets of components that may be executed in parallel, its reliability is calculated as follows:

$$r_\theta = \sum_{j \subset S} \left(\rho_j \cdot \prod_{i \in j} r_i \right) \tag{3}$$

5.3 Reputation

5.3.1 Importance Weight of Components

Each component v_i in a composition has a weight ω_i based on its importance to the reputation of the composition $\omega_i \in \{\omega_1, \ldots, \omega_l\}$ where $0 \leq \omega_i \leq l$, l is the number of component services excluding alternatives in exclusive choice constructs and $\sum_{i=1}^{l} \omega_i = l$. We consider components in an exclusive choice construct as a single unit in terms of their weight ω_θ and its calculation, where θ is the service construct. For example, consider the case where a requirement may be satisfied by only one of two services $\{v_1, v_2\}$ and the trustworthiness of v_1 is more than that of v_2 but its capacity is limited to a certain quantity. When v_1 becomes fully in use, v_2 is invoked. Therefore, a common weighting value is used.

The weighting of the components is used in the calculation of the CS reputation, such as in the case of a Sequence with n components:

$$p_\theta = \prod_{i=1}^{n} p_i^{\omega_i} \tag{4}$$

5.3.2 Aggregation of Reputation Per Construct

1. *Sequence, Synchronized Parallel and Unordered Sequence.* The reputation is calculated as a product of that of constituent services taking the service importance into consideration as in Eq. (4).
2. *Loop.* The reputation of a Loop containing n iterations of a service v_i is the same as a Sequence of n copies of v_i i.e. $p_\theta = p_i^{n \cdot w_i}$.
3. *Exclusive Choice.* Each service v_i among the alternative services in an Exclusive Choice has a probability ρ that it will be executed and $\sum_{i=1}^{n} \rho_i = 1$. As described earlier an Exclusive Choice is considered one unit in the CS component reputation weights. Therefore, we aggregate the reputation of the construct as the sum of the component reputations multiplied by their probabilities of executions.

$$p_\theta = \sum_{i=1}^{n} (\rho_i \cdot p_i)^{\omega_i} \tag{5}$$

4. *Unsynchronized Parallel.* Since all component services are executed, the reputation takes all services into consideration as in Eq. (4).
5. *Multi-choice with Synchronized Merge.* In this construct, the execution of each subset j of all possible subsets of the set S of construct services ($j \subset S$) is associated with a probability ρ_j that it will be executed where $\sum_{j \subset S} \rho_j = 1$. The construct reputation considers both the probability of execution and weighting of the component services:

$$p_\theta = \sum_{j \subset S} \left(\rho_j \cdot \prod_{i \in j} p_i^{\omega_i} \right) \tag{6}$$

6 Component Trustworthiness

6.1 Proposed Framework

Figure 2 illustrates our proposed framework for the exchange of trustworthiness data between CS providers. The CSs are also shown in Fig. 4. The CSs share component services which are orchestrated by each CS provider and executed by its *Execution Management* module. The selected components for each CS are indicated by the corresponding surrounding dotted circle. Consumers can execute a CS and submit a trustworthiness report that indicates either success or failure of the execution and their satisfaction level. Each CS has a *Trustworthiness Agent* that exchanges trustworthiness data with corresponding agents and computes component trustworthiness.

An execution of a CS either succeeds or fails and provides variable satisfaction to the consumer. Therefore, following each execution, the CS is given a reliability rating $R_k \in \{0, 1\}$ and a satisfaction rating $0 \leq P_k \leq 1$, where k is the ratings index; $1 \leq k \leq m$. CS reliability is measured through the number of successful executions to the total executions m over a specified period of time. For example, 50 failures out of 1000 executions in the specified period mean that the CS reliability is 0.95. The failure indicates that a component service has failed. Since components are shared between CSs we use the bidirectional relationship to infer the reliabilities and reputations of components.

Figure 3 describes the procedure for updating the trustworthiness of a CS and its components after a consumer request. A failure or low reputation reported by a consumer triggers collection of trustworthiness statistics from peer CS providers. The statistics, which contain the CSs trustworthiness levels and component graphs, are used to compute components reliabilities and reputations and detect the component causing the failure or bad reputation.

6.2 Example Computation of Component Trustworthiness

Consider the four simple CSs illustrated in Fig. 4 as an example which share multiple components each indicated by a number. This example can be generalised for different constructions of CSs, trustworthiness properties and varying levels of component sharing. All CSs in Fig. 4 consist of only Sequence components. The reliability r_{csj} of a composite service cs_j with n components can be calculated as product of that of its components as in Eq. (1) where j is a numeric identifier for the CS. The failure of a component results in the failure of the whole CS because of the dependencies between the components. The following are the equations that determine the reliability of each CS in the figure:

$$r_{cs1} = r_1 \cdot r_2 \cdot r_3$$

$$r_{cs2} = r_1 \cdot r_4$$

$$r_{cs3} = r_2 \cdot r_4$$

$$r_{cs4} = r_1 \cdot r_3 \cdot r_4$$

For instance, suppose that according to the consumer reports, we have the following reliability scores for CS1 to CS4 respectively; $\{0.78, 0.98, 0.99, 0.82\}$.

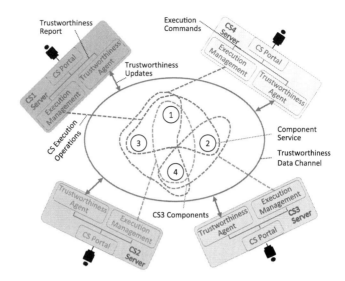

Fig. 2 Trustworthiness data exchange framework

Fig. 3 Procedure of
determining failed
components

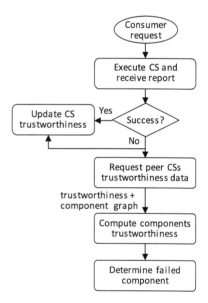

Now we need to know the reliability of each of the shared components. In the case
of these CSs, we have four equations and four variables. Therefore, there is one
solution to the problem. For a CS with n sequence components based on Eq. (1):

$$\sum_{i=1}^{n} \log(r_i) = \log(r_{csj}) \tag{7}$$

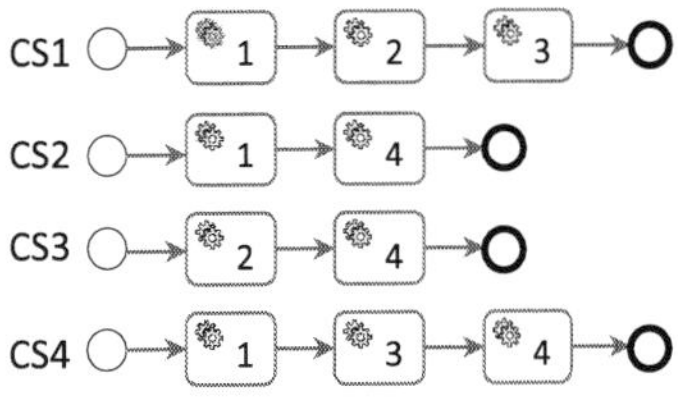

Fig. 4 Simple CSs sharing
components

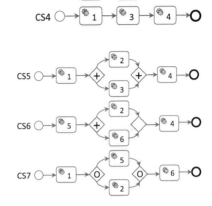

Fig. 5 More complex CSs
sharing components

Accordingly, we can calculate the reliabilities of the components as follows:

$$
\begin{pmatrix} 1 & 1 & 1 & 0 \\ 1 & 0 & 0 & 1 \\ 0 & 1 & 0 & 1 \\ 1 & 0 & 1 & 1 \end{pmatrix} \cdot \begin{pmatrix} \log(r_1) \\ \log(r_2) \\ \log(r_3) \\ \log(r_4) \end{pmatrix} = \begin{pmatrix} 0.78 \\ 0.98 \\ 0.99 \\ 0.82 \end{pmatrix}
$$

This results in the following values for the component reliabilities $\{0.96, 0.97, 0.84, 1.0\}$ for r_1 to r_4 respectively.

The variations of the case discussed above include other service properties and CSs with other types of constructs and different sharing levels of components between CSs (i.e. number of constraints vs. variables). Since the calculation of reliability in CSs containing Parallel and Unordered Sequence constructs is the same as a Sequence-only CS (as discussed in Sect. 5.2), the approach described above applies to such cases as well.

The calculation of reputation of components follows the same approach but with consideration of the weights of components in each CS. CSs such as those in Fig. 5 are more complex containing Unsynchronised Parallel and Multi-choice with Synchronised Merge respectively which require more elaborate procedure to calculate component trustworthiness levels. The procedure is not detailed due to limited space.

7 Conclusion

The paper described an approach to collaboratively infer the trustworthiness of shared components of CSs in distributed services environments. The approach takes into consideration practical requirements for determining the component trustworthiness in those environments. The trustworthiness of a CS is based on consumers' feedback on the reliability and their satisfaction with the service.

In the future we will extend this work to include other properties that may affect the trustworthiness of CSs such as available resources and execution times.

Acknowledgments The research leading to these results has received funding from the EU Seventh Framework Programme (FP7/2007-2013) under grant no 257930 (Aniketos) [17].

References

1. Malik, Z., Bouguettaya, A.: Trust Management for Service Oriented Environments. Springer, New York (2009)
2. Singhal, A., Wingrad, T., Scarfone, K.: NIST guide to secure web services. National Institute of Standards and Technology, Gaithersburg (2007)
3. Takabi, H., Joshi, J., Ahn, G.: Security and privacy challenges in cloud computing environments. IEEE Secur. Priv. **8**(6), 24–31 (2010)
4. Hwang, S., Wang, H., Tang, J., Srivastava, J.: A probabilistic approach to modeling and estimating the QoS of web-services-based workflows. Info. Sci. J. **177**(23), 5484–5503 (2007)
5. Cardoso, J., Sheth, A., Miller, J., Arnold, J., Kochut, K.: Modeling quality of service for workflows and web service processes. J. Web Semant. **1**(3), 281–308 (2004)
6. Hwang, S.-Y., Lim, E.-P., Lee, C.-H., Chen, C.-H.: Dynamic web service selection for reliable web service composition. IEEE Trans. Serv. Comput. **1**(2), 104–116 (2008)
7. Grassi, V., Patella, S.: Reliability prediction for service-oriented computing environments. Internet Comput. **10**(3), 43–49 (2006)
8. Nepal, S., Malik, Z., Bouguettaya, A.: Reputation propagation in composite services. In: Proceedings of 7th IEEE International Conference on Web Services, July 2009
9. OASIS WS-Trust: http://docs.oasis-open.org/ws-sx/ws-trust/v1.4/ws-trust.html, Feb. 2009
10. Elshaafi, H., McGibney, J., Botvich, D.: Trustworthiness monitoring and prediction of composite services. In: Proceedings of 17th IEEE Symposium Computers and Communications (2012)
11. Elshaafi, H., Botvich, D.: Aggregation of trustworthiness properties of BPMN-based composite services. In: Proceedings of 17th IEEE International Workshop on Computer-Aided Modeling Analysis and Design of Communication Links and Networks (2012)
12. Object Management Group: Business process model and notation (BPMN) 2.0. http://www.omg.org/spec/BPMN/2.0
13. Activiti Business Process Management Platform: http://activiti.org
14. Hoffman, K., Zage, D., Nita-Rotaru, C.: A survey of attack and defense techniques for reputation systems. ACM Comput. Surv. **42**(1), 1–31 (2009)
15. A. Jøsang, J. Golbeck: Challenges for robust trust and reputation systems. In: Proceedings of 5th international workshop on security and trust management (2009)
16. Workflow Patterns Initiative: http://www.workflowpatterns.com
17. Aniketos (Secure and Trustworthy Composite Services): http://www.aniketos.eu

Enhanced Middleware for Collaborative Privacy in Community Based Recommendations Services

Ahmed M. Elmisery, Kevin Doolin, Ioanna Roussaki
and Dmitri Botvich

Abstract Recommending communities in social networks is the problem of detecting, for each member, its membership to one of more communities of other members, where members in each community share some relevant features which guaranteeing that the community as a whole satisfies some desired properties of similarity. As a result, forming these communities requires the availability of personal data from different participants. This is a requirement not only for these services but also the landscape of the Web 2.0 itself with all its versatile services heavily relies on the disclosure of private user information. As the more service providers collect personal data about their customers, the growing privacy threats pose for their patrons. Addressing end-user concerns privacy-enhancing techniques (PETs) have emerged to enable them to improve the control over their personal data. In this paper, we introduce a collaborative privacy middleware (EMCP) that runs in attendees' mobile phones and allows exchanging of their information in order to facilities recommending and creating communities without disclosing their preferences to other parties. We also provide a scenario for community based recommender service for conferences and experimentation results.

Keywords Privacy · Clustering · Community Recommendations · Middleware

A. M. Elmisery (✉) · K. Doolin · D. Botvich
TSSG, Waterford Institute of Technology-WIT-Co, Waterford, Ireland
e-mail: ahmedkhmais2001@yahoo.com

I. Roussaki
National Technical University of Athens, Athens, Greece

S.-S. Yeo et al. (eds.), *Computer Science and its Applications*,
Lecture Notes in Electrical Engineering 203, DOI: 10.1007/978-94-007-5699-1_32,
© Springer Science+Business Media Dordrecht 2012

1 Introduction

With the popularity of social networks in the last few years, users are incited to build profiles containing their preferences, join different groups and utilize various services provided within the social platform. Community based recommender service (CRS) is a service running on social media platform and aims at providing end-users referrals to join certain sub-communities out of large number of communities that are relevant for a given end-user's interests. This service is based on the assumption that end-users with similar preferences have the same interests. CRS generates referrals based on end-user profiles containing, for each one, personal data and interests. The CRS is usually accessible and open to all attendees. However, this flexibility brings forward new threats and problems such as malicious behaviors against different participants from both service provider and other participants. For instance, malicious users may get one another's private information, such as current and previous occupations, age and relationship status, even if for the user the information is not supposed to be exposed publicly.

Several strategies have been proposed to control the disclosure of private information. The most popular approach is to permit users to maintain a set of privacy rules, according to which a decision is performed whether to release or not certain preferences in owner profile. However, these approaches are either rather coarse-grained, or require a deep understanding of the privacy control system, any change of one privacy setting may result in unwanted or unexpected behaviors. Moreover, these approaches are based on the logic of either to allow or deny releasing certain preferences in users' profiles. Once, the data is released the user have no control over it and users will be vulnerable for the privacy breaches since released pieces of users' information is often interleaved, adversaries may be able to infer other private information using inference techniques. For example work in [1] shows that private information can be inferred via social relations, and the stronger the relationships people have in the network, the higher inference accuracy can be achieved.

In this paper, we lay out recommending and creating communities functions within user-side, this privacy architecture will help foster the usage and acceptance of our proposed protocols and eliminates the risk of possible privacy abuses as the sensitive data is only available to the owner but not to any other parties. However, as a consequence of applying our protocols, the structure in data is destroyed. In order to facilitate processing of such data, our protocols maintain some properties in this data which is suitable for the required computation. In rest of this work, we will generically refer to attendees' preferences as interests. This paper is organized as follows. In Sect. 2, related works are described. Section 3 presents the proposed middleware enhanced middleware for collaborative privacy (EMCP) used in this work. Section 4 introduces some definition required for this paper. The proposed protocols that are used in EMCP are introduced in details in Sect. 5. In Sect. 6, the Results from some experiments on the proposed mechanisms are reported. Finally, the conclusions and recommendations for future work are given in Sect. 7.

2 Related Works

The majority of the literature addresses the problem of privacy on social recommender services, due to it being a potential source of leakage of private information shared by the users as shown in [2]. In [3] a theoretical framework is proposed to preserve the privacy of customers and the commercial interests of merchants. Their system is a hybrid recommender system that uses secure two party protocols and public key infrastructure to achieve the desired goals. In [4, 5] a privacy preserving approach is proposed based on peer to peer techniques using users' communities, where the community will have a aggregate user profile representing the group as a whole but not individual users. Personal information is encrypted and communication done between individual users but not servers. Thus, the recommendations are generated on the client side. Storing users' profiles on their own side and running the recommender system in a distributed manner without relying on any server is another approach proposed in [6].

3 The Proposed Middleware

In the scope of this work, we aim to achieve privacy by empowering an individual or group to seclude themselves or information about themselves thereby reveal themselves selectively or based on levels. We seek to achieve privacy by implementing a privacy by design approach [7] where we consider a middleware that governs data collection and processing during community building process such that attendees don't have to reveal private interests in their profiles. This will help them to control what they share with various communities and to join specific subcommunity with a customized profile that access only to a subset of their interests. The intuition behind our solution stems from the fact that safest way to protect sensitive profiles data is to not publish them online, but keep them at user side. However, in order to gain most of PCRS's functionalities, attendees disclose their private data in some way to enable PCRS's functionalities.

EMCP is implemented as a middleware running on top of attendees' mobile phones [8–13]. EMCP consists of different agents each of which has a certain task, but their co-operation is required to attain the whole functionality. The local obfuscation agent creates a public profile that is used as an input to encryption agent. The encryption agent is responsible for executing two cryptographic protocols; first one is private community formation (PCF) protocol which builds general communities based on attendees' profiles, while the other one is private sub-community discovery (PSD) protocol that help to discover sub-communities inside each community. These protocols act as wrappers that conceal interests before they are shared with any external entity. EMCP requires attendees to be organized into virtual topology which may be a simple ring topology or hierarchical topology, this ordering enables them to participate in multi-party computations as well. However,

PCRS (private community based recommender service) is the server that initiates the process to extract different communities and sub-communities. The scenario we are considering here is the one introduced in [8] it can be summarized as following based on conference various themes, research strategies and specific topics, the organizers setup a list of available communities on PCRS which act as interaction space that supports any interactions between attendees. Each attendee configures his EMCP to build a public profile that discloses some information about their general interests that are related to conference topics for the purpose of networking and collaboration. Attendees seek to hide from the public their specific expertise, previous conference engagements, details of their research domains and problems in hand, current and previous funded projects, sessions and presentations they are planning to attend and finally their arrival/departure times. Other Private information such as names, company, etc, by default is protected by the privacy protection laws. If attendees already belonging to previously created group, they can form a sub-community inside the conference community such that they can participate in discussions and have access to the already exchanged opinions. EMCP provides referrals to suitable sub-communities and sessions for attendees based on their interests.

3.1 Threat Model

The proposed solution is secure in an honest-but-curious model. Where, every party is obliged to follow the protocol but they are curious to find out as much as possible about the other inputs. The adversaries we consider here are untrusted CRS and malicious attendees that aim to collect other attendees' interests in order to identify and track them. Moreover we do not assume CRS to be completely malicious. This is a realistic assumption because CRS needs to accomplish some business goals and increase its revenues. Intuitively, the system privacy is high if CRS is not able to reconstruct the real attendees' private interests.

4 Problem Formulation

In the following section we outline important notions used in our previous solution in [8] and required in this work, attendees' profiles can be represented in two categories public profiles and private profile. Public profiles is a set of hypernym terms in the same semantic categories for the interests in attendee's profile [8], it represent general information that attendee configures his/her EMCP to disclose, while private profile represents the "hidden" interests that attendee does not want to disclose publically to others. Our goal is to protect private participants' profiles when formulating communities and recommending sub-communities since these are the information that attendees wish to keep private against both PCRS and third parties. The notion of community in this work can be defined:

Definition 1 A community is the set $C = \{c_1, c_2, \ldots, c_n\}$, where n is the number of sub-communities in C, has the following properties: (1) Each $\forall_{i=1}^n c_i \in C$ is a 3-tuple $c = \{I_c, V_c, d_c\}$ such that $I_c = \{i_1, i_2, \ldots, i_l\}$ is a set of generalized interests, $V_c = \{v_1, v_2, \ldots, v_k\}$ is a corresponding set of attendees, and $d_c \in I_c$ is the main-interest of c. (2) For each attendee $\forall_{i=1}^l v_i \in V_c$, v have the interests V_c. (3) d_c is the frequent interest in V_c profiles, and it represents the "core-point" of sub-community c. (4) For any two sub-communities c_a and $c_b (1 \le a, b \le n$ and $a \neq b)$, $V_{c_a} \cap V_{c_b} = \emptyset$ and $I_{c_a} \neq I_{c_b}$.

5 Proposed Privacy Enhanced Protocols for EMCP

In our architecture, privacy is attained using EMCP middleware which is hosted in attendees' mobile phones and equipped with two cryptography protocols which are private community formation protocol (PCF) and private sub-community discovery protocol (PSD) that build communities and sub-communities. EMCP allows the formation of attendees' communities; such that attendees share the same experience can engage in discussions and exchange experiences. An important requirement for our solution is the ability of an attendee to search for and join various sub-communities in private way.

5.1 Private Community Formation Protocol

Our aim is to cluster attendees' profiles into different communities. There are two challenges in identifying these communities: first one is representation of community, i.e., good intra-community similarity and inter-community separation. And the second one is the protection of private profiles in the process of community identification. In order to do so, attendees build public profiles using global information supplied by PCRS (e.g. concept taxonomy and term vocabulary) independently of their profile content, then local obfuscation agent at attendees side start mapping their profiles into this global information space to get public profiles as proposed in [8].

After building public profiles, EMCP invokes the encryption agent to execute PCF protocol that is responsible for clustering attendees into general communities, such that each general community contains various attendees who share similar interests in their profiles. An attendee can belong to multiple communities, thus allowing the separation between public profiles from his/her private profiles. Our novel secure multi-party computation protocol ensures participants privacy when forming communities and matching participant public profile with the list of available communities. PCF is executed in distrusted manor; it first creates a bag of interests representations of each attendee using their profiles data. Then, the extracted interests (words) are stemmed and filtered using domain-specific dictionary; these interests associated with a user V_c are used to create a word vector $V_c = (e_c(w_1), \ldots.. e_c(w_m))$, where m is the total number of distinct words in his/her is profile, and $e_c(w_1)$ describes the degree of importance of user V_c in interest w_1 (weighted frequency). The further computation proceeds to calculate term frequency inverse profile frequency [14] as following:

$$Term - frequency_{V_c}(w_i) = \#w_i \ in \ V_c \ profile / \#words \ in \ V_c \ profile$$

$$inverse - profile - frequency_{V_c}(w_i) = log(\#user / \#profiles \ contain \ word \ w_i)$$

$$e_c(w_1) = Term - frequency_{V_c}(w_i) * inverse - profile - frequency_{V_c}(w_i)$$

The similarity function between two attendees' profiles data should adequately capture the similarity of attendees' interests, and should be easy to calculate in a distributed and private fashion. Specifically, we leverage the Dice similarity for this task. Let $V_c(V_d)$ be the two word vectors for attendees C and D then:

$$UsersSimilarity(V_c, V_d) = 2|V_c \cap V_d| / |V_c|^2 + |V_d|^2$$

Intuitively, this means that two attendees C and D would be considered similar if they share many common words in their associated profiles, and even more so if only a few users share those words. Users have high similarity in set of interests will be clustered into the same community. To protect user privacy, an attendee's interests are stored locally and are not disclosed to other parties including the PCRS. Therefore, a secure multi-party computation mechanism is needed to compute the similarity between every two attendees. We present in the next subsection the similarity calculation procedure in PCF protocol as follows:

1. For any attendee $C, D \in V$ and a set of word vectors $e_c(w_i)$ and $e_d(w_i)$, the similarity is calculated in two steps first, it computes the numerator $|V_C \cap V_D|$ between attendee C and D and then it computes the denominator $|V_C|^2 + |V_D|^2$.
2. After selecting a super-peer as the root for computations, a virtual ring topology between attendees is employed for calculating the numerator between every two participants. Each public profile is associated with certain interests that need to be compared with other participants' public profiles then they submit similarity values to super-peers. Both attendees C and D apply a hash function h to each of their word vectors to generate $V_c = h(e_c(w_i))$ and $V_d = h(e_d(w_i))$. EMCP at attendee C generates an encryption E and decryption U keys then it submit the encryption key E to D.
3. Encryption agent at attendee D hides V_d by $B_d = \{e_d(w_i) \times r^D | w_i \in V_d\}$ where r is a random number for each interest w_i, and send B_d to C.
4. Encryption agent at attendee C signs B_d and get the signature S_d, then sends S_d to D again with the same order it receives. EMCP at attendee D reveals set S_d using the set of r values and obtains the real signature SI_d, then it applies hash function h on SI_d to produce $SIH_d = H(SI_d)$.

5. Encryption agent at attendee C signs the set V_c and gets signature SI_c then applies same hash function h on SI_d to produc e $SIH_c = H(SI_c)$ and submits this set to D.

6. Encryption agent at attendee D compares SIH_d and SIH_c using the knowledge of V_d, D gets the intersection set $IN_{C,D} = SIH_c \cap SIH_d$ that represent $|V_c \cap V_D|$. EMCP at D applies hash function h on $IN_{C,D}$ then it encrypts this value along with $|V_D|$, $|V_C|$ and attendees pseudonyms identities using super-peer public key and forwards them to super-peer of this group.

7. Super-peer collects all these results and decrypts them with its private key. Then it starts to cluster participants into communities, such that each community contains participants who share similar interests. Super-peer performs S-seeds [8] clustering algorithm as follows first, randomly select S attendees' profiles as clusters representatives. Then, it calculates the distance between these S seeds and each data point as specified in PCF protocol. Then, assigns each point to the community with the closest seed. Inside each community, choose the point with the smallest average distance to other data as the new seed. Finally, repeat last two steps until the S-seeds do not change. In S-seeds clustering, only the distance calculations among data points are required to identify the communities without disclosing attendees' profiles.

The above protocol performs it computations on m hashed values held by m parties without exposing any of the inputs values. This protocol is based on secure multi-party computation (SMPC), which was studied first by Yao in his famous Yao's millionaire problem [15].

5.2 Private Sub-Community Discovery (PSD) Protocol

Encryption agent in EMCP executes PSD protocol on the proximate general communities extracted from PCF protocol, PSD protocol determines in a bilateral manor the associated interests within attendees' public profiles, then the final results is used in building sub-communities. PSD protocol is adapted from the work in [16, 17] with the intuition that many frequent interests of attendees should be shared within a sub-community (group) while different sub-communities should have more or less different frequent interests. However, there are no predefined sub-communities yet inside these communities; hence PSD should operate with the available bounded prior domain knowledge and full dimensional profiles.

Definition 2 (Frequent interests) Frequent interests is a notion similar to frequent itemsets in association rule mining, it represent a set of interests that occur together in some minimum fraction of attendees' profiles. For example, let's consider two frequent interests, "libraries" and "C". Profiles containing the interest "libraries" may relate to digital archiving services and Profiles that contain the interest "C" may relate to Healthcare services. However, if both interests occur together in many profiles, then a specific interest sub-community related to C-programming should be identified.

Definition 3 (Global Frequent Interests) Global frequent interests is a set of interests that appear together in more than a minimum fraction of the whole attendees 'profiles in community C; a minimum community support is specified for this purpose. If this set contains k-interests, it called global frequent k-interests such that each interest that belongs to this set is called global frequent interest. Global frequent interest is frequent in sub-community c_i if this interest is contained in some minimum fraction of attendees' profiles; a minimum sub-community support is specified for this purpose.

The attendees are arranged in hierarchical topology in order to compute sub-communities, PSD protocol can be summarized as follows:

1. The initialization process of PSD protocol is invoked by PCRS, whereas attendees form groups then after they negotiate with each other to elect a peer who will act as a "super-peer" for each group. Super-peers distribute a list of 1-candidate frequent interests; therefore, different group members run concurrently a local algorithm to generate local frequent interests using their local support and closure parameters. we use the algorithm presented in [18] to find global & local frequent interests for each group.

2. After local extraction of frequent interests at each member $\forall_1^n P_i$, member P_i encrypts this local list with his own key and send it to member P_{i+1}, such that each member successively sends both his local and received lists to next neighbor. Last member in the group P_{n-1} send collected message to the super-peer. Super-peers now, have a set of local supports and closures of candidate frequent interests; generating global support is done by making the sum of these local supports. The global closure is calculated using intersection of the collected local closure. In the same way, repeating the previous steps, super-peer can generate the candidates of higher size. In order to decrypt the final results, the super-peer encrypts and sends global supports & closures lists to member P_{n-1} in arbitrary order. Member P_{n-1} decrypts his encryption from these lists using his own private key, and then sends this list to the next member P_{n-2} in arbitrary order. When super-peer receives these lists back, these lists will be encrypted with his own key only, which enables him/her to obtain final results.

3. For each adjacent set of global frequent interests at super-peer side, we setup an initial sub-community that includes all attendees' profiles that contain these interests, such that all profiles in this sub-community contain all these global frequent interests. These initial sub-communities are overlapped because each profile may contain multiple global frequent interests. PSD will use these global frequent interests as a sub-community representative. Then after, for each attendee's profile V_i, encryption agent determines the best initial sub-community c_i using the following score function:

$$SimilarityScore(c_i \leftarrow V_i) = \left[\sum_{w_i} e_r(w_i) * sub - community\ support(w_i)\right]$$
$$- \left[\sum_{w'_i} e_r(w'_i) * community\ support(w'_i)\right]$$

where w_i is a global frequent interest in profile r and this interest is also frequent in sub-community c_i while w'_i is a global frequent interest in profile r and is not frequent in sub-community c_i. $e_r(w_i)$ and $e_r(w'_i)$ are the weighted frequency of w_i and w'_i in profile r, which already calculated during the execution of PCF protocol. After this scoring, each attendee's profile belongs to exactly one sub-community.

4. For each community, super-peer organizes sub-communities in hierarchical structure using global frequent k-interests in each sub-community as representatives. In that case, PSD treats all attendees' profiles in each sub-community as single conceptual profile. The sub-community with k-interests will appear at level k in this structure, while the parent sub-community at level k-1 must be a subset of its child sub-community's representatives at level k. The selection of the potential parent for each child sub-community is done using scoring function presented in previous step. After that, super-peers exchange discovered sub-communities with each other to efficiently remove the overly sub-communities based on inter sub-community similarity. The same frequent interests might be distributed over multiple small sub-communities obtained from different super-peers' results, thus merging every two sub-communities into one general sub-community occurs only if they are very similar to each other. Inter sub-community similarity is similar to scoring function presented before with the only difference is that this similarity value should be normalized to remove the effect of varying number of attendees in each sub-community, it is measured using the following functions:

$$SubcommunitySimilarity(c_i \leftarrow c_j) = \left[SimilarityScore\left(c_i \leftarrow \forall_{x=1}^{n} V_x \in c_j\right)\right/$$

$$\left[\sum_{w_j} e(w_j) + \sum_{w'_j} e(w'_j)\right]] + 1$$

Then, *Intersubcommunity similarity*$(c_i \leftrightarrow c_j)$
$$= \left[SubcommunitySimilarity(c_i \leftarrow c_j) * SubcommunitySimilarity(c_j \leftarrow c_i)\right]$$

where c_i and c_j are two sub-communities; $\forall_{x=1}^{n} V_x \in c_j$ stands for single conceptual profile for sub-community c_j. w_j represents a global frequent interest in both c_i and c_j while w'_j represent a global frequent interest in c_j only but not in c_i. $e(w_j)$ and $e(w'_j)$ are the weighted frequency of w_j and w'_j sub-community c_j

5. Finally, for a new attendee, in order to privately recommend suitable sub-communities for him/her, EMCP obtains a list of sub-communities representatives then it generalizes his/her host interests and extract frequent interests for

this generalized profile. EMCP encrypts these frequent interests and measure their similarity with sub-communities' representatives in order to build a list of similar sub-communities. Finally EMCP assigns his/her host to the sub-community with the highest similarity.

6 Experiments

In this section, we describe the implementation of our proposed solution. The experiments are run on 2 Intel® machines connected on local network, the lead peer is Intel® Core i7 2.2 GHz with 8 GB Ram and the other is Intel® Core 2 Duo™ 2.4 GHz with 2 GB Ram. We used MySQL as data storage for the participants' profiles that is acquired by learning agent. PCRS has been implemented and deployed as a web service while *EMCP* has been deployed as an applet to handles the interactions between its owner, PCRS and other participants; it uses the implementation of the MPI communication standard for distributed memory implementation of our proposed protocols to mimic a distributed reliable network of peers. Our proposed protocols implemented using Java and boundycastle© library, RSA key length is set to 512 for the experimental scenario. The experiments were conducted using a dataset pulled from a recruiter network in Denmark (Manpower Professional) in period of 1990–1997. It contains registration data and information related to different participants that attend exhibitions organized by this agent which held concurrently with various scientific conferences. This data set is comprised of approximately 67,000 users and contains various details about them. Each of those details fell into one of several categories: affiliation, expertise, domains, projects, activities, publication and awards, etc. Due to the lack of a reliable subject authority, some other categories were discarded from all experiments. To generate the public profiles from these profiles we use same method proposed in [8].

In the first experiment, we want to measure the execution time for PCF protocol, from first step to last step at each attendee (excluding the time required to generate RSA keys). We divided our dataset into approximately same number of records and distribute then between 20 participants, then we run this experiment 7 times, so each point in the Fig. 1 is the mean value of the 7 runs. Additionally, we performed two other experiments in our dataset in which data was not divided into parts of same number of records. The first experiment, one client got 60 % of total number of records and the rest of records were divided to other clients as parts of approximately same number of records. While, in the second one, one client got 40 % of total number of records, other clients got the rest. The results of these experiments are summarized in Fig. 1. The results indicate the performance benefits of our protocol, as it is not sensitive to the number of shared interests (Fig. 2).

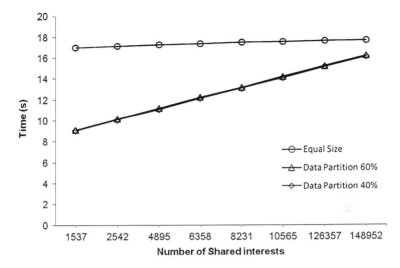

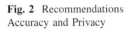

Fig. 1 Execution Time for PCF Protocol

Fig. 2 Recommendations
Accuracy and Privacy

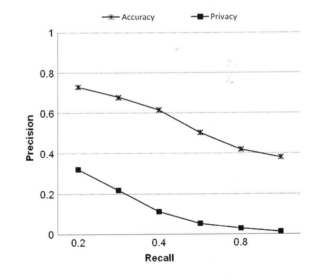

In the next experiment, we need to measure the accuracy of extracted sub-communities using PSD protocol. In order to evaluate the accuracy of our results, we apply hierarchical agglomerative clustering in our dataset in order to indentify natural sub-communities from attendees' private profiles. These sub-communities are utilized for measuring the accuracy of the results produced by PSD protocol. Each cluster represents a sub-community which is constructed from a set of attendees' private profiles who share the same specific interests about the same topic. To measure the goodness of our results, we considered two error metrics

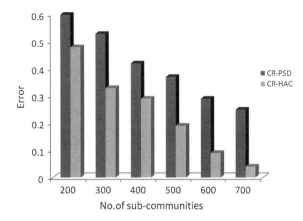

Fig. 3 Grouping Error (GR) of PSD Protocol

defined in [19] which are grouping error (GR) and critical error (CIE). The first one, the grouping error (GR), takes into account the number of attendees' profiles included in a sub-community, but belonging to a topic different from the dominant topic in that sub-community. The second one, the critical error (CIR) measures the number of attendees' profiles belonging to a topic that is not the dominant one in any sub-community. The graphs in Fig. 3 and 4, contain both GR and CIE values for the results obtained from both hierarchical clustering and PSD protocol for different number of sub-communities. This experiment is performed on two versions of our dataset; attendees' generalized profiles are utilized by our PSD protocol, while hierarchical agglomerative clustering utilizes attendees' private profiles that should kept private in our scenario.

We can deduce that both GR and CIE for PSD decrease with the increase in no. of sub-communities till reaching natural number of sub-communities. This indicates that achieving privacy is feasible and does not severely affect the accuracy of the generated sub-communities.

In the last experiment on PSD protocol, we want to measure the overhead of the execution time when applying PSD protocol to preserve attendees' privacy. We divided our dataset into different number of records from 30,000 to 67,000, such that each party held approximately the same number of records. We recorded the execution time when applying our PSD with encryption and without encryption on this data, then we calculated the percentage as following: $percentage = \left(time\ without\ encryption/_{time\ with\ encryption} \right) * 100$. The graph in Fig. 5 shows time comparison of our PSD protocol with and without encryption for different sizes of our dataset. From the results, we can find that the proposed PSD protocol has a reasonable performance and the privacy preserving nature has marginal impact on the execution time in comparison with non encryption option.

In order to measure the correctness of our solution to capture correlated interests between attendees. We extracted sample data from conference proceeding

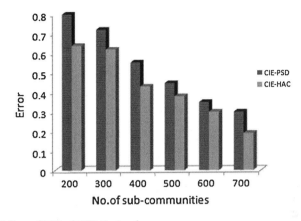

Fig. 4 Critical Error (CIE) of PSD Protocol

related to 500 authors and co-authors. We crawled authors' website to create public profiles for them. Our aim here is to determine if our proposed solution can group attendees in the same sub-community and help them to find the right people to communicate or work with. For every sub-community recommendation for each participant in the conference, we need to test whether or not participants knew each other in this sub-community from previous work and if this recommendation accurate or not. Figure 6 shows a breakdown of the results by our protocols, the percentage of unknown attendees recommended by EMCP are shown above the horizontal center line and the percentages of co-authors below. The chart also shows the percentage of accurate versus inaccurate in two different colors. PCF algorithm recommends other participants than the co-authors, which is not surprising because it mostly creates communities considering only similar interests without take in considerations the correlations between these preferences. In contrast, applying PCF and PSD extract sub-communities for people that are likely similar as sub-communities relies heavily on associations between preferences. These results confirm our intuitions that the more associations between participants' preferences, the more accurate sub-communities are produced.

In the last experiments, we evaluated the proposed solution from different aspects: privacy achieved and accuracy of results. We used precision and recall metrics proposed in [8] to measure privacy and accuracy of the results, the results are shown in Fig. 2. As we can see, a good quality is achieved due to: identifying communities that involve different sub-communities enables accurate recommendations to the attendees who share the same interests. Also, the effect of each interest inside the community can be easily measured, which enables to detect and remove outlier values that are very different than the general interests. We also evaluated the leaked private interests of different attendees when running our solution. We consider users, who published portion of their real interests in their public profiles, for each of these users; we tried the attack procedure proposed in threat model to reveal other hidden interests in their profiles based on the sub-

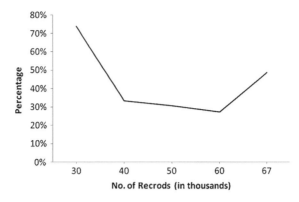

Fig. 5 Percentage Time for PSD Protocol

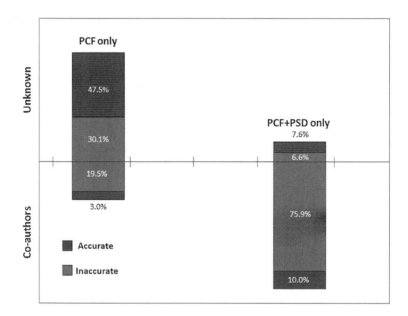

Fig. 6 Co-authors versus Unknown, Accurate versus Inaccurate

community they belong. The obtained interests are quantified using our proposed metrics and the results are shown in Fig. 2. As we can see, our solution manages to reduce privacy leakages for exposed attendees' private interests, However, the revealed interests are only a hashed hypernym terms for attendees private interests.

7 Conclusion and Future Work

In this paper, we presented our attempt to develop an enhanced middleware for collaborative privacy for community based recommender service in conferences or exhibitions. We gave a brief overview of EMCP architecture and proposed protocols. We tested the performance of the proposed protocols on a real dataset. The experimental and analysis results show achieving privacy in recommending sub-communities is feasible under the proposed middleware without hampering the accuracy of the recommendations. A future research agenda will include utilizing game theory to better formulate user groups, sequential preferences release and its impact on privacy of whole profile.

Acknowledgments This work partially supported by the European Comission via the ICT FP7 SOCIETIES Integrated Project (No. 257493). Also it was partially supported from the Higher Education Authority in Ireland under the PRTLI Cycle 4 Programme, in the FutureComm Project (Serving Society: Management of Future Communications Networks and Services).

References

1. He, J., Chu, W.W., Liu, Z.: Inferring privacy information from social networks. Proceedings of the 4th IEEE international conference on Intelligence and security informatics. Springer, San Diego, 154–165 (2006)
2. McSherry, F., Mironov, I.: Differentially private recommender systems: building privacy into the net. Proceedings of the 15th ACM SIGKDD international conference on Knowledge discovery and data mining. ACM, Paris, France 627–636 (2009)
3. Esma, A.: Experimental demonstration of a hybrid privacy-preserving recommender system. In: Gilles, B., Jose, M.F., Flavien Serge Mani, O., Zbigniew, R. (eds.), Vol 0 161–170 (2008)
4. Canny, J.: Collaborative filtering with privacy via factor analysis. Proceedings of the 25th annual international ACM SIGIR conference on research and development in information retrieval. ACM, Tampere, Finland 238–245 (2002)
5. Canny, J.: Collaborative filtering with privacy. Proceedings of the 2002 IEEE symposium on security and privacy. IEEE Computer Society 45 (2002)
6. Miller, B.N., Konstan, J.A., Riedl, J.: PocketLens: toward a personal recommender system. ACM Trans. Inf. Syst. **22**, 437–476 (2004)
7. Rubinstein, I.: Regulating privacy by design. Berkeley Technol. Law J., Forthcoming (2011)
8. Elmisery, A., Doolin, K., Botvich, D.: Privacy Aware community based recommender service for conferences attendees. 16th International conference on knowledge-based and Intelligent information & engineering systems. IOS Press, San Sebastian, Spain (2012)
9. Elmisery, A., Botvich, D.: Privacy Aware Recommender Service using Multi-agent Middleware- an IPTV Network Scenario. Informatica 36 (2012)
10. Elmisery, A., Botvich, D.: Enhanced middleware for collaborative privacy in IPTV recommender Services. J. Convergence **2**:10 (2011)
11. Elmisery, A., Botvich, D.: Privacy aware recommender service for IPTV networks. 5th FTRA/IEEE international conference on multimedia and ubiquitous engineering. IEEE, Crete, Greece (2011)
12. Elmisery, A., Botvich, D.: Multi-agent based middleware for protecting privacy in IPTV content recommender services. Multimed Tools Appl 1–27 (2012)

13. Elmisery, A., Botvich, D.: Privacy aware obfuscation middleware for mobile jukebox recommender services. The 11th IFIP conference on e-business, e-service, e-society. IFIP, Kaunas, Lithuania (2011)
14. Sebastiani, F.: Machine learning in automated text categorization. ACM Comput. Surv. **34**, 1–47 (2002)
15. Yao, A.C.: Protocols for secure computations. Proceedings of the 23rd annual symposium on foundations of computer science. IEEE Computer Society 160–164 (1982)
16. Beil, F., Ester, M., Xu, X.: Frequent term-based text clustering. Proceedings of the eighth ACM SIGKDD international conference on knowledge discovery and data mining. ACM, Edmonton, Alberta, Canada 436–442 (2002)
17. Fung B.C.M.: Hierarchical document clustering using frequent item sets. Master's thesis, Simon Fraser University (2002)
18. Cheung, D.W., Han, J., Ng, V.T., Fu, A.W., Fu, Y.: A fast distributed algorithm for mining association rules. Proceedings of the fourth international conference on on parallel and distributed information systems. IEEE Computer Society, Miami Beach, Florida, United States 31–43 (1996)
19. Cuesta-Frau, D., Pérez-Cortés, J.C., Andreu-Garcia, G.: Clustering of electrocardiograph signals in computer-aided Holter analysis. Comput. Methods Programs Biomed. **72**, 179–196 (2003)

A New k-NN Query Processing Algorithm Using a Grid Structure for Mobile Objects in Location-Based Services

Seungtae Hong and Jaewoo Chang

Abstract In telematics and location-based service (LBS) applications, because moving objects usually move on spatial networks, their locations are updated frequently, leading to the degradation of retrieval performance. To manage the frequent updates of moving objects' locations in an efficient way, we propose a new distributed grid scheme which utilizes node-based pre-computation technique to minimize the update cost of the moving objects' locations. Because our grid scheme manages spatial network data separately from the Point of Interests (POIs) and moving objects, it can minimize the update cost of the POIs and moving objects. Using our grid scheme, we propose a new k-nearest neighbor (k-NN) query processing algorithm which minimizes the number of accesses to adjacent cells during POIs retrieval in a parallel way. Finally, we show from our performance analysis that our k-NN query processing algorithm is better on retrieval performance than that of the existing S-GRID.

Keywords Distributed grid scheme · Query processing algorithm · Road network · Moving objects

1 Introduction

With the advancements on GPS and mobile device technologies, it is required to provide location-based services (LBS) to moving objects which move into spatial networks, like road networks. Several types of location-dependent queries are

S. Hong · J. Chang (✉)
Department of Computer Engineering,
Chonbuk National University, Chonju, Chonbuk 561-756, Jeonju, South Korea
e-mail: jwchang@jbnu.ac.kr

S. Hong
e-mail: dantehst@jbnu.ac.kr

S.-S. Yeo et al. (eds.), *Computer Science and its Applications*,
Lecture Notes in Electrical Engineering 203, DOI: 10.1007/978-94-007-5699-1_33,
© Springer Science+Business Media Dordrecht 2012

significant in LBS, such as range queries [1], k-nearest neighbor (k-NN) queries [1–3], reverse nearest neighbor queries [4], and continuous queries [5]. Among them, the most basic and important queries are k-NN ones. The existing k-NN query processing algorithms use pre-computation techniques for improving performance [6–8]. However, when POIs need to be updated, they are inefficient because distances between new POIs and nodes should be re-computed. To solve it, S-GRID [9] divides a spatial network into two-dimensional grid cells and pre-compute distances between nodes which are hardly updated. However, S-GRID cannot handle a large number of moving objects which is common in real application scenario. As the number of moving objects increases, a lot of insertions and updates of location data are required due to continuous changes in the positions of moving objects. Because of this, a single server with limited resources shows low performance for handing a large number of moving objects. To the best of our knowledge, there exists no work to consider a distributed processing technique using multiple serves for spatial networks. Therefore, we, in this paper, propose a distributed grid scheme which manages the location information of a large number of moving objects in spatial networks. Based on our grid scheme, we propose new k-NN query processing algorithm which minimize the number of accesses to adjacent cells during POIs retrieval in a parallel way.

The rest of the paper is organized as follows. In Sect. 2, we present related works. In Sect. 3, we describe the details of our distributed grid scheme. Section 4 presents a new k-NN query processing algorithm based on our grid scheme. In Sect. 5, we provide the performance analysis of our k-NN query processing algorithm. Finally, we conclude this paper with future work in Sect. 6.

2 Related Work

In this section, we describe some related works on k-NN query processing in spatial networks. First, VN3 [6], PINE [7], and islands [8] were proposed to pre-compute the distance between POIs and nodes (or border points) in road networks. However, when POIs need to be updated, they are inefficient because distances between new POIs and nodes should be re-computed. To resolve the problem of the VN3, PINE and Island approaches, Huang et al. [9] proposed S-GRID (Scalable Grid) which represents a spatial network into two-dimensional grids and pre-computes the network distances between nodes and POIs within each grid cell. To process k-NN query, they adopt the INE algorithm [1] which consists of inner expansion and outer expansion. The inner expansion starts a network expansion from the cell where a given query point is located and continues processing until the shortest paths to all data points inside the cell have been discovered or the cell holds no data points. Whenever the inner expansion visits a border point, the outer expansion is performed from that point. The outer expansion finds all POIs in the cells sharing the border point. This process continues until k nearest POIs are found. In S-GRID, the updates of the pre-computation data are local and POI independent. However, S-GRID have a critical problem that it is not efficient in handling a large number of moving objects, which are common in real application scenario, because it focuses on a single server environment. That is, when the number of moving objects is great, a lot of insertions and updates of location data are required due to continuous changes in the positions of moving objects. Thus, a single server with limited resources shows bad performance for handing a large number of moving objects.

3 Distributed Grid Scheme

To support a large number of moving objects, we propose a distributed grid scheme, by extending S-GRID. Our new grid scheme employs a two-dimensional grid structure for a spatial network and performs pre-computations on the network data, such as nodes and edges, inside each grid cell. In our distributed grid scheme, we assign a server to each cell for managing the network data, POIs and moving objects. Each server stores the pre-computed network data and manages cell-level two indices, one for POIs and the other for moving objects. To assign a unique ID(identifier) to each cell, we define CellID as follows.

Definition 1   Let assume a spatial network is partitioned into n*n two-dimensional grid structure. A unique ID of a cell being located in i-th row and j-th column, CellIDi, j, is defined by $CellID_{i,j} = (i-1)*n + j$.

Figure 1 shows an overall structure of our distributed grid scheme. Each cell consists of seven data structures: a cell table (Cell Table), border point table (BP table), POI R-tree, MO index, Vertex-Edge component, Vertex-Border component, and Cell-Border component.

4 New K-NN Query Processing Algorithm

In this paper, we propose multicasting-based cell expansion (MCE) algorithm which sends a query at once to all the servers for managing the cells where the k nearest POIs are located. Our algorithm sends a query to the servers assigned to cells to be visited and finds the k nearest POIs by performing outer expansion in a parallel way. Our MCE algorithm consists of 2 phases. The first phase is to create a list of boundary cells to find the k nearest POIs. The second phase is to send a query to the boundary cells and to find POIs by using outer expansion. First, to create a list of the boundary cells, we compute the number of expected POIs to be retrieved by visiting the adjacent cells. For this, our MCE algorithm finds the number of POIs in each cell by accessing each cell table. Then, it creates a network of border points by using the cell-border component of other cells and expands the network until the sum of POIs within the expanded cells is k. The total number of

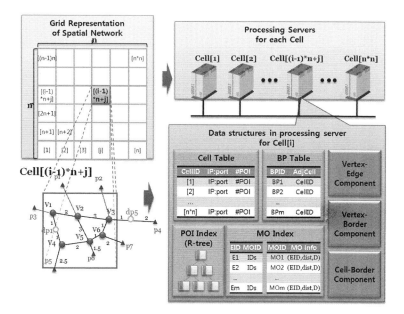

Fig. 1 Overall structure of our distributed grid scheme

POIs within the cells can be calculated by using following Eq. (1). Here, a spatial network consists of n*n grid cells, a query is located in the cell of i-th row and j-th column in the grid, CellIDi,j is the identifier of a query cell, and 'hop' is the number of expansion. Also x and y are the relative row value and the column value between a query cell and a cell visited during the hop-th expansion, respectively. '#POI(CellIDi, j + n*x + y)' is the number of POIs in the cell which is away from a query cell by x rows and y columns.

$$#POI_{hop} = \sum_{x=-hop}^{hop} \sum_{y=-|hop-|x||}^{|hop-|x||} #POI(CellID_{i,j} + n * x + y) \qquad (1)$$

Figure 2 presents a cell list creation algorithm by using the cell-border component of the cells within a boundary. First, the algorithm stores into 'nCandidate' the number of POIs in the cell where a query is located, and it inserts the border points of the cell into Qv (line 1–2). Secondly, for a border point from Qv, the algorithm checks whether the cell being shared with the border point is within the boundary or not. If the cell is not within the boundary, the algorithm calls the existing inner and outer expansion algorithms of of S-GRID (line 5–10). Otherwise, it computes the distances from the query to the other border points in the cell and inserts them into Qv (line 11–13). Next, the algorithm stores the cell ID and the distances into the Celllist (line 14) and updates the number of POIs in the cell from the 'nCandidates' (line 15–17). Finally, the process is repeated until all the cells within the boundary are expanded or 'nCandidates' is k.

```
CreateCellList Algorithm(q, k, CellList) // Qv=φ , Qdp=φ
01. qCell=Celli=findCell(q);   nCandidates = Celltable.getnumofPOI(Celli)
02. for each bp∈Celli.BP Qv.update(bp, dist(q, bp))
03. if nCandidates>=k      dMax=Qv.Dmax()
04. else                   dMax=∞
05. do
06.      bpx=Qv.deque, mark bpx as visited; Cellj=findAdjCell(bpx, Celli)
07.      if Celli is not within range R
08.           CellList.update(qCell, Cellj, bpy, dist(q,bpx)+dist(bpx,bpy))
09.           CellList.callExpansion (Cellj)
10.           continue
11.      for each non-visited bpy bp∈Cellj.BP
12.           if(dist(q,bpx)+dist(bpx,bpy)<dMax)
13.                Qv.update(bpy, dist(q,bpx)+dist(bpx,bpy))
14.                CellList.update(qCell, Cellj, bpy, dist(q,bpx)+dist(bpx,bpy))
15.                if(Cellj is not visited)
16.                     nCandidates = Celltable.getnumofPOI(Cellj)
17.                     mark Cellj as visited
18.      if nCandidates>=k      dMax=Qv.Dmax()
19. while nCandidates<k && Qv≠φ
```

Fig. 2 Cell list creation algorithm

To perform outer expansion, our MCE algorithm creates a list of cells by using the above 'Create CellList' algorithm and sends a query to the relevant cells by multicasting. The cells receiving the query find POIs by using outer expansion. Our MCE algorithm returns the result to the cell where a query is originated. When servers perform outer expansion, they compute the distance between each border point and the nearer node of the edge with POI by using vertex-border component. However, if there are more POIs within a cell than k, the cost of executing outer expansion is higher than that of the network expansion. To solve this problem, we compute the cost of both network expansion and outer expansion and choose one with lower cost to find POIs. The cost of each method can be computed as follows. First, the network expansion expands the spatial network starting from a border point to adjacent nodes by using the vertex-edge component, and it retrieves POIs lying on the expanded edge. Therefore, the cost of network expansion can be computed by $COST_{network\ expansion} = k * E/N$. Here, N and E are the total number of POIs and edges in a cell, respectively, and k is the number of POIs to be retrieved. Secondly, the cost of outer expansion can be computed by $COST_{outer\ expansion} = B * N * 2$. Here, B is the total number of border points associated with the query and N is the total number of POIs in the cell.

Figure 3 shows the outer expansion algorithm. First, the algorithm computes the cost of network expansion, COSTn, and the cost of outer expansion, COSTo. Secondly, if the value of COSTo is lower than that of COSTn, the algorithm finds

```
OuterExpansion Algorithm(q, k, BPlist) // Qdp=φ , Qv=φ
01. COSTne=calculateCOSTne(k, #_Edge, #_POI)
02. COSToe=calculateCOSToe(BPlist.#_BP, #_POI)
03. if(COSToe<=COSTne)
04.     for each bpi∈BPlist
05.         for each bpj∈Cell-Border Component
06.             if (bpi≠bpj)    Qv.update(bpj, dist(q,bpi)+dist(bpi+bpj))
07.             for each POI∈myCell    Qdp.update(POI, dist(q,bpi)+dist(bpi+POI))
08.         dMax=Qdp.dist(k), bp=Qv.deque
09. else
10.     for each adjacent vertex vx of bp∈BPlist
11.         Qv.update(vx, dist(q, bp)+dist(bp, vx))
12.     do
13.         vx=Qv.deque, mark vx as visited
14.         if (vx is a vertex)
15.             for each adjacent vertex vy of vx∈Vertex-Edge Component
16.                 for each POI∈findPOI(ex,y)
17.                     Qdp.update(POI, dist(q,vx)+dist(vx,POI))
18.                 Qv.update(vy, dist(q,vx)+dist(vx,vy))
19.                 if (all POI is discovered or Qdp.maxdist()<dist(q,vx))      break
20.             else    BPList.update(vx, dist(q,vx))
21.             dMax=Qdb.dist(k)
22.     while( d(q, vx) < dMax && Qv≠φ )
23. return POIs in Qdp, bps in Qv
```

Fig. 3 Outer expansion algorithm

POIs by computing the distances between each border point from the BPlist and all of the POIs in the cell (line 4–8). Otherwise, it removes all border points of the BPlist and inserts them into Qv (line 10–11). Thirdly, the algorithm selects a node (or a border point) with the shortest distance from Qv (line 13). Fourthly, if a node is selected, the algorithm inserts adjacent nodes into Qv and stores POIs lying on the adjacent edges into Qdp (line 14–18). Otherwise, it inserts border points into BPlist (line 21). Fifthly, the algorithm repeats the above steps until the number of retrieved POIs is k or all the POIs in the cell are found (line 19, 23). Finally, the algorithm returns the retrieved POIs and BPlist to the coordinator handling the cell where the query is given.

Figure 4 shows our MCE algorithm. First, our MCE algorithm creates a list of cells by using CreateCellList algorithm and sends a query to the cells (line 1–2). Secondly, our MCE algorithm finds POIs within the cell holding a query by using inner expansion. Thirdly, our MCE algorithm integrates the partial results from all the cells receiving the query by the coordinator and inserts the integrated result into Qdp (line 4–6). Fourthly, if there is a border point with a shorter distance than k-th POI, our MCE algorithm creates a new list of cells and sends the query to the servers to perform outer expansion(line 7–10). Finally, our MCE algorithm repeats 3–4 steps until we obtain the k nearest POIs and there is no cell in the cell list.

MCE Algorithm(q, k, Query, Result) // Qv=φ , Qdp=φ , CellList=φ
01. CreateCellList(q, k, CellList)
02. for each celli∈CellList   call OuterExpansion(q, k, celli.BPlist)
03. InnerExpansion(q, k, Qv, Qdp, NULL)
04. while(CellList is not empty)
05.   for each celli∈CellList   ReceiveResult(celli, dplist, celli.BPlist)
06.   for each POIi∈dplist   Qdp.update(POIi)
07.   dMax=Qdp.dist(k)
08.   for each vy∈celli.BPlist and dist(q,vy) < dMax   Cellj=findCell(vy)
09.   if (Cellj≠Celli)   CellList.update(myCell, Cellj, vy, dist(q,vy))
10.   for each celli∈CellList   call OuterExpansion(q, k, celli.BPlist)
11. return POIs in Qdp

Fig. 4 MCE algorithm

5 Performance Analysis

We present performance analysis of k-NN query processing algorithm for our grid scheme. We implement our grid scheme by using visual studio 2003 under HP ML 150 G3 server with Intel Xeon 3.0 GHz dual CPU, 2 GB memory. In our experiments, we used multiple processes in a single server and each process manages a single cell. To provide an environment appropriate to a distributed grid scheme, we let each process use a different port number to communicate with other processes by using TCP/IP protocol. For spatial network data, we use San Francisco Bay map consisting of 220,000 edges and 170,000 nodes, and generate four sets of POIs (i.e., 2200, 4400, 11000, 22000) by using Brinkhoff algorithm [10]. These POIs are indexed by using R-trees. Moreover, we randomly select 100 nodes from San Francisco Bay map as query points. To measure the retrieval performance of k-NN queries, we average response times for all the 100 query points. Because the existing works VN3 [6], PINE [7], island [8] are very inefficient for the update of POIs due to their POI-based pre-computation techniques, they are not appropriate for dealing with a large number of mobile objects in spatial networks. Thus we compare our algorithm with S-GRID algorithm in terms of POI retrieval time.

Figure 5a first shows the performance of k-NN query processing with the different number of grid cells when k = 20 and POI density = 0.01. The performance of our algorithm is better than that of S-GRID when the number of grid cells is more than 10*10. This is because our algorithm performs outer expansion in a parallel way. Figure 5b shows the retrieval time of k-NN query with the varying value of k when the density of POI is 0.01 and the number of grid cells equals 20*20. It is shown that as the value of k increases, the retrieval times of the two algorithms are increased because when the number of disk I/Os of the cell-border and vertex-border components is increased to visit adjacent cells. We can

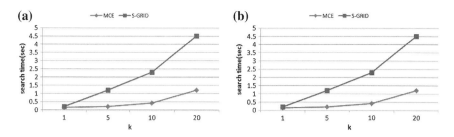

Fig. 5 Retrieval performance **a** with different number of grid cells **b** in terms of k

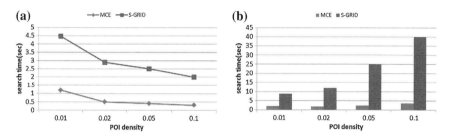

Fig. 6 Retrieval performance **a** in terms of the density of POIs **b** after updating POIs

say form the performance result that our MCE algorithm is better because it can reduce a query propagation step by sending a query to a list of cells at a time.

Figure 6a shows the retrieval time of k-NN query with the varying density of POIs where the number of grid cells equals 20*20 and k = 20. As a result, we can reduce the cost of inner expansion within a cell and the number of adjacent cells to be visited. Figure 6b shows the retrieval time of k-NN query after updating POIs. For this experiment, we measure the search time of k-NN query when the 10 % of POIs is updated. In the case of S-GRID, the retrieval time is exponentially increased as the density of POIs increases. This is because S-GRID uses one R-tree to index all the POIs of the network and so the update of POIs in a cell affects the whole system. Whereas, because our grid scheme uses a separate R-tree per each grid cell to index POIs within it, the update of POIs in a cell does not affect all the grid cells globally. As a result, even though the number of updated POIs increases, the retrieval performance of our grid scheme is not dramatically increased.

6 Conclusion and Future Work

In this paper, we proposed a grid scheme to manage the location information of a large number of moving objects in spatial networks. Our grid scheme makes use of a node-based pre-computation technique so that it can minimize the update cost of the moving objects' locations. Our grid scheme splits a spatial network into two-dimensional grid cells so that it can update network data locally. Based on our grid scheme, we proposed a new k-NN query processing algorithm. Our algorithm improves the retrieval performance of K-NN queries because it decreases the number of adjacent cells visited by transmitting a query to all the shared border points. Our experimental results show that our algorithm is better on retrieval performance than that of S-GRID. As a future work, we need to extend our grid scheme to handle a spatial network with dense and sparse regions in an efficient manner by using non-uniform grid cells.

Acknowledgments This research was supported by Basic Science Research Program through the National Research Foundation of Korea (NRF) funded by the Ministry of Education, Science and Technology (2010-0023800).

References

1. Papadias, D., Zhang, J., Mamoulis, N., Tao, Y.: Query processing in spatial network databases. In: Proc. VLDB, 802–813 (2003)
2. Shahabi, C., Kolahdouzan, M.R., Sharifzadeh, M.: A road network embedding technique for K-nearest neighbor search in moving object databases. In Proc GeoInformatica 7(3), 255–273 (2003)
3. Jensen, C.S., Pedersen, T.B., Speicys, L., Timko, I.: Data modeling for mobile services in the real world. In Proc. SSTD, 1–9 (2003)
4. Rimantas, B., Christian, S., Jensen, Gytis. K., Simonas, Š.: Nearest and reverse nearest neighbor queries for moving objects. In Proc. VLDB, 229–250 (2006)
5. Huang, YK., Chen, C-C., Lee, C.: Continuous K-nearest neighbor query for moving objects with uncertain velocity. In Proc. GeoInformatica 13(1) 1–25 (2009)
6. Kolahdouzan, M.R., Shahabi, C.: Voronoi-based nearest neighbor search for spatial network databases. In Proc. VLDB, 840–851 (2004)
7. Safar, M.: K nearest neighbor search in navigation systems mobile information systems. Mob. Inf. Syst. 1(3), 207–224 (2005)
8. Huang, X., Jensen, C.S., Saltenis, S.: The Islands approach to nearest neighbor querying in spatial networks. In Proc. SSTD LNCS 3633, 73–90 (2005)
9. Huang, X., Jensen, C.S., Lu, H., Saltenis, S.: S-GRID: a versatile approach to efficient query processing in spatial networks. In Proc. SSTD LNCS 4605, 93–111 (2007)
10. Brinkhoff, T.: A framework for generating network-based moving objects. In Proc. GeoInformatica, 153–180 (2002)

An Immune System-Inspired Byte Permutation Function to Improve Confusion Performance of Round Transformation in Symmetric Encryption Scheme

Suriyani Ariffin, Ramlan Mahmod, Azmi Jaafar, Muhammad Rezal and Kamel Ariffin

Abstract In data encryption, the security of the algorithm is measured based on Shannon's confusion and diffusion properties. This paper will proposed the Levinthal's paradox and protein structure essential computation elements on the basis of diversity property of immune systems that satisfy with confusion property of symmetric encryption scheme. This paper measures and analysis the confusion property of the permutation function of a block cipher using the correlation coefficient statistical analysis to identify whether it satisfies Shannon's confusion property. From the analysis carried out, the permutation function block cipher increased the performance of the confusion property, hence, indicating a high non-linear relationship between plaintext and ciphertext in symmetric encryption scheme.

Keywords Symmetric encryption · Block cipher · Protein structure · Levinthal's paradox · Immune system · Permutation · Confusion

S. Ariffin (✉) · A. Jaafar · M. Rezal · K. Ariffin
Universiti Teknologi MARA, Shah Alam 40450 Selangor, Malaysia
e-mail: suriyani@tmsk.uitm.edu.my

A. Jaafar
e-mail: azmi@fsktm.upm.edu.my

M. Rezal
e-mail: rezal@putra.upm.edu.my

R. Mahmod
Universiti Putra Malaysia, Serdang 43400 Selangor, Malaysia
e-mail: ramlan@fsktm.upm.edu.my

S.-S. Yeo et al. (eds.), *Computer Science and its Applications*,
Lecture Notes in Electrical Engineering 203, DOI: 10.1007/978-94-007-5699-1_34,
© Springer Science+Business Media Dordrecht 2012

1 Introduction

In many cryptographic algorithm [1–13] which are based on the Advanced Encryption Standard (AES) symmetric block cipher [14], Shannon's confusion property [15] by byte permutation is being generated using some rounds iterated to produce a secure data transmission [16]. Permutation function is one of the components that is commonly used in symmetric block ciphers to ensure that the ciphers are efficient. The function involved processing of input data through a finite number of iterative operation or repeated functions for every round. However, the byte permutation become more complex since they span over the whole bit of block, and either registers or memories are required in the transformation process to temporarily store the data bytes. It would be more effective if the confusion property can be achieved even when the number of iteration rounds is reduced.

The proposed block cipher consists of a combination of several components of AES symmetric encryption block cipher including a new byte permutation operation. This permutation operates by reading the input data bytes into a $4 \times 4 \times 4$ matrix. The new component, based on immune systems approaches, is proposed to replace the 4×4 matrix *shiftRows* in the diffusion layer and maintain the function in the confusion layer of AES symmetric encryption block cipher components. With the designing and development of the new permutation model within three-round cipher of 128 bits $\times$ 4 (512bits) with 128 bits key, it will reduce the iteration round and increase the level of parallel in the processing bytes of the symmetric encryption block cipher. This paper is organized as follows: the second section reviews the related work on the symmetric block cipher and the reason why immune system is a suitable metaphor model to be use in designing the permutation function, the third section describes the permutation function design, the fourth section measures the confusion based on the correlation coefficient, and the last section covers the conclusions and future works of the paper.

2 Related Works

Numerous examination of the problem since has yielded similarities to the nature of a human immune system. A number of studies have been done [17–20] and it is becoming more common to apply immune system [21–25] in the development of an algorithm or model in different applications. However, there is no research that utilizes the immune system as a basis for designing symmetric encryption block cipher. The Levinthal's paradox and protein structure are possible metaphors models that can be adopted in constructing a transformation operation of symmetric key encryption that satisfies Shannon's confusion properties. To measure it, a security analysis on the components of the block cipher should be done to satisfy Shannon's confusion property.

The symmetric block cipher must possess the functions of diffusion and confusion as mentioned in [26]. The confusion and diffusion defined by [15] are the two basic properties for obscuring the redundancies in a plaintext message. However, this paper will focus on the performance of the confusion property only. A confusion property obscures the relationship between the plaintext and the ciphertext. A long block of plaintext is substituted for a different block of ciphertext and the mechanics of the substitution change with each bit in the plaintext or key. Confusion serves to hide any relationship between the plaintext, the ciphertext and the key. It is one of the important aspects of the security of block ciphers which deals with the dependency of each of the output bits on the input bits. Good confusion produces a random ciphertext and creates a non-linear relationship between plaintext and ciphertext.

The proposed 3D permutation function will be adopted from the process and structure of generating the protein in immune systems to generate random ciphertext from new block cipher. The diversity property of protein structure that inspire the design of the new block cipher are: (a) sequence of amino acids, (b) each protein folds into a unique 3D structure from the combination of amino acid sequence, (c) information of the geometry of protein including distances, angles and dihedral angles in 3D structure and (d) have torsion angles which can rotate freely for every bond at the sequence of amino acids. This structure also is supported by Levinthal's paradox [27] and the problem formulation by [28] and [29] if we take 100 amino acids or residue protein and each residue can have only three positions or conformational states. From this number of residue proteins and positions, it will generate 3^{100} or about 10^{48} possible states of amino acids sequences. The protein can explore a new state of configuration and if each configuration takes about $10-11$ s, there are about 10^8 s in a year or about 10^{25} years which is longer than the estimated universe. In other circumstances, if we drop a necklace pearl to the ground many times, it is impossible to get the same conformational states of the sequence of pearls. Table 1 shows the mapping of the corresponding elements between protein structures in immune systems and ciphertext in block cipher. From the table, it shows that a new algorithm can be proposed based on the randomness of the elements in the protein structure which correspond to the randomness of the ciphertext. The elements of Levinthal's paradox and protein structure are applicable in the permutation function of the block cipher that would satisfy Shannon's confusion property.

3 Design of Permutation Function

This section describes the design principal of the 3D permutation function of block cipher as shown in Fig. 1.

Table 1 Mapping elements between protein structures and characteristics of block cipher

Protein structure	Ciphertext
Sequence of molecule in amino acids	Sequence of bits in a block
Different molecule sequence of amino acids will produce random amino acids sequence	Different sequence of bits in a block will produce random bits sequence
Defined elements in 3D structure	Defined elements in array or matrix
Will create unique sequence by having torsion angle which can rotate or permute freely for every bonds in the sequence of amino acids	Will create unique sequence by permutation or shift bit position
Generate other antibodies via the process of mutation between antigen and antibodies	Generate new bit sequence by substitution and XORed from other bit sequence

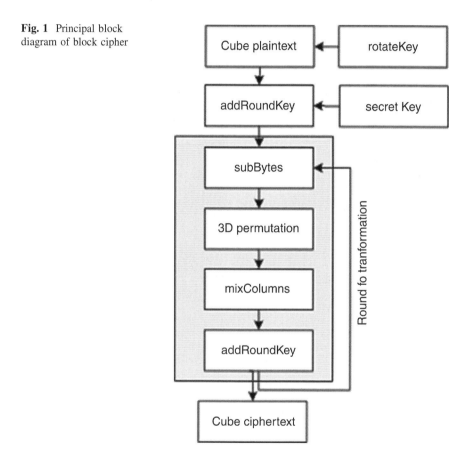

Fig. 1 Principal block diagram of block cipher

3.1 Array of Bytes

The storage of the plaintext in the algorithm's operations is performed on a 3D array (4 × 4 × 4 matrix) of bytes called *Cube* as illustrated in Fig. 2. The *Cube* is mapped to the 3D structure from the combination of amino acid sequence of

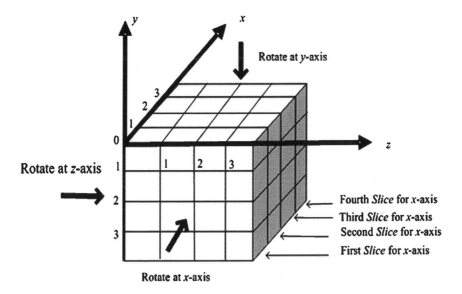

Fig. 2 Storage format in *Cube*

protein with finite length of 64 bytes (512 bits). The permutation process includes rotation at x-axis, y-axis and z-axis called *rotationKey*. The *Cube* for a 64-byte data block, is denoted in (1) with bytes inserted column wise.

$$
Cube, A =
\begin{vmatrix}
a_{000} & a_{001} & a_{002} & a_{003} \\
a_{010} & a_{011} & a_{012} & a_{013} \\
a_{020} & a_{021} & a_{022} & a_{023} \\
a_{030} & a_{031} & a_{032} & a_{033}
\end{vmatrix}
\begin{vmatrix}
a_{100} & a_{101} & a_{102} & a_{103} \\
a_{110} & a_{111} & a_{112} & a_{113} \\
a_{120} & a_{121} & a_{122} & a_{123} \\
a_{130} & a_{131} & a_{132} & a_{133}
\end{vmatrix}
$$

$$
\times
\begin{vmatrix}
a_{200} & a_{201} & a_{202} & a_{203} \\
a_{210} & a_{211} & a_{212} & a_{213} \\
a_{220} & a_{221} & a_{222} & a_{223} \\
a_{230} & a_{231} & a_{232} & a_{233}
\end{vmatrix}
\begin{vmatrix}
a_{300} & a_{301} & a_{302} & a_{303} \\
a_{310} & a_{311} & a_{312} & a_{313} \\
a_{320} & a_{321} & a_{322} & a_{323} \\
a_{330} & a_{331} & a_{332} & a_{333}
\end{vmatrix}
$$

$$(1)$$

The basic unit for processing in the proposed block cipher algorithm is a byte which is referred to a sequence of eight bits treated as a single entity in the AES block cipher. The 3D arrays of bytes will be represented in (1) which refers to the index number:

$$byte_0, \ byte_1, \ byte_2, \ byte_3, \ byte_4, \ldots byte_{63}$$

The bytes and the bit ordering within bytes are derived from the 512 bits input sequence

$$input_0, \; input_1, \; input_2, \; input_3, \; input_4, \ldots input_{511},$$

as follows:

$$byte_0 = input_0, \; input_1, \; input_2, \; input_3, \; input_4, \; input_5, \; input_6, \; input_7$$
$$byte_1 = input_8, \; input_9, \; input_{10}, \; input_{11}, \; input_{12}, \; input_{13}, \; input_{14}, \; input_{15}$$
$$\vdots$$
$$byte_{63} = input_{504}, \; input_{505}, \; input_{506}, \; input_{507}, \; input_{508}, \ldots input_{511}$$

so that:

$$byte_n = input_{8n}, \; input_{8n+1}, \; \ldots, \; input_{8n+7}. \tag{2}$$

3.2 Input and output

The input and output of the proposed block cipher are considered to be 3D arrays of 8-bit bytes or $4 \times 4 \times 4$ of cube. The input of the encryption function is a plaintext block, a rotation key and a secret key to produce the output, ciphertext block. The input of the decryption function is the ciphertext block to produce the output, plaintext block. The cipher-state is the results of every round of the transformation steps. The cipher-state can be illustrated as cube of bytes, with four rows and four column that will be explained in the next section. The number of columns in the state is denoted by N_b, the number of slice is denoted by N_s, and is equal to the block length divided by 64. Let the plaintext block be denoted by:

$$p_0, p_1, p_2, p_3, \ldots, p_{(4.N_b-1)+(16.N_s-1)}$$

where, p_0 denotes the first byte and $p_{(4.N_b-1)+(16.N_s-1)}$ denotes the last byte of the plaintext block. Let the ciphertext block be denoted by:

$$c_0, c_1, c_2, c, \ldots, c_{(4.N_b-1)+(16.N_s-1)}$$

where, c_0 denotes the first byte and $c_{(4.N_b-1)+(16.N_s-1)}$ denotes the last byte of the ciphertext block. Let the cipher-state be denoted by:

$$a_{i,j,k}, 0 \le i \le N_s, 0 \le j \le 4, 0 \le k \le N_b$$

where, $a_{i,j,k}$ denotes the byte in slice i, row j and column k. Also define in (1), the input bytes are mapped onto the state bytes in the order:

$$a_{0,0,0}, a_{0,1,0}, a_{0,2,0}, a_{0,3,0}, a_{0,0,1}, a_{0,1,1}, a_{0,2,1}, \ldots a_{3,3,3}.$$

For encryption function, the input is a plaintext block and the mapping is:

$$a_{i,j,k} = p_{16i+j+4k}, \quad 0 \le i \le N_s, 0 \le j \le 4, 0 \le k \le N_b. \tag{3}$$

For decryption function, the input is a plaintext block and the mapping is:

$$a_{i,j,k} = c_{16i+j+4k}, \quad 0 \leq i \leq N_s, 0 \leq j \leq 4, 0 \leq k \leq N_b. \tag{4}$$

The rotation key is the additional key to rotate the cube at x-axis, y-axis and z-axis in the permutation process. The number of rotation key is denoted by N_t and is equal to three axis. Let the rotation key be denoted by:

$$q_1, q_2, q_3$$

then:

$$Q = q_i, 0 < i < N_t. \tag{5}$$

The cipher key derived from the secret key is illustrated as a rectangular array with four rows and four columns and mapped onto a one dimensional cipher key. The number of columns of cipher key is denoted by N_k and is equal to the key length divided by 32. The bytes of the key are mapped onto the bytes of the cipher key in the order

$$k_{0,0}, k_{1,0}, k_{2,0}, k_{3,0}, k_{0,1}, k_{1,1}, k_{2,1}, \cdots, k_{3,3}.$$

Let the key be denoted by:

$$s_0, s_1, s_2, s_3, \ldots, s_{(4.N_b-1)}$$

then:

$$k_{i,j} = s_{i+4j}, 0 \leq i \leq 4, 0 \leq j \leq N_b. \tag{6}$$

3.3 Permutation Function

The key-iterated block ciphers uses the same round transformation from the key-iterating block cipher structure. As refer to Fig. 1, let the boolean permutation, denoted as $B[k]$ from the number of rounds by r, from k^r to k^0 be:

$$B[k] = \sigma[k^r]^\circ p^r \sigma[k^{r-1}]^\circ \cdots^\circ \sigma[k^1]^\circ p^1 \sigma[k^0] \tag{7}$$

where $\sigma[k^r]$ is the key addition (*addRoundKey* function), p^r is the rth round of the round transformation of the block cipher based on the Wild Trail Strategy [30], k^r is the rth round key and $^\circ$ is the input to $p^2 \sigma[k^2]$ that is derived from the output of $p^1 \sigma[k^1]$ and so on. The boolean permutation $B[k]$ is a sequence-dependent transformation of $p^i \sigma[k^i]$. Let the round transformation be:

$$\rho = \theta^\circ \lambda^\circ \gamma$$

where γ (*subBytes* function) is the non-linear function or substitution function of the slice at round function and θ (*mixColumns* function) is based on the components from the AES block cipher and operated on columns of four bytes each. The

λ (3D permutation function) is the permutation function of the slice at the round function that is inspired from the immune system. From the Eq. (7), the boolean permutation has:

$$B[k] = \sigma[k^r]^\circ \theta^\circ \lambda^\circ \gamma^\circ \sigma[k^{r-1}]^\circ \ldots^\circ \sigma[k^1]^\circ \theta^\circ \lambda^\circ \gamma^\circ \sigma[k^0] \tag{8}$$

where all rounds of the cipher use the same round transformation, which means that the proposed block cipher exhibits is a key-iterated block cipher. Other than a secret key, a rotation key is used as an additive effect to the confusion property in the block cipher. With regards to the permutation function, there are different index numbers for every slice in different axis as the rotation key from Eq. (5). For the purpose of evaluation and testing a new structure, the value of the rotation key is only at value 1 or rotates at x-axis.

4 Performance Evaluation

This section measures and analysis non-linearity between two variables, which is between plaintext, p and ciphertext, c. The correlation values can determine the confusion effect of the proposed block cipher. It is one of the important aspects of the security of block ciphers which deals with the dependency of each of the output bits on the input bits.

4.1 Process and Data Description

Laboratory experiment activities were conducted on Windows Operating System. All data of the 64-byte block of plaintext and the 16-byte key were generated and evaluated offline as similar in [31, 32]. These values were based on data generated using the Blum–Blum–Shub (BBS) pseudo-random bit generator. The BBS was chosen in this experiment because it has been shown to be a cryptographically secure pseudo-random bit generator as mentioned in [16, 33]. For the purpose of evaluation and testing, the value of the rotation key is at value 1 or rotates at x-axis. The 128 sequences of 512 bits with 128 bits keys were generated and converted to ciphertext, which will be incorporated into the laboratory experiment activities.

4.2 Test Description

The correlation values can determine the confusion effect of the block cipher. The correlation coefficient takes on values ranging between +1 and −1. The following values are the accepted range for interpreting the correlation coefficient:

- 0 indicates a non-linear relationship.

- +1 indicates a perfect positive linear relationship: as p increases in its values, c also increases in its values via an exact linear rule.
- −1 indicates a perfect negative linear relationship: as p increases in its values, c decreases in its values via an exact linear rule.
- Values between 0 and 0.3 (0 and −0.3) indicate a weak positive (negative) linear relationship via a unstable linear rule.
- Values between 0.3 and 0.7 (0.3 and −0.7) indicate a moderate positive (negative) linear relationship.
- Values between 0.7 and 1.0 (0.7 and −1.0) indicate a strong positive (negative) linear relationship.

The correlation coefficient functions are used as follows:

$$E(c) = \frac{1}{s} \sum_{i=1}^{s} P_i \tag{9}$$

where the series of s measurements of p and c are written as p_i and c_i where $i = 1$, 2,..., s, c is bits value of change bits (ciphertext), s is the total number of bits, $E(c)$ is the mathematical expectation of c:

$$D(p) = \frac{1}{s} \sum_{i=1}^{s} [p_i - E(p)]^2 \tag{10}$$

$$r_{pc} = \frac{E\{[p - E(c)][c - E(c)]\}}{\sqrt{D(p)}\sqrt{D(c)}} \tag{11}$$

where p is the bits value of the original bits (plaintext), $D(p)$ is the variance of p and r_{pc} is the related coefficients.

To examine the correlation coefficient of the block cipher, the experiment was divided into two different types of testing that is (a) test on permutation function only and (b) test on all functions in the block cipher. The correlation analysis from those laboratory experiments will be presented in the following section.

4.3 Correlation Coefficient on Permutation Function

To measure the diffusion property, the laboratory experiment was conducted on the permutation function only, which is between the 3D permutation function in the proposed block cipher and the *shiftRows* function in the AES block cipher.

The scatter chart of the results is presented in Fig. 3, in which the correlation value for the plaintext before and after the *shiftRows* in the AES block cipher, denoted as *oldF* and for the plaintext before and after the 3D permutation function in the proposed block cipher, denoted as *newF* are recorded. It shows that 21 out of 128 sequences from the *shiftRows* function in the AES block cipher recorded correlation values between 0.7 to 1.0, which indicate a strong positive (or

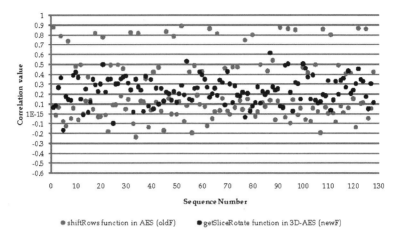

Fig. 3 Scatter chart of the correlation test results on permutation function only

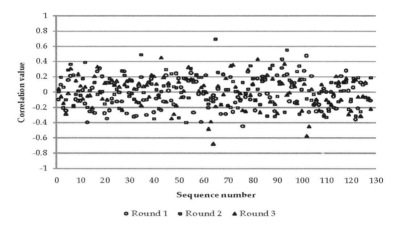

Fig. 4 Scatter chart of the correlation test results on proposed block cipher

negative) linear relationship. However, there is no sequence from the 3D permutation function in the proposed block cipher which recorded the correlation values between 0.7 to 1.0. It also reported that only 25 out of 128 sequences (19 %) from the *shiftRows* function in the AES block cipher recorded the correlation values between 0.3 to 0.7. However, 45 out of 128 sequences (35 %) from the 3D permutation function in the proposed block cipher are within the same ranging of the correlation value. There is an increasing percentage of correlation from a weak positive (negative) non-linear relationship of the AES block cipher to a moderate non-linear relationship of the proposed block cipher. From the results of the analysis, it can be deduced that the 3D permutation function in the proposed block cipher has an increased confusion performance.

4.4 Correlation Coefficient on All Functions

The experiment examined all functions of the proposed block cipher. The scatter chart of the results is presented in Fig. 4, in which the correlation value for every round for each sequence is recorded. It shows that most of the correlation value, at different rounds during the proposed block cipher algorithm implementation, are near to 0, which indicate a strong positive (or negative) non-linear relationship. Only 0.01 % are near to +1 or −1, which indicate a weak positive (or negative) non-linear relationship. From the results of the analysis, it can be concluded that the proposed block cipher has an increased confusion performance between the plaintext and the ciphertext.

5 Conclusion

The correlation coefficient was performed to identify the confusion property of the proposed block cipher. In comparison to the correlation coefficient of the *shift-Rows* in AES block cipher, the permutation function of the proposed block cipher increased the impact of the non-linear relationship between the plaintext and the ciphertext. In comparison to the popular four-round AES block cipher, the proposed block cipher increased the performance of the confusion operations with only three rounds in the round transformation process. From the results of the analysis, it justifies the high non-linearity and sensitivity of the ciphertext generated in round three from this block cipher. Future work can include deeper analysis of the cost evaluation criteria of the proposed block cipher that includes licensing requirements, computational efficiency or speed on various platforms and memory requirements.

Acknowledgments This work is supported by the Fundamental Research Grant Scheme (FRGS) provided by the Ministry of Higher Education Malaysia, under the Grant Number FRGS/1/11/SG/ UPM/02/4 and the Bumiputera Academic Training Scheme (SLAB) Malaysia.

References

1. Ali, F.H.M., Mahmod, R., Rushdan, M., Abdullah, I.: A faster version of Rijndael cryptographic algorithm using cyclic shift and bit wise operations. Int. J. Cryptol. Res. **1**(2), 215–223 (2009)
2. Biryukov, A.: Analysis of involutional ciphers: Khazad and Anubis. In: Johansson, T. (ed.) Fast software encryption, lecture notes in computer science, vol. 2887, pp. 45–53. Springer Berlin/Heidelberg (2003)
3. Daemen, J., Knudsen, L., Rijmen, V.: The block cipher square. In: Biham, E. (ed.) Fast software encryption. Lecture notes in computer science, vol. 1267, pp. 149–165. Springer Berlin/Heidelberg (1997), 10.1007/BFb0052343

4. Elumalai, R., Reddy, A.R.: Improving diffusion power of AES Rijndael with 8x8 MDS matrix. Int. J. Comp. Sci. Eng. **3**(1), 246–253 (2011)
5. Lim, C.H.: Crypton: A new 128-bit block cipher—specification and analysis (1998)
6. Mahmod, R., Ali, S.A., Ghani, A.A.A.: A shift column with different offset for better Rijndael security. Int. J. Cryptol. Res. **1**(2), 245–255 (2009)
7. Mathur, C., Narayan, K., Subbalakshmi, K.: High diffusion cipher: Encryption and error correction in a single cryptographic primitive. In: Zhou, J., Yung, M., Bao, F. (eds.) Applied cryptography and network security, lecture notes in computer science, vol. 3989, pp. 309–324. Springer Berlin/Heidelberg (2006)
8. Nakahara, J.: 3D: A three-dimensional block cipher. In: Franklin, M., Hui, L., Wong, D. (eds.) Cryptology and network security, lecture notes in computer science, vol. 5339, pp. 252–267. Springer Berlin/Heidelberg (2008)
9. Nakahara Jr, J.: New impossible differential and known-key distinguishers for the 3D cipher. In: Feng, B., Jian, W. (eds.) Information security practice and experience 7th International Conference, ISPEC 2011, Guangzhou, China, May 30 June 1, 2011. Proceedings, Lecture Notes in Computer Science, vol. 6672/2011, pp. 208–221. Springer Berlin/Heidelberg (2011)
10. Rijmen, V., Daemen, J., Preneel, B., Bosselaers, A., De Win, E.: The cipher shark. In: Gollmann, D. (ed.) Fast software encryption. Lecture Notes in Computer Science, vol. 1039, pp. 99–111. Springer Berlin/Heidelberg (1996)
11. Simplicio, M., Jr., Barreto, P.S.L.M., Carvalho, T.C.M.B., Margi, C.B., Nslund, M.: The CURUPIRA-2 block cipher for constrained platforms: Specification and benchmarking (2007)
12. Suri, P.R., Deora, S.S.: 3D array block rotation cipher: An improvement using lateral shift. Glob. J. Comp. Sci. Technol. **11**(19), 17–23 (2011)
13. Suri, P.R., Deora, S.S.: Design of a modified Rijndael algorithm using 2d rotations. IJCSNS Int. J. Comp. Sci. Netw. Secur. **11**(9), 141–145 (2011)
14. NIST: Fips197: Advanced encryption standard (AES), FIPS PUB 197 Federal information processing standard publication 197. Technical Reports National Institute of Standards and Technology (2001)
15. Shannon, C.: Communication theory of secrecy systems. Bell Syst. Tech. J. **28**(4), 656–715 (1949)
16. Menezes, A.J., Oorschot, P.C.V., Vanstone, S.A.: Handbook of applied cryptography. CRC Press, Boca Raton (1997)
17. de Castro, L.N., Timmis, J.: Artificial immune systems: A new computational intelligence approach. Springer, New York (2002)
18. Forrest, S., Hofmeyr, S.A., Somayaji, A.: Computer Immunology. Commun. ACM **40**, 88–96 (1997)
19. Harmer, P., Williams, P., Gunsch, G., Lamont, G.: An artificial immune system architecture for computer security applications. Evolutionary computation, IEEE transactions on 6(3):252–280 (2002)
20. Marhusin, M., Cornforth, D., Larkin, H.: Malicious code detection architecture inspired by human immune system. In: Software engineering, artificial intelligence, networking, and parallel/distributed computing, 2008. SNPD'08. Ninth ACIS international conference on. pp. 312–317 (2008)
21. Dasgupta, D.: Advances in artificial immune systems. Computational intelligence magazine. IEEE **1**(4):40–49 (2006)
22. Dasgupta, D., Forrest, S.: Artificial immune systems in industrial applications. In: Intelligent processing and manufacturing of materials, 1999. IPMM'99. Proceedings of the second international conference on. vol. 1, pp. 257–267 vol. 1 (1999)
23. Somayaji, A., Hofmeyr, S., Forrest, S.: Principles of a computer immune system. In: Proceedings of the 1997 workshop on new security paradigms. pp. 75–82. NSPW'97, ACM, New York (1997)
24. Timmis, J.: Artificial immune systems today and tomorrow. Nat. Comput. **6**, 1–18 (2007)

25. Timmis, J.: Artificial immune systems. In: Sammut, C., Webb, G.I. (eds.) Encyclopedia of machine learning. pp. 40–44. Springer, New York (2010)
26. Knudsen, L.R., Robshaw, M.J.: Introduction. In: The block cipher companion, Information security and cryptography, pp. 35–64. Springer Berlin Heidelberg (2011)
27. Levinthal, C.: How to fold graciously. Mssbaun spectroscopy in biological systems proceedings, University of Illinois. Bulletin **67**(41), 22–24 (1969)
28. Karplus, M.: The Levinthal paradox: yesterday and today. Fold Des **2**(Supplement 1), 69–75 (1997)
29. Dobson, C.M., Karplus, M.: The fundamentals of protein folding: bringing together theory and experiment. Curr. Opin. Struct. Biol. **9**(1), 92–101 (1999)
30. Daemen, J., Rijmen, V.: AES and the wide trail design strategy. In: Knudsen, L. (ed.) Advances in cryptology EUROCRYPT 2002. Lecture notes in computer science, vol. 2332, pp. 108–109. Springer Berlin Heidelberg (2002)
31. Ariffin, S., Mahmod, R., Jaafar, A., Ariffin, M.R.K.: Immune systems approaches for cryptographic algorithm. In: Bio-inspired computing: theories and applications (BIC-TA), 2011 sixth international conference on, pp. 231–235. (2011)
32. Ariffin, S., Mahmod, R., Jaafar, A., Ariffin, M.R.K.: Byte permutations in block cipher based on immune systems. In: International conference on software technology and engineering, 3rd (ICSTE 2011). ASME Press, New York (2011)
33. Stallings, W.: Cryptography and network security: principles and practice. Prentice Hall, Upper Saddle River (2011)

New Bilateral Error Concealment Method of Entire Depth Frame Loss for 3DTV and Virtual 3D Videoconferencing Systems

Fucui Li, Gangyi Jiang, Mei Yu, Xiaodong Wang, Feng Shao and Zongju Peng

Abstract In three-dimensional television and 3D videoconferencing systems, depth entire frame loss will degrade the video quality. In this paper, a bilateral error concealment algorithm of the depth entire frame loss is proposed, which utilizes strong temporal correlation and movement correlations between the depth video and its corresponding 2D color video. Experimental results show that the proposed algorithm can conceal the lost entire depth frame and provide better subjective and objective quality of the reconstructed images.

Keywords Three-dimensional television · Depth · Error concealment · Bilateral error concealment

F. Li (✉) · G. Jiang (✉) · M. Yu · X. Wang · F. Shao · Z. Peng
Faculty of Information Science and Engineering, Ningbo University, Ningbo 315211, China
e-mail: lifucuinbu.edu.cn

G. Jiang
e-mail: jianggangyinbu.edu.cn

M. Yu
e-mail: yumeinbu.edu.cn

X. Wang
e-mail: xangxiaodongnbu.edu.cn

F. Shao
e-mail: shaofengnbu.edu.cn

Z. Peng
e-mail: pengzonjunbu.edu.cn

S.-S. Yeo et al. (eds.), *Computer Science and its Applications*,
Lecture Notes in Electrical Engineering 203, DOI: 10.1007/978-94-007-5699-1_35,
© Springer Science+Business Media Dordrecht 2012

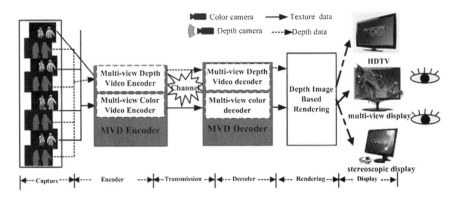

Fig. 1 The 3DTV system with DIBR technique

1 Introduction

Three-dimensional television (3DTV) has recently received an increasing attention since it can represent real world events well [1]. To reduce the complexity of synthesizing a virtual view, depth-image-based rendering (DIBR) technique has been proposed for 3DTV system. Multi-view video plus depth (MVD) is used as the data format for 3D representation of real scene. Both of the color videos and theirs corresponding depth videos should be compressed and transmitted to user end in 3DTV system as shown in Fig. 1 [2, 3]. During transmission of compressed MVD data, despite of application of sophisticated found error protection techniques, packet losses may cur. A single erroneous bit may result in loss of a slice that is often set as a frame. Therefore, decoder error concealment techniques are required in order to provide possibly acceptable image content to the lost frames.

An efficient joint multi-view video coding (JMVC) technique always is used for 3D video [4]. JMVC adopts the hierarchical B prediction (HBP) structure, which uses the inter-view prediction as well as the intra-view prediction to increase coding efficiency [5]. The prediction structure makes compressed bit streams very sensitive to transmission errors since an entire frame loss in one view point will trigger a chain of errors not only in the temporal depth frames but also in other view points that use it as a reference. Thus, it is very important to research the error concealment for entire depth frame loss in 3DTV and virtual 3D Video-conferencing systems.

Some approaches had been proposed to protect the qualities of color sequences against transmission errors [6–8]. It is not suitable to apply the 2-D video error concealment algorithms directly to the depth loss because the depth frame has homogeneous value inside the object, and the value changes a little within the same object between the two consecutive depth frames. Only few algorithms were proposed for robust reconstruction of depth sequences. Liu et al. in [9] introduced an error concealment method for the depth lost block. But this work focused on the block loss not on the frame loss. Hewage et al. proposed that both image and depth frame are transmitted and employ a special scheme of motion vectors sharing between the views [10]. Unfortunately, some modifications of the coding format are required in their methods. These error concealment algorithms may be not applicable to the entire depth frame loss in 3D videos based on HBP coding structure.

In this paper, a bilateral error concealment algorithm of entire depth frame loss for 3D video transmission is proposed, which utilizes the strong bilateral temporal correlations based on the HBP structure and the movement correlations between the depth map and its corresponding 2-D video. The rest of this paper is organized as follows. Section 2 describes the proposed algorithms and Sect. 3 evaluates its performance. Finally, we conclude this paper in Sect. 4.

2 The Proposed Algorithm

In MVD-based 3DTV scheme there are high movement correlations between color video and its corresponding depth video. And the temporal correlations exist between the lost depth frame and its bilateral reference frames. Here, we will exploit the two correlations for error concealment of the entire depth frame loss.

2.1 Motion Vector Mapping Between Color Video and its Depth Video

In MVD-based 3DTV, the color motion information can be used as candidate motion information for the depth video. The analysis of motion correlation of color-plus-depth sequences is described in [11]. So the motion information of the color video can be used as the candidate motion information of the depth video in the case of the depth frame loss. In MVD-based 3DTV scheme, every B frame has the temporal preceding and subsequent frame as its reference frames. This means that the bilateral temporal correlation can also be selected as the information for concealment.

Let D_t^s denote the t-th frame and the s-th view, MV_{D1} be the foreword motion vector (MV) in its subsequent reference frame D_{t+k}^s, and MV_{D2} be the backward motion vector of the block in its preceding reference depth frame D_{t-k}^s. When the depth frame D_t^s losses, motion information of each macroblock B_{lost} in D_t^s is also lost. To recover the lost macroblock B_{lost} with the bilateral reference depth frames which have correctly decoded, we need estimate the motion vectors including MV_{D1} and MV_{D2}. But in general the depth frame is more flat than color texture one. It has homogeneous value inside the object, and the value changes a little within the same object between the two consecutive depth frames. So the MV in depth stream reflect the real motion less accurately than the one in video stream because the video stream can provide more texture information. Thus, the lost macroblock B_{lost} can share the motion vector with the co-located block in its corresponding color video. If the bilateral motion vectors of the co-located block in the color video are the $\hat{v}_{c1}$ and $\hat{v}_{c2}$, the estimated motion vectors of the depth block B_{lost} are $\hat{v}_{d1}$ and $\hat{v}_{d2}$, motion vector mapping between color video and depth video is illustrated in Eqs. (1) and (2)

$$\hat{v}_{d1} = \hat{v}_{c1} \tag{1}$$

$$\hat{v}_{d2} = \hat{v}_{c2} \tag{2}$$

The movement correlations between color frames and depth frames at the decoder will be used to conceal the entire depth frame loss that occurs during the transmission of the compressed MVD.

2.2 Temporal Bilateral Error Concealment

When a depth frame is lost, each block's motion vectors are not available at the decoder. Here, the motion vectors of the lost depth block are estimated by using the motion information of its corresponding color video. When a depth frame D_t^s is detected as lost, it is assumed that the corresponding color frame C_t^s, the previous color frame C_{t-k}^s, the subsequent color frame C_{t+k}^s, the previous depth frame D_{t-k}^s and the depth subsequent frame D_{t+k}^s are available for the depth frame conceal-ment. The bilateral motion vectors of each block $B_d(i)$ in the lost depth frame can be estimated using the motion information of the co-located $B_c(i)$ block in C_t^s. If the object maintains a constant velocity in the time interval between C_{t-k}^s and C_{t+k}^s, the forward motion vector $\hat{v}_{c1}$ of the block $B_c(i)$ can be estimated as the one which minimizes the sum of absolute difference (SAD) between the candidate block in the reference frame C_{t+k}^s and the block $B_c(i)$ in the frame C_t^s. $\hat{v}_{c1}$ is computed as Eqs. (3) and (4)

$$SAD_i(v_{c1}) = \sum_{p \in B_c(i)} \left| C_t^s(p) - C_{t+k}^s(p + v_{c1}) \right| \tag{3}$$

$$\hat{v}_{c1} = \arg \underset{i}{\mathrm{Min}}(SAD_i(v_{c1})) \tag{4}$$

where $B_c(i)$ is the set of pixel coordinates in the i-th block, and $C_t^s(p)$ denotes the value of the pixel p in the frame C_t^s. Calculation of the back vector $\hat{v}_{c2}$ of the block $B_c(i)$ from C_t^s to C_{t-k}^s is similar to $\hat{v}_{c1}$, it can be shown as Eqs. (5) and (6).

$$SAD_i(v_{c2}) = \sum_{p \in B_c(i)} mid; C_t^s(p) - C_{t-k}^s(p + v_{c2}) \tag{5}$$

$$\hat{v}_{c2} = \arg \underset{i}{\mathrm{Min}}(SAD_i(v_{c2})) \tag{6}$$

Then, using the sharing scheme of motion vectors between color frames and depth frames, the motion vector $\hat{v}_{c1}$ of co-located $B_c(i)$ in the corresponding color frame can be used as the forward motion vector $\hat{v}_{d1}$ of the current lost depth block $B_d(i)$, and $\hat{v}_{c2}$ as the backward motion vector $\hat{v}_{d2}$ of $B_d(i)$. After the estimation of the motion vectors $\hat{v}_{d1}$ and $\hat{v}_{d2}$ for the lost block $B_d(i)$, each pixel in the lost block is concealed by using the corresponding pixels:

$$\hat{D}_t^s(p) = \eth_1 D_{t-k}^s(p - \hat{v}_{d1}) + (1 - \eth_1) D_{t+k}^s(p + \hat{v}_{d2}) \quad p \in B_d(i) \qquad (7)$$

with processing each block of the lost depth frame, the lost depth frame is recovered. This concealment mode is called temporal bilateral error concealment based on sharing of motion vectors between color video and depth video in the 3DTV.

2.3 The Proposed Algorithm

The proposed algorithm is described as follows. To recover the block of the entire lost depth frame, it is needed to evaluate the forward motion vector v_{c1} and the backward motion vector v_{c2} of the co-located block in the corresponding color frame. And then $\hat{v}_{c1}$ and $\hat{v}_{c2}$ is served as the motion vectors $\hat{v}_{d1}$ and $\hat{v}_{d2}$ of the depth lost block $B_d(i)$. Finally the block $B_d(i)$ is concealed by using motion compensation with $\hat{v}_{d1}$ and $\hat{v}_{d2}$ in the depth reference frames. The block diagrams of the proposed error concealment method of entire depth frame loss can be seen in Fig. 2.

3 Experimental Results and Discussions

To verify the proposed method in terms of both the objective and subjective qualities, it is integrated into JMVC8.3. Test sequences are Dog color sequences, Pantomime color sequences and their corresponding depth sequences which are created using depth estimation reference software (DERS) [12]. The color and depth frames of the Pantomime test sequences with 8 views are shown in Fig. 3. The resolution is $1,280 \times 960$ each test sequence. In the experiments, we assume that no error in the color video and only the entire depth B frame loss is considered. Each sequence is encoded with JMVC. The length of group of picture (GOP) is 8 and the basis quantization parameter (QP) is set to 32, 27 or 22. After a lot of experiments, the parameter $\eth_1$ in Eq. (5) is set to 1/2.

In order to evaluate the effectiveness of the proposed algorithm, we compare the performances between no error (NOERR), frame copy (FC), which recovers the lost frame by copying the preceding reference frame directly, and backward error concealment (BEC),which conceals the lost frame only using the backward motion information from the lost frame to the preceding map.

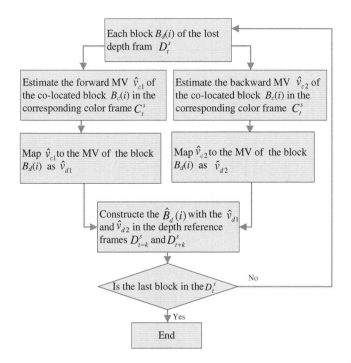

Fig. 2 The block diagram of the proposed algorithm

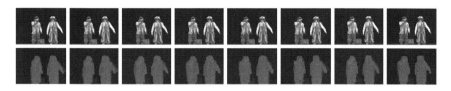

Fig. 3 The color and depth frames of the Pantomime test sequence with 8 views

In MVD-based 3DTV and virtual 3D Videoconferencing systems, depth frames are used to synthesize the virtual view, not to display directly. So, to evaluate the performance of the proposed algorithm, PSNR of the synthesized views based on the recovery of lost depth frame and its corresponding color frame, not of the concealed depth frame, is considered. The smaller ΔPSNR, which denotes the difference between the PSNR of NOERR and other three methods, the closer reconstruct image to the error free depth image. Table 1 shows that proposed method has the best PSNR performance among three models. It provides the smallest ΔPSNR, which is about 0.3–0.5 dB.

Figure 4 also demonstrates that the proposed algorithm can provide the better image subjective quality. In the synthesized views of FC and BEC, there are some artifacts around the arms and hats as shown in Fig. 4b and c. On the other hand, the proposed method suppresses those artifacts and provides a higher synthesized view quality.

Table 1 Comparison of the average PSNR performances (dB)

QP	Sequences	NO ERR	FC		BEC		Proposed	
		PSNR	PSNR	ΔPSNR	PSNR	ΔPSNR	PSNR	ΔPSNR
22	Pantomime	32.18	27.06	−5.12	29.12	−3.06	31.69	−0.49
	Dog	33.59	31.33	−2.26	33.02	−0.57	33.29	−0.30
27	Pantomime	32.16	26.94	−5.22	29.09	−3.07	31.64	−0.52
	Dog	33.61	31.39	−2.22	33.07	−0.54	33.31	−0.30
32	Pantomime	31.95	26.85	−5.10	28.88	−3.07	31.34	−0.61
	Dog	33.45	31.46	−1.99	33.00	−0.45	33.27	−0.18

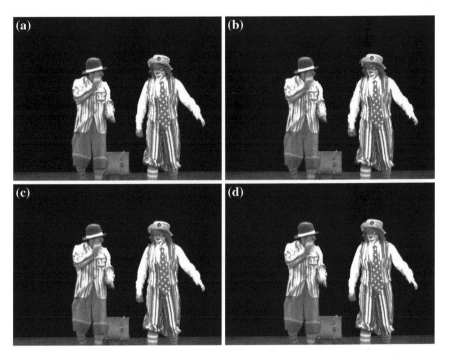

Fig. 4 The subjective quality of the synthesized view of Pantomime sequence (QP = 27). **a** The synthesized view of NOERR PSNR = 32.16 dB **b** The synthesized view of FC, PSNR = 26.94 dB. **c** The synthesized view of BEC, PSNR = 29.09 dB. **d** The synthesized view of Proposed PSNR = 31.64 dbB

4 Conclusions

Multi-view video plus depth (MVD) scheme is used as the main data format for 3DTV. In this scheme there is the high movement correlation between color video and corresponding depth video. So the more accurate motion vectors provided by the color stream with more texture information are mapped as the candidate motion vectors of lost depth frame. Moreover, in the HBP structure, there are bilateral reference frames for every B frame. So a novel temporal bilateral error concealment for entire depth frame loss based on sharing of motion vectors between the color video and the depth video in 3DTV is proposed in this paper. Experimental results demonstrate that the proposed algorithm can conceal the lost entire depth frame and provide better subjective and objective quality of the reconstructed images.

Acknowledgments This work was supported by the Natural Science Foundation of China (Grant Nos. 60832003, 61071120, 61171163), the Zhejiang Provincial Natural Science Foundation of China (Grant No. Y1101240), and Natural Science Foundation of Ningbo (2011A610200).

References

1. Gurler, C., Gorkemli, B., Saygili, G., Tekalp, A.: Flexible transport of 3-D video over networks. Proc. IEEE **99**(4), 694–707 (2011)
2. Merkle, P., Müller, K., Wiegand, T.: 3D video: Acquisition, coding, and display. IEEE Trans. Consumer Electron. **56**(2), 946–950 (2010)
3. Vetro, A., Tourapis, A., Muller, K., Chen, T.: 3D-TV content storage and transmission. IEEE Trans. Broadcast. **57**(2), 384–394 (2011)
4. ISO/IEC JTC1/SC29/WG11 and ITU-T SG16 Q.6, "Draft reference software for MVC", JVT-AE207, London, June (2009)
5. Merkle, P., Smolic, A., Müller, K., Wiegand, T.: Efficient prediction structures for multiview video coding. IEEE Trans. Circuits Syst. Video Technol. **17**(11), 1461–1473 (2007)
6. Pang, L., Yu, M., Yi, W., Jiang, G., et al.: Relativity analysis-based error concealment algorithm for entire frame loss of stereo video, IEEE Int. Conf. Signal Proc. pp. 16–20, China (2006)
7. Yang, S., Wang, S., Zhao Y., Chen, H.: Error Concealment for Stereoscopic Video Using Illumination Compensation, IEEE International Conference on Consumer Electronics (ICCE), (2011)
8. Chen,Y., Cai, C.: Stereoscopic video error concealment for missing frame recovery using disparity-based frame difference projection, International Conference on Image Processing, pp. 4289–4292, (2009)
9. Liu, Y., Wang, J., Zhang, H.: Depth image-based temporal error concealment for 3-D video transmission. IEEE Trans. Circuits Syst. Video Technol. **20**(4), 600–604 (2010)
10. Hewage, C., Worrall, S., Dogan, S., Kondoz, A. M.: Frame concealment algorithm for stereoscopic video using motion vector sharing, IEEE International Conference on Multimedia and Expo, pp. 485–489, (2008)
11. Oh, H., Ho, Y.: H.264-Based Depth Map Sequence Coding Using Motion Information Of Corresponding Texture Video, Advances in Image and Video Technology, vol. 4319, pp. 898–907 (2006)
12. Tanimoto, M., Fujii, T., Suzuki, K., et al.: Reference softwares for depth estimation and view synthesis, in *ISO/IEC JTC1/SC29/WG11 MPEG*, Doc. M15377, Archamps, France (2008)

A REST Open API for Preventing Income Tax Over-Payment by Auditing Year-End Tax Settlement

Min Choi and Sang-Soo Yeo

Abstract RESTful web service everlopers to easily deploy their web service through HTTP protocol. Even though almost all WTO countries' tax law explicitly states the tax incentive to subsidies for research and development, many companies often incorrectly classify the subsidies for research as a taxable income not as a non-taxable income, resulting in over payment of tax for employee. Therefore, we introduce a RESTful Open API for checking whether earned income year-end tax settlement is correct or not. This API automatically audits the tax settlement by some required numbers from the earned income withholding receipt.

Keywords RESTful web service · Year-end tax settlement · Smartphone application development · REST Open API

1 Introduction

There are various categories of incomes for employees: interest income, dividend income, real estate rental income, business income, and earned income. But we are focusing on the earned income. Year-end tax settlement for the earned income is that National Tax Service adjust the amount in case that you have excess or

S.-S. Yeo (✉)
Department of Computer Engineering, Mokwon University, 52, Doan-dong,
Seo-gu, Daejeon 360-213, Republic of Korea
e-mail: ssyeo@mokwon.ac.kr

M. Choi
School of Information and Communication Engineering, Chungbuk National University,
52 Naesudong-ro, Heungdeok-gu, Cheongju, Chungbuk 361-763, Republic of Korea
e-mail: mchoi@cbnu.ac.kr

S.-S. Yeo et al. (eds.), *Computer Science and its Applications*,
Lecture Notes in Electrical Engineering 203, DOI: 10.1007/978-94-007-5699-1_36,
© Springer Science+Business Media Dordrecht 2012

deficiency after contrasting your income tax and residents tax deducted from your monthly paychecks with your income tax and residents tax. The tax liabilities of employee's wage and salary for the relevant taxable year are finalized through the year-end tax settlement. A withholding agent who pays the wage and salary shall make income deduction from the employee's wage and salary for the taxable year, based on report on income deduction filed by the employee. Taking the tax base, the withholding agent should calculate the income tax, and, from there, take off an allowable amount for tax exemption and tax credit as well as the amount of income taxes withheld at source during the relevant tax year. And the balance between the tax paid and tax payable shall be collected as tax or refunded to the taxpayer.

If the taxpayer retires in the middle of the taxable year, the worker's income tax due is finalized for the wage and salary received until the month of retirement based on the report of exemption and deduction from income submitted by the employee. In case where an employee retires during the year, the withholding agent conducts year-end tax settlement for the attributable year of retirement in obligation and issue the receipt for the wage and salary income taxes withholding. If the determined tax resulted in refund, the withholding agent put the return into employee's registered bank account. In case of prepaid tax is lower than final tax liability the employee has to pay the taxes.

Actually, we show in this paper the error case of W company's wrong year-end tax settlement. Likewise, there are often mistakes in the processing of year-and settlement from the withholding agent, so we need to check and make sure the tax settlement is correct or not. This research proposes a public API to audit automatically whether the tax settlement is correct. Even though almost all WTO countries' tax law explicitly states the tax incentive to subsidies for research and development, many companies often incorrectly classify the subsidies for research as a taxable income not as a non-taxable income, resulting in over payment of tax for employee. Therefore, we introduce a RESTful Open API for checking whether earned income year-end tax settlement is correct or not. This API automatically audits the tax settlement by some required numbers from the earned income withholding receipt.

In this research, we designed and implemented the framework for auditing the year-end tax settlement. Thus, this RESTful Open API prevents tax withholders from wrong earned income tax settlement. This service is available on any platforms such as smartphone through REST Open API interface.

The rest of this paper is organized as follows: Section. 2 describes related works on this research. Section 3 focuses on the details of management server architecture for REST web services. Section 4 shows experimental results. Finally, we conclude our work and present future research directions in Sect. 5.

2 Related Work

2.1 WSDL/SOAP Web Service Composition

In the past decade, much research has been put on automated approaches to WSDL/SOAP web service composition. Since WSDL/SOAP web services and REST web service adopt differing styles and view the services from different perspectives, the automated composition problem of these two kinds of web services are very different.

WSDL/SOAP web service composition predominately uses AI planning approaches, and these approaches focus on functional composition of individual web services. That is, how to compose a new functionality out of existing component functionalities. However, REST web services model the system from the perspective of resources. The composition of REST web services focuses on the resource composition and state transfer between candidate web services.

2.2 RESTful Web Service Composition

As shown in the above, whereas there have been many researched SOAP based web service composition, REST web service composition is untouched field. In Zhao [1]'s paper, it demonstrates automated RESTful web service composition in the context of service-oriented architecture (SOA). The author proposed a formal model for describing individual web services and automating the composition. This paper suggests a method to describe RESTful web service by ontology based conceptual model. This model is used to build the automated composition framework. By establishing the model for describing RESTful web services as ontology resources and "state transfer" of ontology resources, the authors form a conceptual model which can be used to facilitate automated composition of RESTful web services. So, the authors present a situation calculus based state transition system (STS) to automate the composition process of RESTful web services.

Zhao [2] proposed a two-stage linear logic based program synthesis approach to automatic RESTful web service composition. The linear logic theorem proof is applied at both resource and service invocation method levels, which greatly improves the searching efficiency and guarantees the correctness and completeness of the service composition. Furthermore, the process calculus is used as formalism for the composition process, which enables the approach to be executable at the business management level.

Pautasso identifies a set of requirements for REST web service composition and extends BPEL to accommodate the REST architecture, which aims to enable composition of both traditional web services and REST web service within the same process-oriented service composition language. Moreover, the work also allows publication of BPEL processes as REST web services. This work is summarized as a BPEL extension for REST that is, so far, the most mature approach for REST web service composition.

Alarcon propose a hypermedia-driven approach based on the Resource Linking Language and Petri Nets. Resource Linking Language focuses on the hypermedia characteristics and serves as a description language for RESTful services. It not only allows resources to be annotated explicitly with domain semantics but also enables machine clients to automatically retrieve the web resources, their domain semantics and the navigation mechanisms. The service Net approach introduced navigation

3 Tax Over Payment Case by Classifying the Subsidies for Research as Taxable Income

In this section, we show a case of wrong tax settlement of a W company. The W Company insists that the mistake occurs due to the error of their computer system. But the process of tax settlement involves a large amount of manual processing, employees have to check and inspect whether the tax settlement processing is correct or not. We need to check tax over payment by classifying the subsidies for research as taxable income.

This case study shows an example of an employee who has worked for the W company from March 2010 to August 2011. The employee moved to working for another C company from September 2011 to now.

Usually, Tax law in almost all WTO countries specifies the allowances for the compensation of actual expenses as non-taxable wage and salary income. Actually in Korea tax law, there are statements of tax incentive to subsidies for research and development. This is because the Korea government wishes to encourage companies and employees that invest research and development. Direct and indirect state aid ensures that companies and employees that carry out research and development activities can do this under the best possible conditions. Actually, the amount of such level as to compensate actual expenses, such as pay for day duty, night watch, or business travel (including amounts not exceeding ₩200,000 per month from among the incurred expenses received by an employee pursuant to payment criteria stipulated by the regulations, etc. of relevant enterprises, in lieu of receiving reimbursement of the actual travel expenses incurred during a business trip within the city using his/her own car to perform his/her duties).

In tax law of Korea, there are such a statement that non-taxable income (Income tax act article 12) can be calculated up to ₩100,000 per month. Amount not exceeding ₩200,000 out of subsidies for research or research activity expenses provided to a person who falls under any of the following items.

- Teaching staff of a school under the early childhood education act, the elementary and secondary education act, the higher education act, and a school equivalent thereto (including an educational institution under a special act)
- A person directly engaging in, or providing direct support for research activities (limited to a person holding qualifications equivalent to that of teaching staff of universities and colleges) in a research institute governed by the support of specific research institutes act, a government-funded research institute established pursuant to a special act, or a local-government-funded research institute established pursuant to the act on the establishment and operation of local government-invested research institutes
- A person who directly engages in researching activities in a research institute attached to a small or medium enterprise or a venture enterprise or three of the enforcement decree of the technology development promotion.

W company has correctly calculated the subsidies for research ₩2,000,000 (₩200,000 × 10 months) up to earned income withholding year-end tax settlement at 2010. C company has also correctly calculated the subsidies for research ₩800,000 (₩200,000 × 4 months) up to earned income withholding year-end tax settlement at 2011. But, the W company has incorrectly calculated the subsidies for research ₩0 up to earned income withholding year-end tax settlement at 2011, resulting in tax over payment for employee of ₩237,607.

W Company incorrectly classified the subsidies for research as a taxable income not as a non-taxable income, resulting in over payment of tax for employee. The W company calculated the research subsidies up to 0 in the earned income withholding receipt. However, the value of subsidies for research in Fig. 1 and Table 1 should be replaced by ₩1,600,000. Because tax law in almost all WTO countries and in Korea also apparently states that non-taxable income (Income tax act article 12) can be calculated up to ₩100,000 per month. Amount not exceeding ₩200,000 out of subsidies for research or research activity expenses provided to a person who falls under any of the following items.

The procedure to check and audit the year-end tax settlement is as the following. First, we have to calculate the tax base which is given by gross wage annual wage excluding non-taxable income (non-taxable income should be already subtracted for generating this gross wage)—deduction for wage and salary income(this is the withholding tax so it was already credited at every month)—Personal deduction basic, additional, multiple children. etc.—pension contribution (this is also credited at every month) national pension, Korea teachers pension—special deduction insurance (such as national health insurance premium), medical, education etc.—other deduction credit card usage, cash usage etc. After getting the tax base (taxable income), we multiply by the tax rates which is determined as from 6 to 35 % depending on the total amount of tax base. Then, we subtract from the above tax base to the tax credit for all salary earners which is commonly applied credit for all earned income employees, especially ₩500,000 in Korea. This is the final tax liability and then we finally subtract the prepaid tax (the withheld at every months), resulting in the additional tax payment or refundable

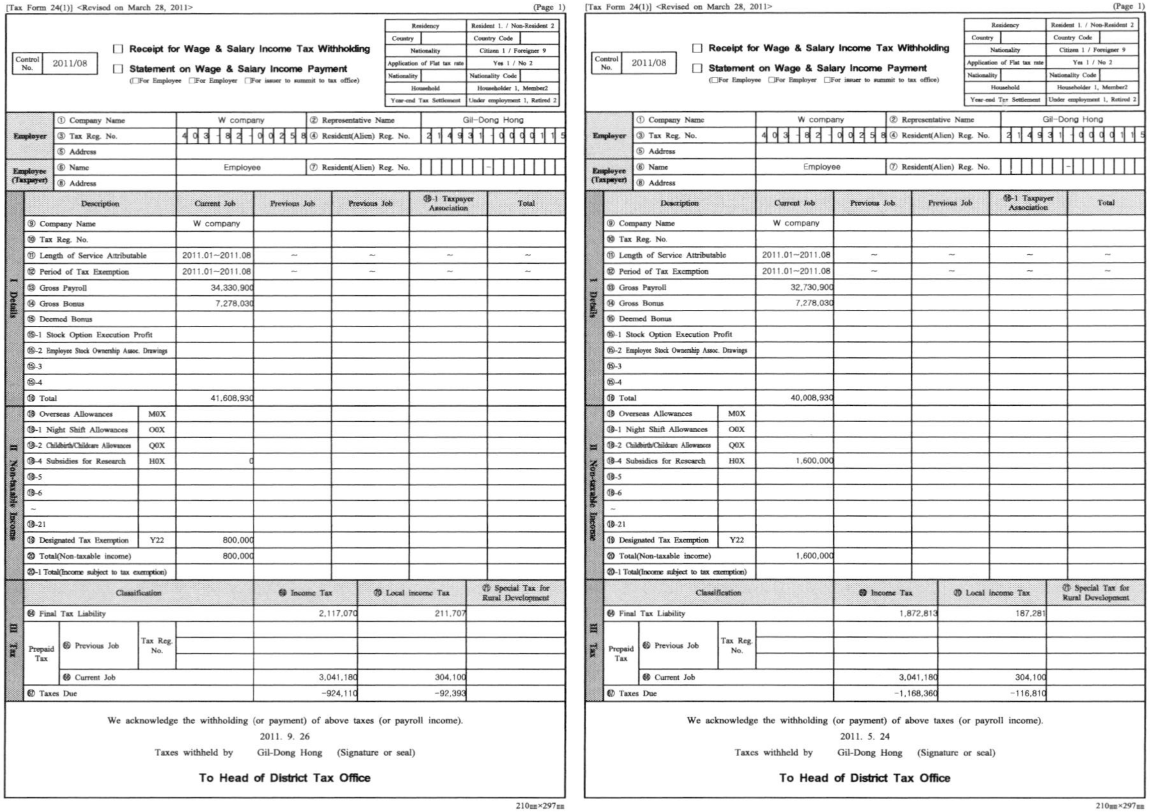

(a)

[Tax Form 24(1)] <Revised on March 28, 2011> (Page 1)

Control No. 2011/08

☐ Receipt for Wage & Salary Income Tax Withholding
☐ Statement on Wage & Salary Income Payment
(☐For Employee ☐For Employer ☐For issuer to summit to tax office)

Residency	Resident 1. / Non-Resident 2
Country	Country Code
Nationality	Citizen 1 / Foreigner 9
Application of Flat tax rate	Yes 1 / No 2
Nationality	Nationality Code
Household	Householder 1, Member2
Year-end Tax Settlement	Under employment 1, Retired 2

Employer	① Company Name	W company	② Representative Name	Gil-Dong Hong
	③ Tax Reg. No.	403-82-00258	④ Resident(Alien) Reg. No.	214931-1000115
	⑤ Address			
Employee (Taxpayer)	⑥ Name	Employee	⑦ Resident(Alien) Reg. No.	-
	⑧ Address			

	Description		Current Job	Previous Job	Previous Job	⑯-1 Taxpayer Association	Total
I Details	⑨ Company Name		W company				
	⑩ Tax Reg. No.						
	⑪ Length of Service Attributable		2011.01~2011.08	~	~	~	~
	⑫ Period of Tax Exemption		2011.01~2011.08	~	~	~	~
	⑬ Gross Payroll		34,330,900				
	⑭ Gross Bonus		7,278,030				
	⑮ Deemed Bonus						
	⑮-1 Stock Option Execution Profit						
	⑮-2 Employee Stock Ownership Assoc. Drawings						
	⑮-3						
	⑮-4						
	⑯ Total		41,608,930				
II Non-taxable Income	⑱ Overseas Allowances	M0X					
	⑱-1 Night Shift Allowances	O0X					
	⑱-2 Childbirth/Childcare Allowances	Q0X					
	⑱-4 Subsidies for Research	H0X	0				
	⑱-5						
	⑱-6						
	~						
	⑱-21						
	⑲ Designated Tax Exemption	Y22	800,000				
	⑳ Total(Non-taxable income)		800,000				
	⑳-1 Total(Income subject to tax exemption)						

	Classification			(69) Income Tax	(70) Local income Tax	(71) Special Tax for Rural Development
III Tax	(64) Final Tax Liability			2,117,070	211,707	
	Prepaid Tax	(65) Previous Job	Tax Reg. No.			
		(66) Current Job		3,041,180	304,100	
	(67) Taxes Due			-924,110	-92,393	

We acknowledge the withholding (or payment) of above taxes (or payroll income).
2011. 9. 26
Taxes withheld by Gil-Dong Hong (Signature or seal)

To Head of District Tax Office

210㎜×297㎜

(b)

[Tax Form 24(1)] <Revised on March 28, 2011> (Page 1)

Control No. 2011/08

☐ Receipt for Wage & Salary Income Tax Withholding
☐ Statement on Wage & Salary Income Payment
(☐For Employee ☐For Employer ☐For issuer to summit to tax office)

Residency	Resident 1. / Non-Resident 2
Country	Country Code
Nationality	Citizen 1 / Foreigner 9
Application of Flat tax rate	Yes 1 / No 2
Nationality	Nationality Code
Household	Householder 1, Member2
Year-end Tax Settlement	Under employment 1, Retired 2

Employer	① Company Name	W company	② Representative Name	Gil-Dong Hong
	③ Tax Reg. No.	403-82-00258	④ Resident(Alien) Reg. No.	214931-1000115
	⑤ Address			
Employee (Taxpayer)	⑥ Name	Employee	⑦ Resident(Alien) Reg. No.	-
	⑧ Address			

	Description		Current Job	Previous Job	Previous Job	⑯-1 Taxpayer Association	Total
I Details	⑨ Company Name		W company				
	⑩ Tax Reg. No.						
	⑪ Length of Service Attributable		2011.01~2011.08	~	~	~	~
	⑫ Period of Tax Exemption		2011.01~2011.08	~	~	~	~
	⑬ Gross Payroll		32,730,900				
	⑭ Gross Bonus		7,278,030				
	⑮ Deemed Bonus						
	⑮-1 Stock Option Execution Profit						
	⑮-2 Employee Stock Ownership Assoc. Drawings						
	⑮-3						
	⑮-4						
	⑯ Total		40,008,930				
II Non-taxable Income	⑱ Overseas Allowances	M0X					
	⑱-1 Night Shift Allowances	O0X					
	⑱-2 Childbirth/Childcare Allowances	Q0X					
	⑱-4 Subsidies for Research	H0X	1,600,000				
	⑱-5						
	⑱-6						
	~						
	⑱-21						
	⑲ Designated Tax Exemption	Y22					
	⑳ Total(Non-taxable income)		1,600,000				
	⑳-1 Total(Income subject to tax exemption)						

	Classification			(69) Income Tax	(70) Local income Tax	(71) Special Tax for Rural Development
III Tax	(64) Final Tax Liability			1,872,813	187,281	
	Prepaid Tax	(65) Previous Job	Tax Reg. No.			
		(66) Current Job		3,041,180	304,100	
	(67) Taxes Due			-1,168,360	-116,810	

We acknowledge the withholding (or payment) of above taxes (or payroll income).
2011. 5. 24
Taxes withheld by Gil-Dong Hong (Signature or seal)

To Head of District Tax Office

210㎜×297㎜

Fig. 1 Earned income withholding receipts. **a** Incorrect version. **b** Correct version

Table 1 Value of subsidies for research

Company	Monthly research funding (non-taxable income)	Etc. (non-taxable income)
W Company 2010.03.01–2010.12.31(10 months)	₩2,000,000 = ₩200,000 × 10 months	₩1,000,000 = ₩100,000 × 10 months
W Company 2011.01.01–2011.08.31(8 months)	₩0	₩800,000 = ₩100,000 × 10 months
C Company 2011.09.01–2011.12.31(4 months)	₩800,000 = ₩200,000 × 4 months	₩400,000 = ₩100,000 × 4 months

depending on the amount of prepaid tax. If the prepaid tax is larger than the final tax liability, then employee gets refund. But if the prepaid tax is smaller than the final tax liability, then employee should pay the rest of them.

4 RESTful Open API for Auditing the Year-End Tax Settlement

With a set of computers connected on a network, there is a vast pool of CPUs and resources, and you have the ability to access files on a cloud. In this paper, we propose a novel approach that realizes the mobile cloud convergence in transparent and platform-independent way. Since we are targeting on OS independent platform, web service is the best fit for the framework that is not depending on a certain smart-phone OS platform. REST web service is a lightweight approach for the provision of services on the web. Unlike WSDL-based web services, the set of operations is reduced, standardized, with well known semantics, and changes the resource's state in REST web service. To this end, we propose the management server architecture for REST web services. By this way, complex business logics and computations will be offloaded by cloud computing platforms. With this mobile cloud computing framework you will be able to enjoy all such application only if you can access web through your cell phone.

REST web service is core technology for smartphone application development. This is because REST web service is the most appropriate way for accessing information through internet. Usually, a smartphone application needs information from several sources of (one or more) REST web services [1]. So, we need to utilize two or more REST web services composition to realize a target application [3, 4]. In this paper, we propose a server architecture for managing REST web services. This server is for managing web services so as to provide web server maintenance, especially on composition, deployment, and management of REST web services. It enables service developers to conveniently develop, deploy, upload, and run their composed web services with the use of general OOP languages.

In this section, we introduce our details of the RESTful Open API for auditing year-end tax settlement. The ③ represents an example of the actual usage of our REST Open API for auditing the year-end tax settlement. The reason why values of some items are 0 is because this case is from the simplified year-end tax settlement. The employee is moved from W company to another C company during the year.

- API description: Open API for auditing year-end tax settlement
- API interface

AuditTaxSettlement (Key, Year, SubsidiesForResearch, Total, EmploymentInsurance, NationalHealthInsurance, QualifiedInsurance, MedicalExpenses,

EducationExpenses, Donations, PrivateSchoolTeachersPension, TaxCreditfor AllSalaryEarners);

- Sample URL (An example case from Fig. 1)

http://embed.cbnu.ac.kr/AuditTaxSettlement?Key=OAuthKey&Year=2012& SubsidiesForResearch=1,600,000&Total=40,008,930&NationalHealthInsurance= 1,148,380&Donations=40,000&QualifiedInsurance=0&MedicalExpenses=0& PrivateSchoolTeachersPension=2,050,900&TaxCreditforAllSalaryEarners= 500,000

- Responses

Results data represent the response for example. This research provides responses as XML format either to make use of the Open API to Object conversion as described in Sect. 3 or to make easily parse/extract the data part that they want.

```
<ResultSet>
  <info>
  <type>Paper Content</type>
  <lastBuildDate>datetime : date of result generated</lastBuildDate>
  <total>integer : total number of documents</total>
  </info>
  <item>
  This is correct earned income withholding receipt!!
  </item>
</ResultSet>
```

- Error messages

The REST web server responses to the Open API request as an error message when there are some errors during the process of Open API execution. The error messages are one of the followings as in the following Table 1.

000	System error
010	Your query request count is over the limit
011	Incorrect query request
020	Unregistered key
021	Your key is temporary unavailable
100	Invalid target value
101	Invalid display value
102	Invalid start value
110	Undefined sort value
200	Reserved
900	Undefined error occurred

5 Concluding Remarks

We proposed a server architecture for managing REST web services. This server is for managing web services so as to provide web server maintenance, especially on composition, deployment, and management of REST web services. It enables service developers to conveniently develop, deploy, upload, and run their composed web services with the use of general OOP languages. The REST web service management server is useful for clients such as smartphone applications. This is because simply by uploading their web service package onto our system, web service developers can operate their service without physical server. Object caching significantly improves performance of our web service composition and management system, especially on sudden batch requests within a short period time. Experimental results show that our REST web service composition management server improves performance up to 35.35 % compared to conventional usage of web services.

Acknowledgments This work is supported by the Basic Science Research Program through the National Research Foundation of Korea (NRF) funded by the Ministry of Education, Science and Technology (2011-0027161). Corresponding author of this paper is Min Choi (mchoi@cbnu.ac.kr).

References

1. Choi, M.: REST web service composition. In: 2nd International Workshop of Mobile Platform, Computer Applications, 2012
2. Choi, M.: A platform-independent smartphone application development Framework. In: 1st International Workshop of Mobile Platform, Computer Applications, 2011
3. Zhao, H., Doshi, P.: Towards automated RESTful web service composition. In: International Conference on Web Services (ICWS), 2009
4. Zhao, X., Liu, E., Clapworthy, GJ., Ye, N., Lu, Y.: RESTful web service composition: extracting a process model from linear logic theorem proving. In: IEEE International Conference on Next Generation Web Service Practice (NWeSP), 2011

BAT: Bimodal Cryptographic Algorithm Suitable for Various Environments

Jesang Lee, Kitae Jeong, Jinkeon Kang, Yuseop Lee, Jaechul Sung, Ku-Young Chang and Seokhie Hong

Abstract In this paper, we propose a new bimodal cryptographic algorithm BAT. BAT provides a hash function BAT-H and a block cipher BAT-B. Moreover, according to some parameters, it is possible to combine BAT-H and BAT-B to one integrated module. Thus, the algorithm is suitable for various environments, such as RFID and USN, where a hash function and a block cipher are required simultaneously. From our implementation results, our integrated module is more efficient than the case of the combination of known dedicated hash functions and block ciphers.

Keywords Block cipher · Hash function · Integrated module · BAT

J. Lee · K. Jeong · J. Kang · Y. Lee · S. Hong (✉)
Center for Information Security Technologies, Korea University, Seoul, Korea
e-mail: shhong@korea.ac.kr

J. Lee
e-mail: jesang.lee@gmail.com

K. Jeong
e-mail: kite.jeong@gmail.com

J. Kang
e-mail: jinkeon.kang@gmail.com

Y. Lee
e-mail: Yusubi@korea.ac.kr

J. Sung
Department of Mathematics, University of Seoul, Seoul, Korea
e-mail: jcsung@uos.ac.kr

K.-Y. Chang
Cyber Security-Convergence Research Department, Electronics and Telecommunication Research Institute, Daejeon, Korea
e-mail: jang1090@etri.re.kr

S.-S. Yeo et al. (eds.), *Computer Science and its Applications*,
Lecture Notes in Electrical Engineering 203, DOI: 10.1007/978-94-007-5699-1_37,
© Springer Science+Business Media Dordrecht 2012

1 Introduction

Recently, the research on lightweight cryptographic primitives has been received considerable attention. Since these primitives can be efficiently implemented under restricted resources such as low-cost, low-power and lightweight platforms, they are applicable to low-end devices such as RFID tags, sensor nodes and smart devices. So far, many lightweight cryptographic algorithms (e.g., block ciphers such as PRESENT [1], LED [2], HIGHT [3] and Piccolo [4], and hash functions such as QUARK [5], H-PRESENT [6], SPONGENT [7] and PHOTON [8]) have been proposed. However, most of them are cryptographic algorithms which are dedicated only to a block cipher or a hash function. Therefore, in order to construct a cryptosystem providing both confidentiality (e.g., block ciphers) and integrity (e.g., hash functions), we should consider several cryptographic algorithms simultaneously.

So far, there have been several attempts to design a multi-purpose cryptographic algorithm (e.g., ARMADILLO [9], DM-PRESENT [6] and H-PRESENT [6]). Most of them are based on the modes of operation. In detail, DM-PRESENT and H-PRESENT are hash functions based on block cipher PRESENT, and ARMADILLO can be operated as PRNG based on the modes of operation. Thus, the construction of a hash function and a block cipher in one integrated module is one of the interesting topics. To our knowledge, this approach had not been studied until now.

In this paper, we propose a new bimodal cryptographic algorithm BAT (Bimodal cryptographic AlgoriThm). BAT includes a hash function BAT-H and a block cipher BAT-B. BAT-H is the hash function generating 128/256/384-bit hash values. According to the length of hash value, we denote it BAT-H128/256/384, respectively. BAT-B is the block cipher with 64/128/192-bit block and 64/128/192-bit secret key, respectively. According to the length of data block, we denote it BAT-B64/128/192, respectively. The parameters of BAT are shown in Table 1. Comparing with known cryptographic algorithms, the superior property of BAT is that it is possible to combine BAT-H and BAT-B for some parameters by using a multiplexer. In detail, we can construct an integrated module BAT_i as follows ($i = 1, 2, 3$).

- BAT_1: BAT-H128 + BAT-B64
- BAT_2: BAT-H256 + BAT-B128
- BAT_3: BAT-H384 + BAT-B192.

Thus, our algorithm can be applied to various environments, such as RFID and USN, where a hash function and a block cipher are required simultaneously. Note that we do not consider the modes of operations and hash functions based on block ciphers, since these approaches are different from ours. Thus, we compare our algorithm with known dedicated hash functions and block ciphers.

Table 2 presents the area requirements for BAT and the combination of known dedicated hash functions and block ciphers where the parameters of each ciphers are similar to BAT_i, respectively. From our hardware implementation results, the area requirements for BAT_i are 1,762, 3,489 and 5,199 gate equivalents, respectively. From this table, the area requirements for BAT_1 and BAT_3 are smaller than

Table 1 The parameters of BAT

BAT-H	Hash value (bits)	# of steps	BAT-B	Block (bits)	Secret key (bits)	# of rounds
BAT-H128	128	48	BAT-B64	64	64	24
BAT-H256	256	48	BAT-B128	128	128	24
BAT-H384	384	48	BAT-B192	192	192	24

Table 2 The comparison between BAT and other combinations

The combination	Area (GE)
BAT₁ (BAT-H128 + BAT-B64)	**1,762**
SPONGENT-128 + LED-64	2,026 (= 1,060 + 966)
PHOTON-128 + LED-64	2,342 (= 1,122 + 1220)
BAT₂ (BAT-H256 + BAT-B128)	**3,489**
SPONGENT-256 + Piccolo-128*	3,284 (= 1,950 + 1,334)
PHOTON-256 + LED-128	3,442 (= 2,177 + 1,265)
SHA-256 + AES-128	10988 (= 8,588 + 2,400)
BAT₃ (BAT-H384 + BAT-B192)	**5,199**
SHA-384 + ARMADILLO2-D)	49,884 (= 43,330 + 6,554)

* With registers for storing the secret key

other combinations, respectively. In the case of BAT_2, it has the similar efficiency to (SPONGENT-256 + Piccolo-128) and (PHOTON-256 + LED-128). It means that BAT is the most suitable for even constrained environments, such as RFID and USN, where a hash function and a block cipher are required simultaneously. Note that the area requirement for Piccolo-128 is measured without registers for storing the secret key. Thus, for fairness, we considered a version of Piccolo-128 with registers for storing the secret key (758 GE + 576 GE). In fact, the area requirements for these combinations can be reduced by sharing state registers in a block cipher and a hash function. However, since each result is based on serialized implementation dedicated to the target algorithm, sharing state registers will cause overhead for control logic. Hence, we just added the area requirements for other combinations.

The rest of this paper is organized as follows. In Sect. 2, we introduce the specification of BAT. In Sect. 3, the design rationale of BAT is presented. Security analysis and implementation results are provided in Sects. 4 and 5, respectively. Finally, we give our conclusion in Sect. 6.

2 Specification

In this section, we introduce the structure of BAT. Since both BAT-H and BAT-B are based on the G function, we first present the G function. Then BAT-H and BAT-B are presented. At last, we introduce the method to construct an integrated module.

Table 3 S-box of the G function

w	0x0	0x1	0x2	0x3	0x4	0x5	0x6	0x7	0x8	0x9	0xA	0xB	0xC	0xD	0xE	0xF
$Sbox(w)$	0x1	0x2	0x4	0xB	0xD	0xE	0xA	0x5	0xF	0x8	0x9	0x7	0x6	0x3	0x0	0xC

2.1 G Function

The G function consists of the S-box layer S-layer and the diffusion layer D-layer: $G(W) \mapsto D(S(W))$. S-layer is a substitution function which satisfies the confusion property on each 4-bit word. A 32-bit input word W is divided into eight 4-bit words $(w_0, \cdots, w_7)$ and then each 4-bit word goes through a bijective S-box. S-layer is defined as follows: $S(w_0, \cdots, w_7) \mapsto (w_0', \cdots, w_7')$, where $w_i' = Sbox(w_i)$ $(i = 0, \cdots, 7)$. The 4×4 S-box $Sbox$ is shown in Table 3.

The diffusion layer D-layer is a permutation which satisfies the diffusion property. A 32-bit input word W is split into eight 4-bit words $(w_0, \cdots, w_7)$ and D-layer mixes all 4-bit words. It is the same as the P function of Camellia [10].

$$
\begin{pmatrix} w_0' \\ w_1' \\ w_2' \\ w_3' \\ w_4' \\ w_5' \\ w_6' \\ w_7' \end{pmatrix}
=
\begin{pmatrix}
0 & 1 & 1 & 1 & 1 & 0 & 0 & 1 \\
1 & 0 & 1 & 1 & 1 & 1 & 0 & 0 \\
1 & 1 & 0 & 1 & 0 & 1 & 1 & 0 \\
1 & 1 & 1 & 0 & 0 & 0 & 1 & 1 \\
0 & 1 & 1 & 1 & 1 & 1 & 1 & 0 \\
1 & 0 & 1 & 1 & 0 & 1 & 1 & 1 \\
1 & 1 & 0 & 1 & 1 & 0 & 1 & 1 \\
1 & 1 & 1 & 0 & 1 & 1 & 0 & 1
\end{pmatrix}
\begin{pmatrix} w_0 \\ w_1 \\ w_2 \\ w_3 \\ w_4 \\ w_5 \\ w_6 \\ w_7 \end{pmatrix}
$$

2.2 Hash function BAT-H

Algorithm 1 BAT-H($128 \cdot i$)(M)
Input: t padded message blocks $M = (M_0, M_1, \cdots, M_{t-1})$
Output: ($128 \cdot i$)-bit hash value (H_0, H_1, H_2, H_3)
1. $S = (S_0, \cdots, S_{4i-1}) = (0, 0, \cdots, 0, 128 \cdot i)$;   //initialization
2. $\text{Perm}_i(S)$;
3. For $j = 0$ to $t - 1$ //absorbing phase
   − For $k = 0$ to $i - 1$
   • $S_{4k+3} = S_{4k+3} \oplus M_{j, k-1}$;
   − $\text{Perm}_i(S)$;
4. $H_0 = \text{SQUEEZE}(S, i)$;   //squeezing phase
5. For $k = 1$ to 3
   − $\text{Perm}_i(S)$;
   − $H_k = \text{SQUEEZE}(S, i)$;

BAT-H uses the hermetic sponge strategy as depicted in Fig. 1. According to the digestive length, BAT-H is denoted by BAT-H128/256/384 or BAT-H($128 \cdot i$) for brevity ($i = 1, 2, 3$). The bit rate and the capacity of BAT-H($128 \cdot i$) are $32 \cdot i$ and $96 \cdot i$, respectively. Thus, the width of BAT-H128/256/384, that is the size of internal state, is 128, 256 and 384 bits, respectively. BAT-H($128 \cdot i$) processes a padded message $M = (M_0, M_1, \cdots, M_{t-1})$ as shown in Algorithm 1. Here, $S = (S_0, \cdots, S_{4i-1})$ is an internal state of BAT-H($128 \cdot i$). Note that each S_k and $M_j = (M_{j,0}, \cdots, M_{j,i-1})$ are 32-bit and ($32 \cdot i$)-bit values, respectively. The SQUEEZE function used in squeezing phase is defined as follows.

$$\text{SQUEEZE}(S, i) = \begin{cases} S_3, i = 1; \\ S_3 \parallel S_7, i = 2; \\ S_3 \parallel S_7 \parallel S_{11}, i = 3. \end{cases}$$

An input message is padded and parsed into ($32 \cdot i$)-bit blocks. Our padding method is as follows. Let l be the total length of message bits, a bit 1 is appended to the end of the message, followed by k zero bits, where k is the smallest non-negative integer such that $l + 1 + k = 0 \bmod (32 \cdot i)$.

Perm_i is a permutation of BAT-H($128 \cdot i$), which is defined as shown in Algorithm 2. Here, the parameters of left circular rotations are $rot_0 = 19$, $rot_1 = 1$ and $rot_2 = 14$. As a concrete example, Fig. 3a presents Perm_1 of BAT-H128.

Algorithm 2 $\text{Perm}_i(S)$
1. For $r = 0$ to 47
    − For $k = 0$ to $i − 1$
- $S_{4k+3} = S_{4k+3} \oplus r$;
- $S_{4k} = S_{4k} \oplus S_{4k+1}$;
- $S_{4k+2} = S_{4k+2} \oplus S_{4k+3}$;
- $S_{4k} = S_{4k} \oplus G(S_{4k+2})$;
- $S_{4k+2} = S_{4k+2} \oplus (S_{4k} < \ll rot_k)$;
    − $Temp = S_{4i-1}$;
    − For $k = 4i - 1$ to 1
- $S_k = S_{k-1}$;
    − $S_0 = Temp$;

2.3 Block Cipher BAT-B

BAT-B has the generalized Feistel-like structure. According to the length of data block, we denote it BAT-B64/128/192 or BAT-B($64 \cdot i$) for brevity ($i = 1, 2, 3$). That is, BAT-B($64 \cdot i$) generates a ($64 \cdot i$)-bit ciphertext $C = (C_0, \cdots, C_{2i-1})$ by using a ($64 \cdot i$)-bit plaintext $P = (P_0, \cdots, P_{2i-1})$ and the ($64 \cdot i$)-bit secret key $K = (K_0, \cdots, K_{2i-1})$. Note that each P_i, C_i and K_i are 32-bit values.

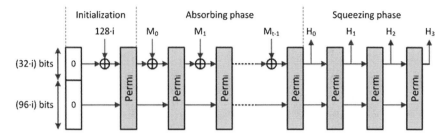

Fig. 1 BAT-H with the modified sponge construction

We focus on the encryption process because the decryption process is similar to the encryption process. Algorithm 3 shows the encryption process of BAT-B($64 \cdot i$). Here, the parameters of left circular rotations are $rot_0 = 19$, $rot_1 = 1$ and $rot_2 = 14$. $S = (S_0, \cdots, S_{2i-1})$ is an internal state of BAT-B($64 \cdot i$).

Algorithm 3 BAT-B($64 \cdot i$)(P, K)
Input: ($64 \cdot i$)-bit plaintext P, ($64 \cdot i$)-bit secret key K
Output: ($64 \cdot i$)-bit ciphertext C
1. For $k = 0$ to $2i - 1$
    $- S_k = P_k$;
    $- RK_k = K_k$;
2. For $r = 0$ to 23
    $-$ Keyschedule($64 \cdot i$)(RK, $2r$);
    $-$ For $k = 0$ to $i - 1$
  $S_{2k+1} = S_{2k+1} \oplus RK_{2i}$;
  $S_{2k} = S_{2k} \oplus G(S_{2k+1})$;
  $S_{2k+1} = S_{2k+1} \oplus (S_{2k} <\ll rot_k)$;
    $-Temp = S_{2i-1}$;
    $-$ For $k = 2i - 1$ to 1
  $S_k = S_{k-1}$;
    $- S_0 = Temp$;
3. For $k = 0$ to $2i - 1$
    $- C_k = S_k$;

The keyschedule of BAT-B($64 \cdot i$) generates total twenty four ($32 \cdot i$)-bit round keys. As shown in Algorithm 4, the structure of it is similar to Algorithm 3 except the use of a round counter Ctr instead of a round key RK. See Fig. 3b for the structure of BAT-B64.

Algorithm 4 Keyschedule ($64 \cdot i$)(RK, Ctr)
1. For $k = 0$ to $i - 1$
    $- RK_{2k+1} = RK_{2k+1} \oplus Ctr$;
    $- RK_{2k} = RK_{2k} \oplus G(RK_{2k+1})$;
    $- RK_{2k+1} = RK_{2k+1} \oplus (RK_{2k} <\ll rot_k)$;
2. $Temp = RK_{2i-1}$;
3. For $k = 2i - 1$ to 1

(a) **(b)** **(c)**

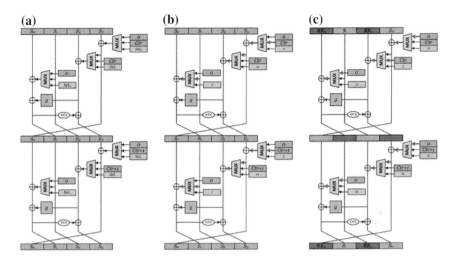

Fig. 2 An integrated module BAT$_1$ (BAT-H128 + BAT-B64)

$$- RK_k = RK_{k-1};$$
4. $RK_0 = Temp;$

2.4 Construction of an Integrated Module

Now, we introduce the method to construct BAT-H and BAT-B in one integrated module. Recall that the structures of BAT-B and Perm of BAT-H are similar as shown in Algorithm 2 and Algorithm 3. This fact enables us to construct an integrated module by combining BAT-B with BAT-H.

To achieve our goal, we first should determine how to deal with internal states. In the case of block ciphers, we should consider internal states used in both the encryption process and the keyschedule independently. However, in hash functions, all internal states are updated interactively. To solve this problem, we use a multiplexer. In electronics, a multiplexer (or mux) is a device that selects one of several analog or digital input signals and forwards the selected input into a single line. In detail, we use the *MUX* function which is defined as follows.

$$MUX(X_0, X_1, Sel) = \begin{cases} X_0, Sel = 0; \\ X_1, Sel = 1. \end{cases}$$

As a concrete example, Fig. 2a presents an integrated module BAT$_1$, the combination of BAT-H128 and BAT-B64 (i.e. BAT-H128 + BAT-B64). BAT$_1$ uses two 1-bit parameters (Sel_0, Sel_1). According to these parameters, BAT$_1$ can be operated as BAT-H128 or BAT-64 optionally. To operate BAT$_1$ as BAT-H128,

(Sel_0, Sel_1) is assigned to $(1, 0)$ as shown in Fig. 2b. In the figure, the bold lines indicate the value that the *MUX* function selects.

In the case of BAT-B64, we first divide the internal state into two parts; one is used for the encryption process and the other is used for the keyschedule. In detail, an internal state (S_0, S_1, S_2, S_3) is assigned to (RK_0, S_1, RK_1, S_0). Since BAT-B64 consists of the encryption part and the keyschedule part, different parameters are used in odd and even rounds, respectively. In odd rounds where the keyschedule part is conducted, (Sel_0, Sel_1) is assigned to $(0, 1)$. On the other hand, (Sel_0, Sel_1) is assigned to $(0, 0)$ in even rounds of the encryption part.

Other versions such as BAT_2 (BAT-H256 + BAT-B128) and BAT_3 (BAT-H384 + BAT-B192) can be constructed similarly. These integrated modules can be also constructed by two parameters.

3 Design Rationale

3.1 G Function

As basic operations, we chose a S-box-based approach rather than an ARX-based approach. Generally, the latter allows simpler and more compact implementations. However, we believe that the former is more secure than the latter. Also, it is well known that the security of an algorithm based on S-boxes can be evaluated more thoroughly.

In the case of S-box, we use a single 4×4 S-box. Since the implementation of 4×4 S-boxes is typically more compact than 8×8 S-boxes, this is a direct consequence of our pursuit of hardware efficiency. Our S-box satisfies the following design criteria. These criteria are based on the design rationale of S-box used in PRESENT [1]. Here, the Fourier coefficient of S is defined as $S_b^W(a) = \sum_{x \in \mathbb{Z}_2^4} (-1)^{<b,S(x)> + <a,x>}$.

1. For any fixed non-zero input difference $\Delta_I \in \mathbb{Z}_2^4$ and any fixed non-zero output difference $\Delta_O \in \mathbb{Z}_2^4$, we require

$$\#\{x \in \mathbb{Z}_2^4 | Sbox(x) + Sbox(x + \Delta_I) = \Delta_O\} \leq 4$$

2. For any fixed non-zero input difference $\Delta_I \in \mathbb{Z}_2^4$ and any fixed output difference $\Delta_O \in \mathbb{Z}_2^4$ such that $wt(\Delta_I) = wt(\Delta_O) = 1$, we have

$$\#\{x \in \mathbb{Z}_2^4 | Sbox(x) + Sbox(x + \Delta_I) = \Delta_O\} \leq 4$$

3. For all non-zero $a \in \mathbb{Z}_2^4$ and non-zero $b \in \mathbb{Z}_2^4$, it holds that $\left| S_b^W(a) \right| \leq 8$.

We checked all 4×4 S-boxes satisfying the above design criteria. From simulation results, we obtained about $2^{21.08}$ 4×4 S-boxes. And then, we ranked them with respect to the number of operations (AND, XOR) (since the cost of NOT operations is generally much smaller than costs of the other operations, we did not consider the number of NOT operations). As a result, we got 18432 4×4 S-boxes. They consist of the same number of operations and have the same security. Thus, we chose one of them.

Second, we selected for the P function of Camellia as a diffusion layer. The 8×8 MDS code provides the best diffusion property, but it is not suitable for the limited implementation environment because it requires more operations when it is implemented in hardware environments. We tried to find an efficient 8×8 MDS code, but we could not find any efficient MDS code with properties found in AES or Whirlpool. Since the binary matrix multiplication provides both efficiency on hardware implementation and sufficient avalanche effects, we chose the P function of Camellia whose security and efficiency has been verified.

Finally, S-box and a binary diffusion matrix are nibble-oriented operations. Thus, in order to maximize the avalanche effect, we use non-nibble rotation parameters. Therefore, through these three operations (a 4×4 S-box, a binary diffusion matrix and a left circular rotation), we expect that BAT has the enough security.

3.2 The Hermetic Sponge Strategy in BAT-H

The sponge function has proven that the sponge construction has the preimage resistance, the second preimage resistance, the collision resistance, and the in-differentiability under the assumption that the compression function is a random permutation [11, 12]. In contrast with HAIFA or Wide-Pipe, the sponge function uses small internal states and does not require any additional message expansion. Hence, the sponge construction method is more suitable for limited environments like low-end devices.

However, it is difficult to construct the sponge function based on a random permutation. So, the hermetic sponge strategy is generally considered [13]. In this strategy, we should only an underlying permutation that should not have any structural distinguishers.

4 Security Analysis

4.1 Security Analysis of BAT-H

The security claims of the sponge function are shown in Table 4. Here, c is the capacity and r is the bit rate.

Table 4 Security claims of the sponge function

Security	Sponge	BAT-H128	BAT-H256	BAT-H384
Collision	$\min(2^{c/2}, 2^{n/2})$	2^{48}	2^{96}	2^{144}
Preimage	$\min(2^{c}, 2^{n-r})$	2^{96}	2^{192}	2^{288}
Second preimage	$\min(2^{c/2}, 2^{n})$	2^{48}	2^{96}	2^{144}

Differential cryptanalysis on Perm Since Perm consists of left circular rotations and the G function, finding differential characteristics with high probability seems to be hard. So, we checked some particular types of differential characteristics of $Perm_i$, such as iterative differential characteristics and differential characteristics which passed minimal active G functions. We first attempted to find iterative differential characteristics of $Perm_i$ ($i = 1, 2, 3$). From simulation results, for each $Perm_i$, there exist several 1-round iterative differential characteristics with a probability of $2^{-11 \cdot i}$ (see Table 5). It means that an attacker cannot distinguish Perm and a random permutation by using $r (\geq 12)$-round differential characteristics from these 1-round iterative differential characteristics. Also, we found that iterative differential characteristics on 2 or more rounds have a lower probability than 1-round iterative differential characteristics.

Second, we considered differential characteristics where input differences of the G function in many rounds are zero. In the case of BAT-H128, we found several 3-round differential characteristics with a probability of 1 and we could extend them to 8-round differential characteristics with a probability 2^{-24} (this differential characteristic includes four active G functions). However, we could not apply them to more rounds. Similarly, we found differential characteristics including a small number of G functions on small round of BAT-H256/384. However, it is not possible to apply them to more rounds, either.

Integral attack and algebraic attack To show the strength of $Perm_i$ against an integral attack, we considered all possible 4-bit word input patterns that can be used in integral attacks. For the convenience of analysis, we considered a tweaked version of $Perm_i$. In detail, the rotation parameters (19, 1, 14) are modified to (20, 4, 16). Since a tweaked version consists of only nibble-based operations differently from the original version, it is reasonable that a tweaked version is less secure than the original version against an integral attack. As a result, there is no integral property in a 10-round tweaked version. Thus, we expect that $Perm_i$ has the enough resistance to an integral attack.

In algebraic attacks, we first derive an over-defined system of algebraic equations. In the case of BAT-H, the S-box has the degree of 3 as a vector Boolean function. However, a diffusion layer, a left circular rotation and XOR operation give the superior diffusion property, and the number of rounds is enough to reach a sufficient degree. Therefore, it is impossible to convert any equation system in $Perm_i$ into an over-defined system.

Other cryptanalysis The slide attack was recently applied to sponge-like hash functions [14]. This attack is to exploit the self-similarity of a permutation. In the case of a permutation level, since all steps of $Perm_i$ use different constants, it is

Table 5 1-round iterative differential characteristics of Perm

	Input difference	Probability
Perm$_1$	$(\alpha, 0, 0, \alpha)$	2^{-11}
Perm$_2$	$(\alpha, 0, 0, \alpha, \alpha, 0, 0, \alpha)$	2^{-22}
Perm$_3$	$(\alpha, 0, 0, \alpha, \alpha, 0, \alpha, \alpha, 0, 0, \alpha)$	2^{-33}

impossible to apply a slide attack to it. In the case of a hash mode level, since the padding rule of BAT-H prevents the last message block to be zero, BAT-H has the enough strength against a slide attack.

The cryptanalysis for ARX primitives is newly introduced and applied to several SHA-3 candidates [15]. This attack uses the evolution of a rotated variant of some input words through the round process. Recall that Perm$_i$ uses a left circular rotation operation and consists of the G function, which is based on the S-box and matrix multiplication, instead of addition operations. Thus, any rotational property is not remained by the application of the G function. Therefore, this attack cannot be applied to Perm$_i$.

A rebound attack is one of the most popular methods to analyze hash functions based on AES-like permutations. Recently, a rebound attack against 11-round Feistel structure is introduced [16]. Recall that BAT-H has basically a generalized Feistel-like structure. However, our algorithm includes several additional operations such as XOR operations between some branches, which enable to have the more avalanche effects. Moreover, the number of steps in Perm$_i$ is 48. Thus, we expect that BAT-H has the enough security against a rebound attack.

5 Security Analysis of BAT-B

Differential and linear cryptanalysis First, the resistance of a block cipher against differential cryptanalysis depends on probabilities of differential characteristics, which are the paths from a plaintext difference to a ciphertext difference. We first tried to find the best differential characteristics of BAT-B64. However, since it is impossible to find all differential characteristics of BAT-B64 for the given 2^{64} possible input differences, we simulated to find the best differential characteristics of it when all possible 2^{32} input differences $(\alpha, 0)$ are given. From simulation results, we found thirteen 11-round differential characteristics $(\alpha, 0) \rightarrow ((\alpha << 19), \alpha)$ with a probability of 2^{-60}, where $\alpha \in \{$0x201d0e20, 0x20202c2c, 0x2cc000 cc, 0x82101046, 0x84400045, 0x90600fe7, 0xb0026f30, 0xb0062f70, 0xc9400094, 0xc9800098, 0xe6200062, 0xe6600066, 0xec6000c6$\}$. Applying the above method to BAT-B128, we found one 7-round differential characteristic $(\beta, 0, \beta, 0) \rightarrow ((\beta << 1), \beta, (\beta << 19), \beta)$ with a probability of 2^{-99}, where $\beta = $ 0x005ea276. In the case of BAT-B192, we could not find any differential characteristic where a input difference is $(\gamma, 0, \gamma, 0, \gamma, 0)$. However, we checked that a 6-round differential characteristic $(\gamma_1, 0, \gamma_2, 0, \gamma_3, 0) \rightarrow (\gamma_1, 0, \gamma_2, 0, \gamma_3, 0)$ which includes at least 63 active S-boxes. It means that the probability of this characteristic is less than or equal to 2^{-126}. Therefore, considering that the number of rounds of BAT-B is 24, we expect that these algorithms have the enough security against differential cryptanalysis.

With respected to linear cryptanalysis, we found several 11-round linear approximations with $\varepsilon^2 = 2^{-62}$. They were constructed by iterating linear approximations on small rounds. In the case of BAT-B128/192, we could not find any more good linear approximation. However, we checked that they have the similar property to BAT-B64. Thus, it is reasonable for us to expected that BAT-B has the enough security against linear cryptanalysis.

Integral attack and slide attack To evaluate the security against an integral attack, we considered a tweaked version of BAT-B similarly to the case of BAT-H. In detail, the rotation parameters (19, 1, 14) are modified to (20, 4, 16). As a result, there is no integral property in a 10-round tweaked version. Thus, we expect that BAT-B has the enough resistance to an integral attack.

A slide attack makes the number of rounds in a cipher obsolete. The iterated ciphers with an identical round function are known to be vulnerable to this attack. However, since BAT-B uses a different constant for each round, it is secure against a slide attack.

Related-key and weak-key cryptanalysis In a related-key attack, it is assumed that an attacker has an ability to control the condition of the secret key. The simple keyschedule makes it easy to construct the specific differential patterns of round keys by using related-key differences. If finding differential characteristics of the keyschedule is as hard as finding them of the round function, it is known to be hard to apply a related-key attack to the target algorithm. Because the keyschedule and round function of BAT-B have the same structure, it is hard to construct differential patterns of each round key. Various cryptanalytic results including differential cryptanalysis and linear cryptanalysis on BAT-B prove these facts. Hence, BAT-B is resistant against the attacks involving related-key cryptanalysis.

A weak key is a key that makes a specific cipher behave in some undesirable way. But it takes only modest portion of the overall key space. Because the keyschedule of BAT-B has a strong non-linearity and an avalanche effect, it has the sufficient randomness. Thus, it is hard for an attacker to find a weak key for BAT-B. Even if he finds a weak key, this weak key cannot be used in other attacks.

6 Implementations Results

6.1 Hardware Implementation

We implemented BAT in Verilog-HDL and used Mentor Graphics ModelSim SE 6.2c for the functional simulation. The functional simulation was performed by using test vectors generated by the software implementation for test benches. In order to evaluate area and throughput, the code was synthesized by Synopsys Design

Compiler which is based on the TSMC 0.25 µm CMOS technology. One gate equivalent (GE) is equivalent to the area of a NAND gate. We used Synopsys Power Compiler to estimate the power consumption with a typical voltage of 2.5 V. For synthesis and power estimation, we advised the compiler to use a clock frequency of 100 kHz, a typical operating frequency of cryptographic modules for RFID tags.

Tables 6 and 7 show the comparison between hardware implementations of BAT and several known dedicated hash functions and block ciphers, respectively. Note that we do not consider the modes of operations and hash functions based on block ciphers by using hashing modes such as DM (Davies-Mayer), MMO (Matyas-Meyer-Oseas) and MP (Miyaguchi-Preneel). Because the main goal of BAT is to implement a hash function and a block cipher in one cryptographic module, these approaches are different from ours. Thus, we compare our algorithm with the combinations of known dedicated hash functions and block ciphers.

In the hardware implementation for lightweight application, the area requirement is one of the important measures. Thus, if the area requirement of BAT is smaller than that of the combination of known block cipher and hash function, it is reasonable to say that our algorithm is more efficient than it. From Table 6 and 7, we selected one hash function and one block cipher where the parameters are similar to BAT_i, respectively ($i = 1, 2, 3$). Then, we compute the area requirements for BAT_i with it. The comparative result is shown in Table 2.

6.2 Software Implementation

Software implementations are optimized by using table lookups and the measurements were performed on an Intel (R) Core (TM) i7 CPU 950 clocked at 3.07 GHz. As a simulation result, BAT-H128/256/384 run at 106, 126 and 154 Mbps, respectively. We checked that the software performances of BAT-H are similar to that of known lightweight hash functions such as PHOTON. In the case of BAT-B64/128/192, they run at 430, 508 and 604 Mbps, respectively. From these results, BAT-B has a half software performance of AES (about 1.2 Gbps).

7 Conclusion

In this paper, we proposed a bimodal cryptographic algorithm BAT. BAT provides a hash function BAT-H and a block cipher BAT-B. A remarkable feature of BAT is that it is possible to combine BAT-H and BAT-B to one integrated module. Thus, this algorithm is suitable for various environments, such as RFID and USN, where a hash function and a block cipher are required simultaneously. From hardware implementation results, BAT_1 and BAT_3 are more efficient than all possible combinations of known hash functions and block ciphers including lightweight cryptographic algorithms. In the case of BAT_2, it has the similar efficiency to them. In the case of the software performance, BAT-H has an almost similar performance to lightweight hash functions and BAT-B has a half

Table 6 Hardware implementation results of some hash functions

Hash function	DigestSize (bits)	Block size (bits)	Cycles per block	Throughput at 100 kHz (kbps)	Logic process (m)	Area (GE)	FOM
BAT-H128	**128**	**32**	**48**	**66**	**0.25**	**1,605**	**258.80**
H-PRESENT-128	128	128	559	11.45	0.18	2,330	21.09
[6]	128	128	32	200	0.18	4,256	110.41
ARMADILLO2-	128	64	256	250	0.18	4,353	13.19
B [9]	128	64	64	1000	0.18	6,025	27.55
MD4 [17]	128	512	456	112.28	0.18	7,350	20.78
MD5 [17]	128	512	612	83.66	0.18	8,400	11.86
U-QUARK [5]	128	8	544	1.47	0.18	1,379	7.73
	128	8	68	11.76	0.18	2,392	20.56
PHOTON-128	128	16	996	1.61	0.18	1,122	12.78
[8]	128	16	156	10.26	0.18	1,708	35.15
SPONGENT-128	128	8	2,380	0.34	0.13	1,060	2.99
[7]	128	16	70	11.43	0.13	1,687	40.16
BAT-H256	**256**	**64**	**48**	**133**	**0.25**	**3,193**	**130.78**
ARMADILLO2-	256	128	512	25	0.18	8,653	3.34
E [9]	256	128	128	100	0.18	11,914	7.05
SHA-256 [18]	256	512	490	104.48	0.25	8,588	14.17
BLAKE [19]	256	32	816	72.79	0.18	13,575	0.21
Grostl [20]	256	64	196	261.14	0.18	14,622	1.53
PHOTON-256	256	32	156	3.21	0.18	2,177	6.78
[8]	256	32	156	20.51	0.18	4,362	10.17
SPONGENT-256	256	16	9,520	0.17	0.13	1,950	0.44
[7]	256	16	140	11.43	0.13	3,281	10.62
BAT-H384	**384**	**96**	**48**	**200**	**0.25**	**4,753**	**88.53**
SHA-384 [21]	384	1,024	84	1219.04	0.13	43,330	6.49

Table 7 Hardware implementation results of some block ciphers

Block cipher	Key size (bits)	Block size (bits)	Cycles per block	Throughput at 100 kHz (kbps)	Logic process (m)	Area (GE)	FOM
BAT-B64	**64**	**64**	**48**	**133**	**0.25**	**1,423**	**658.46**
LED-64 [2]	64	64	1,248	5.1	0.18	966	55
KLEIN-64 [22]	64	64	207	30.9	0.18	1,220	207.72
BAT-B128	**128**	**128**	**48**	**266**	**0.25**	**2,804**	**339.17**
AES [23]	128	128	226	56.6	0.13	2,400	98
HIGHT [3]	128	64	34	188.0	0.25	3,048	202.61
LED-128 [2]	128	64	1,872	3.4	0.18	1,265	21
PRESENT-128 [1]	128	64	32	200.0	0.18	1,886	562
ARMADILLO2-B [9]	128	192	64	300	0.18	6,025	82.64
	128	192	256	75	0.18	4,353	39.58
Piccolo-128 [4]	128	64	528	12.1	0.13	1,334(1)	68.11
	128	64	33	193.9	0.13	1,773(2)	616.94
BAT-B192	**192**	**192**	**48**	**400**	**0.25**	**4,171**	**229.92**
ARMADILLO2-D [9]	192	288	96	300	0.18	8,999	37.04
	192	288	384	75	0.18	6,554	17.46

performance of AES. From these results, it is reasonable to say that BAT has the enough security, efficiency and applicability simultaneously.

Acknowledgments This work was supported the IT R&D program of MKE, Korea [Development of Privacy Enhancing Cryptography on Ubiquitous Computing Environment].

Appendix: Figure

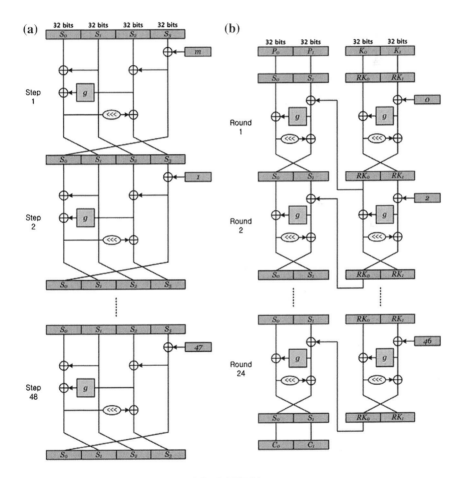

Fig. 3 The structures of (a) Perm₁ and (b) BAT-B64

References

1. Bogdanov, A., Knudsen, L., Leander, G., Paar, C., Poschmann, A., Robshaw, M., Seurin, Y., Vikkelsoe, C.: PRESENT: An ultra-lightweight block cipher. In: Paillier, P., Verbauwhede, I. (eds.) CHES 2007, LNCS, vol. 4727, pp. 450–466. Springer, Berlin (2007)
2. Guo, J., Peyrin, T., Poschmann, A., Robshaw, M.: The LED block cipher. In: Preneel, B., Takagi, T. (eds.) CHES 2011, LNCS, vol. 6917, pp. 326–341. Springer, Berlin (2011)
3. Hong, D., Sung, J., Hong, S., Lim, J., Lee, S., Koo, B., Lee, C., Chang, D., Lee, J., Jeong, K., Kim, H., Kim, J., Chee, S.: HIGHT: A new block cipher suitable for low-resource device. In: Goubin, L., Matsui, M. (eds.) CHES 2006, LNCS, vol. 4249, pp. 46–59. Springer, Berlin (2006)
4. Shibutani, K., Isobe, T., Hiwatari, H., Mitsuda, A., Akishita, T., Shirai, T.: Picollo: An ultra-lightweight blockcipher. In: Preneel, B., Takagi, T. (eds.) CHES 2011, LNCS, vol. 6917, pp. 342–357. Springer, Berlin (2011)
5. Aumasson, J., Henzen, L., Meier, W., Naya-Plasencia, M.: Quark: A lightweight hash. In: Mangard, S., Standaert, F.-X. (eds.) CHES 2010, LNCS, vol. 6225, pp. 1–15. Springer, Berlin (2010)
6. Bogdanov, A., Leander, G., Paar, C., Poschmann, A., Robshaw, M., Seurin, Y.: Hash functions and RFID tags: Mind the gap. In: Oswald, E., Rohatgi, P. (eds.) CHES 2008, LNCS, vol. 5154, pp. 283–299. Springer, Berlin (2008)
7. Bogdanov, A., Knežević, M., Leander, G., Toz, D., Varıcı, K., Verbauwhede, I.: SPONGENT: A lightweight hash function. In: Preneel, B., Takagi, T. (eds.) CHES 2011, LNCS, vol. 6917, pp. 312–325. Springer, Berlin (2011)
8. Guo, J., Peyrin, T., Poschmann, A.: The PHOTON family of lightweight hash functions. In: Rogaway, P. (ed.) Crypto 2011, LNCS, vol. 6841, pp. 222–239. Springer, Berlin (2011)
9. Badel, S., Dagtekin, N., Nakahara, J., Ouafi, K., Reffe, N., Sepehrdad, P., Susil, P., Vaudenay, S.: ARMADILLO: A Multi-purpose cryptographic primitive dedicated to hardware. In: Mangard, S., Standaert, F.-X. (eds.) CHES 2010, LNCS, vol. 6225, pp. 398–412. Springer, Berlin (2010)
10. Aoki, K., Ichikawa, T., Kanda, M., Matsui, M., Moriai, S., Nakajima, J., Tokita, T.: Camellia: A 128-bit block cipher suitable multiple platforms—design and analysis. In: Stinson, D. R., Tavares, S. (eds.) SAC 2000, LNCS, vol. 2012, pp. 39–56. Springer, Berlin (2000)
11. Bertoni, G., Daemen, J., Peeters, M., Assche, G.: On the indifferentiability of the sponge construction. In: Smart, N. (ed.) EUROCRYPT 2008, LNCS, vol. 4965, pp. 181–197. Springer, Berlin (2008)
12. Bertoni, G., Daemen, J., Peeters, M., Assche, G.: Sponge-based pseudo-random number generators. In: Mangard, S., Standaert, F.-X. (eds.) CHES 2010, LNCS, vol. 6225, pp. 33–47. Springer, Berlin (2010)
13. Bertoni, G., Daemen, J., Peeters, M., Assche, G.: Cryptographic sponge functions. Available at http://sponge.noekeon.org/CSF-0.1.pdf. (2011)
14. Gorski, M., Lucks, S., Peyrin, T.: Slide attacks on a class of hash functions. In: Pieprzyk, J. (ed.) ASIACRYPT 2008, LNCS, vol. 5350, pp. 143–160. Springer, Berlin (2008)
15. Khovratovich, D., Nikolic, I., Rechberger, C.: Rotational rebound attacks on reduced skein. In: Abe, M. (ed.) ASIACRYPT 2010, LNCS, vol. 6477, pp. 1–19. Springer, Berlin (2010)
16. Sasaki, Y., Yasuda, K.: Known-key distinguishers on 11-round feistel and collision attacks on its hashing modes. In: Joux, A. (ed.) FSE 2011, LNCS, vol. 6733, pp. 397–415. Springer, Berlin (2011)
17. Feldhofer, M., Rechberger, C.: A case against currently used hash functions in RFID protocols. In: Meersman, R., Tari, Z., Herrero, P. (eds.) OTM 2006, LNCS, vol. 4277, pp. 372–381. Springer, Berlin (2006)
18. Kim, M., Ryou, J., Jun, S.: Efficient hardware architecture of SHA-256 algorithm for trusted mobile computing. In: Yung, M., Liu, P., Lin, D. (eds.) Inscrypt 2008, LNCS, vol. 5487, pp. 240–252. Springer, Berlin (2009)

19. Henzen, L., Aumasson, J., Meier, W., Phan, R.: LSI characterization of the cryptographic hash function BLAKE. Available at http://131002.net/data/papers/HAMP10.pdf. (2010)

20. Tillich, S., Feldhofer, M., Issovits, W., Kern, T., Kureck, H., Mhlberghuber, M., Neubauer, G., Reiter, A., Kofler, A., Mayrhofer, M.: Compact hardware implementations of the SHA-3 candidates ARIRANG, BLAKE, Gröstl, and Skein. Cryptology ePrint Archive, Report 2009/349. Available at http://eprint.iacr.org/2009/349. (2009)

21. Lee, Y., Chan, H., Verbauwhede, I.: Iteration Bound Analysis and Throughput Optimum Architecture of SHA-256 (384, 512) for Hardware Implementations, WISA 2007, LNCS, vol. 4867, pp. 102–114. Springer, Berlin (2007)

22. Gong, Z., Nikova, S., Law, Y.-W.: KLEIN: A new family of lightweight block ciphers. In: Juels, A., Paar, C. (eds.) RFIDSec 2011, LNCS, vol. 7055, pp. 1–18. Springer, Berlin (2012)

23. Moradi, A., Poschmann, A., Ling, S., Paar, C., Wang, H.: Pushing the limits: A very compact and a threshold implementation of AES. In: Paterson, K. G. (ed.) EUROCRYPT 2011, LNCS, vol. 6632, pp. 69–88. Springer, Berlin (2011)

Applying Forensic Approach to Live Investigation Using XeBag

Kyung-Soo Lim and Changhoon Lee

Abstract The law enforcement agencies in the worldwide are confiscating or retaining computer systems involved in a crime/civil case at the preliminary investigation stage, even though the case does not involve a cyber-crime. They are collecting digital evidences from the suspects's systems and using them in the essential investigation procedure. It requires much time, though, to collect, duplicate and analyze disk images in general crime cases, especially in cases in which rapid response must be taken such as kidnapping and murder cases. It is efficient and effective to selectively collect only traces of the behavior of the user activities on operating systems or particular files in focus of triage investigation in live system. On the other hand, if we just acquire essential files from target computer, it is not suitable forensically soundness. Therefore, we need to use standard digital evidence container to prove integrity and probative of evidence from various digital sources. In this article, we describe a forensic approach to live investigation using Xebeg, which is easily able to preserve collected digital evidences selectively for using general technology such as XML and PKZIP compression technology, which is satisfied with generality, integrity, unification, scalability and security.

Keywords Digital forensics · Live investigation · Incident response · Digital evidence container

K.-S. Lim
Convergence Service Security Research Laboratory, Electronics and Telecommunications Research Institute (ETRI), Daejeon, South Korea
e-mail: lukelim@etri.re.kr

C. Lee (✉)
Department of Computer Science and Engineering, Seoul National University of Science & Technology, Seoul, South Korea
e-mail: chlee@snut.ac.kr

S.-S. Yeo et al. (eds.), *Computer Science and its Applications*,
Lecture Notes in Electrical Engineering 203, DOI: 10.1007/978-94-007-5699-1_38,
© Springer Science+Business Media Dordrecht 2012

1 Introduction

Currently, computer forensic paradigm is in the midst of change, from the conventional disc-image-based to evidence-content-based investigation. As digital forensic technology has been enlarged to conventional civil/criminal cases, the scope of seizure has enlarged to digital evidence by a warrant. On the other hand, an efficient investigation methodology is required because a single hard disc volume of normal computers reaches more than 1 terabyte. Accordingly, methodologies for the selective collection and analysis of evidences are becoming increasingly important [1–3].

The conventional methodology, wherein disc images are collected and analyzed, requires huge time to investigate. In cases which are needed rapid response, it is more appropriate to select and collect relevant evidence. Therefore, it is efficient to selectively collect only traces of the behavior of the user on the target devices.

These contents-based digital investigations cannot be easily presented in the court currently, because there is still no standard digital evidence container for containing and describing particular digital evidence such as a word document, image file etc. Most recent studies of forensic community on digital evidence storage formats are related to the improvement of disc image format. A digital evidence storage format is required to meet the requirements of the selective digital evidence investigation technique. Therefore, we presented a concept of portable digital evidence format in IEEE International Conference on Control System, Computing and Engineering 2011 (ICCSCE 2011); we called XeBag [1]. XeBag is easily able to preserve collected digital evidences selectively for using general technology such as XML and PKZIP compression technology, which is satisfied with generality, integrity, unification, scalability and security. In this study, we describe a forensic approach to live investigation using XeBag, which is to store selectively collected the digital evidence in live system [4, 5].

2 Related Works

2.1 Digital Evidence Containers

Earlier studies on digital evidence container originated from the DD image, which is created by the bit-stream duplicate and has the same size and type as those of the original disc drive. These studies have been enlarged, and disc image file format in each disc browsing solutions, which are currently being used commercially. Most representative example is the EnCase image file. In the EnCase image file format, the entire disc is divided into specific volume and it is compressed, and the integrity of disk image is proven using CRC check and cryptography hash function.

According to digital forensic academia in earlier period, digital evidence container called the disc image format to which an additional function or a compression technique is added, e.g. the EnCase image file format.

It means the concept of DEC (we called DEC which is the abbreviation of 'digital evidence container') is changing from a digital container for the disk image to the format of digital evidence container. It means the concept of the evidence collection bag or physical evidence container can be applied to the digital evidence. Therefore, metadata, including information and timestamp from cases and hash values for verifying the data integrity, can be included, and the collected evidence data can be stored in the DEC format. This concept was first presented in Philip Tuner's paper in DFRWS in 2005 as the 'digital evidence bag'. And he had shown several research paper to enlarge his ideas such as selective imaging and live data [3, 6, 7].

The 'digital evidence bag' is a comprehensive storage format wherein digital evidences can be stored, or sets of evidence objects with a hierarchical architecture that include multiple digital evidences. This design concept aims to develop a standard format in which digital evidences from diverse devices, including computers, mobile devices, and network systems, can be stored. The standard format will enable consistent processing with a unified format, and will simplify the data processing procedure. It will also be suitable for distributed or parallel processing environments. This format is designed to include the metadata created during the investigation, including the audit information for the collected evidence data. All the metadata are stored in plain text and all the data are recorded in the raw binary format to ensure their universal use. The concept of DEC was originally that of the simple image format, but it has been developed into the multifunctional format, beyond merely a device to reduce the image volume or to ensure the image integrity [6].

2.2 Selective Evidence Collection for Triage Investigation

The cyber forensic field triage process model (CFFTPM) methodology was developed to allow quick action on cases in the field. The CFFTPM was developed to enable quick investigation of kidnapping, hostage, and murder cases in as short a time as possible. It focuses on the selection, collection, and investigation of files in the field that are required for handling of cases, based on the files instead of the disc images. In conclusion, it is a process model that selectively collects the required evidence data in terms of their importance and priority, and investigates them in the field [2].

At the first stage of the CFFTPM model, the investigation is planned for a specific case. The next stage is the evidence selection (triage) stage, in which the evidence data are selectively collected according to priority. In the analysis of user usage profiles, the directories in which the user's data are mostly stored are investigated (e.g., 'My documents' and 'Desktop'), as well as the file properties, including the file creation, modification, and access data and the file types. The user's traces in the system are examined by investigating the system, network, and user traces stored in the registry. The chronology timeline analysis enables the analysis of a user's actions and patterns by time-line sorting the collected data in order. The Internet usage analysis collects and analyzes the personal data related to the Internet, including the records of the Web browser, e-mail, and messenger. The case-specific process of analyzing data is as follows.

The CFFTPM methodology was firstly realized to a forensic toolkit by Guidance Soft's EnCase Portable Kit. The EnCase Portable Kit is a solution that combines a 4 GB USB drive equipped with the EnCase Portable software with a 16 GB USB drive for additional storage space, USB hub, and dongle key [8].

The main feature of EnCase Portable allows the user to select information for investigation from the target computer, for automatic collection. To use the EnCase Portable Kit, the system is booted using the EnCase Portable software, and the investigator selects the required data type to be automatically collected and stored in the USB. These collected data include document files, Internet usage history, image files and the imaging of the entire disc. The collected data are stored using the EnCase logical evidence format.

Thus, EnCase Portable is realizing the selective digital evidence collection technique. But, the DEC format for EnCase Portable is not opened for forensic community and it is focused on live investigation on personal computer system only. So these cannot apply other digital forensic area such as enterprise forensics and mobile forensics.

3 Applying XeBag to Live Investigation

3.1 Design of XeBag

We proposed a concept of new digital evidence storage format, the 'XML and Compression-file based digital evidence Bag (XeBag)' in ICCSCE 2011. XeBag is easily able to preserve collected digital evidences selectively for using general technology such as XML and PKZIP compression technology, which is satisfied with generality, integrity, unification, scalability and security. The XML and zip compression formats are used in OpenXML, which is applied to the Microsoft Office 2007 product group. The basic structure is based on PKZip or WinRAR to store digital evidence, and it includes XML document for describing forensic metadata and log history for provenance. In the XeBag format, the basic data format is based on PKZip, and the data for expression in the forensic view are described and stored using XML. Because PKZip is widely used as a compression technology, the increase in the volume due to the evidence format when adding metadata can be compensated for by the compression technology. Moreover it can be applied other compression file format such as WinRAR [1].

The use of XML technology increases the technical contents of digital evidences, but the basic scalability of XML can handle this issue. This technology is also highly flexible because the existing format can be extended and continuously used when the evidence format changes. The XML infrastructure has already been distributed, so the generality for diverse sectors is satisfied. The existing XML security solutions can be directly applied to the privacy and security problems involving collected data. In addition, the integrity of the data can be ensured by defining together the hash values or metadata of the digital evidences in XML documents.

In the XeBag structure, the collected evidence files or data sets are stored based on PKZip, and the main forensic data for the evidence data are represented in the XML format. The basic XeBag file format complies with that of the zip file format (Fig. 1). In the zip file format structure, the file entries are those that are compressed when the compression software is used. In the XeBag structure, the main evidences or files that are collected via the digital evidence selection technique are added as file entries one by one. The XeBag structure defines the file entry as the evidence object.

In the XML documents that are used in the XeBag structure, the XML document is created, which consists of the file element of each object and the root element in which basic data on the case are stored, to allow the description of the evidence object in the forensic view. The root element manages the file element stored in the XeBag structure and stores the basic information from the computer. The stored file elements include the MAC time or hash value of the collected files. Thus, the XML document has a hierarchical structure with a file element for describing each evidence object and a root element for managing it. This is similar to the hierarchical structure set that is suggested in the digital evidence bag. When many XeBag files with tens of stored files are analyzed, they do not have to be extracted because XML documents alone allow the selection or reading of evidence objects, which leads to efficient investigation (Fig. 2).

In the XeBag structure, the XML document for each evidence object is not created by simply creating the XML file and adding it to the file entry, but in the manner in which the extra areas that are available in the zip file structure can be used. This allows the general compression software or libraries to read and extract the evidence objects. In addition, this design is not intended to add files created by the evidence bag to the collected evidence object list. Following these concepts of XeBag design provides portability and generality to use various digital forensic areas.

3.2 XeBag Framework

The XeBag framework describes Fig. 3, which is how to apply XeBag in the general live forensic toolkit. In case of live data, the results of data collection are written a XML document by XML Writer. It will be compressed and archived by

Fig. 1 The cyber forensic
field triage process model [2]

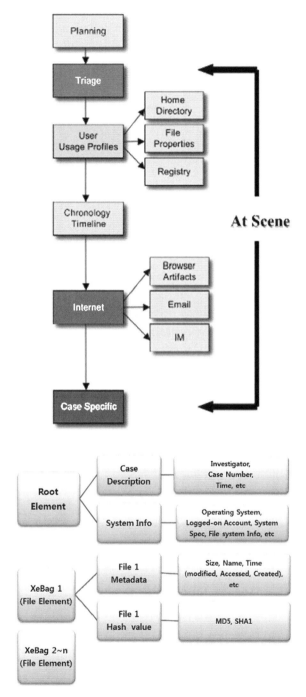

Fig. 2 The structure of
XeBag format

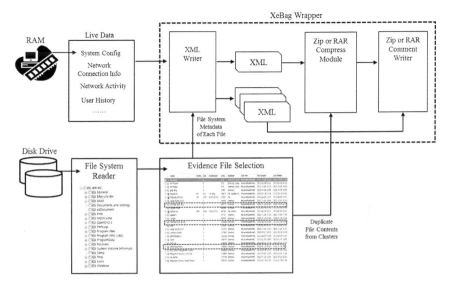

Fig. 3 XeBag framework in live investigation

compress module as single evidence file, metadata of this evidence insert to comment area of Zip format. In addition to single file collection, file system reader accesses and parses file system metadata such as $MFT table of NTFS file system. Each MFT entry has location list of clusters in disk drive, which is corresponding with the file. The evidence file selection module duplicates bit-stream contents of the target file from these clusters; it makes an identical file compared to original file. The record of a MFT entry has timestamp such as file size, cluster location, MAC time and so on. This information will be created XML document by the XML writer module. Finally, the compress module creates and saves a single PKZip or WinRAR file format and the comment writer will be added original metadata information of a MFT entry.

The most difference of XeBag compared with 'Digital Evidence Bag (DEB)' is forensic approach in live investigation. Recent research paper by Turner has shown that DEB can be just presented live data including system configuration, network connections and others by using the wrapper tool [3]. But his approach does not consider triage investigation such as CFFTPM and EnCase Portable. The forensic approach of target system still is same as disk-image based investigation except for original strength of DEB format. But, XeBag framework is based on triage investigation in live system for rapid response. It provides selective collection and presentation of digital evidence using file system reader and file duplication module. An investigator can be browsed, searched, selected and collected evidence files in the target computer using XeBag framework with satisfying with forensic soundness. And this approach can be enlarged various forensic area such as database forensic, CCTV forensic, embedded system forensic etc.

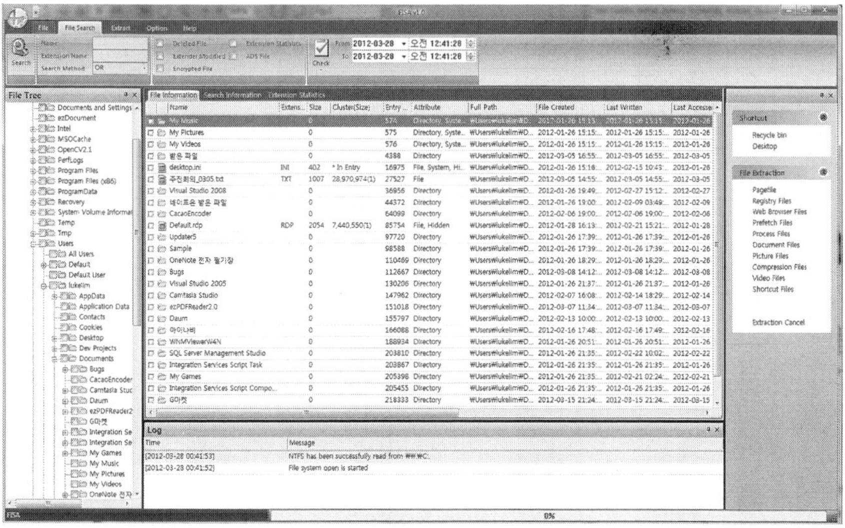

Fig. 4 File System Analyzer (FISA) by 4 and 6 Tech

3.3 Implementation

Figure 4 shows FISA forensic tool, which is a file system analyzer base on live system investigation. FISA is an easy and effective live forensic tool to examine file system metadata and collect potential evidence files on Windows system in forensically manner. The latest version of FISA supports file-extraction feature with evidence container for saving collected files, which is based on XeBag framework [9].

4 Conclusion and Future Works

In this study, we describe how to use XeBag in live investigation with forensic toolkit. XeBag is a new evidence storage format that is satisfied with the requirements of the digital evidence container, which are generality, scalability, security and integration. XeBag structure is advantageous because it can be applied to the existing file-unit-based forensic tool by merely using the existing extraction tool. When the data are extracted using the general extraction tool and used as entries for the general reader, a certain level of security is ensured because XML still has the time information of the original data and the XeBag itself can be individually stored. And portability and unification of it can be used various forensic areas. Further studies will be performed to show possibility in various forensic areas to apply it such as database forensic, CCTV forensic and so on.

Acknowledgments This research was supported by Basic Science Research Program through the National Research Foundation of Korea (NRF) funded by the Ministry of Education, Science and Technology (grant number 2012-0003832).

References

1. Lim, K.S., Park, J., Lee, C., Lee, S.: A New proposal for a digital evidence container for triage investigation. In: ICCSCE' 11 Nov 2011
2. Rogers, M.K., Goldman, J., Mislan, R., Wedge, T., Debrot, S.: Computer forensics field triage process model. Conference on Digit Forensics, Security and Law, 2006
3. Turner, P.: Applying a forensic approach to incident response, network investigation and system administration using Digital Evidence Bags. Digit. Investig. **4**(1), 30–35 (2007)
4. Lim, K.-S.: XFRAME: A XML framework for digital forensic investigation. Thesis of Master Course Degree, Graduate School of Information Security, Korea University (2008)
5. Lim, K.-S., Lee, S.B., Lee, S.: Applying a stepwise forensic approach to incident response and computer usage analysis. In: 2nd International Conference on Computer Science and its Application (CSA 2009)
6. Turner, P.: Unification of digital evidence from disparate sources (Digital Evidence Bags). Digit. Investig. **2**(3), 223–228 (2005)
7. Turner, P.: Selective and intelligent imaging using digital evidence bags original research article. Digit. Investig. **3**(1), 59–64 (2006)
8. Gudiance Soft. EnCase Portable, http://www.guidancesoftware.com/encase-portable.htm
9. Four&Six Tech FISA: File System Analyzer, http://www.4n6tech.com/pro_eng/info/info.php?pn=1&sn=1&dn=1

The Development of an Internet-Based Knowledge Management System for Adapted Physical Education Service Delivery

Jong-Jin Bae, Jung-Chul Lee, Min-Woo Cheon and Seung-Oh Choi

Abstract The purpose of this study is to develop an internet-based system for the actual implementation and practice of knowledge management to support adapted physical education service delivery by using distributed knowledge accumulated through exchanging and sharing case studies among different schools and educational institutions over the Internet, and to evaluate the effectiveness of the newly developed system by asking the adapted physical education teachers. The results are as follows: This system makes it easy to archive and share classroom teaching information, and also supports collaboration among people in different schools and educational institutions. It also permits new teachers who lack expertise and actual classroom experience to work on case studies in collaboration with experienced APE specialists from other schools, and thereby improve their skills in teaching children with disabilities.

Keywords Adapted physical education · Video consultant · Web-based system · Teaching method · Autism

J.-J. Bae
Department of Physical Education, Gongju National University, 56 Bonghwang-dong, Gongju-si, Chungcheongnam-do, Republic of Korea314-701,

J.-C. Lee (✉)
Department of Exercise Prescription, Dongshin University, 252 Daeho-dong, Naju, Jeonnam, 520-714, Republic of Korea
e-mail: channel365@hanmail.net

M.-W. Cheon
Department of Biomedical Science, Dongshin University, 252 Daeho-dong, Naju, Jeonnam, 520-714, Republic of Korea

S.-O. Choi
Department of Sport Science, Hannam University, 133 Ojung-dong, Deadeok-gu, Daejeon, 306-791, Republic of Korea

S.-S. Yeo et al. (eds.), *Computer Science and its Applications*, Lecture Notes in Electrical Engineering 203, DOI: 10.1007/978-94-007-5699-1_39, © Springer Science+Business Media Dordrecht 2012

1 Introduction

Adapted physical education (APE) is changing. This is more than just a simple change of nomenclature, because the target population for adapted physical education service has expanded to individuals with many developmental disabilities such as learning disabilities, attention deficit disorder, and high-functioning autism that were not previously covered in the service delivery system. A number of school-aged children are affected by these more broadly defined conditions. Some progress has been made to accommodate and support these children in inclusive physical education classes, but it has also been shown that appropriate support and teaching practices are not being provided in many regular physical education classrooms [1]. One factor that has contributed to this lack of support is that there are simply not enough teachers with APE expertise to teach these children. Another factor is that the very concept of APE is so new that, aside from APE specialists, there are no clear guidelines on how to deal with the behavioral issues of children with mild developmental disabilities [2]. Instead of trying to assign new regular physical education teachers or APE specialists, a new approach can be considered, in which regular physical education teachers collaborate with APE specialists in other institutions to develop case studies and teaching strategies to deal with behavioral problems [3]. Such an approach is called an expended collaborative consultation model. In order to pursue such a collaborative approach, it is necessary to archive and share detailed information about specific cases and teaching strategies to deal with specific behavioral problems [4].

The purpose of this study is to develop an Internet-based system for the actual implementation and practice of knowledge management to support adapted physical education service delivery by using distributed knowledge accumulated through exchanging and sharing case studies among different schools and educational institutions over the Internet, and to evaluate the effectiveness of the newly developed system by asking the adapted physical education teachers.

2 Methods

2.1 Development Tools

The Internet web page where the specialist group and the teacher group will be directly connected uses allocated space at the laboratory in the school.

Although some people use client-oriented X-internet web pages to avoid overload on the server, this study develops conventional Hyper Text Markup Language (HTML)-based web pages since the direct overload usually only appears in the direct download and transmission of video files [5]. The developmental environment and developing tool are shown in Table 1.

Table 1 Development environments and tools

	Tool name	Manufacturer
Main sever	Unix solaris 8.0	Sun microsystems
Database	Oracle 9i	Oracle korea
Client computer	Windows XP	Intel, samsung Ltd., microsoft co.
Html	Web studio 2010 standard	SEJOONGNAMO INC
Script	Edit Plus v. 2.21	ES-Computing
Web browser	Internet explorer 6.0	Microsoft co.
Ftp protocol	All FTP 6.0	East soft.

Common gateway interface (CGI) programs, such as message boards and guest books, were designed using professional html preprocessor (PHP) with Edit Plus. By using free CGI programs available on the web, the developing time was reduced and the design could be modified in accordance with the intentions of this study.

2.2 User Interface

2.2.1 Adapted Physical Education Teacher and General Teacher

By posting various examples that help people rightly extract videos from video cameras, people could see and follow these examples. After completing the video, we selected consultants to examine the profiles of the specialist committee or the video analysts who delegate their analytical points. The designed video files and texts of other questions or difficulties from physical activities were sent to the homepage. The analyzed video and text information were displayed separately on the web to explain how to setup the playback for teachers' convenience. People can watch the analyzed video and practice the corrections contained in them, and produce another video to be analyzed by specialists. Since the repetitive examination can give advantages of comparing the later analysis with one's previous video, we tried our best to motivate people to use the reexamination system.

2.2.2 APE Specialist

The specialist group that analyzes the students' videos and returns them with notes and guidance consists of professors in special physical education and PE teachers with at least 10 years of experience.

The procedure for using the webpage by the specialist group is shown in Fig. 1. After login, the specialist member check sent videos and texts, and resend them to the video analysts with analytical points and guidance. The video analysts download the requested video and the reference video for comparison. Video analysts use dartfish[TM] software to intensively analyze the problematic parts of the

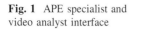 **Fig. 1** APE specialist and video analyst interface

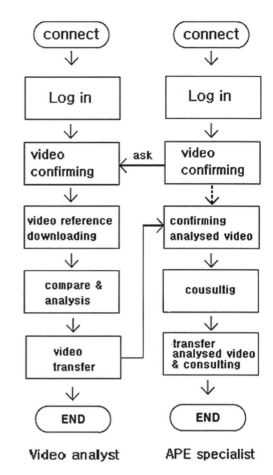

video advised by the specialists. When the analysis completes, the analysts send the analyzed video to the specialists. The reason for this process is to save time in analyzing the videos. Since there could be various subjective points of analysis even if the videos are all the same, specialists take the responsibility and delegate the analysis to the video analysts to avoid any confusion. Also, the specialists were allowed to advise the requests without any video attached, and to advise only with consultants without video analysis at their discretion.

2.3 System Structure

The internal structure of the homepage system is shown in Fig. 2. When special PE teachers or typical teachers sign up for the membership and select the specialist they want comments from, one message board will be generated automatically for each member teacher, allocated in the database of the laboratory. In this system,

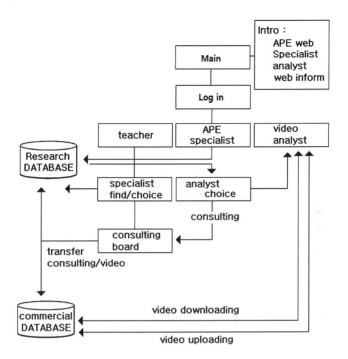

Fig. 2 Structure of internet-based knowledge management system for APE

the video file among the contents uploaded by the member teacher will be saved in the commercial database, and only texts will be saved in the message board. When the consulting contents are saved in the message board generated in the laboratory, the individual page for the specialists will notify it. The specialist will access the client's message board to check the requested contents, and if necessary, the video will be sent to the video analyst to analyze the contents.

The video analysts will receive the messages about the video analysis from the specialists, and access to the commercial web disk to download the requested video and the example video to analyze them to meet the demand from the specialists, and send them back to the commercial web again. At the same time, when they notify the specialists of the completion of analysis, their responsibility is finished. After receiving the message of completion from the video analyst, the specialist accesses the client's message board to give advice about the analyzed video and the example video.

2.4 Video Capturing and Processing

The knowledge management system consists of webcams for capturing the problematic behavior of children with disabilities and a Web database. The webcam is shown in Fig. 3. When the teacher wants to capture behavioral problems on video

Fig. 3 Webcam for
capturing

Table 2 Table structure of video information

Field	Type	Null	Key	Default	Description
ID	int(8)		PRI	0	Number of video characteristic
Name	char(16)			NULL	Teacher name who made video
APEN	char(256)			NULL	Information of students characteristic
APE	char(256)	YES		NULL	Problematic behavior
Place	int(1)	YES		NULL	Recording place (field, gym, class)
Time	time	YES		NULL	Teaching day
Favor	char(16)	YES		NULL	Class of physical education
Pres	memo(2048)	YES		NULL	education chart and a bulletin board

during class, the timing is set by pressing a button on the mouse. Since behavioral problems often occur very suddenly, the webcam begins to record the incident 10 s prior to the timing set by the teacher. The incident recording time is set at 3 min.

After the class is over, the teacher chooses footage from the camera that had the best angle in capturing the behavioral incident, and the video is archived in the Web database. Figure 3 is a screenshot showing how a typical case is displayed in the Web database. The database screen consists of three parts: a video record, an education chart, and a bulletin board section. Table 2 shows database structure of video information.

3 Results

3.1 Video Information

The video clips show the actual problematic behavior that is recorded during the classes by the webcam system described earlier. Behavioral problems of children with developmental disabilities were defined as behavior meeting two criteria: (a) behavior that is either too frequent or too intense or not frequent enough for the age of the child in question; (b) behavior that is inappropriate either for the setting or place, or behavior that is inappropriate from the standpoint of the prevailing social code [6].

The authors will develop a classification of behavioral problems including 20 types based on close observation of many incidents involving children with learning disabilities (LD) and attention deficit hyperactivity disorders (ADHD), and 20 types based observation of many behavioral incidents involving children with high-functioning autism and Asperger's syndrome. Data for the behavioral problem video clips was based on this classification. The way the teacher attempts to deal with the behavioral problems was also included in the video clips. Note that it was difficult to rigorously differentiate high-functioning autism from Asperger's syndrome [7], so the teaching response and record maintenance were handled similarly.

The education chart information corresponds to the particular behavioral problem that could be seen in the video clip. All the descriptive information on the Web was represented in the database as a single education chart. The descriptive information in the education chart was organized as a template based on ABC analysis. ABC analysis is stands for antecedents (A), behavior (B), and consequences (C). Applied behavioral analysis attempts to comprehend behavioral problems from the standpoint of interaction between the individual (i.e., the child) and the environment [8].

3.2 Knowledge Management Practice

The behavioral problems and teaching strategies stored in the Web database were stored in separate child-specific archives on the Web, and only those people with access to the child's information could view the child's data and exchange and share information about the child over the Internet. This teaching example data stored in the Web database could be shared by a diverse range of schools and educational institutions using the system and used as the basis of collaborative case studies with many different participants contributing the knowledge and expertise.

All discussions regarding teaching ideas were conducted in the bulletin board area, and the teaching plan development process was recorded as a case study record. Participants choose a comment category and post their comments. When comments are posted to the bulletin board, the content of the comments are sent to all registered PCs and cell phone email addresses by the system. The teaching plans were created from data entered by all of the participants, and were provided as information along with the video clips on the Web. This information was the basic material of the case studies, and through collaboration within and between teams including the involvement of the consultants team, this distributed knowledge that was so useful for developing the special-needs teaching skills of teachers could be archived in the Web database.

Based on the video clip and record of an incident, several teachers from widely dispersed schools defined comment categories and post comments from their various perspectives. Cases were displayed in the database in sequential order for each school day, under the title of recording place and teaching day, and the behavioral problems recorded during teaching have multiple links. Currently as of 2012, there are 35 video clip records of behavioral problems, bulletin board discussion relating to the video records, 166 education charts created based on discussions, and more records continue to be archived as multimedia data.

4 Discussion

A proposed online system features simplified teacher administration, visualization of child data and checklists, and enables all the people involved to collaborate and participate in creating teaching plans. Internet-based systems that permit different schools and institutions scattered in remote locations to exchange and share information have become a practical reality, and this kind of system is highly effective for sharing information among teachers and other people who are concerned about children with disabilities. However, it is hard to describe the behavioral problems of children with disabilities and the range of teaching solutions, because they are so diverse. In addition, it is also difficult to archive and share teaching information when it is represented as text alone. Rather, it would be preferable to make a video record of the problematic behavior that actually occurs in the classroom and how the teacher deals with the behavior, and share that information with the people involved.

However, one of the problems with this approach is that editing the video in the classroom would create an enormous amount of extra work for the teacher. This calls for a system in classrooms and APE facilities in which the behavioral problem occurs. Teaching information such as video records can be more easily archived and shared [9]. The challenges APE teachers are facing include: (a) lack of enough teachers with adapted physical education experience and expertise in the schools; (b) a need for collaborative teaching that involves multiple schools and educational institutions; and (c) a need for the capability to easily archive and share teaching material such as video records [10].

To address these issues and challenges, we have proposed a powerful knowledge management system that archives teaching strategies and solutions for dealing with behavioral problems in the classroom, permits sharing of information among widely dispersed schools and educational institutions, and supports collaborative development of teaching plans. More specifically, the system consists of two basic elements—webcams for recording behavioral problems in the classroom, and a Web database. The system has the following key features: (a) video records are created online during class by the teacher that has behavioral problems and information on how the teacher attempts to deal with the problems, which are easily archived on the Web as information that can be shared with other teachers and APE specialists; (b) archived data including video records, template formatted education charts, and bulletin board records, allowing teachers and APE specialists around the country to share teaching examples that occur in classrooms on a daily basis over the internet; and (c) participants collaborate over the internet to develop case studies and teaching plans based on the shared teaching examples, and new APE teachers or regular physical education teachers can access and use the archived case studies and plans.

5 Conclusion

This system makes it easy to archive and share classroom teaching information, and also supports collaboration among people in different schools and educational institutions. It also permits new teachers who lack expertise and actual classroom experience to work on case studies in collaboration with experienced APE specialists from other schools, and thereby improve their skills in teaching children with disabilities.

References

1. Bruce, A.K.: Towards inclusion of students with special educational needs in the regular classroom. Support Learn **14**(1), 3–7 (2003)
2. Michael, J.G.: The nature and meaning of social integration for young children with mild developmental delays in inclusive settings. J Early Interv **22**(1), 70–86 (1999)
3. Adelman, H., Taylor, L.: Conduct and behavior problems related to school aged youth. Center for Mental Health in Schools at UCLA, LA (2008)
4. Bos, C.S., Vaughn, S.: Strategies for teaching students with learning and behavior problems fifth edition. A Pearson Education Company, Boston (2002)
5. Bob, O.: 'The X-internet' Connecting the Physical World with the Cyber World, Motorola, Inc, Illinois (2006)
6. Nagamori, M., Nagasawa, M., Ueno, M.: Webcam-based knowledge management system for special needs education. In Luca, J., Weippl, E. (eds.) Proceedings of World Conference on Educational Multimedia, Hypermedia and Telecommunications, pp. 3500–3509 (2008)
7. Raviola, G., Gosselin, G. J., Walter, H. J., DeMaso, D. R.: Pervasive developmental disorders and childhood psychosis. In: Kliegman, R. M., Behrman, R. E., Jenson, H. B., Stanton, B. F.(eds.) Nelson Textbook of Pediatrics, 19th edn. Saunders Elsevier, Philadelphia, 28 (2011)
8. Alberto, A.P., Troutman, C.A.: Applied Behavior Analysis for Teachers. Prentice Hall College Division, New York (2002)
9. Naeyc.: Technology and Interactive Media as Tools in Early Childhood Programs Serving Children from Birth through Age 8, National Association for the Education of Young Children, Fred Rogers Center for Early Learning and Children's Media, Washington (2012)
10. Kelly, L.E.: Designing and Implementing Effective Adapted Physical Education Programs. Sagamore Publishing LLC, IL (2011)

Intangible Capital, Opportunity Exploitation and Institutional Endorsement in Emerging IT Industry

Choi Youngkeun

Abstract. In early stage of entrepreneurial process, an entrepreneur makes the best use of his/her human and social capital to overcome the liability of newness. In this study, we define human/social capital aspects of an entrepreneur's personal resources "intangible capital". We investigate how intangible capital of an entrepreneur influences their ability to exploit business opportunity in a promising industry and to induce venture capital investment. The data come from Korean startups in information technology industry. An empirical analysis of 357 firms gone public in the KOSDAQ market shows that an entrepreneur with prior relevant work experience and network in related firm more likely exploit business opportunity in a promising industry and induce venture institutional endorsement. The managerial background of an entrepreneur turns out to be a more important factor than education.

Keywords Information Technology · Intangible capital · Opportunity Exploitation · Institutional Endorsement

1 Introduction

The entrepreneurial process is essentially interactive between an entrepreneur and business environments [1–3]. It can be defined as a process in which the feasibility of a business concept is verified. It consists of series of events to be passed through to overcome the liability of newness [4, 5]. In this process, the entrepreneur sits in

C. Youngkeun (✉)
Sangmyung University, 20, Hongjimun 2-gil, Jongno-gu, Seoul 110-743, Korea
e-mail: penking1@smu.ac.kr

S.-S. Yeo et al. (eds.), *Computer Science and its Applications*,
Lecture Notes in Electrical Engineering 203, DOI: 10.1007/978-94-007-5699-1_40,
© Springer Science+Business Media Dordrecht 2012

the driver's seat in the initial absence of key resources [6, 7]. Here, we define human/social aspects of an entrepreneur's personal resources "intangible capital."

In the early stage, an entrepreneur should depend on his/her personal intangible capital heavily [1]. In the early stage, both human and social capital of an entrepreneur may affect opportunity exploitation because they contribute to his/her unique information corridor and knowledge set [8]. In later stages, venture capitalists tend to regard social background and relations of an entrepreneur as a key criterion in judging the company's growth potential [4].

Since the early 1990s, Korean small and medium venture firms have developed with 2 features [9]. First, with help from emerging new technologies mainly information technologies, founding venture firms were activated. Second, the more start-ups were created in information technology industry, the more venture capital investments were activated. The research background of this study is the process of Korean small and medium venture firm's development. Our purpose is to analyze the effects of intangible capital in the entrepreneurial process, that is, (1) the relationship between the intangible capital of an entrepreneur and opportunity exploitation in a promising industry, and (2) the contribution of intangible capital to venture capital financing as institutional endorsement of an emerging company.

The paper proceeds as follows: we review theory and previous research on human/social capital and entrepreneurial process. This leads to the generation of four hypotheses to be tested. We then describe the methods we have used for data collection and analysis. Following that, we present the results of our analysis. This paper concludes with interpretation of our results and states their implications.

2 Theoretical Background and Hypothesis

2.1 Intangible Capital

2.1.1 Human Capital

The human capital perspective suggests that human capital is an important variable that has an influence on entrepreneurial opportunity [10]. Human capital includes an entrepreneur's knowledge which has been acquired through education and experience. The knowledge set plays a critical role in intellectual performance, which may help business development. It also assists in cumulating and integrating new knowledge as well as in adapting to new situations [11].

Formal education is an element of human capital assisting the accumulation of explicit knowledge and useful skills [12, 13]. It provides entrepreneurs with increased abilities both in evaluating opportunities and mobilizing resources [14, 15]. Human capital is not only the result of formal education, but also includes work experience that represents tacit knowledge. Entrepreneurs having different functional backgrounds in a prior job are expected to have somewhat different cognitive bases [16].

2.1.2 Social Capital

Experiences in prior affiliations help build social capital. An entrepreneur's social capital includes networks both in relevant firms and in schools. A college degree from an elite institution often becomes a basis for later friendships and contacts that not only allow companionship, but networks that are at times beneficial to one's later career [17]. And the degree from an elite institution is an indicator of potential networking ability [18, 19].

The social network built in a related firm is a major means of opportunity exploitation and resources mobilization [8, 20]. Social networking creates opportunities for knowledge acquisition and exploitation. Especially, social networking with customers creates both customer and brand royalties, and increases sales (Park and Luo 2001). Good relationships with suppliers also provide quality raw materials, superior service, and reliable deliveries [20]. The failure rate of emerging companies is quite high, because they cannot establish stable relationships with customers and suppliers [7, 21].

2.2 Opportunity Exploitation

Emerging companies experience the inherent liability of newness [7]. The liability of newness is compounded by the lack of a clearly defined success model. This increases the level of uncertainty in technological and market changes, as potential entrants face the added cost of recognizing and evaluating information. An entrepreneur's mental construct plays a central role in the process of opportunity evaluation. It is not the knowledge itself, but the way a human being applies knowledge, that is crucial in evaluating opportunities [22]. For an entrepreneur to benefit from an opportunity, he/she must evaluate its value potential. Knowledge provides an entrepreneur with increased evaluative abilities [23–25]. Therefore, entrepreneurs with a higher level of human capital should have a better ability in opportunity exploitation.

Hypothesis 1   The more human capital an entrepreneur has, the more likely he/she is to exploit a new business opportunity in a promising industry than in other industries

Market information is distributed unevenly throughout society. This creates situations where some entrepreneurs have better access to information [26, 27]. As results, some entrepreneurs pursue attractive opportunities while others do not. This explains why an entrepreneur's networks are important to opportunity exploitation. Entrepreneurs, who have extended networks, can recognize significantly more opportunities than isolates (Hills et al. 1997).

Hypothesis 2   The more social capital an entrepreneur possesses, the more likely he/she is to exploit a new business opportunity in a promising industry than in other industries

2.3 Institutional Endorsement

There are imperfections in capital markets that render external financing expensive and constrain investment decisions [28]. Emerging companies suffer most from these capital market imperfections because most of their assets are intangible and firm specific such that they have little collateral value [29]. Thus literature suggests that the prior experience of an entrepreneur is one of the most important considerations in venture capitalists' investment decision as the emerging company lacks enough track records [30]. Unfortunately, however, previous studies have not paid much attention to the influence of an entrepreneur's prior experience in accessing venture capital money.

We suggest that an entrepreneur's prior experience as intangible capital enhances the chances of getting venture capital support. Generally, it is difficult for emerging companies to obtain adequate external financing, and thus they tend to rely exclusively on personal capital. In turn, these financial constraints prevent high potential companies from growing as fast as they would with adequate financing at the early stage of growth [29]. In this context, venture capital is a critical financial resource for the growth of emerging companies. Although they are investment experts in emerging companies, venture capitalists have trouble in investment decision making due to the lack of track records as well as collateral for those companies.

So, venture capitalists depend largely upon the expected abilities of an entrepreneur for decision making [29], 1987). Under this situation, an entrepreneur's intangible capital provides them with some confidence in the people side. New technology-based firms established by individuals with greater human capital tend to possess better capabilities [30]. Studies also suggest that social capital embodied in social networks with external entities affects a firm's competitive advantage and performance [20]. Overall, although there is some study not confirming consistent results [15], we can still expect a positive relationship between this intangible capital of entrepreneurs and firm performance though although some studies examining this relationship have not yielded consistently strong results.

Hypothesis 3   The level of an entrepreneur's human capital positively increases the likelihood of obtaining venture capital support

Hypothesis 4   The level of an entrepreneur's social capital positively increases the likelihood of obtaining venture capital support

3 Methods

3.1 Data

Our sample consists of 357 firms listed in the KOSDAQ (Korea Securities Dealers Automated Quotation) market. These firms represent all the companies with a founder as CEO at the initial public offering that went public from July 1, 2000 to December 31, 2005, except those that were an affiliate of an existing company. We first collected corporate disclosure information from an electronic disclosure system called DART (Data Analysis, Retrieval and Transfer System), and then supplemented the database with press releases, publications, corporate homepages and phone calls.

3.2 Variables

3.2.1 Dependent Variables

We first define starting up in information technology as opportunity exploitation in a promising industry, considering the fact that information technology industry was the highest growth sector in Korea during the period. Regarding venture capital support, we checked if the firm attracted venture capital investment before going public. We excluded investments by banks, securities companies, and insurance companies.

3.2.2 Intangible Capital

Human capital is measured with the formal educational level and managerial background of an entrepreneur. Formal education has two categories, i.e., undergraduate degree or lower, and (2) master's or doctoral degree. Managerial background is divided into output functions (such as R&D, and sales/marketing) and throughput functions (such as operations, engineering, and accounting) according to [16]. We define attending a well-known elite college and having a prior job experience in a relevant firm as a basis of social capital.

3.2.3 Controls

We controlled for gender and age of an entrepreneur. The gender of an entrepreneur may influence founding and firm performance [15]. Younger people can make more commitment both physically and mentally, and thus have a greater entrepreneurial propensity [31, 32].

Table 1 Pearson correlations between the various variables used in the regression models

	Mean	SD	1	2	3	4	5	6	7
Entrepreneur's gender	0.98	0.13							
Entrepreneur's age	38.96	6.34	0.05						
Level of education	0.32	0.47	-	0.00	−0.04				
Output function	0.67	0.47	0.05	−0.19***	0.01				
N/W in Elite school	0.29	0.45	0.04	−0.11*	0.29***	−0.03			
N/W in relevant firm	0.73	0.44	0.07	−0.26***	0.06	0.46***	0.17**		
Opportunity exploitation	0.66	0.47	0.05	−0.24***	0.02	0.94***	−0.05	0.48***	
Obtaining VC support	0.69	0.46	0.01	−0.05	0.07	0.24***	0.07	0.26***	0.25***

$N = 357$; * $p < 0.05$, ** $p < 0.01$, *** $p < 0.001$

4 Results

Table 1 shows the descriptive statistics and Pearson correlations of our research variables. On average, 98 % of entrepreneurs were male (standard deviation 13 %) and age of entrepreneur was 38.96 (std. dev. 6.34). 32 of entrepreneurs had master's degree (std. dev. 47), 67 had output background (std. dev. 47), 29 are from elite school (std. dev. 45), and 73 % had prior work experience at related firm (std. dev. 44 %). 66 of venture firms were founded in information technology industry (std. dev. 47), and 69 % of firms obtained venture capital investment (std. dev. 46 %). Table 1 further shows that all correlations between independent variables are high. Although not reported here because of space limitations, we also tested for multicollinearity by calculating tolerance values, variance inflation factors, condition indices, and variance proportions. All values were far below acceptable thresholds for multivariate analysis as reported by [33] and showed that multicollinearity is not a problem in our data set.

The two regression models, business opportunity exploitation and institutional endorsement, are each constructed in a three-stage process. The first stage contains the control variables, while the second stage contains control and human capital variables and the final stage contains control, human and social capital variables. Table 2 contains the results of the regression analyses. Sub-models 1 through 3 are for business opportunity exploitation in a promising industry, and sub-models 4 through 6 are for institutional endorsement. Models 1 and 4 contain only the control variables, models 2 and 5 the control variables and the human capital-related variables, while models 3 and 6 contain the control variables and the social capital-related variables.

Model 1 shows that only entrepreneur's age is statistically significant, younger entrepreneurs are better in business opportunity exploitation in a promising

Table 2 Regression results of intangible capital on opportunity exploitation and institutional endorsement

	Opportunity exploitation			Institutional endorsement		
	Model 1 (control model)	Model 2 (human capital)	Model 2 (Social capital)	Model 3 (control model)	Model 4 (human capital)	Model 5 (Social capital)
Constant	2.73*	1.10	1.04	1.28	−0.15	0.05
Gender	0.99	0.54	0.60	0.16	−0.25	−0.12
Age	−0.08***	−0.06*	−0.16**	−0.02	0.01	0.00
Level of education		−0.94**			0.20	
Functional background		2.46***			1.23***	
N/W in elite school			1.22			0.35
N/W in relevant firm			7.90***			1.22***
-2Log Likelihood	421.01	349.73	78.35	440.16	416.58	414.22
Cox & Shell R^2	0.05	0.24	0.64	0.00	0.07	0.07

N = 357; * p < 0.05, ** p < 0.01, *** p < 0.001

industry than older ones ($\beta = -0.08$, p < 0.001). If we add the human capital-related variables (Model 2), we see that both level of education and functional background are significant predictors. As expected, entrepreneur's functional background has a positive influence on his/her exploiting a new business opportunity in a promising industry ($\beta = 2.46$, p < 0.001). What is not expected, however, is the negative effects of level of education ($\beta = -0.94$, p < 0.01). This means that the higher level of education he/she has, the less able he/she will be in exploiting business opportunity in promising industry. Hypothesis 1 is only partly confirmed by the data. Finally, adding the social capital-related variables to the basic model have the following results (see model 3). The control variable 'entrepreneur's gender' has no effect. The 'network in elite school' is not significant variable, but 'network in relevant firm' is significant one ($\beta = 7.90$, p < 0.001). This implies that the more constrained a relevant firm is, the better he/she is in exploiting business opportunity in promising industry. Hypothesis 2 is only partly confirmed by the data. The results of the regression analyses regarding institutional endorsement are presented in models 4 through 6. Table 2 shows that the control variables (model 4) are not statistically significant. Model 5 adds the human capital-related variables. As expected, functional background is significant predictor of institutional endorsement ($\beta = 1.23$, p < 0.001). What is not expected, however, that the entrepreneur's level of education has no effect on institutional endorsement. The human capital model partly confirms Hypothesis 3. Finally, adding the social capital-related variables to the basic model have the following results (see model 6). As expected, the 'network in relevant firm' is significant variable ($\beta = 1.22$, p < 0.001). What is not expected, however, is that the 'network in elite school' is not a significant variable. This means that Hypothesis 4 is partly confirmed. To summarize, among the entrepreneur's intangible capital, functional background at prior work and relevant work experience are influential on business opportunity exploitation in a promising industry and institutional endorsement, However, entrepreneur's level of education an network in elite school are not. We come back to this unexpected result in the discussion.

5 Discussions

This study attempts to analyze influence of entrepreneur's intangible capital on entrepreneurial process. Specially, we defined entrepreneurial process as the process of verifying an entrepreneur's business idea in reality and we explained it with two events; establish a new firm and obtain venture capital. An entrepreneur experiences lack of resources during entrepreneurial process [7], and to overcome this problem he/she depends on his/her intangible capital. Therefore, we maintain that intangible capital plays an important role in exploitation new business opportunity and leading an institutional endorsement in emerging companies.

Regarding business opportunity exploitation in information technology and obtaining venture capital investment, among human capital variables, functional background effects more than educational background does, and among social capital variables, network in related firm effects more than network in elite school does. Information technology industry contains rapid change in technologies and markets. To exploit a business opportunity in information technology industry, entrepreneurs are required to have cognition ability and network [16, 34, 35]. The result show that output background that explains cognition ability in technologies and markets is more effective than the level of education that explains general cognition ability. And network in related firm which is field-specific social network is more advantageous than network in elite school which is general social network. For organization-information technology industry embeddedness, corporate is more effective than university. So, among variables that explain entrepreneur's intangible capital, only variables from managerial background have positive relation with business opportunity exploitation in information technology industry. Dahlstrand [36] compared founders form corporate with ones firm university. Founders from university have advantage of technology development, while ones from corporate have strengths in creating profitability and sales growth. That is the reason why venture capitals are interested more in realization of financial performance than in ability of technology development. Therefore, Korean venture capitals consider the managerial background as being more important than educational background when they evaluate entrepreneur's intangible capital. Venture capitals are professional investment organizations. They invest their resources in venture firms with their capital collected by attracting investors to the funds they are operating. Venture capitals should collect their invested capital in investment before the fund expiration through initial public offerings (IPO) of their investment or merger and acquisitions, because they receive evaluation from fund investors with investment earning rate.

Research contributions in this study are summarized as the follows. First, this study is the first to examine how intangible capital of an entrepreneur influences on his business opportunity exploitation in emerging companies. Literatures studies business opportunity identification in successful enterprises or examined if founding a firm was successful or not [15]. We maintained that high level of intangible capital is required to exploit a business opportunity in information technology industry of high uncertainty and empirically tested. Second, this study suggests a relationship between the social capital of entrepreneurs and obtaining VC. Although some study suggests that human capital of entrepreneurs is more positively associated with the likelihood of obtaining venture capital investment [14], there is no empirical study that examines a relationship between the social capital of entrepreneurs and obtaining venture capital investment. This study provides an integrative understanding of venture capital investment decision-making by testing both the social and human capital of the entrepreneur in obtaining venture capital investment. Finally, we find that considering the formation process of intangible capital, entrepreneur's managerial background is more important than educational background. It implies that to exploit a new business opportunity in emerging industry, recognition ability and network in technologies and markets can be built up more effectively in managerial background than in educational background. Venture capitals consider the managerial background as being more important than educational background when they evaluate entrepreneur's intangible capital, because they are evaluated by fund investors with investment earning rate.

We hope that future research will continue along the lines of this study by addressing its limitation. The limitation is that this study measured independent variables dichotomously. We use binary variables to measure intangible capital therefore we have limitations in understanding and interpreting effect size of intangible capital.

References

1. Archer, M.: Structure, Agency, and the Internal Conversation. Cambridge University Press, New York (2003)
2. Archer, M.: Realist Social theory: The Morphogenetic Approach. Cambridge University Press, New York (1995)
3. Sarason, Y., Dean, T., Dillard, J.: Entrepreneurship as the nexus of individual and opportunity: A structuration view. J. Busin. Ventur. 21(3) 286–305 (2006)
4. Block, Z., MacMillan, I.: Milestones for Successful Venture Planning. Harvard Business Review, September-October, pp 184–196 (1985)
5. Galbraith, J.: The stages of growth. J. Busin. Strateg. 3(4), 70–79 (1982)

6. Archer, M.: For Structure: Its Reality, Properties and Powers: A Reply to Anthony King. Sociological Review, 48/3, 464-472 (2000)
7. Stinchcombe, A.: Social structure and organizations. In: March, J. (ed.) Handbook of Organization, pp. 163–178. Rand McNally, Chicago (1965)
8. Shane, S., Venkataraman, S.: The promise of entrepreneurship as a field of research. Acad. Manag. Rev. **25**(1), 217–226 (2000)
9. Chung, S., Choi, Y.K., Lee, J., Park, S., Shin, H.: Policy intervention in the development of the Korean venture capital industry. In: Phan, P.H., Venkataraman, S. (eds.) Entrepreneurship in Emerging Regions around the World, Theory, Evidence and Implications, pp 206–238. Edward Elgar Publishing, Massachusetts (2008)
10. Dimov, D., Shepard, D.: Human capital theory and venture capital firms: Exploring "home runs" and "strikes outs". J. Busin. Ventur. **20**(1) 1–21 (2005)
11. Weick, K.: Drop your tools: an allegory for organizational studies". Admin. Sci. Quart. **41**(2) 301–314 (1996)
12. Honig, B. Education and self employment in jamaica. Comp. Education Rev. **40**(2) 177–193 (1996)
13. Reynolds, P., White, S.: The Entrepreneurial Process: Economic Growth, Men, Women, and Minorities: Quorum Books, Westport, Connecticut (1997)
14. Colombo, M., Grill, L.: On Growth Drivers of High-tech Start-ups: Exploring the Role of Founders' Human Capital and Venture Capital. Journal of Business Venturing forthcoming (2009)
15. Davidsson, P., Honig, B.: The role of social and human capital among nascent entrepreneurs. J. Busin. Ventur. **18**(3), 301–331 (2003)
16. Hambrick, D., Mason, P.: Upper echelons: The organization as a reflection of its top managers. Acad. Manag. Rev. **9**(2), 193–206 (1984)
17. Zweigenhaft, R.L., Domhoff, G.W.: Blacks in the White Establishments?. A Study of Race and Class in America. Yale University Press, New Haven (1991)
18. Belliveau, M., O'Reilly, C., Wade, J.: "Social capital at the top; Effects of social similarity and status on CEO compensation". Acad. Manag. J. **39**(6), 1568–1593 (1996)
19. Useem, M., Jerome, K.: Pathways to top corporate management. Am. Sociol. Rev. **51**/2, 184–200 (1986)
20. Peng, M., Luo, Y.: Managerial ties and firm performance in a transition economy: The nature of a micro-macro link. Acad. Manag. J. **43**(3), 486–501 (2000)
21. Hannan, M., Freeman, J.: Structural inertia and organizational change. Am. J. Sociol. **49**(2) 149–164 (1984)
22. Yu, T.F.: Entrepreneurial alertness and discovery. Rev. Austrian Econ. **14**(10) 47–63 (2001)
23. Becker, G.: Human Capital: A Theoretical and Empirical Analysis with Special Reference to Education. University of Chicago Press, Chicago (1964)
24. Mincer, J.: Schooling. Experience and Earnings. Columbia Univ. Press, New York (1974)
25. Schultz, T.: Investment in man; An economist's view. Soc. Serv. Rev. **33**(2), 69–75 (1959)
26. Cooper, A., Gimeno-Gascon, J., Woo, C.: Initial human and financial capital as predictors of new venture performance. J. Busin. Ventur. **9**(5), 371–395 (1995)
27. Kirzner, I.: Entrepreneurial discovery and the competitive market process: an Austrian approach. J. Econ. Lit. **35**(1) 60–85 (1997)
28. Jaffee, D., Russel, T.: Imperfect information, uncertainty and credit rationing. Quart. J. Econ. **90**(4) 650–666 (1976)
29. Carpenter, R., Peterson, B.: Capital market imperfections, high-tech investment, and new equity financing. Econ. J. **112**(477), 54–72 (2002)
30. MacMillan, I., Siegel, R., Narasimha, S.: Criteria used by venture capitalists to evaluate new venture proposals". J. Busin. Ventur. **1**(1), 119–128 (1985)
31. Child, J.: Managerial and organizational factors associated with company performance. J. Manag. Stud. **11**(1), 13–27 (1974)
32. Stevenson, H, Roberts, M., Grousbeck, I.: New Business Ventures and the Entrepreneur. Irwin, Homewood, IL (1985)

33. Hair, J. F., Anderson, R. E., Tathum, R. L., Black, W. C.: Multivariate Data Analysis Macmillan, New York (1998)
34. Roure, J., Keeley, R.: Predictors of success in new technology based ventures. J. Busin. Ventur. **5**(4) 201–220 (1990)
35. Vesper, K.: New Venture Strategies. Prentice-Hall, Englewood Cliffs, NJ (1980)
36. Dahlstrand, L.: Entrepreneurial Origin and Spin-off Performance: A Comparison between Corporate and University Spin-offs. European Commission Report. John Wiley and Sons, New York (2001)

A Virtualization Security Framework for Public Cloud Computing

Jong Hyuk Park

Abstract Cloud computing is very efficient and useful technology for many organizations with its dynamic scalability and usage of virtualized resources, infrastructure, software and platform as a service through the Internet. Also, virtualization is a technology that has been widely applied for sharing the capabilities of physical computer's resources and software. Therefore, in order to increase the usage performance of cloud computing resources, virtualization was used. This research aims to make a virtualization framework for cloud computing. The framework was considered security service process with virtual machines and used in public cloud computing.

Keywords Cloud computing · Virtualization framework · Cloud security

1 Introduction

The recent years have witnessed the continuing development of the Internet from its original communication purpose (e.g., email) and content provision (e.g., Web) to an application deployment platform, where increased computing and storage capabilities are constantly being made available to end users. In parallel, an unprecedented number of personal computers are deployed worldwide according to a recent Gartner report, as worldwide PC shipments have reached 82.9 million units just in the second quarter of 2010, representing a 20.7 % increase from the second quarter of 2009. Recently, cloud computing paradigm has emerged as an

J. H. Park (✉)
Department of Computer Science and Engineering, Seoul National University of Science and Technology (SeoulTech), 172 Gongreung 2-dong, Nowon-gu, Seoul 139-743, South Korea
e-mail: jhpark1@seoultech.ac.kr

S.-S. Yeo et al. (eds.), *Computer Science and its Applications*,
Lecture Notes in Electrical Engineering 203, DOI: 10.1007/978-94-007-5699-1_41,
© Springer Science+Business Media Dordrecht 2012

energy efficient approach which enables ubiquitous, on-demand network accesses to a shared pool of flexibly reconfigurable computing resources including networks, servers, storage, applications, and services that can be rapidly deployed with minimal management effort or service provider interactions [1]. Thus, the cloud computing model is expected to become the next-generation computing infrastructure. In a cloud system, virtualization is an essential tool for providing resources flexibly to each user and isolating security and stability issues from other users. Because users' services and applications run on separate virtual machines (VMs), virtualization helps cloud platforms to accurately monitor and control the amount of resources being provided to users. Virtualization also facilitates the guarantee of service-level agreements (SLAs) between users and cloud platform providers by resolving QoS crosstalk between services and applications. Moreover, separation of the execution context and data between VMs is supported by processors and hypervisors so that users use their VMs as if they own their dedicated server hardware. Owing to these benefits, many commercial cloud systems including Amazon EC2 employ virtualization so that users can freely configure their virtual servers from kernel to application layers [2].

Additionally To advance cloud computing, series of critical problems should be solved and security is a top priority. Internet information system under cloud computing environment faces numerous challenges: (1) No identical standard. The deployment of existing cloud computing platform is scattered, the major manufacturers have set up their own cloud platform, each with a strong computing power, but interactivity among systems is not accessible because there is no uniform standard. (2) Security risks in the cloud. Potential security risks emerge simultaneously as users choose cloud computing service such as leak of privacy, invasion of information assets, security and audit ability of data, credibility of the cloud services platform and errors of large-scale distributed system. (3) Security risks whose origins are traditional Internet. (4) The relevant policies and regulations are not sound [3].

This research aims to make a framework for a virtualization framework for security model in public cloud computing environment. The rest of the paper is organized as follows: Sect. 2 discusses about the cloud computing environment and their virtualization, Sect. 3 describes the proposed model for virtualization framework in cloud computing and Sect. 4 discusses about the Conclusion.

2 Public Cloud Computing and their Virtualization Technology

2.1 Public Cloud Computing

Cloud computing is of interest for the computational science community as a promising alternative and extension of the grid computing model [4]. It has three main categories as below [5].

- **Infrastructure as a Service (IaaS)** provisions hardware, software, and equipments to deliver software application environments with a resource usage-based pricing model. Infrastructure can scale up and down dynamically based on application resource needs. Typical examples are Amazon EC2 (Elastic Cloud Computing) Service and S3 (Simple Storage Service) where compute and storage infrastructures are open to public access with a utility pricing model. This basically delivers virtual machine images to the IaaS provider, instead of programs, and the Machine can contain whatever the developer want.
- **Platform as a Service (PaaS)** offers a high-level integrated environment to build, test, and deploy custom applications. Generally, developers will need to accept some restrictions on the type of software they can write in exchange for built-in application scalability.
- ┤|**Software as a Service (SaaS)**. User buys a Subscription to some software product, but some or all of the data and codes resides remotely. Delivers special-purpose software that is remotely accessible by consumers through the Internet with a usage-based pricing model. In this model, applications could run entirely on the network, with the user interface living on a thin client.

In each of the models outlined above, the underlying infrastructure (operating systems, network, servers, operating systems, storage facilities) is usually in the control of the provider (although not always e the provider may well reserve the right to sub-contract any aspect of the service it provides to any sub-contractor anywhere in the world), although the seller may permit the customer a certain degree of control over selected networking components, such as firewalls, for instance. Each of these service models in turn is controlled and run in a variety of ways such as private cloud, community cloud, public cloud and hybrid cloud. Especially, a 'public cloud', where a provider owns the infrastructure and makes it available to anybody that wishes to pay for the service. The way each provider deals with the rise and fall in demand will affect how data is dealt with under this model. In essence, the providers act in a similar way as an electricity grid: they will trade between each other to buy and sell capacity to process data or store data, or both process and store data [6].

2.2 Security for Cloud Computing

In cloud computing environment, customers can then access these services to execute their business jobs in a pay-as-you-go fashion while saving huge capital investment in their own IT infrastructure. However, customers often have concerns about whether their privacy can be protected when facilitating services in the cloud, since they do not have much control inside the cloud. Correspondingly, privacy protection has become a critical issue and one of most concerning. Without it, customers may eventually lose the confidence in and desire to deploy cloud computing in practice [7]. In general Security means, focus will be giving

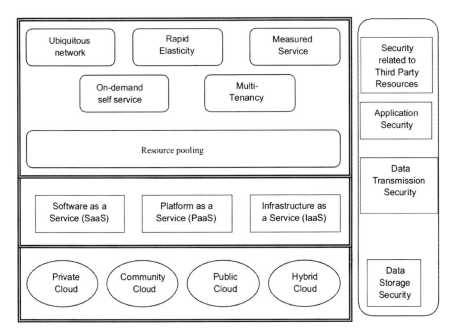

Fig. 1 Complexity of security in cloud environment

attention on confidentiality, Integrity, Availability. But will that be sufficient? Cloud Computing is providing services Such as Infrastructure as a Service, Platform as a Service, Software as a Service, or Anything as a Service through internet based as pay per usage model like utility computing [5]. Though cloud computing is targeted to provide better utilization of resources using virtualization techniques and to take up much of the workload from the client, it is fraught with security risks. The complexity of security risks in a complete cloud environment is illustrated in Fig. 1. In Fig. 1, the lower layer represents the different deployment models of the cloud namely private, community, public and hybrid cloud deployment models. The layer just above the deployment layer represents the different delivery models that are utilized within a particular deployment model. These delivery models are the SaaS, PaaS and IaaS delivery models. These delivery models form the core of the cloud and they exhibit certain characteristics like on-demand self-service, multi-tenancy, ubiquitous network, measured service and rapid elasticity which are shown in the to player. These fundamental elements of the cloud require security which depends and varies with respect to the deployment model that is used, the way by which it is delivered and the character it exhibits. Some of the fundamental security challenges are data storage security, data transmission security, application security and security related to third-party resources [8].

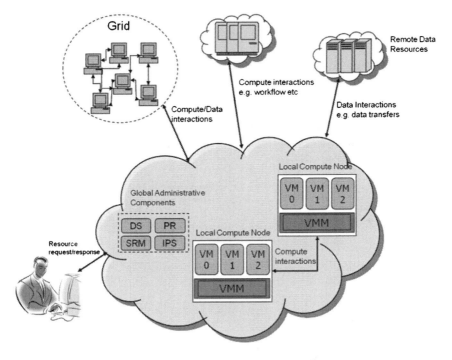

Fig. 2 A Cloud computing system with VM

2.3 Virtualization Technology in Cloud Computing

Traditionally, virtualization is a technology that has been widely applied for sharing the capabilities of physical computers by splitting the resources among OSs. The journey started back in 1964 when IBM initiated a project that shaped the meaning of the term Virtual Machines (VM) [9]. Cloud uses virtualization as its key technology. When end user submits their requirement, a separate Virtual Machine is created to run their specific application. In a single host machine itself multiple Virtual Machines can be run to utilize the resources [5]. A cloud computing system environment with VM is shown in Fig. 2. In this architecture, a system virtual machine, as described in this definition, serves as the fundamental unit for the realization of a Cloud infrastructure and emulates a complete and independent operating environment [10]. Virtualization helps multiple instances of same application can be run on one or more cloud resources. It automatically provides scalability when more number or user wants to run their application. It gives each user that their application is running on a single virtual machine. Here end user cannot see other user's data. Proper isolation of Virtual machine is important [5].

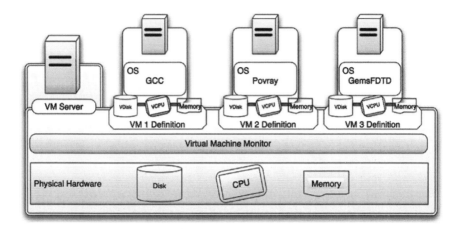

Fig. 3 Consolidation of heterogeneous VMs in a cloud node [2]

And Kim et al. [2] proposed the structure of heterogeneous VMs for cloud computing as shown in Fig. 3.

3 Framework for Security Model with Virtualization

In this research, the security model framework with virtualization in cloud computing was proposed (see Fig. 4).

This system architecture consists of main two parts; private cloud and public cloud. The private cloud system is process for supporting the public cloud system. In their system, *Business and Application Service* has their own business strategy, therefore the service is able to get their application to combine the VM according to business boundary. The public cloud system has many their own service and application that are to provide them to user. Especially, the *On Demand CPU* service is connected to *VM Server*. The *VM Server* deals with perform to manage and control the system resources such as CPU, memory and disk (storage devices). In order to provide privacy preserving authorization service, proposed system used obligation process. The Obligations are actions that must be performed when a certain event occurs. When the event is an authorization decision, then the obligations are actions that must accompany this decision. Enforcement of obligations, such as sending an email to a user when his/her data is accessed, or writing to an audit trail, or deleting data of a user after a certain time, is very important for a privacy preserving system. Some obligations need to be performed "before" the decision is enforced, some "after" the decision has been enforced, and some along "with" the enforcement of the authorization decision [11].

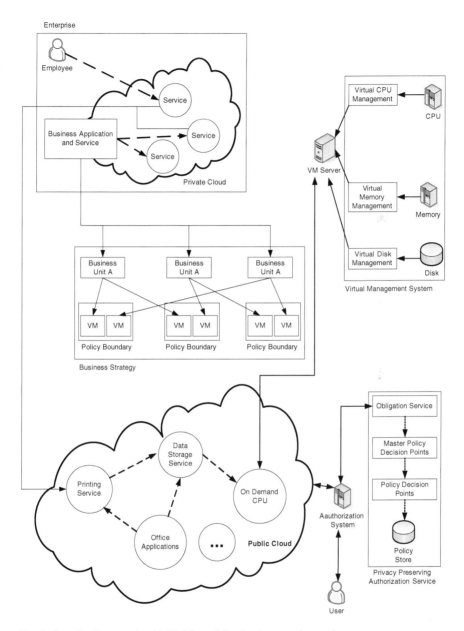

Fig. 4 Security framework with VM for public cloud computing environment

4 Conclusion

Cloud computing is a new paradigm to provide infrastructure, platform and software as a service that has emerged as a result of distributed Information Technology (IT) resources over the Internet. Also security becomes a hot issue in cloud computing cause of accessing the service over the internet. It can provide access to heterogeneous IT resources, which can be physical or virtual, as a service using the Internet. And the other technology, virtualization of hardware and software resources, enables clouds to efficiently adapt service provisioning. This research aimed to make a security framework for public cloud computing. To apply the virtualization technique, this research was considered VM according to business strategies in private cloud computing to support their service to public cloud. In order to provide security process to the user, the framework used *Authorization System* process. And virtualization technique for hardware resources in this framework was used *VM Server*.

Acknowledgments This work was supported by NIA, KOREA under the KOREN program (1295100001-120010100).

References

1. Li, J., Li, B., Wo, T., Hu, C., Huai, J., Liu, L., Lam, K.P.: CyberGuarder: a virtualization security assurance architecture for green cloud computing. Futur. Gener. Comput. Syst. **28**, 379–390 (2012)
2. Kim, N., Cho, J., Seo, E.: Energy-credit scheduler: an energy-aware virtual machine scheduler for cloud systems. Futur. Gener. Comput. Syst. (2012). doi:10.1016/j.future.2012.05.019
3. Liu, P., Liu, D.: The new risk assessment model for information system in cloud computing environment. Procedia Eng. **15**, 3200–3204 (2011)
4. Malawski, M., Meizner, J., Bubak, M., Gepner, P.: Component approach to computational applications on clouds. Procedia Comput. Sci. **4**, 432–441 (2011)
5. Loganayagi, B., Sujatha, S.: Enhanced cloud security by combining virtualization and policy monitoring techniques. Procedia Eng. **30**, 654–661 (2012)
6. Mason, S., George, E.: Digital evidence and 'cloud' computing. Comput. Law Secur. Rev. **27**, 524–528 (2011)
7. Zhang, G., Yang, Y., Chen, J.: A historical probability based noise generation strategy for privacy protection in cloud computing. J. Comput. Syst. Sci. (2012). doi:10,1016/j.jcss.2011.12.020
8. Subashini, S., Kavitha, V.: A survey on security issues in service delivery models of cloud computing. J. Net. Comput. Appl. **34**(1), 1–11 (2011)
9. Rodr'ıguez-Haro, F., Freitag, F., Navarro, L., Hern'andez-S'anchez, E., Far'ıas-Mendoza, N., Guerrero-Ib'ãnez, J.A., Gonz'alez-Potes, A.: A summary of virtualization techniques. Procedia Technol. **3**, 267–272 (2012)
10. Arshad, J., Townend, P., Xu, J.: A novel intrusion severity analysis approach for Clouds. Futur. Gener. Comput. Syst. (2011). doi:10.1016/j.future.2011.08.009
11. Chadwick, D.W., Fatema, K.: A privacy preserving authorization system for the cloud. J. Comput. Syst. Sci. (2012). doi:10.1016/j.jcss.2011.12.019

Environmental Awareness in Green Supply Chain and Green Business Practices: Application to Small and Medium-Sized Enterprises

Se-Hak Chun, Ho Joong Hwang and Young-Hwan Byun

Abstract This study investigates how the green business activities can be affected by awareness of green supply chain management (SCM) and existing cost reduction behaviors. To this end, this study conducts an empirical research of small and medium-sized enterprises based on a questionnaire survey method. This study shows green business practices are affected by types of industry, but not by the awareness of green SCM and current cost reduction activities.

Keywords Green SCM · Green business practices · Climate change · Environmental awareness

1 Introduction

As global warming has made the public aware of problems such as climate change, peak-oil, high volume of greenhouse gas emissions, loss of biodiversity, potable water pollution, growing landfill areas, and population pressure, firms also consider green management as their business strategies [1]. Many leaders understand that sustainability is now a critical part of the core value of a company because

S.-H. Chun · H. J. Hwang
Department of Business Administration, Seoul National University of Science
and Technology, Kongneung-gil 138 Nowon-gu, Seoul 139-743, Republic of Korea
e-mail: shchun@seoultech.ac.kr

H. J. Hwang
e-mail: intostar@korea.com

Y.-H. Byun (✉)
Department of Business Administration, Hallym University,
Chuncheon-si, Republic of Korea
e-mail: yhbyun@hallym.ac.kr

S.-S. Yeo et al. (eds.), *Computer Science and its Applications*,
Lecture Notes in Electrical Engineering 203, DOI: 10.1007/978-94-007-5699-1_42,
© Springer Science+Business Media Dordrecht 2012

returns from launching green projects like positive cash flow, reduced energy, material, and operating costs can make or break a company today [2]. Today almost all companies face the challenges to get the balance between their activities and the environment [3]. Although market becomes highly competitive in global environment, yet is a market which draws much attention to the aspects of environmental protection. Consequently, companies face the need to change the philosophy of their actions [3]. Thus, recently firms including small medium sized firms consider green SCM as a tool to cope with rapid changeable business environment.

The term 'supply chain' describes the network of suppliers, distributors and consumers [4]. With increased awareness to corporate responsibility and the requirement to meet the terms with environmental policy, green supply chain management (GSCM) is becoming increasingly important. Companies that have adopted GSCM practices with a focus on distribution activities have successfully improved their business and environmental performance on many levels. The major four activities of the green supply chain management are green purchasing, green manufacturing, green marketing and reverse logistics [5]. Activities of "green" supply chain can be addressed to the four areas found in the traditional supply chain, such as upper and lower flows, and the activities occurring within the organization and logistics processes [6].

Although in the twenty-first century developing a global supply chain strategy is necessary, however, advances in technology has opened previously closed markets to small and medium enterprises (SMEs) that otherwise lacked the resources to compete globally. In the past SMES were not under pressure to engage in environmental initiatives, since SMEs have little individual impact on the environment. However, SMEs supplying large organizations are now under pressure from those organizations, seeking to green entire supply chains, to conform to environmental standards. Laws and large customers are starting to demand that smaller firms reduce environmental impact. There are some barriers when SMEs Adopts green initiatives because many lack the resources (people, money, or knowledge). The lack of information focusing on green supply chain management in SMEs acts as a deterrent, especially for SMEs with limited resources [7]. Additional reasons SMEs may not engage in green supply chain management include: lack of environmental awareness, belief that environmental practices cost more than it pays, belief that it is time consuming to implement green initiatives. As larger firms often invest in the environmental capacity of smaller suppliers, it becomes crucial for SME suppliers to be involved in green initiatives, because a green supply chain is hard to achieve without full participation [7].

While much research has been done on environmental management in supply chains, little research has focused on green supply chain management in SMEs. This study investigates how the green business activities can be affected by awareness of green supply chain management (SCM) and existing cost reduction behaviors. The paper is organized as follows. In Sect. 2, we describe research model and the data used in the study. In Sect. 3, we show the results of our analyses. Section 4 concludes this study.

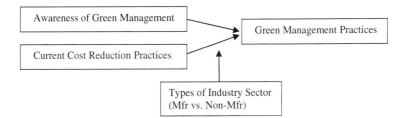

Fig. 1 Research model

2 Research Methodology and the Data

We investigate how green management practices are related to current cost reduction activities and green awareness according to types of industry sector as shown in Fig. 1. We hypothesizes that green management activities are related to these factors such as existing cost reduction activities, green awareness and types of industry sector as follows;

H1: As green awareness of SMEs is higher, there are more green management practices in SMEs.
H2: As SMEs have current cost reduction activities more, there are more green management practices in SMEs.
H3: Manufacture sector has more green management practices than non-manufacture sector.

Questionnaire survey is widely used in business and management research [8] (Saunders et al. 2009) and is known as helpful to obtain straightforward information from the respondents [9] (McIntyre 2005). A total of One hundred and thirty companies were selected personally and among them 75 copies of the questionnaire were returned and the data output was then processed. The statistical analyzing software, SPSS is used to analyze the data collected.

Table 1 lists the distribution of respondents in terms of their types of industry, company size and experience years. The percentage of respondents consists of manufacture sector (44.0 %) and non-manufacturer sector (56.0 %). Regarding the size of respondents ranged from under 20 to over 300 employees which found that respondent's companies are mainly from less than 20 as shown in Table 1.

3 Results of the Study

3.1 Factor Analysis

This research tries to identify how Korean SMEs in the manufacturing industry and non manufacturing industry are aware of green SCM and act green practices. First we did factor analysis because factor analysis is proposed to put some

Table 1 Demograhpy of respondents

Items	%
Types of industry	
Manufacture sector	44.0
Non-manufacture sector	56.0
Size (employees)	
>300	6.1
100–300	6.1
50–100	15.2
20–50	15.2
<20	57.6
Job position	
CEO	23.5
Director	13.6
General manager and below	62.9

Table 2 KMO and Bartlett's test

Kaiser–Meyer–Olkin measure of sampling		
Adequacy		0.786
Bartlett's test of sphericity	Approx. Chi-Square	803.499
	df	91
	Sig.	0.000

correlated factors together into one factor, helping to reduce the factors and to provide clearer evaluations.

Table 2 shows KMO and Bartlett's test result. The test results of KMO show that the compared value is 0.786, significantly exceeding the suggested minimum standard of 0.5 required for conducting factor analysis [10]. The Kaiser–Meyer–Olkin Measure of Sampling Adequacy is 0.786, indicating that the variables are suitable for factor analysis. The significance is 0.000 thus concluding that the strength of the relationship among variables is strong, and that the variables are suitable for factor analysis. We performed factor analysis to extract factors in accordance with the eigenvalues of discontinuity which is greater than 1 [11].

Table 3 shows results of factor analysis based on 14 items which factor loading value are over 0.600 and are categorized in three factors such as Awareness of Green Management, Current Cost Reduction Activities Green Practices. As shown in Table 3 most of companies do not show green practices although most of them show higher value of 3.0 in two factors such as awareness and cost reduction activities.

3.2 Regression Result

Table 4 presents the results of pairwise *t*-tests between selected variables of manufacturing and non-manufacturing sectors. Manufacturing sector is significantly different in green practices from non-manufacturing sector. However, other

Table 3 Factor analysis

Factor	Items	Average	Factor loading
Awareness of green management	Awareness green consuming	3.25	0.872
	Environment regulatory policy (RoHS)	2.91	0.844
	Regulation reduction CO_2	3.00	0.783
	Awareness needs of green SCM	2.84	0.604
Current cost reduction activities	Use of recycling products	3.08	0.877
	Reducing electronic power	3.27	0.863
	Use of renewal papers	3.71	0.793
	Turn off power and plug off in offices	3.23	0.767
	Not using one time usage	2.93	0.721
Green practices (dependent)	Measure of CO_2	1.92	0.932
	Reduction CO_2 plan	2.01	0.893
	Attending seminar and education about green management	2.24	0.800
	Investment for renewable energy	2.40	0.761
	Green information sharing with partner	2.31	0.744

Table 4 Results of pairwise t-tests

Variables (Mfr versus Non-Mfr)	Mean	Variance	t-statistic	P-value
Awareness	0.138 versus −0.108	0.760 versus 1.151	1.063	0.291
Cost reduction activities	−0.008 versus 0.006	1.088 versus 0.938	−0.068	0.946
Green practices	0.447 versus −0.351	0.956 versus 0.896	**3.723**	**0.000**

Table 5 Model Summary

	Coefficient	Standard error	t-statistic	P value
Constant	1.266	0.358	**3.534**	**0.001**
Awareness	−0.050	0.109	−0.458	0.649
Cost reduction activities	0.003	0.109	0.030	0.976
Types of industry sector	−0.812	0.219	**−3.705**	**0.000**
	$R^2 = 0.162$ F $= 4.577$ Signif F $= 0.005$			

two variables have not significant results although manufacturers are more seen to be aware of green management than non-manufacturers.

Next, we investigate how performances of green technologies can be affected by other factors such as awareness of green management, current cost reduction activities and types of industry sector. For this, we assume a regression framework as follows:

$$Green\ Practices = \alpha + \beta_1 * Awareness + \beta_2 * Cost\ Reduction\ Activities$$
$$+ \beta_3 * Types\ of\ Industry +$$
$$\varepsilon\alpha : constant\ \beta : standardized\ coefficient\ \varepsilon : residuals$$

Table 5 shows the results of the regression of green practices and three variables. The dependent variable is the green practices. Table 5 indicates that two factors such as awareness of green management and current cost reduction activities are not statistically significant and the types of industry sector is significant at level $p < 0.01$.

4 Conclusion

This study investigates how the green business practices are related to other factors such as awareness of green management, current cost reduction behaviors and types of industry. This study shows that green business practices are affected by types of industry, but not by the awareness of green management and current cost reduction behaviors. Also there are no differences in green awareness and cost reduction behaviors between manufacturing sector and non-manufacturing sector. However, this study shows the difference in green SCM practices between manufacturing sector and non-manufacturing sector.

This study has limitation in that the analysis was based on relatively less amount of data set and many respondents has less information about green SCM, which may lead to low reliability to test practical green management practices.

Acknowledgments This study was partially supported by Seoul National University of Science and Technology.

References

1. Wills, B.: The Business Case for Environmental Sustainability (Green). A 2009 HPS White Paper. (2009)
2. OECD. Towards Green ICT Strategies: Assessing Policies and Programmes on ICT and the Environment. (2009)
3. Beamon B.M., Designing the green supply chain. Logist. Inf. Manag. **12**(4), 332–342 (1999)
4. Sabari, R.P., Xavier, M.J., Israel, D.: Green purchasing practices: a study of E-procurement in B2B buying in Indian small and medium enterprises. J. Supply Chain Oper. Manag. **10**(1), 13–23 (2012)
5. Research Journal of Recent Sciences Vol. 1(6), 77–82, June (2012) Res. J. Recent Sci. International Science Congress Association 77. An Overview of Green Supply Chain Management in India. Nimawat Dheeraj1 and Namdev Vishal2, Department of Mechanical Engineering, Singhania University, Pacheri Bari, Jhunjhunu, Rajasthan, INDIA Government Engineering College Jhalawar, Rajasthan, INDIA

6. Emmett, S., Sood, V.: Green Supply Chains—An Action Manifesto? Wiley, New York (2009)
7. http://www.articlesbase.com/environment-articles/smes-and-green-supply-chain-management-5890224.html
8. Saunders, M., Lewis, P., Thornhill, A.: Research Methods for Business Students, 5th edn. Prentice Hall, London (2009)
9. McIntyre, L.J.: Need to Know: Social Science Research Methods. McGraw-Hill, New York (2005)
10. Hair, J.F. Jr., Anderson, R.E., Tatham, R.L., Black, W.C.: Multivariate Data Analysis, 3rd edn. Macmillan Publishing Company, New York (1995)
11. Tabachnick, B.G., Fidell, L.S.: Using Multivariate Statistics. Harper & Row, New York (1989)

Computational Analysis of the Bargaining Power and Channel Strategies in Supply Chain Relationship

Se-Hak Chun

Abstract This paper analyzes hybrid channel strategies of a manufacturer when it considers an online store using analytical and computational simulation methods, and discusses some strategic implications from the perspective of market transaction costs and the portion of online customers.

Keywords Bargaining power · Channel management · Direct online channel · Electric commerce · Computational analysis

1 Introduction

The prevailing popularity of the Internet has led thousands of manufacturers, such as IBM, Hewlett-Packard, Pioneer Electronics, Sony, Kodak, Minolta, Panasonic, Cisco, the former Compaq, Mattel, Estee Lauder and Nike, to add direct online channels to their existing retail networks [1–6]. However, some manufacturers such as JVC, NEC, 3M, Nikon, Cannon, Olympus, Samsung and LG have used the Internet as a mere medium to provide information about their products without selling via their websites [3]. These examples show that the adoption of the Internet channel seems to be very complicated. Why do those companies use different channel strategies?

This paper focuses on analyzing a manufacturer's optimal channel using analytical and computational simulation methods, and discusses some strategic

S.-H. Chun (✉)
Department of Business Administration, Seoul National University of Science and Technology, Kongneung-gil 138 Nowon-gu, Seoul 139-743, Republic of Korea
e-mail: shchun@seoultech.ac.kr

S.-S. Yeo et al. (eds.), *Computer Science and its Applications*,
Lecture Notes in Electrical Engineering 203, DOI: 10.1007/978-94-007-5699-1_43,
© Springer Science+Business Media Dordrecht 2012

implications from the perspective of market transaction costs and the portion of online customers. This paper shows that manufacturers choose a dual channel strategy using both offline and online channels simultaneously when convenience associated with the online purchase becomes larger (smaller a), inconvenience associated with the offline purchase becomes larger (larger t) and its offline customers are over a certain range.

The remainder of this paper is organized as follows: Sect. 2 presents the basic model and analyzes the hybrid channel model. Section 3 discusses results of the study with strategic implications. In Sect. 4, conclusions and future research are discussed.

2 The Model

I analyze a model when a manufacturer launches a direct online channel in competition with the traditional offline retailer. I choose the commonly used spatial competition model of Hotelling [7] and extend Chun and Kim's model [8, 9].

I considered a linear city of length s where there is a retailer (or manufacturer) at the end of the city. There is also a direct online channel of the manufacturer that sells the same good with no physical location. The unit production cost of both a retailer and a direct online channel of the manufacturer is equal to c. Consumers are distributed uniformly along the city. In other words, letting denote s the distance from the retailer, which is located at 0, is uniformly distributed on $[0, s]$. The distance here can represent different preferences such as offline product service preference, the opportunity cost of time, the implicit cost of inconvenience, as well as the real cost of travel [8, 10]. We assumed that the location of a retailer is at 0. Thus a consumer located at 0 has maximum valuation of the good, V, and others preferences are decreasingly differentiated according to s.

Each consumer consumes one or zero units of the good. A fraction of consumers at each location point, m, have a strong preference on offline services and they are served from distributors or retailers' product services. They are a group of "service sensitive group" [11]. The other group, $1 - m$, can have access to the Internet and they are "price sensitive group" who are easily switch from offline to online when they compare two channel's transaction costs and prices. Thus, this group of consumer may buy the good from the offline retailer, in which case the consumer has to travel to the retail store and pay transportation cost ts, where t is the transportation cost per unit of length. If the consumer buys the good from the online firm, the consumer incurs cost, a, which may be search cost and other costs related to quality uncertainty, security risk, and delivery cost. These costs represent online specific inconvenience costs, incurred when consumers buy from online channel, and can affect the consumer's willingness to purchase from online channel. We assume that consumers are homogeneous with respect to these costs, though they can be heterogeneous. Then, the utility of a consumer with the Internet access (or a group of "price sensitive", easily switchable consumer group from offline to online) located at s is:

$$\begin{cases} V - p_r - ts & \text{if he buys from the offline firm} \\ V - p_d - a & \text{if he buys from the online firm} \\ 0 & \text{if he does not buy,} \end{cases}$$

where p_r and p_d are the prices charged by the offline and online firms. Letting $\hat{s} = \frac{(p_d + a - p_r)}{t}$, consumers with $s < \hat{s}$ prefer the offline channel while consumers with $s > \hat{s}$ prefer the online channel. Thus, the demand functions are given by

$$D_{off} = \frac{(1-m)(p_d + a - p_r)}{t} + \frac{m(V - p_r)}{t}$$

$$D_{on} = (1-m)\left(\bar{s} - \frac{(p_d + a - p_r)}{t}\right).$$

2.1 The Basic Model

In this section, I analyze when a manufacturer does not use an online channel. In order to compare the performance of the model where a manufacturer uses direct online channel, I find optimal wholesale and retail prices of the model when it uses only the traditional channel. I denote this case by the subscript, B, to identify prices and resulting profits of the manufacturer and retailer. When a manufacturer acts as Stackelberg-leader, in the second period, given wholesale price, w, the retailer finds the optimal retail price maximizing its profit as follows;

$$\Pi_R^B = (p_r - w)\left(\frac{V - p_r}{t}\right)$$

From this, the retail price is derived as

$$p_r^B = \frac{1}{2}V + \frac{1}{2}w.$$

Anticipating the retailer's price, in the first period, the manufacturer chooses the profit maximizing wholesale price from its profit function as follows;

$$\Pi_M^B = (w - c)\left(\frac{V - p_r}{t}\right)$$

From this, wholesale price and retail prices are derived as

$$w^B = \frac{1}{2}V + \frac{1}{2}c, \quad p_R^B = \frac{3}{4}V + \frac{1}{4}c.$$

Thus, profits of the retailer and manufacturer are derived as follows;

$$\Pi_M^B = \frac{(V-c)^2}{8t}, \quad \Pi_R^B = \frac{(V-c)^2}{16t}$$

2.2 The Hybrid Model

Next, I consider a situation when the manufacturer uses an online channel. The manufacturer takes the retailer's reaction function in consideration in order to find wholesale price, w, and a direct online price, p_d, maximizing its profit. The retailer profit function is

$$\Pi_R^H = m(p_r - w)\left(\frac{V-p_r}{t}\right) + (1-m)(p_r - w)\left(\frac{p_d + a - p_r}{t}\right).$$

In the first period, using the retailer's reaction function, the manufacturer finds wholesale price and its direct online price from its profit function as follows;

$$\Pi_M^H = m(w-c)\left(\frac{V-p_r}{t}\right) + (1-m)(w-c)\left(\frac{p_d + a - p_r}{t}\right) + (1-m)(p_d$$
$$-c)\left(1 - \frac{p_d + a - p_r}{t}\right).$$

From these profit functions, optimal prices are derived as follows;

$$w^H = \frac{t}{2m} + \frac{V+c-t}{2}, \quad p_D^H = \frac{t}{2m} + \frac{V+c-a}{2}, \quad p_R^H$$
$$= \frac{t}{2m} + \frac{V+c-t}{2} + \frac{a}{4} + \frac{V-c-a}{4}m$$

3 Results of the Study

I compare two profits of the basic model and hybrid model using maple's mathematical software for simulation. Table 1 shows the result of profit differences, $\Pi_M^H - \Pi_M^B$, between the two models. The scenario is when $V = 30$, $c = 15$, $t = 2.5$ for example of manufacturer's profit.

The none-area means outside cases when optimal solutions cannot be derived because assumptions are violated. The blue-area denotes conditions when a manufacturer should add its new direct online channel to exploit profit while red-area when it should not. As shown in Table 1, apparently when m and a are lower the manufacturer tends to launch its online business at the same time. Figures 1, 2, 3 and 4 show the results of sensitive analysis through simulation.

Table 1 Profit differences between the hybrid model, Π_M^H and Π_M^B when $V = 30$, $c = 15$, $t = 2.5$

m	a						
	0.01	1	2	3	4	5	6
0.01	None	None	None	None	None	None	None
0.1	None	None	None	None	None	None	None
0.2	1.29	0.71	0.21	−0.19	−0.49	−0.70	−0.81
0.3	1.19	0.58	0.05	−0.39	−0.74	−0.99	−1.16
0.4	None	0.77	0.24	−0.21	−0.58	−0.86	−1.06
0.5	None	None	0.46	0.03	−0.34	−0.63	−0.84
0.6	None	None	None	0.22	−0.11	−0.38	−0.59
0.7	None	None	None	None	0.05	−0.18	−0.37
0.8	None	None	None	None	0.13	−0.04	−0.19
0.9	None	None	None	None	None	0.02	−0.06
1	None	None	None	None	None	0.00	0.00

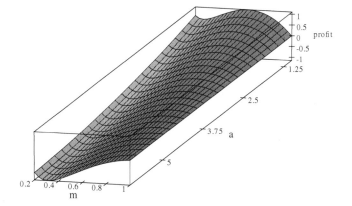

Fig. 1 The effect of a and m on the profit of the manufacturer when $V = 30$, $c = 15$, $t = 2.5$

Figure 1 supports the results of Table 1, which seemingly implies that as m and a are relatively lower the profit of manufacturer with additional online channel are higher than that of manufacturer without an online channel.

Figure 2 compares profits of two models. The dotted curve denotes the profit of hybrid model when a manufacturer uses both channels while the blue line represents the profit when a manufactures use only the offline channel. As shown in Fig. 2, as a increases the profits decreases, which means as online customers feel more convenience the manufacturer tends to launch its online business.

To see how m can affect on the profits of the hybrid model, Fig. 3 compares profits of two models when a is fixed at 3. The dotted curve denotes the profit of hybrid model while the blue line represents the profit of basic model when a manufactures use only the offline channel. Figure 3 shows that a manufacturer uses

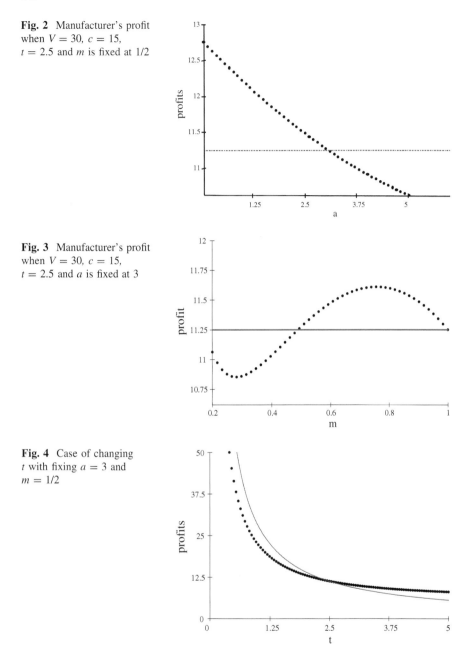

Fig. 2 Manufacturer's profit when $V = 30$, $c = 15$, $t = 2.5$ and m is fixed at 1/2

Fig. 3 Manufacturer's profit when $V = 30$, $c = 15$, $t = 2.5$ and a is fixed at 3

Fig. 4 Case of changing t with fixing $a = 3$ and $m = 1/2$

online at the same time when its offline customers are over a certain range. Also its maximum profits can be attained when m is in medium range.

To see the effect of t Fig. 4 compares profits of two models when a and m are fixed at 3 and 1/2 respectively. The dotted curve denotes the profit of hybrid model

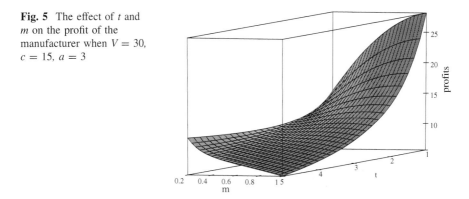

Fig. 5 The effect of t and m on the profit of the manufacturer when $V = 30$, $c = 15$, $a = 3$

while the blue curve represents the profit of basic model. Figure 4 is closely related to Fig. 2 in a perspective of each market transaction costs and shows that as t increases the profits of the hybrid model increases, which means as offline customers feel less convenience the manufacturer tends to launch its online business at the same time.

Figure 5 is closely related to Fig. 1 and shows apparently when m and t are lower the manufacturer tends to launch its online business at the same time.

4 Conclusion

In this paper, I analyzed a manufacturer's channel strategy when it considers an online store. Through analytical and computational simulation methods this paper has drawn some important implications. First, the manufacturer enters an online market with existing its offline market when convenience associated with the online purchase becomes larger (smaller a) and inconvenience associated with the offline purchase becomes larger (larger t). Second, the manufacturer uses online at the same time when its offline customers are over a certain range and its maximum profits can be attained when m is in medium range. For future study we anticipate conducting research more on bargaining power between manufacturers and retailers and strategic interactions.

Acknowledgments This work was supported by the National Research Foundation of Korea Grant funded by the Korean Government (NRF-2012-016673).

References

1. Chiang, W., Chhajed, D., Hess, J.: Direct marketing, indirect profits: a strategic analysis of dual-channel supply chain design. Manag. Sci. **49**, 1–20 (2003)
2. Hua, G., Wang, S., Cheng, T.C.E.: Price and lead time decisions in dual channel supply chains. Eur. J. Oper. Res. **205**, 113–126 (2010)

3. Kumar, N., Ruan, R.: On complementing the retail channel with a direct online channel. Quant. Mark. Econ. **4**(3), 289–323 (2006)
4. Tedeshi, B.: What a tangled Web they weave: the rush to e-commerce pits brick-and-mortar retailers against unlikely competition, they own supplies. National Post March 3 C16 (1999)
5. Tedeshi, B. Compressed data; big companies go slowly in devising net strategy: New York Times (March 27) (2000)
6. Tsay, A.A., Agrawal, N.: Channel conflict and coordination in the e-commerce age. Prod. Oper. Manag. **13**(1), 93–110 (2004)
7. Tirole, J.: The Theory of Industrial Organization. The MIT Press, Cambridge (1995)
8. Chun, S.-H., Kim, J.-C.: Pricing strategies in B2C electronic commerce: analytical and empirical approaches. Decis. Support Syst. **40**(2), 375–388 (2005)
9. Chun, S.-H., Kim, J.-C.: Analysis of price competition and strategic implications for heterogeneous market structure. Int. Rev. Econ. Finance **14**, 455–468 (2005)
10. Strader, T.J., Shaw, M.J.: Characteristics of electronic markets. Decis. Support Syst. **21**, 185–198 (1997)
11. Chun, S.-H., Rhee, B.-D., Park, S.Y., Kim, J.-C.: Emerging dual channel system and manufacturer's direct retail channel strategy. Int. Rev. Econ. Finance **20**, 812–825 (2011)

New Hybrid Data Model for XML Document Management in Electronic Commerce

Eun-Young Kim and Se-Hak Chun

Abstract XML has been known as a document standard in representation and exchange of data on the Internet, and is also used as a standard language for the search and reuse of scattered documents on the Internet. The issues related to XML are how to model data on effective and efficient management of semi-structured data and how to actually store the model data when implementing a XML contents management system. This paper considers a relational data system to store XML document efficiently and propose a hybrid data model for XML document management.

Keywords XML document management · Relational database · Objective database · Multi-format information retrieval · Electronic commerce

1 Introduction

Extensible Markup Language (XML) is a simple and flexible markup language. As XML has been used for data transaction in EDI from 1998, it supports all kinds of electronic commerce transaction [1]. It provides a method for finding information which users want and XML's abundant data representative method enables users

E.-Y. Kim
Department of Multimedia Contents, Sin Ansan University,
671 Chosi-dong, Ansan City, Kyunggi-do 425-792, Republic of Korea
e-mail: key@sau.ac.kr

S.-H. Chun (✉)
Department of Business Administration, Seoul National University of Science and Technology, Kongneung-gil 138 Nowon-gu, Seoul 139-743, Republic of Korea
e-mail: shchun@seoultech.ac.kr

S.-S. Yeo et al. (eds.), *Computer Science and its Applications*,
Lecture Notes in Electrical Engineering 203, DOI: 10.1007/978-94-007-5699-1_44,
© Springer Science+Business Media Dordrecht 2012

to do business on the web by making intellectual mechanism. Therefore, it has received a lot of attention and it has been utilized on almost all fields on the web such as on-line banking, push technology, search engine, web based control system and agent and so on. In addition, application areas are expanding rapidly, and XML documents are made possible to be reused from returned XML documents through the search engine, another existing XML documents, or transmitted XML documents users can draw out data that they want and process their own data structure and store it. Approaches to manage XML documents data can be classified into using new database system for XML documents and using existing database system such as relational database system. Using new database system for XML documents is based on a new data model that represents semi-structured XML data. Also, when users take the existing relational database system or the object system in consideration, mapping model that fits in a relational database system or an object system is required.

Individual data structure was defined according to the XML documents type in the previous researches in data model of XML documents. This poses some problems because the new data structure should be defined for XML documents that have new structures and data; it is difficult to expand the domain and the data structure that needs to be defined in a different way every time according to the object to apply and top-down access is the only available way to search for information. Both the data and structure view of XML's original documents are lost in the data model to map the existing relational database system or object system, so XML documents cannot be generated from the stored data. The range of the application can be expanded easily when the data model for XML documents has a unique data structure regardless of data and structures of XML documents. Also, in terms of searching for information on XML documents, bottom-up access and left–right access should be possible as well as top-down access. The search should also be possible without previous knowledge on the XML document structure.

In this paper, we suggest a data model that supports all these requirements and is applicable to new database system and existing relational database system for the XML document and represents the data and the structure view of the XML document. The paper is organized as follows. In Sect. 2, we describe research motivation. In Sect. 3, we propose the hybrid data model for XML document management. Section 4 concludes this study.

2 Research Motivation

An XML document describes data and structures but it cannot be regarded as a database system simply because it describes data. Even though it contains data, it is just a general text file that cannot perform any function without additional software that manages data. However, once an XML document, a related XML tool and XML's various functions are united, it can be regarded as a database system because it contains storage, schema, query language, programming interface (factors of database system) and so forth. But, it cannot support effective storage, index, security, transaction, data perfection, multi-user access, trigger, and query on multiple documents that the existing database system supports. Therefore, if the amount of data is not large and the number of users is small and the circumstances require just general capabilities, the XML document can be used as a database, but if there are many users and the circumstances require data perfection and advanced capabilities, the XML document can be used as a database. In this case, the database that stores the XML data and structure is necessary. At present, there are two ways to store data of the XML document; one is to store it as a new data model in the new database system only for the XML document and another is to store the data by mapping the data of the XML document into an existing relational or object system.

The new database system only for the XML document requires a new model that can store, represent and query data of the XML document effectively, while the data model for the existing relational database system or object system requires a data model that coincides with these database system models naturally.

3 New Proposed Hybrid Data Model

This paper proposes a data model that is needed when a new XML-native database system for only XML is designed and is applicable to an existing relational database system. Based on the basic structure of Lore's XML Data Model and Edge Labeled Graph Model [2, 3], proposed hybrid xml data model is renewed and extended to support requirements shown below.

(1) Document in global domain is the object.
(2) Data model doesn't depend on the type of database system chosen.
(3) Both structure and data of the document are represented.
(4) Structure of the document and data change is applicable flexibly.
(5) Mapping document from data model reversely, all element orders of document are preserved.
(6) Top-down, bottom-up, left-right, right-left search about query should be possible.

3.1 Basic Model of XML

It is the most natural to represent the XML document as a graph because the element structure of the XML document can be represented as a tree and mutual reference among elements is added. When representing a basic structure of the XML document data as a graph, each element of the XML document covers a graph that takes root on the node having its own tag name. Therefore, another element included to element is represented as a sub-graph in the element graph

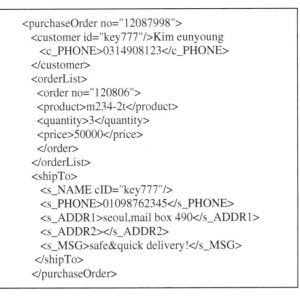

```
<purchaseOrder no="12087998">
  <customer id="key777"/>Kim eunyoung
    <c_PHONE>0314908123</c_PHONE>
  </customer>
  <orderList>
    <order no="120806">
    <product>m234-2t</product>
    <quantity>3</quantity>
    <price>50000</price>
    </order>
  </orderList>
  <shipTo>
    <s_NAME cID="key777"/>
    <s_PHONE>01098762345</s_PHONE>
    <s_ADDR1>seoul,mail box 490</s_ADDR1>
    <s_ADDR2></s_ADDR2>
    <s_MSG>safe&quick delivery!</s_MSG>
  </shipTo>
</purchaseOrder>
```

Fig. 1 XML document

containing itself. Also, each node is linked by edges and has level or represents reference meanings.

In the basic model, the XML document is represented as a graph that all nodes and edges are labeled and that is ordered among nodes and edges and that edges are directed on. When the XML document is represented as a graph $G = \{V,E,A\}$ where V is a vertex, i.e. set of nodes, E, a set of edges, and A, a set of attributes defined on start tag of element. Figure 1 shows an example of XML document for electronic commerce and Fig. 2 represents it with graph.

An element in the XML document means all contents from the equivalent start tag to finish tag are related to the start tag, and each element of the XML document is represented as a sub-tree/graph that takes the tag of an equivalent element as a root node. The label of node V is a tag value in the case of a middle node and it corresponds with the value (contains NULL and EMPTY) in the case of a leaf node. The label of the edge Ei from node Vi corresponds with relation between node Vi and arriving child node Vj. If child node Vj is another element, it is labeled as CHILD that indicates child element information of node Vi and if the child node Vj is a value of element, it is labeled as VALUE that manifests Vi value information.

3.2 Data Extended Model Considering DTD

DTD means a mutual agreement on the XML document transmitted when XML documents are exchanged. Attributes referring elements in DTD are presented by defining the type of IDREF or IDREFS. Also the value of attributes of IDREF or

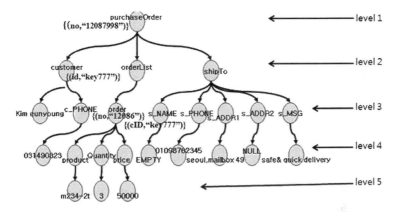

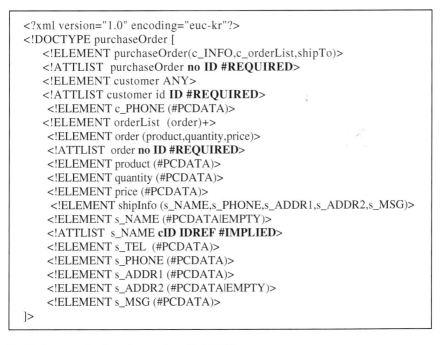

Fig. 2 XML document represented as a graph of the data model

```
<?xml version="1.0" encoding="euc-kr"?>
<!DOCTYPE purchaseOrder [
    <!ELEMENT purchaseOrder(c_INFO,c_orderList,shipTo)>
    <!ATTLIST  purchaseOrder no ID #REQUIRED>
    <!ELEMENT customer ANY>
    <!ATTLIST customer id ID #REQUIRED>
     <!ELEMENT c_PHONE (#PCDATA)>
    <!ELEMENT orderList  (order)+>
     <!ELEMENT order (product,quantity,price)>
     <!ATTLIST  order no ID #REQUIRED>
     <!ELEMENT product (#PCDATA)>
    <!ELEMENT quantity (#PCDATA)>
    <!ELEMENT price (#PCDATA)>
     <!ELEMENT shipInfo (s_NAME,s_PHONE,s_ADDR1,s_ADDR2,s_MSG)>
    <!ELEMENT s_NAME (#PCDATA|EMPTY)>
    <!ATTLIST s_NAME cID IDREF #IMPLIED>
    <!ELEMENT s_TEL  (#PCDATA)>
    <!ELEMENT s_PHONE (#PCDATA)>
    <!ELEMENT s_ADDR1 (#PCDATA)>
    <!ELEMENT s_ADDR2 (#PCDATA|EMPTY)>
    <!ELEMENT s_MSG (#PCDATA)>
]>
```

Fig. 3 An example of purchase order with IDFEF type

IDREFS must be identical to the value of the specified property in ID type of different type. Figure 3 shows an example of purchase order with IDFEF type. Figure 4 shows the graph of the XML document with reference information between elements by using DTD.

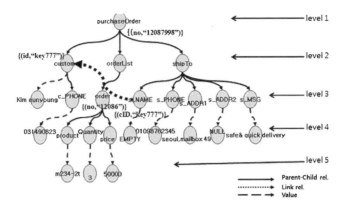

Fig. 4 XML document represented as a graph of the data model with reference information

3.3 Implication

The users see XML documents as a set of one or more objects. These objects can be vertex, edge, attribute, or reference edge objects. Each vertex object has a "label", which users can look at to perceive the hierarchy structure of XML documents. Each vertex can also contain links (called "Edge Object") to other attribute object in the system; the relationships between attribute objects can be displayed as connections (called "Reference Edge Object"). Under certain complex conditions, the user can expand and discern a data model for XML documents.

XML has been employed as a critical means for exchange enterprises' information. The design objectives of XML documents must be clearly and easily understood at the outset and must be defined in terms of the business requirements. For those purposes, the techniques for modeling structures and data of XML documents have been pointed out (e.g., a graph of a data model). The most important reason to build a database of XML documents is to improve the usability of XML data from the user's perspective.

In spite of this usefulness, the hybrid XML data model has a weakness related to automatic problems. For higher efficiency of the data model, it is necessary to implement the automatic capability for obtaining XML documents in two ways: one is to make an automatic coordination possible with a Computer-Aided Software Engineering (CASE) tool which supports development of the graph of the hybrid XML data model and the other is to make possible automatic storages and reconstruction of XML documents.

4 Conclusion

This paper proposes the new data model for the XML document that is suggested as a document standard to represent the data on the Internet and to exchange data mutually. The hybrid data model in this paper is an applicable model in the case of designing a new database system for the XML document and in the case of using an existing database system like a relational database system. Because the data view and the structure view of the original XML document are lost in the data model for mapping to the existing relational database system [4–6] or the object-oriented system, not only cannot XML generate again from the stored data, but also XML sub-graph corresponding to element as search result cannot be returned [7].

However, in the proposed model, this problem is solved because the data view and structure view of the original XML document are stored. Therefore, since the proposed model can form a new XML document from the search result of the XML document, it supports reusability as another great feature. The search about the XML document in the proposed model is possible on the condition of all the factors in the XML document, and not only top-down search but also bottom-up search is possible. Besides, as it has the order information between child elements that have same parents, left–right search is possible and the proposed model preserves order of same level as well as hierarchy structure between the elements of the XML document. For a future study, we expect to apply this modeling technique to XML based Request For Proposal (RFP) or Request For Quotation (RFQ) in business to business electronic commerce and other contents management system (CMS).

References

1. Wills, B.: The Business Case for Environmental Sustainability (Green). A 2009 HPS White Paper. (2009). MicroSoft, "XML Scenarios". http://msdn.microsoft.com/xml/scenario/inro.asp
2. Shanmugasundaram, J., Gang, H., Tufte, K., Zhang, C., DeWitt, D.J., Naughton, J.F.: Relational databases for querying XML documents: limitations and opportunities. In: Proceedings of 25th International Conference on Very Large Data Bases, VLDB'99. Edinburgh, Scotland, pp. 302–304 (1999)
3. Florescu, D., Kossman, D.: A performance evaluation of alternative mapping schemes for storing XML data in a relational database. Technical Report 3684, INRIA, March 1999
4. Du, F., Amer-Yahia, S., Freire, J.: ShreX: managing XML documents in relational databases. In: Proceedings of the 30th VLDB Conference, Toronto, Canada (2004)
5. Kappel, G., Kapsammer, E., Retschitzegger, W.: X-ray-towards integrating xml and relational database systems. Technical Report, July 2000
6. Choi, R.H., Wong, R.K.: Efficient date structure for XML keyword search. DASFAA, pp. 549–554 (2009)
7. Atay, M., Chebotko, A., Liu, D., Lu, S., Fotouhi, F.: Efficient schema based XML-to-relational data mapping. Inf. Syst. (IS) **32**(3), 458–476 (2007)

A Study on Advanced Penetration Testing and Defensive Schemes for Web Service Vulnerability Analyses

Ji Soo Park, Chang-Hyun Mun, Chul Ho Shin and Jong Hyuk Park

Abstract In modern society, due to the popularity of web service, the risk of information leakage has increased. To reduce web service security incidents, web service providers must understand the possible vulnerabilities and the necessary responses. A number of web services exist that are not patched for security about open vulnerabilities. Thus, because of the possibility of new hacking techniques, a new security solution study is required. In this thesis, we suggest the advanced penetration testing and defensive Scheme for Web service (ASW) security of which was reinforced. This ASW model offers improved penetration testing and defensive schemes to analyze vulnerabilities to respond to new hacking techniques.

Keywords Web service · Penetration testing · Vulnerability analyses

1 Introduction

As the use of the internet has increased, users collect and share a great deal of information. On the web, we can find multi-media content such as games, travel information, movies, music, stocks, and real estate. Above this, much personal

J. S. Park · C.-H. Mun · C. H. Shin · J. H. Park (✉)
Department of Computer Science and Engineering, Seoul National University of Science and Technology (SeoulTech), 172 Gongreung 2-dong, Nowon-gu, Seoul 139-743, Korea
e-mail: jhpark1@seoultech.ac.kr

J. S. Park
e-mail: jisoo08@seoultech.ac.kr

C.-H. Mun
e-mail: ckdgus2482@seoultech.ac.kr

C. H. Shin
e-mail: sch1992@seoultech.ac.kr

S.-S. Yeo et al. (eds.), *Computer Science and its Applications*,
Lecture Notes in Electrical Engineering 203, DOI: 10.1007/978-94-007-5699-1_45,
© Springer Science+Business Media Dordrecht 2012

information and data are required for web services. This information is sensitive, and the risk of information leaks has increased. Because new attacks, such as the zero day attack, continue to occur, real-time security solution updates are an increasingly important part of web services. People find web service vulnerabilities using penetration testing, and apply these findings to security solutions or to continuous web service setting updates. However, it is difficult to respond to new attack schemes. In this thesis, we investigate the newest web service security issues, web service security trends, and the existing penetration testing plans. We suggest a service model to respond to new, real-time attack techniques and to reinforce the security of web services through improved defensive schemes and penetration testing to solve web service security problems. This thesis is organized as follows: related works are discussed in Sect. 2, the web service model is proposed in Sect. 3, and Sect. 4 offers a conclusion.

2 Related Works

In this section, we look into the security issues of web services, web service models, and penetration testing and defensive schemes.

2.1 Web Services Security Issues

The Open Web Application Security Project (OWASP) found ten main security issues of web applications provided by web services, and offered attack scenarios and counterplans for each. The ten issues were as follows: Injection, Cross Site Script (XSS), Broken Authentication and Session Management, Insecure Direct Object References, Cross Site Request Forgery (CSRF), Incorrect Configuration for Security Reasons, Insecure Password Storage, Failure to Restrict URL Access, Insufficient Transport Layer Protection, and Unvalidated Redirects and Forwards. Incorrect Configuration for Security Reasons and Unvalidated Redirects and Forwards were added to the list in 2010 because they had a high probability of causing serious damage [1] (Table 1).

In Korea, security issue's monthly, itemized statistics on the web are could confirm through the research paper published by Korea Internet and Security Agency (KISA). We can confirm that malicious code infection and the spread of malicious DDoS attacks leading to bot are the most critical issues for web services, according to survey results in 2010 [2].

2.2 Web Services Models

Web service is defining models that is associated with data expression and transfer system, and progressing standardization for user interface (UI) provided to data transmission and user. Studies on web service models are indispensable because

Table 1 OWASP Top 10

No	Risk	Exploitability	Impact
1	Injection	Easy	Severe
2	Cross-site scripting (XSS)	Average	Moderate
3	Broken authentication and session management	Average	Severe
4	Insecure direct object references	Easy	Moderate
5	Cross-site request forgery (CSRF)	Average	Moderate
6	Security misconfiguration (NEW)	Easy	Moderate
7	Insecure cryptographic storage	Difficult	Severe
8	Failure to restrict URL access	Easy	Moderate
9	Insufficient transport layer protection	Difficult	Moderate
10	Unvalidated redirects and forwards (NEW)	Average	Moderate

web service attacks are mostly based on data. Recently, most web services use XML.

- **Simple Object Access Protocol (SOAP)**. SOAP is a kind of directory service. A provider in this web service descript own service function and register WEDL to universal description, discovery, and integration (UDDI). A user can easily search the web service, which is registered through UDDI, and can communicate with the service provider using web service description language (WSDL)'s client creation [3].
- **Web Service Description Language**. WSDL defines the message schema to use web service by using web service description script language based on web XML. It is composed of a port and binding, which is delivered from WSDL. In the binding, information about abstract messages, operations, port types, and how to map the return protocol are included [3].
- **Universal Description, Discovery and Integration**. UDDI is a kind of directory service. A provider in this web service owns a service function and registers WEDL to UDDI. A user can easily search the web service, which is registered through UDDI, and can communicate with the service provider using the WSDL's client creation [3, 4].

2.3 Penetration Testing and Defensive Schemes

- **Penetration Testing**
  Penetration testing to the general software development test process like proceeding in a similar testing.
  Penetration testing involves three stages. However, in the case of a software-based document with development specification or source code, vulnerabilities are expected.
  Penetration testing goes through three phases to collect basic information: the information gathering phase, the second part of the expected vulnerability of

vulnerability for the analysis phase, the third is expected to do about the vulnerabilities, vulnerability to attack is the utilization stage. And analysis and will repeat the attack.

In particular, an automated scanning tool can be used to collect information in the information gathering phase. Some of the vulnerabilities and exploits for the vulnerability analysis can be performed automatically [5].

A web service in a running state, when performing dynamic penetration testing using automated tools such as SQL Injection and Cross Site Script to detect and to fix security vulnerabilities, is easy, and user input validation on the data alone and a significant vulnerability part can be resolved.

However, dynamic analysis is to note that at the time. If vulnerability is not found, an analysis of the real and legal sanctions and exposure of sensitive information could be important. It agreed in advance in accordance with the scope and plan if progress is to avoid the risk of many parts [6, 7].

- **Defensive Scheme** Attack for web service in order to defend on web server and service on running need to check and to deal with it when the incident occurred is required for the procedures and methods. The web service details, web server vulnerabilities, network vulnerabilities, DB vulnerabilities, and application vulnerabilities need to be checked.

Of these, the web server, network, and DB administrator can reduce the administrator's settings, such as using the official version, applying the latest update and authentications, and setting an access control policy.

Applying an SQL Injection, XSS, and File Upload, and so the user web service abnormal transmission of data using an attack is possible to check, which the administrator's security settings with each application for input data validation through an abnormal value of the input block of the file signature Check the web shell and upload the same file can be prevented [8].

Reacting to a security accident involves incident preparation prior to an incident occurring, incident detection, the initial reaction, the systemized reaction strategy, data collection, data analysis, and report writing. Of these, data collection and analysis on effective accident and the remaining log record to the server and run the equipment or process information was previously run by the grasp will react and perform restoration of [9].

3 ASW: Advanced Penetration Testing and Defensive Scheme for Web Service

3.1 ASW Architecture

Penetration testing and defensive schemes for web services should be considered for the next service model. The service model suggested consists of a web service server to strengthen security and make the service server independent; a penetration test server for vulnerability analysis and attack; and a response

Fig. 1 Service model architecture

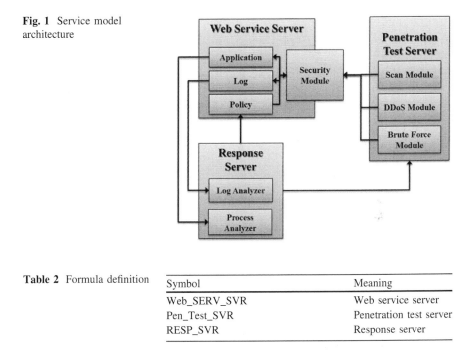

Table 2 Formula definition

Symbol	Meaning
Web_SERV_SVR	Web service server
Pen_Test_SVR	Penetration test server
RESP_SVR	Response server

server to the update security policies and applications of the web service server. The web service server consists of a security module that responds to outside attacks, a log that includes the server log and attack log, a policy that contains the security policy in the security module, and Application that Application information provided for Web Service is saved in. The penetration test server consists of detailed modules to check various vulnerabilities, a scan module that analyzes vulnerability, SQL Injection or XSS, CSRF, etc., a DDos module that creates a DDos packet to analyze the defense ability of the web service server, and a brute force module that inputs certification key fulfills brute-force attack. Lastly, the response server is in charge of reacting to an invasion and consists of a log analyzer that calls in and analyzes log records and a log analyzer that calls in and analyzes running processes and processes that have been run (Fig. 1, Table 2).

3.2 Use Case Scenario

Figure 2 shows the use case scenario of the service model. It is a cycle type process that performs penetration testing from Pen_Test_SVR to Web_-SERV_SVR and analyzes the result in RESP_SVR to update the security policy of Web_SERV_SVR and delivers the checked data to Pen_Test_SVR.

Fig. 2 Use case scenario

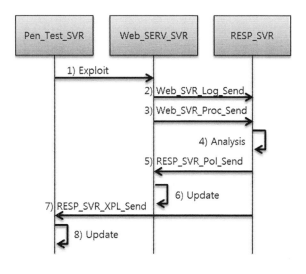

The concrete running process consists of the following eight steps:

Step 1   Pen_Test_SVR → Web_SERV_SVR: Exploit
           Pen_Test_SVR tries to attack Web_SERV_SVR to find security
           vulnerability.

Step 2   Web_SERV_SVR → RESP_SVR: Web_SVR_Log_Send
           Web_SERV_SVR sends Web_SVR_Log_Send, server access log to
           RESP_SVR.

Step 3   Web_SERV_SVR → RESP_SVR: Web_SVR_Proc_SendWeb_SERV_
           SVR sends Web_SVR_Proc_Send, application's process log to RESP_
           SVR.

Step 4   RESP_SVR, AnalysisRESP_SVR analyzes Web_SVR_Proc_Send sent,
           and saves the information to update Web_SERV_SVR and Pen_Test_
           SVR.

Step 5   RESP_SVR → Web_SERV_SVR: RESP_SVR_Pol_SendRESP_SVR
           sends RESP_SVR_Pol_Send to Web_SERV_SVR to update security
           policy.

Step 6   Web_SERV_SVR, Update
           Web_SERV_SVR fulfills the security policy update RESP_SVR_Pol_-
           Send sent.

Step 7   RESP_SVR → Pen_Test_SVR: RESP_SVR_XPL_Send
           Send RESP_SVR_XPL_SEND information according to attack success
           from RESP_SVR to Pen_Test_SVR.

Step 8   Pen_Test_SVR, Update
           Pen_Test_SVR fulfills check vulnerability through RESP_SVR_XPL_
           SEND sent and fulfills update.

3.3 ASW Model Analysis

The suggested models consist of a penetration testing server, web service server, and response server. Each part fulfills an independent function and checks the results of the penetration testing. They later fulfill functions that the update web service server and penetration server. When the response server does a vulnerability analysis, the administer can fulfill analysis about some vulnerability itself. But we can undertake effective penetration testing and identify results through cycle on three server. We can also fulfill the conative reaction through immediate updates when vulnerability factors occur.

4 Conclusion

In this thesis, we discussed penetration testing and defensive schemes, and we examined improved penetration testing and defensive schemes for web service vulnerability analysis. Penetration testing that analyzes web vulnerability was used to solve the security vulnerability element. To protect against vulnerability, the administer must update the policy set-up and web application vulnerable point about web service server. When penetration testing occurs, we can strengthen security and enhance business efficiency. If we fulfill attack and analysis reaction continually and update policies, information and web service server during each server connected.

In this thesis, the proposed penetration testing and defend the security of Web services by using the technique in order to strengthen elements of the security issues and constantly updated network penetration in the inner parts as well as regional factors outside the network part test in considering on research is needed.

Acknowledgments This study was financially supported by Seoul National University of Science & Technology.

References

1. The Open Web Application Security Project (OWASP). https://www.owasp.org/
2. Korea Internet & Security Agency (KISA): 2010 Hacking Virus Present Condition and Response. KISA-RP-2010-0051, 2010
3. Lee, S.-H.: An analysis on trend of web service and bio-information based security enhancement method. J. Korean Inst. Inf. Technol. **10**(3), 213–218 (2012)
4. Cho, P.-Y., Lee, N.-Y., Lee, C.-K.: Design and implementation of UDDI to provide the user-side quality of web service. J. Korea Inst. Intell. Transp. Syst. **8**(4), 102–112 (2009)
5. Bacudio, A.G., Yuan, X., Chu, B.-T.B., Jones, M.: An overview of penetration testing. Int. J. Netw. Secur. Its Appl. **3**(6), 19–38 (2011)

6. Sreenivasa Rao B, Kumar N, Web application vulnerability detection using dynamic analysis with penetration testing. Int. J. Enterp. Comput. Bus. **2**(1), 16–40 (2012)
7. J. Bau, E. Bursztein, D. Gupta, J. Mitchell, in *State of the Art: Automated Black-Box Web Application Vulnerability Testing, 2010 IEEE Symposium on Security and Privacy*, pp. 332–345, 2010
8. Korea Internet & Security Agency (KISA), Web Server Deployment Security Examination Guide, KISA Guide, No.2010-9, 2010
9. Korea Internet & Security Agency (KISA), Security Incident Analysis Procedures Guide, KISA Guide, No.2010-8, 2010

Orthogonal Unified Buffer with Memory Efficiency

SangHyun Seo and JaeMoon Choi

Abstract In this paper, we introduce an Orthogonal Unified Buffer (OUB), which is an extension of the Layered Depth Cube (LDC) as an alternative to the Orthogonal Frame Buffer (OFB). Compared to the OFB, the OUB achieves significant improvement in memory efficiency, which enables a high-resolution representation of 3D models that have various depth-complexities. Similar to the OFB, the OUB resamples the surface nearly uniformly while conserving data locality. As a grid-based texture mapping method, the OUB shows satisfactory memory efficiency and is independent of an object's shape or topology. Our method is built and handled on the GPU, thus it is well suited for real-time rendering.

Keywords Orthogonal unified buffer (OUB) · Layered depth cube (LDC) · Orthogonal fragment buffer (OFB) · Texture mapping

1 Introduction

In the past decade, scalable graphics algorithms have received considerable attention among the graphics community, especially because of developing trends in high definition display devices, and recently, there is a way to represent Ultra-High Definition images with a single device. For 3D rendering that is well suited for such devices, a texture mapping method that provides both high

S. Seo (✉)
Liris Laboratory, University Claude Bernard Lyon 1, Lyon, France
e-mail: shseo75@gmail.com

J. Choi
InusTech Inc, Seoul, Korea
e-mail: invisible@cglab.cau.ac.kr

S.-S. Yeo et al. (eds.), *Computer Science and its Applications*,
Lecture Notes in Electrical Engineering 203, DOI: 10.1007/978-94-007-5699-1_46,
© Springer Science+Business Media Dordrecht 2012

resolution and uniform quality over the surface must first be attained. Unfortunately, few research efforts have considered such a data structure.

Shade et al. proposed the Layered Depth Image (LDI) [1], which uses a single camera view to keep multiple pixels residing in the line of sight. The Layered depth cube (LDC) [2] uses three mutually orthogonal directions to keep data elements on the surface. Compare to the LDC, the Orthogonal Frame Buffer (OFB) [3] achieves fast construction of data structure and improvements in performance by using current graphics hardware. However, because the OFB saves data onto a sample-based data structure (k-buffer) [4], it may suffer from memory efficiency issues for expression of models that have a high depth complexity. Similar to LDC and OFB, our method, OUB, re-samples the sparse data onto a nearly-uniform grid.

Our goal is to make a data structure with satisfactory memory efficiency and performance which are problems that have arisen in previous grid-based approaches. To achieve this, we propose an Orthogonal Unified Buffer (OUB) as a new grid-based surface representation. Since data in OUB are sampled along a uniform grid, it is free from mapping distortion and seam discontinuity while providing satisfactory performance and scalability. In this paper, we adopt a per-pixel linked list introduced by Jason et al. [5], for rapid generation of our OUB.

2 Related Work

There are a number of technologies that map the texture of 3D surfaces, such as 2D parameterized texture space for 3D Mesh, grid-based representation of surfaces and so on. Parameterization of a 3D surface onto a 2D texture space (aka texture atlas) is also common in texture mapping. Many researchers have studied automatic generation of texture atlases [6–9], and some others have proposed techniques that guarantee one-to-one mapping [10–12], and though much research has been conducted to achieve techniques that produce a satisfactory parameterization, the existing methods are not completely free from mapping distortion and seam discontinuity

Problems with 2D parameterization of a 3D surface can be addressed by grid-based representation of a 3D surface. A high-resolution grid requires a considerable number of sequential indirection for accessing the elements in the finest detail. Thus, it shows insufficient performance for high resolution rendering in real-time. Lefebvre and Hoppe proposed Perfect Spatial Hashing [13], which reconstructs spatial data onto a hash. This method shows good memory efficiency. However, construction of a perfect hashing function can be considered an optimization problem, and this procedure requires considerable amounts of time. Moreover, it is not guaranteed to work for any type of surface. BÜGER et al. proposed Orthogonal Frame Buffer (OFB) [3], which is an extension of the Layered Depth Cube (LDC) [2].

The OFB samples data by three mutually orthogonal directions and provides nearly-uniform distribution over the surface, depending on the sorting of

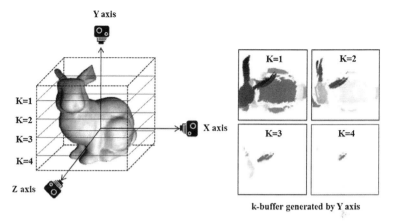

Fig. 1 K-buffer (K = 4) of orthogonal fragment buffer (OFB) sampled by Y-axis

the structure. For the construction of the OFB, stencil routing is used to create a data structure. Specifically, his stores all surface samples onto a sample-based data structure (k-buffer), and it tends to be memory inefficient, generally because of the insufficient data occupancy of a k-buffer.

3 Orthogonal Unified Buffer

3.1 Memory Efficiency of Orthogonal Fragment Buffer

The Orthogonal Fragment Buffer (OFB) is a data structure that stores surface samples at a nearly uniform distribution over the surface. The data access operations have a complexity that is logarithmic in the depth complexity of the surface. Memory efficiency is closely related to data occupancy, which is a proportion of practically used data compared to allocated memory. As shown in Fig. 1, data occupancy in the OFB tends to dramatically decrease with each progressing depth layers (k-buffer). White area in each buffer is not object and unnecessary memory usage.

3.2 Construction of Orthogonal Unified Buffer

The proposed Orthogonal Unified Buffer (OUB) samples the surface data along three mutually orthogonal sampling directions. For each direction, the OUB performs orthographic rendering and stores pixels and attributes onto a corresponding unified buffer (UB), which is one buffer stored all K-buffer of OFB. Since the sampling is performed on three orthogonal directions, redundant sampling of data

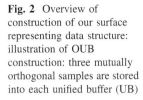

UB 1 :Y axis

UB 2: X axis

UB 3: Z axis

occurs. To manage this problem, we divide the mesh into three chunks in the geometry shader. Each of those chunks consists of triangles that have the lowest angle between their normal direction and the sampling direction, as shown in Fig. 2. Each cell in the start address buffer keeps an address that indicates recently drawn pixels in the data buffer.

Figure 3 shows the comparison of memory usage between OFB and proposed OUB. The data occupancy in the OFB tends to decrease with each progressing depth layers. However, the OUB shows near-maximal data occupancies, sacrificing fixed amounts of overheads in a start address buffer. In general, overheads in a start address buffer are often considered less than overheads in a data buffer, because it saves only offset data. OUB requires a single start buffer and a single data buffer for each direction. All dotted areas are unused memory and dashed areas are saved memories that are not required by the OUB.

Each data buffer stores color or surface attributes like normal, curvature, visibility, etc. In addition, each node requires an address for the next node of K-buffer. Upon the creation of the data structure, a number of pixels may fall into the same cell, which results in the addition of a new depth layer (K). For the insertion of the new node, the address in the start offset buffer to the most recently drawn pixel is first updated, and then a link between the new node and the last occupied node in the linked list is made. Basically, the OUB can be constructed by rasterization or ray casting. For implementation of the OUB, we used the per-pixel linked list introduced by Jason et al. [5], which is based on generic atomic operations available in OpenGL 4.0 and DirectX 11. In our research, we implemented the OUB with the UAB in DirectX 11.

4 Experimental Result

We discuss the memory efficiency of OUB, which are general issues of concern among the existing grid-based surface representation methods. We have tested the OUB on a system that consists of the Intel E6400 (2.14 GHz), NVidia GeForce GTX460 with (2 GB), 4 GB RAM, and a 1280 × 1024-viewport resolution.

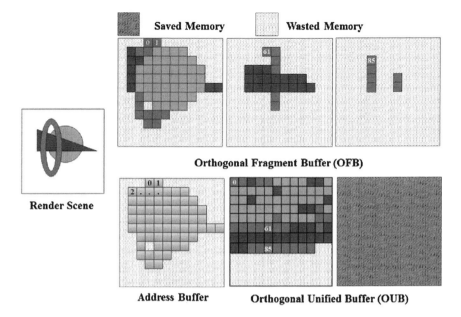

Fig. 3 Comparison of memory occupancy between OFB and OUB

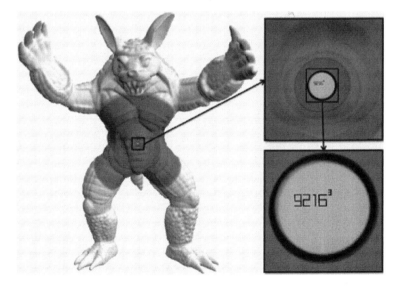

Fig. 4 Rendering image using OUB

Because the OUB is constructed based on the linked list, we needed to allocate memory space for the start address buffer. In spite of this, the OUB generally shows better memory efficiency than the OFB, because the OUB can maintain the maximal data occupancy in most conditions. Figure 4 shows an armadillo model

Table 1 Comparison of memory consumption between OUB and OFB

Model	Layer depth			Memory usage (MB)		OFB/UOFB (%)
	X	Y	Z	OFB	OUB	
Box	2	2	2	192	336	57
Bunny	4	5	4	416	146	284
Virus	47	46	53	4672	1240	377
Armadillo	8	6	5	608	116	523
Ventricle	11	12	13	1152	172	669
Lucy	9	8	10	864	90	957

textured using OUB with 9216^3 spatial domain and 120.6 million color elements. Left images show the zoomed parts of original scene.

We also demonstrate the relationship between depth and memory efficiency in Table 1. A test was performed on the 2048^3 spatial domain of OUB, and each element's size was 12 bytes (depth, color, and link), and the OFB was 8 bytes (depth and color), which implies that the link overhead is the maximal and worst element size in terms of memory consumption. Memory efficiency will become greater in cases where each element has to carry additional information or the start address buffer's occupancy is high.

5 Conclusions

We have introduced the OUB, a simple data structure for keeping the surface attributes of 3D polygonal models. We have demonstrated that the OUB shows plausible memory efficiency. The OUB guarantees near-maximal data occupancy. The OUB is easily implemented and used on the GPU. The OUB shares the same general strong and weak points of the grid-based surface representation method. In addition to this, the proposed technique can be applied to 3D surface painting and curvature estimation techniques based. Because data structure storing sparse 3D data are common in graphics, we expect that our approach can be applicable for various applications.

Acknowledgments This work was supported by the Korea Research Foundation Grant funded by the Korean Government (KRF-2011-357-D00202).

References

1. Shade, J., Gortler, S., He, L.-W., Szeliski, R.: Layered depth images. In: Proceeding of SIGGRAPH'98, pp. 231–242 (1998)
2. Lischinski, D., Rappoport, A.: Image-based rendering for nondiffuse synthetic scenes. In: Proceeding of Rendering Techniques'98, pp. 301–314 (1998)
3. Buerger, K., Kruger, J., Westermann, R.: Sample-based surface coloring. IEEE Trans. Vis. Comput. Graph. **16**, 763–776 (2010)

4. Myers, K., Bavoil, L.: Stencil routed a-buffer. In Proceeding of SIGGRAPH'07 Sketches (2007)
5. Yang, J.C., Hensley, J., Grn, H., Thibieroz, N.: Real-time concurrent linked list construction on the GPU. J. Comput. Graph. Forum **29**(4), 1297–1304 (2010)
6. Ma, S.D., Lin, H.: Optimal texture mapping. In: Proceeding of Eurographics'88, pp. 421–428 (1988)
7. Bennis, C., V'ezien, J.-M., Igl'esias, G.: Piecewise surface flattening for non-distorted texture mapping. In: Proceeding of SIGGRAPH'91, vol. 25, pp. 237–246 (1991)
8. Maillot, J., Yahia, H., Verroust, A.: Interactive texture mapping. In Proceeding of SIGGRAPH'93, pp. 27–34 (1993)
9. Zhang, E., Mischaikow, K., Turk, G.: Feature-based surface parameterization and texture mapping. ACM Trans. Graph. **24**, 1–27 (2005)
10. Hormann, K., Greiner, G.: MIPS: an efficient global parameterization method. In: Laurent, P.-J., Sablonniere, P., Schumaker, L.L. (eds.) Curve and Surface Design: Saint-Malo 1999. Vanderbilt University Press, Nashville, pp. 153–162 (2000)
11. Sander, P.V., Snyder, J., Gortler, S.J., Hoppe, H.: Texture mapping progressive meshes. In: Proceeding of SIGGRAPH'01, pp. 409–416 (2001)
12. Floater, M.S.: Mean value coordinates. J. Comput. Aided Geom. Des. **20**, 19–27 (2003)
13. Lefebvre, S., Hoppe, H.: Perfect spatial hashing. In: Proceeding of ACM SIGGRAPH'06, pp. 579–588 (2006)

The Study on Smart Sensor Network Based Production Management Service Design

Hangbae Chang

Abstract The small and medium-sized manufacturers, which produce a small amount of different items, have a problem in terms of cost and time that is difficult to introduce automated production process control system for manufacturing and management of new products. In addition, companies' ERP systems could not cope with real-time errors arisen from production processes dealing with actual logistics. And companies recently focus on introduction of Smart sensors for managing production processes, however, they have difficulty in optimizing deployment suitable to the corresponding company's process by prioritizing only the technological introduction. Accordingly, this paper analyzed the production process currently used for manufacturing, and designed and implemented a production process control system that could deploy optimized Smart sensors and connect with them based on the analysis.

Keywords Smart sensor · Production monitoring service

1 Introduction

The automated production process control system delivers data required between different systems and sends information needed for managers or producers to the monitoring system in real time in most manufacturers. And it draws information such as estimations of production amount for each process, process status of

H. Chang (✉)
Division of Business Administration,
College of Business, Sangmyung University, Seoul, Korea
e-mail: hbchang@smu.ac.kr

S.-S. Yeo et al. (eds.), *Computer Science and its Applications*,
Lecture Notes in Electrical Engineering 203, DOI: 10.1007/978-94-007-5699-1_47,
© Springer Science+Business Media Dordrecht 2012

certain products, status of defective products etc. based on the received data for efficient manufacturing of products. In addition, because the cost of production could be lowered by increasing efficiency of raw materials' supply/demand and inventory control through the automated production process control system, it could increase the small and medium enterprises' production competitiveness [1]. However, the general small and medium manufacturers, which produce a small amount of different items, have a problem in terms of cost and time that is difficult to build the system for manufacturing and management of new products [2].

In addition, even though companies have introduced the Enterprise Resource Planning (ERP) system for managing resources efficiently, it is a system for managing resources and is difficult to cope with real-time errors arisen from the production process [3–5]. Recently, companies have focused on introduction of Smart sensors for managing production processes, however, they have difficulty in optimizing deployment suitable to the corresponding company's process by prioritizing only the technological introduction. Accordingly, this study would like to design and implement a production process control system that could introduce optimized Smart sensors and be connected with them by analyzing the production process in manufacturing for general manufacturers. In detail, it would like to design and implement a production monitoring and control system based on Smart sensors.

2 Concepts and Features of USNs and Smart Sensors

Ubiquitous Sensor Network (USN) sensors could detect, process and store information about contexts of objects and their surrounding environment by attaching electronic tags or sensors etc. to the objects, deliver the collected information in real time via wire and wireless networks, and utilize it. In addition, it enables the person-to-person, person-to-object and object-to-object communications, and provides safety and convenience throughout industries and life including distribution, logistics, medicine and environment as a basic infrastructure realizing a ubiquitous society. The Smart sensors mean an active sensor environment to realize a ubiquitous environment based on such a USN.

The Smart sensor is divided into 4 layers as Fig. 1. The information resource layer is one to collect information about objects and environment, which includes physical elements such as sensor and sink nodes, and logical elements such as sensor networks of a topology form. The BcN backbone and access network layer has broadband Broadband convergence Networks (BcN) networks including gateways accessing data. Data accessed by this layer is sent to the middleware via various schemes of networks. The middleware layer carries out integration and processing of the received data, and offers context-aware services or event functions etc. for providing appropriate functions suitable to application services. In addition, it also carries out processing for reducing the database's load and enhancing its reliability. The application service layer utilizes the processed data

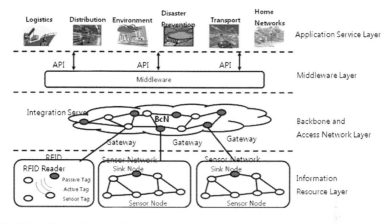

Fig. 1 Hierarchical diagram of smart sensors

through queries. The application service is currently operated on a trial basis and served in a diversity of areas such as logistics, distribution, environment, disaster prevention, transport and home networks etc.

3 Design and Implementation of a Smart Sensor-Based Production Monitoring and Control System

3.1 Features of the Production Monitoring and Control Service Process

Computerization for production sites in general manufacturing has an effect of optimized productivity, quality increase, supply chain synchronization, efficient production facilities management by digitalizing production activities. In general, a manufacturing process is composed of production facilities that make products directly, programmable logic controllers that generate basic data and control the facilities, and an ERP that manages an actual output and quality etc. However, it does not contribute to productivity or quality improvement because there are not sufficient systems to manage information on production sites in real time. Therefore, in order to collect information (temperature, humidity, pressure etc.) generated from production sites in real time, it could not use RFIDs that simply recognize only chips, so that it is essential to acquire information through Smart sensors. Consequently, it would like to build a service to achieve improvement of quality and productivity in manufacturing by collecting, analyzing and context-recognizing associated data (temperature, pressure, humidity etc.) of facilities and devices on production manufacturing sites in real time to quickly take action when there is an abnormality (Fig. 2).

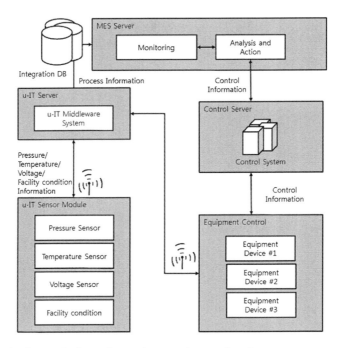

Fig. 2 Production monitoring and control process in manufacturing

3.2 Analysis on Production Monitoring and Control Process

Analyzing problems of production monitoring and control services built currently in general manufacturing, they are as follows. Efficient management of facilities is important due to properties of manufacturing industries, however, it carries out manually without systematic management for them, so that there are causes of lowering quality and productivity due to failure of facilities. And, because a lot of data is generated in real time on production sites, it is difficult for people to manage it manually, so that it could not analyze without building an additional system. Therefore, because there are many factors that could affect communication environment like electric impulses in production sites, it is required to build wireless environment that could be operated stably in industrial environment.

Even though the recently introduced production facilities provide a method to acquire information through sensors, production facilities used in most companies are so outdated that great expense is required to add a sensor system. Because the current facilities could not collect and analyze quality information generated from production sites in real time, follow-up management is costly due to quality problems (Fig. 3).

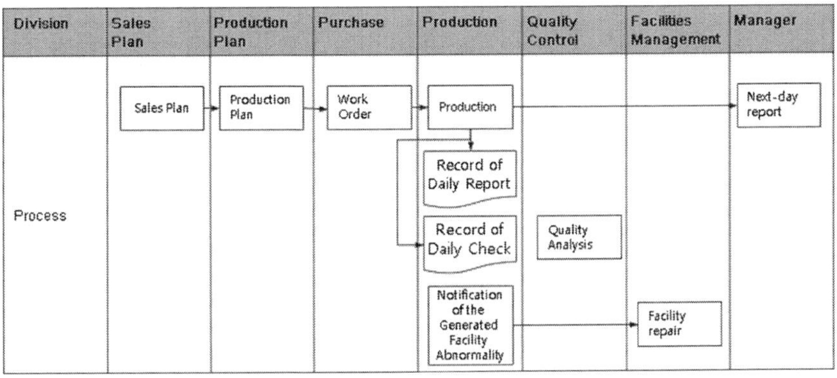

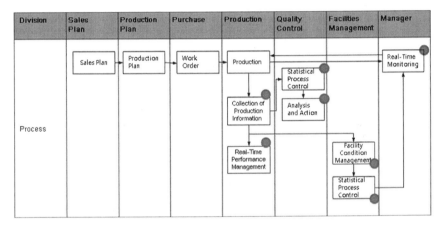

Fig. 3 Schematic diagram of recognizing and sending refractory's locations through forklifts

Fig. 4 Schematic diagram of recognizing and sending refractory's locations through forklifts

3.3 Design of a Smart Based Production Monitoring and Control Process

A design of the future monitoring and control service to solve problems of the production monitoring and control service currently built in general manufacturing is as follows. Building a real-time manufacturing environment with Smart sensor technologies, it collects data (temperature, pressure, humidity etc.) of facilities and devices on production manufacturing sites in real time. And, it monitors facility conditions and operating environment to build optimum production environment, and enables to rapidly support action through facility control when a failure occurs (Fig. 4).

It utilizes Smart sensors to collect information by using pressure and temperature sensors attached to production facilities, and sends information about

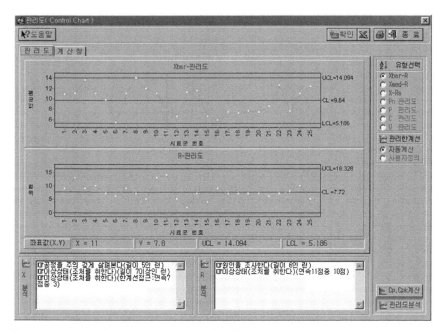

Fig. 5 Smart sensor based system to analyze facilities and context-information on the production manufacturing site

operating status of facilities to a production process management system via wireless networks. And, the production process management system analyzes abnormalities by statistical process management techniques, and sends it to programmable logic controllers based on the analysis results to control operation of the production facilities.

4 Implementation and Expected Effects of the Smart Based Production Monitoring and Control System

4.1 Implementation of the Smart Based Production Monitoring and Control System

The production monitoring and control service for general manufacturing was implemented as Fig. 5. It would contribute to improvement of quality and productivity by applying sensor networks to production manufacturing process sites to collect information in real time. And, it could secure stability and competitive advantages in manufacturing by removing obstacles and waste factors through online/offline integration of the production manufacturing sites.

4.2 Expected Effects of Smart Based Production Monitoring and Control System

The production monitoring and control system, which is improved by utilizing Smart sensors through the analysis of general manufacturing's processes used inefficiently, has the following expected effects. Qualitatively expected effects are improvement of quality control ability through real-time quality control, enabling systematic quality control by tracking the products' process history, quick responding to failures through facilities monitoring and control, reducing risk factors of work sites by preparing safety measures for high-temperature facilities and moving vehicles, improvement of worker's work conditions by measuring a pollution level in the plant etc. Quantitatively expected effects are management of process information in real time, detection of facilities and quality abnormality, reduction of a defect ratio (1 %) and improvement of productivity, improving accuracy of quality information, facilities operating time, production results information etc.

5 Conclusion

An automated production process management system could increase companies' competitiveness. However, small and medium-sized manufacturers, which produce a small amount of different items, have a problem in terms of cost and time that is difficult to build the system for manufacturing and management of new products. In addition, even though they have introduced the ERP system for their efficient resource management, it is difficult to cope with real-time errors arisen from production processes dealing with actual logistics. And, companies recently focus on introduction of Smart sensors for managing production processes, however, they have difficulty in optimizing deployment suitable to the corresponding company's process by prioritizing only the technological introduction.

Accordingly, this paper analyzed the production process currently used for manufacturing. Based on the analysis, it designed and implemented a production process control system that could deploy optimized Smart sensors and connect with them. It is expected that manufacturers' competitiveness could be improved by building a real-time manufacturing status management environment for lots of small and medium manufacturers that do not systematically manage information generated from production sites because production sites are not interworked with office systems.

In the future study, it would like to design and implement a system with Smart sensor networks related to company's entire business processes from production processes to logistics and distribution.

Acknowledgments This research was supported by a 2012 Research Grant from Sangmyung University.

References

1. Bhattacharyya, D., Kim, T-h., Pal, S.: A comparative study of wireless sensor networks and their routing protocols. Sensors **10**(12), 10506–10523 (2010)
2. Jing, L., Cheng, Z.: Functional safety problems in the ubiquitous environment. In: Advanced Information Networking and Applications Workshops, AINAW, vol. 2 pp. 1035–1040 (2007)
3. Lee, S., Kim, H., et al.: A study on the multidimensional service scenario evaluation methodology for ITSM construction considering ubiquitous computing technology. J Soc e-Business Stud **12**, 155–194 (2007)
4. Jennifer, Y., Biswanath, M., Dipak, G.: Wireless sensor network survey. Comput. Netw. **52**(12), 2292–2330 (2008)
5. Antoine, B., Marco, Z., Gordon, I., Simon, S., David, G.: Ubiquitous sensor networking for development (USN4D): an application to pollution monitoring. Sensors **12**(1), 391–414 (2010)

R&D Intensity and Productivity: Evidence from IT Firms versus Non-IT Firms in KOSDAQ Market

SungSin Kim

Abstract We investigate the effects of R&D intensity on efficiency of productivity in non-IT firms and IT firms, in a sample taken from the KOSDAQ market over the period 2000–2011. The main empirical result is as follows: First, by comparing production efficiency between IT firms and non-IT firms, we find that R&D intensity has a positive effect on production efficiency and plays an important role in production growth. Second, R&D intensity of Non IT firm is relatively smaller than IT firms compared. Third, performance (Tobin's Q or ROA) of IT firms is more sensitive to productivity of R&D intensity than that of non-IT firms, and stock market is favorable to firms with high level of productivity of R&D intensity.

Keywords R&D intensity · Productivity · IT firms · Non IT firm

1 Introduction

Since the year of 2000, thanks to the support of government policies, the IT industries in Korea have been developing rapidly and many have been listed in KOSDAQ. The recent rapid growth of IT industries has reinforced the competitiveness of Korean export industry and this phenomenon has been regarded as an important role in the improvement of economic growth in Korea.

S. Kim (✉)
Sangmyung University, 20, Hongjimun 2-gil, Jongno-gu, Seoul 110-743,
Korea
e-mail: scolass@smu.ac.kr

S.-S. Yeo et al. (eds.), *Computer Science and its Applications*,
Lecture Notes in Electrical Engineering 203, DOI: 10.1007/978-94-007-5699-1_48,
© Springer Science+Business Media Dordrecht 2012

This paper investigates the effect of R&D intensity of IT firms and non-IT firms listed in KOSDAQ on the value added, from the sample of 243 in KOSDAQ market in period from 2000 to 2010.

A nation's scientific technology acts as a medium to boost the industrial competitiveness of a nation, and advanced technology presents itself as a very critical position in the field. IT is not only highly essential in the information-oriented society of the twenty-first century, but also has a crucial effect on the great value added and the on the socio-economic ripple effect. Therefore, IT is considered to be a very important field in the economic and industrial development. Also, we can forecast that a new technology will lead the world market in the next 10 years.

Knowledge can be created and accumulated through R&D efforts of firms or industries, and will become available to product innovations [1]. Indeed, advanced countries including Korea have invested substantial expenditures on R&D part. In addition to the conventional role of stimulating innovation, R&D enhances technology transfer by improving the ability of firms to learn about advances in the leading edge [2].

Wang and Tsai [3] suggest that R&D investment is a significant determinant of firm productivity growth during the second half of the 1990s, and high-tech and other firms are statistically significantly different in R&D elasticity.

Adopting the model of R&D intensity, Clark and Griliches [4, 5], and [6], the rates of return on R&D are between 10 and 39 percent in U.S. manufacturing firms.

Kafouros [7] shows that comparing the impact of R&D on productivity growth of high-tech firms with the corresponding impact on productivity growth of low-tech firms, the R&D-elasticity is considerably high for high-tech sectors, but statistically insignificant for low-tech sectors. Palia and Litchenberg [8] suggest that the role of productivity in firm performance is of a fundamental importance to the U.S. economy and market value of a firm increases with increasing production efficiency. Griffith et al. (2004) find evidence that R&D is statistically and economically important in this catch up process as well as stimulating innovation directly. In addition, Human capital also plays an major role in productivity growth, but we only find a small impact of trade.

Our main purpose is that we investigate the effects of R&D intensity on efficiency of productivity in non-IT firms and IT firms, in a sample taken from the KOSDAQ market. In general, we expect that the proportion of R&D expenditure spent on basic research in non-IT firms has been rather small. We also expect that the efficiency of productivity for non-IT firm is worse than that of productivity for IT firms.

Our empirical results demonstrate that the effect of R&D intensity on productivity is positive and statistically significant. Additionally, IT firms are significantly more productive and profitable than non-IT firms compared. Overall, IT firms are more efficient in the productivity of R&D intensity than non IT firms. The paper is organized as follows. In Sect. 2, we develop a main model in this paper and present extended production function, based on Cobb-Douglas function. Furthermore, we investigate the relationship between the productivity of R&D capital and its performance. In Sect. 3 we describe data and statistics and in Sect. 4, we show results of empirical analysis. Finally, Sect. 5 concludes.

2 Model

2.1 Extended Production Function

To estimate R&D capital and productivity growth, we adopt the extended Cobb-Douglas production function model.

$$Y_{it} = AK_{it}^{\alpha}L_{it}^{\beta}R_{it}^{\gamma} \tag{1}$$

where Q, K, L, and R respectively represent output (value added), physical capital, labor, and R&D capital. A is a constant and α, β and γ are the elasticity of output with respect to the physical capital, labor, and R&D capital. And constant elasticity of substitution has been assumed with the respect to labor and physical capital. Taking logarithms in Eq. (1), model of log form give us as follow.

$$\ln Y_{it} = A + \alpha \ln K_{it} + \beta \ln L_{it} + \gamma \ln R_{it} + \varepsilon_{it} \tag{2}$$

γ is the impact of R&D on productivity and εit is the error term. α and β are the output elasticity of physical capital and labor.

Specially, to examine the effect of R&D capital on productivity for non-IT firms and IT firms, we obtain output elasticity of R&D capital from Eq. (3) as follows.

$$\varepsilon = \frac{\partial(\ln Y)/(\ln Y)}{\partial \ln R/\ln R} = \gamma \tag{3}$$

It equals the ratio of the marginal product of capital to the average product of capital. If $\gamma < 1$, then the next small change in R&D capital makes less change in the output than the average capital per unit R&D capital, i.e., there is diminishing returns on R&D capital.

Because of expecting more sensitive to the R&D intensity in the case of IT firms, we hypothesize that λ of IT firms is larger than that of non-IT firms.

2.2 Productivity of R&D Capital and Performance

To examine whether the relationship between productivity of R&D and performance differ across R&D intensity, we adopt the approach that was suggested by [8]. First, by rewriting Eq. (2), we get the productivity residuals from Eq. (4).

$$\ln Y_{it} = \varpi_i + \beta \ln K_{it} + \alpha \ln L_{it} + \gamma \ln R_{it} + \delta_t + \varepsilon_{it} \tag{4}$$

where ϖi is the firm effect and δ_t is the year effect. Y_{it} is output(value added) and defined as sales for firm i at the end of the year. K_{it} is defined as net property and plants. L_{it} is the number of employees. R_{it} is research and development expenditure.

Second, we estimate Tobin's Q and ROA residuals from Eq. (5)

$$Perforamce_{it} = \varpi_i + \beta_1 \frac{R_{it}}{SALES_{it}} + \beta_2 \frac{D_{it}}{TA_{it}} + \beta_3 \ln TA_{it} + \varepsilon'_{it} \qquad (5)$$

where performances are Tobin's Q or ROA. We measure Tobin's Q as sum of market value of equity, book value of debt, and the book value of preferred stock divided by the book value of total assets. ROA is return on total assets and earnings before interest tax divided by the book value of total assets. $\frac{R}{SALES}$ is R&D intensity and R&D capital to sales. $\frac{D}{TA}$ is debt ratio and debt to total assets. TA is total assets.

To control the firm effect and year effect, we estimate productivity residual and performance residual by means of panel analysis. Then, we regress performance residuals ε'_{it} on productivity residuals ε_{it}.

3 Data and Statistics

In this paper, we examine the impact of R&D on productivity, from the sample of 243 firms listed in KOSDAQ market over the period from 2000 to 2010. The data of financial statements is collected from Fn-Guide. KOSDAQ market in Korea consists of IT firms and non-IT firms. We exclude the firms in financial sector and the firms that are impaired of capital. We also exclude the firms that the fiscal year isn't in December. The sample numbers for IT firms and non-IT firms are 134 and 109. Non-IT firms in this paper are included in the manufacturing sector.

Table 1 shows descriptive statistics of all firms used in this paper. Panel A shows the results for non-IT firms and Panel B shows the results for IT firms. SALES is sales for firm in the end of the year and PPE is net property and plants (fixed asset). Labor is average number of workers employed and R&D is research and development expenditure. ROA is return on total assets. Specially, Table 1 reveals that mean of R&D intensity for IT firms is 0.0368 and that of non-IT firms is 0.0156. These statistics indicate that R&D intensity of IT firms is substantially larger than that of non-IT firms. Consequently, R&D intensity varies across the industry sector. Furthermore, we find that mean PPE for non-IT firms and IT firms are 70,107 and 42,642 respectively. However, non-IT firms have more capital intensity than IT firms.

4 Empirical Test and Results

To examine the effects of R&D capital on productivity of IT firms and non-IT firms, we estimate Eq. (4). Table 2 presents our results of expanded production function. We estimate the parameters of production function, by the method of

Table 1 Descriptive statistics

Panel A: Non IT Firm (firm that included in manufacturing industries)

Variable	Mean	std	Min	max
Sales	191,558	299,949	5,080	3,696,371
PPE	70,107	104,323	846	1,060,553
Labor	293	306	15	2,135
R&D	1,706	2,454	1	15,867
ROA	0.0434	0.0680	-0.2264	0.2971
Tobin' s Q	1.0376	0.4366	0.3969	3.5679
R/SALES	0.0156	0.0242	0.0000	0.2711
D/TA	0.5363	0.1788	0.0564	0.9526
TA	190,138	297,240	24,760	3,671,790

Panel B: IT Firm

Variable	Mean	std	min	max
Sales	153,600	164,857	4,642	1,528,172
PPE	42,642	46,199	188,059	465,983
Labor	316	282	9	1,811
R&D	4,011	5,474	1	50,973
ROA	0.0350	0.0886	-0.2254	0.2998
Tobin' s Q	1.1376	0.5367	0.3689	3.5887
R/SALES	0.0368	0.0568	0.0000	0.6650
D/TA	0.4663	0.1838	0.0638	0.9672
TA	137,303	109,376	25,455	1,171,548

ordinary least squares (OLS). Panel A shows the results for cross sectional regression for non-IT firms and Panel B shows the results for IT firms. This paper shows that the two groups (non-IT firm and IT firms) have distinct difference. As can be noted from the Table 2, a positive relationship between R&D capital and production can be assumed. The coefficient estimate for lnR for non-IT firms and IT firms is 0.0328 and 0.111, which is statistically significant at the 1 % level. The firms that have larger R&D capital show more efficient in their production than the other firms. If R&D expenditure increases by 1 %, sales for non-IT firms and IT firms increase by 0.0328 and 0.111 %, respectively. Overall, these results suggest that value added of IT firms is more sensitive to R&D capital than that of non-IT firms. Interestingly, IT firms have larger elasticity of labor than non-IT firms. A labor (human capital) is substantially important in the enhancement of output to the field of IT industry. In the comparisons of R&D capital and labor, coefficient estimate of physical capital for non-IT firms is larger than that for IT firms. This paper explains that output in IT firms with high R&D intensity is created mainly through R&D efforts.

Figure 1 shows R&D intensity of Non IT firm and IT firm by year. It is shown that R&D intensity of IT firm decreases sharply after 2001 and becomes shape of concave function. As reported in previous literature, R&D intensity of Non IT firm is relatively small.

Table 2 Analysis of expanded production function

Variable	Panel A: Non IT Firm	Panel B: IT Firm
Alpha	5.3758***	
		10.4849***
	(12.41)	
		(25.78)
lnK	0.6531***	
		0.2241***
	(25.46)	
		(8.24)
lnL	0.2386***	
		0.468***
	(7.06)	
		(12.57)
lnR	0.0328**	
		0.1111***
	(2.02)	
		(5.61)
obs	586	
		623
F-statistic	427.44	
		249.03
Adj. R^2	68.6	
		54.4

***, ** Significance at the 1 and 5 % level

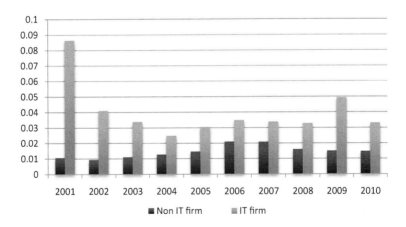

Fig. 1 R&D intensity of Non IT firm and IT firm

We test whether the relationship between productivity of R&D intensity and performance exists and how it differs across characteristics of industry. Palia and Litchtenberg [8] and [9] claim that firms have strong positive relationship between productivity and Tobin's Q.

Table 3 The relationship between productivity of R&D intensity and performance

Panel A: Non IT Firm

Variable	Tobin's Q	ROA
Productivity	0.0665[**]	0.0452[***]
	(1.97)	(7.55)
F-statistic	3.89	56.95
Adj. R^2	0.01	0.09

Panel B: IT Firm

Variable	Tobin's Q	ROA
Productivity	0.1143[***]	0.0636[***]
	(2.64)	(7.67)
F-statistic	6.95	58.87
Adj. R^2	0.01	0.09

***, ** Significance at the 1 and 5 % level

To examine this topic, we adopt the approach suggested by [8]. First, we obtain the productivity residuals from Eq. (4). Then, estimate performance residuals from Eq. (5). To control the firm effect and year effect, this paper applies panel regression with fixed effect model. To test whether fixed or random effect model is appropriate, we use Hausman test. Finally, we regress performance residuals (Tobin's Q or ROA) on productivity residuals.

To examine this topic, we adopt the approach suggested by [8]. First, we obtain the productivity residuals from Eq. (4). Then, estimate performance residuals from Eq. (5).

To control the firm effect and year effect, this paper applies panel regression with fixed effect model. To test whether fixed or random effect model is appropriate, we use Hausman test. Finally, we regress performance residuals (Tobin's Q or ROA) on productivity residuals. To measure performance of firms, we use Tobin's Q and ROA as suggested in [9]. The results of analysis are given in Table 3.

In the column 1, we find that in the case of Tobin's Q, the coefficient estimates of productivity residuals for non-IT firms and IT firms are 0.0665 and 0.1143, which are statistically significant at the 5 and 1 % level respectively. These results suggest a strong positive relationship between productivity and Tobin's Q. These results are consistent with the works by Palia and Litchtenberg [8] and [9]. Overall, these results suggest that stock market rewards firms when these firms increase their level of productivity of R&D intensity [8]. Moreover, performance of IT firms is more sensitive to productivity of R&D intensity than that of non-IT firms, and stock market is favorable to firms with high level of productivity of R&D intensity. In the column 2, the results of empirical test using ROA are similar to those of Tobin's Q.

5 Conclusion

In this paper, we investigate the relationship between R&D intensity and productivity of IT firms and non IT firms in KOSDAQ market over the period 2000–2011. First, by comparing production efficiency between IT firms and non-IT firms, we find that R&D intensity has a positive effect on production efficiency and plays an important role in production growth. Further, output in IT firms with high R&D intensity is created mainly through R&D efforts. Second, R&D intensity of Non IT firm is relatively smaller than IT firms compared. Third, performance (Tobin's Q or ROA) of IT firms is more sensitive to productivity of R&D intensity than that of non-IT firms, and stock market is favorable to firms with high level of productivity of R&D intensity.

References

1. Mansfield, E.: Rates of return from industrial R&D. Am. Econ. Rev. **55**, 863–873 (1965)
2. Griffith, R., Redding, S., Reenen, J.: Mapping the two faces of R&D: productivity growth in a panel of OECD industries. Rev. Econ. Stat. **86**, 883–895 (2006)
3. Wang, J.C., Tsai, K.H.: Productivity growth and R&D expenditure in Taiwan's manufacturing firms. National Bureau of Economic Research vol. 13. (2004)
4. Clark, K.B., Griliches, Z.: Productivity growth and R&D at the business level: Results from the PIMS data base, in R&D, patents, and productivity, University of Chicago Press (1984)
5. Griliches, Z.: Productivity, R&D, and basic research at firm level in the 1970s. Am. Econ. Rev. **76**, 141–154 (1986)
6. Lichtenberg, F., Siegel, D.: The impact of R&D investment on productivity: new evidence using linked R&D-LRD data. Econ. Inq. **29**, 203–229 (1991)
7. Kafouros, M.I.: R&D and productivity growth: evidence from the UK. Econ. Innovation New Technol **14**, 479–497 (2005)
8. Palia, D., Lichtenberg, F.: Managerial ownership and firm performance. A re-examination using productivity measurement. J. Corp. Financ. **5**, 323–329 (1999)
9. Martikainen, M., Nikkinen, J., Vähämaa, S.: Production functions and productivity of family firms: evidence from the S&P 500. The Quarterly Rev. Econ. Financ. **49**, 295–307 (2009)

An Improved CRT-based Broadcast Authentication Scheme in WSNs

Yunjie Zhu and Yu Shen

Abstract As wireless sensor networks (WSNs) are increasingly widespread, probability of being under attack also will soar, so the security of WSNs has raised more concern. Broadcast communication plays an important role in WSNs due to the existence of a large number of sensor nodes and the broadcast nature of wireless communications. Therefore, the security of broadcast communication directly relates the safety of the entire network. Authentication is one of the basic security services needed to construct a practical WSNs. In this paper, we present a high-security broadcast authentication protocol. Our proposal combines time synchronization with the Chinese Remainder Theorem (CRT) to implement dual authentication. As a result, this scheme has greatly improvement to some existed schemes in term of security.

Keywords Wireless sensor networks · Broadcast authentication · Chinese remainder theorem

1 Introduction

Wireless sensor networks (WSNs) are composed of a large number of tiny nodes which resources, communication ability and computing power are extremely limited [1]. In recent years, it has been widely applied to battlefield management, medical monitoring, environmental monitoring and so on. Broadcast communication plays an important role in WSNs due to the existence of a large number of

Y. Zhu (✉) · Y. Shen
East China Normal University, Shanghai, China
e-mail: csdn322@gmail.com

S.-S. Yeo et al. (eds.), *Computer Science and its Applications,*
Lecture Notes in Electrical Engineering 203, DOI: 10.1007/978-94-007-5699-1_49,
© Springer Science+Business Media Dordrecht 2012

sensor nodes and the broadcast nature of wireless communications. Therefore, the security of broadcast communication directly relates the safety of the entire network, especially nodes being deployed in the harsh conditions and lack of supervision. In order to ensure the security of WSNs, many security mechanisms are proposed, and broadcast authentication mechanism is one fundamental and essential of them. It can save WSN bandwidth and reduce the communication delays. In a broadcast authentication mechanism, the base station generally broadcasts the network nodes commands or sends data packets. When receiving the broadcast data from the base station, nodes need verify the authenticity of source, integrity, freshness of the packet. Due to the limited computing power of the sensor nodes, broadcast authentication mechanism generally does not use public key digital signature technology, instead of using symmetric scheme with time synchronization and hash function.

TESLA is a remarkable mature scheme using broadcast authentication mechanism in sensor network [2, 3]. μTESLA [1], proposed by Perrig and Szewczyketc, is based on TESLA protocol. It introduces asymmetry by delaying the disclosure of symmetric keys which generates through one way hash chain. Liu and Ning have proposed Mulit-level μTESLA [4, 5] reduced overheads of key chain commitment distribution. However, these cannot immediately authenticate these messages. To address this situation, Zhang et al. [6] proposed a Chinese Remainder Theorem-based broadcast authentication scheme—CRTBA. CRTBA scheme associates the authenticating procedure of the authentication key with MAC of the broadcast message. However, this scheme can be attacked easily if an attacker captured a node. So we present an improved scheme based on CRTBA. We first budget the time delay between node and station and find the max delay as the threshold. Then, adding time slice ensures that the time interval is much larger than delay and maintains the freshness of the broadcast messages.

The rest of the paper is organized as follows. In the next section, we present review of the CRTBA scheme. In section III, we discuss security weakness in the CRTBA scheme. In section IV, we introduce our scheme and analysis. Finally, we conclude the paper in section V.

2 Review of the CRTBA Scheme

This section reviews the CRTBA scheme. Table 1 presents the list of notations used in this paper.

Participants in the CRTBA scheme include a base station and multiple sensor nodes. In the pre-deployment phase, the base station (BS) first chooses a random K_n and repeatedly uses one-way function F to generate a series of keys as:

$$K_u = F(K_{u+1}), \ 0 \leq u \leq n - 1$$

Table 1 Some notations used in the paper

Notation	Description
BS	Base station
MAC	Message authentication code
M	Value of the function
m	The message
F	One-way function
U	CRT solution
BT_s	BS start operation time
ct_s	BS current time
ct_i	Node current time
PT_i	The most recent time interval from last updating key time
p_0, p_1	Global prime numbers
n	Length of key chain
$K_u(u = 0, 1, 2 \ldots n)$	Key chain
K_0	Commitment of key chain
K_i	Current key
TL	The length of time interval
T_i	Pre-set transmission delay threshold between nodes and BS
t_i	Actual transmission delay threshold between nodes and BS
T_{max}	The maximum value of transmission delay threshold between nodes and BS
Q	Message queue
ΔTC	Minimal time interval of two messages

Then BS chooses two large prime numbers p_0 and p_1. Finally, BS installs p_0,p_1, and K_0 on each sensor node.

If BS wants to broadcast a message, denote by m, in i-th times, it broadcasts the following advertisement message to all the sensor nodes:

$$\langle m, U, \rangle$$

where U is a CRT solution of congruent equations

$$U \equiv M mod p_0$$

$$U \equiv K_i mod p_1$$

and M is the value of the MAC function (i.e. $M = MAC(m, K_i)$).

when a node receives $\langle m, U, \rangle$, it first computes $K_i = U \mod p_1$, and confirms if K_j is equal to $F^{i-j}(K_i)$ where K_j is the last authentic key. If it is confirmed, then the node verifies the integrity of message m by testing whether M is equal to $MAC(m, K_i)$. If all are authentic, the message is accepted.

3 Security Weakness in the CRTBA Scheme

In this section, we point out an inherent design flaw in the CRTBA scheme [6]. We then demonstrate that this design flaw results in the CRTBA scheme being vulnerable to an impersonation attack (i.e., an attacker can easily impersonate *BS* to successfully broadcast message on sensor nodes), in violation of its security claims.

Recall that in the CRTBA scheme, the parameter K_0 needs to be pre-installed on all sensor nodes before deploying them, and subsequently K_i will be placed by $(K_{i+1}(0 \le i \le n-1))$ with all sensor nodes whenever they accept a correct advertisement message. Thus, K_i acts as a dynamic secret key shared by *BS* and all sensors nodes in the CRTBA scheme. This is, however, a design flaw as it enables an attacker to obtain K_i (together with p_0 and p_1) once he/she has compromised a sensor node. After all, it has been shown that compromised node to be revealed. Therefore, any secure broadcast authentication inside the compromised node to be revealed. Therefore, any secure broadcast authentication scheme should be designed to be robust against node compromise. We now show how this design flaw in the CRTBA scheme enables an attacker to impersonate *BS* to broadcast message on sensor nodes.

Assume that A is an attacker who wants to broadcast his/her own message m' on sensor nodes. Assume also that A has obtained the parameters K_i, p_0 and p_1 from a compromised node and intercepted the latest broadcast message $\langle m, U, \rangle$, where U is a solution of the congruent equation

$$U = MAC(m, K_{i+1}) mod p_0$$

$$U = K_{i+1} mod p_1.$$

It is important to note that, similar to other networks, the communication medium is insecure in WSNs: A is able to eavesdrop on, inject, delay, and intercept messages. As a result, it is reasonable to assume that the existing nodes fail to receive the above broadcast message $\langle m, U, \rangle$.

Next, A can generate an advertisement message $\langle m', U' \rangle$ as follows and successfully impersonate *BS* to broadcast it to all sensor nodes:

- Recover the key K_i by

$$K_{i+1} = U mod p_1$$

- Computes the *MAC* value M' as

$$M' = MAC(m', K_{i+1})$$

- Obtains U' by solving the congruent equation

$$U' = M' mod p_0$$

$$U' = K_{i+1} \bmod p_1$$

The forged message $\langle m', U' \rangle$ will pass the verification tests by all sensor nodes as shown below:

- Since K_{i+1} (originally from the BS) is not changed, the verification in Step-1 is easily performed.
- The integrity verification of message m' will succeed because the value of $MAC(m', K_{i+1})$ is clearly equal to $U' \bmod p_0$.
- Consequently, all sensor nodes will accept the message $\langle m', U' \rangle$ and subsequently replace the key K_i by K_{i+1}.

4 Our Proposal

In this section we propose an improved scheme based on the above scheme CRTBA.

Our scheme requires that the base station and nodes be loosely time synchronized and each node knows the upper bound on transmission delay threshold between nodes and BS. To send an authenticated message, the base station computes a MAC on message with current key and constructs the congruent equation to get CRT solution U. Because of adding time slice and having delay, it is possible that the node receives the message from BS in a next time interval.

Therefore, the node will calculate invalid key and cannot pass authentication. In consideration of this, we introduce a mechanism that each time interval is divided into the security segment and insecure segment. And we design a message queue Q in BS. When $c_t - BT$ in $(aTL, (a+1)TL - T_{max})$ is in the security segment, BS can broadcast messages in turn at ΔTC. When $c_t - BT$ in $((a+1)TL - T_{max}, (a+1)TL)$, the message is put into Q until the next time. In order to reduce the possibility of the sending and receiving time respectively in two time interval, BS needs to check the current time before it sends a message. When a node gets a message, it can recover M and Ki from U and confirms if $F^{v+1}(K_i) = K_0$, where $v = \lfloor (cti - BTi)/TL \rfloor$. If yes, the node continues to verify message m. If the equation holds well, m is accepted. If any of these checks fails, the packet is considered illegal.

It hypothesizes that the transmission delay between node and BS obey the normal distribution. The threshold (T_i) is set as $T_i = \mu + 1.96\sigma$, namely the probability that the actual transmission delay between node and BS less than T_i is around 97.25% . Before deployment, BS does a lot of off-line tests to get the data and calculate the value of T_i. The value of the time interval TL should be far longer than T_{max}.

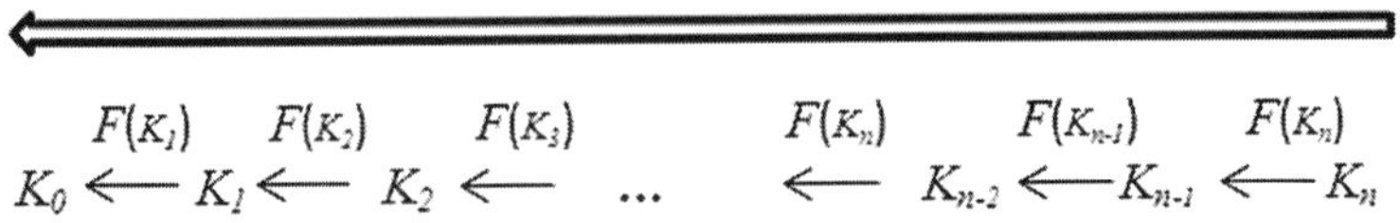

$$F(K_1) \quad F(K_2) \quad F(K_3) \qquad\qquad F(K_n) \quad F(K_{n-1}) \quad F(K_n)$$
$$K_0 \leftarrow K_1 \leftarrow K_2 \leftarrow \quad \cdots \quad \leftarrow K_{n-2} \leftarrow K_{n-1} \leftarrow K_n$$

Fig. 1 Generating keys

In the following section, there is a detailed description of our scheme. This scheme has three phases: *BS* setup, Broadcasting Authenticated Messages and Message Authentication.

4.1 BS setup

Before the deployment of sensor nodes, the base station first needs a one-way key chain. To generate it, *BS* first chooses a random K_i as the last one, and then repeatedly uses one-way function F. The final key chain has form: $K_0, K_1, ..., K_n$. Because F is a one way function, anyone can compute the previous keys using $K_i = F(K_{i+1})(0 \leq i \leq j)$. However, no one can computer backward. Then, *BS* randomly chooses two different prime numbers (p_0, p_1), and stores p_0, p_1, K_0 into each sensor memory (Fig. 1).

4.2 Broadcasting Authenticated Messages

Time is divided into uniform time intervals and *BS* associates each key of the one-way key chain with one time interval. In consideration of time delay, each time interval is divided into the security segment and no-secure segment.

When $c_t - BT$ in $(aTL, (a+1)TL - T_{max})$, it is in the security segment. Otherwise, $c_t - BT$ in $((a+1)TL - T_{max}, (a+1)TL)$, it is in the insecure segment. If *BS* needs to broadcast a message m in time interval i, it first computes a *MAC* value M of m using current key K_i. Then *BS* can get the solution U of the following congruent equations by CRT (refer to Fig. 2):

$$U \equiv M mod p_0$$

$$U \equiv K_i mod p_1$$

Finally, *BS* broadcast the message $\langle m, U, \rangle$ to all the sensor nodes. Because of the time delay, *BS* should broadcast $\langle m, U, \rangle$ in a security segment of time interval i to reduce unnecessary error. A FIFO (first-in, first-out) message queue Q, it should check whether the current time is in the security segment. If checked, *BS* broadcasts the messages in turn at every ΔTC, otherwise, the message is in the Q until the next time.

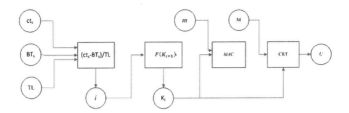

Fig. 2 Message constructing

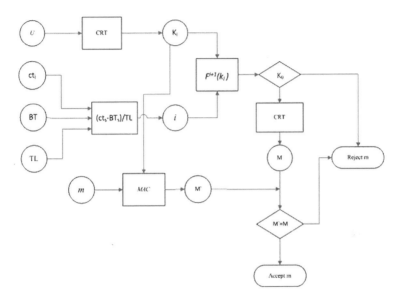

Fig. 3 Message authentication

4.3 Message Authentication

After a node receives the message m, U, it need to verify the message's authenticity. The node recovers the K_i and M as

$$K_i = U mod p_0$$

and

$$M = U mod p_1$$

The node confirms if $F^{v+1}(K_i) = K_0$, where $v = \lfloor (cti - BTi)/TL \rfloor$. If yes, the node continues to verify message m by checking if $M = MAC(m, K_i)$. If the equation holds well, m is accepted and BT and K_0 are replaced by $BT + v$ and K_i, respectively. If any of these checks fails, the packet is considered illegal.

In our proposed scheme, the receiver is able to get the authentication key and broadcast message in a certain time interval. It is assumed that tampering with a packet takes some time and after the time the key becomes invalid. At the same time, it uses CRT to authenticate the message again, therefore achieving dual authentication. As a result, our proposed scheme has higher security than the original CRTBA scheme (Fig. 3).

5 Conclusion

We point out an inherent design flaw in the CRTBA scheme, which enables an attacker to easily impersonate the base station to install his/her preferred program on sensor nodes and gain control over the network. We then presented a modified scheme to prevent such an attack based station on a restricted WSN.

References

1. Perrig, A., Szewczyk, R., Wen, V., Culler, D., Tygar, J.: Spins: security protocol for sensor networks. In: Proceedings of Seventh Annual International Conference on Mobile Computing and Networks, July 2001
2. Perrig, A., Canetti, R., Tygar, J. D., Song, D.: The TESLA broadcast authentication protocol. In: CrytoBytes, Summer/Fall 2002
3. Perrig, A., Canetti, R., Tygar, J.D., Song, D.: Efficient authentication and signing of multicast streams over lossy channel. In: IEEE Fr Symposium on Security and Privacy, May 2000
4. Liu, D., Ning, P.: Efficient distribution key chain commitments for broadcast authentication in distributed sensor networks. In: Proceedings of the 10th Annual Network and Distributed System Security Symposium (NDSS'03), February2003, pp. 263–276
5. Liu, D., Ning, P.: Multi-level μTESLA: broadcast authentication for distributed sensor networks. In: ACM Transactions in Embedded Computing Systems (TECS), vol. 3, no. 4 2004
6. Zhang, J., Yu, W., Liu, X.: CRTBA: Chinese remainder theorem-based broadcast authentication in wireless sensor networks. In: International Symposium on Computer Network and Multimedia Technology, 2009. CNMT 2009. 18–20 January 2009

Evaluation and Exploration of Optimal Deployment for RFID Services in Smart Campus Framework

Yao-Chung Chang

Abstract With the application of new technologies such as pervasive wireless computing and RFID, the theoretical foundation and technical support of smarter campus are evolved rapidly. Especially, the campus security in children surveillance has been regarded as a critical issue. This work establishes a smart campus framework based on innovative RFID services. The Tabu search strategy and optimal deployment mechanism are used to find the better locations of deploying readers. In order to keep the safety of campus, this framework analyzes and designs five innovative scenarios toward children surveillance, including campus gate management service (CGMS), student temperature anomaly management service (STAMS), hazardous area management service (HAMS), campus visitor management service (CVMS) and campus equipment management service (CEMS). Then, the pervasive RFID services of five scenarios are implemented in the campus according to the demand from parents' and teachers' point of view. The middleware of the proposed framework adapts to collect both active/inactive RFID tags deployed around the campus, processes and filters the necessary information and events generated by the RFID reader from different locations or user rules. Finally, this work is successfully deployed and currently in use by the National Taitung University Laboratory Elementary School with more than 150 students and 30 administration staffs and faculties. Compared to traditional dense mechanisms, this mechanism has a 36 % lower deployment cost.

Y.-C. Chang (✉)
Department of Computer Science and Information Engineering,
National Taitung University, Taitung, Taiwan, Republic of China
e-mail: ycc@nttu.edu.tw

S.-S. Yeo et al. (eds.), *Computer Science and its Applications*,
Lecture Notes in Electrical Engineering 203, DOI: 10.1007/978-94-007-5699-1_50,
© Springer Science+Business Media Dordrecht 2012

Keywords Smart campus · Pervasive RFID service · Application framework · Tabu search · Optimal deployment · Campus Gate Management Service (CGMS) · Student Temperature Anomaly Management Service (STAMS) · Hazardous Area Management Service (HAMS) · Campus Visitor Management Service (CVMS) · Campus Equipment Management Service (CEMS)

1 Introduction

In recent years, student kidnappings, phone scams and phone threats have become common, and students' safety has become an important issue for schools, parents and teachers. As elementary school students do not have totally independent behavior capability, and teachers cannot take care of individuals all the time, students' security on campus is questionable.

In order to successfully implement the research program, this study first obtained the support of the teachers and students of National Taitung University Laboratory Elementary School. Both parents and teachers were asked to describe the current school situation, in order to determine the needs related to campus security as criteria to develop the key technologies for RFID and its integrated technologies [1–3]. In addition, when implementing the information system, the researcher paid attention to the development of foreign and domestic RFID technologies and adjusted the system implementation in order to result in a suitable system for campus security [4–7].

This study presents RFID readers deployment mechanism, which uses mobile RFID readers to find out the better locations in pipelines [8–10], that is able to read more tags and decreases the interference by other readers. Figure 1 illustrates the proposed smart campus framework environment. This framework is divided into multiple regions and every region contains immovable RFID tags and information gateway. The information gateway stores the unique ID data of RFID tags in the connection table.

The remainder of this paper is organized as follows: Sect. 2 outlines the optimal deployment agent and Tabu search strategy for RFID applications. Five innovative RFID scenarios for campus security are discussed in Sect. 3. Section. 4 addresses key frameworks for the RFID smart campus framework. Section. 5 presents the results and discussion. Finally, conclusions of this paper are drawn in Sect. 6.

2 Optimal Deployment Mechanism

Figure 2 illustrates the deployment agent for RFID applications, which mainly can be divided into two parts: superior mechanism in RFID reader deployment and reading speed adjustment for anti-collision. The first mechanism utilizes the virtual

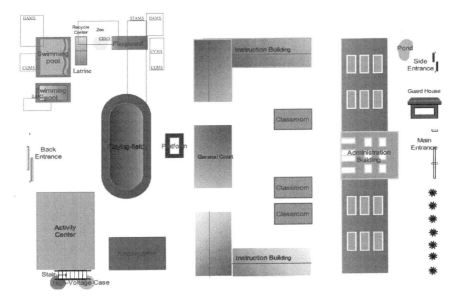

Fig. 1 Campus RFID services

reader model to estimate the backscatter power and the Tabu search [11] to find the better locations of deploying readers. In the procedure of Tabu search, it dynamically uses the adaptive power control approach to correct power coverage of readers, so that the interference of reader-to-reader collisions can be avoided and Query-Hit-Ration (QHR) value can be enhanced [12]. Assume RFID readers have been deployed into the target area. Sometimes it may cause Reader-to-Reader collision because these readers should cover border tags. The parameters of the optimal deployment mechanism are defined as follows:

$$wd_x + ID_x = \rho_{max} \cdot MRC(wd_x)$$

$$ld_y + ID_y = \rho_{max} \cdot MRC(ld_y)$$

$$MRC(\mu) = \begin{cases} [\mu], & if \ [\mu]\%2 = 0 \\ [\mu] + 1, & otherwise \end{cases} \quad (1)$$

where wd_x, ld_y denotes the width distance and length distance, ID_x, ID_y denotes the Increased Distance for wd_x, ld_y. MRC (μ) function is making increased width and length distance that can be multiple of Maximum Read Coverage. This mechanism selects new unit area for deploying RFID readers without Reader-to-Reader collision.

This mechanism also computes how many RFID readers can be deployed in the target area by utilizing following Eq. 2, where IS_n denotes deployment number of RFID readers in Initial State of system, RC_{max} denotes maximum radius of Reader Coverage, and $|TA|$ denotes actual measure of Target Area. This mechanism uses

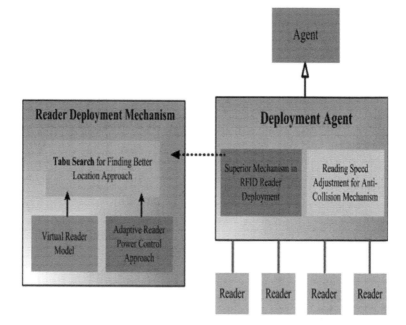

Fig. 2 Agent architecture

circular formula for computing IS_n value, and deploys IS_n number of RFID readers to search out better locations.

$$IS_n \times \left\lfloor \frac{1}{2} RC_{max}^2 \times \pi \right\rfloor = |TA| \tag{2}$$

IF_i denotes the interference value from other readers to $reader_i$, and can be estimated by Eq. 3 where TP_j denotes *Transmission Power* of RFID reader. The channel loss from $reader_i$ to $reader_i$ is considered as CL_{ij},

$$IF_i = \sum_{j \neq i} CL_{ij} \times TP_j + \varphi \tag{3}$$

The R_d demotes the desired radius of reader coverage, and R_{actual} denotes the actual radius value of reader coverage. The R_{actual} can be calculated by utilizing Eq. 4 where T_{snr} denotes the threshold value of SNR (Signal-to-Noise Ratio). T_{snr} is dependent with noise power, bit rate, radio channel bandwidth and he Bit-Error-Rate. SNR_i denotes received SNR value in $reader_i$ at unit of time.

$$R_{actual} = R_d \left(\frac{SNR_i}{T_{SNR}} \right)^{1/4q} \tag{4}$$

QHR denotes the Query-Hit-Ratio value for RFID readers, which represents the quality level of read RFID tags, and can be calculated with Eq. 5, where QR_s is the successful number of query and reading. QR is the number of query and reading.

$$\text{QHR} = \frac{QR_s}{QR} \tag{5}$$

In other words, that is incapable for covering all of Tags. To solve the above problem, first of all, this mechanism calculates if there is more space for deployment, and then determines that how many readers can be raised. Secondly, this method increases power coverage of readers which locating at the neighbor for the unread tags if there still has some tags can't be covered. Finally, the optimal deployment mechanism can utilize lowest RFID readers to reach the better readers deployment in RFID applications.

3 Innovative RFID Scenarios for Smart Campus

Regarding campus security, this study implemented a framework for teachers and parents to control students' situations in school. The advantages of the system are shown in the following programs under various pervasive RFID services:

3.1 Campus Gate Management Service (CGMS)

When the front entrance reader senses a student entering or leaving, it will transmit the information to the student security system, which delivers the said information to the parents. In addition, students' entering and leaving the entrance can be checked by the system.

3.2 Student Temperature Anomaly Management Service (STAMS)

The sensor will regularly report body temperatures. When the student health system recognizes an abnormal body temperature, it will immediately inform medical personnel for follow-up treatment.

3.3 Hazardous Area Management Service (HAMS)

When the system senses a student entering dangerous areas such as remote corners on campus, attics, swimming pools and high voltage boxes, the sensor will transmit the information to the student security system, which will immediately inform the guard and teachers.

3.4 Campus Visitor Management Service (CVMS)

When visitors enter the school, they must sign in with the guard and wear an active tag. The information transmitted by the tag can immediately show the areas entered by the visitors. Once the visitors enter a forbidden area, the system will immediately inform the guard.

3.5 Campus Equipment Management Service (CEMS)

(1) On campus, active RFID asset tags are hung on important assets.
(2) In order to borrow important assets, students must reserve the assets using their RFID student cards.
(3) If a student does not reserve the assets and randomly use them, the system can detect the abnormal use of the RFID card. The system will record the event and e-mail the homeroom teacher and parents.

As to the reporting measures of the service platform, when students have activities in the four areas of home, front entrance, classrooms and campus, the RFID reader provides three services to three different subjects using RFID middleware and the campus security platform, and informs teachers and parents using an electronic contact book. The student/device positioning and dynamic area control system informs the security company and school manager, and GSM informs the security company.

4 Smart Campus Framework

This study constructed an RFID campus security service platform in order to guarantee student safety on campus using RFID. The service platform provided an in-school report service, an abnormal student body temperature management service, a dangerous area management service, a campus visitor management service and a campus asset management service, as well as an active/passive RFID sensor and a campus. The information platform provided an open program development interface for the integration of RFID campus security and the student protection platform as well as the development of new situations in order to incorporate a user interface on service platform. The devices and the specifications of the programs in the information platform were as shown below.

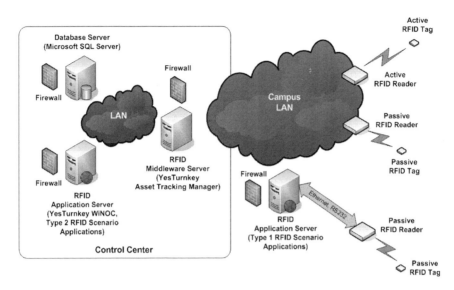

Fig. 3 Network framework

4.1 Network Framework

The service platform network framework is shown in Fig. 3. The devices required for the system included a PC server, an active/passive RFID reader, RFID tags and converters and a router for connecting the network.

4.2 Software Framework

The information service platform is constructed using the following programs, and the program framework of the information system is shown in Fig. 4. The program components are executed using an RFID application program server at the front, an RFID medium program server at the back, a database server at the back, and an RFID application program server at the back.

5 Implementation Results and Discussion

Before the integration of the system, the functions of the system components had to be confirmed and the system test and test validation had to be incorporated with the requirement documents and the design documents. The system demand specifications and system design documents described the necessary integration tests. According to different platform construction stages, the following steps were implemented to validate the feasibility of the system:

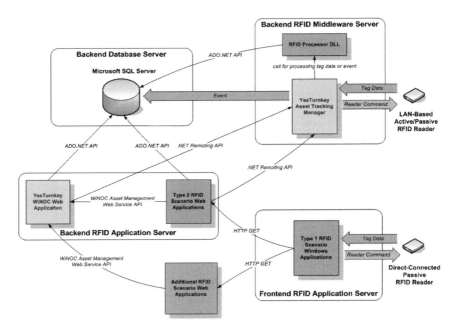

Fig. 4 Software framework

(1) Network test: this tested the normal installation and wiring of a LAN-based RFID reader and set the TCP/IP network configuration. The step confirmed the precise routes of the related servers and the reader.

(2) RFID tag sensor test: this tested the RFID tag that can be detected by the RFID reader and the information that could be collected by the YesTurnkey Asset Tracking Manager [13]. The step confirmed the correct installation and setting of the YesTurnkey Asset Tracking Manager and the normal operation of all RFID tags.

(3) RFID database writing test: this tested if the RFID information collected by the asset tracking manager could be correctly written to the back-end database server. This step confirmed the normal operation of the VPN between the front and back-end servers.

(4) YesTurnkey WiNOC test: this tested if the RFID reader and the RFID tag information could be presented on the YesTurnkey WiNOC webpage user interface. This step confirmed the normal installation and setting of the Yes-Turnkey WiNOC.

(5) Situation program test: before the installation of situation programs on the server, a unit test, integration test, validation test, and system test were conducted, according to the program engineering development.

(6) System integration test: this tested if the system operation met the plan. This step confirmed the system standards.

By integrating the tests and the system framework indicated above, this study implemented a service platform combining RFID R&D and applications for teachers and parents to check the students' in-school situation and guarantee campus safety. The classes and administrative units that participated in this experiment are shown as follows: The classes using active RFID were: grade 2 class 1, grade 2 class 3, and grade 2 class 5 (45 participants in total). The classes using passive RFID were: grade 4 class 1, grade 4 class 2 and grade 4 class 6 (82 participants in total). There were 127 participants. The administrative units included the office of student affairs, the office of general affairs, the office of academic affairs (information division), the health center and the guardroom.

As mentioned above, in order to demonstrate the completeness of the system, this study uses a large number of tags to continuously test the reader and ensure that all tags are read by the reader and were precisely transmitted to the database. As to the pressure test, 35 passive tags are adopted to continuously test the reader. Each tag is detected for about 2 s. The reading of the tags is judged using RFID middleware and the information is acquired from the SensedAssetLog of the database AssetDB. The continuous sensing of the passive tags begins at 13:38:12, and it finished at 13:39:49. There were 35 passive tags detected, requiring 1 min and 37 s. This demonstrated good detection of the passive tags. Regarding the pressure test, 123 active tags were used for immediate testing of the reader. A total of 123 active tags were used to test the reader. The tag readings were judged using RFID middleware and the information was acquired from the SensedAssetLog of the database AssetDB. The test results showed that it was in operation from 9:03:48 to 9:06:27, taking 2 min and 39 s. All 123 active tags were detected. This showed the good detectability of the active tags.

6 Conclusion and Future Work

This study employs a virtual reader model and Tabu search mechanism to optimize the deployment of RFID readers. Experimental results indicate that the proposed system is more efficient than current systems when used in large deployments with numerous tags. Besides, the RFID services in smart campus framework developed by this paper not only enhances campus security, but also provides the additional services of tracking students before school and after school, measuring and managing abnormal student body temperatures, and managing dangerous areas, which are all concerns of parents. Furthermore, this framework can protect the students' personal safety in the school, and integrate with information communication, such as e-mail or mobile phones. Parents can be informed of their children's situations at school when they are at work. It is expected that the framework in this study can be applied to campus security and combined with wireless sensors and cloud technologies. As successful cases, experiences and effects are accumulated, it can be applied to other instructional domains.

Acknowledgments The author would like to thank Prof. Jiann-Liang Chen and Yu-Shan Lin, Mr. Ci-Jhih Fong and Miss Jing-Rong Wang for this work, especially the Ministry of Education under Contract No. 98TH01-AA8 and National Science Council of Taiwan under Contract No. NSC 100-2221-E-143-003 for financially supporting this research.

References

1. RFID Weblog. http://www.rfid-weblog.com
2. Shepard, S.: RFID Radio frequency identification. McGraw-Hill, New York (2005)
3. Chen, H.Y.: Introduction to RFID systems—radio frequency identification system. Kings Information Corporation, Twain (2004)
4. Lee, R.K., Yu, C.H., Liang, M.S., Feng, M.W.: An approach to children surveillance with sensor-based signals using complex event processing. In: IEEE International Conference on e-Business Engineering, IEEE Press, Macau, China (2009)
5. Zangroniz, R., Pastor, J.M., Dios, J.J.D., Garcia-Escribano, J., Morenas, J., Garcia, A.: RFID-based traceability system for architectural concrete. In: 2010 European Workshop on Smart Objects: Systems, Technologies and Applications, Ciudad, Spain (2010)
6. Dimitriou, G. Antonis, Bletsas, Aggelos; Polycarpou, C. Anastasis, Sahalos and John N.C.: On efficient UHF RFID coverage inside a room. In: Fourth European Conference on Antennas and Propagation, Barcelona, Spain (2010)
7. Liu, C.L., Xie, Z.F Peng, P.: A discussion on the framework of smarter campus, In: Third International Symposium on Intelligent Information Technology Application, Nanchang, China (2009)
8. Chen, N.K., Chen, J.L., Lee, C.C.: Array-based reader anti-collision scheme for highly efficient RFID network applications. Wirel. Commun. Mobile Comput. **9**, 976–987 (2009)
9. Kuo, C.H., Chen, H.G.: The critical issues about deploying RFID in healthcare industry by service perspective. In: IEEE Conference on System Sciences, Hawaii (2008)
10. Birari, S.M., Iyer, S.C.: Mitigating the reader collision problem in RFID networks with mobile readers. In: IEEE Conference on Networks, pp. 463–468, Malaysia (2005)
11. Glover, F.: Tabu search: a tutorial. Instit. Manag. Sci. **20**(4), 74–94 (1990)
12. Chang, Y.C., Fong, C.J., Lin, Y.S., Chen, J.L., C.: Optimal deployment strategy of RFID networks using tabu search mechanism. In: 12th International Conference on Advanced Communication Technology, Korea (2010)
13. YesTurnkey Technology, Inc. http://www.yesturnkey.com

Towards Analyzing Family Misconfiguration in Tor Network

Xiao Wang, Jinqiao Shi and Guo Li

Abstract As one of the most popular low-latency anonymous communication systems in the world, Tor has become the research hotspot recent years. However, disproportionately little attention has been paid to Tor family design, a feature that plays an important and irreplaceable role in Tor network. This paper gives an analysis of family misconfiguration in Tor network. Two types of family misconfigurations together with the corresponding methods that discover and correct them are presented. The experiment over live Tor network data confirms the pervasiveness of family misconfigurations in Tor network. Besides, families misconfigurations discovered in the experiment as well as methods used to determine and correct them may not only help Tor operators to make family declarations correctly, but also help Tor client to avoid potential risk when constructing circuits.

Keywords Tor · Family declaration · Misconfiguration · Anonymity · Similarity

X. Wang (✉)
Institute of Computing Technology, CAS, Beijing, China
e-mail: wangxiao@software.ict.ac.cn

X. Wang · J. Shi · G. Li
Institute of Information Engineering, CAS, Beijing, China
e-mail: shijinqiao@software.ict.ac.cn

G. Li
e-mail: guoli@software.ict.ac.cn

X. Wang
Graduate University, CAS, Beijing, China

X. Wang · J. Shi · G. Li
National Engineering Laboratory for Information Content Security Technology, Beijing, China

S.-S. Yeo et al. (eds.), *Computer Science and its Applications*,
Lecture Notes in Electrical Engineering 203, DOI: 10.1007/978-94-007-5699-1_51,
© Springer Science+Business Media Dordrecht 2012

1 Introduction

Tor [1] is one of the most popular low-latency anonymous communication systems in the world. It provides anonymity service by forwarding user's traffic through an anonymous communication channel consisting of Tor nodes all around the globe. Tor was originally designed, implemented and deployed based on the onion routing project of the U.S. Naval Research Laboratory. Currently, in Tor network, there are over 2,000 volunteer servers simultaneously and hundreds of thousands of users from all around the world [2]. It has attracted users with a wide variety of purposes ranging from military, journalists, law enforcement, officers to activists and so on [3].

Tor family plays an important and irreplaceable role in Tor network. A Tor node family is a set of Tor nodes that are under the same administrative control of a volunteer. As specified in Tor path specification [3], Tor never chooses more than two nodes from the same family when constructing circuits. This constraint ensures that Tor clients will always choose its first hop and last hop from Tor nodes provided by different volunteers, which are important to the anonymous service provided by Tor. Besides, Tor never choose more than two nodes from the same/16 subset in a circuit, which can avoid the end-to-end correlation attack committed by an/16 subnet-level adversary. Edman et al. [4] even enhanced this constraint to AS-level, with the purpose of avoiding AS-level observers. Both the/16 subnet-level constraint and AS-level constraint, even a country-level constraint cannot prevent Tor clients from selecting Tor nodes from the same family, for Tor nodes within a family always scatter across different/16 subnets, different ASes and different countries. For example, Tor nodes provided by "perfect-privacy" organization[1] (The PPrivCom family) distribute widely around the world, in over a dozen countries.

In this paper, the family design as well as its implementation is examined. We notice that, family declarations made by volunteers are not always correct. Two types of family misconfigurations are defined in this paper. Methods that discover and correct these misconfigurations are presented too. We validated these methods using experiment over the live Tor network data. The experiment shows that: there exist many family misconfigurations in Tor, which should arouse attention from both Tor node providers and Tor users. The methods presented in this paper as well as family misconfigurations discovered in the experiment can not only help Tor operators to examine their family declarations, but also help Tor clients to make a proper selection when constructing circuits.

The remainder of this paper is organized as follows. Section 2 gives a brief introduction to the background and related work in Tor family. Section 3 examines the family design as well as its implementation in the live network. Two types of family misconfigurations as well as methods that discover and correct them are presented in Sect. 4. Section 5 gives an analysis of family misconfigurations over

---

[1] http://www.perfect-privacy.com

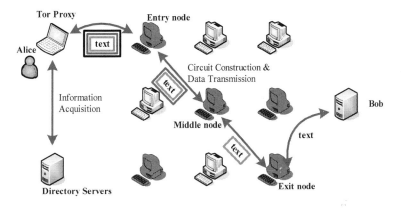

Fig. 1 Tor network's basic components and its working process

the live Tor network data. Section 6 summarizes this paper and discusses the limitations of this study as well as a few possible future research problems.

2 Background and Related Work

2.1 Tor

Tor is a popular low-latency anonymous communication system that provides anonymity service for TCP applications. As shown in Fig. 1, there are three basic components in Tor network: Tor directory servers, Tor proxy and Tor relay nodes. Tor directory servers are the core of Tor network. Information about the whole Tor network is gathered and distributed by Tor directory servers. Tor relay nodes form the basis of Tor network. All traffic in Tor network from the sender to the receiver is forwarded by Tor relay nodes. Tor proxy is the client-side component of Tor network. It serves as an interface between user applications and the Tor network.

Figure 1 also illustrates how these components work together to provide anonymity service. The process is divided into three phases as follows:

- Information Acquisition. Tor proxy acquires an overview of Tor network's current status (called network consensus) from Tor directory servers. The detailed information (called node descriptor) of each node will be further downloaded by Tor proxy soon after consensus.
- Circuit Construction. A circuit is an anonymous communication channel created by Tor proxy through a few (3 by default) randomly chosen Tor relay nodes. Nodes in the circuit are called as entry node, middle node and exit node respectively.

- Data Transmission. In this phase, Tor proxy encrypts user traffic and injects it into the circuit; it also decrypts traffic from Tor circuit and passes it to user applications.

2.2 Related Work

Recent years, Tor has become the research hotspot in the anonymous communication area. Methodologies in this field include theoretical analysis, simulation, proof-of-concept experiments over private Tor network or even an actual deployment in the live Tor network. Research in this field mainly focuses on a few topics such as censorship resistance [5, 6], privacy enhancement [4, 7, 8], performance improvement [9] and scalability [10, 11]. Most of these researches are related to the Tor path selection algorithm specified in Tor path specification.

Tor path selection algorithm is one of the most essential algorithms in the circuit construction phase presented in Sect. 2.1. By default, Tor selects three relay nodes to construct a circuit and these three nodes are selected in the order of exit node, entry node and then middle node. In order to ensure the performance of Tor service, Tor weights node selection according to node bandwidth. In order to ensure anonymity, Tor never uses more than two nodes from the same/16 subnet or family in a circuit.

As mentioned above, Tor family plays an irreplaceable role in Tor network for the/16 subnet-level constraint, even an AS or country-level constraint cannot prevent Tor clients from selecting Tor nodes from the same family. To the best of our knowledge, there exists no in-depth analysis of this important feature in Tor, Tor family design. This has motivated our research in examining Tor family design and analyzing family misconfigurations in Tor network.

3 Understanding Tor Family in Depth

3.1 Tor Family

Let V to be volunteers that contribute Tor nodes and provide anonymous service in Tor network. N is used to denote all Tor nodes in Tor network. $N(v_i)$ is used to denote a set of Tor nodes that provided by volunteer v_i and $V(n_i)$ is used to denote the volunteer that operates Tor node n_i.

According to the description in Tor path specification and Tor FAQ [3], a Tor node family is a set of Tor nodes that are under the same administrative control of a volunteer. We take $N(v_i)$ as a node family (i.e., $N(v_i) \in F$) iff $|N(v_i)| \geq 2$. In general, for a given set of Tor nodes S, we say S is a F **family** (i.e., $S \in F$) iff it meets the following two conditions: i. $\forall n_i, n_j \in S$, $V(n_i) = V(n_j)$; ii. $\forall n_i \in S$, $n_j \notin S$, $V(n_i) \neq V(n_j)$. The first constraint is the straight requirement for family

Fig. 2 An example of Tor
node families

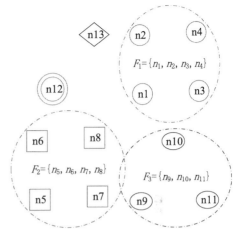

determination. The second constraint ensures that a node family is as large as possible, with no family members excluded from it.

As an example, Fig. 2 presents 13 nodes operated by five different volunteers. We use different node shapes to represent different volunteers, i.e., nodes with the same shape are provided by the same volunteer. According to the definition of node family, three node families can be observed from Fig. 2, $F = \{F_1, F_2, F_3\}$, where $F_1 = \{n_1, n_2, n_3, n_4\}$, $F_2 = \{n_5, n_6, n_7, n_8\}$ and $F_3 = \{n_9, n_{10}, n_{11}\}$.

3.2 Family Declarations

In practice, Tor relies on Tor nodes' self-announced family declarations to implement their family design. It's family declaration that makes family design a practical feature in Tor network. As specified in Tor FAQ [3], a volunteer who runs more than one relay nodes should set "MyFamily" option in the configuration file of each node, listing all the other relay nodes that are under his control. Only mutual agreements among family members result in a valid node family.

Let $F(n_i)$ to be Tor nodes that are listed in n_i's "MyFamily" configuration, i.e., n_i announces all nodes in $F(n_i)$ to be in the same family with it. The relations between two Tor nodes n_i and n_j can be classified into the following three types:

-
$$n_i \cong n_j, \text{ iff } n_i \in F(n_j) \text{ and } n_j \in F(n_i)$$

-
$$n_i \simeq n_j, \text{ iff } n_i \in F(n_j) \text{ or } n_j \in F(n_i)$$

-
$$n_i \approx n_j, \text{ iff } \exists n_{k_1}, n_{k_2}, \ldots, n_{k_m} \in N \text{ that } n_i \simeq n_{k_1}, n_{k_1} \simeq n_{k_2}, \ldots, n_{k_m} \simeq n_j$$

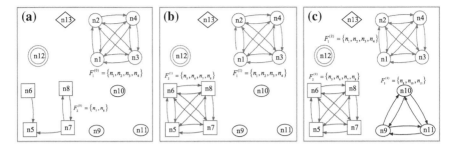

Fig. 3 Tor nodes and their original family declarations, Type I misconfiguration corrected family declarations and Type II misconfiguration corrected declarations; **a** Original, **b** Type I corrected, **c** Type II corrected

As implied by the definitions of these relations, we have $n_i \cong n_j \Rightarrow n_i \simeq n_j \Rightarrow n_i \approx n_j$. And vice versa, $n_i \not\approx n_j \Rightarrow n_i \not\simeq n_j \Rightarrow n_i \ncong n_j$.

Though the formal definition of node family given in Sect. 3.1 is based on the volunteer-node relationship, there exists no direct information about who is operating which nodes in the live Tor network. The self-announced family declarations are the most useful information that we can rely on when determining node families. However, we noticed that, many volunteers make their family declarations incorrectly or even fail to make any family declarations where they are necessary. As a result, Tor families determined on the basis of family declarations may have differences from the definition presented in Sect. 3.1. $F^{(0)}$ **family** is defined over family declarations corresponding to the former presented volunteer-node relationship based family definition. For a given set of Tor nodes S, we say S is a $F^{(0)}$ family iff it meets the following two conditions: i. $\forall n_i, n_j \in S$, $n_i \cong n_j$; ii. $\forall n_i \notin S$, $\exists n_j \in S$, that $n_i \ncong n_j$. The first constraint is the straight requirement for family determination that each node within a node family should declare the others as in the same family with it. The second constraint ensures that a node family is as large as possible.

Figure 3a shows 13 Tor nodes together with their family declarations. The relation $n_j \in F(n_i)$ is represented by an arrow from n_i to n_j in this figure. According to the definition of $F^{(0)}$ family, two $F^{(0)}$ families can be observed from Fig. 3a, $F^{(0)} = \left\{ F_1^{(0)}, F_2^{(0)} \right\}$, where $F_1^{(0)} = \{n_1, n_2, n_3, n_4\}$ and $F_2^{(0)} = \{n_7, n_8\}$. We can notice that $F^{(0)} \neq F$ due to the existence of family misconfigurations.

4 Family Misconfiguration

In this paper, we deal with two types of family misconfigurations: Type I misconfiguration and Type II misconfiguration. For each misconfiguration type, we give its definition as well as the probable causes in the following discussion.

Ways of correcting them and the corresponding variants of family definitions are also presented.

4.1 Type I Misconfiguration

Tor node n_i is considered to make a **Type I misconfiguration** iff there exists a node n_j that, $n_i \approx n_j$ but $n_j \notin F(n_i)$. Type I misconfiguration are probably caused by volunteers who know Tor family design and misconfigure some of their Tor nodes by accident. For example, we noticed many unidirectional (instead of bidirectional) family declarations in Tor network, which are the main causes of Type I misconfiguration. $N^{(1)}$ is used to denote all Tor nodes that have a Type I misconfiguration.

Type I misconfiguration of n_i can be corrected by adding additional nodes into its family declaration, listing each node n_j in the "MyFamily" configuration iff $n_i \approx n_j$. Using the same definition of $F^{(0)}$ family, we can draw a few families based on these Type I misconfiguration-corrected family declarations. Families determined by this way are defined as $F^{(1)}$ **families**. A few conclusions can be drawn that:

-
$$\forall F_i^{(1)}, F_j^{(1)} \in F^{(1)}, F_i^{(1)} \bigcap F_j^{(1)} = \emptyset$$

-
$$\forall F_i^{(0)} \in F^{(0)}, \exists F_j^{(1)} \in F^{(1)} \text{that } F_i^{(0)} \subseteq F_j^{(1)}$$

-
$$\left| \bigcup F_i^{(0)} \right| \leq \left| \bigcup F_j^{(1)} \right|$$

Figure 3b presents the 13 Tor nodes as well as their Type I misconfiguration-corrected family declarations. It can be observed from the figure that, there exists two $F^{(1)}$ families, $F^{(1)} = \left\{ F_1^{(1)}, F_2^{(1)} \right\}$, where $F_1^{(1)} = \{n_1, n_2, n_3, n_4\}$ and $F_2^{(1)} = \{n_5, n_6, n_7, n_8\}$. $F_2^{(1)}$ is larger than $F_2^{(0)}$ due to the discovery of Type I misconfigurations and the corresponding revision of family declarations.

4.2 Type II Misconfiguration

Tor node n_i ($\forall n_k \in N, n_i \not\approx n_j$) is considered to make a **Type II misconfiguration** iff there exists a node n_j that the similarity between n_i and n_j is greater than the threshold δ (i.e., $S(n_i, n_j) > \delta$). By definition, only Tor nodes that neither list others nor be listed by other in the "MyFamily" configuration can cause a Type II misconfiguration. Type II misconfiguration are probably caused by volunteers that are not familiar with Tor family design, i.e., they provide more than one Tor node and never make any family declarations at all. $N^{(2)}$ is used to denote all Tor nodes that have a Type II misconfiguration.

Unlike it is in the definition of Type I misconfiguration, Type II misconfiguration is defined based on the similarity between two Tor nodes which is independent of Tor nodes' self-advertised family declarations. The similarity between two Tor nodes can be determined by their configurations such as nickname, contact information and so on. Type II misconfigurations can be corrected in the following way: first, each Tor node in $N^{(2)}$ should add a "MyFamily" line into its configuration file and list nodes that have a similarity greater than δ in it; then, the family declarations should be checked again to correct Type I misconfigurations caused either by the original family declarations or newly added family declarations. Using the same definition of $F^{(0)}$ family, we can draw a few families based on these Type II misconfiguration-corrected family declarations. Families determined by this way are defined as $F^{(2)}$ **families**. A few conclusions can be drawn from the description above:

- $$\forall F_i^{(2)}, F_j^{(2)} \in F^{(2)}, F_i^{(2)} \bigcap F_j^{(2)} = \emptyset$$

- $$\forall F_i^{(1)} \in F^{(1)}, \exists F_j^{(2)} \in F^{(2)} \text{that } F_i^{(0)} = F_j^{(1)}$$

- $$\left| \bigcup F_i^{(1)} \right| \leq \left| \bigcup F_j^{(2)} \right|$$

Figure 3c presents the 13 Tor nodes as well as their Type II misconfiguration-corrected family declarations. We suppose that $S(n_9, n_{10}) > \delta$ and $S(n_{10}, n_{11}) > \delta$, so four extra arrows ($n_9 \rightarrow n_{10}$, $n_{10} \rightarrow n_9$, $n_{10} \rightarrow n_{11}$ and $n_{11} \rightarrow n_{10}$) will be added into the original family declaration figure shown in Fig. 3a. Then, all Type I misconfigurations in the revised family declarations are corrected. Finally, the Type II misconfiguration-corrected family declarations are presented in Fig. 3c, from which we can find three $F^{(2)}$ families, $F^{(2)} = \left\{ F_1^{(2)}, F_2^{(2)}, F_3^{(2)} \right\}$, where

$F_1^{(2)} = \{n_1, n_2, n_3, n_4\}$, $F_2^{(2)} = \{n_5, n_6, n_7, n_8\}$ and $F_3^{(2)} = \{n_9, n_{10}, n_{11}\}$. $F_3^{(2)}$ is found due to the discovery of Type II misconfiguration and the corresponding revision of family declarations.

5 Experiments and Results

5.1 Experiments

Data Set. We examined all 2,451 Tor nodes listed in the Tor consensus file announced at the 23:00 of Dec. 31, 2011. In order to obtain some detail about each node such as family declaration and contact information, we also checked the corresponding server descriptor files of these nodes. All these can be fetched from Tor metrics website [2].

Node families. These 2,451 Tor nodes as well as the corresponding family declarations are viewed as a vertex-edge graph. Each node is represented by a vertex and every bidirectional edge between two nodes is represented by an undirected edge in the graph. All unidirectional edges between two nodes are ignored. According to the definition of $F^{(0)}$ node family, each family corresponds to a maximal clique [12] in this graph. With the help of a free Java graph library JGraphT,[2] we can find all maximal cliques from the figure, i.e., all node families.

Type I misconfigurations and $F^{(1)}$ node families. In order to discover Type I misconfigurations, we can also view these Tor nodes and family declarations as a vertex-edge graph. However, this time, each edge between two nodes, no matter it is bidirectional or unidirectional, is represented by an undirected edge in the graph. Based on the graph, JGraphT can help to discover a few connected sets. Each pair of Tor nodes from the same connected set have a relation that $n_i \approx n_j$. As a result, each pair of Tor nodes from the same connected set should have bidirectional family declaration between them. Tor nodes that fail to do so should be taken as making a Type I misconfiguration. These Type I misconfigurations can be corrected accordingly. Based on the Type I misconfiguration-corrected family declarations, we can draw a few $F^{(1)}$ families out using the same method as $F^{(0)}$ families. It's not hard to understand that, every $F^{(1)}$ family we find will correspond to a connected set mentioned above.

Type II misconfigurations and $F^{(2)}$ node families. The nickname, contact info and software version are taken into consideration when calculating the similarity between two Tor nodes. We use Jaro-Winkler similarity [13] to measure the nickname similarity and contact info similarity between two Tor nodes. We sorted Tor versions based on their release time and assigned a sequence number to each version. The version similarity between software versions of two Tor nodes is

---

[2] http://jgrapht.org/

Table 1 There are 85 $F^{(0)}$ families, 88 $F^{(1)}$ families and 115 $F^{(2)}$ families in Tor network, covering 10.4, 12.4 and 15.7 % of all Tor nodes at that moment

	Number of families	Number of Tor nodes
$F^{(0)}$	85	256 (10.4 %)
	88	305 (12.4 %)
$F^{(1)}$	115	386 (15.7 %)
$F^{(2)}$		

determined based on the distance between two versions' sequence numbers. Then, the similarity between two Tor nodes is calculated by the weighted sum of nickname similarity, contact info similarity and version similarity with a weight of 0.4, 0.4 and 0.2 respectively. The threshold δ is set to be 0.9, a very high value that only Tor nodes with almost the same configuration can exceed. Tor nodes with a similarity over this threshold are considered to have a Type II misconfiguration. We add extra family declarations for these misconfigured Tor nodes and correct Type I misconfigurations as mentioned before. Then, $F^{(2)}$ families can be found accordingly.

5.2 Results

Tor nodes with family misconfigurations. $|N^{(1)}| = 94$ and $|N^{(2)}| = 81$ Tor nodes are found to be making Type I and Type II misconfigurations respectively. Type I misconfigured Tor nodes compose 3.8 % of all Tor nodes in the network while Type II misconfigured Tor nodes compose 3.3 % of that. Compared with the number of all Tor nodes with a family declaration [343 (14.0 %)] in Tor network, the fraction is really large enough to cause our attention.

Family numbers and Coverage. Table 1 shows the number of $F^{(0)}$ families, $F^{(1)}$ families and $F^{(2)}$ families we draw from these 2,451 Tor nodes. The coverage of these families is also given in the table. It can be observed from the table that $|F^{(0)}| < |F^{(1)}| < |F^{(2)}|$ and node families cover about 10–16 % nodes in Tor network.

Examples of misconfigured families. As an example of Type I misconfiguration, we found a node family $F_i^{(0)} = \{$SKYNETDE1, SKYNETDE2$\}$ and its corresponding $F^{(1)}$ family $F_j^{(1)} = \{$SKYNETDE1, SKYNETDE2, SKYNETUS1$\}$. We looked into the family declarations that cause this result and found that: $F($SKYNETDE1$) = \{$SKYNETDE2, SKYNETUS1$\}$, $F($SKYNETDE2$) = \{$SKYNETDE1, SKYNETUS1$\}$ and $F($SKYNETUS1$) = \emptyset$. The Type I misconfiguration made by this volunteer in node SKYNETUS1 causes the difference between its $F^{(0)}$ family and the corresponding $F^{(1)}$ family. Most $F^{(1)}$ families that we found are caused by a similar reason as this one.

As for the Type II misconfiguration, we found a $F^{(2)}$ family $F_k^{(2)} = \{$Caldron, Caldron2, Caldron3$\}$ that all nodes in this family have the same contact address of "tor@liebt-dich.info", similar nicknames with the same prefix "Caldron" and software versions of 0.2.1.32 and 0.2.2.35 (which are released on the same day, Dec. 16, 2011). We looked into the descriptor files of these nodes and found that, none of them made any family declarations. Most $F^{(2)}$ families that we found are caused by a similar reason as this one.

The experiment over live Tor network data validated the formal analysis presented in Sects. 3 and 4. It also confirmed the existence of both Type I and Type II family misconfigurations in Tor network.

6 Conclusion and Future Work

In this paper, an analysis of family misconfigurations in Tor network is presented. According to the probably different causes, family misconfigurations are classed into two types. The formal definitions of both types of family misconfigurations are given, together with methods to discover and correct them. The experiment over live Tor network data shows that family misconfigurations prevail in Tor network. Both Tor operators and Tor users should pay attention to the family misconfigurations revealed in this paper. Beside, methods presented in this paper may not only help Tor operators to make family declarations correctly, but also help Tor clients to avoid potential risk when constructing circuits.

Both Type I and Type II misconfigurations studied in this paper are caused by Tor operator failing to list enough (if any) nodes in the "MyFamily" configuration. Misconfigurations caused by Tor operators listing more nodes in "MyFamily" than necessary are not considered. Actually, a malicious node who lists all other nodes in the "MyFamily" line may cause serious availability problem in Tor network. A sophisticated malicious family declaration can even take advantage of the "no multiple nodes from the same family" constraint of Tor path selection algorithm and induce Tor clients to select Tor nodes they provide with a high probability, proposing potential threats to Tor users' anonymity. We will further our research to include family misconfigurations of other types and to analyze family misconfigurations' influence on Tor network performance, availability and anonymity.

Acknowledgments This work is supported by National Natural Science Foundation of China (Grant No.6100174), National Key Technology R&D Program (Grant No.2012BAH37B04) and High Technology Research and Development Program of China, 863 Program (Grant No.2011AA010701).

References

1. Dingledine, R., Mathewson, N., Syverson, P.: Tor: The second-generation onion router. In: Proceedings of the 13th Conference on USENIX Security Symposium, vol. 13, pp. 21–21. USENIX Association, Berkeley, CA, USA (2004)
2. The Tor Project: Tor metrics portal: https://metrics.torproject.org/
3. The Tor Project: Tor Project: Anonymity Online. https://www.torproject.org/
4. Edman, M., Syverson, P.: As-awareness in tor path selection. In: Proceedings of the 16th ACM Conference on Computer and communications security, pp. 380–389. CCS'09, ACM, New York, NY, USA (2009)
5. Dingledine, R., Mathewson, N.: Design of a blocking-resistant anonymity system. Technical report, The Tor Project (2006)
6. Vasserman, E., Jansen, R., Tyra, J., Hopper, N., Kim, Y.: Membership-concealing overlay networks. In: Proceedings of the 16th ACM Conference on Computer and communications security, pp. 390–399. CCS'09, ACM, New York, NY, USA (2009)
7. Bauer, K., McCoy, D., Grunwald, D., Kohno, T., Sicker, D.: Low-resource routing attacks against tor. In: Proceedings of the 2007 ACM Workshop on Privacy in Electronic Society, pp. 11–20. WPES'07, ACM, New York, NY, USA (2007)
8. Murdoch, S.J., Danezis, G.: Low-cost traffic analysis of tor. In: Proceedings of the 2005 IEEE Symposium on Security and Privacy, pp. 183–195. SP'05, IEEE Computer Society, Washington, DC, USA (2005)
9. Tang, C., Goldberg, I.: An improved algorithm for tor circuit scheduling. In: Proceedings of the 17th ACM Conference on Computer and communications security, pp. 329–339. CCS'10, ACM, New York, NY, USA (2010)
10. McLachlan, J., Tran, A., Hopper, N., Kim, Y.: Scalable onion routing with torsk. In: Proceedings of the 16th ACM Conference on Computer and communications security, pp. 590–599. CCS'09, ACM, New York, NY, USA (2009)
11. Mittal, P., Olumofin, F.G., Troncoso, C., Borisov, N., Goldberg, I.: PIR-tor: Scalable anonymous communication using private information retrieval. In: USENIX Security Symposium. USENIX Association (2011)
12. Samudrala, R., Moult, J.: A graph-theoretic algorithm for comparative modeling of protein structure. J. Mol. Biol. **279**(1), 287–302 (1998)
13. Winkler, W.E.: String comparator metrics and enhanced decision rules in the fellegi-sunter model of record linkage. In: Proceedings of the Section on Survey Research, pp. 354–359 (1990)

Intelligent Heterogeneous Network Worms Propagation Modeling and Analysis

Wei Guo, Lidong Zhai, Yunlong Ren and Li Guo

Abstract As many people rely on Internet and Mobile network communications for business and everyday life, the integration of these two networks called heterogeneous network is unavoidable in the future. In this paper, we analysis an intelligent worms in Heterogeneous network called H-worm which is unlike scanning worms such as Code Red or Slammer. It spreads through transferring files with regulating computing resource consuming (CRC) to avoid user security awareness. To well study the H-worm and find the difference between H-worm and traditional worms, we present an H-worm simulation model, divided into connection probability, opening probability, host-defense and CRC. From the results, we give an empirical rule to keep a high propagation speed and some advises on defense H-worm.

W. Guo · L. Zhai (✉) · Y. Ren · L. Guo
Institute of Information Engineering, Chinese Academy of Sciences, Beijing, China
e-mail: nakusakula@gmail.com; zhailidong@cte.ac.in

W. Guo
e-mail: Guowei_apple@hotmail.com

Y. Ren
e-mail: renyunlong@iie.ac.cn

L. Guo
e-mail: guoli@iie.ac.cn

W. Guo
Beijing University of Posts and Telecommunications, Beijing, China

Y. Ren
Beihang University, Beijing, China

S.-S. Yeo et al. (eds.), *Computer Science and its Applications*,
Lecture Notes in Electrical Engineering 203, DOI: 10.1007/978-94-007-5699-1_52,
© Springer Science+Business Media Dordrecht 2012

1 Introduction

In the past, some theories used in Internet viruses can be found in Mobile network and have caused great damage. For example, Knysz et al. [1] form a mobile WiFi botnets on smartphone that can support rapid command propagation, with commands typically reaching over 75 % of the botnet only 2 h after injection sometimes. As Mobile network became more popular, viruses and worms absorbing the technology form Internet quickly evolved the ability to spread through the Mobile network by various means such as file downloading, e-mail, exploiting security holes in software, and so forth.

Currently, most serves of Mobile network could run on Internet. The integration of Internet and Mobile network defined as heterogeneous network is unavoidable in the future. Since the two networks user combined into one network, the intelligent Heterogeneous network worms (H-worm) will constitute one of the major network security problems. There has been an example of H-worm, Stuxnet which propagates by infecting removable drives and copying itself over the network using a variety of means in Internet and finally attacks the computer with industrial control systems (ICS) in Industry Network. Although Stuxnet does not strictly work on the Heterogeneous network we defined, it is able to propagate in two different networks. In order to avoid the detection and infect more nodes within a limit time, H-worm in this paper mainly spread through transferring files and regulate the computing resource consuming (CRC) to control the propagation speed to avoid detection.

In order to understand how H-worms works, we define H-Worm as a file attach with a piece of malicious code have the ability controlling CRC and propagates through transferring files by any serves the user may use on the heterogeneous network, such as www, email, Msn, Bluetooth. A user will be infected if he or she opens the abnormal files attached with H-worm and is unaware of the anomaly caused by H-worm. Once the user is infected, H-worm will regulate CRC to copy itself and attach other files with H-worm.

2 Related Work

In this section, we first introduce the modeling of worm propagation. Then, we explore work related to intelligent H-worm model.

Due to substantial damage caused by worms in past years, there have been significant efforts on modeling and detecting worms. As we know, there are mainly 5 models. First, the two-factor worm model [2] considers the two factors: Human countermeasures and Decreased infection rate. Second, The WAW model is a worm anti worm model (WAW) [3]. The model assumes that there are two kinds of worms occurring on the network, worm A and worm B. Worm A is malicious worm, and Worm B is anti-worm [4]. Second, to model the epidemic spreading on topological networks, Pastor Satorras and Vespignani [5] presented a differential equation for an

SIS model by differentiating the infection dynamics of nodes with different degrees. In this paper, we model the H-worm described in the fifth model with matrix and consider the human countermeasures in the two-factor worm model.

All these studies focus on traditional worms, while our study focuses on modeling the propagation of H-worm that could avoid the detection intelligently. There are two types of systems for worm detection: host-based detection and network-based detection. H-worm is primarily to against host-based detection. There are already some intelligent worms. The "Self-stopping" worms investigated in [6] are special cases of self-disciplinary worms [7]. Self-disciplinary worms could partition into two categories, namely static and dynamic self-disciplinary worms. "Self-stopping" is the static one. H-worm belongs to this dynamic intelligent worm. Although Self-disciplinary could be a dynamic worm, it only give the model of static one and left a lot of work on the dynamic intelligent worm. In this paper, we give the model of dynamic intelligent worm.

3 Modeling of the H-Worm

3.1 H-Worm

H-Worm refers to a document attach with a piece of malicious code that propagation through transferring files by any way the user may use on the heterogeneous network, such as www, email, Msn, Bluetooth.

The process of infecting also can be generally divided into four steps showed in Fig. 1. After a user receiving a file, it is scanned by anti-worm software. If the file doesn't pass the scanning, it will be deleted directly. When the passed file opened by user, user will adjudge the computing resource consuming. If it is suspected by the user, the new progress of the worm will be killed. Otherwise, we consider that the user trust this progress. Even another abnormal file opened on an infected node and suspected by the user, the user only kills the new progress and keeps the trust progresses. If the opened file is a normal file, it will continue receive file. On the other hand, H-worm runs with infected file opened and it will begin to control the computing resource. Finally, it continue receive file too. The process will be repeated in the whole network until all H-worm removed.

3.2 Connection Probability

In our paper, the topology of the heterogeneous network plays a critical role in determining the propagation dynamics of H-worm. Moreover, there is no an existent topology to describe this network which may change with time. In this paper, we first create a new model to describe the changing heterogeneous topology.

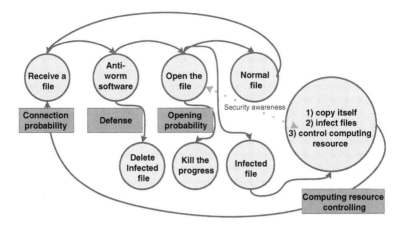

Fig. 1 Process of infecting

Considering the heterogeneous network consisting of internet and mobile network, and worm propagation exhibits significant difference between internet and mobile network, terminals in heterogeneous network is classified into two categories: Fix node (computer) and Mobile node (smartphone). Let diagonal matrix $M_{n \times n}$ denotes the Mobile node and diagonal matrix $F_{n \times n}$ denotes the Fix node.

We introduce a network proportion parameter called Np. The network proportion Np can be expressed as

$$Np = \begin{cases} \dfrac{\sum\limits_{i=1}^{N} M_i}{N} = \rho \\ 1 - \dfrac{\sum\limits_{i=1}^{N} M_i}{N} = 1 - \rho \end{cases}$$

where $0 \leq \rho \leq 1$ is the proportion of computer in heterogeneous network.

For the Internet, Recently Faloutsos et al. [8] The exponent α is obtained by performing a linear regression on P(k) when plotted 2002 AS-level topology where $\alpha = -2.18$. To keep it simple, the Internet in this paper is defined as a scale-free network with the degree distribution $P(k) \sim k^{-2}$ [9]. For the mobile network, Lambiotte al. [10] It is shown that the degree distribution in the mobile network has a power-law degree distribution $P(k) \sim k^{-5}$. Obviously, the degree distribution of heterogeneous network has a power-law exponent $-5 \leq \alpha \leq -2$ and α changed with network proportion Np. We denote that $\alpha = -2 - 3 \times \rho$ is the power-law exponent of heterogeneous network.

In this paper, we use the Generalized Linear Preference (GLP) power-law generator presented in [11]. We choose the GLP power-law network generator instead of other generators because it also has an adjustable power-law exponent η. The following is the formula of η.

$$\frac{2m - \beta(1 - p)}{(1 + p)m} = \eta$$

where $\beta \in (-\infty, 1)$ is a tunable parameter that indicates the preference for a new node (edge) connecting to more popular nodes. The smaller the value of β is, the less preference gives to high degree nodes. Due to the fact that GLP delete self-loops and merge duplicate edges. Then [11] demonstrated that $\eta = \alpha + 1$ approximately. According to the α, $\eta = -1 - 3\rho$.

Among the large number of connected nodes, which node the infected node would like to choose is a significant problem. We assume that there are n different nodes in the network. $L_1(n)$ is the state of node 1 connected to node n. $CN_{1(1 \times n)}$ is state of node 1 connected to all the nodes in the network. k_1 is the degree of node 1. Diagonal matrix $K_{n \times n}$ has all the degrees of the network. $NCP_{n \times n}$ is the network connected probability matrix.

Then, we calculate the connected probability which is different with [5]. We consider a node only transferring files to the other node that it connected. Therefore, the probability of node i connected to node j equals to the degree of node j divide the sum of degree of the nodes connected to node i.

So, $NCP = [CN_1''^T, \cdots, CN_n''^T]^T$. $NCP_{i,j}$ is the connected probability of node i connected to node j and always not the same with node j connected to node i.

3.3 Defense

In this paper, host-based detection systems play an important role on preventing H-worm propagation. As I know, there are already some anti-worm software developed for smartphone on Android and OS. Although most of them transplant the proven technology from computer to smartphone, they won't work as well as in the computer. Actually, it is a great challenge to detect worm rapidly and accurately on smartphone with the limited computing ability, low update frequency, new application such as SMS, MMS, Bluetooth brought by mobile network.

Therefore, we believe that H-worm will infect smartphone easier than computer under the monitoring of anti-worm software. In order to distinguish the anti-work software effects between computer and smartphone, we introduce a parameter Def(n) denoting the node n having a probability of undetected worm, $\alpha_m \geq \alpha_f$. Def is the undetected probability of the whole network.

$$\text{Def(n)} = \begin{cases} \alpha_m \text{ if } M_{n,n} = 1 \\ \alpha_f \text{ if } F_{n,n} = 1 \end{cases}$$

$$\text{Def} = \alpha_m * M + \alpha_f * F$$

3.4 Opening Probability

In this paper, the level of user awareness directly decides the probability of a worm successfully cheating user and infect the system. Therefore, qualifying the user awareness is one of the most significant studies of worm propagation model.

User awareness is a random variable that is very difficult to characterize due to user's individual nature. From the above description, we believe that as the increasing of CRC, user is more like to notice the system anomaly. We had tried some other equation, such as circle equation with radio equals two, quadratic equation and so on. Finally, we found that they are not much difference with linear equation. To keep it simple, Let $op_n(t)$ be the opening probability of node n at time t. $op_n(t) = 1-rs_n(t)$ where $rs_n(t)$ is computing resource consuming and $rs_n(t) \in [0,1)$. OP(t) is the opening probability of whole network, for calculating in the worm propagation model. $OP_{i,1}(t) = op_1(t)$ is the opening probability of node 1.

3.5 Computing Resource Controlling

Computing resource controlling (CRC) is a complex factor that has great impact on H-worm propagation speed. Let diagonal matrix $RS(t)$ be the abnormal files probability of the whole network. $RS_{i,i}(t) \in [0,1]$ is the sum of $rs_t(t)$ from time one to time t divide a constant C. We try to get a high propagation speed by regulating CRC according to user security awareness called greedy method. First, infect a normal node with a relative low CRC to avoid user security awareness. Next, H-worm increases the consuming. We create a rule to solve the problem by testing the user security awareness. The rule is to control the new progress consuming created by opening a new abnormal files equaling to the consuming which the H-worm will increase. If the increasing is noticed, only the new progress will be killed. Let $rs_n(t)=\mu + cou_n(t) \times \varphi$ where μ is the initial computing resource consuming, $cou_n(t)$ is the time of an infected opening the abnormal files, φ is the increasing computing resource consuming.

3.6 SIS Model

The most simple and popular differential equation model is the SIS epidemic model in complex network, which has been used by many papers [37]. However, the traditional model is not suitable for our model.

We present a matrix model, let $EINF_{n \times n}$ be expectation of infected probability shown below. $EINF_{i,j}$ is the probability that node j infect node i. If node i is infected, then $\sum_{j=1}^{j=n} EINF_{i,j} \geq 1$. If node i is cured, then $\sum_{j=1}^{j=n} EINF_{i,j}=0$. Let $IT_{n \times n}$

be the infected state of the whole network. If node 1 is infected, then $IT_{1,j}=1$, j is from 1 to n. So, $EINF(t + 1) = EINF(t) + EINF(t) \cdot NCP \cdot RS(t) \cdot OP(t) \cdot IT$

4 H-Worm Simulation Studies

To find the characters of H-worm propagation in heterogeneous network, we simulate the propagation on Omnet. We compare the H-worm spreading simulation results with different parameters. We make a default parameters list of heterogeneous network. N = 20,000, Np = 0.5, $\beta = 3.5$, m = 1, p = 0.5, $\mu = 0.1$ $\varphi = 0.05$, $\alpha_m = 0.1$, $\alpha_f = 0.1$.

4.1 Network Topology

We try to find the difference of worm propagation on Internet, mobile network and heterogeneous network. We assume that there are 10,000 nodes in the Internet and mobile network respectively and the node in heterogeneous network containing the all the fix nodes and mobile nodes is 20,000.

From the result showed in Fig. 2, it seems that worm propagation in internet is a little faster than in mobile network, but, it is not much difference between them. However, worm propagation in heterogeneous network is much faster than anyone. Although the number of heterogeneous network node is sum of internet and mobile network, worm propagation speed is faster than the sum of the propagation in single internet and mobile network as shown in the Fig. 2.

4.2 Greedy Method

The greedy method is decided by μ and φ. We try to work out a right μ and φ to keep the H-worm propagation at a high speed. Hence, our simulation can be divided into two steps: first, find which μ has the highest propagation speed when the φ keep at a very low value; Next, from the first step results, test the different φ to find the right μ and φ with the highest H-worm propagation speed.

First, we simulate H-worm propagation with μ from 0.2 to 0.8. The result showed in Fig. 3, when μ equals 0.4, 0.5 and 0.6, they are almost the same and have the higher propagation speed than others. Then we call them highest area $Area_\mu \in [0.4, 0.5, 0.6]$. If μ is less than the minimum of $Area_\mu$, the propagation speed will increase when μ grow. In addition, if μ is greater than the maximum of $Area_\mu$, the propagation speed will reduce sharply going with μ increasing.

Fig. 2 H-worm in heterogeneous network compared with single network

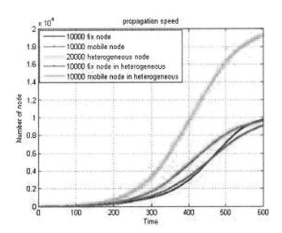

Fig. 3 H-worm propagation with various initial computing resource consuming μ

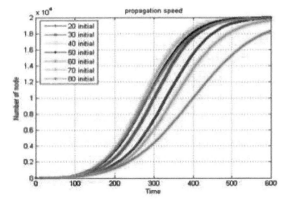

Fig. 4 H-worm propagation μ equals 0.3, and $\varphi \in [0.05, 10, 20, 30]$

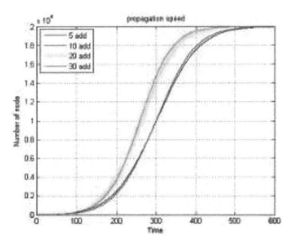

Fig. 5 H-worm propagation
μ equals 0.4, and
$\varphi \in [0.05, 10, 20, 30]$

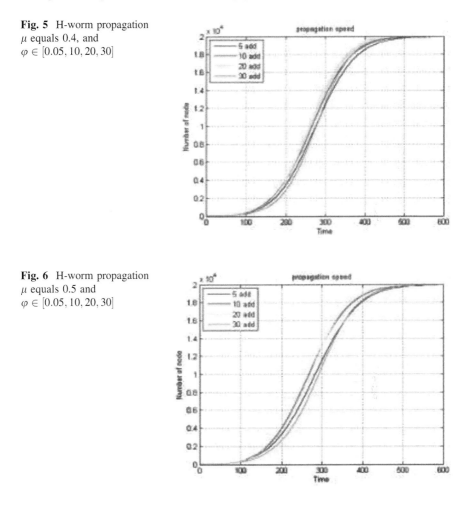

Fig. 6 H-worm propagation
μ equals 0.5 and
$\varphi \in [0.05, 10, 20, 30]$

Second, we test various φ with different μ. We only give the remarkable results of μ from 0.3 to 0.5 and φ from 0.05 to 0. 3. From Figs. 4, 5 and 6, we can draw the conclusion that $\mu + \varphi = \text{Max}(Area_{\mu})$ has the highest propagation speed in each Figure. Also, we can see that the propagation speed became faster when μ decrease. However, we limit the minimum of $\mu = 0.3$, because, user will be more sensitive to the computing resource consuming suddenly changing. Hence, we consider that H-worm with $\mu = 0.3$ $\varphi = 0.3$ has the highest propagation speed.

From the above description, first, no matter what is the user awareness equation, find $Area_{\mu}$; second, set a minimum μ and calculate φ by $\mu + \varphi = \text{Max}(Area_{\mu})$.

5 Conclusion

In this paper, we get three conclusions. First, a worm like H-worm outbreak in heterogeneous network, the worm will cause greater damage than traditional one. Second, we get an empirically rule to pick up CRC parameters with relative highest propagation speed.

Acknowledgments This work is partially supported by 863 National Hi-tech Research and Development Program (2011AA01A103).

References

1. Knysz, M., Xin, H., Yuanyuan, Z., Shin, K.G.: Open WiFi networks: Lethal weapons for botnets? In: Proceedings IEEE INFOCOM, 2012, pp. 2631–2635 (2012)
2. Zou, C.C., Towsley, D., Gong, W.: On the performance of Internet worm scanning strategies. Perform Eval **63**, 700–723 (2006)
3. Pastor-Satorras, R., Vespignani, A.: Epidemic dynamics and endemic states in complex networks. Phys. Rev. E **63**, 066117 (2001)
4. Yu, Y., Liqiong, W., Fuxiang, G., Wei, Y., Ge, Y.: A WAW model of P2P-based anti-worm. In: IEEE International Conference on Networking, Sensing and Control, ICNSC 2008, pp. 1131–1136 (2008)
5. Pastor-Satorras, R., Vespignani, A.: Epidemic spreading in scale-free networks. Phys. Rev. Lett. **86**, 3200–3203 (2001)
6. Ma, J., Voelker, G.M., Savage S.: Self-stopping worms. In: Presented at the Proceedings of the 2005 ACM workshop on Rapid malcode, Fairfax, VA, USA (2005)
7. Wei, Y., Nan, Z., Xinwen, F., Wei, Z.: Self-disciplinary worms and countermeasures: modeling and analysis. In: IEEE Transactions on Parallel and Distributed Systems, vol. 21, pp. 1501–1514 (2010)
8. Faloutsos, M., Faloutsos, P., Faloutsos, C.: On power-law relationships of the Internet topology. SIGCOMM Comput. Commun. Rev. **29**, 251–262 (1999)
9. Albert, R., Barabási, A.-L.: Topology of evolving networks: Local events and universality. Phys. Rev. Lett. **85**, 5234–5237 (2000)
10. Lambiotte, R., Blondel, V.D., de Kerchove, C., Huens, E., Prieur, C., Smoreda, Z., Van Dooren, P.: Geographical dispersal of mobile communication networks. Physica A **387**, 5317–5325 (2008)
11. Tian, B., Towsley, D.: On distinguishing between Internet power law topology generators. In: Twenty-First Annual Joint Conference of the IEEE Computer and Communications Societies. Proceedings IEEE INFOCOM 2002, vol. 2, pp. 638–647 (2002)

The Design of Remote Control Car Using Smartphone for Intrusion Detection

Chang-Ju Ryu

Abstract This paper proposes the use of remote control car for intrusion detection. The scheme concerns the use of a camera, remote control car, TCP server and smartphone. The arrangement allows us to observe any given location in real time, through a camera mounted on the RC car. The RC car is controlled by a computer via Bluetooth and the commands to the computer can be sent using a smartphone application through a TCP network. As an example application the arrangement was used to observe a residence by obtaining images from there at specified intervals. Application of Bluetooth for computer to RC car communication has allowed the achievement of maximum communication speed and we have also shown that the proposed scheme has less BER.

Keywords Smartphone · Remote control · Bluetooth · BER

1 Introduction

Current security systems employ video cameras for observing remote locations through a designated security centre [1–3]. They can be used to periodically take pictures of the location for assessing the situation, or to know if there have been any intrusions. However, in most such systems the camera is fixed and so has a limited field of vision [4–6]. There are always areas which lie outside the cameras range. Some implementations circumvent this problem by using multiples cameras, each

C.-J. Ryu (✉)
Department of Information and Communication Engineering,
Chosun University, Gwangju, Korea
e-mail: rcjlove@naver.com

S.-S. Yeo et al. (eds.), *Computer Science and its Applications*,
Lecture Notes in Electrical Engineering 203, DOI: 10.1007/978-94-007-5699-1_53,
© Springer Science+Business Media Dordrecht 2012

Fig. 1 Recently used camera for crime prevention

one observing different regions. But this can be very expensive, as it requires more equipment. This study proposes the use of a mobile RC Car for removing the problem of cost. A camera is mounted on top of an RC Car so that it can take security pictures and send it to a remote server. Since the field of vision of the camera is limited, for those areas that lie outside the camera's range, the RC Car itself is moved so that the camera points at the desired location. The RC car communicates with the server through Bluetooth technology in the 2.4 GHz band range. The camera on the RC Car can take pictures at designated intervals of time, and send to be stored in the remote server via bluetooth. The remote server is also connected to a smartphone through a TCP network, so that the image may be viewed in the smartphone as well. Furthermore the smartphone can be used to send commands to the RC Car via the remote server, through a designated application installed on the smartphone. A BER (Bit Error Rate) analysis performed has also shown that the BER of this setup is very low and is in the acceptable range of 10^{-5}. In this paper, Sect. 2 gives an introduction to the related preliminary study. Section 3 describes the proposed scheme in detail, including the hardware and the software parts. Section 4 gives a performance evaluation of this scheme while Sect. 5 concludes the paper.

2 Related Study

At present, camera based crime prevention modules have a general setup as shown in Fig. 1.

Although commercial camera modules can be rotated, the range of rotation is not very large. Also, since the camera is immobile, there will always be the presence of a region that the camera cannot cover. Such a region is called a blind spot. To eliminate such blind spots, many camera modules can be set up, each one observing a different direction. However, such a setup will be more costly.

Real time monitoring of these cameras by the client through a PC or notebook is also possible, but the drawback is that computer needs to be around the client, which may not always be the case.

The system proposed in this paper has been specifically designed to overcome such limitations. Unlike current CCTV systems which utilizes a fixed camera, the proposed system uses a camera attached to a mobile RC Car in order to avoid blind spots. Moreover as this allows a single camera to cover multiple locations the cost of implementation is also reduced.

In addition, real-time monitoring through a Windows based client PC or laptop can be replaced by monitoring through a smartphone. This has the advantage that the client will be able to monitor the camera from any location and need not have access to a computer. In other words, the proposed system gives mobility to the client as well. The RC Car is further equipped with ultrasonic sensors for rear end collision detection which helps to improve its safety.

3 Proposed Scheme

3.1 System Architecture

The system basically consists of two parts: the hardware part and the software part. The hardware part consists of the RC Car, it's sensors, camera, micro controller module. It is mainly concerned with initialization of camera which sets the image size and resolution, the actual image capture as well as transfer of this captured image wirelessly through bluetooth.

The second part of the system is the software part. It mainly consists of the TCP server running on a Windows Computer and the GUI provided to the client through a smartphone application. The software part basically concerns itself with receiving the image data from the bluetooth module as well as transfer of control signals from the smartphone client to the RC Car.

The flowchart in Fig. 2 shows the general outline of the hardware operation. Similarly Fig. 3 shows the block diagram of the overall system architecture.

3.2 Hardware Part

The camera module is connected to an AVR micro controller through an I2C bus. It is first initialized, after which image data can be obtained through eight lines from the camera module. The eight lines are connected to eight AVR GPIOs. Figure 4 shows an outline of the general structure of the camera module and its connection with the AVR micro controller.

Physically, the AVR micro controller board, motors and ultrasonic sensors are located at the rear end of the RC car. If AVR sends trigger signal to ultrasonic module, it generates a high frequency sound which gets reflected back from surrounding obstacles. This high frequency echo is detected by the ultrasonic sensors. The time duration between the transmission of the signal and the reception

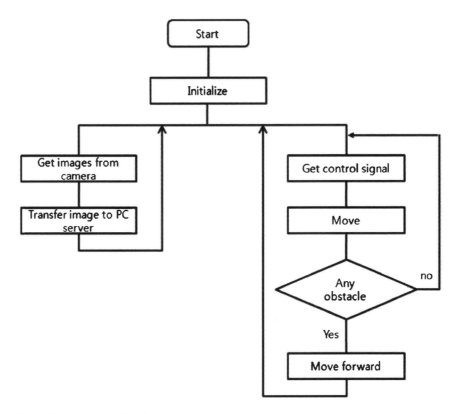

Fig. 2 Hardware operation flowchart

of the echo can be analyzed by the AVR micro controller to estimate the distance between the RC Car and the obstacle. The RC Car utilizes two stepper motors for its motion. AVR interfaces with the stepper motor using two L297 Stepper motor driver ICs. When the micro controller sends control signals and clock signal (PWM) to L297, it generates 4 phase signals and sends them to the stepper motor operation module which in turn drives the stepper motors. The image data from the camera is sent to bluetooth module through an 8-bit bus which is received by the AVR micro controller. Figure 5 shows the block diagram of the connection between the AVR micro controller, bluetooth module and the stepper motors.

The size of a single image taken from the camera is given by Eq. 1 while the total time required for transferring this image via bluetooth is given by Eq. 2.

$$
\begin{aligned}
input\ data &= 120 \times 160 \times 60 \\
&= 307,200\ \text{bits}
\end{aligned}
\tag{1}
$$

$$
\begin{aligned}
response\ time &= 307,200 \div 230,400\ \text{bps} \\
&= 1.33\ \text{sec}
\end{aligned}
\tag{2}
$$

Fig. 3 System architecture

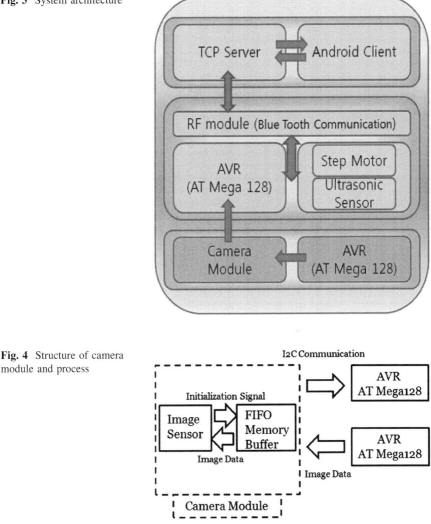

Fig. 4 Structure of camera module and process

3.3 Software Part

The software section of this project consists of TCP server and android client which communicate with each other as shown in Fig. 6. The TCP server is constructed using MFC and is used to send image data from the camera module to the smartphone as well as receive control signals for the RC Car from the smartphone. Data transfer between the TCP server and the android client is achieved through TCP/IP socket communication. The server is a Windows application based on dialog boxes. The GUI (Graphical User Interface) consists of sections for viewing image and a Log printing the details of communication and transmission [7, 8].

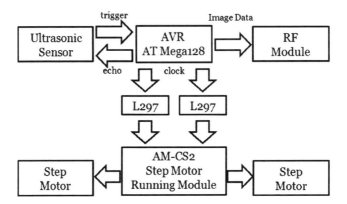

Fig. 5 Structure of AVR, motor and process

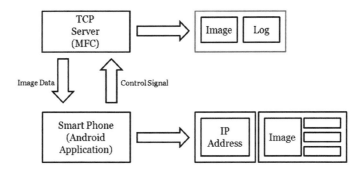

Fig. 6 Structure of software

The Android Client application consists of two activities. The first activity takes as input the IP address of the server, and opens the second activity. The second activity GUI shows the transmitted image as well as buttons for sending control signals for the movement of the RC Car.

After the server obtains images from the RC Car camera through bluetooth, it sends them to the client through TCP/IP socket communication. The socket communication transmission procedure can be summarized through the skeleton code given below:

```
if(readStart){
POSITION pos;
pos = m_ListenSocket.m_ptrClientSocketList.GetHeadPosition();
CClientSocket* pClient = NULL;
while(pos ! = NULL)
{
pClient = (CClientSocket*)
m_ListenSocket.m_ptrClientSocketList.GetNext(pos);
```

Fig. 7 Remote RC car for intrusion detection

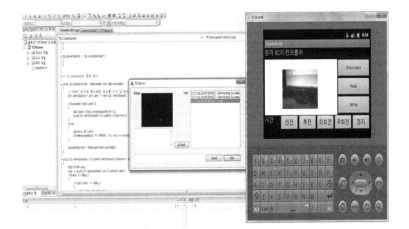

Fig. 8 Using MFC server

```
if(pClient ! = NULL)
{
pClient- > Send(ImageData, sizeof(ImageData));
```

4 Performance Evaluation

Figures 7, 8 Errors in the data packets may occur during data transmission between bluetooth devices while other wireless communication systems in the surrounding area can interfere with the data transmission as well. This paper does not consider the other interruptive agents in real life environment such as other RF devices that may be operating nearby. However a detailed analysis of the Bit Error Rate (BER)

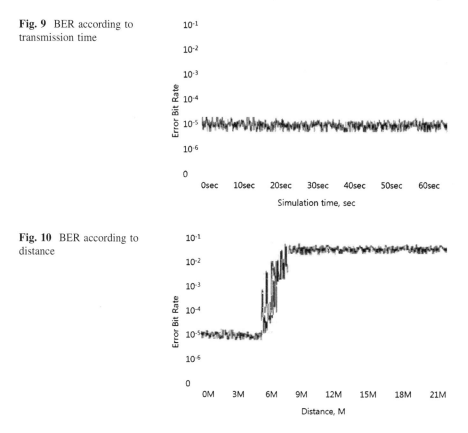

Fig. 9 BER according to transmission time

Fig. 10 BER according to distance

observed during data transmission was performed. Figure 9 shows the graph of BER against transmission time in ideal condition and Fig. 10 shows BER against distance between RC Car and TCP Server.

As we can see from Fig. 9 the value of BER is 10-5 on average throughout the sample transmission time of 60 s. This implies that the system did not encounter any significant loss of data. Again, from Fig. 10 we can see that as long as the separation between the RC Car and the TCP server is less than 6 m, the BER remains almost constant at the acceptable value of 10-5. However as the separation is increased the BER also increases rapidly.

5 Conclusion

The main advantage of this system is that it gives mobility to the security camera, as it is mounted on top of a mobile RC Car. As present systems utilize cameras that stay at a fixed position, they cannot remove presence of blind spots completely. But in this system, the camera can be used to observe any direction as long as the

RC Car has freedom to move. Since a single camera can be used to monitor many different directions, this system is also much cheaper than setups that require the use of multiple cameras to improve the field of vision. The communication between the RC Car and the server was achieved wirelessly through Bluetooth which makes the data transfer significantly faster. Image transfer between camera hardware and the server via bluetooth was seen to be achieved at a high speed of 230,400 bps. At this rate it was possible to transfer one image in 1.33 s, which is a relatively good rate. BER analysis was also done on the basis of distance of the RC Car from the remote server. An acceptable value of BER is 10–5, and it was seen that this was achieved as long as the separation was not more than 6 m. However stable image could still be obtained for a separation of up to 9 m. The system also allowed for efficient remote control of the RC Car. Instead of having to be at a computer terminal for operating the car, a designated smartphone App is used, which allows the client to control the RC Car through the server, as well as view the security images transmitted by the camera. This has the advantage of making not only the camera mobile, but the client mobile as well. Hence the system is cheaper and more convenient to implement.

Needless to say, as with any new system, this proposed system has its disadvantages as well. For example the paper does not consider environmental interferences in bluetooth communication. Various devices operating around the vicinity of the RC Car may be interfering with its communication with the server. So as a future improvement, further study can be done to reduce the environmental interferences on the system.

References

1. ETS 300 328: Radio Equipment and Systems (RES); Wideband transmission systems; Technical characteristics and test conditions for each transmission equipment operating in the 2.4 GHz ISM band and using spread spectrum modulation techniques, 1996
2. Burgeoning Bluetooth, IDC, April 2000
3. Bluetooth: Update on the Wireless Link for Mobile Computer, May 8 2000
4. Khan J.Y., Wall, J.: Bluetooth-based wireless personal area network for multimedia communication. IEEE Computer Society 2002, pp. 47–51
5. Kim, B-K.: A method to support high data transmission rate in ad-hoc networks based on bluetooth. ITC-CSCC **3**, 507–5008
6. Bluetooth Specification 1.0B, http://www.bluetooth.com
7. Bluetooth Special Interest Group, Specification of the Bluetooth system Version 1.1B, Specification Vol. 1&2, Profiles, February 2001
8. Wireless Portable Devices : World Market for 2G, 2.5G, and 3G Devices and Connectivity to the Wireless Internet, Allied Business Intelligence, 1Q 2001

Part II
Computerized Applications for the Sustainable Development of Aviation Industry

The Impact of Systems Engineering on Program Performance: A Case Study of the Boeing Company

Samuel K. Son and Sheung-Kown Kim

Abstract This study provides clear empirical evidence that Systems Engineering (SE), when implemented by Program Management (PM), drives improvements in program performance. During a three-year interval, SE processes were systematically integrated into The Boeing Company's Program Management Best Practices (PMBP) on the multi-purpose airlifter program. Given SE processes providing a foundation for high-quality PM, program managers heavily rely upon project team members to consistently implement SE processes throughout the project lifecycle to ensure successful program performance. The study's results indicate that a strong and positive correlation exists between the implementation of SE processes and program performance improvements.

Keywords Systems engineering · Program management · Program management best practices · Technical performance measurement · Integrated product team

1 Introduction

Program management (PM) is the centralized coordinated management of a program to achieve the program's strategic objectives and benefits [1]. It is the formal process of developing, managing, and administering processes and resources while

S. K. Son
Department of Systems Engineering, The Boeing Company, Huntington Beach, CA, USA
e-mail: samuel.k.son@boeing.com

S.-K. Kim (✉)
Graduate School of Management of Technology, Korea University, Seoul, Korea
e-mail: kimsk@korea.ac.kr

S.-S. Yeo et al. (eds.), *Computer Science and its Applications*,
Lecture Notes in Electrical Engineering 203, DOI: 10.1007/978-94-007-5699-1_54,
© Springer Science+Business Media Dordrecht 2012

utilizing the pertinent infrastructure to ensure efficient, effective, and compliant program execution [2].

Originally, The Boeing Company identified the PMBP (14 best practices) as a formal means of assessing process documentation and process implementation for program management as well as sharing program experiences with Integrated Product Teams (IPTs). These IPTs are cross-functional (i.e., multi-disciplinary) groups of people who are tasked with delivering a product either internally to their organizations or externally to the customer. In this way, each organization is provided the authority to proactively design its method of collaboration [3]. The IPT approach requires a team-based organizational structure characterized by open communication among team members, access to evolving system design information, clear and concise communication channels to PM, and commitment to the program rather than commitment to the functional office [4].

Program management best practices (PMBP) independent assessment results provide PM with one of the most powerful analysis tools to gain a better understanding of how well a program is currently functioning and identifying areas for future improvement [2]. In this case, the assessment provided the program manager with a "fresh pair of eyes" for looking at the inner workings of the program. By utilizing an assessment built on the PMBP, independent reviewers provided feedback on how consistent the practices were applied, what worked well, and where improvements could directly lead to opportunities to increase business results.

2 SE Elements in PMBP

SE and PM are the two groups within these program environments that ensure that the program's infrastructure is correctly defined and implemented in a healthy way. Systems engineers are also critical to making sure that the programs meet their customer needs. From this perspective, the main SE objective is to complete the requirements analysis and then allocate the requirements to the various parts of the system [5–8]. Therefore, it is important for program managers and IPT leaders to understand the SE shared vision and implement the SE processes.

The SE elements provide a framework to support several elements of the PMBP. The SE's *Requirements Analysis* element supports the *Requirements Definition* and the *Baseline Management* best practices of the PMBP to help the program team identify customer needs, define requirements, and manage requirement changes. The *Requirements Analysis* element establishes a means to manage requirements development, allocation, traceability, and verification throughout the program's lifecycle to include the customer and suppliers' requirements. The clear flow-down of all applicable requirements to suppliers is a critical part of the *Requirements Definition* and the *Baseline Management* best practices of the PMBP. The SE's *Synthesis and Integration* element supports the *Supplier Integration* best practice of the PMBP through the development of suppliers' teaming agreements, establishment of early commitments to maximize flexibility, and the implementation of interface requirements management [9]. Early alignment with suppliers' strategies and incentives assists the program manager to better manage performance issues and identify supplier risks. Suppliers participate in the Configuration Management and the Change Management Control processes to document, control, and maintain accurate subcontract baselines.

Finally, the SE's *System Analysis and Control* element integrates the *Risk, Issue and Opportunity (RIO) Management* [10–12] best practice with the *Integrated Plan and the Integrated Schedule* best practices of the PMBP, and utilizes the TPM process to conduct program performance measurements [13]. The RIO processes draw from team members across the entire organization, which include the customer and suppliers, to identify and manage risks, issues, and opportunities.

3 Results and Discussion

3.1 PMBP Data Collection and Assessment

The Boeing Company required all divisions to conduct the PMBP assessment yearly and focused on two assessment criteria: "Process Documented" and "Process Implemented." The best practice score was defined in accordance with the following criteria:

Score 1: Processes are not documented and not implemented sufficiently to ensure basic program execution.

Score 2: Processes are partially documented and partially implemented with no improvement plans.

Score 3: Processes are partially documented and partially implemented but with effective improvement plans.

Score 4: Processes are appropriately documented, but implementation needs improvement.

Score 5: Processes are documented and implemented to the extent necessary to ensure successful program performance.

Prior to conducting the PMBP assessment, it was essential for the assessment team to have a detailed understanding of the program. About 3 weeks prior to the assessment, the assessment team researched and analyzed program documentation and plans, such as the Program Execution Plan (PEP), Team Execution Plans (TEPs), Integrated Business Plan (IBP), and RIO Management Plan.

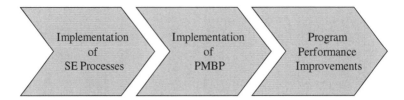

Fig. 1 Contingent effects of implementing the SE processes in the PMBP on the program performance

3.2 Improvements of Program Performance by Implementation of SE

One of the main purposes of implementing the SE processes in the PMBP is to reinforce Boeing standard program management procedures.

As shown in Fig. 1, the effects of SE on program performance are contingent upon and mediated by the PMBP. In this model, SE support is defined as the number of SE processes contained in the PMBP, with the strength of SE support increasing in direct proportion to their number. Therefore, one could hypothesize that the improvement in program performance should be proportional to the number of SE processes implemented in the PMBP.

In practical terms, a stronger *Requirements Analysis* element increased SE involvements with the customer as well as requirements traceability with the suppliers, as shown in Table 1. Before 2007, PM mainly concentrated on requirements definition with design engineers in system development with little SE involvement. Needless-to-say, this led to the ineffective implementation of program's requirements management including requirements development, allocation, and change management. Similarly, increasing the *Synthesis and Integration* element resulted in a greater focus on interface management. Finally, the traditional *System Analysis and Control* element tended to emphasize scheduling, risk management, and design reviews. The assumption was that the requirements were already established and merely had to be managed. This element actually lessened in importance over this time interval, as it was discovered that the assumption was faulty, and stronger requirements and traceability processes were needed.

3.3 Results

Figure 2 compares system reliability and maintainability performance improvements for the pre- and post-PMBP intervals. The regression equations of reliability in the pre-PMBP period are mean time between maintenance, inherent (MTBMI) = $-0.0264T + 52$, mean time between maintenance, corrective (MTBMC) = $-0.1569T + 314$, and mean time between removal (MTBR) = $-$

Table 1 Percentage (%) changes of the SE attributes from 2007–2009

SE Element	% Change of SE attribute		
	2007 (%)	2009 (%)	Delta (%)
Requirements analysis	18	44	26
Synthesis and integration	21	33	12
System analysis and control	61	23	−38

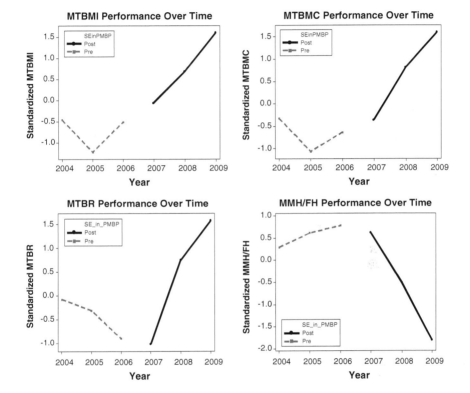

Fig. 2 Regression graphs for pre- and post-PMBP

0.4129T + 827, while the regression equation of maintainability in the pre-PMBP period is maintenance man-hours per flight hour (MMH/FH) = 0.2428 T–486. In the other hand, the regression equations of reliability in the post-PMBP period are MTBMI = 0.83209T–1670, MTBMC = 0.9804T–1968, and MTBR = 1.2976T–2605, while the regression equation of maintainability in the post-PMBP period is MMH/FH = −1.214T + 2437. The pre-PMBP slopes are statistically indistinguishable from zero, but the post-PMBP slopes are statistically significant. Therefore, the system design improvements driven by the Program Management had a positive influence on system reliability and maintainability. The results are summarized in Table 2.

Table 2 Regression analyses for pre- and post-PMBP

Mean time between maintenance inherent			Mean time between maintenance corrective		
PMBP	β	p	PMBP	β	p
Pre-PMBP	−0.0264	0.961	Pre-PMBP	−0.1569	0.725
Post-PMBP	0.83209	0.041	Post-PMBP	0.9804	0.073
Mean time between removal			Maintenance man hour per flight hour		
PMBP	β	p	PMBP	β	p
Pre-PMBP	−0.4129	0.154	Pre-PMBP	0.2428	0.121
Post-PMBP	1.2976	0.132	Post-PMBP	−1.214	0.024

Table 3 T test results for pre- and post-PMBP slopes

Mean time between maintenance, inherent (MTBMI)					
SE status	Intercept	Slope	SE(Slope)	t	p
Pre-PMBP	52.00	−0.026	0.427	1.994	0.092
Post-PMBP	−1670.00	0.832	0.053		
Mean time between maintenance, corrective (MTBMC)					
Pre-PMBP	314.00	−0.157	0.340	3.176	0.043
Post-PMBP	−1968.00	0.980	0.113		
Mean time between removal (MTBR)					
Pre-PMBP	827.00	−0.413	0.102	5.888	0.014
Post-PMBP	−2605.00	1.300	0.272		
Maintenance man-hours per flight hour (MMH/FH)					
Pre-PMBP	−486.00	0.243	0.047	21.986	0.001
Post-PMBP	2437.00	−1.210	0.047		

A *t* test [14–18] was performed to assess whether these pre- and post-PMBP slopes were significantly different. The test results are summarized in Table 3, and indicate that the differences were significant at the 0.10 level. In Table 3, the standard error for the beta coefficient (SE slope) is included as an estimate of precision. This T test supports that there were significant changes in the rates at which system reliability and maintainability were improving.

3.4 Discussion

The implementation of SE in the PMBP shows that it has a strong positive influence on program performance. In order to achieve these results, three key lessons about PM were required.

The first key lesson is that the program-level TPMs must be the main communication means between the program manager and the IPTs. The program-level TPMs are considered fundamental to PM specifically because they enable the program manager and the IPTs to focus on the most important contractual

Table 4 Percentage (%) changes of the PMBP attributes from 2007–2009

Attribute	% Change of Attribute		
	2007 (%)	2009 (%)	Delta (%)
Requirements-related	28.4	37.7	9.30
Cost-related	8.2	19.9	11.70
Schedule-related	3.0	11.1	8.10
Other	60.4	31.3	−29.10

requirements. In the case of the multi-purpose airlifter program, MTBMI, MTBMC, MTBR, and MMH/FH TPMs constituted the primary contractual requirements. In 2008, the Requirements Definition best practice was added and became a key best practice of the PMBP for the systematic requirements flow-down to all of the relevant suppliers.

The next key lesson is that the Integrated Plan and the Integrated Schedule best practices have been integrated with the RIO Management best practice. By being received emphasis of the process integration, they enabled increased visibility and risk management in PM.

The final key lesson is the Leadership & Accountability best practice that has been added in the 2008–2009 time frame. In 2007, the PMBP contained no Leadership best practice. The Leadership & Accountability best practice received a perfect assessment score, which reflects the involvement of the Boeing Leadership Team (LT). The Boeing LT's involvement became a thrust that improved all of the other best practices and provided the foundation for the performance improvements described in this paper.

In addition, the most prominent contribution to the program performance improvements in PM was an increase of requirements-related attributes. For example, the 2008 PMBP Model emphasized the need to better understand customer and supplier-based requirements and the allocation of system-level requirements down to the subsystem-level and component-level. The new model also indicated the importance of developing a requirements allocation process and requirements traceability matrices to be presented to the customer at design reviews.

Some may think that the effect of cost-related and/or schedule-related attributes is equal to the effect of requirements-related attributes regarding program performance. However, Table 4 shows that the cost-related attributes increased from 8.2 to 19.9 %, and the schedule-related attributes increased from 3.0 to 11.1 %. Although the cost and schedule-related attributes increased in the PMBP, these attributes could not have contributed to the increase of technical performance improvements. On the other hand, the attributes in the "Other" category decreased from 60.4 to 31.3 %, which included: business plan, organization, work break-down structure, change management, supplier management, import/export compliance, and communication. These "Other" category attributes could not contribute to the technical performance improvements either. Therefore, the results of the study indicated that a positive correlation existed between the increase of requirements-related attributes and program performance improvements.

4 Conclusion

This paper has demonstrated that the implementation of SE in PM exerts a strong positive effect on the program performance. The paper showed significant reliability improvements in mean time between maintenance, inherent (MTBMI), mean time between maintenance, corrective (MTBMC), and mean time between removal (MTBR) and also significant maintainability improvement in maintenance man-hours per flight hour (MMH/FH). The results of the study proved the hypothesis that implementing the SE processes in the PMBP improves the program performance.

These system reliability and maintainability improvement measurements represent technical performance measurements of the program. However, since schedule and cost also measure the program performance, this paper should be expanded in the future to include them as well. Specifically, how have the SE processes affected the schedule and cost performance of the program after they were implemented into PM?

Acknowledgments The authors thank Boeing's Department of Public Relations/Communications for permission and use of the multi-purpose airlifter program as a case study, MDC10K0041. The work of the 2nd author was supported by the Korea University Grant.

References

1. Project Management Institute: A Guide to the Project Management Body of Knowledge (PMBOK Guide), 4th edn. Global STANDARD (2008)
2. The Boeing Company: Program Management: A Reference Guide Written From the Experience of Program Manager (2009)
3. Zhao, X., Liu, C., Yang, Y., Sadiq, W.: Aligning collaborative business processes-an organization-oriented perspective. IEEE Trans. Sys. Man Cybern. A Sys. Hum. **39**(6), 1152–1164 (2009)
4. Roe, C.L.: Project management and systems engineering in an IPD environment. In: 6th Annual International Symposium of the NCOSE (1996)
5. Blanchard, B.S.: System Engineering Management, 4th edn. Wiley (2008)
6. Blanchard, B.S., Fabrycky, W.J.: Systems Engineering and Analysis, 2nd edn. Prentice-Hall, Inc. (1990)
7. Grady, J.O.: System Requirements Analysis. McGraw-Hill, Inc. (1993)
8. Huang, C., Kusiak, A.: Modularity in design of products and systems. IEEE Trans. Sys. Man, Cybern. A Sys. Hum. **28**(1), 66–77 (1998)
9. Qiu, R.G., Zhou, M.: Emerging approaches to integrating distributed, heterogeneous, and complex systems. IEEE Trans. Sys. Man Cybern. A Sys. Hum. **36**(1), 2–4 (2006)
10. Chapman, C.B., Ward, S.C.: Project Risk Management: Processes, Techniques, and Insights. Wiley, Chicester (1997)
11. Grabowski, M., Merrick, J.R.W., Harrald, J.R., Mazzuchi, T.A., René van Dorp, J.: Risk modeling in distributed, large-scale systems. IEEE Trans. Sys. Man, Cybern. A Sys. Hum. **30**(6), 651–60 (2000)
12. Larson, N., Kusiak, A.: Managing design processes: A risk assessment approach. IEEE Trans. Sys. Man Cybern. A Syst. Hum. **26**(6), 749–759 (1996)

13. Jiang, J.J., Klein, G.: User evaluation of information systems: By system typology. IEEE Trans. Sys. Man Cybern. A Sys. Hum. **29**(1), 111–116 (1999)
14. Casella, G., Berger, R.L.: Statistical Inference, 2nd edn. Duxbury (2002)
15. Mendenhall, W., Sincich, T.: A Second Course in Statistics: Regression Analysis, 6th edn. Pearson Education, Inc. (2003)
16. Montgomery, D.C.: Design and Analysis of Experiments, 6th edn. Wiley (2005)
17. Paternoster, R., Brame, R., Mazerrolle, P., Piquero, A.: Using the correct statistical test for the equality of regression coefficients. Criminology **36**(4), 864–866 (1998)
18. Larsen, R.J., Marx, M.L.: An Introduction to Mathematical Statistics and Its Applications, 3rd edn. Prentice Hall (2001)

IT Framework and User Requirement Analysis for Smart Airports

Sei-Chang Sohn, Kee-Woong Kim and Chulung Lee

Abstract An airport is a physical site where aircrafts take off and land, and passengers head to their destinations through a series of arrival and departure processes. But it takes considerable time to go through the immigration process, security checks, and other airport-associated processes. As a result of reviewing airport processes in order to increase airport productivity and improve the passenger convenience, we divide passenger-related processes into three sub-processes, namely u-Fast Passenger, u-Guidance, and mobile services. To better explain a series of improvements in the process, a new paradigm of "Smart Airport Service Framework" is presented. This delivers a faster and more convenient service for passengers and improves the work process and resource utilization. Consequently, passengers are enabled to enjoy and be entertained by the airport facilities with this way of saving time, while the airport authority can enjoy additional profit.

Keywords Business process · Smart airport · Quality function deployment

S.-C. Sohn
Incheon International Airport Corporation, Incheon, Korea and Aviation Business
Administration, Korea Aerospace University, Gyeonggido, Korea
e-mail: scsohn@airport.kr

K.-W. Kim (✉)
Aviation Business Administration, Korea Aerospace University, Gyeonggido, Korea
e-mail: scsohn@yahoo.com

C. Lee
Division of Industrial Management Engineering and Graduate School
of Management of Technology, Korea University, Seongbuk-gu, Seoul 136-701, Korea
e-mail: leecu@korea.ac.kr

S.-S. Yeo et al. (eds.), *Computer Science and its Applications*,
Lecture Notes in Electrical Engineering 203, DOI: 10.1007/978-94-007-5699-1_55,
© Springer Science+Business Media Dordrecht 2012

1 Introduction

An airport is the place where airlines' passengers come to reach or return from a destination through a range of processes such as being permitted to depart or enter the physical location through the landing and departing of airplanes. It takes considerable time for those processes to be enacted. Therefore, many airports are making an effort to simplify the procedures of entry and departure and thereby provide passengers with not just convenience but enjoyable experiences at the airport.

The most important thing in improving the airport business is to streamline those processes that are related to the passengers. Further, current immigration procedures need improvements. By streamlining these processes, it will be possible for passengers to pass through the procedure promptly, thereby providing them with faster and more convenient service including guidance and improves the business process of the airport and maximizes the utilization of the resources. To do so, it is necessary to introduce and apply the Smart Airport framework, which is a comprehensive plan built around cutting-edge smart technology that improves the customer satisfaction and enhances the competitiveness of the airport. This method can be considered as one way for the airport industry, which is in the midst of keen competition, to have a Sustainable Creating Competitive Advantage.

The management of the airport is most efficiently operated when associated companies including governmental organizations, airline companies, travel agencies, and private companies are connected and operated organically. To further this cooperation, information technology (IT) is increasingly being implemented, although mostly only as regards partial aspects [1], such as dealing with self-check-in kiosks. Clearly, as regards the entire airport, this is insufficient. Therefore, in this study, we transform and establish the integrated view about the airport passenger service, and develop the concept of using smart technology in the airport [2] into the concept of the Smart Airport, in respect of passenger service and airport resource management.

This study, by suggesting the Smart Airport Framework, which will not only improve the process of fulfilling the airport's mission, but also be helpful for both airport and passengers, will develop the way to overcome the possible obstacles during the process and apply up-to-date technology.

2 Analysis of Airport Business Processes

An airport is a facility designed for the processing of airplanes, passengers, and freight, which is illustrated in Fig. 1 in the form of an airport process model. Focusing on the international component of this process, the main processes are composed of departure and arrival processes.

In the case of passenger arrival, the airplane lands and has been guided to the boarding gate, after the passengers have deplaned, the airplane is either taking off

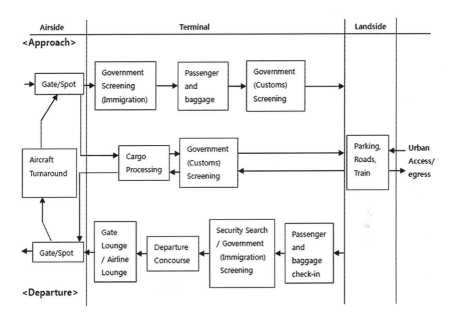

Fig. 1 Airport process model and the flows of the baggage

or making a turnaround [3]. Passengers, after going through the governmental screening such as immigration, retrieving the baggage from the baggage reclaim, and getting out of the terminal, then finally head to their final destination using surface transportation.

In the case of the departure process, after arriving at the terminal by the ground transportation, passengers obtain their boarding passes at the check-in counter, handle their baggages, and then enter into the screening zone, and they pass through the security. After that, the passengers pass through the governmental screening such as quarantine, customs, and immigration, go through the duty-free shopping area and get on a plane at a gate and then depart to their destination.

The improvement of a given process is attained by removing from it any duplications and by automating it to increase the efficiency of the business. Toward this end, the purpose of Business Process Management (BPM) is to build the system to manage this kind of process improvement [4]. The target for the improvement is where a high number of manual activities or many interactions between fields exist, which require to be automated or to cooperate with other fields so the process can be improved [5]. Accordingly, our study will examine those parts. When examining the airport business, the passenger service process is regarded as the front-end business and the plane handling process as the back-end business. Thus, this study examines the passenger process from the perspective of the customer.

Considering the current main process of the airport (Fig. 1), the main obstacle is the constraints of time and space that occur as part of handling the airport business process. Also, the passengers' flows are identified through manual activities and

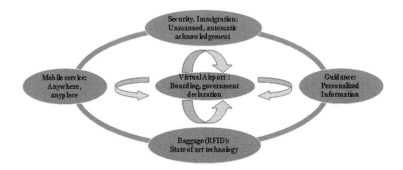

Fig. 2 Smart airport: conceptual design

baggage is processed by barcodes, which are used to identify the flow of baggage. In parallel with that, when departing, security search and immigration are performed respectively and the immigration screening is made by face-to-face meeting. As each process is handled, the location changes constantly, which makes the guidance information system even more important. To solve the obstacles of the restraints in the airport process such as constraints of time and space, security search/face-to-face screening in Immigration, and the management of baggage by barcodes, we present some measures in Fig. 2.

3 Smart Airport Service Framework

To achieve these improvements, this study suggests the new paradigm of the airport management, the Smart Airport Service Framework, in the field of airport passenger service, which provides a faster, more convenient, and secure journey for the passengers by actualizing the simplification of the departing procedure using state-of-the-art technology, and gives a convenient experience with simplified boarding procedures, Security Search, and Immigration through utilizing the ubiquitous technology and integration of services. This framework will minimize the unnecessary time in the airport and generate spare time in the airport for the creation of added value. The concept of a smart service system, which provides information proactively whenever the user needs it, is applied to the two categories, the guidance and the fast passing through. Therefore, this concept is embodied into an information system that supplies information automatically and speeds up the system. More than that, it is developed not into a blueprint concept with only partial application but into a more enhanced system in the form of the framework. The following are the consequences of reviewing the application of the framework.

As a result of examining the areas for improvements, a u-cyberterminal processes the online items in advance, which previously have been done at the airport. Subsequently, the only aspect of checking into be dealt with at the airport

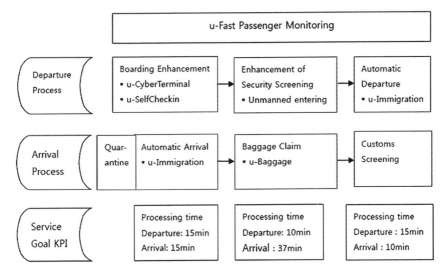

Fig. 3 Process management of u-fast passenger

is the Security Search. And the other domains consist of u-SelfCheckin, which is possible for all airlines, and is operated unmanned; u-immigration, providing an unmanned immigration system; u-Baggage, offering real-time location service using the radio-frequency identity (RFID) technology; u-Board, which offers personalized information such as flight times, weather, immigration, parking, etc.; u-signage, providing customized information automatically for each user; and Mobile service including the supply of individualized information and all kinds of process functions without the constraints of location or space. These three, taken together, are comprehended as the services for improvements.

In respect of the process, it can be divided into u-Fast Passenger, u–Guide, and mobile service. U-Fast Passenger supports the process by conducting the prompt passing of the passengers onward and is composed of the process covering self-service and the non-face-to-face paperless system shown in Fig. 3.

Also, u-Guide includes the domain that supports the information-related process and is a generally complex portal, which consists of sections applying additional mobile technology that must be developed constantly.

The International Civil Aviation Organization (ICAO) has the recommended standard for immigration, which sets sixty minutes for departure and forty-five minutes for arrival. Considering that time, the service goal key performance indicator (KPI) [6] is set and managed on the process that requires more time. The improved processes such as boarding enhancement (handling airport business in advance + u-SelfChec-kin), enhancement of security screening (unmanned entering-departing zone) and automatic departure (u-Immigration) in the departure process, and quarantine, automatic arrival (u-immigration), unmanned entering-departing zone, baggage claim (u-Baggage), and Customs screening in the arrival process should be managed so as to promote the improvement in the passenger service level.

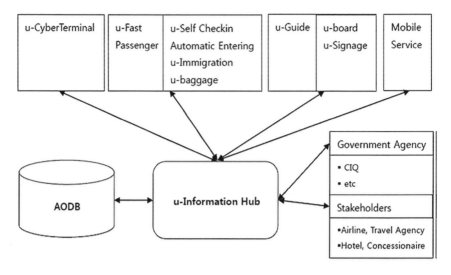

Fig. 4 Smart airport service framework

The most important part of this process improvement is the u-Information Hub, which is equivalent to the Airport Operation Database (AODB), which centralizes all the information in the airport operation domain. The main concept of this u-Information Hub is to collect and provide the information based on the AODB, government agencies, and airport-related companies, and to enable the management of real-time passenger information. Therefore, it is connected to the u-cybertermi-nal, which offers the government agency an information service, airline service, and travel agencies service, duty-free service, roaming service, insurance service, and additional service of all kinds. These are utilized in pre-travel service and later integrated into each level of security and immigration information service.

Figure 4 is the Framework as the blueprint [7] applying each corresponding process. This stage, first of all, will represent the realization of the Virtual Airport (Cyber Terminal), which handles the boarding procedure and delivers all kinds of governmental declarations online at once. Secondly, the simplification and auto-mation of Immigration and boarding procedures using bio-data and passport, and integration and automation of the departing procedure is also realized. Thirdly, with the collaboration and reinforcement of the main information regarding passengers, it becomes possible to utilize the shared information among the related organizations.

4 Analysis of Customer Requirements for Smart Technologies

With the expansion of smart devices, the smart device users demand various kinds of applications. Accordingly, many companies offer users applications for smart devices. Banks have started to offer banking service for smart devices, while libraries, cinemas, restaurants, etc. are also presenting their applications and thus various kinds of applications are pouring onto the market. However, for Incheon International Airport, the development of smart device application is not fully completed yet. Currently, most applications are those provided by each airline company, while the application offered by the airport itself is just a link to the airport webpage. And since mobile service part is urgently needed in the Airport Service Framework, an analysis of smart device applications for the airport is required.

It is important to capture the smart device user's needs to increase the ease of use. As a means to grasp these needs, the quality function deployment (QFD) technique [8, 9] is employed to comprehend the diverse demands of smart device users at the airport and to try to create a high-quality smart device application that satisfies their demands. Also, the claims from customers gained through the QFD will be reflected thoroughly onto the new development of applications by the systematization and establishment of the database.

The survey was conducted with 269 people by questionnaire about the necessary airport-related applications. Two hundred and fifty-nine valid voice-of-customer (VOC) questionnaires were collected; among them, 197 were male and 62 were female, the age range was 189 people younger than 20, 58 in the 20s, and 6 in the 30s, so most of them often use smart devices. As regards whether they have smart devices or not and the intent of purchasing one, 91 % of respondents expected to have smart devices. Also, as a result of asking whether they used airport applications or had the intent of using, 195 respondents, which were around 75 % of the sample, responded that they have used it and 37 people, around 14 %, said they are going to use it, so overall, 89 % of this population is expected to use the application. Through the source information stage such as this VOC survey and the development of SCENE, we made the customer requirements' development table including the detailed items and decided the customers' important needs by using AHP [10, 11]. Lastly, the application for development is drawn from the quality characteristics development and computing the weight of the plan attributes. Through KJ law [12, 13] using the VOC survey deducted from the survey, the weights of the primary items were drawn as 5 for quality, 5 for function, and 4 for design through the 5 points of the Likert scale survey. The weights of the secondary items are determined by AHP law (9-point Likert scale); as a consequence, it is shown that in the design part, an interface with great legibility is the most important, in the quality part the supply of accurate information is the most important, in the function aspect, the flight time table and the internal/external facilities of the airport are the most important (see Table 1).

The functions necessary to the smart device application for the customers were the flight timetable as the highest, the location information such as internal airport facilities information (Airlines, Duty-free, Q&A, etc.) and external airport facilities information (Accommodations, surrounding areas, the sights) came second, followed by E-ticket and transportation information. Unlike the existing information

Table 1 "Voice of the customer" analysis

Primary	Secondary	Evaluation by customers[a]
Design	1.1 Interface with great legibility	(4*4*0.31) = 7.44
	1.2 Easy operation	6.48
	1.3 Larger operating button	4.32
	1.4 Etc.	5.76
Quality	2.1 Quick supply of the information	7.35
	2.2 Provide accurate information	12.25
	2.3 Minimization of program capacity	4.9
	2.4 Proper number of functions(4–6)	4.9
	2.5 Etc.	5.6
Function	3.1 E-ticket	6.4
	3.2 Flight timetable	19.2
	3.3 Internal airport facilities information (Airlines, duty-free, Q&A, etc.)	11.2
	3.4 External airport facilities information (accommodations, surrounding areas, the sights)	11.2
	3.5 Airport-connected transportation information	16
	3.6 Information guide for the airport	5.6
	3.7 Information about the destination (the airport, weather, tour, etc.)	4.8
	3.8 Real-time location service and airport service navigation	4
	3.9 Interpretation service customized to the destination	8.8
	3.10 Etc.	12.8

[a] (Weights of primary items × Number of secondary items × Weights of secondary items) = Evaluation by customers

system, a smart device application requires a more direct expansion of convenience, for instance, the flow of the floating population in the interior space, figuring out of the customer's location and the guide for the best way to get to the amenities the customer wants.

5 Conclusion

After examining the departure, arrival, and transfer processes, which comprise the main airport passenger processes in order to increase the airport profit through enhancing the efficiency of the airport and improving the customer satisfaction, we find some restrictions regarding the constraints of time and space or the face-to-face screening in all kinds of screenings. The aspect of speeding up and improving the efficiency in passenger service should also be considered. Among these airport processes, we find the area requiring improvement and optimization. The requirements of customers associated with IT system are defined as the Smart Airport Service Framework.

Also, by limiting the application of the up-to-date technology to the smart device, which is the dominant technology nowadays, related application requirements are analyzed by QFD. The resulting analysis found that flight information, location information, and transportation information are the main applications. This has great meaning in that it defines the Smart Airport Service Framework, which is important for competitiveness in the airport industry and deduces the requirements of the latest technology in general, and demonstrates it is necessary to develop the system and prove the results based on these consequences. Since Smart Airport consists of smart airport service and smart airport operation, it is considered that further study is needed for establishing the Smart Airport Operation Framework to operate as the system coordinating the whole airport operation.

Acknowledgments This research was supported by Basic Science Research Program through the National Research Foundation of Korea(NRF) funded by the Ministry of Education, Science and Technology(2009-0076365).

References

1. Jarrell, J.: Self-service Kiosks: Museum pieces or here to stay? J. Airpt. Manage. **2**(1), 23–29 (2007)
2. Massink, M., Harrison, M., Latella, D.: Scalable analysis of collective behaviour in smart service systems. In: SAC '10 Proceedings of the 2010 ACM Symposium on Applied Computing, pp. 1173–1180, New York (2010)
3. Lindh, A., Andersson, T., Varbrand, P., Yuan, D.: Intelligent air Transportation—a resource management perspective. In: Proceedings of the 14th World Congress on Intelligent Transport Systems (2007)
4. Crich, N., Waterston, M.: Developing an IT strategy for a growing regional airport. J. Airpt. Manage. **3**(4), 328–336 (2009)
5. Smith, H., Neal, D., Ferrara, L., Hayden, F.: The emergence of business process management. Technical report, CSC's Research Services (2002)
6. IIAC & KMA consultants Inc.: Something different IIAC. KMAC (2010)
7. Rohloff, M.: An integrated view on business—and IT-architecture. In: SAC '08 Proceedings of the 2008 ACM Symposium on Applied Computing, pp. 561–565, New York (2008)
8. Carnevalli, J.A., Miguel, P.C.: Review, analysis and classification of the literature on QFD—types of research, difficulties and benefits. Int. J. Prod. Econ. **114**(2), 737–754 (2008)
9. Chan, L.-K., Wu, M.-L.: Quality function deployment: A literature review. Eur. J. Oper. Res. **143**(3), 463–497 (2001)
10. Triantaphyllou, E., Mann, S. H.: Using the analytic hierarchy process for decision making in engineering applications: some challenges. Int. J. Ind. Eng. Appl. Pract. **2**(1), 35–44 (1995)
11. Saaty, T.L.: Decision making with the analytic hierarchy process. Int. J. Serv. Sci. **1**(1), 83–98 (2008)
12. Cheng, Y.-M., Leu, S–.S.: Integrating data mining with KJ method to classify bridge construction defects. Expert Sys. Appl. **38**(6), 7143–7150 (2011)
13. Scupin, R.: The KJ method: A technique for analyzing data derived from Japanese ethnology. Hum. Org. **56**(2), 233–237 (1997)

The Framework Development of a RFID-Based Baggage Handling System for Airports

Chang-gi Kim, Kee-woong Kim and Youn-Chul Choi

Abstract This paper introduces a study of a Baggage Handling System (BHS) interface solution and the expected effects of applying a new Radio Frequency Identification (RFID) technology in the Incheon airport baggage handling facilities. In addition, this study suggests that the RFID application could improve the use of human and material resources. For example, this could lead to a decrease in the amount of delayed baggage by increasing the tag-read rate, improvements in convenience by providing baggage information, improvements in security screening efficiency, and a reduction in Minimum Connecting Time (MCT).

Keywords RFID (Radio frequency identification) · BHS (Baggage handling system)

1 Introduction

The Baggage Handling System (BHS) is a cutting-edge complex automated logistics system that transports and sorts passengers' departure/transfer baggage according to flight destination by recognizing bag tags automatically. The aviation

C. Kim
Incheon International Airport Corporation 424-47 Gonghang-gil,
Jung-gu, Incheon 400-700, Korea
e-mail: changkkr@yahoo.co.kr

K. Kim
Aviation Business Administration, Korea Aerospace University, Gyeonggido, Korea
e-mail: kimkw@kau.ac.kr

Y.-C. Choi (✉)
Division of Aeronautical Studies, Hanseo University, Seosan city,
Chungcheongnamdo 356-705, Korea
e-mail: pilot@hanseo.ac.kr

S.-S. Yeo et al. (eds.), *Computer Science and its Applications*,
Lecture Notes in Electrical Engineering 203, DOI: 10.1007/978-94-007-5699-1_56,
© Springer Science+Business Media Dordrecht 2012

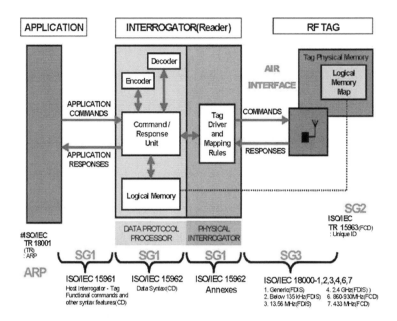

Fig. 1 RFID system configuration

industry is making a steady effort to prevent bag mishandling through the use of advanced technology, new BHSs, and process improvement. Radio Frequency Identification (RFID) is applied to BHS to automate the baggage handling process by attaching RFID tags to passengers' baggage at check-in time. RFID technology can reduce baggage handling errors, because it has a very high recognition rate for baggage tags compared to barcodes, which are commonly used in airports.

The introduction of RFID is a good solution to reduce delayed baggage and reduce the number of baggage handlers required. This will contribute to reducing airline costs, helping them remain competitive through business improvements, and improving the credibility of Incheon Airport.

2 Implementation of RFID Technologies for the Airport BHS

RFID is used to identify and recognize individual items by electronic devices using radio frequencies or magnetic fields. RFID consists of three parts: the tag (the identification device attached to the tracked item), the reader (which recognizes and reads data from the tag), and the software (which assigns meaning to the data from the reader) as shown in Fig. 1.

Table 1 Comparison of RFID and barcode technologies

Identification method	RFID tag	Barcode
Max. data volume	Kbytes	Bytes
Max. communication distance	0–100 m^3	50 cm
Security ability	Difficult to duplicate/falsify	Easy to duplicate/falsify
Tolerance for contamination	High	Low
Batch reading	Easy	Difficult
Cost (chip)	0.5 USD~	Little or nothing
Standardization	In progress	International
Recycling	Possible	Impossible
Penetration	Possible	Impossible

Two major identification technologies, RFID and Barcodes, are analyzed, as shown in the following Table 1.

The biggest difference between barcodes and RFID is the method of information readout. Scanners have to read all the barcode information, and information input must be done manually. In contrast, RFID can recognize a large amount of information, and input can be automated. In addition, barcodes must be operated at close range. However, because RFID is able to recognize information using various forms of radio frequency, efficient processes and improved productivity can be realized. In addition, RFID can modify or re-enter input information, while barcodes cannot. In terms of quantity of input and output information, RFID can process dozens of kilobytes, while barcodes can process only limited information. One weakness of RFID is the expense. However, when barcodes and RFID are compared in terms of re-use, damage, utilization, and lifecycle, RFID is more economical. Reading labels from a barcode scanner and recognizing tags from RFID are physically similar, but a review of the two shows that there are many differences in functionality [1].

2.1 Airport Baggage Handling Process

At the airport, checked-in baggage goes through security checks, is transported to Early Baggage Storage (EBS[1]) if necessary, and is sorted according to final destination (lateral) in time for departure. Transfer baggage is entered from the transfer input station and sorted to the final destination after security checks. In the case of arrival baggage, it is normally transported from the input station to the reclaim carousel using the point-to-point method. The general baggage transportation route is presented in Fig. 2 [2].

---

[1] EBS: At Incheon Airport, bags checked in BHS more than three hours earlier than ETD (Estimated Time of Departure) are stored in EBS. When ETD becomes less than three hours, bags are automatically sorted to the final destination (lateral).

Fig. 2 General baggage transportation route

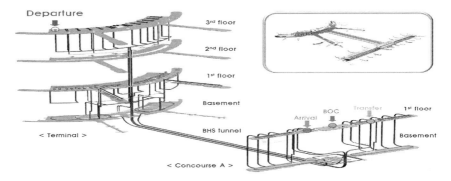

Fig. 3 Baggage-handling system at Incheon International Airport

2.2 Overview of the Incheon International Airport BHS

At the Incheon International Airport, as shown in Fig. 3, BHS is now made up of Phase 1 BHS facilities in the main terminal and Phase 2 in the concourse and connecting tunnel. Phase 1 is made up of a conveyor system and a tilt-tray sorter system. Phase 2 consists of a conveyor system and a high-speed system. The total length of BHS is 88 km with a commercial speed of 420 m/min (7 m/s).

2.3 Baggage Handling Process Using RFID Technologies

The baggage handling process using RFID is illustrated in the following Fig. 4.

When passengers leave baggage at the counter of the airport when departing, a bag tag for baggage treatment with the RFID reader is issued. This bag tag collects tracking information through the RFID reader installed in the baggage handling process. RFID is embedded in the bag tag, and RFID printers issue bag tags and

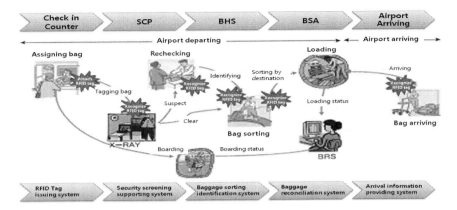

Fig. 4 Baggage handling information delivery concept by RFID [3]

record representative baggage numbers. This part is done in conjunction with the airline transportation system and shares information about baggage owners. After the passenger checks in the baggage, the baggage goes through x-ray inspection equipment installed in the baggage input counter's end-conveyor to inspect dangerous goods in baggage. For baggage for which classification is completed, the RFID reader once again checks for accuracy by recognizing the bag tag when loading baggage to containers to the baggage destination. When the aircraft arrives at the airport, the baggage is transported to the baggage claim where passengers are waiting to take the baggage. In this area, by installing RFID readers, passengers are informed of whether the baggage has arrived.

2.4 RFID Interface with the Existing BHS

The interface between the RFID tags and the existing BHS is illustrated in Fig. 5.

The RFID system linked to the existing system can be divided into three parts: hardware, network, and database. In order to review the introduction of RFID systems, linkages with existing systems for continued operation must be considered. In the case of the hardware, this is simple if the server used or existing products employ the same equipment as the RFID system construction. However, if the equipment used to build the RFID system differs from the existing equipment, the equipment connected via middleware for heterogeneous linkages in the system must be secure. In addition, when using RFID systems to collect data, the data must be organized and utilized. Depending on the role of existing information systems, if it does not work between the existing database and the data obtained by applying the RFID system, it will be possible to associate through middleware by creating a view type of existing data RFID system that must be built.

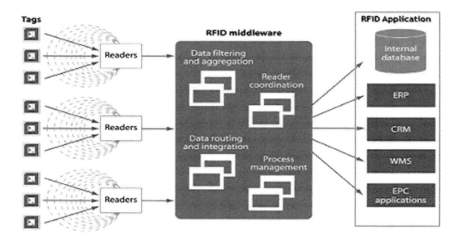

Fig. 5 RFID middleware concept

Table 2 Overview of RFID trial read rates

RFID trial	Date	Read-rate (average)	Read-rate range (%)
Kuala Lumpur Airport	2005–2006	With Gen 2 : 100 % With Class 0 Gen 1 : >98 %	
Kansai Airport-Hong Kong Airport	2005	98.78 %	94.25–100
Asiana-Korean Airport Corporation	2004–2005	97.00 %	–
TSA World-wide Trial	2004–2005	∼99 %	96–100
Narita Airport (HF)	2004	–	92–95
British Airways at Heathrow T1	1999	96.40 %	95.4–99.4

2.5 Previous Implementation Results at Other Airports

The tag-recognition rate of RFID technology is higher than that of barcodes used in airports, as shown in Table 2. Therefore, the likelihood of incorrect baggage handling is lowered significantly. The introduction of RFID at Lisbon Airport in Portugal increased air transfer baggage handling processing speed 66 % and reduced transfer baggage handling processing time to 10 min compared to 30 min in the past. In addition, baggage handling mistakes were reduced 50 %. In the case of Chek-Lap-Kok Airport, the baggage read rate accuracy previously averaged 80 %, but since the introduction of RFID tags, the accuracy has increased to 96 %. In terms of improving work efficiency, it results in 5 % more baggage handling, and in terms of processing baggage automatically, it reduces handling time. Below are the test results of RFID read rates [4].

2.6 Advantages From the Application of RFID

- **Rise in tag recognition rate**

The RFID tag-recognition rate is much higher than that of scanners. Thus, compared with barcodes, RFID can reduce time in treatment [e.g., unrecognized tag Manual Encoding System (MES)]. Therefore, it can shorten the processing time, thereby reducing the number of aircraft loading baggage mistakes, and baggage tracking can significantly reduce the rate of lost luggage.

- **Reduced misclassification errors during operations**

If the stub tag is not removed from past baggage or the stub tag is attached from a baggage conveyor belt during transport, this may result in unclassified baggage. Otherwise, because of the high recognition rate, RFID does not require separate stub tags. Through RFID tags and the reader attached ULD, the data can be transmitted to the server whether the baggage is sent or not without ground handling identification, and remote data inquiries are possible.

- **Additional customer services**

Through the tracking technology, RFID enables customers to obtain baggage handling information directly from their mobile phone. Accurate information about baggage handling delivery relieves passengers' anxieties.

- **Improved security efficiency and enhanced tracking for customs purposes**

By installing RFID readers at arrival entry gates, it is possible to pinpoint the position of suspicious baggage and the location of passengers when baggage arrives in the arrivals hall, thereby increasing the efficiency of personnel management in the arrivals hall.

- **Reduced MCT with rapid baggage handling**

Transfer-baggage handling is one of the most important factors in reducing MCT[2]; a great deal of transfer baggage from the airport of departure has a very bad state of the barcode. In addition, in the process of the transfer, the Baggage Source Message (BSM[3]) or tags of some baggage may be deleted or lost. Such luggage is

---

[2] MCT: The shortest time interval needed to transfer from one flight to a connecting flight (Source: OAG).

[3] BSM: Information used for baggage sorting. Airline's host server sends it to BHS server at the time of check-in.

put into the MES and handled manually by employees. RFID adoption is considered a solution to these problems.

- **Streamlining of maintenance personnel and ground handling staff**

At the Incheon Airport baggage handling facilities, 98 out of 520 members (18.8 %) of the maintenance staff are MES operators, which is a high ratio. Due to labor-management issues and labor costs, it is a huge burden on business. In addition, after baggage is sorted, the ground handlers read the bag tags and check to ensure the information from check-in matches. If we introduce RFID and increase the read rate, we will enable smooth baggage handling in spite of minimal operating MES. In addition, we will be able to remotely manage airborne information without a BHS operation, which could lead to the streamlining of maintenance personnel and ground handling staff.

2.7 Future Research Issues for the Optimal RFID-Based BHS Design

It is necessary to examine investment costs of the introduction of RFID in the current situation. In addition, we need to review the incidental benefits of RFID adoption. The cost of replacing barcode scanners unit is a good example. Barcode scanners unit is very expensive ($74,000 unit) and last 5 years. Thus, replacing 75 barcode scanners unit installed at Incheon Airport comes at a huge cost. The introduction of RFID will reduce replacement costs.

At Incheon Airport, airport authorities contend that introducing RFID will lead to $2.26 million in annual cost savings and that the total investment of $6.3 million will be recouped within approximately 2.6 years. RFID tags currently cost about $0.174 (and this price is decreasing). However, they are still more expensive than barcode price tags (i.e., $0.052) [1]. Although Incheon Airport is fully adopting RFID, it is expected to lead to high operating costs for airlines. In the case of Hong Kong Airport, passengers pay for the RFID tags, and a reasonable method that considers the long-term perspective needs to be devised.

3 Conclusion

The technical preparation required, benefits, and limitations of the new RFID technology for BHSs have been discussed.

First, an analysis of the differences between RFID and the barcode scanners currently being used at Incheon Airport BHS clarified the benefits of RFID. RFID systems last longer than barcode scanners, require less maintenance and replacements, and involve a shorter baggage processing time because of the high tag-recognition rate of RFID.

Second, effects other than those associated with shortened process time were analyzed. These included preventing misclassification due to errors in operation, providing additional customer services, enhancing customs security and tracking features, and streamlining maintenance personnel and ground handling staff.

Third, an analysis of the limitations of RFID adoption led to a potential future strategy. Airlines' unwillingness to support RFID and the high price of RFID tags were presented as problems to be solved as soon as possible.

Despite the many advantages of RFID, RFID adoption in the airport industry has only begun recently. The main reason is that RFID adoption relies on the plans of the airlines, and the standardization of the technology has not yet been established. In order to introduce RFID successfully, alliances between airlines and between airports need to be in place. Technical standards and a decline in RFID tag prices are also required.

Although current RFID tag prices are declining gradually, they are still more expensive than barcode tags. Thus, if RFID is introduced, it is expected to lead to high operating costs for airlines.

References

1. Jung, D.-W.: A Study on the RFID adoption for efficiency improvement of baggage handling at Incheon International Airport. Korea Aerospace University Graduate School of Aerospace Management (2012)
2. SITA: Baggage report 2010. Technical report (2011)
3. Chae, G.-M.: A Study on the RFID applied to the shipping. Konkuk University Graduate School of Information and Communication (2007)
4. IATA: http://www.iata.org/pressroom

A Modified Carbon Calculator for Enhanced Accuracy, Reliability and Understandability

Gun-Young Lee, Kwang-Eui Yoo, Kee-Woong Kim and Bo-Myung Kim

Abstract Devices for calculating the carbon dioxide emissions from flights have been developed by several agencies. However, these carbon calculators introduce many assumptions to simplify the calculation process. As a result, the pre-existing calculators have deficiencies in general to be used as consumer references. This study assesses carbon calculators for aviation emissions and suggests a modified calculation methodology using the pre-existing computer reservation system. The new methodology is more accurate and sophisticated to support consumer participation in the carbon offset program or selection of more environmentally friendly airlines or routes in online reservation.

Keywords Carbon calculator · Emissions · Climate change · Flight fuel burn

G.-Y. Lee
Ministry of Land, Transport and Maritime Affairs, Graduation School, Business Administration, Korea Aerospace University, Gyeonggido, Korea
e-mail: airsafe@korea.kr

K.-E. Yoo (✉)
Korea Aerospace University, Gyeonggido, Korea
e-mail: keyoo@kau.ac.kr

K.-W. Kim
Korea Aerospace University, Gyeonggido, Korea
e-mail: kimkw@kau.ac.kr

B.-M. Kim
Hanyoung Foreign Language High School, Seoul, Korea
e-mail: kathbmk@gmail.com

S.-S. Yeo et al. (eds.), *Computer Science and its Applications*,
Lecture Notes in Electrical Engineering 203, DOI: 10.1007/978-94-007-5699-1_57,
© Springer Science+Business Media Dordrecht 2012

1 Introduction

Air travel has become an essential part of global society. Air transportation is not one of the main contributors of greenhouse gases (GHG); however, the expectation of growth of emissions in the medium and long term is problematic. The Kyoto Protocol, which was signed in 1997 and entered into force in 2005, represents the most complete framework for a global emissions trading scheme. The Kyoto Protocol, however, excludes international air transportation from its targets; hence, many stakeholders in international aviation, such as individual states, groups of states (e.g., the EU), and international organizations (e.g., ICAO, IATA), have been arguing and struggling with the issues of carbon reduction and trade, especially represented as an EU-ETS [1]. Accordingly, there are many fundamental issues to be solved relating to the reduction of aircraft's CO_2 emissions; these issues need to be discussed in the political and diplomatic arenas. However, the present study is focused on a carbon calculation methodology in online reservation systems. Also, this study proposes a modified aircraft's CO_2 emission calculator and an associated methodology for carbon calculation.

2 Carbon Emissions Calculating Methodology

2.1 Function of the Carbon Emission Calculator

The actual amount of carbon emissions can be measured after the flight by multiplying the conversion factor with fuel burned. This number will be utilized for market-based measurement schemes like the EU-ETS; however, there are some countermeasures to reduce carbon emissions in a proactive manner. The aviation carbon offset program is one of them. The IATA believes that many flying public want to know what the industry is doing to mitigate its climate change impacts, and, increasingly, they want to know what they can do help [2]. It is necessary to provide carbon emissions information to the public who want to neutralize it by contributing required funds to offset program provider when they purchase flight tickets. As a result, consumers can have pertinent data about their planned flights and use it when they choose airlines for their travel. To implement such an important program, carbon calculation in advance without actual fuel burning figures is required.

2.2 Factors Influencing Carbon Emissions

There are various factors that influence carbon emissions: for example, meteorological conditions such as headwinds or tailwinds; flight distance variations due to

Table 1 ICAO GCD correction factor

Great circle distance (GDC) (km)	Correction to GCD (km)
Less than 550	+50
Between 550 and 5,500	+100
Above 5,500	+125

Source ICAO [3]

vectoring to avoid bad weather; and air traffic control matters like holding patterns, mass of aircraft load, type of aircraft, type of engines and seating configurations. Conventional emission calculators utilize almost identical same methodology to cover such factors.

2.2.1 Distance Flown

The distance can be calculated by using the geographical coordination of the two airports. Once the Great Circle Distance (GCD) is calculated, it needs to be corrected. The relationship between emissions and distance flown is not linear; this is because of the take off part of the flight. A short flight has much higher emissions per km flown because of a greater proportion of fuel burning in takeoff. Mathematically, emissions can be represented as a function of distance in one of three ways.

The simplest methodology is:

$$y = ax \, (\text{y is fuel burn, x is distance flown}) \tag{1}$$

The more sophisticated methodology is:

$$y = ax + b. \tag{2}$$

The more accurate representation methodology (polynomial formula) is:

$$y = ax^2 + bx + c \tag{3}$$

The ICAO has also developed the GCD correction factor in conjunction with stopovers, and diverts from straight line of the GCD, as shown in Table 1.

2.2.2 Aircraft Type

The ICAO calculator identifies the scheduled aircraft from the database and maps it into one of the fifty equivalent aircraft types existing in the aircraft fuel consumption database. The problem is that the aircraft types that cannot be mapped are excluded from the calculations, and the ICAO asserts that it would only have a minor influence on the final result. Other conventional calculators utilize similar databases or estimation methods.

2.2.3 Passenger Load Factors and Passenger to Cargo Factor

Deduction of flight emissions associated with the freight and mail carried on the flight from the total is necessary. This calculation needs to utilize a historic database. The calculation is as follows:

$$\text{Total mass} = [((\text{No. Passengers} * 100 \text{ kg}) + (\text{No. of seats} * 50 \text{ kg}))/1,000(\text{tones})] + \text{Freight (tones)} + \text{Mail (tones)} \tag{4}$$

2.2.4 Cabin Class

The traveler is responsible solely for the carbon emissions for the seat they occupy. Premium seats occupy more space, so it seems natural to allocate a greater share of load per passenger. In case of the ICAO, the organization employs a simplified approach by using two cabin class factors (economy and premium) when allocating emissions to passengers, with a ratio of 1:2.

2.3 Pre-existing Emissions Calculators

2.3.1 DEFRA Emissions Calculator Methodology

The Department for Environment, Food, and Rural Affairs (DEFRA) of the United Kingdom has been responsible for CO_2 emissions reporting in the UK and developed its own calculating methodologies, which have subsequently been adopted by other agencies [4]. The DEFRA methodology uses a series of emission factors for short, medium, and long haul flights. Fuel burn data are calculated for typical aircraft over illustrative trip distances. Emissions are allocated in the proportions of the respective weights of passengers and freight. Passenger load factors are also allocated for domestic (66.3 %), short-haul (81.2 %) and long-haul (78.1 %) flights.

2.3.2 ICAO Carbon Emissions Calculator

The ICAO developed a general methodology to estimate the amount of carbon emissions generated by a passenger in a flight [3]. The ICAO methodology employs a distance-based approach to estimate the individual's aviation emissions using data currently available on a range of aircraft types. The ICAO uses publicly–available data regarding fuel consumption to continuously monitor and seek improvement in the data used in order to obtain better estimation. Figure 1 describes the ICAO carbon emissions calculation procedure.

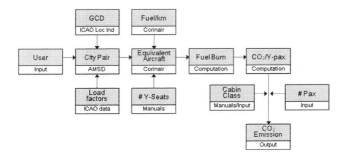

Fig. 1 ICAO emissions calculation procedure (*Source*: ICAO [3])

2.3.3 The Sabre Holdings Model

Sabre is a company that provides a computer reservation system to airlines, railways, hotels, and travel agents. The Sabre database already has critical information for carbon emissions calculation, like date of travel, airline, departure and destination airport, and aircraft model and configuration. Main engines of the Sabre model are the Passenger Name Record (PNR) and the System for assessing Aviation's Global Emissions (SAGE) [5]. The PNR contains information about individual flights, the point of origin, destination, airline, and plane type, and can access seating configuration. The SAGE was developed by the FAA Office of Environment and Energy. The SAGE can provide fuel burn data at the individual flight level based on the aircraft type using crude average and assumptions. The SAGE model has extensive fuel burn data for more than 200 aircraft types.

3 Analyses

3.1 Aviation Carbon Offset Programs

Offsetting climate change is behavior that companies or individuals undertake to compensate for greenhouse gas emissions. The offset can be equivalent to part or all of the associated emissions. Offsets can either be bought from within the international compliance system under the Kyoto Protocol or in the voluntary market. Offsetting in aviation emissions can be done either by airlines or by their customers. Such efforts to offset emissions can be sourced from various types of activities, like forestation or renewable energy projects, and emissions can be purchased from offset providers or carbon brokers. The buyer then receives a certificate or a record of amount of CO_2 reduced. If an emissions calculator is to be made for the flying public, it should be integrated with the ticket purchasing system, because it allows easy calculation of the CO_2 emissions associated with a ticket purchase. The calculator generally can provide electronic data on CO_2 footprint of all the sectors flown by the airline.

Table 2 Carbon dioxide emissions calculation (CDG-JFK, economy, round trip)

Carbon calculator	International organization (ICAO)	Airlines (AIRFRANCE)	NGO (ClimateCare)
Distance (round trip)	11,658 km	12,250 km	11,666 km
Mass of emission	838 kg/CO$_2$	1,049 kg/CO$_2$	1,620 kg/CO$_2$
Deviation	100 %	125 %	193 %

3.2 Limitations of Pre-existing Carbon Calculators

3.2.1 Inaccuracy

To evaluate the accuracy of pre-existing calculators, let us show a case of a flight between Paris (CDG) and New York (JFK) trip. The carbon emission during that flight was calculated, and Table 2 shows the result of carbon dioxide emission calculation by interested entities.

The airline's calculator indicates 25 % more CO$_2$ than that of the international organization. The private carbon offset company shows almost double compared to ICAO calculation. This simple comparison shows that it is difficult to find consistency between different methods.

3.2.2 Assumptions

The pre-existing calculators use assumptions in many calculating steps. For example, it is necessary to define the type of aircraft for carbon footprint calculation; however, in order to avoid inconvenience to link the actual aircraft type data, many calculators use the aircraft type matrix date, which is outdated and does not incorporates new type of aircraft, such as the Airbus 380 or Boeing B787. In addition, in terms of the passenger to freight ratio, seat capacity, and passenger load factors, many assumptions are applied without efforts to update data.

3.2.3 Accessibility of the Calculator

To calculate carbon emissions of a flight, users have to visit the website of the ICAO or airlines (in case the airline has a carbon calculator). With this poor accessibility of carbon calculators, active participation of the offset program or selection of greener airline is not to be expected in most cases. It is necessary to make carbon calculators available on more convenient position.

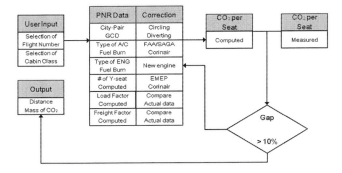

Fig. 2 New methodology for the proposed carbon calculator

3.2.4 Neglect of an Important Factor

The aircraft engine is one of the important factors to be considered. Engine manufacturers are developing greener engines for the next generation. In case of the General Electric, they said that the GE's next generation engine family offers an overall 15 % less CO_2 emissions. Pratt and Whitney, one of the GE's competitors, says their new PW1000G engine burns 16 % less versus today's best engine. Such savings are too big to be neglected. Furthermore, the recognition of new advanced engines that burn less fuel will boost the policy of airlines for investment in new engines.

4 Improved Methodology for Carbon Calculators

To avoid inaccuracy in the calculation of carbon dioxide emissions, this study proposes a methodology using advanced IT. This new methodology utilizes actual data from the PNR database during the ticket purchasing phase. When a traveler is browsing the trip, a city pare is selected. Consequently, the GCD is calculated. The advantage of this calculator comes from the usage of data on actual aircraft and engine types. Fuel burning of each type of aircraft and engines are mapped based on the pre-existing database. Seating configuration can be calculated based on actual data, and freight passenger ratio and load factors can be derived from the pre-existing database. Next, the computed emission per passenger number is randomly evaluated with the actual fuel burn data. If the difference is less than 10 %, it can be said that the system is operating as anticipated. If the difference is greater than 10 %, the database and correction factors can be adjusted. Figure 2 illustrates the newly proposed methodology.

If each airline introduces this calculator on its reservation system, it would be a good reference for consumers to select more environmentally friendly flights. Moreover, the new calculator could provide more accurate mass of carbon (less

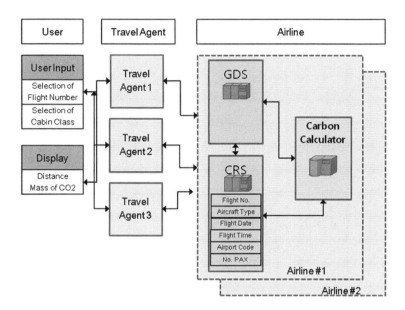

Fig. 3 Framework on data flow for flight reservation and carbon emissions display

than 10 % deviation from the actual emission) data, and it will be an accurate reference for the carbon offset program.

Figure 3 illustrates the application of the new carbon calculator to a Computer Reservation System (CRS). When travelers browsing available flights, computed distance and mass of carbon emissions can be displayed (if each airlines introduce the carbon calculator in conjunction with existing computer reservation systems). The CRS with the new carbon calculator linkage can provide a new criterion for choosing airlines and specific flights.

5 Conclusions

Carbon dioxide emissions calculation can provide an important reference for consumers regarding selection of environmentally friendly airlines and participation in the carbon offset program. Through review and comparison, this study finds that the pre-existing calculators are not sufficiently accurate and useful for the flying public. To provide more accurate and useful information on carbon emissions, a new calculation methodology is proposed and explained. The new methodology utilizes actual data instead of computed data, and assumptions based on the outdated matrix can be avoided with use of the modified process. The ICAO carbon calculator is the first meaningful step of carbon emissions calculation; however, it does not solve the underlying causes of inconsistency. The new methodology has the potential to address such inconsistencies by providing emissions calculated from actual data for specific flight. Further studies on comparison of calculated and measured carbon emissions would be of interest to extend the scope of the current study.

References

1. Pache, E.: On the compatibility with international legal provisions of including greenhouse gas emissions from international aviation in the EU emission allowance trading scheme as a result of the proposed changes to the EU emission allowance directive (2008)
2. IATA Aviation Carbon Offset Programmes: IATA guidelines and toolkit. Technical report (2008)
3. ICAO: ICAO Carbon Calculation Emissions Calculator Ver 3. Technical report (2010)
4. DEFRA: Act on CO_2 Calculator Ver 1.2: Data, methodology and assumptions. Technical report (2008)
5. Sabre Holdings: Carbon model description—description of Sabre Holding's emissions model for air. Technical report (2008)

A Framework for Developing Internet-Based Global Integrated Logistics Management System

Ho-Seon Hong, Ki-sung Hong, Kee-woong Kim and Chulung Lee

Abstract This paper introduces a differentiated framework for an internet-based global integrated logistics management system (IGILMS) which can access and mediate cargo transportations among small- and medium-sized domestic/international logistics corporations as regards cargo, registration and search for space, public tender and negotiation, contract, payment and safety-guarantee, transportation through the internet, wireless communication means, electronic data interchange (EDI), or smart phone etc. In addition, we describe a model to provide customized services in the B2B e-market for IGILMS and the structure and the control of its fulfillment process, and provide technical architecture of the IGILMS for an internet-based global logistics management system.

Keywords Global logistics · Global network · Internet-based · Brokerage system · Consolidation · B2B e-market · Safety guarantee

H.-S. Hong
Corea International Logistics Co. Ltd., Paju-si, Gyeonggido, Korea
e-mail: cilceo@hanmail.net

H.-S. Hong · K. Kim
Aviation Business Administration, Korea Aerospace University,
Goyang, Gyeonggido, Korea
e-mail: kimkw@kau.ac.kr

K. Hong
Graduate School of Information Management and Security, Korea University, Seoul, Korea
e-mail: justlikewind@korea.ac.kr

C. Lee (✉)
Division of Industrial Management Engineering and Graduate School of Management
of Technology, Korea University, Seongbuk-gu, Seoul 136-701, Korea
e-mail: leecu@korea.ac.kr

S.-S. Yeo et al. (eds.), *Computer Science and its Applications,*
Lecture Notes in Electrical Engineering 203, DOI: 10.1007/978-94-007-5699-1_58,
© Springer Science+Business Media Dordrecht 2012

1 Introduction

Due to globalization and the economy tending to be based on e-business, international transport is changing rapidly into a value-added international multimodal transport business where swiftness and value-added services are emphasized. The reason for such changes can be said to be derived from being multi-dimensional, rapid development, the diversification of international logistics management systems, and the rapid increase of such services, which follow from the changes to international logistics management systems, etc.

Following such changes, global integrated logistics management systems are rapidly being reorganized to focus on specialized logistics corporations (e.g., global integrated logistics corporations), and international transport is shifting into international multimodal transport linking inland, maritime and air transport systemically. With the spread of global network strategies of globalized corporations, the international economic environment demands a higher-grade and more precise global logistics management system. Moreover, with the trend of cargoes becoming smaller in volume and being transported to multiple locations, the demand for international multimodal transport through global integrated logistics corporations is increasing.

With the acceleration of the growth of the global logistics market and tougher competition in domestic and foreign markets, it is necessary to develop international multimodal transport corporations to strengthen the global competitiveness of national logistics industries. For a successful international multimodal transportation company, it is, above all, necessary to construct an integrated logistics network across the world. However, it is not easy to construct such a network because it requires a huge capital investment.

Large logistics corporations perform in the international logistics market by investing huge capital in means of inland, maritime, and air transport. On the other hand, in the area of multimodal transport through mixed loading suitable for small- and medium-sized international logistics corporations, those companies fail to load cargoes to their capacities, and the proportion of cargoes not fully loaded (20–30 % of the maximum cargoes) is increasing. Thus, the cost competitiveness factor works as an obstacle for small- and medium-sized international logistics corporations to cultivate and enter foreign markets.

To respond properly to the changes in international logistics markets and to develop international logistics corporations, small- and medium-sized international logistics corporations need to enter the global logistics market through a differentiated alliance network. To build such an alliance network of small- and medium-sized international logistics corporations working in a global market, it is necessary to build a brokerage system which can mediate cargo transport among alliance partners.

Kim et al. [1] suggested a brokerage system for truckload freights and Jung and Jo [2] proposed a two-layered multi-agent framework for brokerage between buyers and sellers. However, they do not consider the characteristics of international logistics such as surety insurance which can guarantee safe cargo transportation.

In this paper, we propose a framework for an Internet-based global integrated logistics management system which can mediate cargo transport among small- and medium-sized international logistics corporations.

The paper is organized as follows: Section 2 presents a conceptual model for an Internet-based global integrated logistics management system, and Sect. 3 describes the B2B e-market model or Internet-based global integrated logistics management system. In Sect. 4, we provide the technical architecture for an Internet-based global integrated logistics management system. The final section provides conclusions.

2 Internet-Based Global Integrated Logistics Management System

2.1 Requirements on the System

This paper analyzed the requirements for the system applying the scenario-based requirements analysis that is one of the analysis methods for the system requirements and the drawn requirements are as follows.

1. It should be available to inquire into the information about cargo or space.
2. It should be available to exchange the opinions for negotiations on freight costs.
3. The credit information on the other party of the contract should be provided.
4. Transportation contract should be secured by the system.

In this paper, the framework of the system is proposed, which the drawn four requirements are reflected.

2.2 Framework of the System

The internet-based global integrated logistics management system (IGILMS) in this study involves a system center, Forwarders which have cargo (FC), and Forwarders which have space (FS) as shown in Fig. 1. An FC that has joined the IGILMS-safety guarantee module (IGILMS-sgm) should deposit a total amount of transportation costs ensuring the fulfillment of a contract in the way of an ESCROW in a bank. For the case of an FS, it should get surety insurance which can guarantee safe cargo transportation with a specific sum of money fixed by the IGILMS-sgm. By doing this, the FC and FS enter into a contract. The procedure in which the FC and FS are matched with transport tasks is as follows: the FC registers the system center "Cargo Information and Transportation Information",

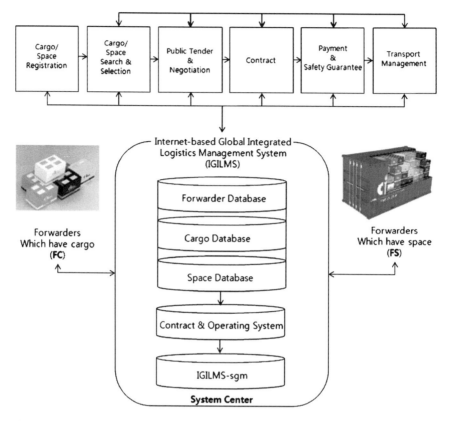

Fig. 1 Conceptual model for IGILMS

i.e. Volume of cargo, Pickup and Delivery Location, Estimated time of departure (ETD), Estimated time of arrival (ETA), and other special conditions about the transportation service as well as "a deadline of estimated value" through the Internet, a wireless communication means, electronic data interchange (EDI), or a smart phone. The system center administers cargo information, conditions and a deadline of estimated value that the FS can access and participate in open bidding.

The FS accesses the web site of the system and posts up estimates and service conditions etc. to participate in open bidding. The FC and FS select their counterpart that can generate maximum profits and negotiate prices to enter into a contract. After the conclusion of a contract, the FS provides the FC with a transportation schedule and transportation documents such as Bill of Landing (B/L), Container Load Plan (CLP), Cargo Insurance etc. as well as a Cargo Tracking Service. Before entering a port, the FS forwards customs information to the FC and consults with the FC to clear customs. Then, if the shipment is sent to its final destination, the FS reports and gives notice of payment permit to the ESCROW bank through the IGILMS-sgm. However, if a problem occurs from one

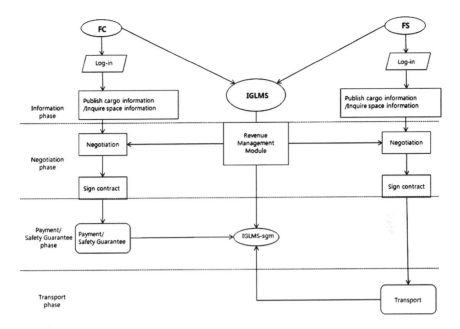

Fig. 2 The transaction process of IGILMS

side, the IGILMS-sgm can manage the risk of member companies by controlling the ESCROW bank and cargo transportation insurance.

3 The B2B e-Market Model for IGILMS

The B2B e-market model for IGILMS consists of a four transaction process. Transactional processing is slightly different from the definition in the four phase models (information, agreement, settlement and communication) proposed by Schmid and Lindemann [3], Selz and Schubert [4] and Zhao et al. [5]. Communication occurs at every stage of the transactional process in the e-market; thus, it is not seen as a separate phase. Therefore, in our model, the four phases of the transactional process are identified as: information, negotiation, payment/safety guarantee. The transaction process of IGILMS is shown in Fig. 2.

3.1 Personalized and Customized Services

In the information phase, the IGILMS provides a "Member Company Page", which enables forwarders to manage their business information. In a Member Company Page, forwarders provide a "Self Operating" function that can always

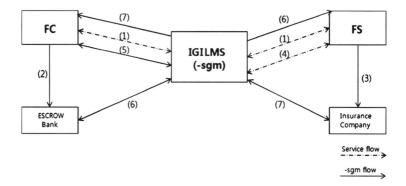

Fig. 3 IGILMS's fulfillment process

revise or register their business information, including registration, issuance, contract signing, etc.

In the negotiation phase, the IGILMS supports a forwarder's price decision through the revenue management module. A cargo space is a perishable item like an airline seat. That is, the value of a cargo space changes with time. The revenue management module calculates an optimal price of a cargo space at a given time and provides alternative price information to the forwarder.

3.2 Structure and Control of the Fulfillment Process

The IGILMS's payment/safety guarantee process is intertwined with the transport phase. The structure of the fulfillment process involves FC, ESCROW bank, FS, FS cargo transportation insurance, and IGILMS-sgm. The IGILMS's fulfillment process is as follows (see Fig. 3):

1. IGILMS -> FC and FS enter into a contract through mediation of the IGILMS
2. FC -> Deposit total amount of transportation cost (ESCROW way)
3. FS -> Get cargo transportation insurance
4. FS -> Completion of transportation service or Report a problem
5. FC -> Notify approval of payment or disapproval of payment
6. IGILMS-sgm -> Inform ESCROW bank and FS of FC's willingness to pay
7. IGILMS-sgm -> Forward a compensation application for FC's damage caused by FS's fault or bankruptcy to insurance company and FC.

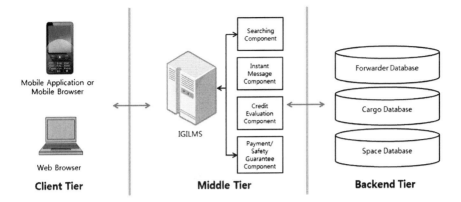

Fig. 4 IGILMS GEM technical architecture

4 Technical Architecture of IGILMS

The IGILMS is designed to provide a global electronic marketplace (GEM) between small- and medium-sized international logistics corporations. The IGI-LMS is not a technologically innovative platform on GEM but an innovative customer-oriented e-commerce model. The technical architecture of the IGILMS can be described in Fig. 4.

The IGILMS platform is a three tier architecture; client tier, middle tier and backend tier. The client tier consists of a Mobile Application (or Mobile Browser) and Web Browser. The Web Browser (by real-time communication) can execute functions of cargo and space search and sending and receiving trade information. The middle tier applies the IGILMS as middleware responsible for coordinating messaging with the Web Browser and communicating with the backend tier (database).

The IGILMS platform consists of four components; searching component, instant message component, credit evaluation component, and payment/safety guarantee component.

4.1 Searching Component

Discovery of the forwarders, cargo and space information is always the first step of a transaction process. The searching component of the IGILMS provides a dynamic network-wide search using a robust, efficient and adaptable search algorithm based on well-designed indexes.

4.2 *Instant Message Component*

The IGILMS has to collaborate and integrate with instant message component for real-time communication during the transaction process. This service combines FCs and FSs directly for a pricing conversation, order conversation, and credit conversation when negotiation is needed. The basic messages for these conversations include a request, a response, and the requester's reply (e.g. rejecting or accepting a price) and so on.

4.3 *Credit Evaluation Component*

It is of paramount importance to identify/verify the credit-worthiness of the participants engaging in a transaction process due to the legally and financially binding nature of business [6]. The Credit Evaluation Component determines whether the credit of a participant is available for transportation using credit management.

4.4 *Payment/Safety Guarantee Component*

The Payment Component indicates the amount, date and receiver/payer. For a safe transaction, the FC deposits the total amount of the transportation cost in an ESCROW way and, after getting a report of transportation completion, it permits transportation payment to the FS through the IGILMS-sgm. However, if a problem occurs, it puts a freeze on the deposit until the problem can be resolved.

5 Conclusions

To respond properly to the changes in the international logistics markets and to develop international logistics corporations, small- and medium-sized international logistics corporations need to enter the global logistics market through a differentiated alliance network. To build such an alliance network of small- and medium-sized international logistics corporations, it is necessary to build a safety guarantee global integrated logistics management system which can mediate cargo transport among alliance partners.

This paper introduces a framework for an Internet-based safety guarantee global integrated logistics management system which can mediate cargo transport among international logistics corporations. We describe a B2B e-market model and provide technical architecture for an Internet-based safety guarantee global integrated logistics management system.

Small- and medium-sized international logistics corporations can reduce the proportion of cargoes not fully loaded and increase their cost competitiveness using the proposed system.

Acknowledgments This research was supported by Basic Science Research Program through the National Research Foundation of Korea (NRF) funded by the Ministry of Education, Science and Technology (2009-0068528).

References

1. Kim, K.H., Chung, W.J., Hwang, H., Ko, C.S.: A distributed dispatching method for the brokerage of truckload freights. Int. J. Prod. Econ. **90**, 150–161 (2005)
2. Jung, J.-J., Jo, G.-S.: Brokerage between buyer and seller agents using constraint satisfaction problem models. Decis. Support Syst. **28**, 293–304 (2000)
3. Schmid, B.F., Lindemann, M.A.: Elements of a reference model for electronic markets. In: Proceedings of the Thirty-First Hawaii International Conference on System Sciences, 1998, vol. 4, pp. 193–201
4. Selz, D., Schubert, P.: Web assessment—a model for the evaluation and the assessment of successful electronic commerce applications. In: Proceedings of the Thirty-First Hawaii International Conference on System Sciences, 1998, vol. 4, pp. 222–231
5. Zhao, J., Wang, S., Huang, W.V.: A study of B2B e-market in China: e-commerce process perspective. Inf. Manag. **45**, 242–248 (2008)
6. Fan, J., Ren, B., Cai, J.-M.: Design of customer credit evaluation system for e-commerce. In: 2004 IEEE International Conference on Systems, Man and Cybernetics, 2004

The Implication of Environmental Costs on Air Passenger Demand for Airline Networks

Baek-Jae Kim and Kwang-Eui Yoo

Abstract The aviation sector has developed dramatically in recent decades. However, environmental and social concerns are gradually posing limitations on the growth of the air transport industry. Air travel contributes to climate change, and causes significant environmental damage. It is therefore subject to environmental impact charges, which will intensify over time. These environmental charges will influence air travel. This research aims to investigate the impact of environment costs by examining their influence on air passenger demand within the context of two major airline network models: the hub-and-spoke network and the point-to-point network. The additional costs, caused by environmental rehabilitation, will affect both the networks and profit structures of airlines, and these structures will need to be amended in time. This study may contribute to the South Korean aviation industry, especially to the airlines that will be affected by the CO_2 constraints imposed by the European Union Emissions Trading System from 2012 onwards.

Keywords Environmental costs · EU-ETS · Airline networks

B.-J. Kim
Department of Aviation Business Administration, Korea Aerospace University,
Hwajeon-dong, Deogyang-gu, Gyeonggi-do, Goyang 412-791, Korea
e-mail: baekjae1004@hotmail.com

K.-E. Yoo (✉)
Department of Air Transportation, Transportation and Logistics, Korea Aerospace
University, Hwajeon-dong, Deogyang-gu, Gyeonggi-do, Goyang 412-791, Korea
e-mail: keyoo@kau.ac.kr

S.-S. Yeo et al. (eds.), *Computer Science and its Applications*,
Lecture Notes in Electrical Engineering 203, DOI: 10.1007/978-94-007-5699-1_59,
© Springer Science+Business Media Dordrecht 2012

1 Introduction

Like other transportation methods, air travel contributes to climate change, and causes environmental and economic damage through CO_2 emissions. The contribution of the aviation sector to total CO_2 emissions is only about 3 % at this point in time, whereas 19 % are caused by road transport. However, the real impact of air travel CO_2 emissions may be twice as high, or even higher, compared with other CO_2 emissions since a significant proportion of aircraft emissions are at high altitude. These emissions give rise to important environmental concerns regarding their global impact and effect on air quality at ground level.

Aviation demand has grown rapidly in recent decades, and it is expected that this growth will continue [1]. At the same time, the fuel consumption and CO_2 emissions of the aviation sector have grown by about 2–4 % and are expected to grow further [2]. The demand for air transport has increased steadily over the years. Passenger numbers have grown by 45 % over the last few decades, and have more than doubled since the mid-1980s. In addition, freight traffic has increased even more rapidly, by over 80 % on a tonne-kilometer performed basis over the last decade, and almost three-fold since the mid-1980s [3]. Expected technological innovations cannot prevent an increase in CO_2 emissions from the aviation sector due to the high increase in demand. The Intergovernmental Panel on Climatic Change (IPCC) completed a report on the effects of aviation on the global climate, and this is a much cited reference in this field [4].

In July 2008, the European Council and the European Parliament agreed to incorporate international aviation into the existing European Union Emissions Trading Scheme (EU-ETS) in order to limit CO_2 emissions. The Directive came into force in February 2009 [5]. Aircraft operators are obliged to surrender allowances for all commercial flights landing at and departing from any airport in the EU from 2012 onwards. Within this regulatory process, the EU-ETS will not only impact airlines based in the EU, but in fact all carriers operating at airports in the EU. Therefore, flights from South Korea to the EU will also be affected. The EU Directive contains the following provisions for the inclusion of aviation into the existing emissions trading scheme:

- The emission trading scheme will cover all flights departing from or arriving at EU airports from 2012 onwards.
- Aircraft operators will be the responsible entities for holding and surrendering allowances for CO_2 emissions.
- Regulations for emission monitoring and reporting took effect in 2010 while an emission cap for all aircraft operators will be introduced in 2012.
- The EU Directive for the period 2013–2020, agreed by the Council of the European Union in December 2008, aims to improve and extend the greenhouse gas emission allowance trading system of the Community.

The cost effects for passengers will depend to a large extent on the ability of the airlines to pass through cost increases to the demand side. Similar to the imposition of fuel surcharges in recent years due to high oil prices, it can be expected that airlines will pass a large extent of the cost increase to the passenger. Even if airlines are able to pass through only the acquisition costs of allowances to be purchased, in addition to the free allocation, passengers can expect an average cost increase per ticket. The cost increase for long-haul flights can be substantial: if the complete trip is subject to the EU-ETS and acquisition as well as opportunity costs, then these will be passed on to the passenger. The direct operating cost of long-haul route airlines to EU airports will be affected more than medium-haul and short-haul route airlines. Between the Far East Asia and EU routes, transisting via the Middle East is more favorable than direct operation since CO_2 emission charges are calculated on the basis of flying distance from departing airport to EU airports. Hence it implies a cost disadvantage for long-haul routes operating direct flights.

2 Air Traffic Demand Forecast

Forecasting demand is the most critical area of airline management. An airline forecasts demand in order to plan the supply of services required to meet the demand. This is dependent on external factors such as the economic situation, exchange rates, tourism trends, etc. Most external factors are unpredictable, and result in much of the uncertainty in airline forecasting. Air traffic demand is also dependent on many internal factors and changes [6].

There are numerous qualitative methods and techniques available to forecast air traffic demand. Executive judgment is most widely used to modify and adapt other, more mathematical forecasts. It is based on the insight and assessment of a person who has specialist knowledge of the market in question. A wide range of market research techniques can be used by airlines in order to analyze the characteristics of demand. These techniques include attitudinal and behavioral surveys of passengers. The Delphi approach is also used for a consensus forecast based on the views of individuals who are considered to have sufficient expertise to be able to anticipate future trends. Most econometric forecasts of air traffic tend to be based on simple or multiple regression models where traffic is a function of one or more independent variables. The two variables most frequently used are airfares and some measure of per capita income.

$$T = f(F, Y, t) \tag{1}$$

where
T is the annual number of passengers traveling between two points;
F is the average fare in real terms;
Y is an income measure; and
t is an underlying time trend

The choice of fare is critical. Due to the addition of environmental costs and the concomitant airfare increases, passenger demand for air travel will be affected. The responsiveness of demand to price or fare changes can be measured in terms of an elasticity coefficient:

Price elasticity = % change in demand / % change in price (fare)

Price elasticity is always negative since price and demand must move in opposite directions. If the fare goes up, demand is expected to fall and vice versa, as it is invariably a negative sign in the equation. Price elasticity will be dependent on the value of the environmental costs, however it can be expected that general air traffic demand will be negatively affected. Furthermore, these costs will predominantly influence current airlines networks.

3 Hub-and-Spoke Network and Point-to-Point Network

There have been many changes in the routing structure of airlines in the past decades. Since deregulation in the United States and elsewhere, the hub-and-spoke network has emerged as the dominant network design. The hub-and-spoke network is a system of connections arranged like a chariot wheel, in which all traffic moves along spokes connected to the hub at the center. A small number of routes generally lead to more efficient use of aircraft, and new routes can be created easily.

An airline that is able to develop and operate a hub-and-spoke system with a series of complexes can demonstrate potential advantages. The increase in city-pair coverage, which can be obtained as a result of hubbing, is beneficial. If three point-to-point direct links from cities A to B, C to D, and E to F are replaced by six direct services from each of these six airports to a new hub at an intermediate point (G), the number of city-pair markets that can be served increases from 3 to 21. This advantage increases in proportion to the square of the number of routes or spokes operated from the hub. Therefore, if a hub has n spokes, the number of direct links is n to which must be added $n(n-1)/2$ connecting links via the hub.

The progressively greater impact of adding more links through a hub can be seen in Table 1. The ability to reach a large number of destinations from any one origin gives the airline operating the hub system considerable market appeal. An effective hub-and-spoke network generates substantial volumes of additional traffic—and revenue—and most of the traffic is transferred through the hub airport. In spite of some negative aspects of the hub-and-spoke network, such as extra landing charges, additional fuel consumption, etc., this network has been the dominant airline network for many decades. However, it may lose favor when environmental costs, such as the ETS charge, emission charges, noise charges, and other various environmental taxes, come into play, as these can offset the economic benefits of the hub-and-spoke network. On the other hand, the financial success of carriers offering direct services has been noted. The point-to-point

Table 1 Impact of hub-and-spoke network on the number of city-pairs serviced [6]

Number of spokes from the hub n	Number of points connected via the hub $n(n-1)/2$	Number of points linked to the hub by direct flights n	Total city-pairs served $n(n+1)/2$
2	1	2	3
6	15	6	21
10	45	10	55
50	1,225	50	1,275
75	2,775	75	2,850
100	4,950	100	5,050

network is normally used by low cost carriers (LCC) as it can minimize the number of connections and travel time. The point-to-point network is a transportation system where a plane travels directly to its destination rather than going through a central hub.

In order to maximize profits, an airline needs to design a network and schedule that will minimize airline and passenger costs. This is because changes to the network schedule that decrease passenger costs can be captured by the airline through price increases. Depending on the gradual increment of environmental costs, the point-to-point network will become more favorable than the hub-and-spoke network at some point. It is beneficial for airlines to reduce the number of landings and take-offs depending on the various environmental cost increases. However, other operating cost reductions should also be considered by airlines in order to maximize their profits.

4 Influence of CO_2 Constraints on Passenger Behavior

When researching the influence of the EU-ETS CO_2 constraints and the resultant effect on passenger behavior, we found some important results. Major variables identified were the nationality of the airlines, air journey time, and airfare. The logit model was calibrated against these variables to test the hypothesis that travelers prefer national carriers, short journey times, and low airfares, and the degree of importance these variables have on flight choice was identified. In this utility function, the expected EU-ETS charges were added as an additional variable. The EU-ETS assigns an economic value to carbon dioxide emissions. Under such market-based schemes, the price is both variable and strongly influenced by supply and demand. The EU-ETS is currently trading at 10 Euros per ton of carbon dioxide, and this actual data and the calculation of the volume of CO_2 produced on a flight is based on actual operational performance rather than an industry average.

The utility function of the model could be written as:

$$U = ß0 + a_1 NATION + a_2 JT + a_3(FARE + EU - ETS) \qquad (2)$$

where

NATION   nationality of airlines (0 for Korean, 1 for Foreign);
JT   air journey time (min);
FARE   airfare (KRW);
EU-ETS   environmental costs (KRW); and
a_1, a_2, a_3,   coefficients to be estimated

Based on the results of the regression utility function, it could be written as:

$$U = 20.636 - 21.797 \text{ NATION} - 3.435 \text{ JT} + 1.736(\text{FARE} + \text{EU} - \text{ETS})$$
$$(3)$$

As we have negative numbers for the nationality of airlines and air journey time, these factors will be affected by the passengers' behavior when other conditions remain the same. However, we have concluded that airfares reflecting the EU-ETS CO_2 emission charge are not severely affected by passenger choice of airline. Our analysis determined that the EU-ETS charge is not sufficient to affect passenger behavior, and, in addition, environment issues are not yet considered important in the Korean aviation market. But, the situation will become more intensified as environmental costs increase rapidly, and this will likely affect air passengers and therefore current airline networks.

The emerging carbon market shows evidence of increasing activity levels, and a number of new players will enter the market in the near future. In addition to the EU-ETS charges, other environmental charges will be levied to airlines, and this will affect air ticket prices directly. The impact and influence of these factors will be different for each airline as it is dependent on each airline's network. Further study is required to identify the actual impact.

5 Conclusions

Given the constraints of external environmental costs, such as the ETS charge, emission charges, noise charges, and other environmental taxes, the current major airline networks, hub-and-spoke and point-to-point, will be affected, and the impact will become more intensified as time passes.

In recent decades, the hub-and-spoke network has emerged as the dominant network design. However, the financial success of carriers offering direct services, i.e., airlines using the point-to-point network, has been noted. LCCs operate in this way as it can minimize the number of connections required as well as actual travel time.

Given the new environmental conditions, we need to establish which airline network will be able to operate more favorably, or if both networks will be affected by the additional costs. We need to build into this the fact that environmental costs can be expected to increase for a certain period. Essentially, it will depend on the environmental cost level, but from this vantage, the point-to-point network will be more favorable than the hub-and-spoke network since the latter will be more costly due to a higher landings frequency, inefficient fuel consumption, etc. Again, at this stage we do not know the actual costs of the environmental factors, and therefore further studies will be required once the actual environmental charges are in place.

References

1. The Boeing Company 2007 Annual Report (2007)
2. Eyring, V., Ivar S.A., Berntsen, I.T., Collins, W.J., et al.: Transport impacts on atmosphere and climate: Shipping, atmospheric environment (2009)
3. ICAO: Report on Voluntary Emissions Trading for Aviation (VETS). Technical report (2007)
4. Penner, J.E., Lister, D.H., Dokker, D.J., MacFarland, M.: Aviation and the global atmosphere; a special report of IPCC working groups I and III, Cambridge University Press, Cambridge (1999)
5. Council of the European Union: Council adopts climate-energy legislative package 8434/09 (2009)
6. Doganis, R.: Flying Off Course: Airline Economics and Marketing, 4th edn. Routledge, London (2010)

An Empirical Study on the Design Peak Hourly Traffic at a Major International Airport

Sang-kyu Lee, Kee-Woong Kim and Youn-Chul Choi

Abstract Airports need to accommodate peak-time traffic, especially as the demand for air travel has increased dramatically in recent years. In small airports, the gap between hourly volume and design peak hourly traffic has little economic effect; however, it has a greater impact in larger airports. This study on the design peak hourly traffic in large airports showed that the value of the typical peak hour passenger (TPHP) is very similar to the value of inflection point that is used in the road design for estimating the design peak hourly traffic. It is expected that in large airports, the value of the TPHP and the cumulative traffic between 2.5 and 3.5 % would be preferable.

Keywords Design peak hourly traffic · Standard busy rate (SBR) · Typical peak hour passengers (TPHP) · Busy hour rate (BHR)

S. Lee
Incheon International Airport Corporation, Incheon, Korea and Aviation Business Administration, Korea Aerospace University, Gyeonggido, Korea
e-mail: LSK@airport.kr

K.-W. Kim
Aviation Business Administration, Korea Aerospace University, Gyeonggido, Korea
e-mail: kimkw@kau.ac.kr

Y.-C. Choi (✉)
Division of Aeronautical Studies, Hanseo University, Chungcheongnamdo, Korea
e-mail: pilot@hanseo.ac.kr

S.-S. Yeo et al. (eds.), *Computer Science and its Applications*,
Lecture Notes in Electrical Engineering 203, DOI: 10.1007/978-94-007-5699-1_60,
© Springer Science+Business Media Dordrecht 2012

1 Introduction

The improvement of the air transportation industry and the growing number of large airplanes have led to the increase of developing of large airports capable of satisfying the demands of more than one hundred million passengers worldwide [1]. But as airports expand, the cost for airport development becomes more extravagant. Thus, it is crucial to consider the costs for planning and constructing airports of reasonable size.

Airport construction must compute the facility size to meet the future air traffic demands. According to demand satisfaction, the results of economics and service are shown contrarily. The calculation of airport demands is done by setting usual peak hours as the standard. In the case of setting the maximum peak hour (the 1st peak hour demand) as the standard, the level of service (LOS) is increased but the economic value is decreased. On the other hand, in the case of setting the minimum peak hour as the standard, it is advantageous in economic aspects; however, because of the decrease in the LOS, the competitiveness is significantly diminished. Thus, when setting an appropriate standard of construction, design peak hourly traffic (or demand) is a crucial planning factor for determining demands. Design peak hourly traffic used in airports means that a given LOS can accommodate the traffic volume. In Korea, the standard busy rate (SBR) and typical peak hour passenger (TPHP) are commonly used as design peak hourly demand. These are known to have similar traffic volumes; however, because the terms generally are based on small airports with little traffic volume, the opinion dominates that such terms cannot be applied to complex large airports. Consequently, further research on design peak hourly traffic applicable for large airports is needed. The following study conducts an actual research of design peak hourly traffic based on the data from Incheon International Airport.

2 Background of Theory

2.1 Definition of Design Peak Hourly Traffic

The estimation of design peak hourly traffic has the same background theory as the design peak hourly traffic of the roads, which is defined as the sudden changing point (inflection point) on the curve that is designed based on the annual data of 8,760 h by arranging the traffic volume from the most to the least. With this notion, the Bureau of Public Roads in the USA suggests the idea of the 30th traffic volume as a design hourly traffic for the roads [2]. A reason behind such a suggestion is that since the break point, the curve representing the traffic volume per hour becomes flatter.

Table 1 Relationship between annual traffic and peak hourly traffic (according to the FAA)

Annual passengers (in millions)	PHF
Over 20	0.030
10–20	0.035
1–10	0.040
0.5–1	0.050
0.1–0.5	0.065
Under 0.1	0.210

2.2 Definition of Design Peak Hourly Traffic in Airports

The design peak hourly traffic in airports is usually applicable to small airports; thus, assuming that there are not many differences between peak hourly traffic volumes, the following are the most commonly used methods in Korean airports.

2.2.1 Standard Busy Rate

The SBR is commonly used in Great Britain airports as determining design peak hourly traffic. In addition, this method has been a prevailing way of designing the highway as a design peak hourly traffic. The SBR measures the hourly traffic volume (365 days × 24 h) for the year and uses the 30th traffic volume out of measurements of the year. There is no apparent relationship between highest peak hour demand and the SBR. However, the SBR generally is known to have a correlation of about 1.2 × SBR (or peak hour demand × 0.8) [3], and this number is known to increase up to about 0.9 as traffic volume increases [4].

2.2.2 Typical Peak Hour Passenger

The TPHP is defined as an hourly demand of the average day of the peak month, which is a method the Federal Aviation Administration (FAA) uses. The TPHP calculates the design peak hourly traffic using the peak hour factor (PHF), or the ratio of annual and hourly traffic.

Table 1 shows the PHF according to the FAA; however, it is flawed because it applies the same concentration coefficient in large airports with twenty million passengers. Still, the PHF is known to be useful in airports where small, centered, domestic flights dominate.

Fig. 1 Annual traffic pattern curve (*Source* Leslie J. Hempsey et al., 1999)

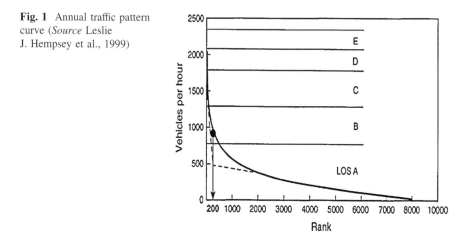

2.2.3 Busy Hour Rate

The BHR is the top 5 % accumulative demand taken care of during the busiest hour of the year. It can be computed by arranging the hourly demand by size and adding the accumulated 5 % of the yearly demand.

2.3 Estimation of Inflection Point in Roads

To find out the design hourly traffic volume, it is crucial to determine the break point of the curve appropriately (see Fig. 1).

The break point is where the slope of the curve changes dramatically; it resembles the meeting point of two tangent lines. Therefore, it is defined as the point with the longest distance between x-axis; the points are on the connected line between point with a slope of 1.0×10^{-5} and the 1st rank point [2].

2.4 Further Examination of Design Peak Hourly Traffic

In 1999, Wang and Pitfield [5] conducted an empirical examination of design peak hourly traffic considering large airports, specifically about 48 Brazilian airports. Wang and Pitfield [5] used 200 data of hourly traffic volume pattern and investigated the appropriateness of design peak hourly traffic by dividing the hourly standard deviation of five years by the mean traffic volume.

The study used the standard deviation of hourly traffic in order to deduct the break point. Overall, the study is meaningful because it empirically analyzed the design peak hourly traffic of large airports.

Table 2 Annual passenger performance

Years	Annual passengers (in millions)	Years	Annual passengers (in millions)
2007	30.8	2010	32.9
2008	29.6	2011	34.5
2009	28.1		

Source Incheon International Airport

In the case of Sao Paulo International Airport analyzed in the study, the 80th point was chosen as the design peak hourly traffic because it showed stability in the given traffic volume since the initial fluctuations.

3 Analysis Results

3.1 Research Planning

Most airports in Korea usually use the TPHP or SBR for the design peak hourly traffic to plan airport facilities. In its First Airport Development program, the national level of master plan, a peak hourly demand was used FAA's PHF. On its second and third Development planning, it had corrected the PHF coefficient to fit with the characteristics of local airports [6]. This study estimated the appropriateness of design peak hourly traffic in large airports according to the recent records regarding Incheon International Airport's international passengers over a five-year period (2007–2011; see Table 2).

The results of estimated design peak hourly traffic were compared with Wang's 1999 study results as well as the SBR, TPHP, and BHR (see Fig. 2).

3.2 Empirical Analysis Using Data from a Major International Airport

3.2.1 Design Peak Hourly Traffic by Inflection Point Approach

According to the results obtained by the inflection point method, the yearly inflection point formed at about a little over the 100th point (see Table 3). The accumulation value is about 3.5 % of the yearly demand and resembles similar results presented in Wang's 1999 study of Brazilian airports. The annual regression model was analyzed using the annual traffic performance statistics such as aircraft and passengers of 8,760 h (365 days × 24 h) from the Airport Operation Management System (AOMS) of the Incheon International Airport.

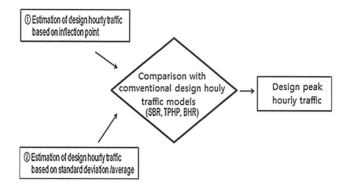

Fig. 2 Hybrid approach to obtain the design peak hourly traffic

Table 3 Estimation of inflection points

Years	Equation of regression	R²	Inflection point	Accumulation (%)
2007	y = 0.125−0.0070 ln(x)	0.993	131st	3.5
2008	y = 0.127−0.0073 ln(x)	0.991	143rd	3.8
2009	y = 0.131−0.0076 ln(x)	0.984	128th	3.6
2010	y = 0.126−0.0060 ln(x)	0.971	101st	2.6
2011	y = 0.118−0.0062 ln(x)	0.978	128th	3.3

Table 4 Comparison of estimation results in design peak hourly traffic

Classification	Rank	Average accumulation (%)
Using inflection point	125th	3.4
Using standard deviation	90th	2.5

3.2.2 Estimation of Design Peak Hourly Traffic Using Standard Deviation

At about the 90th point, according to the results of estimation on design peak hourly traffic for the last five years using standard deviation, the value was stable and accounted for 2.5 % of accumulation (see Table 4).

3.2.3 Estimation of Design Peak Hourly Traffic with Air Traffic Demands

The results of estimating design peak hourly traffic using aviary demands are shown in Table 5. Whereas the TPHP and BHR show similar values to one another, the SBR shows a great difference from the two. Thus, it is judged that using the SBR or 30th yearly traffic volume is inappropriate.

Especially, the SBR tends to approach near the highest peak hourly traffic. Thus, the SBR should be used at an appropriate distance to separate with the

Table 5 Design peak hourly traffic via definition of air traffic demands

Years	SBR	TPHP	BHR (5 %)
2007	30th	184th	194th
2008	30th	160th	191st
2009	30th	157th	185th
2010	30th	116th	199th
2011	30th	150th	201st

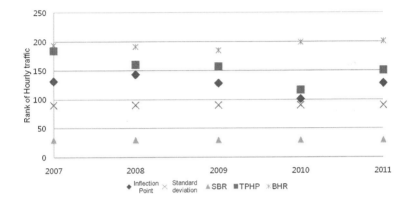

Fig. 3 Comparison of design peak hourly traffic estimation results

highest peak hour demand in order to sustain the purpose of design peak hourly traffic.

3.2.4 Estimation of Appropriate Degree of Design Peak Hourly Traffic

Figure 3 shows the results of the above estimations. The estimation of design peak hour traffic using inflection point shows similarities with the TPHP value's rank.

In addition to examining the degree of difference between the highest peak hourly traffic and appropriateness, the percentage ratios between the highest peak hour and design peak hourly traffic were examined. By estimating according to inflection points as seen in Fig. 4, the TPHP and BHR with an inflection point showed very similar results.

Compared with the estimation of using inflection points, there is a possibility to overestimate by using the SBR. Figure 5 shows the accumulation of the above values. This, as well, shows great similarities between estimations based on inflection points and the TPHP. In the case of the design peak hourly traffic using inflection points, the yearly traffic volume is about 2.5–3.5 %.

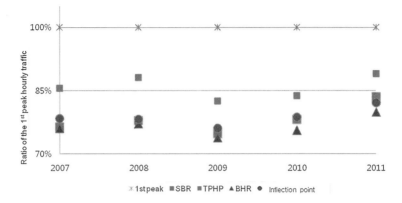

Fig. 4 Comparing ratios of the highest peak hour and design peak hourly traffic

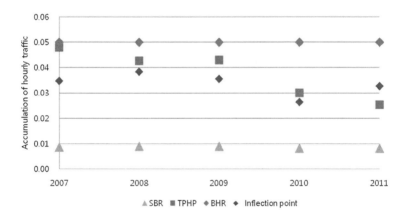

Fig. 5 Comparison with accumulation of design peak hourly traffic

4 Conclusions

As the demand for air traffic increases, airport size should increase as well. In the case of small-airport construction, planning demands do not have much of an economic effect because the absolute traffic volume is small. However, for large airports, the influence is huge indeed. This study estimated design peak hourly traffic by setting Incheon International Airport as the subject and searching for inflection points.

As a result, the study estimated that the TPHP used in many airports showed similar values. In addition, applying such results in reality, the SBR—the 30th value of the yearly demand—had the possibility of being overestimated. Looking at the case of Incheon International Airport, the accumulation of design peak hourly traffic in Korea is between 2.5 and 3.5 %. Thus, when constructing large international airports, it is advised that the TPHP should be used as design peak hourly demand, but a reference should be added to place accumulated traffic volume between 2.5 and 3.5 % according to the characteristics of each airport.

References

1. ICAO: Airport planning, design, operation and safety, 5th edn, technical manual (2009)
2. KOEX (Korea Express Corporation): The study of design hourly factor(K) for proper roads lane (2007)
3. Yang, S.S.: Airport planning and operation, easy books (2001)
4. IIAC (Incheon International Airport Corporation) : Incheon international airport master plan (after the 2nd phase) for the next generation (2008)
5. Wang, P.T., Pitfield, D.E.: The derivation and analysis of the passenger peak hour. J. Air. Transp. Manage. **5**(3), 135–141 (1999)
6. KOTI (Korea Transport Institute): The study of design criteria for standardization of airport by size; Appendix 1. 4 the space requirements by airport size 4.9 (2005)

Review of CRS in the Airline Industry: New Categorization with Previous Literatures

Sun Oh Bang, Jaehwan Lee, Kee-woong Kim and Chulung Lee

Abstract This paper was written to act as a new foundation of the Computer Reservation System (CRS) study approach. Although previous studies focused on various areas of the CRS industry, the short history of CRS and the change of business environments made previous researches unable to provide good intuition for later researchers. This paper conducted literary reviews to offer new categories based on the recent trends and activities in the CRS industry, and we believe the new categories can provide more valuable issues to study to future researchers.

Keywords Computer reservation system (CRS) · Global distribution system (GDS) · GDS new entrants (GNE) · Airline distribution · Categorize · Travel agency · Review

S. O. Bang
TOPAS Co., Ltd, 19F Hanjin New Bidg., 51, Seoul, Korea
e-mail: sobang@topas.net

S. O. Bang · K. Kim
Aviation Business Administration, Korea Aerospace University, Gyeonggi-do, Korea
e-mail: kimkw@kau.ac.kr

J. Lee
Department of Industrial Management Engineering, Korea University, Seoul, Korea
e-mail: lee-jh1012@korea.ac.kr

C. Lee (✉)
Division of Industrial Management Engineering and Graduate School of Management of Technology, Korea University, Anam-dong Seongbuk-gu, Seoul 136-701, Korea
e-mail: leecu@korea.ac.kr

S.-S. Yeo et al. (eds.), *Computer Science and its Applications*,
Lecture Notes in Electrical Engineering 203, DOI: 10.1007/978-94-007-5699-1_61,
© Springer Science+Business Media Dordrecht 2012

1 Introduction

The history of Computer Reservation system (CRS) started in the 1960s. At the time, CRS was a feasible airline reservation system. As it was originated from airlines' system, the early stage of CRS had been set to display biased results for giving advantages for the owner's company. However, it caused complaints from travel agencies as well as small/medium sized airlines that were not able to operate CRS. The complaints were finally addressed by DOT's (US Department of Transportation) CRS rules, which imposed regulations on the CRS business. After the rule was imposed, CRS became independent companies separated from airlines. During the 1990s, CRS became a Global Distribution System (GDS) thanks to the alliance between companies in the same or different fields and they extended their business territory to other countries [1–15] (see Fig. 1).

As CRS developed, it had to build relationships with other companies resulting in a more complex CRS business environment.

Unfortunately, the previous studies had no concerns over the environmental changes including current trends and updates. Therefore, this study shows the issues of previous studies based on conventional categories. It also analyzes the recent trends and builds new categories for future studies.

2 Previous Studies in Historical Order

The previous studies searched in "Google scholar (http://scholar.google.co.kr/)" were with keywords such as "Airline" and "Computer Reservation System" or "CRS" without "patent". The period was set at default.

In the result of the search, the previous studies are separated into two parts. One was papers focusing on the airline business, and the other was CRS-oriented papers. In this research, we cover the CRS-oriented papers only.

In the 1980s, there are papers and researches which focused on regulations and the abuse of CRS. Mietus [16] studied the impact of CRS on the air transportation industry and compared CRS regulations in the US and Europe. This article is based on regulation and distribution strategies.

In the 1990s, there was a boom in the CRS study. Ellig [17] studied CRS as a new distribution method of creatively destructing the market structure, which can provide more service to customer. In this paper, he emphasized the role of CRS as a distribution channel in deregulation circumstances. Truitt et al. [18] studied European CRS and CRS in the USA to benchmark the application of the system and policy issue in response to the regulation of CRS. Kleit [19] argued the importance of CRS in the deregulation environment and the importance of players who form the business environment as a distribution channel. Chrismar and Meier [20] studied a model of competing inter-organizational systems and its application to airline reservation systems with the change of cost and its impact.

	1970's		1980's		1990's~
Stage	Infant years	Growth years	Expansion	Consolidation	alliances
Action(s)	As an ARS	Launching competing GDSs	Search for new revenue resources	Network investments, global strategic concerns	Remaining GDS players seek global alliances.
Rationale	Airline automation tool	Competitive tool : selling host airline product	Globalization : network expenditures, economics of scale, international coverage, and etc.		
Events	Birth of Sabre and Apollo	1978 : Airline Deregulation Act	1984 : release CRS rule		

Fig. 1 The history of CRS

Their approach was clear and useful to decide the position in a competitive environment. However, the study was limited to the providers' side only. Archdale [21] studied to examine the main use of CRS and identify possible policy issues for public tourist offices. This study focused on the relation between CRS and tourist offices as a customer. In addition, this study analyzed the status of CRS and proposed distribution policies to benefit CRS owners. Duliba et al. [22] studied the indirect value of CRS ownership. It is not only an attempt to evaluate indirect benefits, but also the change of CRS core roles. They concluded that only major airlines can develop and operate CRS because CRS businesses require great deals of investment. However, it is no longer valid as IT cost is cheaper and technology is more advanced than before.

In the 2000s, there were big differences in the topics. In the late 1990s, Information & Computer Technology (ICT) was drastically advanced. Pemberton et al. [23] studied competing with CRS-generated information. This paper emphasized the importance of CRS-generated information as a competitive tool within the airline industry. In other words, airlines should be supported by CRS to survive and CRS stood at the center of such studies. Granados et al. [24] conducted a case study of "Orbitz", the online travel agency created by airlines to bypass CRS.

Gasson [25] conducted a case study of "SABRE (American Airline's CRS)" with Porter's five-force model of the industry competitive forces. Also, she concluded with comparison of the impact on information technologies between U.S. and European travel industries. Shon et al. [26] studied ticketing channels with CRS systems and they concluded that virtual channels are advantageous to dominate the market and traditional channels are useful in some specific segments. Alford [27] proposed a framework for mapping and evaluating business process costs in the tourism industry supply chain. He considered the weakness of ICT, which is frequently advocated as a business process enabler, and verified which organizations are fit to be implemented. Sengupta and Wiggins [28], Vukmirovic et al. [29], Lu [30], Kim et al. [31] studied the pricing of tickets, strategies for pricing, and distribution through CRS. In these studies they were concerned over agent-based environments and there are many other researches for this issue. However, most of them have same or similar approaches.

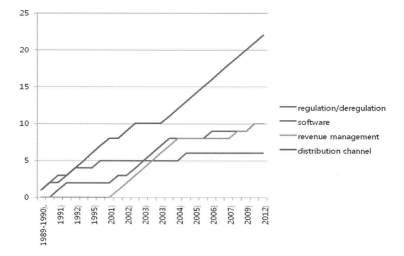

Fig. 2 Cumulative trend about the number of study

Figure 2 shows the cumulative trend of previous studies based on traditional categories. The categories are composed of four issues; CRS regulation/deregulation, software application, technological infrastructure, and distribution strategy.

In this figure, we can find that the CRS regulations are studied only during the early stages of CRS history. It is because most of the regulations in the US and Europe were deregulated by the government as the CRS environment changed. We can also find that the innovation of ICT had an effect on CRS studies, especially on the software application and technological infrastructure of CRS. Studies on distribution strategies are kept up continuously as it is an essential part of the CRS industry.

3 Recent Trends of the CRS Industry

Recently, there were many upheavals in the CRS industry. According to Quinby [32], GDS and CRS play central roles in the online travel agency, which is growing very fast. And there are new players who entered the CRS business which is based on new technologies. Those companies (which are called GDS New Entrants or GNE) can be a strong challenge to the traditional players. For example, one of the GNEs called "ITA Software" was purchased by Google, and now Google started to use ITA's technology to create new flight search tools. Apple has applied for a patent on an application called "iTravel", which provides travel planning, travel contents searching, reviewing, booking flights/hotels/cars and other transportation. Considering Google and Apple's strong influence on the IT industry and their brand power, it will be a threat to existing CRS sooner or later.

Airlines also try to increase their direct distribution channels through the internet, as part of their strategy to save costs and increase customer loyalty. All of the above factors show that there is huge change in the traditional distribution strategy, and the battle to win the new distribution network is underway.

Furthermore, Offutt [33] also shows that the development of technology such as location-based service, smartphone applications, as well as semantic search service is newly approaching customers. In addition, heavy and expensive legacy systems became light and flexible open systems. It is related with software applications, but also related with the innovation of the technological infrastructure.

Of course, there are changes in the internal side of the CRS business as well. GDSs are becoming large and diversifying their services by alliance with other contents providers like cruises, insurances, travel packages and low cost carrier (LCC)s. So, the business is naturally expanding. Aside from the contents expansion, some GDSs started unconventional businesses. Amadeus' system outsourcing service for airlines, which covers the airline's inventory management to the departure control service, is now one of the major business areas of Amadeus along with its distribution business. Sabre also succeeded to diversify its business to non-distribution fields, namely the online travel business. Travelocity, Sabre's subsidiary company, is now one of the top three online travel agencies in the U.S. All of the business diversification was the result of GDS's effort to survive, and we will see that more and more diversified services will be provided by them.

Mergers and acquisitions (M&A) are activated by local CRS as well as GDS. Galileo and Worldspan, two of the top 4 GDSs, were merged into one company (TravelPort), while local CRS like AXESS in Japan and TOPAS in Korea are enhancing their partnership with GDS. Furthermore, TravelSky, the dominant CRS in China, has been targeted for M&A by GDSs for a long time. If TravelSky chooses any one of the GDSs as its system partner, it will change the current landscape of the CRS industry.

Those are examples of new trends in the CRS industry and there will be more news or trends to come. However, when we try to apply the previous CRS categories to these new trends, we can find that some of the trends do not match well. We need to create new categories to cover them all, while we also need to drop some of the previous categories like CRS Regulation/Deregulation.

4 New Categories and Previous Categories

The previous studies fall into four categories. (1) CRS regulation/deregulation, (2) software application, (3) technological infrastructure, and (4) distribution strategy. As we commented above, the current trends do not fit into traditional categories. Therefore, new categories are needed.

The recent trends can be distinguished based on four categories.

Distribution Strategy needs to be changed into Distribution Network Expansion in order to accommodate the appearance of new players such as GNE and Google.

Fig. 3 Change in the categorization

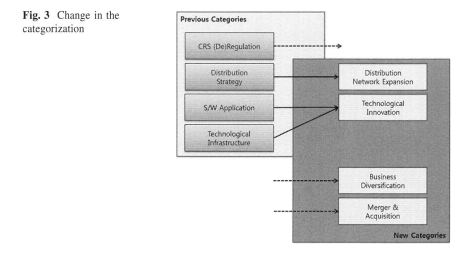

Software Application and Technological Infrastructure needs to be combined into Technological Innovation, as it has a common background in technology. Location-based service and semantic search can be part of this category. Business Diversification and M&A can be additional categories to cover new trends like GDS's unconventional businesses and GDS/CRS' strategic partnerships.

Figure 3 shows the relationship between previous and new categories. As we can see, some categories are excluded or merged, while new categories need to be created.

5 Future Study Issues

Each of the categories gives us items and issues to consider, and future studies need to be involved in the following areas. First, regarding "Distribution Network Expansion", it is meaningful to study more about the competition between new and traditional players. Competition can change the landscape of the CRS industry, and predicting the winners and their strategy will give us a glimpse of the future travel industry.

"Technological Innovation" also gives us items to study. The innovation that GDSs are making would bring customers more value and benefits and it is worth studying what is the best way to support customers from GDS's point of view.

For "Business Diversification", more study needs to be conducted to analyze the current GDS business model. All GDSs are looking for new sources of revenue in different areas, and it will be very interesting which areas and models are more lucrative. Future researchers can also study what is the new business model they can suggest GDSs to invest their resource and knowledge.

Lastly, for "Mergers and Acquisitions", it is worth looking into who is working with whom. Mergers between Galileo and Worldspan will not be the last M&A we will see. To survive or to succeed, GDSs will do their best to find strategic partners to bring higher synergy effects to them. It can be between GDS and CRS, or it can be with other players from a completely different industry.

6 Conclusion

This paper was written to act as a new foundation of the CRS study approach. We proposed new categories based on new trends in the CRS industry.

The new categorization that this paper offers is generated in relation with environmental changes, trends of the CRS industry, and development of technology. Since the business environment has always followed the trends, new categories would provide good intuition for researchers in the future.

Acknowledgments This research was supported by Basic Science Research Program through the National Research Foundation of Korea (NRF) funded by the Ministry of Education, Science and Technology (2009-0068528).

References

1. Marin, P.L.: Competition in European aviation: pricing policy and market structure. J. Ind. Econ. **43**, 141–159 (1995)
2. Smith, B.C., Gunther, D.P., Rao, B.V., Ratliff, R.M.: E-Commerce and operations research in airline planning, marketing, and distribution. Interfaces **31**, 37–55 (2001)
3. Jarach, D.: The digitalisation of market relationships in the airline business: the impact and prospects of e-business. J. Air. Transp. Manag. **8**, 115–120 (2002)
4. Riege, A.M., Perry, C., Go, F.M.: Partnerships in international travel and tourism marketing: a systems-oriented approach between Australia, New Zealand, Germany and the United Kingdom. J. Travel. Tour. Mark. 59–77 (2002)
5. Ma L., Yan L., Liu J.Y.: An analysis of current development of China's GDS system. Tour. Sci. **3**, 33–36 (2003)
6. Buhalis, D.: eAirlines: strategic and tactical use of ICTs in the airline industry. Inf. Manag. **41**, 805–825 (2004)
7. Ravich, T.M.: Deregulation of the airline computer reservation systems (CRS) industry. 69. J. Air L. & Com. **26**, 387–412 (2004)
8. Tahayori, H., Moharrer, M.: E-tourism : the role of ICT in tourism industry, innovations and challenges, http://hdl.handle.net/123456789/778 (2006)
9. Kozak, N: Transformation of tourism distribution channels: implications of e-Commerce for Turkish travel agencies. J. Hosp. Leis. Marketing, 95–119 (2007)
10. Okulski, R.R.: The role of ICT reservation systems for operational management of air transportation companies. In: Computer sciences and convergence information technology. ICCIT '09. Fourth international conference, 1493–1498 (2009)
11. Rebezova, M., Sulima, N., Surinov, R.: Development trends of air passenger transport services and service distribution channels. Transp. Telecommun, 159–166 (2012)

12. Quinby, D.: An uneasy peace, the PhoCusWright snapshot, PhoCusWright Inc., Nov (2006)
13. AOL Inc., http://techcrunch.com/2010/07/01/google-ita-700-million/
14. tnooz,   http://www.tnooz.com/2012/04/23/news/travelport-agrees-major-partnership-with-japanese-gds-axess-jal-says-goodbye-to-sabre/'
15. Travel tech consulting, http://www.traveltechnology.com/2010/04/apple-itravel/
16. Mietus, J.R.: European community regulation of airline computer reservation systems. Law. Policy. Int. Bus. **26**, 93–118 (1989–1990)
17. Ellig, J.: Computer reservation systems, creative destruction, and consumer welfare: Some unsettled issues. 19 Transp. L. J. 287–308 (1990)
18. Truitt, L.J., Teye, V.B., Farris, M.T.: The role of computer reservations systems international implications for the travel industry. Tour. Manag. **12**, 21–36 (1991)
19. Kleit, A.N.: Computer reservations systems: Competition misunderstood. Antitrust Bull. 833–862 (1992)
20. Chrismar, W.G., Meler, J.: A model of competing interorganizational systems and its application to airline reservation systems. Decis. Support Syst. **8**, 447–458 (1992)
21. Archdale, G.: Computer reservation systems and public tourist offices. Tour. Manag. **14**, 3–14 (1993)
22. Duliba, K.A., Kauffman, R.J., Lucas, H.C.: Appropriability and the indirect value of CRS ownership in the airline industry. In: center for digital economy research stern school of business working paper. Stern, New York (1996)
23. Pemberton, J.D., Stonehouse, G.H., Barber, C.E.: Competing with CRS-generated information in the airline industry. J. Strateg. Inf. Syst. **10**(1), 59–76 (2001)
24. Granados, N., Gupta, A., Kauffman, R.J.: Orbitz, online travel agents and market structure changes in the presence of technology-driven market transparency. Inf. Decis. Sci., Citeseer (2003)
25. Gasson, S.: The impact of e-commerce technology on the air travel industry. In: Annals of cases on information technology. IRMA & Idea Group, 234–249 (2003)
26. Shon, Z.Y., Chen, F.Y., Chang, Y.H.: Airline e-commerce: the revolution in ticketing channels. J. Air. Transp. Manag. **9**, 285–295 (2003)
27. Alford, P.: A framework for mapping and evaluating business process costs in the tourism industry supply chain. Inf. Commun. Technol. Tour. Springer, Vienna (2005)
28. Sengupta, A., Wiggins, S. N.: Airline pricing, price dispersion and ticket characteristics on and off the internet. NET Institute Working Paper, 06–07 (2006)
29. Vukmirovic, M., Ganzha, M., Paprzycki, M.: Developing a model agent-based airline ticket auctioning system. Springer-Verlag, Heidelberg (2006)
30. Lu, W.: An analysis of airline e-Commerce strategies in ticket distribution. In: International conference on service systems and service management, **1–5**, 9–11 (2007)
31. Kim, H.B., Kim, T.G., Shin, S.W.: Modeling roles of subjective norms and eTrust in customers' acceptance of airline B2C eCommerce websites. Tour. Manag. **30**, 266–277 (2009)
32. Quinby, D.: The Role and value of the global distribution systems in travel distribution. PhoCusWright Inc., Nov (2009)
33. Offutt, B.: Travel innovation and technology trends: 2010 and beyond. PhoCusWright Inc., March (2010)

Simulation Analysis for the Design of the Airport Security System

Yoon-tae Sim, Sang-beom Park and Youn-chul Choi

Abstract As the number of air passengers is rapidly increasing worldwide, airports have become more crowded, impairing their efficiency. One of the main reasons why airports are congested is the time required in the security check processes. This paper estimated the time required at the security check process at Incheon International Airport in 2015 based on a simulation; plans and strategies for the improvement are provided in light of the findings. It was concluded in the research that if the gates and search lines go through reforms, there will be a reduction 27 % of the time required at the time of the study.

Keywords Air passenger · Security check · Security search line · Simulation · Airport efficient

Y. Sim
Incheon International Airport Cooperation, 424–47 Gonghang-gil,
Jung-gu, Incheon 400-700, Republic of Korea
e-mail: ytsim01@hanmail.net

S. Park
Korea Aerospace University, 76 Hanggongdaehak-ro, Deogyang-gu,
Koyang-si, Geonggi 412-791, Republic of Korea
e-mail: psb@kau.ac.kr

Y. Choi (✉)
Hanseo Universty, 46 Hanseo-ro, Haemi-myeon, Seosan-si,
Chungcheongnamdo 356-706, Republic of Korea
e-mail: pilot@hanseo.ac.kr

S.-S. Yeo et al. (eds.), *Computer Science and its Applications*,
Lecture Notes in Electrical Engineering 203, DOI: 10.1007/978-94-007-5699-1_62,
© Springer Science+Business Media Dordrecht 2012

1 Introduction

The annual growth rate of Worldwide Gross Domestic Product (GDP) is due to be 3.3 % from 2011 to 2030, according to the data of Boeing. On the other hand, the annual growth rate of air traffic volume, air passengers, air cargo and airplane numbers, is expected to exceed the annual growth rate of Worldwide GDP. As shown in Table 1, over the same period, the rise in the number of global air passengers is expected to be 4.2 % annually. From 2011 to 2030, the annual growth rate increase of worldwide air traffic volume is expected to be 5.1 %; notably, the Asia–Pacific area is expected to take about 28.5 % of global air traffic volume in 2030 [1].

Incheon International Airport (IIA), which opened in 2001, has gone through a continuous increase in its number of passenger: in 2011, 34.5 million people used the airport; this number is predicted to increase to 57.8 million people by 2020 (a 67 % increase from current number). Since the more passengers use airport, the more demands are made on airport processes, the airport authorities act to increase facilities or improve procedures by forecasting the passenger traffic to optimize performance [2].

The security search line takes a lot of time at most airports. As many passengers undergo congested conditions, total time for an aviation security search is delayed and affects the overall organization in the airport. At worst, such shortcomings may lead to airplane delays [3]. Thus, it is crucial to minimize the time it takes for aviation security searches. The easiest way to reduce the lead time for aviation security searches is to increase the number of security search lines, increasing process efficiency: however such increase in number of lines, considering Peak time, causes decrease in cost efficiency of the airport [4].

It costs a significant amount of human and physical resources if a single search space and search team[1] is added. To avoid this cost, process development is needed towards prompt security activity while maximizing the use of the existing facilities at the airport. The present study analyzed efficient operation of the security system at IIA to increase passengers under a simulation [5, 6].

2 Security Check Process and Principle of IIA

Security activities at airports are essentially a responsibility of airports concerning the safety of passengers, as mention in the International Civil Aviation Organization, Annex 17. Airport security search is defined as an act for detection of things that are usable as weapons or items that can cause an illegal disturbance at airport facilities or on airplanes, or jeopardize the safety of passengers, according to the "Aviation Safety and Security Act" Article 2 regarding security. Thus, the

---

[1] Usually one team requires five members including team leader.

Table 1 Prediction of future global air traffic volume

Item	Annual-average-growth-rate (%)
Worldwide GDP	3.3
Global air traffic volume	5.1
Global air passengers	4.2
Global air cargo volume	5.6
Worldwide number of airplanes	3.6

Source Boeing current market outlook 2011–2030 (2011)

security search performed in airports can be explained by a justifiable act of detecting any goods that may interfere in safety aviation, principally done through physical searches of passengers and baggage.

Currently, as terrorist activities have been increasing internationally not to mention domestically, early search to prevent such activities is also strengthened, increasing time for security screening of passengers and impacting entire operational process of airport. On the other hand, as most passengers prefer easy and prompt security check, their complaints over the lead time increase is getting bigger. It is at the peak time that most complaints are made. If the number of gates and security search lines were increased, security searches would be processed promptly since passengers pass through security search lines after the departure gate. However, the fact that peak time at airport stretches resources remains problematic.

Generally, in airports, there are certain times when passengers are particularly concentrated (according to characteristics and conditions at specific airports); it is during such times that confusion takes place at IIA, around 2–3 times per day for about 5 h. Therefore, if the security search lines are operated with considering of the maximum passenger of peak time, the passenger processing time decreases. However, the usage rate of an airport will diminish rapidly during the off-peak portion of the day (approximately 19 h). Consequently, as it is difficult to increase security search lines under such circumstances; instead, efficient airport development and operating plan are required.

Usually, in the case of IIA, 2–3 h prior to departure, passengers arrive at the airport and finish embarkation procedures at the airline check-in counters. Then they move into one security area from among the four in an airport. Once passengers pass through the security gate, they move to the security search lines. There are 10 security search lines at each security checkpoint, and it is possible to adjust the number of lines depending on the number of passengers. After putting hand baggage on an X-ray machine, passengers pass by through a metal detector. During this process, if chemicals or metals are found, machine alarms are activated and secondary screening is processed. Passengers receiving additional searches will go through Pat-down searches, and the process will not stop until the alarm is turned off. Usually, it takes 10 min on average to go through this process, unless a passenger violates aviation security. In case of IIA, the flight security search has a form as shown in Fig. 1.

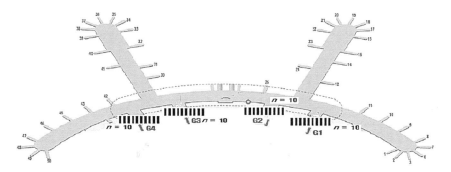

Fig. 1 IIA security gate and security search line

Passengers move to the search gate after the ticketing at check-in counters. IIA has installed four gates (G1, G2, G3 and G4). At each search gates, 10 search lines are installed. That is, a total of 40 search lines are operating at IIA. According to the statistics, confusion rarely happens at the 4 gates in IIA: however, it sometimes happens at peak time. Specifically, it was found that the confusion was high at Gate 1 and Gate 4 at peak time. Still, such delays correspond to international standards in general, and thus do not cause passengers to feel inconvenience during the delayed time.

3 Simulation Methodologies for Security System

3.1 Passenger Demand Forecasting

As the number of passengers increases, it is estimated that passenger processing delays will be a big problem. Considering this, IIA is currently underway for a third-phase expansion to solve the problem regarding increased passengers.

The biggest issue here is the airport operation in 2015, that will bring a number of passengers that will approach airport capacity. For related issues, the best solution of problems is to facilitate more efficient passenger processing [7]. Thus, this study analyzed air passenger processing under conditions anticipated for IIA in 2015. The number of passengers in each time was estimated through a passenger forecasting system. The variance data of passengers was calculated using the average number of passenger notice data collected on August 2011.

According to the analysis results, as shown in Fig. 2, the airport security search experienced the most delay during the peak time (18:00–20:00 and 8:00–10:00). Maximum hourly passenger numbers was 2,648 people at (9:00 in Gate 4) maximum passengers appeared.

On the other hand, it showed that at 19:00, both Gates 1 and 4 processed about 200 passengers, while at Gates 2 and 3, one thousand people entered, thus showing a big difference.

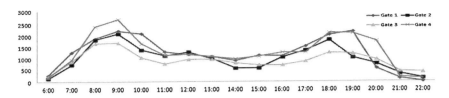

Fig. 2 2015 volume of passenger/time period

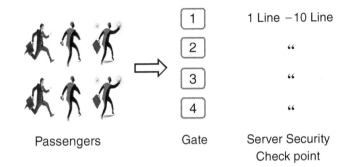

Fig. 3 Security search M/M/s queuing model

3.2 Simulation Model

This section presents the M/M/s queuing model to determine the optimal number of security search lines to be opened in given period. The M/M/s queuing model is suitable for the airport security system management problem as security search lines are designed as multi queuing lines and multi servers [8, 9]. An explanation regarding queuing system is as shown in Fig. 3.

In this study, an M/M/s model shows the following assumptions:

- Arrival rate obeys the Poisson distribution, which is fluid per time.
- Service process follows the Triangular Distribution.
- The number of service is convertible as whole number from 1 to 10.
- The number of people who can be waiting in a single line is 150.
- Security service, is based on "First come first serve" measures. If service of one passenger is over, the next passenger enters the system.
- The possible number of people allowed to enter the queue system is limitless.

M/M/s model and the related symbols and equations are as follows:

Parameters

λ_t	The average arrival rate per hour
μ_t	The average screening time per hour
REQ_{dt}	The necessary number of security checkpoints for each time period.

The following formulas are used for determining the required security line for each time period in security queue.

$$P_0 = \left[\sum_{n=0}^{s-1} \frac{(\lambda_t/\mu_t)^n}{n!} + \frac{(\lambda_t/\mu_t)^s}{REQ_{dt}!} \left(\frac{1}{1-p} \right) \right]^{-1} \tag{1}$$

Formula (1) shows the probability that no customer exist in the system.

$$L_q = \frac{P_0(\lambda_t/\mu_t)^s p}{REQ_{dt}!(1-p)^2} \tag{2}$$

Formula (2) represents the average number of customer waiting at security lines.

$$w_q = L_q/\lambda_t \tag{3}$$

Formula (3) represents the average time spent waiting at the security search lines.

$$w = w_q + \frac{1}{\mu_t} \tag{4}$$

Formula (4) represents the average time spent in the system, including service.

$$L = \lambda_t W \tag{5}$$

Formula (5) represents the average number of passenger in the system.

$$\text{Prob}(W > 0.05) \leq 0.25 \tag{6}$$

Formula (6) represents that the security search should satisfy probability within 95 %, allowing for times given in service systems.

The objective of determining model of security search line is to minimize the number of required security search lines.

Minimize REQ_t

subject to

$$\text{Prob}(W \geq 0.25416) \leq 0.05 \tag{7}$$

Constraint (7) represents the average time spent in the system, including service, and it has to be less than the maximum allowance level (15 min and 15 s, in our case) for 95 % of all passengers.

$$\text{Prob}(W_q \geq 0.25) \leq 0.05 \tag{8}$$

Constraints (8) represents the average waiting time before entering the system, which has to be less than the maximum allowance level (15 min and 15 s, in our case) by 95 % the average waiting.

$$REQ_t \geq 0 \tag{9}$$

Table 2 Number of security check line by simulation result at time periods

Time	6:00	7:00	8:00	9:00	10:00	11:00	12:00	13:00	14:00	15:00	16:00	17:00	18:00	19:00	20:00	21:00	22:00
Gate 1	1	2	6	8	10	9	6	6	5	4	5	5	7	9	10	3	1
Gate 2	1	1	4	7	9	6	5	6	5	3	3	5	6	8	5	4	2
Gate 3	1	1	4	8	8	5	4	5	5	4	4	4	4	6	6	5	2
Gate 4	1	1	5	10	10	8	5	6	5	5	5	6	6	10	10	8	1
Sum	4	5	19	33	37	28	20	23	20	16	17	20	23	33	31	20	6

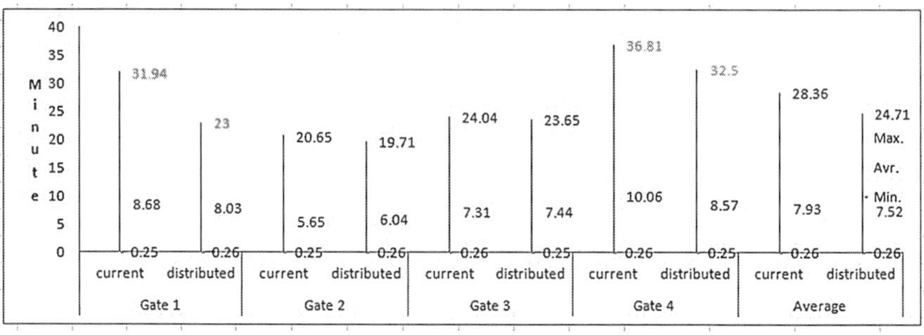

Fig. 4 Simulation results for total security time

Constraints (9) represents the number of required security search lines for each time period.

To resolve waiting personnel in the security check, a simulation model was made using Java program, assuming the improvement conditions; in the simulation of the improving condition: first, a scenario was designed to increase the security search lines before peak-time in order to avoid bottleneck. Simulation for the increase in security checkpoints was modeled per hour; peak times are 8:00–10:00 and 18:00–20:00. By applying real-time monitoring system to the passenger, the simulation assumed that the passengers can be led to gates with less waiting time at the security checkpoint by letting them know the delay times for each departure.

4 Simulation Results

The operating optimum plan analysis results of a security search by each gate derived through simulation are as follows. We can get the largest effect to operate current facilities by operating optimum gates and adjusting the number of security search lines in gates, according to modeling result of this study. The analysis results of optimum operating method of security search line by each gate derived in simulations are as follows. We also assume that passengers disperse to neighborhood gates following the forecast. Therefore, it is required to lead the passengers to disperse to each gate following the forecast. According to the modeling results of this study, in order to resolve such problems, it is most effective to adjust the number of security search lines within each gate and operate the gates optimally.

As shown in Table 2, it was computed that, at most, 10 security search lines have to be operated per day at 6 different places. During peak hour, It was computed to be advisable to operate one or two additional nearby security checkpoints.

According to Fig. 4, since the processing time of each security checkpoint is operated around 5.65–10.06 min, it was expected that 95 % of the time, the security search place will be operated with little confusion. In the case of Gate 4, the average processing time was shown to be between 10.06 and 36.81 min, which clearly requires improvement. After improvement of the average processing time from 7.93 to 7.52 min, the maximum processing time was reduced from 28.32 to 24.71 min. In case of Gate 4, which experienced the heaviest congestion, the processing time was reduced from 10.06 to 8.57 min, which comprises about 13 % improvement. Maximum time in the simulation ranged from 36.81 to 32.50 min. Gate 1 has been reduced from 8.68 to 8.03 min for average time, and 31.4 to 23.0 min for maximum time, which comprises about 27 % improvement.

The range of average time, especially at the two peak times from (5.65–10.06 min) to (6.04–8.57 min).

5 Conclusions

This study was done to help preclude passenger-processing delays caused by excessive security search times predicted at IIA from 2015 and on. That is, by applying the predicted number of passengers expected in 2015 to the current aviation security system, we analyzed the expected lead time at each gate in terms of passenger and security search lines. This study proposed that the optimum security search operation system under current conditions and with equipment that will direct passengers to each gate based on the computed data of M/M/s simulation will reduce the waiting time for each passenger.

As a result of simulations, it was determined that if the reformed gate system run side by side of security search lines, there will be a maximum 27 % decrease in waiting time for aviation security searches. These measures are predicted to optimize the operation efficiency of an aviation security system.

The simulation results are calculated considering only the processing time of the security search lines. Thus, the airport authority will have to naturally lead passengers to voluntarily disperse to appropriate gates through the means of electric bulletin boards, which will inform them of the line times at each gate; this way, the airport will become a "smart airport", optimizing use of resources.

References

1. Boeing: Current market outlook 2011–2030 (2011)
2. Ioanna, E.M., Konstantinos, G.Z.: Assessing airport terminal performance using a system dynamics model. J. Air Transp. Manag. **16**, 86–93 (2010)
3. Gkritza, K., Niermeier, D., Mannering, F.: Airport security screening and changing passenger satisfaction: an exploratory assessment. J. Air Transp. Manag. **12**, 213–219 (2006)
4. George, P.: Some practical considerations on multi-server queues with multiple poisson arrivals. Omega **6**(5), 443–448 (1981)
5. Michel, R.G., Simon, W.W.: Analysis and simulation of passenger flow in an AIRPORT terminal. Proceedings of the 1999 winter simulation conference (1999)

6. Brusco, M., Jacobs, L., Bongiorno, J., Lyons, V., Tang, B.: Improving personnel scheduling at airline stations. Oper. Res. **43**, 741–751 (1995)
7. Jacobson, S., Virta, J., Bowman, J., Kobza, J., Nestor, J.: Modeling aviation baggage screening security systems: a case study. IIE Trans. **35**, 259–269 (2003)
8. Gilliam, R.: An application of queuing theory to airport passenger security screening. Interfaces **9**, 117–122 (1979)
9. Seo, S., Choi, S., Lee, C.: Security manpower scheduling for smart airports. Computer science and convergence, Lecture Notes in Electrical Engineering **114**:519–527(2012)

Measuring Asia Airport Productivity Considering the Undesirable Output

Sangjun Park and Chulung Lee

Abstract It becomes important for airports to consider environmental aspects such as the concept of sustainable growth, especially for Asian airports growing and competing aggressively. Thus, it requires incorporating the environmental factor into the productivity indicator of airport and is important to understand how the efforts impact to their current performance and future competition. This paper used Data Envelopment Analysis (DEA) method including the two undesirable outputs to assess the productivity of 11 Asia airports. Inclusion in the analysis of airport operations with the undesirable effects leads to greater airport's efficiency scores today and shed a further insight in a competitiveness and risk among Asia airports, when each Asian government extends their environment regulation and tax.

Keywords Data envelopment analysis · Airport · Efficiency · Environment

1 Introduction

There are more efforts to improve an environmental performance from airports in according to the recent movement in a government regulation and cost increase on the environmental factors like noise and carbon emission. The most of airlines

S. Park
Department of Industrial Management Engineering, Korea University, Anam-dong
Seongbuk-gu, Seoul 136-701, Korea
e-mail: edmond@korea.ac.kr

C. Lee (✉)
Department of Industrial Management Engineering and Graduate School of Management
of Technology, Korea University, Anam-dong Seongbuk-gu, Seoul 136-701, Korea
e-mail: leecu@korea.ac.kr

S.-S. Yeo et al. (eds.), *Computer Science and its Applications*,
Lecture Notes in Electrical Engineering 203, DOI: 10.1007/978-94-007-5699-1_63,
© Springer Science+Business Media Dordrecht 2012

flying to EU became complying to the agreement of European carbon emissions [1] and some of Asia country governments are planning to apply this regulation gradually. The city area near Incheon international airport is under the close monitoring and controlling by a new Korean environment regulation against airport noise since 2010 and the $2M claim against Korean airport noise was accepted by Korea court recently. Such an environment dispute with a legal and financial claim has been kept arising from major cities near the airport who showed a rapid expansion recently such as Seoul [2] and Shanghai [3]. The expansion in Asian air transportation market causes a concern on an environmental issue by increasing the undesirable factors such as noise and carbon emission, since "Sustainable growth" has been emerged [6] and emphasized globally with the reinforced environment policy and regulation in recent years. However, there are rarely studies including an environment factor across international borders. Yu [7, 8] had included the airline noise of Taiwan local airport performance and Lin [9] compared 20 international airport performance using desirable factors. But among the existing DEA approaches to the airport performance, there was no research considering the environment factors including the both, a noise and a carbon emission across Asia airports in a significant growth with competing in same region. This paper is approaching to measure Asia airport performance including both desirable and undesirable factors and to compare each airport performance trend for a further business insight in a competiveness and risk, when each Asian government extends their environment regulation and tax. This paper is organized as follows. Section 2 provides a brief background of Asia airport market and the literature. Section 3 introduces the performance model including the environmental factors. Section 4 describes the selection of variables and data. Section 5 gives the empirical results and Sect. 6 concludes this study.

2 Background and Literature Review

Under the increasing requirements and regulations on the noise and carbon emission control from airports, some Asia airports are proactively responding to such a demand. Airport Authority Hong Kong (AAHK) and 40 business partners pledged to make Hong Kong International Airport (HKIA) the world's greenest airport [10]. Incheon airport announced invest $65 M in the next 5 years for the CO2 emission reduction of 50,400 tons in their annual report, 2011. In the last decade, the operation performance of airport has been studied in many papers using a nonparametric approach called data envelopment analysis (DEA). Gillen and Lall [11] analyzed US 21 main airports with DEA and Parker [12] showed the performance change of UK 22 airports by DEA before and after their liberalization. However, then certain outputs are undesirable, they may yield biased measures of efficiency if only "goods" are valued and "bads" are ignored [8]. Therefore it requires considering not only favorable factors, but also unfavorable factors. In spite of the importance of environmental factors in an airport performance review, there are not many studies including the undesirable factors up to now. Yu [7, 8] performed the DEA analysis for Taiwan's airports including the noise and tt showed that more airports moved closely to the efficient frontier when removing the noise factor. This paper analysis and compare the airport performance across Asia including the environmental factors, Noise and Carbon emission.

3 Methodology

3.1 Directional Distance Function

Coelli et al. [14] explained the capacity in a specific industry as the maximized output with Input Non-negative value ($\neq X$) and the technology. Chambers et al. [15] defined the capacity with the below function. The production function is composed of desirable and undesirable output which is made from the input vector.

$$P(x) = (y, b) : x \text{ can produce}(y, b) \tag{1}$$

- $Y \in R^{m+}$: Desirable output, Vector ($m = 1, 2, ..., M$)
- $b \in R^{j+}$: Undesirable output, Vector ($j = 1, 2, ..., J$)
- $x \in R^{n+}$: Input output, Vector ($n = 1, 2, ..., N$).

Färe et al. [16] and Färe and Grosskopf [17] explained the optimized production function $P(x)$ with the below conditions.

(1) $P(x)$ is convex compact, (2) $P(0) = (0,0)$, (3) If $(y, b) \in$ and $P(x) b = 0$, then $y = 0$ (4) then $\leq \leq$ If $(y, b) \in P(x)$ for y' y, $(y', b) \in P(x)$, and for x' x, $(y', b) \in P(x) \in P(x')$ and $0 \leq \theta \leq 1$ then (θ_y, θ_b) and (5) If $(y, b) \in P(x) \in P(x)$

A production function $P(x)$ was described with Fig. 1 with the above 5 conditions. $P(x)$ can be obtained from the Eq. (2) with DEA.

$$P(x)[(y, b) :$$

$$\sum_{k \in K} \lambda_k y_{km} \geq y_{km}, \quad m = 1, 2, ..., M$$

$$\sum_{k \in K} \lambda_k b_{kj} = b_{kj}, \quad j = 1, 2, ..., J \tag{2}$$

$$\sum_{k \in K} \lambda_k x_{kn} = x_{kn}, \quad n = 1, 2, ..., N$$

$$\lambda_k \geq 0, k = 1, 2, ..., K]$$

The directional distance function (DDF) can consider the both, a desirable output and an undesirable output under the fixed input. Chamber et al. [15], Färe and Grosskopf [18] introduced Directional Output Distance Function (DODF).

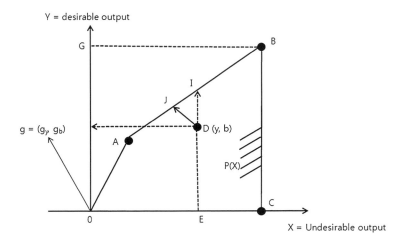

Fig. 1 Construct of production possibility set and the DDF

$$D_0 = (x, y, b; g_y, -g_b) = \max \beta : (y + \beta g_y, b - \beta g_b) \in P(x) \tag{3}$$

In the Eq. (3), $g = (g_y, g_b)$ as the vector of the desirable output and undesirable output, $(g_y \in R_+^M, g_b \in R_+^J)$, $g = (y, 0)$ does not include the undesirable output and $g = (1, -1)$ has the same weight on the both, the desirable and undesirable output.

$\beta\beta\beta$ DDF can be applied into other model and the below linear programming (4) was transformed by DEA with a directional DDF to evaluate the production efficiency. The airport is efficient with zero while the airport efficiency is worse when the number increases.

$$\text{Max } \beta_k$$
$$s.t. \sum_{k \in K} \lambda_k y_{km} \geq y_{km} + g_y \beta_k, \quad m = 1, 2, \ldots, M$$
$$\sum_{k \in K} \lambda_k b_{kj} = y_{kj} + g_b \beta_k, \quad j = 1, 2, \ldots, J \tag{4}$$
$$\sum_{k \in K} \lambda_k x_{kn} \leq x_{kn}, \quad n = 1, 2, \ldots, N$$
$$\lambda_k \geq 0, \quad k = 1, 2, \ldots, K$$

- x: Input valve, λ_k: Weight for each variable, y: desirable output value,
- b: undesirable output valve, β_k: Efficiency score of each airport
- g: Weighting between the desirable and undesirable output.

4 The Data Selection

4.1 Variable Selection

It is required to define the input and the output of the airport to evaluate Asia airports. Each research has defined the diverse input and output according to their focused elements. In this paper, they are focused and approached to evaluate the airport operation performance including the main two environment factors, a noise and carbon emission that have been controlled recently by a regulation with a financial cost like EU [1] and Korea [19], as shown in Fig. 2. Thus, the annual Work Load Unit (WLU) capacity and the operation cost are considered as the input and WLU, profit, airport noise and the carbon emission are considered as output for 11 major Asian airports. The facility environment data like the size of a runway, terminal and parking lot are excluded from this paper because their basic facility environment from each Asia airport. It requires to be considered when the airport facility conditions are not equal and the facility environment is out of direct control [7].

4.2 Data Selection

This research selected 11 Asia main international airports which are in a stiff competition as a hub airport in Asia and 5 years data are collected to show their performance trend yearly. It requires enough number of Decision Making Unit (DMU) comparing to the number of input and output for the required degree of freedom because DEA is a model to compare among DMU. Boussofiane et al. [20] suggested the required number of DMU with n m s, while Fitzsimmons [5] did with $n \, 2 \, (m + s)$ and Cooper et al. [21] did $n \, Max\{3 \, (m + s), m \, s\}$. 'n' is the minimum required quantity and 'm' is the number of input and 's' s the number of output. As this analysis is consisting with two input and four output for 11 Asia airports, it does not satisfied with the above suggested conditions. Therefore, the extended 'n-year-window DEA' [4] was used with the time leg with two years in this paper.

4.2.1 Carbon Emission from Airports

Among the guide line of IPCC, the Tier 2 was used with the weighted emission quantity with LTO which is based of Time in Mode (TIM). The level of emission can be calculated with no. of LTO and LTO emission index. The LTO index is different from an airline engine with the Eq. (5).

Production model of airport operation

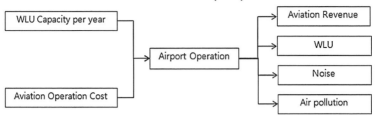

Fig. 2 Production model of airport operation

$$E_{ij} = \sum_k \left(TIM_{jk} \times FF_{jk} \times EI_{ijk} \times N_j \right) \tag{5}$$

- E_{ij}: Carbon emission, i from the aircraft j at 1 LTO
- TIM_{jk}: Averaged time of the air, j at the mode k
- FF_{jk}: Fuel quantity from the aircraft j at the mode k
- EI_{ijk}: Carbon emission, i in the air, j at the mode k
- N_j: No. of engine from the aircraft j.

This research considered the carbon emission, NOx, CO and HC which impact mainly in the cost. Also they have different impacts per each carbon gas emission [13].

$$API_a = \sum_{i=1}^{4} C_i \times E_{ia} \tag{6}$$

- API_a: Carbon emission from airport a. C_i: Cost of the carbon emission,
- E_{ia}: Quantity of the carbon emission from the airport a.

4.2.2 Noise from Aircraft at Airports

The airport noise is decided by the aircraft type and the number of movement. The noise can be recorded at three points, Flyover, Lateral and Approach. Also the Effective Perceived Noise Level (EPNL) was considered and calculated with the index data from European Aviation Safety Agency (EASA).

$$WECPNL = \overline{EPNL} + 10 \, \log N - 39.36 \tag{7}$$

$N = N_1 + 6N_2 + 10N_3, \quad N_1 = \text{Movement from 7 to 19}$
$N_2 = \text{Movement from 19 to 22}, \quad N_3 = \text{Movement from 22 to 7}$
$\overline{WECPNL}1$ annual average can be shown by the function (8)

Table 1 Input and output factors per case

	Case 1	Case 2	Case 3a	Case 3b
Input	WLU capacity, operation cost	WLU capacity, operation cost,	WLU capacity, operation cost,	WLU capacity, operation cost,
Output	WLU, profit	WLU, profit, carbon emission, noise	WLU, profit, carbon emission	WLU, profit, noise

Table 2 Airport performance comparison (traditional vs. analysis with undesirable factors)

	Case 1	Case 2	Case 3a	Case 3b
Input	WLU capacity,	WLU capacity,	WLU capacity,	WLU capacity,
	Operation cost	Operation cost,	Operation cost,	Operation cost,
Output	WLU, profit	WLU, Profit,	WLU, Profit,	WLU, Profit,
		Carbon emission,	Carbon emission,	Noise
		Noise		

Fig. 3 Airport rank under a categorization with environment factors

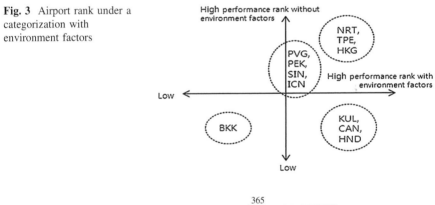

$$\overline{WECPNL} = 10LOG\left[(1/365)\sum_{i=1}^{365} 10^{0.1 \times WEGPNL_i}\right] \qquad (8)$$

$WECPNL_i$ = WECPNL amount for i date.

5 Result

The empirical result is shown under the three cases in the Table 1. The case 1 only includes the desirable factor while the case 2 also includes the undesirable factor. In the case 3, each single output from two unfavorable outputs was deleted each time.

Table 2 and Fig. 3 provide the main results with the score and rank from each airport in the three cases. In general, NRT, HKG and TPE airports got a good performance while BKK and KUL get comparatively lower score in the case 1.

But in the case 2, it showed a different result. CAN airport showed a better score and rank while BKK still did in a low score and rank. The main purpose in testing the case 3 on the implementations of DEA is to review the effects of a particular environmental output on overall efficiencies. In other word, one single input may have significant influence of a unit but this factor may not necessarily play an important role on other units. One important observation is that each airport showed the different result in a performance score and rank by deleting each output factor. In the airport management perspective, it might be questioned how each environmental factor works with their airport performance and this result can provide specifically the productivity information which they have to focus on the environmental factors. In addition, future research would also develop and assess the additional environmental factors and conditions future forecasting work on the productivity based on their past performance data.

6 Conclusion

This paper measured and compared the performance under which two undesirable outputs considered: noise and carbon emission for 11 Asian major airports who are in a stiff completion as the first time by DEA method with the non-parametric directional output distance function (DDF) to assess during 2005–2010. The result shows that the performance of Asian major airports was measured differently when including the undesirable factor comparing to the traditional measurement review only with the desirable factors. Also it shows the different performance trend per year with the assumption linking to the recent proactive efforts on the environmental area from each Asian airport. In other words, the integrated performance could be increased when it is optimized by the various influential factors on airport's productivity performance such as an aircraft selection in a size and a type considering the reduction of noise and carbon emission from Asia airports. The proven inclusion in the analysis of Asian airport operations with the undesirable effects leads to greater airport's efficiency scores today and shed a further insight in a competiveness and risk among Asia airports, when each Asian government extends their environment regulation and tax in future.

Acknowledgments This research was supported by Basic Science Research Program through the National Research Foundation of Korea(NRF) funded by the Ministry of Education, Science and Technology(2009-0076365)

References

1. European Commission: Climate Action on Aviation sector (2012)
2. Choi, J.S.: Increasing law sue against Korea airports, Choson Daily (2012)
3. Xiaoru, C.: Airport protests about serious aircraft noise, Shanghai Daily (2012)

4. Nghiem, H.S., Coelli, T.: The effect of incentive reforms upon productivity. J. Dev. Stud. **39**(1), 74–93 (2002)
5. Fitzsimmons, J., Fitzsimmons, M.: Service Management for Competitive Advantage. McGraw-Hill, New York (1994)
6. Labuschagne, C.: Assessing the sustainability performances of industries. J. Clean. Prod. **13**, 373–385 (2005)
7. Yu, M.M.: Measuring physical efficiency of domestic airports in Taiwan with undesirable outputs and environmental factors. J. Air Transp. Manag. **10**, 295–303 (2004)
8. Yu, M.M., Hsu, S.H.: Productivity growth of Taiwan's domestic airports in the presence of aircraft noise. Transp. Res. Pare E **44**(3), 543–554 (2008)
9. Lin, L.C., Hong, C.H.: Operational performance evaluation of international major airports. J. Air Transp. Manag. **12**(6), 342–351 (2006)
10. Elliott, M.: HKIA pledges to be world's greenest airport, Travel News (2012)
11. Gillen, D., Lall, A.: Developing measures of airport productivity and performance: an application of data envelopment analysis. Transp. Res. E **33**(4), 261–273 (1997)
12. Parker, D.: The performance of the BAA before and after privatization. J. Transp. Econ. Policy **33**, 133–146 (1999)
13. Ding, J.M.W., Wit, R.C.N., Leurs, B.A., Davidson, M.D., Fransen, W.: External Costs of Aviation. Federal Environmental Agency, Umweltbundesamt, Berlin (2003)
14. Coelli, T., Rao, D.S P., O'Donell, C.J. Battese, G.E.: An introduction to Efficiency and Productivity Analysis. Springer, New York (2005)
15. Chambers, R.G., Chung, Y., Färe, R.: Profit, directional distance function and Nerlovian efficiency. J. Optim. Theory Appl. **98**, 351–364 (1998)
16. Färe, R., Grosskopf, S., Lovell, C.A.K., Pasurka, C.: Multilateral productivity comparisons when some outputs are undesirable. Rev. Econ. Stat. **71**, 90–98 (1989)
17. Färe, R., Grosskopf, S,: Productivity and undesirable outputs: a directional distance function approach. J. Environ. Manag. **51**, 229–240 (1997)
18. Färe, R., Grosskopf, S.: Theory and application of directional distance functions. J. Prod. Anal. **13**(2), 93–103 (2000)
19. Ministry of Land, Transport, and marine time affairs: Regulation on airport noise (2010)
20. Boussofiane, A., Dyson, R.G., Thanassoulis, E.: Applied data envelopment analysis. Eur. J. Oper. Res. **52**, 1–15 (1991)
21. Cooper, W.W., Seiford, L.M., Tone, K.: Data Envelopment Analysis. Kluwer Academic Publishers, Boston (2000)

A Study on Aviation Technology Forecast for Sustainable (Green) Aviation Using Patent Analysis

Hyejin Kwon and Chulung Lee

Abstract For efficient technology development, prediction of directions for technology development in the future is important. In the present study, trends of changes in development of green aviation technology will be analyzed through patent analysis, and directions for development in the green aviation technology in the future will be predicted. The numbers of countries' filed patents per year will be compared through an analysis of trends in the case numbers among quantitative analysis methods, and what technology areas are developed, with emphasis on the country to be analyzed through an analysis of companies with multiple filings. Finally, after preparing summary lists and worksheets with a qualitative analysis method, trends of technology development, along with directions for development of green aviation technologies in the future, will be predicted through preparation of a techno-map based on such results.

Keywords Sustainable aviation · Green aviation · Aviation technology · Patent map · Patent analysis

H. Kwon
Department of Information Management Engineering,
Korea University, Seongbuk-gu, Seoul 136-701, Korea
e-mail: hjkwon87@korea.ac.kr

C. Lee (✉)
Division of Industrial Management Engineering and Graduate School of Management
of Technology, Korea University, Seongbuk-gu, Seoul 136-701, Korea
e-mail: leecu@korea.ac.kr

S.-S. Yeo et al. (eds.), *Computer Science and its Applications*,
Lecture Notes in Electrical Engineering 203, DOI: 10.1007/978-94-007-5699-1_64,
© Springer Science+Business Media Dordrecht 2012

1 Introduction

The amount of greenhouse gases generated by the aviation industry accounts for about 3 % of the total generated amount in the world. Compared with the 10 % by road-related areas and 13 % by agricultural areas, this seems to be a relatively small number [1]. However, since the greenhouse gases caused by aviation are exhausted at a high altitude, they are less apt to be absorbed by trees and plants. Thus, the greenhouse gases exhausted from airplanes have direct influence on the greenhouse effects. For this reason, despite the relatively small emissions, 13 % of the overall greenhouse effects were shown to be caused by greenhouse gases by airplanes [1]. Therefore, in a reality with increasing carbon cost, investments in technologies for reducing a carbon exhaust from the aviation industry are expected to alleviate carbon exhaust costs and enable further sustainable business by airline companies. For efficient technology development, prediction of directions for technology development in the future is important.

Agarwal [2], Jung [3], and Lee [4] offer useful references for green aviation technology prediction. Agarwal [2] is describes each technology as classifying and organizing contents related to green aviation technology. Jung [3] analyzes technologies with specialized statuses utilizing RTA (Revealed Technological Advantage) among the patent analysis index, and CII (Current Impact Index) is utilized for technology influence analysis. Lee [4] analyzed development forecast and trends of future air traffic ways. These studies have limits since they are simply describing green aviation technology and analyzing only current technology levels. However, this study is differentiated from those studies in analyzing current technology level through patent analysis and predicting future technology development direction based on this.

This study writes a patent map (PM) through patent analysis related to green aviation technology in the world, analyzes green aviation technology at each technology field based on the written PM, and predicts future research development direction of Green Aviation technology considering relations between technologies.

2 Patent Analysis

A sequence of preparing a PM is broadly divided into stages of data survey and classification, and analysis and PM preparation [5]. In the stage of data survey and classification, themes and directions of patents to be searched are set and strategies are established. After that, keywords are set as per the area of a patent theme, followed by a search. After the patent search, abstracts of the texts are extracted, and reviewed. This is followed by setting of secondary keywords through supplementary surveys, and the previous stages will be repeated. Subject data for a PM are extracted based on patented technologies from the supplementary surveys,

Table 1 Keywords of green aviation technology

Category I	Category II	Category III	Keyword
Hardware improvement (1)	Alternative fuels (A)	Green fuels (i)	Bio fuel, Fuel cell, Hydrogen-powered aircraft, Solar-powered aircraft, Hydrocarbon fuel
	Technological innovation in aircraft design and engines (B)	Airframe type and design (ii)	Composite metal, Laminar flow wing, Aircraft recycling, Supersonic aircraft, Coating material. Aircraft noise
		Engine type and design (iii)	Geared Turbofan, Open-rotor engine, Electric battery aircraft, Combustion engines, Energy accumulation engine
Software improvement (2)	Operational improvement (C)	Flight operatin system (i)	Close formation flying, Wireless cabin, Aircraft navigation, Airway control system, Refueling in-flight
		Airport management system (ii)	Electric/hybrid ground vehicle, Sustainable airport

with the data analyzed and fabricated. Finally, patent data from fabrication and analysis go through stages of being converted into diagrams and media convenient to use for completion of a PM. Through the PM, what level of green aviation technology is reached at present and how it may be changed in the future can be identified. Detailed contents per stage for preparation of a PM in this study are as follows.

2.1 Keyword Selection and Data Collection Stage

In the present article, a PM is prepared for the green aviation technology, which may be largely divided into Categories I, II, and III, as shown in Table 1. Patented technologies are searched after keywords are selected according to subdivisions. Keywords for each subdivision were selected after a literature survey for the green aviation technology [2, 6–10].

Keyword extraction is performed through repeat of verification process, including the keyword draw regarding green aviation technology through the document survey, followed by a secondary keyword draw by additional analysis of the patent documents found by patent search and verification through another patent search. Categorized results are shown in Table 1.

2.2 Data Analysis and PM Preparation

In general, quantitative analysis and qualitative analysis are employed for patent analysis [5]. In addition, both analysis methods are applied for this study.

With the quantitative analysis, patent information is analyzed simply with the number of patents. However, the number of patents is the concrete evidence of technology development, so that various hidden situations and flows and technology become visible when the trend on technology development is understood. Through the qualitative analysis, basic patents in research project technology area, important patents establishing foundation of technology area, and problematic patents with possibilities of right problems on a company's product design can be drawn.

2.2.1 Quantitative Analysis

For this study, the case number trend analysis and the multiple applicant analysis of the qualitative analysis method are applied. The case number trend analysis is an analysis based on application dates to check changes of various patent applications according to a course of time.

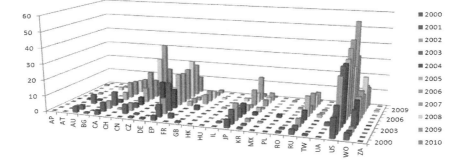

Fig. 1 Trends of patent number of green aviation technology classified by nation

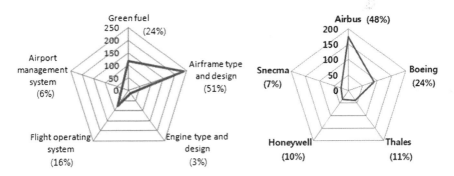

Fig. 2 Radar map of application numbers regarding technology areas and applicants

According to Fig. 1, the U.S. is shown to be applying for patents regarding green aviation technology more than any other nation, while the number of applications is continuously growing. It also shows that Europe has relatively higher application numbers. In analyzing the reasons for the high number of patent in the USA and Europe, a main factor is aerospace companies with headquarter located in the USA or Europe—such as AERION, HONEYWELL, and BOEING in the USA, SNECMA and AIRBUS in France, etc—were most active in the field related to green aviation technology according to number of patents by applicants.

Next, the distribution of application case numbers for each technological area is reviewed by applying the multiple applicant analysis. The multiple applicant analysis is conducted to identify which areas are the focus in related patented technologies.

Figure 2 presents accumulation of application numbers regarding five technology areas in the radar map. According to the radar map, it is found that airframe type and design technology have more applications than other technology areas do. In other words, in green aviation technology, developments are found to be concentrated on the airframe type and design area, including aircraft design for fuel saving.

The reason why many patents were registered in the area of airframe type and design is because there were many patents in the aircraft noise and supersonic aircraft area. The numbers of patent in aviation noise are increased because of increased corresponding legal regulations, since jet flight causes direct and considerable damages to adjacent areas. In addition, supersonic aircraft were developed by jet engine development, and their importance is increased in the role of combat and reconnaissance. Accordingly, every country concentrates their research on the technology to increase air speed and spur supersonic aircraft development. Also, it is found that patents on supersonic passenger plane technology are consistently increased since needs for them rapidly growing along with growing transportation demands in the longest distances between US, European, and Asian countries.

According to a radar map for patent applications as per applicant, it can be seen that the top 5 businesses with patent applications for green aviation technology were Airbus, Boeing, Thales, Honeywell and Snecma. Airbus is producing even research outcomes in airframe type and design, flight operating system and green fuel. In the case of Boeing, patent applications are done mainly in flight operating system and airframe type and design parts. In the case of Honeywell and Snecma companies, patent applications are made primarily in flight operating system and airframe type and design areas, respectively. Finally, with Thales company, there are many applications related to navigation among flight operation system, showing that their emphasis is placed on development of navigation-related technology.

2.2.2 Qualitative Analysis

Patented technologies corresponding to the keywords are collected and analyzed to establish a techno map showing a technology development chart that displays technological development status in the time series. The technology development chart is necessary to understand technology areas' technology development flows. Next is the technology development chart of green aviation technology from 2000 to 2010.

According to the technology development chart of green aviation technology, it is shown that fuel cell and noise technology areas have been continuously developed. In 2000, a patent on noise reduction on aircraft's wings was applied for. From 2003 to 2010, noise reduction patents on engines and landing gear are shown to be continuously applied for. According to applications submitted for patents for fuel cell technology by Boeing Corporation, Ion America Corporation, and AIRBUS GMBH in 2002, 2005, and 2006, respectively, it is possible to assume that interests on the fuel cell technology are have increased in the aviation industry.

In fuel cell technology, patents are consistently registered for application to aircraft energy sources from early 2000 to 2010. Airbus registered patents on fire protection systems and portable water inside aircrafts by utilizing water recycled from fuel cell in 2005 and 2006, respectively. In addition, ION AMERICA registered patents on energy, heat, water, and oxygen generation by utilizing fuel cell systems in 2006. In utilizing fuel cells in the aircraft like this, the application technology of fuel cell for functions other than power is developed from of the mid-2000 s.

In addition, many patent applications for geared turbofan and open rotor engine started to be submitted from 2009; thus, these technology areas are found to be relatively recently developed compared with other technology areas.

According to the Society of Japanese Aerospace Companies (SJAC), a government support program is underway for a highly efficient engine development, including an open rotor engine, for fuel efficiency improvement as a measure against global warming and fuel cost increase in the USA and Europe.

The geared turbofan was recently successfully implemented in technology development over 20 years with a one trillion US$ budget since 1998 by Pratt & Whitney in the USA. The flight test results showed that engine noise and nitrogen dioxide generation were reduced by 50 %, and carbon dioxide generation was also reduced by 15 % in the geared turbofan engine. Supersonic aircraft technology is one of the main research topics of aerospace industry. The fact that this area registered the most patents from 2000 to 2010, and the consistent increase in the number of applications, indicates this tendency.

3 Green Aviation Technology Forecasting

As shown in Fig. 3, the noise technology area is more focused in technology development than other areas (as noise regulations to aircrafts around airports are reinforced), and is expected to be developed continuously in the future as well. In addition, following the developments made in the noise reduction technology area, the related technology areas for engine and body design are expected to be developed. In addition, the patents regarding fuel cells as an alternative fuel are continuously submitted, indicating that the related technology development will be continuously developed in the future. While patents regarding noise and fuel cell are continuously submitted, patents on aviation engines and designs are low in number. However, patents regarding new aviation engines and designs such as geared turbofan and open rotor engine are being submitted recently to indicating that new aviation engines and designs will be developed as a new technology area of green aviation.

Also, as shown in Fig. 3, supersonic aircraft technology is constantly increasing. The needs for supersonic aircraft are increased as transportation demand is rapidly increased in the longest distance between US, European, and Asian countries. However, there are many current technical difficulties and problems to commercializing supersonic aircraft. First of all, an improvement of aviation technology is necessary to overcome environmental restrictions such as noise, ozone layer destruction, etc. since fast fueling and breathing systems are required.

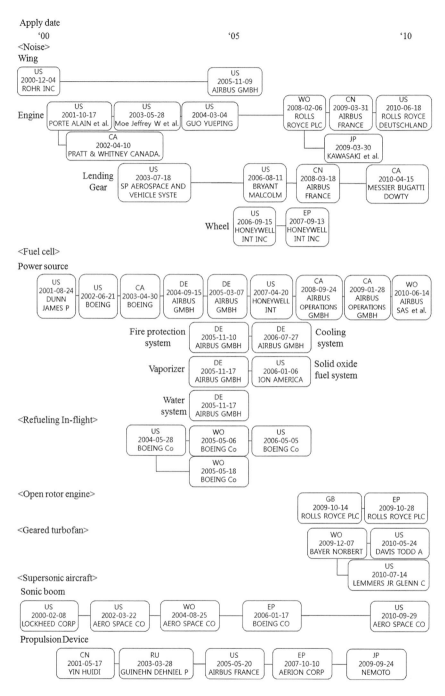

Fig. 3 Technology development chart of green aviation technology

For supersonic aircraft, severe noise together with sonic booms occur at takeoff and landing. Also, whereas general aircraft fly in the troposphere below 20,000 m, supersonic aircraft fly higher, in the stratosphere. Presently, ozone layer existing in the stratosphere is destroyed by catalysis of nitrogen oxide emitted from aircraft engines. Accordingly, technology development to reduce supersonic aircraft noise and nitrogen oxide is necessary.

In 2010, Boeing and Lockheed Martin proposed a new supersonic aircraft research development to NASA. The main research aim was reducing crashing sound and engine roar. Also, in 2011, the European Aerospace, Defense, and Space Industry (EADS) and Japan presented the Jest concept of environmental friendly supersonic passenger aircraft. Jest can reduce Nitrogen Oxide up to 75 % by using bio fuel. However, it has the limitation of a passenger capacity only up to 100 persons. Accordingly, it is expected that supersonic engine development will be continued for large aircraft design development to reduce noise and for Nitrogen Oxide reduction as demand for long range airline is increasing.

Technology relating to fuel cells is proceeding to two directions. As shown in Fig. 3, studies on utilization of fuel cell for aircraft power source are continuing, and application studies on utilizing fuel cell for other technology other than power source have been performed since of the mid-2000 s.

Fuel cell technology has been available to use for long period of time due to high energy density, and many patents were registered for aircraft power sources due to low noise. However, for large aircraft, a fuel cell is used only as an auxiliary power device. This is because fuel cells alone cannot drive enough thrust for large aircraft, and does not presently have economic feasibility due to high cost. Accordingly, it is expected that, for utilizing fuel cells as a main power source for large aircraft, technology development for improving manufacturing methods to reduce manufacturing cost and improve thrust is necessary in the future.

4 Conclusions

Currently, the aviation industry promotes investments in green aviation technology development to reduce green house gas emission from aircrafts by reducing the carbon emission and reducing the gigantic cost incurred in using aircraft fuels. This type of trend is expected to be continued in the future [11]. For efficient technology development, prediction of directions for technology development in the future are important.

In this study, PM establishment is conducted to forecast the technology flow of green aviation technology and the future technology development direction. The results show that technology developments for fuel cell and noise area in the green aviation technology area were continuously performed in the 2000 s. Also, recently, as patents for the geared turbofan and open rotor engine technology area were submitted, new aircraft engines are expected to be focused on for development as a new technology for future green aviation.

Acknowledgments This research was supported by Basic Science Research Program through the National Research Foundation of Korea (NRF) funded by the Ministry of Education, Science and Technology (2009-0068528)

References

1. Lewis, M.J.: Military aviation goes green, Serospace Ameriga, 24–31 (2009)
2. Agarwal, R.K.: Review of technologies to achieve sustainable (Green) aviation. Recent Advances in Aircraft Technology **19**, 427–464 (2012)
3. Jung, H.G.: The Technological competitiveness analysis of aircraft-based industries using patent information, the Korean operations research and management science society, 111–127 (2008)
4. Lee, K.S.: Development prospect and opportunity of the future aviation transportation: revolve around personal air vehicle (PAV), Aerospace Industry Research Institute, 70–93 (2008)
5. Suzuki, S.I.: Introduction to patent map analysis, Japan Patent Office (2011)
6. Kronenberg, E., White, J., Dickinson, R., Ramanathan, R.: The future of green aviation, Booz & Company (2008)
7. NASA.: Green aviation: a better way to treat the planet. National Aeronautics and Space Administration (2010)
8. NAST.: Revolutionizing aviation in the 21st century. A proposal for research for NASA, NAST (2008)
9. Agarwal, R.K.: Sustainable air transportation, recent researches in environment. Energy Systems and Sustainability, 130–139, (2012)
10. OECD.: Green growth and the future of aviation (2012)
11. GBI Research: Green aviation market to 2020-stringent regulations to drive investment in green technologies, GBI Research

Part III
Signal Processing for Image, Video, and Networking

A Vision-Based Universal Control System for Home Appliances

Chaur-Heh Hsieh, Ping S. Huang, Shiuh-Ku Weng, Chin-Pan Huang, Jeng-Sheng Yeh and Ying-Bo Lee

Abstract Based on computer vision techniques, this paper presents a universal remote control system for home appliances. This system consists of three major components: a paper control panel, a web camera wore on the user's chest, and a laptop computer. The system operates as follows. First, the user points his finger tip to a virtual button on the paper panel to select a specific appliance. Second, a function button is pointed at to operate an assigned function. The user's hand image is captured by the camera and the virtual button pointed by the fingertip is detected by computer vision techniques. Then a specific infrared code is emitted to control the corresponding appliance for a specific function. The advantage of this universal remote control system is that several remote controllers can be integrated together to simplify the operations of different appliances. The system performance is shown in the experimental results.

Keywords Computer vision · Remote control system · Home appliances

C.-H. Hsieh · C.-P. Huang · J.-S. Yeh · Y.-B. Lee
Department of Computer and Communication Engineering,
Ming Chuan University, 333 Taoyuan, Taiwan
e-mail: hsiehch@mail.mcu.edu.tw

C.-P. Huang
e-mail: hcptw@mail.mcu.edu.tw

J.-S. Yeh
e-mail: jsyeh@mail.mcu.edu.tw

P. S. Huang (✉)
Department of Electronic Engineering, Ming Chuan University, 333 Taoyuan, Taiwan
e-mail: pshuang@mail.mcu.edu.tw

S.-K. Weng
Department of of Computer Science and Information Engineering, Chung Cheng Institute
of Technology, National Defense University, 335 Taoyuan, Taiwan

S.-S. Yeo et al. (eds.), *Computer Science and its Applications*,
Lecture Notes in Electrical Engineering 203, DOI: 10.1007/978-94-007-5699-1_65,
© Springer Science+Business Media Dordrecht 2012

1 Introduction

To improve the quality and convenience of modern human life, more and more new consumer electronic devices are rapidly developed and released. However, each device almost comes with its own remote controller. Therefore, to operate those complicated devices, remote controllers appear nearly everywhere in our house and many users are frequently confused with a variety of controllers and buttons. This makes the usage of home appliances becomes annoying for us, especially for senior citizens. For instance, when we are watching movies in the living room, we may have to control the television, the DVD player, and the hi-fi system with three individual remote controllers at the same time. And this sometimes takes a while for us to figure out the correct control sequence. As such, to facilitate the usage of home appliances, a unified system is needed to integrate all control functions of home appliances into one single interface and this interface is responsible for the interaction between humans and the computer. Moreover, this computer drives the operations of selected home appliances.

In recent years, due to the prosperous development of ubiquitous computing, current traditional user interfaces become insufficient to fulfill the control requirements of humans and the computer. Furthermore, research efforts seeking to develop intuitive user interfaces which provide more human-centered methods of interacting with devices have gained many interests [1–7]. Motivated by those promising research results, in this paper, we focus on designing a multifunctional (universal) virtual remote controller via hand gesture recognition. The video of hand gesture motion is captured by a web camera wearing on the user's chest and the problem of hand gesture recognition is tackled by using computer vision techniques. After the hand gesture is recognized and the correct position of the finger tip pointed on a paper control panel is located, the corresponding home appliance is then activated and associated functions are operated. This paper control panel is actually used as a virtual human computer interface. The rest of this paper is organized as follows. Section 2 describes the system design. Experimental results are demonstrated in Sect. 3, prior to the conclusions in Sect. 4.

2 System Design

The scenario of proposed system is illustrated in Fig. 1. There are three major units in the system: a web camera wore on the user's chest, a paper control panel, and a computer controller. The user points a finger tip to a certain virtual button on the paper panel to control home appliances.

The system flowchart is demonstrated in Fig. 2 and three phases are included: selecting a home appliance, recognizing the touch of a virtual button, and generating an infrared control signal. Before the fingertip recognition, in the first phase, the system needs to detect the region of the paper panel from the input

Fig. 1 The scenario of
proposed system

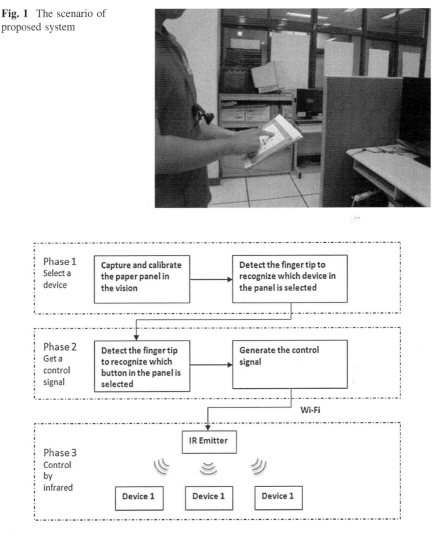

Fig. 2 The system flowchart

image scene. And, it is very possible that the paper panel is held in a tilted and rolled condition and this will increase the recognition error of the pressed buttons. Thus, it is necessary to do alignment for the panel image before recognition. After the system has detected and aligned the paper panel, then the user fingertip is detected using the skin color feature. Note that the skin color model is learned online. According to the detected position of the fingertip on the panel, the corresponding control signal is used and transmitted to drive the IR emitter and control the home appliance. The details are described in the following paragraphs.

Fig. 3 The layout of paper
control panel

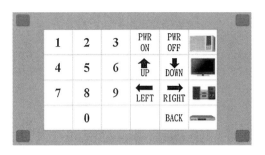

2.1 Detection and Alignment of the Paper Control Panel

To let the user manipulate the system easily, several factors of designing the
control panel need to be considered including the size and the number of buttons,
the gap between two buttons, the orientation alignment of the panel, and so on. In
this system, the control panel is made of a paper board and there are 32 (4 × 6)
cells on the panel. As shown in Fig. 3, every cell represents a virtual button with a
certain function. The green area and four red rectangles located at the corners of
the panel are used as the alignment patterns. For panel detection and alignment, the
green color and the red color must be trained beforehand and the training algorithm is listed as follows.

Step 1: Print out a piece of green (or red) paper and take 20 paper images using a
        camera with different directions and under variable illumination
        conditions.
Step 2: Calculate the histogram of Cb and Cr color components (h(Cb,Cr)) for
        those 20 images and find the bin with the largest frequency, denoted as
        (Cb_{Max}, Cr_{Max}).
Step 3: Set the green (or red) color range as
        $Cb_{Max} - 10 \leq Cb \leq Cb_{Max} + 10$ and $Cr_{Max} - 10 \leq Cr \leq Cr_{Max} + 10$.

    After the color ranges of green and red are decided, they are applied to detect
and align the panel and the algorithm is described as follows.

Step 1: Detect green color pixels of the panel using the range obtained during the
        training process. If the number of green color pixels detected is greater
        than a preset threshold value, go to **Step 2**; otherwise repeat **Step 1**.
Step 2: Using the region growing method [8] to connect the detected green pixels
        and form a green region.
Step 3: Apply the Sobel edge detector [8] on the green region and extract the
        region contour. Then find the four corner points of the contour in upper
        left, upper right, lower right, and lower left directions. Using those 4
        points, we can build a search rectangular area for **Step 4**.

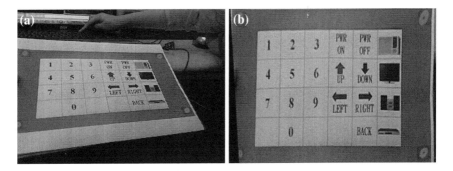

Fig. 4 Paper panel detection and alignment

Step 4: Detect red color pixels in the search area. If the number of red color pixels found in the search area is greater than a preset threshold value, go to **Step 5**; otherwise go to **Step 1**.

Step 5: Use the method of K-means clustering to classify the detected red points into 4 classes and use them to separate the panel image into 4 quadrants.

Step 6: Find the center position of the red points in every quadrant. The perspective transforms [Eqs. (1) and (2)] are used to map the 4 red center points to the predefined points in a 2D plane as shown in the right picture of Fig. 4. In Eq. (1), the (x_1, y_1), (x_2, y_2), (x_3, y_3) and (x_4, y_4) are 4 red center points in the source image (in the image plane, Fig. 4a) and (x_1', y_1'), (x_2', y_2'), (x_3', y_3') and (x_4', y_4') are the predefined corresponding 4 points in the target image (Fig. 4b). Using those points, the transformed parameters h_{ij} in Eq. (1) can be obtained. Then, Eq. (2) is used to convert all pixel positions on the source image into their corresponding coordinates on the aligned image. Now the source image is aligned.

$$
\begin{bmatrix}
x_1 & y_1 & 1 & 0 & 0 & 0 & -x_1x_1' & -y_1x_1' & x_1 & y_1 & 1 & 0 & 0 & 0 & -x_1x_1' & -y_1x_1' \\
0 & 0 & 0 & x_1 & y_1 & 1 & -x_1y_1' & -y_1y_1' & 0 & 0 & 0 & x_1 & y_1 & 1 & -x_1y_1' & -y_1y_1' \\
x_2 & y_2 & 1 & 0 & 0 & 0 & -x_2x_2' & -y_2x_2' & x_2 & y_2 & 1 & 0 & 0 & 0 & -x_2x_2' & -y_2x_2' \\
0 & 0 & 0 & x_2 & y_2 & 1 & -x_2y_2' & -y_2y_2' & 0 & 0 & 0 & x_2 & y_2 & 1 & -x_2y_2' & -y_2y_2' \\
x_3 & y_3 & 1 & 0 & 0 & 0 & -x_3x_3' & -y_3x_3' & x_3 & y_3 & 1 & 0 & 0 & 0 & -x_3x_3' & -y_3x_3' \\
0 & 0 & 0 & x_3 & y_3 & 1 & -x_3y_3' & -y_3y_3' & 0 & 0 & 0 & x_3 & y_3 & 1 & -x_3y_3' & -y_3y_3' \\
x_4 & y_4 & 1 & 0 & 0 & 0 & -x_4x_4' & -y_4x_4' & x_4 & y_4 & 1 & 0 & 0 & 0 & -x_4x_4' & -y_4x_4' \\
0 & 0 & 0 & x_4 & y_4 & 1 & -x_4y_4' & -y_4y_4' & 0 & 0 & 0 & x_4 & y_4 & 1 & -x_4y_4' & -y_4y_4'
\end{bmatrix}
\begin{bmatrix}
h_{11} \\ h_{12} \\ h_{13} \\ h_{21} \\ h_{22} \\ h_{23} \\ h_{31} \\ h_{32}
\end{bmatrix}
=
\begin{bmatrix}
x_1' \\ y_1' \\ x_2' \\ y_2' \\ x_3' \\ y_3' \\ x_4' \\ y_4'
\end{bmatrix}
$$

$$(1)$$

$$\begin{bmatrix} x' \\ y' \\ 1 \end{bmatrix} = \frac{\begin{bmatrix} h_{11} & h_{12} & h_{13} \\ h_{21} & h_{22} & h_{23} \\ h_{31} & h_{32} & 1 \end{bmatrix} \begin{bmatrix} x \\ y \\ 1 \end{bmatrix}}{\begin{bmatrix} h_{31} & h_{32} & 1 \end{bmatrix} \begin{bmatrix} x \\ y \\ 1 \end{bmatrix}} \tag{2}$$

2.2 Online Skin-Color Model Learning

For skin color detection, the YCbCr color space is used in this paper. Usually, the range of the standard skin color is $97.5 \leq Cb \leq 142.5$ and $134 \leq Cr \leq 176$ [9]. However, this range is only suitable for normal environmental conditions and it is not good enough for the application in varying conditions such as different human races or lighting conditions. Therefore, we propose an online skin-color model learning method to generate an appropriate skin-color range which can adapt to a particular user and lighting conditions.

In the learning stage of skin color range, the training skin color pixels are firstly collected from 30 effective frames in the initial period of a video. The effective frames refer to those frames with enough number of skin color pixels. Based on the detected skin color pixels from the effective frames, the histogram of (Cb, Cr) pair can be generated. Experimental results show that the highest occurrence frequency of skin pixels is at the center of the histogram, and the least frequency appears at the boarder of the histogram. Therefore, to utilize those characteristics, different weighting values are assigned to various histogram bins. Also, each color component (Cb or Cr) is uniformly divided into five partitions with different weighting values in which the middle partition has the largest weighting value. As illustrated in Fig. 5, the weighting distribution shows that the color values in every bin are multiplied by the assigned weighting values. Then the weighting averages of Cb and Cr can be obtained and given by

$$Cb' = \frac{1}{N} \sum_{K=1}^{N} \sum_{(x,y) \in Ps_k} W_{(Cb_n, Cr_n)} \times Cb_{(x,y)} \text{ and}$$
$$Cr' = \frac{1}{N} \sum_{K=1}^{N} \sum_{(x,y) \in Ps_k} W_{(Cb_n, Cr_n)} \times Cr_{(x,y)} \tag{3}$$

where N is the total number of frames, (x, y) is the pixel location, Ps_k is the skin color set of kth frame and $W_{(Cb_n, Cr_n)}$ is the weighting value of (Cb_n, Cr_n) bin. Currently, the final pair of (Cb', Cr') is calculated from the averaging results of continuous 30 frames (N = 30). The pair of (Cb', Cr') represents the learned skin color under the current environment.

According to the experimental results, the skin color ranges are adjusted to

Fig. 5 The distribution of
weighting values

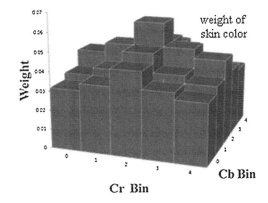

$$Cb' - 10 \leq Cb' \leq + 10 \tag{4}$$

and

$$Cr' - 10 \leq Cr' \leq + 10 \tag{5}$$

respectively. Using those color ranges, the skin color pixels (palm area) can be obtained. Furthermore, to make the palm area more complete and clearer, the operations of erosion and dilation [8] are performed three times to clarify the detected palm area. In summary, the skin color self-learning algorithm is listed as follows.

Step 1: Detect and align the paper panel. Display the image of aligned panel.
Step 2: Put a palm on the paper panel and use the standard skin color ranges (97.5 $\leq$ Cb $\leq$ 142.5 and 134 $\leq$ Cr $\leq$ 176) to count the number of the skin color pixels. If the percentage of skin color pixels is greater than a threshold, go to **Step 3**, otherwise, stay in **Step 2**.
Step 3: Builds the histogram of (Cb, Cr) pair for the detected skin color pixels from continuous 30 frames.
Step 4: Use Eq. (3) to calculate Cb$'$ and Cr$'$ values and adjust the new skin color range by Eqs. (4) and (5).

2.3 Finger Orientation and Fingertip Detection

To find the finger position and decide which virtual button is pressed, firstly, the system needs to distinguish the skin color pixels on the palm from those background pixels by the learned skin color ranges. After that, the image with palm pixels is converted to a binary image. Then, the contour of the palm is extracted by the Canny edge detector [8]. After the contour is extracted, the center of the palm

Fig. 6 The skin color
training and the result of
fingertip detection

and the distance from the center to every point along the contour can be calculated. The point with the longest distance can be considered as the fingertip position.

2.4 Recognizing the Pressed Button

In this process, if the detected fingertip position is located inside the area of a request button, this button is considered as being possibly pressed by the user. After the fingertip has stayed at the same area longer than 1 s (about 30 continuous frames), the system will ensure that this button is actually selected by the user. Then the control signal will be sent out when the finger is moved away this area. The steps of recognizing the pressed button are described below.

Step 1: Scan the image frame and count the total number of skin color pixels N_s.

Step 2: If the N_s is greater than a threshold value, then the system starts to learn the skin color under current environment and go to **Step 3**. Otherwise, go to **Step 1**.

Step 3: Find the center of the user palm and extract the palm's contour. Calculate the distance from the center to each point along the contour. Find the contour point with the longest distance and set it as the fingertip position.

Step 4: Check the movement of fingertip position. If the fingertip stays at one of 4×6 cells (buttons) in the following 30 frames, then the control signal will be sent to the IR emitter when the palm is moved away. Otherwise, repeat **Step 4**.

3 Experimental Results

At first, the system aligns the paper panel to its correct position. According to experimental results, the tilted and rolled degrees of the paper panel are limited in the ranges between -30 and $30°$, respectively.

After the panel alignment is completed, an aligned panel will be popped up on the screen to inform the user to put his finger on the panel for function selection. Then, the system starts the skin color learning process using continuous 30 frames. After the learning process, the skin color is set to the ranges decided by Eq. (4) and Eq. (5) to detect the user's finger. Moreover, the contour of the finger will be extracted to find the fingertip point. Figure 6 illustrates the results of detecting the finger and the fingertip point. According to the detected fingertip position and the panel area, the relative position of the fingertip point on the panel can be found. To assure a virtual button is actually pressed, the system must check if the fingertip has stayed on a virtual button area for continuous 30 frames (1 s). In case a button is pressed, the system will sends a feedback to tell the user that it is ready; meanwhile, the system sends the control signal to the device and makes a correct or an erroneous sound when the finger moves away the button cell.

4 Conclusions

In this paper, a vision-based universal control system for home appliances is implemented. This system is also a multifunctional virtual remote controller. The system hardware components include a webcam, a piece of paper (control panel), wireless network, an infrared emitter and receiver. The costs of all hardware components are very low and affordable. For those developed functions by software programs, they are panel detection and calibration, skin color training method, the finger orientation and fingertip detection, and the pressed button recognition. Experimental results have demonstrated that the system can control the home appliances accurately and it is easy to use.

In the future, we plan to modify this system and let it be self-adjusted to most of users. Therefore, the analysis to users with different ages including children, adults and aged users is needed. This is to improve the robustness of the proposed system.

Acknowledgments This research is supported in part by the National Science Council, Taiwan under the grant NSC 99-2632-E-130-001-MY3.

References

1. Saffer, D.: Designing Gestural Interfaces, p. 95472. O'Reilly Media, Inc., Sebastapool (2009)
2. Wachs, J.P., Kölsch, M., Stern, H., Edan, Y.: Vision-based hand-gesture applications. Commun. Assoc. Comput. Mach. **54**(2), 60–71 (2011)

3. Pang, Y.Y., Ismail, N.A., Gilbert, P.L.S.: A Real Time Vision-Based Hand Gesture Interaction. In: Fourth Asia International Conference on Mathematical/Analytical Modeling and Computer Simulation, pp. 237–242. Borneo, Malaysia (2010)
4. Zabulis, X., Baltzakis, H., Argyros, A. A.: Vision-based hand gesture recognition for human computer interaction. In: The Universal Access Handbook, pp. 34.1–34.30. Lawrence Erlbaum Associates, Inc. (LEA), Series on Human Factors and Ergonomics (2009)
5. Rautaray, S.S., Agrawal, A.: A novel human computer interface based on hand gesture recognition using computer vision techniques. In: First International Conference on Intelligent Interactive Technologies and Multimedia, pp. 292–296. Allahabad, India (2010)
6. Hassanpour, R., Shahbahrami, A.: Human computer interaction using vision-based hand Gesture recognition. J. Comput. Eng. **1**, 21–30 (2009)
7. Mistry, P., Maes, P.: SixthSense: A Wearable Gestural Interface. In: SIGGRAPH Asia 2009, Article 11. Yokohama, Japan (2009)
8. Gonzalez, R., Woods, R.E.: Digital image processing, 3rd edn. Prentice Hall, Upper Saddle River, NJ, United States (2008)
9. Chai, D., Ngan, K.N.: Face segmentation using skin-color map in videophone applications. IEEE Trans. Circ. Sys. Video Technol. **9**(4), 551–564 (2002)

An IMU-Based Positioning System Using QR-Code Assisting for Indoor Navigation

Yih-Shyh Chiou, Fuan Tsai, Sheng-Cheng Yeh and Wu-Hsiao Hsu

Abstract In this paper, a positioning scheme combining inertial measurement unit (IMU) observation with QR code recognition is proposed to improve the location accuracy in an indoor environment. For the location-estimation technique, the proposed positioning scheme based on IMU observations is handled by the dead-reckoning (DR) algorithm; in terms of a QR-code-assisted calibration technique, the proposed approach is an accuracy enhancement procedure that effectively reduces the error propagation caused by DR approach. Namely, with the assisting approach to recognize the locations of the QR-code-reference nodes as landmarks, a DR-based scheme using the landmark information can calibrate the estimated location, and then the error propagation effect is reduced. The experimental results demonstrate that the location based on the proposed approach have much lower location errors in an IMU positioning platform. As compared with the non-QR-code-assisted approach, the proposed algorithm can achieve reasonably good performance.

Keywords Dead reckoning · Error propagation · Location estimation · Inertial measurement unit · QR code

Y.-S. Chiou · F. Tsai
Center for Space and Remote Sensing Research, National Central University,
Jhongli, Taoyuan 32001, Taiwan
e-mail: choice@alumni.ncu.edu.tw

F. Tsai
e-mail: ftsai@csrsr.ncu.edu.tw

S.-C. Yeh (✉) · W.-H. Hsu
Department of Computer Science and Information Engineering,
Ming Chuan University, Guishan, Taoyuan 33324, Taiwan
e-mail: peteryeh@mail.mcu.edu.tw

W.-H. Hsu
e-mail: wuhsiao@mail.mcu.edu.tw

S.-S. Yeo et al. (eds.), *Computer Science and its Applications*,
Lecture Notes in Electrical Engineering 203, DOI: 10.1007/978-94-007-5699-1_66,
© Springer Science+Business Media Dordrecht 2012

1 Introduction

Location-aware services have received great attention for commercial, public-safety, and military applications [1]. In the literatures, a number of positioning schemes have been reported [2]. There are two basic schemes applied in location-estimation systems. One is the radio ranging scheme (absolute scheme) based on wireless network services [2–5], the other is the speed sensing scheme (relative scheme) based on an inertial measurement unit (IMU), a package of inertial sensors (gyroscopes and accelerometers) [6, 7]. However, providing customers with tailored location-based services is a fundamental problem. Consequently, the fusion algorithm between the different devices is considered an important technique to improve location accuracy [8]. In addition, an accurate location can be improved with location tracking algorithms. The Kalman filtering algorithm is considered an optimal tracking algorithm for the linear Gaussian model [2–5].

Recently, for relative positioning scheme, a mobile terminal (MT) with low cost MEMS (micro-electro-mechanical systems) sensors, such as accelerometer, gyroscope, and compass, has made the dead-reckoning (DR) algorithm that becomes an attractive choice for indoor environment [6]. However, a magnetic compass does not function well in an indoor environment. That is, for indoor complexity environments, the magnetic fields will change for different locations, and the low-cost gyro usually suffers from high drift rate [7]. That is, the location error of the DR approach will be enlarged with the inaccuracy heading observations, and it would be useful to develop assisted schemes for improving location accuracy.

Previously, human-recognition systems have used active-tag technologies to detect MTs for location tracking. Lately, the features of quick response (QR) codes have been implemented for human tracking by image-processing and recognizing systems [9]. A QR code is the trademark for a type of a two-dimensional (2-D) code. Compared with a traditional one-dimensional (1-D) code, a QR code has features of fast readability and large storage capacity. Nevertheless, as compared previous active-tag approaches with low-cost passive QR-code approaches, how to maintain the normal function of the active tags with batteries is a heavy burden for location-estimation system. Recently, a smartphone with multi-sensor systems and cooperative capabilities has become widely available. Specifically, a smartphone can offer more advanced computing ability and allows the user to run multi-task applications [10]. Therefore, the portable navigation and tracking system (a smartphone or an MT) can combine with the diverse sensing capabilities. That is, a location-estimation system combining with different devices can be considered an important technique to improve location accuracy. As an MT with the diverse sensors, the image recognizing scheme and speed sensing scheme are natural choices for multi-sensor positioning systems to improve the location estimation of MTs. In other words, for improving location accuracy, it would be useful to develop an IMU-based positioning system using QR-code assisting in indoor navigation systems.

In this paper, an IMU-based positioning system with QR-code assisting is proposed to improve the location accuracy. Specifically, to effectively reduce the error propagation of DR approach in indoor environments, the location-estimation process involves three main steps. The first step is location estimation based on DR approach; the second step is QR-code recognition using pattern matching approach to decode the identification of the QR-code locations; the third step is combined DR approach with QR-code recognition to reduce error propagation in location-estimation system. According to the experiment results, they demonstrate that the proposed scheme can effectively reduce the impact factor of error propagation, and the location accuracy of the proposed scheme is better than that of non-QR-code-assisted schemes.

2 Background

2.1 Dead-Reckoning Algorithm

For location-estimation systems, a DR algorithm is one of popular positioning schemes calculating the current location according to a previously determined location. Namely, the DR algorithm relies on the initial information, it pluses heading and speed (or location) information, and then the next location is estimated for navigation. Consider a system described with dead reckoning as illustrated in Fig. 1. If the system is based on a 2-D coordinate system, the mathematical models for the system at time k can be taken by

$$x_k = x_0 + \sum_{i=0}^{k-1} d_i \cdot \cos \theta_i, \quad k \geq 0 \tag{1}$$

$$y_k = y_0 + \sum_{i=0}^{k-1} d_i \cdot \sin \theta_i, \quad k \geq 0 \tag{2}$$

where x_0 and y_0 are the initial location at time 0; x_k and y_k are the estimation location at time k; d_i and θ_i are the moving distance and the observation (absolute) angle in terms of east (E) at time i, respectively. N and E are north and east in Fig. 1, respectively. However, a DR algorithm based on the estimates of velocity (speed) and direction will cause error propagation for location estimation [11]. As a result, it would be useful to develop an algorithm for improving location accuracy.

2.2 QR Code

A QR-code is a small 2-D barcode image that conveys much more information than the traditional 1-D barcode, and it has been designed to store a large amount

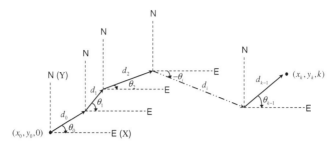

Fig. 1 A DR algorithm using distance information and direction information for location estimation

of data for applications. For the structure of QR codes, the foursquare QR code consists of coding region and functional region [12]. The coding region is described by version, format, and data characters. The functional region is combined with localizing, correcting, and seeking graphs. As illustrated in Fig. 2, QR codes can be read and decoded by smart phones directly. That is, the information of QR code can be extracted with image recognizing (decoding) scheme by smartphones. Afterward, the location information containing in QR codes can be used for the applications of location-based services.

3 The Proposed Algorithm

For location estimation in an indoor environment, the proposed algorithm is based on the 3DM-GX3 IMU developed by Microstrain [7], where the IMU has features as follows: orientation range of 360° about all axes, accelerometer range of ±5 g standard, gyro range of ±300°/sec standard, magnetometer range of ±2.5 Gauss. In addition, the IMU can provide a data rate of 1–1000 Hz. Without loss of generality, this paper only focuses on location-estimation approaches in a 2-D coordinate system, and the extension of the scheme to a three-dimensional model is straightforward.

3.1 Experiment Setup

The experimental platform was located on the sixth floor of a building (CCE, MCU, Taiwan). The floor layout is shown in Fig. 3; the sampled locations are denoted by circle • with 1 m separation between adjacent points; the QR-code locations denoted by

Fig. 2 QR-code recognition (decoder) with a mobile smartphone

are distributed around the hallway. The data sample rate of the IMU was set to 50 or 100 per second of the experiments. Furthermore, the IMU was placed on a box, and it was dragged along the sampled locations of the hallway in Fig. 3.

3.2 Location Estimation Using QR-Code Assisting

In an inertial navigation system (INS), not only will the characteristic of indoor environments affect the heading measurement, but also the error propagation effect impacts on the location estimation. To improve location accuracy, a simple technique employing landmark-assisted scheme is used to overcome the effect of the error propagation caused by the DR algorithm for dramatic time-varying systems. That is, an MT with assisting approach can calibrate and modify its location based on sensing landmark locations. In this paper, QR codes are encoded and decoded based on image characteristics. In terms of the recognition approach, the information of QR-code coordinates is as landmarks for calibrating locations. By calibrating some fixed known locations, the procedure used to avoid the error propagation caused by the DR algorithm is illustrated in Fig. 4. It indicates the

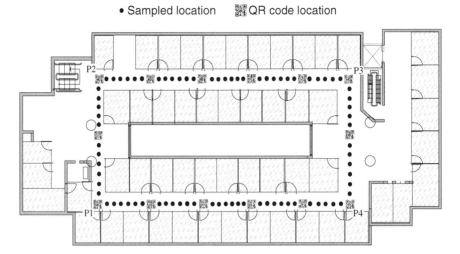

Fig. 3 The floor layout of the experimental environment

process diagram about the QR-code-assisted approach for improving location accuracy. In brief, the procedure of the proposed approach involves three main operations and is described as follows. (1) Location estimation: The computations are implemented with Asus Eee PC, and the 3DM-GX3 IMU is connected PC with an USB connector for extracted IMU information. The upper part of Fig. 4 illustrates the concept of an INS procedure, where the INS is based on IMU information for location estimation. The INS approach used in this paper is based on Ref. [13]. (2) QR-code recognition: As is well known, the encoder (generator) and decoder (reader) of QR codes are widely used in different applications. In order to recognize the QR-code information as landmarks along the test path, this paper combines a QR code decoder with fixed coordinate information for calibrating the estimated locations. The illustrations about the QR-code recognition have been proposed in [9] and [12]. (3) Combination of information sources: In order to reduce error propagation caused by the heading observation and the DR approach, the encoded QR codes are static nodes placed at a known position. As a smartphone detects the information extracted from these QR codes, the well known coordinate information is obtained. And then, the proposed algorithm combined the INS procedure with the QR-code-recognition procedure is used to improve the location accuracy.

4 Experimental Results

In this paper, an Asus Eee PC 1005HA is taken as a smartphone. The experimental investigation combines IMU positioning system with QR-code-assisted approach for location estimation; the estimated results are conducted to demonstrate the

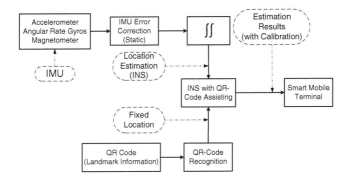

Fig. 4 Location-estimation system based on the QR-code-assisted approach

accuracy of the proposed scheme, where the QR-code information as the land-mark's coordinates is encoded and decoded by image characteristics for the recognition approach.

4.1 Location Estimation Based on DR Approach

In order to verify the performance of the proposed scheme, the models describing the shift of translational motion and the rotational motion models of MTs are given by

$$d_{x,k} = \Delta_k V_{x,k} + \frac{\Delta_k^2}{2} a_{x,k}$$
$$d_{y,k} = \Delta_k V_{y,k} + \frac{\Delta_k^2}{2} a_{y,k}$$

(3)

$$x_{k+1} = x_k + \iint\limits_{t_k \to t_{k+1}} (a_{x,k} \cos\theta - a_{y,k} \sin\theta) d\tau dt$$
$$y_{k+1} = y_k + \iint\limits_{t_k \to t_{k+1}} (a_{x,k} \sin\theta + a_{y,k} \cos\theta) d\tau dt,$$

(4)

respectively, where V, a, and θ are the velocity, the acceleration, and the absolute angle of observation, respectively. Δ_k is the measurement interval between time k (t_k) and time $k + 1$ (t_{k+1}). Furthermore, in this paper, the heading information is the angle of the MT referring to true north. In addition, the heading is the moving direction of the MT, and the units are degrees from north in a clockwise direction. As an MT moved along the test path in Fig. 3, the experimental results of the heading information read from and the velocity information extracted from the 3DM-GX3 IMU are illustrated in Fig. 5, where the measurements of the first 60 s

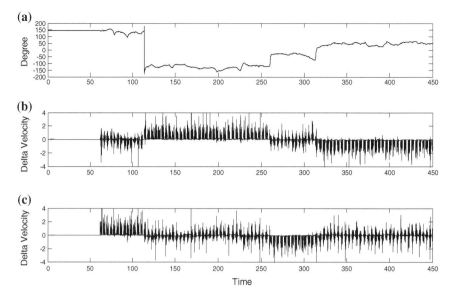

Fig. 5 Heading and delta velocity results in terms of data rate of the IMU set to 50 per second along of the hallway in a clockwise direction in Fig. 3

are used for the static calibration. As illustrated in Fig. 5, in terms of IMU measurements, the performed experiments are used to get the characteristics of the inertial sensor and to verify the performances of a DR approach. Furthermore, the information reading from the MEMS IMU is under the assumption of the same bias through whole experiment.

As the MT moved along the hallway in a clockwise direction, the location-estimation results in terms of the IMU's data rate set to 50 and 100 per second are given in Fig. 6. The results demonstrate that the positioning scheme using the DR algorithm causes error propagation, and the error increases rapidly in this indoor environment. In fact, the property of the DR-based method is based on calculating the current location according to the previously determined information. Namely, it relies on the previous heading and velocity information for estimating the next location. However, the accumulations of location error become inevitable and unbounded with sensor biases and noises. In addition, the heading information based on the accuracy of the electronic compass is easy to be affected in indoor environments. As shown in Fig. 6, the location-estimation approach with the DR-based algorithm causes large location errors in this indoor environment.

4.2 Location Estimation Using QR-Code Assisting

As demonstrated in Fig. 3, the adjacent distance of QR-code locations is set about 7 m around the hallway. Figure 7 indicates the experimental results using

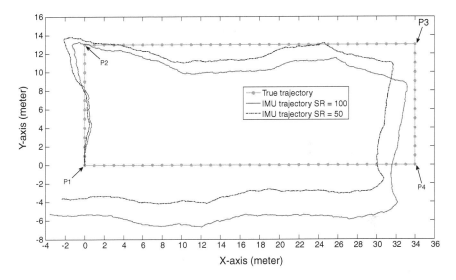

Fig. 6 The DR-based approach for location estimation in terms of IMU observations as the MT moves in a clockwise direction of the hallway from P1 to P1 (P1 → P2 → P3 → P4 → P1) with two sampling rates, 50 and 100, in Fig. 3

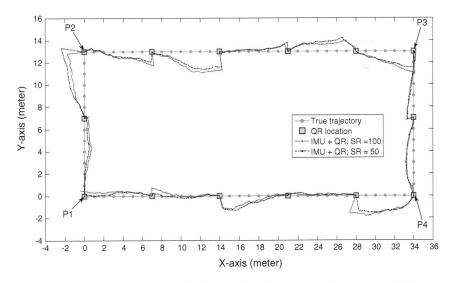

Fig. 7 The DR-based approach with QR-Code assisting for location estimation as the MT moves in a clockwise direction of the hallway from P1 to P1 (P1 → P2 → P3 → P4 → P1) in Fig. 3

QR-code-assisted scheme for location estimation. The results illustrate that the QR-code locations as landmarks can improve the location accuracy very well. As compared with Figs. 6, 7 shows that the IMU-based positioning technique using the QR-code-assisted approaches based on the image recognition approach can reduce error propagation and improve location accuracy. Figures 6, 7 indicate the comparison between the IMU-based method and the IMU-based with the QR-code-assisted method as an MT moves along a test path in Fig. 3. After using QR-code assistance, the experimental results illustrate that the location accuracies of the proposed QR-code-assisted scheme are better than the non-QR-code-assisted scheme. After combining the QR-code-assisted technique with the DR algorithm, Fig. 7 shows that not only can the data sample rate of the IMU set to 100 closely predict the MT's location, but also the data sample rate of the IMU set to 50 can closely predict the MT's location. In brief, with the QR-code-assisted approach, the results show that the proposed algorithm could mitigate and overcome the dramatic time-varying environment of the navigation path more efficiently.

5 Conclusion

In this paper, we have proposed a new scheme for location estimation based on the DR approach with QR-code-assisted scheme in indoor environments, where the QR-code coordinates extracted from image characteristics are as landmarks. According to the experiment results with the DR approach, the MT can not properly track the path in terms of the IMU observations for location estimation. In addition, the heading angle is extracted from an electric compass, and the measurement of heading angle is easily affected in indoor environments. Therefore, the inaccuracy angle would enlarge the error propagation of the DR-based approaches. However, in terms of QR codes encoded and decoded with image characteristics, the DR scheme with QR-code assisting can calibrate the location estimation and alleviate the error propagation. According to the experimental results about investigating the performance of the DR-based scheme under the indoor environment in Fig. 3, we conclude that the proposed scheme demonstrates much better accuracy as compared with the non-QR-code assisted scheme. The proposed scheme has good features of location accuracy. According to experimental results, it has been shown that the estimated locations have error distances less than DR-based approach. Compared with the non-assisted schemes, the proposed location platform combining the QR-code-assisted scheme in an indoor environment is attractive for use in various LBS systems.

Acknowledgements This work was supported in part by the National Science Council of the Republic of China under grant NSC 98-2221-E-008-097-MY2.

References

1. Rantakokko, J., Rydell, J., Stromback, P., Handel, P., Callmer, J., Tornqvist, D., Gustafsson, F., Jobs, M., Gruden, M., Gezici, M.: Accurate and reliable soldier and first responder indoor positioning: multisensor systems and cooperative localization. IEEE Wirel. Commun. **18**(2), 10–18 (2011)

2. Barton, R.J., Zheng, R., Gezici, S., Veeravalli, V.V.: Signal processing for location estimation and tracking in wireless environments. EURASIP J. Adv. Signal Process. **2008**, 1–3 (2008)
3. Chiou, Y.-S., Wang, C.-L., Yeh, S.-C., Su, M.-Y.: Design of an adaptive positioning system based on WiFi radio signals. Elsevier Comput. Commun. **32**, 1245–1254 (2009)
4. Chiou, Y.-S., Wang, C.-L., Yeh, S.-C.: An adaptive location estimator using tracking algorithms for indoor WLANs. ACM/Springer Wirel. Netw. **16**(7), 1987–2012 (2010)
5. Chiou, Y.-S., Wang, C.-L., Yeh, S.-C.: Reduced-complexity scheme using alpha-beta filtering for location tracking. IET Commun. **5**(13), 1806–1813 (2011)
6. Glanzer, G., Bernoulli, T., Wiessflecker, T., Walder, U.: Semi-autonomous indoor positioning using MEMS-based inertial measurement units and building information. In: IEEE WPNC 2009, pp. 135–139. IEEE Press, New York (2009)
7. Microstrain Inc., http://www.microstrain.com/inertial.aspx
8. Fischer, C., Gellersen, H.: Location and navigation support for emergency responders: a survey. IEEE Pervasive Comput. **9**(1), 38–47 (2010)
9. Anezaki, T., Eimon, K., Tansuriyavong, S., Yagi, Y.: Development of a human-tracking robot using QR code recognition. In: FCV 2011, pp. 1–6 (2011)
10. Zheng P., Ni, L. M.: The rise of the smart phone. IEEE Distrib. Syst. Online, **7**(3), art. no. 0603-o3003 (2006)
11. Hardt, H.-J. von der, Wolf, D., Husson, R.: The dead reckoning localization system of the wheeled mobile robot ROMANE. In: IEEE MFI '96, pp. 603–610. IEEE Press, New York (1996)
12. A-Lin, H., Yuan, F., Ying, G.: QR code image detection using run-length coding. In IEEE ICCSNT 2011, pp. 2130–2134 IEEE Press, New York (2011)
13. Britting, K.R.: Inertial Navigation Systems Analysis. Artech House, Massachusetts (2010)

A Visual-Audio Assisting System for Senior Citizen Reading

Yu-Qi Li, Jia-Jiun Liu, Shih-Yu Huang, Chen-Kuei Yang, Chin-Chun Chang, Li-Tien Wang and Kuei-Fang Hsiao

Abstract This paper proposes an augmented reality (AR) platform to solve problems of seniors' eyesight deterioration. A finger detection technique is first employed to retrieve articles in newspaper captured by the AR platform, the extracted article is then enlarged by the bilinear interpolation. Furthermore, an image projection algorithm to recompose the text on the enlarge image thus provide an ideal visual reading interface. In addition, the proposed AR platform designs an audio assisting function for the senior to listen the extracted article in newspapers or magazines. The experiment results show that the proposed AR platform provides a good visual-audio interface for senior citizen.

Y.-Q. Li (✉) · S.-Y. Huang · C.-K. Yang · L.-T. Wang
Department of Computer Science and Information Engineering, Ming Chuan University,
Taoyuan 333, Taiwan, Republic of China
e-mail: am997vivian@gmail.com

S.-Y. Huang
e-mail: syhuang@mail.mcu.edu.tw

C.-K. Yang
e-mail: ckyang@mail.mcu.edu.tw

L.-T. Wang
e-mail: ltwang@mail.mcu.edu.tw

J.-J. Liu · C.-C. Chang
Department of Computer Science and Information Engineering, National Taiwan Ocean
University, Keelung 202, Taiwan, Republic of China
e-mail: t90615@gmail.com

C.-C. Chang
e-mail: cvml@gmail.ntou.edu.tw

K.-F. Hsiao
Department of Information Management, Ming Chuan University, Taoyuan 333, Taiwan,
Republic of China
e-mail: kfhsiao@mail.mcu.edu.tw

S.-S. Yeo et al. (eds.), *Computer Science and its Applications*,
Lecture Notes in Electrical Engineering 203, DOI: 10.1007/978-94-007-5699-1_67,
© Springer Science+Business Media Dordrecht 2012

668 Y.-Q. Li et al.

Keywords Augmented reality · User interface

1 Introduction

Studies have shown that Senior citizens in Taiwan commonly disturbed by vision problem caused by presbyopia. Presbyopia is the deterioration of ability to regulate lens for viewing objects at varying distances, which the root cause is the dysfunction of ciliary muscle. When people see through eyes, light go through the vitreous, and an image on the retina is formed. The focus of lens is adjusted according to the distance of the object. Younger people own more soften lens, so that the ability to regulate the focus is much superior. For aged persons, gradually hardened lens began to affect the flexibility in adjusting focus in different distance. Objects are having problem to be projected clearly on the retina. In particular, blurred objects made reading books and newspapers less comfortable and took away the enjoyment of reading.

In order to tackle the problem of presbyopia, reading glasses has been the classical remedy almost for centuries. Objects in newspapers and magazines will be to enlarge by the optical lenses. However, being facilitated by fast growing computer hardware and software nowadays, advanced technologies are introduced to easy the vision issue for elderly people. Figure 1 is the structure of the proposed AR platform which idea comes from the Sixth Sense device proposed by Pranav Mistry [1]. The gear includes a camera and a micro-projector together with a mirror. The camera captures images from newspapers or magazines, and feeds digital image to the innovated device for signal process. Through the projector, captured information on the digital image will be projected with the Image Super-resolution technology. Seniors citizen can easily recognized information on newspapers or magazines, and even scene in far distance.

Figure 2 is the software flow chart of the proposed AR platform. It has five modules: article detect module (AD module), article confirm module (AC module), article enlarge module (AE module), article recompose module (AO module), and OCR-TTS module. The AD module will detect an entire article from the newspaper captured by the camera of AR platform. A finger detection technique is used in the AC module to confirm the article detected by AD module. The AE module will enlarge the confirmed article by the method of bilinear interpolation. The AO module recomposes the enlarged article to provide a good visual interface for seniors. The OCR-TTS module will recognize the characters in the detected article and transform them to a text file and transfer the text file into audio files by TTS technology.

The rest of this paper is organized. Section 2 presents the details of the five modules in the proposed AR platform. Experimental results are shown in Sect. 3, and a brief conclusion is made finally in Sect. 4.

Fig. 1 The structure of the proposed AR platform

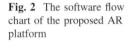

Fig. 2 The software flow chart of the proposed AR platform

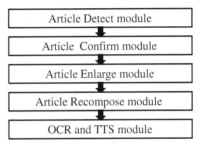

2 The Software Design in the Proposed AR Platform

There are five modules in the proposed AR platform. Details of AD module, AC module, AE module, AO module, and OCR-TTS module are introduced in the following.

2.1 The Design of AD Module

Newspapers or magazines often use a fixed color (e.g. white) as the background. There are text spacing between characters, line spacing between lines, and white space between articles [2–4]. AD module uses white space between articles to detect articles in the newspaper or magazine. The design of AD module is based on a projection technology. Figure 3 give a demonstrative example of the projection technology in AD module. The approach is to first find out the X-axis Projection and Y-axis projection on the image captured by the AR platform, where the X-axis Projection vertically counts the number of white pixels and the Y-axis Projection horizontally counts the number of white pixels. White spaces between articles clearly generate consecutive 0 in the projections respectively. Therefore,

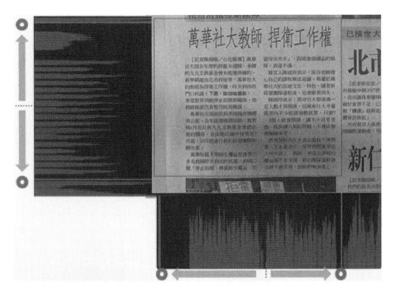

Fig. 3 A demonstrative example of AD module

AC module then finds two positions with the smallest X-projection value from the middle position to both sides in the captured image. The found two positions are the left and right positions of the detected article. Similarly, the top and bottom positions of the detected article are determined by the two positions with the smallest Y-projection value from the middle position to both sides in the captured image. The red rectangle in Fig. 3 gives the article detected by AD module.

2.2 The Design of AC Module

AC module utilizes a finger detection technique to confirm article detected by AD module. Figure 4 give a demonstrative example of AC module, where the article is extracted when the senior use his finger point to the red rectangle. In AC module, the pixels in RGB color space are first transformed into pixels in YCbCr color space to overcome the light problem. Eq. (1) gives the corresponding transform function. A pixel is classified into skin pixel if the corresponding Y value is ranging from 60 to 255, the Cb value is ranging from 97 to 142 and the Cr value is ranging from 134 to 176, respectively [5]. If the number of skin pixels in the red rectangle is larger than a predefined threshold, AC module extracts the article with red rectangle and output it to AE module.

$$\begin{bmatrix} Y \\ Cb \\ Cr \end{bmatrix} = \begin{bmatrix} 16 \\ 128 \\ 128 \end{bmatrix} + \begin{bmatrix} 65.481 & 128.533 & 24.966 \\ -37.797 & -74.203 & 112 \\ 122 & -93.786 & -84.214 \end{bmatrix} \cdot \begin{bmatrix} R' \\ G' \\ B' \end{bmatrix} \quad (1)$$

Fig. 4 A demonstrative example of AC module

Fig. 5 The bilinear
interpolation technique to
enlarge R ratio

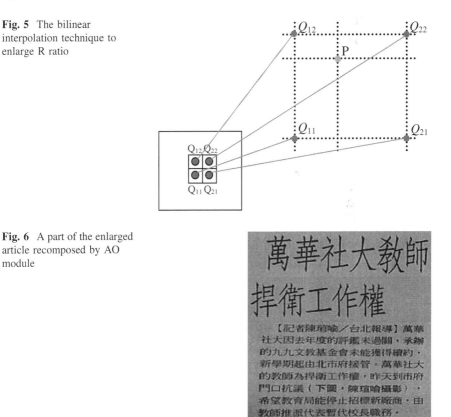

Fig. 6 A part of the enlarged
article recomposed by AO
module

2.3 The Design of AE Module

Using image enlargement technology is very promising to assist senior citizens.
Many existing technologies are proposed in the past two decades [6–9]. Bilinear

萬導九府停萬遭會日師喊羊淑專隊要內讓團王巾駐
華】文接門止華檢終將生口「玲唱進澳不除永、後ｔ
社萬教管口招社調止進權號「表於萬滿少少承學ｔ城
大華基·萬華社大抗樣大「契行益·表一經民受表示怔
教社大議新因約祈促停「部分老師革後定濟弱員及不
師大能《商校長教·校由龍山國中業·萬市府抗議課會
捍去要Ｉ導，係經營如標，教育隊教師重師昨生准益，程澳
衛因能下圓，由教師端團隊名譯氣球，表達不滿·發言人陳
工年度的許鑑，新學期Ｉ導教師端端涉嫌挪用自己的課程無法延續，
作的評約，陳瓊喑攝影）代表暫代公款）Ｌ）Ｌ特色·隨著新經營團
權的許鑑未過關，承辦的新學期起由北市教育局長職務·部分社大
【記者陳堉喃／台北報導】希望暫代校長，去年底到考量區里
者陳堉喃／台北報九市能·陳瓊喑１月31日與）Ｌ文教基金20大
過未約·萬華社大的圓，陳師瓊喑１月31日准益，拒當待宰羔陳
嘖關，萬華社大教師為捍衛工作權，昨天到市府抗議，新的團隊進
堉未萬華社大的圓，長教·校長團多，龍山國標教師昨生８個人就會開課，
台／承辦的九市能的下商校，營如標，教育隊教師重益都不會受根，個人整惆
北的，昨天到市府抗議，新學期起由北市教育局文教基金20大里
報九市能·20萬華社大的圓，陳端團隊名譯重師昨生准益，不應由整惆

Fig. 7 The results of OCR with the input of the article extracted in the above example

Interpolation techniques [9] are used in AE module. Figure 5 shows the schematic view of amplification of Bilinear Interpolation techniques. Assume that the magnification rate is R, and Q_{11}, Q_{12}, Q_{21}, and Q_{22} are values of four Pixels in the Original Block. Four red colored Pixels on the enlarged image are result of Bilinear Interpolation of the four original Pixels Q_{11}, Q_{12}, Q_{21}, and Q_{22}. The coordinate value the green colored Pixel is calculated by the relative distance to all four projected red color Pixels.

2.4 The Design of AO Module

It is not suitable for a text reading in the enlarged image generated by AE module. Similar to the Microsoft Windows environments, the enlarged image has to be rolled to left and right constantly in order to finish the article. The rolling back and forth will simply lead to visual fatigue for elderly people. This problem had already overcome by the projection technique published in [10].

The projection technique can be used not only to detect articles in newspapers, but also to determine the current sizes of each text lines. Each individual text line extracted partially from the original image based on the magnification ratio in order to fit into the enlarged image. The remaining not yet extracted text will be fed to next round of extraction. Hence the typesetting is accomplished line by line. Also, the truncation of each text line should be intelligent. Therefore during the extraction, hyphenation will not be used on the last word. In case there is a need of truncation, the word will be moved to next line automatically. Figure 6 gives a part of the enlarged article recomposed by AO module.

Fig. 8 Two examples of the results of the proposed AR platform. **a** The result of AC module. **b** The result of AE module with enlarged ratio of two. **c** The result of OCR module

2.5 The Design of OCR-TTS Module

Audio assistance is a premium function to help senior reading. Senior can request system to read out the article as audio assistance. The module will first perform OCR function (Optical Character Recognition) to analyze the article extracted by AC module and transform into text file. The proposed AR platform applies OCR engine from Microsoft Office Document Imaging to convert optical text into digital characters. In addition, this platform applies TTS technique from MS Speech SDK to convert text line into audio file. Note that, the TTS doesn't support traditional Chinese characters so they have to transfer to simplified Chinese characters first. Figure 7 gives the results of OCR with the input of the article extracted in the above example in Fig. 4.

3 Experimental Results

We simulate the proposed AR platform and the proposed AD module, AC module, AE module, AO module, and OCR-TTS module on a PC with 3.0 GHz Intel® Core™2 Duo CPU and 2.0 GB memory. Figure 8 gives two examples of the results of the proposed AR platform. It's easily seen that the articles can efficiently be extracted by AC module and the AE module can successfully enlarge the extracted articles. But, the extracted article cannot completely be recognized by OCR. The accuracy rate of the above two examples are 93 and 96 %, respectively. This is the results of the characters captured by the camera are too small to be recognized.

4 Conclusions and Future Works

The contribution of this paper has two. The first one is that an AR platform integrating a camera and a micro-projector together with a mirror is proposed to be a new reading interface for senior citizens. In addition, five software modules named AD module, AC module, AE module, AO module, and OCR-TTS module are designed to provide visuals and audio assisting for senior citizens. The experimental results show that the proposed system is a good reading interface integrating visuals and audio for senior citizens. This paper covers articles with a clean text content only. Other related issues such as skewed articles, untidy image, non-text image, and so forth did not take into account yet in the discussion. These topics can be further studied.

Acknowledgments This research is under the support of Nation Science Council of project with number NSC 99-2632-E-130-001-MY3.

References

1. Mistry, P., Maes, P.: Sixthsense: A wearable gestural interface. In: SIGGRAPH Asia 2009, Article 11. Yokohama, Japan (2009)
2. Dos Santos, R.P., Clemente, G.S., Tsang, I.R., Calvalcanti, G.D.C.: Text line segmentation based on morphology and histogram projection. In: International Conference on Document Analysis and Recognition, 2009
3. Lue, H.-T., Wen, M.-G., Cheng, H.-Y., Fan, K.-C., Lin, C.-W., Yu, C.-C.: A novel character segmentation method for text images captured by cameras. ETRI J. **32**(5), 729–739 (2010)
4. Murguia, M.I.C.: Document segmentation using texture variance and low resolution images. IEEE Southwest Symposium on Image Analysis and Interpretation, April 1998, pp. 5–7

5. Phung, S.L., Bouzerdoum, A. Sr., Chai, D. Sr.: Skin segmentation using color pixel classification: analysis and comparison. IEEE Trans. Pattern Anal. Mach. Intell. **27**(1), 148–154 (2005)
6. http://en.wikipedia.org/wiki/Upsampling
7. http://en.wikipedia.org/wiki/Nearest-neighbor_interpolation
8. http://en.wikipedia.org/wiki/Bilinear_interpolation
9. http://en.wikipedia.org/wiki/Bicubic_interpolation
10. Liu, J.-J., Chang, C.-C., Wang, L.-T., Yang, C.-K., Huang, S.-Y.: A new reading interface design for senior citizens. In: First International Conference on Instrumentation, Measurement, Computer, Communication and Control, 2011

Automated Text Detection and Text-Line Construction in Natural Images

Chih-Chang Yu, Ying-Nong Chen, Wang-Hsin Hsu
and Thomas C. Chuang

Abstract This work develops an automated system to detect texts in natural images captured by the cameras embedded on mobile devices. Unlike former researches which focus on detecting with straight texts, this work proposes a text-line construction algorithm which is able to extract curved text-lines in any orientations. An image operator called the Stroke Width Transform is adopted to exploit connected components which have stroke-like properties. Text components are classified into two types: active and passive. The links of active components are considered the initial orientation of text-lines. Complete text-lines are constructed by linking active and passive components. The system is implemented on the Android platform and the experimental results demonstrate the feasibility and validity of the proposed system.

Keywords Text detection · Stroke width transform · Mobile application

1 Introduction

Employing text detection algorithms on mobile devices assists users in understanding or gathering useful information. Today, many advertisements embed specific QR-codes, allowing users to capture the code using the camera on

C.-C. Yu (✉) · W.-H. Hsu · T. C. Chuang
Department of Computer Science Information Engineering,
Vanung University, 32061 Zhongli, Taiwan, Republic of China)
e-mail: tacoyu@mail.vnu.edu.tw

Y.-N. Chen
Department of Computer Science Information Engineering, National Central University,
32001 Zhongli, Taiwan, Republic of China.)

S.-S. Yeo et al. (eds.), *Computer Science and its Applications*,
Lecture Notes in Electrical Engineering 203, DOI: 10.1007/978-94-007-5699-1_68,
© Springer Science+Business Media Dordrecht 2012

smartphones and then direct them to a specific webpage. Others may ask users to type in keywords on browsers. A system incorporating with text detection and Optical Character Recognition (OCR) techniques provides a more convenient, flexible, and intuitive approach for users. Mobile applications with text detection techniques can greatly assist people. For example, Ezaki et al. [1] proposed a text-to-speech system to assist visually impaired people to understand the content of an image.

Algorithms used for detecting texts are categorized into region-based method and connected component-based method. Region-based methods adopt features such as edges [2], DCT or wavelet coefficients [3], and histograms of oriented gradients [4] are used for analyzing whether the features in a sliding window reveal text-like properties. However, this kind of algorithm is not suitable for applications in mobile devices due to the high computational complexity. Another approach is the connected component-based method. Text candidates are extracted by thresholding firstly. Then, non-text components are filtered out based on certain geometric rules. The connected component-based method is usually computationally inexpensive than the region-based method. Therefore, applying a connected component-based method to mobile devices is a reasonable option. In terms of connected component-based method, Epshtein et al. [5] proposed an image operator called the Stroke Width Transform (SWT). However, the authors in [5] did not explicitly explain how to construct text-lines. Moreover, most text detection algorithms [6, 7] focus on detecting horizontal or straight texts. However, texts may be curved in photographs (see Fig. 3a). Therefore, built upon the effort of SWT, this work proposes a text-line construction algorithm. The advantage of the proposed method is that it can obtain text orientation, regardless of how the text is displayed. Moreover, the proposed method is able to detect texts in different colors. That is, the system does not assume that letters in the same text should have the same color.

2 Proposed System

2.1 Stroke Width Transform

The SWT operator manifests image regions containing parallel edges. Non-edge pixels inside the stroke are filled with the stroke width value. Pixels are then grouped to form components according to their stroke width value. Because the width of strokes should be a fixed value ideally, components which have small variances of SWT values are regarded as text candidates. More detail can be found in [5].

Fig. 1 Minimum rectangle
which encloses letter 'W' and
its properties

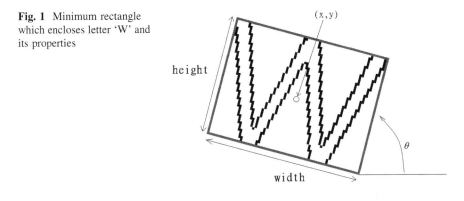

2.2 Text Components Classification

The idea of SWT is very simple, which is to find parallel edges in images.
However, many objects in natural images also contain parallel edges, such as
fences or stripes. These objects often cause false alarms. To remedy this problem,
this work classifies text components into two types, active type and passive type.
First, extracted components after performing SWT are discarded if they meet the
following criteria: (i) the size of the CC is too small. (ii) The variance of SWT
value is larger than a specific threshold. (iii) the CC is overlapped with more than
three other CCs. (iv) The average SWT value of the CC is too large.

After eliminating most non-text components, the rest CCs are then enclosed by
a minimum rectangle as shown in Fig. 1. For each CC, the following properties are
obtained based on the minimum rectangle: the height and width of the rectangle,
the centroid of CC, the orientation θ, and the number of contours.

A CC is discarded if it satisfies the following criteria: (i) The number of
contours is larger than a specific value. This is because the maximum number of
contours of alphanumeric is 3 (e.g., 'B' and '8'). This work set the value to 4 to
prevent some characters may stick together due to imperfect SWT. (ii) The height/
width ratio of CC is too large. (iii) The fill rate, which is the number of pixels of
CC over the area of the rectangle, is too small. CCs which pass the above criteria
are classified into active and passive components. A CC is defined as a passive
component if it meets the following criteria: (i) the number of contour is 1 and the
height/width ratio is larger than a specific threshold. Characters such as 'I' and '1'
will be classified as passive components according to this criterion. (ii) The height/
width ratio of CC is larger than a threshold th (th is set to 4 in our experiments).
(iii) The fill rate is larger than 0.8. A CC which does not satisfy these criteria is
classified as active type. The active components try to connect with other CCs
while the passive components are only connected by active components. A problem arises in estimating the orientation of character "O/o" because the letter is
nearly round, and thus, allows for multiple orientations. Therefore, before estimating the orientation of the rectangle, a CC is first examined if it is character "o"
before computing its orientation. The characteristic of "o" is stated as follows:

(i) the height/width ratio is close to 1. (ii) It contains exactly two contours. (iii) The difference of height/width between the inner contour and outer contour is close to the stroke width. When a component is determined as character "o", it is classified as active component but its orientation is not computed.

In the proposed design, many non-text CCs will be classified as passive components. This work does not discard these components directly because true text components may be broken or merged together due to imperfect SWT extraction. The orientations of active CCs are used to search nearby active CCs.When such things happen; the properties of these components are very similar to that of noises. Therefore, this work classifies them as passive components so that they can be recovered by connecting with active components.

2.3 Text-Line Extraction

According to the orientation of an active component C_i, the directions of two orthogonal rays are considered (see Fig. 2a). Both rays are given scores. The score of the ray is increased if a component C_j that is passed by the ray meet the following criteria: (i) C_i and C_j have similar stroke width. (ii) C_i and C_j have similar orientation or C_j is recognized as character 'O/o'. (iii) The height ratio or width ratio of C_i and C_j do not exceed a threshold. The length of R_1 and R_2 is three times of the diagonal of C_i in this work. The ray with the higher score is considered the potential text-line direction of C_i. C_i will connect to another active component C_j which is closest to C_i according to the potential text-line direction. Figure 2b shows the connection results of active components.

Afterwards, the system starts to extend/merge these text-lines. For each connected set, the end CCs, C_1 and C_2, can be found. According to the link direction of C_1 or C_2, the opposite directions are extended within a certain distance until it encounters another component C_k. C_k is connected to the set if the following criteria are satisfied: (i) C_k is an active component. (ii) C_k is a passive component and the stroke width value of C_k is similar to C_1/C_2. (iii) C_k is a passive component and the height ratio or width ratio of C_k and C_1/C_2 is similar. The main difference to the above step is that the passive component is considered in this step and the orientation information of CCs is discarded. Finally, those connected sets with less than two components are eliminated. The extension results are shown in Fig. 2c.

3 Experiments

To verify the feasibility and validity of the system, the proposed system is developed on the Android platform. 181 images were collected using the smartphone HTC Desire, which has a CPU speed 1 GHz with 576 MB RAM. The resolution of the embedded camera on the smart phone is 500 megapixels. In this

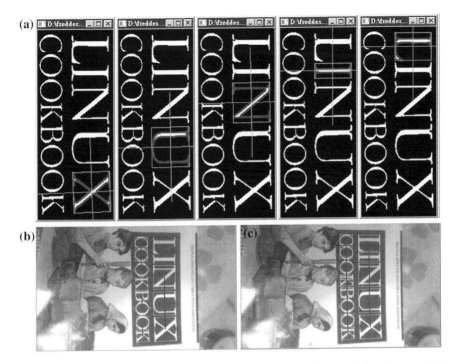

Fig. 2 Example of text-line construction: **a** directions of text-line seeking with each CC, **b** connected active components, **c** constructed text-lines after extension

work, the resolution of the input images is fixed at 640-by-384. Although the camera on HTC Desire is capable of taking higher resolution images, the processing time is too long for practical mobile applications. Test images may contain some backgrounds to test the robustness of the system.

The performance of the proposed system is evaluated using two forms of notation, precision and recall, which are calculated based on the area of the rectangles (see Fig. 3). The definition of precision and recall are listed as follow:

$$\text{Precision} = \frac{C}{E} \tag{1}$$

$$\text{Recall} = \frac{C}{T} \tag{2}$$

E is called the estimate, which is the total area of the rectangles detected by the proposed algorithm. T is the area of the rectangles of true texts, which is manually labeled. C is the intersected area between E and T. In general, the standard f-measure is used for evaluating the quality of the algorithm:

$$f = \frac{1}{\frac{\alpha}{precision} + \frac{1-\alpha}{recall}} \tag{3}$$

Fig. 3 Detection results with different situations. **a** Curved texts, **b** curved texts and horizontal texts, **c** slanted texts, **d** texts with different colors

α is set to 0.5 to give equal weight to the precision and recall. In this work, the precision rate is 89.2 % and the recall rate is 64.14 %, and the corresponding *f*-score is 0.75. Figure 3 shows some detection results. It is noticeable that the proposed system is able to detect text-lines in any direction or curved text-lines even the colors of characters inside a text-line are different.

4 Conclusions

This paper proposes a system which automatically finds texts in natural images. The image operator called stroke width transform is employed to extract possible text components in images. To detect curved text-lines, a two-stage framework is proposed. Text components are classified into two types: active and passive. Active text components are first linked each other based on some rules to form strong text sets. The orientations of these links are regarded as initial text-lines. These text-lines are further extended to link passive components to form the refined text-lines. This work only considers the detection performance of alphanumeric letters. This work can be further improved by detecting characters of other languages (e.g. Chinese) in the future.

References

1. Ezaki, N., Bulacu, M., Schomaker, L.: Text detection from natural scene images: towards a system for visually impaired persons. In: 17th International Conference on Pattern Recognition, vol. II, pp. 683–686 (2004)
2. Chen, X.R., Yuille, A.L.: Detecting and reading text in natural scenes. In: IEEE Conference on Computer Vision and Pattern Recognition, Washington, USA, pp. 366–373 (2004)
3. Gllavata, J., Ewerth, R., Freisleben, B.: Text detection in images based on unsupervised classification of high-frequency wavelet coefficients. In: 17th International Conference Pattern Recognition, pp. 425–428 (2004)
4. Ma, L., Wang, C., Xiao, B.: Text detection in natural images based on multiscale edge detection and classification. In: 3rd International Congress on Image and Signal Processing, pp. 1961–1965 (2010)
5. Epshtein, B., Ofek, E., Wexler, Y.: Detecting text in natural scenes with stroke width transform. In: IEEE Conference on Computer Vision and Pattern Recognition (2010)
6. Ferreira, S., Garin, V., Gosselin, B.: A Text detection technique applied in the framework of a mobile camera-based application. In: Workshop of Camera-Based Document Analysis and Recognition (2005)
7. Subramanian, K., Natarajan, P., Decerbo, M., Castañòn, D.: Character-stroke detection for text-localization and extraction. In: International Conference on Document Analysis and Recognition (2005)

Part IV
Science, Technology and Society

Security Weakness of a Dynamic ID-Based User Authentication Scheme with Key Agreement

Mijin Kim, Namje Park and Dongho Won

Abstract A remote user authentication scheme is a method to confirm the identity of a remote individual login to the server over an untrusted, public network. In 2012, Wen-Li proposed a dynamic ID-based user authentication scheme with key agreement and claimed that their scheme resisted impersonation attack and avoided leakage of partial information However, we find out that Wen-Li's scheme could leak some key information to an adversary and is exposed to man in the middle attack launched by any adversary. In this paper we conduct detailed analysis of flaws in Wen-Li's scheme.

Keywords Dynamic ID · User authentication · Key agreement · Man-in-the-middle attack · Security

1 Introduction

In many areas of computing to provide secure communication over an untrusted, public network, remote user authentication protocols have been developed [1–7], but there are still research problems to ensure security. In 2004, Das et al. [1]

M. Kim · D. Won (✉)
School of Information and Communication Engineering, Sungkyunkwan University,
300 Cheoncheon-dong, Jangan-gu,, Suwon-si, Gyeonggi-do 440-746, Korea
e-mail: dhwon@security.re.kr

M. Kim
e-mail: mjkim@security.re.kr

N. Park
Department of Computer Education, Teachers College, Jeju National University,
61 Iljudong-ro, Jeju-si, Jeju 690-781, Korea
e-mail: namjepark@jejunu.ac.kr

S.-S. Yeo et al. (eds.), *Computer Science and its Applications*,
Lecture Notes in Electrical Engineering 203, DOI: 10.1007/978-94-007-5699-1_69,
© Springer Science+Business Media Dordrecht 2012

proposed a dynamic ID-based authentication scheme to avoid leakage of partial information about the user's login message to the adversary. After that, researchers have proposed improved dynamic ID-based user authentication protocols to remedy weaknesses in the previous authentication protocols. In 2012, Wen-Li proposed a dynamic ID-based authentication scheme with key agreement [7] which remedied security flaws and weaknesses of Wang et al.'s scheme [4]. Session key was used to establish a secure communication channel in Wen-Li's scheme. A secure authentication with key agreement should accomplish both mutual authentication and session key establishment [8]. Wen-Li claimed that their scheme resisted impersonation attack and avoided the leakage of partial information [9–13]. However, we find out that Wen-Li's scheme leaks partial information about communication party's secret parameters and any adversary can exploit the leaked information to deduce session keys. In this paper, we present the vulnerability of Wen-Li's scheme to the man in the middle attack.

The rest of this paper is organized as follows. In Sect. 2, Wen-Li's dynamic ID-based user authentication scheme with key agreement is reviewed. The security weaknesses of Wen-Li's scheme are presented in Sect. 3. Finally, concluding remarks are given in Sect. 4.

2 Wen-Li's Scheme

Wen-Li's scheme consists of four basic phases: registration phase, login phase, authenticated key exchange phase, mutual authentication and key confirmation phase, and three functional phases: revocation phase, off-line password change phase and on-line secret renew phase.

Registration phase. User U_i performs the registration phase to be a legal participant in the scheme. New users have to submit $<ID_i, pw_i>$ to S through a secret channel. The detailed description is given below.

R1. $U_i \rightarrow S <ID_i, pw_i>$
R2. $S \rightarrow U_i <h(\cdot), N_i, n_i>$

   1. S computes $n_i = h(ID_i \| pw_i)$. The unique number n_i is kept by S to check the validity of the smart card.
   2. S computes $m_i = n_i \oplus x, N_i = h(ID_i) \oplus h(pw_i) \oplus h(x) \oplus h(m_i)$, where x is the server's secret number.
   3. S stores some parameters $<h(\cdot), N_i, n_i>$ in the U_i's smart card.
   4. S sends the smart card to U_i through a secret channel.

Login Phase. In this phase, when U_i wants to login the server S, U_i inserts his/her smart card and keys ID_i and PW_i. The smart card performs the following:

L1. $U_i \rightarrow S <M_1>$

1. The smart card computes the needed parameters to create the login request message.

$$A_i = h(ID_i) \oplus h(pw_i),$$
$$B_i = N_i \oplus h(ID_i) \oplus h(pw_i) = h(x) \oplus h(m_i),$$
$$CID_i = h(A_i) \oplus h(h(n_i) \oplus B_i \oplus h(N_i) \oplus T).$$

2. U_i sends the login request message $M_1 = \{CID_i, n_i, N_i, T\}$ to S.

Authentication and key exchange phase. In this phase, when S receives the login request message from U_i, S performs the following steps.

A1. $<S \to U_i> \cdots <M_2>$

1. Upon receiving the login request message at time T', S checks the validity of the timestamp T If $T' - T \leq \Delta T$ holds and n_i is in the registered list, S continues the next step.
2. S computes $m_i = n_i \oplus x, B_i = h(x) \oplus h(m_i), A_i = N_i \oplus B_i = h(ID_i) \oplus h$
(pw_i).
3. S verifies whether the equation $CID_i \oplus h(A_i) = h(B_i \oplus h(N_i) \oplus h(n_i) \oplus T)$ holds.
4. If so, S computes $C_i = h(A_i \oplus T' \oplus h(n_i))$. S can compute the session key $SK = h(A_i \| T \| B_i \| T')$, and key confirmation message $KC' = h(B_i \| SK \| T')$..
5. S sends the replied message $M_2 = \{C_i, KC', T'\}$.

Mutual authentication and key confirmation phase. In this phase, when U_i receives the replication at time T'', U_i performs the following steps.

M1. $U_i \to S <M_3>$

1. U_i checks whether the timestamp T' is valid.
2. If the time interval is valid, U_i computes $h(A_i \oplus T' \oplus h(n_i))$, and verifies if it is equal to C_i.
3. U_i computes $SK = h(A_i \| T \| B_i \| T')$, then checks whether the key confirmation message KC' is correct. If so, U_i computes $KC = h(A_i \| SK \| T'')$
4. U_i sends the message $M_3 = \{KC, T''\}$.
M2. S verifies the message M_3, if the equation $KC = h(A_i \| SK \| T'')$ holds, this scheme is finished.

There are three additional functional phases: revocation phase, off-line password change phase, on-line secret renew phase in Wen-Li's scheme [7].

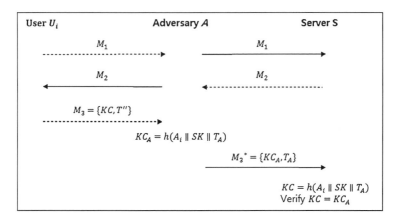

Fig. 1 Man-in-the-middle attack on Wen-Li's scheme

3 Security Analysis of Wen-Li's Scheme

We point out security weakness of Wen-Li's scheme. In Wen-Li's scheme, when U_i sends the login request message M_1 to S through public network, the secret values stored in U_i's smart card $\{N_i, n_i\}$ which is included in the message M_1 can be revealed. The adversary A may exploit these values to achieve offline guessing attack and man-in-the-middle attack.

3.1 Man-in-the-middle Attack

We assume that adversary A interposes the communication between U_i and S. A has intercepted the user U_i's login message $M_1 = \{CID_i, N_i, n_i, T\}$ and authentication and key exchange message $M_2 = \{C_i, KC', T'\}$ between U_i and S. Since the login request, and authentication and key exchange message are sent through public network, any users including illegal ones can intercept them from the public network. The attack scenario is outlined in Fig. 1 where a dashed line indicates that the corresponding message is intercepted by A en route its destination. A more detailed description of the attack is as follows:

1. From the intercepted message M_1 and M_2, the adversary A can get $h(n_i)$, T', and C_i. The leakage of these information is equivalent to compromise the secret A_i.
2. Since $CID_i = h(A_i) \oplus h(h(n_i) \oplus B_i \oplus h(N_i) \oplus T)$, the secret information A_i would be more helpful for the adversary to launch off-line guessing attacks in order to reveal B_i.
3. Therefore, the adversary A can calculate the session key $SK = h(A_i \parallel T \parallel B_i \parallel T')$ which is a secret value between U_i and S.
4. Thereafter, A can impersonate U_i to S without knowing ID_i and pw_i.

5. On the other hand, upon receiving the message M_1 in Wen-Li's scheme, S computes and sends the replied message M_2 to U_i. However, this message is intercepted by the adversary A.

6. The adversary A forwards the message M_2 to U_i. Then U_i operates, as specified in Wen-Li's scheme, and sends the message $M_3 = \{KC, T''\}$ to S. However, this message is intercepted by A and creating a timestamp T_A, the adversary A computes $KC_A = h(A_i \parallel SK \parallel T_A)$, A forges a message $M_3^* = \{KC_A, T_A\}$. Then A sends the forged message M_3^*, as if it originated from U_i.

7. According to Wen-Li's scheme, upon receiving the message M_3^*, S verifies the last key confirmation message, if the equation $KC = h(A_i \parallel SK \parallel T_A) = KC_A$ holds. Since M_3^* is valid, this passes, verifying U_i.

Following this, as described in the above attack, S, U_i, and adversary A share the session key SK. However, S and U_i cannot detect they share the session key with the adversary A. From now, the adversary A could impersonate U_i to S and impersonate S to U_i, only through intercepting the message transmitted in the public channels. Such attack could be serious, such as in the financial fields, the adversary could impersonate the legal user to transfer accounts to somebody. The worse effect could be happens in the government or military departments [7].

4 Conclusion

In 2012, Wen-Li proposed a dynamic ID-based remote user authentication scheme [7] with session key agreement and demonstrated its resistance to impersonation attack and nonleakage of partial information. However, after reviewing Wen-Li's scheme and analyzing its security, man-in-the-middle attack on the scheme is presented. The analyses show that the scheme reveals partial information and is insecure for practical applications. Future work could be undertaken to refer Lv et al.'s scheme [14] and remedy Wen-Li's scheme.

Acknowledgments This research was supported by the KCC (Korea Communications Commission), Korea, under the R&D program supervised by the KCA (Korea Communications Agency) (KCA-2012-12-912-06-003).

References

1. Das, M.L., Saxana, A., Gulati, V.P.: A dynamic ID-based remote user authentication scheme. IEEE Trans. Consum. Electron. **50**(2), 629–631 (2004)
2. Awasthi, A.K.: Comment on a dynamic ID-based remote user authentication scheme. Trans. Cryptol. **1**(2), 15–16 (2004)
3. Ku, W.C., Chang, S.T.: Impersonation attacks on a dynamic ID-based remote user authentication scheme using smart cards. IEICE Trans. **5**, 2165–2167 (2005)

4. Wang, Y.Y., Liu, J.Y., Xiao, F.X., Dan, J.: A more efficient and secure dynamic ID-based remote user authentication scheme. Comput. Commun. **32**(4), 583–585 (2009)
5. Juang, W.S., Wu, J.L.: Two efficient two-factor authenticated key exchange protocols in public wireless LANs. Comput. Electr. Eng. **35**(1), 33–40 (2009)
6. Lee, Y., Kim, S., Won, D.: Enhancement of two-factor authenticated key exchange protocols in public wireless LANs. Comput. Electr. Eng. **36**(1), 213–223 (2010)
7. Wen, F., Li, X.: An improved dynamic ID based remote user authentication scheme with key agreement scheme. Comput. Electr. Eng. **38**(2), 381–387 (2012)
8. Tsaur, W., Li, J., Lee, W.: An efficient and secure multi-server authentication scheme with key agreement. J. Syst. Softw. **85**(4), 876–882 (2012)
9. Park, N., Kwak, J., Kim, S., Won, D., Kim, H.: WIPI mobile platform with secure service for mobile RFID network environment. In: Shen, H.T., Li, J., Li, M., Ni, J., Wang, W. (eds.) APWeb Workshops 2006, LNCS, vol. 3842, pp. 741–748. Springer, Heidelberg (2006)
10. Park, N.: Security scheme for managing a large quantity of individual information in RFID environment. In: Zhu, R., Zhang, Y., Liu, B., Liu, C. (eds.) ICICA 2010, CCIS, vol. 106, pp. 72–79. Springer, Heidelberg (2010)
11. Park, N.: Secure UHF/HF dual-band RFID: strategic framework approaches and application solutions. In: ICCCI 2011, LNCS. Springer, Heidelberg (2011)
12. Park, N.: Implementation of terminal middleware platform for mobile RFID computing. Int. J. Ad HocUbiquitous Comput, **8**(4), 205–219 (2011)
13. Park, N., Kim, Y.: Harmful adult multimedia contents filtering method in mobile RFID service environment. In: Pan, J.-S., Chen, S.-M., Nguyen, N.T. (eds.) ICCCI 2010, LNCS(LNAI), vol. 6422, pp. 193–202. Springer, Heidelberg (2010)
14. Lv, C., Ma, M., Li, H., Ma, J., Zhang, Y.: An novel three-party authenticated key exchange protocol using one-time key. J. Netw. Comput. Appl. (available online, 2012)

Development and Application of STEAM Teaching Model Based on the Rube Goldberg's Invention

Yilip Kim and Namje Park

Abstract STEAM is an acronym of Science, Technology, Engineering, Arts, and Mathematics. Rube Goldberg's Invention requires various inputs and efforts ranging from scientific knowledge, mathematical reasoning, engineering design, to ability for technical operation. It can be an ideal activity for STEAM education that stands for science, technology, engineering, art and mathematics. In this regard, the study identified elements of Rube Goldberg's Invention that could be applied to STEAM education.

Keywords Rube Goldberg · Elementary education · Creativity · STEAM

1 Introduction

The late Steve Jobs, an iconic figure of the 21st-century innovation, pioneered a new IT frontier with iPhone that blended technology, engineering elements and creativity [1–6]. These days, creative thinking and technology are directly related to the

This work was supported by the Korea Foundation for the Advancement of Science and Creativity (KOFAC) grant funded by the Korean Government (MEST).

Y. Kim
Jeju Nam Elementary School, 149 Namsung-ro, Jeju-si, Korea

Y. Kim (✉) · N. Park (✉)
Department of Computer Education, Teachers College, Jeju National University,
61 Iljudong-ro, Jeju Special Self-Governing Province 690-781 Jeju-si, Korea
e-mail: yilip@paran.com

N. Park
e-mail: namjepark@jejunu.ac.kr

S.-S. Yeo et al. (eds.), *Computer Science and its Applications*,
Lecture Notes in Electrical Engineering 203, DOI: 10.1007/978-94-007-5699-1_70,
© Springer Science+Business Media Dordrecht 2012

competitiveness of a society. In coming years, technical innovation will be about reinventing and converging existing technologies, rather than creating something out of nothing. In an era of convergence, leaders will need creativity and artistic sensibilities in addition to scientific knowledge and technical savvy [7–11]. In this circumstance, STEAM(Science, Technology, Engineering, and Mathematics) has emerged as a new educational catchphrase that will enhance students' understanding of mathematics and science. Many educational programs are being developed accordingly [12–16].

The basic idea of Rube Goldberg's Invention is to design the most complicated mechanism to solve the simplest problem. This process involves scientific principles, mathematical intuition, engineering maneuver, creative design and skills. It has all the elements that STEAM education aims to instill in students. This study compares the mechanical elements of Rube Goldberg's Invention with ideas of STEAM, in order to design learning materials that can be introduced to school curricula.

2 Need for STEAM Education

STEAM is an acronym of Science, Technology, Engineering, Arts, and Mathematics. As the low interests and accomplishments of American teenagers in math and science, the STEM education started as an educational solution [17–20]. However, the STEM education was missing a very important piece. This is that Art, a comparatively competitive and innovative field as STEM in creativity, was also needed. In addition, the science education could not keep up with the current changes in science, technology, and engineering and the teenagers who are used to the various advanced technology products were bound to lose interests as well as creating a gap in creativity cultivation in science education during elementary and middle school years.

Therefore, the experts argued for "amicability between science and art" because a dichotomous thought that art is illogical and science is not creative ruined the future and the art and science should be taught together before the concept of STEAM education emerged. In this perspective, the art education is crucial in developing creativity that is highly valued in modern education; therefore, the art education should be added to the education of science, technology, engineering, and mathematics.

3 Rube Goldberg's Invention for STEAM Education

3.1 Introduction to Rube Goldberg's Invention

Students watch video of Rube Goldberg's Invention contest or actual Rube Goldberg's Invention to develop interest and share ideas. By doing so, students are motivated and perceive that scientific activities can be familiar and easily approached.

Fig. 1 Working with an actual Rube Goldberg's invention

3.2 Exploring Scientific and Mathematical Aspects of Rube Goldberg's Invention

Students work with Rube Goldberg's Invention to examine and understand various scientific and mathematical principles involved such as operation of a pulley, conversion of energy on a slope, operation of a lever, and center of gravity. By working with an actual Rube Goldberg's Invention, students intuitively absorb scientific and mathematical ideas, and are encouraged to find examples of Rube Goldberg's Invention in a daily life (Fig. 1).

3.3 Analyzing Mechanical Design of Rube Goldberg's Invention

Students examine each part of Rube Goldberg's Invention to see how they are designed and interconnected.

3.4 Design of Rube Goldberg's Invention

Students work in a team to design Rube Goldberg's Invention. They are encouraged to apply the scientific ideas and principles to each part. Students draw design concepts and apply various scientific and mathematical principles, absorbing the knowledge through the process with an anticipation for the outcome (Fig. 2).

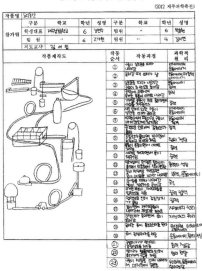

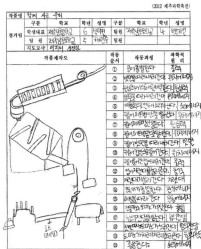

Fig. 2 Design of Rube Goldberg's invention

Fig. 3 Design of Rube Goldberg's invention

3.5 Manufacturing Rube Goldberg's Invention Using Colored Styrofoam and 4D Frame

With the design concept and drawing, students work in a team to construct Rube Goldberg's Invention and make necessary modifications.

3.6 Demonstration of Rube Goldberg's Invention

Students present mechanism and design of their Rube Goldberg's Invention and demonstrate its operation (Fig. 3).

3.7 Peer Review

Students compare and analyze various Rube Goldberg's Inventions and evaluate them.

4 Conclusion

The objective of the paper is to apply Rube Goldberg's Invention to explore how to connect school curricula with invention activities, establish a theoretical structure of invention education to draw lessons from mechanical and scientific principles. Also, it aims to design an invention program for schools. Rube Goldberg's Invention could be a useful tool to revitalize invention contests and programs at school. In the stages of understanding STEAM and Rube Goldberg's Invention, the related activities stimulate students' academic curiosity and interest (A) and help them develop a positive attitude toward science (S). They examine and analyze (ST) Rube Goldberg's Invention to learn its scientific (S) and mathematical (M) principles as well as engineering mechanism (E).

Acknowledgments This work was supported by the Korea Foundation for the Advancement of Science and Creativity(KOFAC) grant funded by the Korean Government(MEST). This paper is extended from a conference paper presented at the second international conference on Computers, Networks, Systems, and Industrial applications (CNSI 2012), Jeju Island, Korea. The author is deeply grateful to the anonymous reviewers for their valuable suggestions and comments on the first version of this paper. The corresponding author is Namje Park (namjepark@jejunu.ac.kr).

References

1. Debra O'Connor.: Application sharing in K-12 education: teaching and learning with Rube Goldberg. TechTrends, vol. 47, pp. 6–13. Springer, Boston (2003)
2. Wolfe, M.F., Goldberg, R.: Rube Goldberg: inventions! Simon & Schuster, New York (2000)

3. Park, N., Kwak, J., Kim, S., Won, D., Kim, H.: WIPI mobile platform with secure service for mobile RFID network environment. In: Shen, H.T., Li, J., Li, M., Ni, J., Wang, W. (eds.) APWeb Workshops 2006. LNCS, vol. 3842, pp. 741–748. Springer, Heidelberg (2006)

4. Park, N.: Security scheme for managing a large quantity of individual information in RFID environment. In: Zhu, R., Zhang, Y., Liu, B., Liu, C. (eds.) ICICA 2010. CCIS, vol. 106, pp. 72–79. Springer, Heidelberg (2010)

5. Park, N.: Secure UHF/HF Dual-band RFID: Strategic framework approaches and application solutions. In: ICCCI 2011. LNCS, Springer, Heidelberg (2011)

6. Park, N.: Implementation of terminal middleware platform for mobile RFID computing. Int. J. Ad Hoc Ubiquitous Comput. **8**(4), 205–219 (2011)

7. Park, N., Kim, Y.: Harmful adult multimedia contents filtering method in mobile RFID service environment. In: Pan, J.-S., Chen, S.-M., Nguyen, N.T. (eds.) ICCCI 2010. LNCS(LNAI), vol. 6422, pp. 193–202. Springer, Heidelberg (2010)

8. Park, N., Song, Y.: AONT Encryption based application data management in mobile RFID environment. In: Pan, J.-S., Chen, S.-M., Nguyen, N.T. (eds.) ICCCI 2010. LNCS(LNAI), vol. 6422, pp. 142–152. Springer, Heidelberg (2010)

9. Park, N.: Customized healthcare infrastructure using privacy weight level based on smart device. In: Communications in Computer and Information Science, vol. 206, pp. 467–474. Springer (2011)

10. Park, N.: Secure data access control scheme using type-based re-encryption in cloud environment. In: Studies in Computational Intelligence, vol. 381, pp. 319–327. Springer (2011)

11. Park, N., Song, Y.: Secure RFID application data management using all-or-nothing transform encryption. In: Pandurangan, G., Anil Kumar, V.S., Ming, G., Liu, Y., Li, Y. (eds.) WASA 2010. LNCS, vol. 6221, pp. 245–252. Springer, Heidelberg (2010)

12. Park, N.: The implementation of open embedded S/W platform for secure mobile RFID reader. J. Korea Inf. Commun. Soc. **35**(5), 785–793 (2010)

13. Park, N., Song, Y., Won, D., Kim, H.: Multilateral approaches to the mobile RFID security problem using web service. In: Zhang, Y., Yu, G., Bertino, E., Xu, G. (eds.) APWeb 2008. LNCS, vol. 4976, pp. 331–341. Springer, Heidelberg (2008)

14. Park, N., Kim, H., Kim, S., Won, D.: Open location-based service using secure middleware infrastructure in web services. In: Gervasi, O., Gavrilova, M.L., Kumar, V., Laganá, A., Lee, H.P., Mun, Y., Taniar, D., Tan, C.J.K. (eds.) ICCSA 2005. LNCS, vol. 3481, pp. 1146–1155. Springer, Heidelberg (2005)

15. Park, N., Kim, S., Won, D.: Privacy preserving enhanced service mechanism in mobile RFID network. In: ASC, Advances in Soft Computing, vol. 43, pp. 151–156. Springer, Heidelberg (2007)

16. Park, N., Kim, S., Won, D., Kim, H.: Security analysis and implementation leveraging globally networked mobile RFIDs. In: PWC 2006. LNCS, vol. 4217, pp. 494–505. Springer, Heidelberg (2006)

17. Kim, Y.: Harmful-word dictionary DB based text classification for improving performances of precesion and recall. Sungkyunkwan University, Ph.D. Thesis (2009)

18. Kim, Y, Park, N., Hong, D.: Enterprise data loss prevention system having a function of coping with civil suits. In: Studies in Computational Intelligence, vol. 365, pp. 201–208. Springer (2011)

19. Kim, Y., Park, N., Won, D.: Privacy-enhanced adult certification method for multimedia contents on mobile RFID environments. In: Proceedings of IEEE international symposium on consumer electronics, pp. 1–4. IEEE, Los Alamitos (2007)

20. Kim, Y., Park, N., Hong, D., Won, D.: Adult certification system on mobile RFID service environments. J. Korea Contents Assoc. **9**(1), 131–138 (2009)

Security Enhancement of User Authentication Scheme Using IVEF in Vessel Traffic Service System

Namje Park, Seunghyun Cho, Byung-Doo Kim, Byunggil Lee
and Dongho Won

Abstract Vessel Traffic System (VTS) is an important marine traffic monitoring system which is designed to improve the safety and efficiency of navigation and the protection of the marine environment. And the demand of Inter-VTS networking has been increased for realization of e-Navigation as shore side collaboration for maritime safety. And IVEF (Inter-VTS Data Exchange Format) for Inter-VTS network has become a hot research topic of VTS system. Currently, the IVEF developed by the International Association of Lighthouse Authorities (IALA) does not include any highly trusted certification technology for the connectors. However, the vessel traffic information requires high security since it is highly protected by the countries. Therefore, this study suggests the certification system to increase the security of the VTS systems using the main certification server and IVEF.

N. Park
Department of Computer Education, Teachers College, Jeju National University,
61 Iljudong-ro, Jeju-si, Jeju Special Self-Governing Province 690-781, Korea
e-mail: namjepark@jejunu.ac.kr

S. Cho · D. Won (✉)
College of Information and Communication Engineering, Sungkyunkwan University,
300 Cheoncheon-dong, Jangan-gu, Suwon-si, Gyeonggi-do 440-746, Korea
e-mail: dhwon@security.re.kr

S. Cho
e-mail: shc@security.re.kr

B.-D. Kim · B. Lee
Electronics and Telecommunications Research Institute (ETRI), 218 Gajeong-ro,
Yuseong-gu, Daejeon 305-700, Korea
e-mail: bdkim@etri.re.kr

B. Lee
e-mail: bglee@etri.re.kr

S.-S. Yeo et al. (eds.), *Computer Science and its Applications*,
Lecture Notes in Electrical Engineering 203, DOI: 10.1007/978-94-007-5699-1_71,
© Springer Science+Business Media Dordrecht 2012

Keywords VTS · Inter-VTS · IVEF · Navigation · Security

This research was supported by the KCC (Korea Communications Commission), Korea, under the R&D program supervised by the KCA (Korea Communications Agency) (KCA-2012-12-912-06-003).

1 Introduction

E-Navigation is a project suggested by the International Marine Organization with the purpose of marine security and safety as well as the marine environment protection in 2006 and also performs the standardization of the technology and fulfillment with the International Association of Lighthouse Authorities (IALA), an international vessel route association with still an active research in the field in 2010 [1–8]. The Vessel Traffic System (VTS) in the e-Navigation is the marine traffic control system that includes the vessel traffic information and marine environment. IVEF is a gateway protocol between the VTS developed by IALA to allow the information exchange between VTS's and is undergoing international standardization. This study suggests a system to resolve the issues in the information exchange between VTS using IVEF.

2 Related Work

2.1 Introduction of VTS System

Generally, the factors that comprise of VTS system are shown in Fig. 1, where the VTS center on land, various sensors (CCTV, Radar, DF, MET, and sensors) and the base station with AIS with the control center with actual VTS utilization are connected in a communication net of ship, satellite, and sensors in the control system [9–13].

The FP project in Europe promoted various e-Navigation studies, including next-generation vessel transportation control related technologies of VTS, VTMIS, and PCS. This is the "e-Navigation" project that will be realized in 201-2020 to provide cruise support and SAR service through safe and efficient information management by collecting the various pieces of information, such as vessel movements and docking, sea area climates, geography, and environment, with diverse equipments [14–18]. In addition, the MarNIS project carried out by the EU studied the wide sea area communication technology for an improved multimedia vessel communication. Especially, the improved sea area control function, vessel multimedia

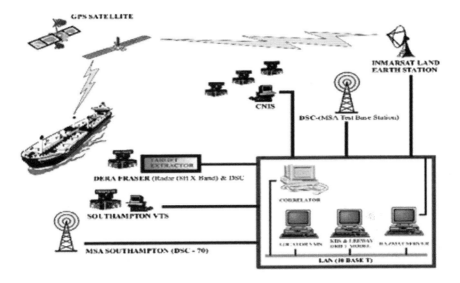

Fig. 1 System architecture of VTS

communication function were studied and the practical service realization and international standards request continuous future research.

2.2 Overview of IVEF Protocol

The VTS Committee of IALA is a framework of methods and services to promote the safety, security, efficiency, and environment protection in all transportable waters that is evolving from a traditional VTS to an e-Navigation service. In other words, this service structure is advanced from the vessel traffic monitor and control to business service and e-Navigation in vessel computing environment as a new service form [19–21].

The various information collection and management in the sea encountered a rapid development in its technology, which aims to provide the information service for the vessels during their voyage, such as sea situations, sea map vessel support, etc. At this time, the vessel information collection/management/production/sharing/provision services should be enabled through the information collection from vessels or trusted information exchange between the land systems.

IVEF service is a gateway service in the currently developing land system structure by IALA-AISM's e-Navigation working group. In other words, the IVEF service can have an external third party system linking structure as the client that requests the service and the mutually trusted network gateway security service is required [21–24]. Figure 2 shows the IVEF information exchange of VTS center with other users and institutions, such as neighboring control center, control

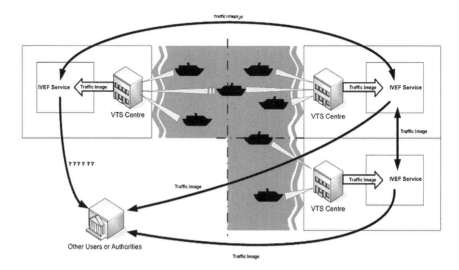

Fig. 2 VTS and IVEF service

institutions within the control area, or other relevant institutions. As shown in the figure, the traffic information provides the necessary information to the nearby system through the IVEF service.

IVEF service should be defined as a mutually linked service between domains. In addition, for a safe IVEF service, the land systems of regional VTS, national VTS, related institutions and companies should be interconnected in a safe structure.

3 Security Enhancement of User Authentication Scheme

IVEF is an open-source SDK for VTS information exchange that is being developed by IALA and is almost complete in its international standardization as a gateway. The official IVEF technology documents provided by IALA specify that the data security except for authentication and authorization is out of the IVEF scope. The IVEF security suggested at this point only codes the user authorization information in an open key method. However, when the physical link is terminated and then reconnected between VTS's, the VTS system may be delayed from temporary traffic overload. This may lead to data leakage. A solution requires studies on the main authorization server.

This chapter suggests the main authorization server for user authentication as shown in Fig. 3.

Figure 3 briefly summarizes the information exchange system after authentication for the Main Authentication Server with the IVEF Client (IC) and IVEF Server (IS). The MAS is comprised of AS (Authentication Server) and TGS (Ticket Granting Server).

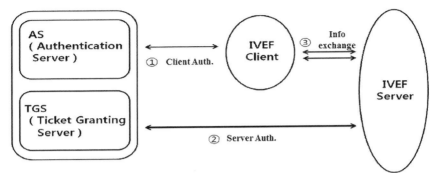

Main Authentication Server

Fig. 3 Main authorization scheme for user authentication in IVEF

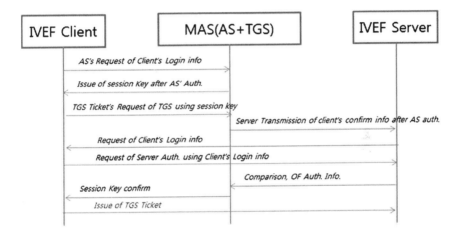

Fig. 4 Secure protocol between IC, IS, and MAS

Figure 4 shows the protocol between IC, IS, and MAS. The step 1 in Fig. 4 shows how IC requires to confirm the user from the AS in MAS using the Login information. Step 2 is the issuing of the session key after the certification of the IC in AS. Step 3 is the request of the TGS ticket issuing from TGS with the issued Session key. TGS ticket holds the Client ID, IP address, Ticket issue time, and ticket validity information. Step 4 sends the IC confirmation certified in AS in MAS to IS. Step 5 requests the Login information from IC by IS directly. Step 6 is the request of the server authentication with the Client Login information. Step 7 is the result delivery after the comparison of the confirmation from AS in step 4 and the client login result. Step 8 confirms the issued session key in step 2. Lastly, step 9 certifies IC an IS by issuing the same TGS ticket if there were no errors in all steps.

Therefore, MAS can authenticate IC and IS at the same time to sense the link termination in certain areas. In addition, the TGS ticket IP address and ticket valid time information can prevent the illegal access and replay attack of the attackers.

4 Conclusion

The e-Navigation technology standardization progress until show shows that IVEF will be highly utilized as the information exchange gateway between VTS but will still have its weaknesses. The existing open key improved the weaknesses, but such a method will lead to another weakness.

This study introduces the MAS concept using the basic IVEF structure and comparison keys to improve the new authentication process and the future studies will present specific key distribution methods in an efficient manner.

Acknowledgments This research was supported by the KCC(Korea Communications Commission), Korea, under the R&D program supervised by the KCA(Korea Communications Agency) (KCA-2012-12-912-06-003). The corresponding author is Dongho Won (dhwon@security.re.kr).

References

1. Arifin B., Ross E., Brodsky Y.: Data security in a ship detection and Identification system. IEEE RAST2011, pp. 634–636 (2011)
2. IVEF Recommendation V-145 on the Inter-VTS Exchange Format (IVEF) Service (2011)
3. Open IVEF, http://www.openivef.org
4. Carter, B., Green, S., Leeman, R., Chaulk, N.: SmartBay: better information—better decisions. IEEE Oceans **2008**, 1–7 (2008)
5. Frejlichowski D., Lisaj A.: Analysis of lossless radar images compression for navigation in marine traffic and remote transmission. IEEE Radar Conference, pp. 1–4 (2008)
6. Garnier B., Andritsos F.: A port waterside security systemic analysis. IEEE WSS Conference 2010, pp. 1–6 (2010)
7. Park, N., Kwak, J., Kim, S., Won, D., Kim, H.: WIPI mobile platform with secure service for mobile RFID network environment. In: Shen, H.T., Li, J., Li, M., Ni, J., Wang, W. (eds.) APWeb Workshops 2006. LNCS, vol. 3842, pp. 741–748. Springer, Heidelberg (2006)
8. Park, N.: Security scheme for managing a large quantity of individual information in RFID environment. In: Zhu, R., Zhang, Y., Liu, B., Liu, C. (eds.) ICICA 2010. CCIS, vol. 106, pp. 72–79. Springer, Heidelberg (2010)
9. Park, N.: Secure UHF/HF dual-band RFID: strategic framework approaches and application solutions. In: ICCCI 2011. LNCS, Springer, Heidelberg (2011)
10. Park, N.: Implementation of terminal middleware platform for mobile RFID computing. Int. J. Ad Hoc Ubiquitous Comput. **8**(4), 205–219 (2011)
11. Park, N., Kim, Y.: Harmful adult multimedia contents filtering method in mobile RFID service environment. In: Pan, J.-S., Chen, S.-M., Nguyen, N.T. (eds.) ICCCI 2010. LNCS(LNAI), vol. 6422, pp. 193–202. Springer, Heidelberg (2010)

12. Park, N., Song, Y.: AONT encryption based application data management in mobile RFID environment. In: Pan, J.-S., Chen, S.-M., Nguyen, N.T. (eds.) ICCCI 2010. LNCS(LNAI), vol. 6422, pp. 142–152. Springer, Heidelberg (2010)

13. Park, N.: Customized healthcare infrastructure using privacy weight level based on smart device. In: Communications in Computer and Information Science, vol. 206, pp. 467–474. Springer (2011)

14. Park, N.: Secure data access control scheme using type-based re-encryption in cloud environment. In: Studies in Computational Intelligence, vol. 381, pp. 319–327. Springer (2011)

15. Park, N., Song, Y.: Secure RFID application data management using all-or-nothing transform encryption. In: Pandurangan, G., Anil Kumar, V.S., Ming, G., Liu, Y., Li, Y. (eds.) WASA 2010. LNCS, vol. 6221, pp. 245–252. Springer, Heidelberg (2010)

16. Park, N.: The implementation of open embedded S/W platform for secure mobile RFID reader. J. Korea Inf. Commun. Soc. 35(5), 785–793 (2010)

17. Park, N., Song, Y., Won, D., Kim, H.: Multilateral approaches to the mobile RFID security problem using web service. In: Zhang, Y., Yu, G., Bertino, E., Xu, G. (eds.) APWeb 2008. LNCS, vol. 4976, pp. 331–341. Springer, Heidelberg (2008)

18. Park, N., Kim, H., Kim, S., Won, D.: Open location-based service using secure middleware infrastructure in web services. In: Gervasi, O., Gavrilova, M.L., Kumar, V., Laganá, A., Lee, H.P., Mun, Y., Taniar, D., Tan, C.J.K. (eds.) ICCSA 2005. LNCS, vol. 3481, pp. 1146–1155. Springer, Heidelberg (2005)

19. Park, N., Kim, S., Won, D.: Privacy preserving enhanced service mechanism in mobile RFID network. In: ASC, Advances in Soft Computing, vol. 43, pp. 151–156. Springer, Heidelberg (2007)

20. Park, N., Kim, S., Won, D., Kim, H.: Security analysis and implementation leveraging globally networked mobile RFIDs. In: PWC 2006. LNCS, vol. 4217, pp. 494–505. Springer, Heidelberg (2006)

21. Kim, Y.: Harmful-word dictionary DB based text classification for improving performances of precesion and recall. Sungkyunkwan University, Ph.D. Thesis (2009)

22. Kim, Y, Park, N., Hong, D.: Enterprise data loss prevention system having a function of coping with civil suits. In: Studies in Computational Intelligence, vol. 365, pp. 201–208. Springer (2011)

23. Kim, Y., Park, N., Won, D.: Privacy-enhanced adult certification method for multimedia contents on mobile RFID environments. In: Proceedings of IEEE International Symposium on Consumer Electronics, pp. 1–4. IEEE, Los Alamitos (2007)

24. Kim, Y., Park, N., Hong, D., Won, D.: Adult certification system on mobile RFID service environments. J. Korea Contents Assoc. 9(1), 131–138 (2009)

Weakness of Tan's Two-Factor User Authentication Scheme in Wireless Sensor Networks

Youngsook Lee, Jeeyeon Kim and Dongho Won

Abstract As wireless sensor networks (WSN) continue to grow, so does the need for effective security mechanisms. Because sensor networks may interact with sensitive data and/or operate in hostile unattended environments, it is important that these security concerns be addressed from the beginning of the system design. So WSN requires main security goal of authenticating among a remote individual, the sensor nodes, and the gateway node. In 2011, Tan proposed a two-factor user authentication scheme suited for WSN environments, in which users can be authenticated using a single password shared with the gateway node. A fundamental requirement for password-based authentication is security against off-line password guessing attack. However, Tan's scheme does not meet the requirement. In this work, we demonstrate this security problem with Tan's user authentication scheme.

Keywords Wireless sensor network · Smart card · Password · Off-line password guessing attack · Two-factor authentication

Y. Lee
Department of Cyber Investigation Police, Howon University, Gunsan, Korea
e-mail: ysooklee@howon.ac.kr

J. Kim · D. Won (✉)
School of Information and Communication Engineering, Sungkyunkwan University,
Suwon-si, Korea
e-mail: jeeyeonkim@paran.com

J. Kim
e-mail: jeeyeonkim@paran.com

S.-S. Yeo et al. (eds.), *Computer Science and its Applications*,
Lecture Notes in Electrical Engineering 203, DOI: 10.1007/978-94-007-5699-1_72,
© Springer Science+Business Media Dordrecht 2012

1 Introduction

A wireless sensor network (WSN) consists of spatially distributed autonomous sensors to monitor physical or environmental conditions, such as temperature, sound, vibration, pressure, humidity, motion or pollutants and to cooperatively pass their data through the network to a main location [1]. Wireless Sensor Network is composed of large number of sensor nodes that are scattered in hostile unattended environments [2]. The inclusion of wireless communication technology also incurs various types of security threats. It is important that these security concerns be addressed from the beginning of the system design. So understanding security of wireless sensor network is important issue. There are so many mechanisms are developed to provide the security to sensor network or node. One of the important issues in security of wireless sensor network is main security goal of authenticating between a remote individual and the sensor nodes, between the sensor node and the gateway node, and between the remote individual and the gateway node [3–11]. Figure 1 depicts an example of such wireless sensor network architecture.

In 2009, Das proposed a two-factor user authentication scheme for WSN environment [12]. However Khan et al. have pointed out that Das's scheme is still vulnerable to a bypassing attack and privileged-insider attack, and so on [13]. To patch these security flaws, Khan et al. have recently presented a modified version of Das's scheme. Their enhanced scheme overcomes the GWN bypassing attack and provides mutual authentication between the gateway node GWN and the sensor node [14, 15]. Unfortunately, in 2011, Tan uncover that Khan et al.'s scheme is still insecure and vulnerable to several attacks, the off-line password guessing attack and the stolen verifier attack [16]. Tan proposed the enhanced scheme on Khan et al's scheme. In its article, it claims that the scheme provides a two-factor authentication which guarantees the security of the scheme when either the user's smart card or its password is stolen, but not both. But unlike the claim, its modification does not achieve a two-factor user authentication. We show this by mounting the off-line password guessing attack on Tan's scheme.

2 Review of Tan's Two-Factor User Authentication Scheme

This section reviews a two-factor user authentication scheme proposed by Tan [16]. The scheme participants include a gateway node, a remote user, and multiple sensor nodes. For simplicity, we denote the gateway node by GWN, the remote user by U_i, and the senssor nodes by SN_1, SN_2,…, SN_n.

Tan's scheme consists of four phases: registration phase, login phase, verification phase, and password change phase. The registration phase is performed only once per user when a new user registers itself with the gateway node. The login and verification phases are carried out whenever a user wants to gain access to

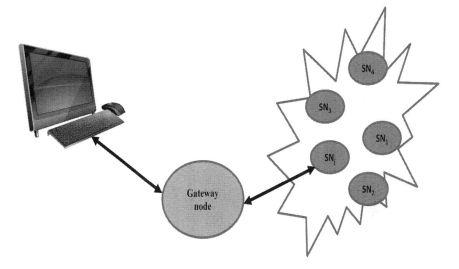

Fig. 1 Wireless sensor network architecture

Table 1 Notation

U_i	i-th User
SN_j	j-th sensor node
GWN	Gateway node
PW_i	Password of an entity U_i
ID_i	Identity of an entity U_i
SID_j	Identity of a sensor node SN_j
T_i	i-th timestamp
X_U	Secret value shared by GWN
X_S	Secret value shared by SN_j and GWN
$h()$	One-way hash function
$\|$	Concatenation operation
$\oplus$	XOR operation

each sensor node. Before the registration phase is performed for the first time, the gateway node *GWN* decides on the following system parameters a one-way hash function *h* and two cryptographic keys X_U and X_S. The key X_U is shared securely with the gateway node and X_S is shared securely with the gateway node and the sensor node. The notation in Table 1 is employed throughout this paper. A high level depiction of the scheme is given in Figs. 2, 3 where dashed lines indicate a secure channel, and a more detailed description follows:

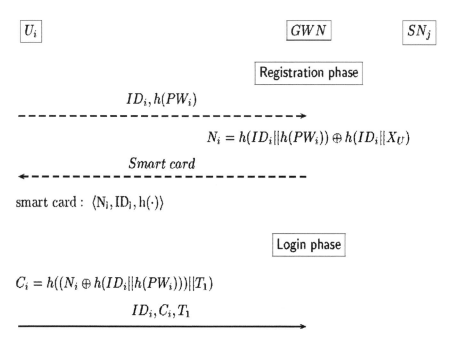

Fig. 2 Tan's registration and login phase

2.1 Registration Phase

This is the phase where a new registration of a user takes place. The registration proceeds as follows:

Step 1. A user U_i, who wants to register with the gateway node GWN, chooses its password PW_i at will. Then U_i computes $h(PW_i)$ and submits a registration request, consisting of its identity ID_i and $h(PW_i)$, to the gateway node GWN via a secure channel.

Step 2. Upon receiving the request $<ID_i, h(PWi)>$, GWN computes $N_i = h(ID_i||h(PW_i)) \oplus h(ID_i|||X_U))$ and issues a smart card containing $<N_i, ID_i, h()>$ to U_i.

2.2 Login Phase

Step 1. When U_i wants to log into the system, he inserts his smart card into a card reader and enters his identity ID_i and password PW_i.

Step 2. Given ID_i and PW_i, the smart card generates the timestamp T_1 and computes $C_i = h((N_i \oplus h(ID_i||h(PW_i)))||T_1)$.

The smart card then sends the login request message $<ID_i, C_i, T_1>$ to the gateway node GWN.

U_i $\boxed{GWN}$ $\boxed{SN_j}$

$$\boxed{\text{Verification phase}}$$

$$T_2 - T_1 \overset{?}{\leq} \Delta T$$

$$D_i = h(h(ID_i\|X_U)\|T_1)$$

$$D_i \overset{?}{=} C_i$$

$$A_i =$$

$$h(h(ID_i\|C_i\|h(h(SID_j\|X_S)\|T_3))$$

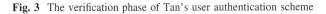

$$ID_i, SID_j, C_i, A_i, T_3$$

$$T_3 - T_4 \overset{?}{\leq} \Delta T$$

$$B_i =$$

$$h(h(ID_i\|C_i\|h(h(SID_j\|X_S)\|T_3))$$

$$B_i \overset{?}{=} A_i$$

$$E_j = h(h(SID_j\|X_S)\|T_5)$$

$$SID_j, E_j, T_5$$

$$T_6 - T_5 \overset{?}{\leq} \Delta T$$

$$F_j = h(h(SID_j\|X_S)\|T_5)$$

$$F_j \overset{?}{=} E_j$$

Fig. 3 The verification phase of Tan's user authentication scheme

2.3 Verification Phase

With the login request message $<ID_i,\ C_i,\ T_1>$, the scheme enters the verification phase during which GWN and SN_j perform the following steps:

Step 1. When the login request arrives $<ID_i,\ C_i,\ T_1>$, the gateway node GWN first chooses the current timestamp T_2 and T_3 and computes $D_i = h(h(ID_i\|X_U)\|T_1)$ and $A_i = h(h(ID_i\|C_i\|h(h(SIDj\|X_S)\|T_3))$ verifies that: (1) ID_i is valid

(2) $T_2 - T_1 \leq \Delta T$, ΔT is the maximum allowed time difference between T_1 and T_2, and (3) D_i equals C_i. If any of these is untrue, *GWN* rejects the login request and aborts the scheme. Otherwise, *GWN* accepts the login request and sends the message $<ID_i, SID_j, C_i, A_i, T_1>^1$ to the sensor node SN_j.

Step 2. After receiving $<ID_i, SID_j, C_i, A_i, T_3>$ from S_j, the sensor node SN_j obtains the new timestamp T_4 and T_5 and computes

$$B_i = h(h(ID_i||C_i||h(h(SID_j||X_S)||T_3)),$$
$$E_j = h(h(h(SID_j||X_S)||T_5)).$$

The sensor node SN_j verifies that: (1) $T_4 - T_3 \leq \Delta T$ where $\leq \Delta T$ is the maximum allowed time difference between T_3 and T_4 and (2) B_i equals A_i. If both of these conditions are not satisfied, SN_j aborts the scheme. Otherwise, SN_j believes that the responding party is the genuine user and *GWN*. Then *SNj* sends $<SID_j, E_j, T_5>$ to the gateway node *GWN*.

Step 3. After that, *GWN* obtains the current timestamp T_6 and computes $F_j = h(h(SID_j||X_S)||T_5)$. *GWN* verifies that (1) $T_6 - T_5 \leq \Delta T$ and (2) F_j equals E_j. If both of these conditions are hold, *GWN* accepts as authentic the sensor node. Otherwise, *GWN* stop the following procedures.

3 Cryptanalysis of Tan's User Authentication Scheme

Here we point out a security problem is that Tan's two-factor user authentication scheme has not password security. We interpret this problem as the vulnerability of the scheme to off-line password guessing attack. A few years ago, we demonstrate the password security of remote user authentication schemes using smart cards [17]. The security of remote user authentication schemes using smart cards need to consider the two-factor security [18]. To guarantee the security of the scheme when either the user's smart card or its password is stolen is the best we can do. In this section we point out that Tan's user authentication scheme suffers from an off-line password guessing attack.

3.1 Off-line Password Guessing Attack

Tan claims that their authentication scheme prevents an attacker from learning some registered user's password via an off-line password guessing attack. But, unlike the claim, Tan's scheme is vulnerable to an off-line password guessing

---

[1] $<ID_i, C_i, A_i, T_1>$ was incorrectly stated as $< ID_i, SID_j, C_i, A_i, T_1 >$ in the seventeenth-to-last line of Sect. 3 of [16].

attack mounted by extracting the secret information from a smart card. Assume that the attacker, who wants to find out the password of the user U_i, has stolen the U_i's smart card or gained access to it and extracted the secret values stored in it by monitoring its power consumption [19, 20]. Now the attacker U_a has obtained the value N_i stored in the U_i's smart card. Then the following description represents our off-line password guessing attack mounted by U_a against U_i's password.

In Tan's user authentication scheme, assume that an attacker has stolen the U_i's smart card or gained access to it and extracted the secret values stored in it by monitoring its power consumption. Now the attacker U_a, has obtained the value N_i, stored in the U_i's smart card. Then the following description represents our off-line guessing attack mounted by the attacker U_a, against U_i's password: The attacker U_a, who wants to find out PW_i, now guesses possible passwords and checks them for correctness.

1. As preliminary step, the attacker U_a, who has obtained N_i stored in U_i's smart card.
2. As the login phase proceeds, the attacker U_a eaves drops on the login request message $<ID_i, C_i, T_i>$ sent from U_i to GWN.
3. Attacker U_a makes a guess PW_i' for the password PW_i and computes $C_i' = h\big(\big(N_{ii} \oplus h\big(ID_i \| h\big(PW_i'\big)\big)\big)\|T_1\big)$.
4. U_a then verifies the correctness of PW_i by checking that C_i' is equal to C_i. If PW_i' and PW_i are equal, the equality C_i' ought to be satisfied.
5. U_a repeats 3 and 4 using another guessed password until a correct password is found.

Acknowledgments This work war supported by Howon University in 2012. The first author is Youngsook Lee (ysooklee@howon.ac.kr).

References

1. Rathod, V., Mehta, M.: Security in wireless sensor network: a survey. GANPAT Univ. J. Eng. Technol. **1**(1), 35–44 (2011)
2. Akyildiz, I.F., Su, W., Sankarasubramaniam, Y., Cayirci, E.: A survey on sensor networks. IEEE Commun. Mag. **40**(8), 102–114
3. Chang, C., Kuo, J.Y.: An efficient multi-server password authenticated keys agreement scheme using smart cards with access control. In: IEEE Proceeding of the 19th International Conference on Advanced Information Networking and Applications **2**, 257–260 (2005)
4. Ku, W.-C., Chang, S.-T., Chiang, M.-H.: Weaknesses of a remote user authentication scheme using smart cards for multi-server architecture. IEICE Trans. Commun. **E88-B**(8), 3451–3454 (2005)
5. Li, L.-H., Lin, I.-C., Hwang, M.-S.: A remote password authentication scheme for multi-server architecture using neural networks. IEEE Trans. Neural Netw. **12**(6), 1498–1504 (2001)

6. Lin, I.-C., Hwang, M.-S., Li, L.-H.: A new remote user authentication scheme for multi-server internet environments. Futur. Gener. Comput. Syst. **19**, 13–22 (2003)
7. Sun, H.-M.: An efficient remote user authentication scheme using smart cards. IEEE Trans. Consumer Electron. **46**(4), 958–961 (2000)
8. Tsai, J.-L.: Efficient multi-server authentication scheme based on one-way hash function without verification table. Comput. Secur. **27**, 115–121 (2008)
9. Tsuar, W.-J.: An enhanced user authentication scheme for multi-server internet services. Appl. Math. Comput. **170**, 258–266 (2005)
10. Tsuar, W.-J., Wu, C–.C., Lee, W.-B.: A flexible user authentication for multi-server internet services, Networking-JCN. LNCS **2093**(2001), 174–183 (2001)
11. Tsuar, W.-J., Wu, C–.C., Lee, W.-B.: A smart card-based remote scheme for password authentication in multi-server Internet services. Comput. Stand. Interfaces **27**, 39–51 (2004)
12. Das, M.L.: Two-factor user authentication in wireless sensor networks. IEEE Trans Wirel. Comm. **8**, 1086–1090 (2009)
13. Khan, M.K., Alghathbar, K.: Cryptanalysis and security improvements of two-factor user authentication in wireless sensor networks. Sensors **10**, 2450–2459 (2010)
14. Park, N., Kwak, J., Kim, S., Won, D., Kim, H.: WIPI mobile platform with secure service for mobile RFID network environment. In: Shen, H.T., Li, J., Li, M., Ni, J., Wang, W. (eds.) APWeb Workshops 2006. LNCS, vol. 3842, pp. 741–748. Springer, Heidelberg (2006)
15. Park, N.: Implementation of terminal middleware platform for mobile RFID computing. Int. J. Ad Hoc Ubiquitous Comput. **8**(4), 205–219, Inderscience Publishers (2011)
16. Tan, Z.: Cryptanalyses of a two-factor user authentication scheme in wireless sensor networks. Adv. Inf. Sci. Serv. Sci. **6**(4), 117–128 (2011)
17. Lee, Y., Won, D.: Security vulnerabilities of a remote user authentication scheme using smart cards suited for a multi-server environment. LNCS, **5593**, 164–172 (2009)
18. Tian, X., Zhu, R.W., Wong, D.S.: Improved efficient remote user authentication schemes. Int. J. Netw. Secur. **4**(2), 149–154 (2007)
19. Kocher, P., Jaffe, J., Jun, B.: Differential power analysis. In: Wiener, M. (ed.) Advances in Cryptology-Crypto'99. Springer, Berlin (1999) pp. 388–397
20. Messergers, T.S., Dabbish, E.A., Sloan, R.H.: Examining smart card security under the threat of power analysis attacks. IEEE Trans. Comput. **51**(5), 541–552 (2002)

Design of Mobile NFC Extension Protocol for Various Ubiquitous Sensor Network Environments

Jun Wook Lee, Hyochan Bang and Namje Park

Abstract The mobile RFID related technologies that offer similar services as NFC were also analyzed to understand the potential linkage to NFC and the requirements for the linkage. To apply the linking method of two technologies suggested in this paper, the dual tag development with marketability and NDEF decoder development as well as additional technology development, standardization, and patents of related technologies are required. Based on these efforts, if the mobile RFID and NFC technologies are linked, the companies can use the existing infrastructures to reduce the initial technological costs.

Keywords Mobile RFID · NFC · UHF · Dual band · U-sensor network

1 Introduction

In the international competition where various advanced technologies are rapidly developing, an in-depth analysis and prediction on the international standardization and domestic application of advanced technologies and international standards

J. W. Lee · H. Bang
Electronics and Telecommunications Research Institute (ETRI), 218 Gajeong-ro,
Yuseong-gu, Daejeon 305-700, Korea
e-mail: junux@etri.re.kr

H. Bang
e-mail: bangs@etri.re.kr

N. Park (✉)
Department of Computer Education, Teachers College, Jeju National University,
61 Iljudong-ro, Jeju-si, Jeju Special Self-Governing Province 690-781, Korea
e-mail: namjepark@jejunu.ac.kr

S.-S. Yeo et al. (eds.), *Computer Science and its Applications*,
Lecture Notes in Electrical Engineering 203, DOI: 10.1007/978-94-007-5699-1_73,
© Springer Science+Business Media Dordrecht 2012

as well as leading the international standardizations for Korea to acquire an international competitiveness. The recently emerged 13.56 MHz substitution no-touch close range wireless communication technology, NFC will be widely applied in practical life, such as mobile transportation card or credit card transactions with the integration with Smartphones in the future [1–11]. Therefore, the nations and companies are attempting to gain the leading power in this technology, especially focusing on the standardizations. When the NFC technology is utilized in the future, it will be connected to various technology standards, such as mobile RFID and it is important to understand the most pressing areas for standardization. Next, the developed standardizations should be understood around the world to find the additional targets and promote the standardizations.

ISO/IEC JTC 1, an NFC international standardization institute has started an international standardization on NFC communication scale in 2003 and has been carrying out standardization tasks to acquire the interoperability with existing no-touch smartcard technology and RFID with much an accomplishment. In addition, NFC forum was established in 2004 to start the attempts to use it commercially in many ways but in 2011, the connection with the Smartphones allowed many service models to emerge to bring another spotlight [12–20].

Korea also needs to reflect such international standardization trends of NFC standardization and also create an appropriate connection platform for Korea through a comparative analysis with the domestic standardization of mobile RFID technology. In addition, the new standardizations need to be reflected internationally in an active manner. For these reasons, this report provides the foundation for domestic distribution of related technologies and standardizations by connecting the two technologies through the analysis of NFC system international standardization activities and trends with mobile RFID technology. This paper analyzed the standard statuses of NFC that is promoted internationally. Also, the mobile RFID related technologies that offer similar services as NFC were also analyzed to understand the potential linkage to NFC and the requirements for the linkage.

2 Necessity of Mobile RFID/NFC Linkage

The mobile RFID and NFC have a common aspect of using the short range wireless communication technology, RFID, but the used frequency range, standardized institution and scope, and the service areas still differ. In detail, the 900 MHz UHF substitution tag and reader-based mobile RFID technology has been standardized by the mobile RFID forum and the Korean Telecommunications Technology Association (TTA) and is appropriate for the distribute services because of its characteristic to recognize over long distances. On the other hand, the NFC technology is based on 13.56 MHz HF substitution and is being standardized internationally with the NFC forum but is only usable in short distances

Table 1 Comparison of mobile RFID and NFC technologies

	Nokia's mobile RFID	KDDI's mobile RFID (Passive)	KDDI's mobile RFID (Active)	NFC (Near field communication)	Korea's mobile RFID
Radio Frequency	13.56 MHz	2.45 GHz	315 MHz	13.56 MHz	860–960 MHz
Reading Range	2–3 cm	∼5 cm	∼10 cm		
Compliant Standards	ISO/IEC 14443 A		ISO/IEC 18092	ISO/IEC 18000-6 B/C	
Feature	HF RF Reader	RF reader	Active RFID Reader	Tag and reader	UHF RF Reader

to be used in transactions and P2P services. The table below shows the characteristics and differences of mobile RFID technology and NFC technology (Table 1).

As shown above, the mobile RFID and NFC technologies hold their appropriate service areas based on their characteristics but the common characteristic of linking to the communication net on a portable communication device allows them to be used for the same purpose. For example, the tag that holds the information on a specific item can be attached to use the mobile RFID and NFC technologies to actively receive the information about the item.

NFC provides 3 types of services. The first type of service is the card emulation mode where the built-in NFC chip acts as the tag within the device. This mode can provide no-touch transaction services, such as the transportation card. The second service type is reading/writing mode. In this mode, the NFC chipset acts as the reader to read the information from the tags attached to items. The third service type is P2P mode to directly exchange the data between the devices. The mobile RFID technology can only be used as the reader using the chipset within the device, unlike NFC. Therefore, it does not support as many service types as NFC [21].

As shown in the figure above, the Read/Write mode where the device acts as a reader among the NFC technology service type can cover the mobile RFID technology service type. However, the two technologies have appropriate use environment based on their physical characteristics and neither technologies alone can provide a wide range of services. In addition, the mobile RFID has already completed the standardization on the information delivery system, such as code system, ODS, or OIS and the infrastructure establishment and commercialization have occurred, which need to be maximally utilized. Therefore, the interoperable linking between mobile RFID and NFC technologies should be studied to increase the compatibility in the code system and information delivery system as well as complementing the related standards to activate the domestic RFID industry to acquire the competitiveness and initiatives in the world (Fig. 1).

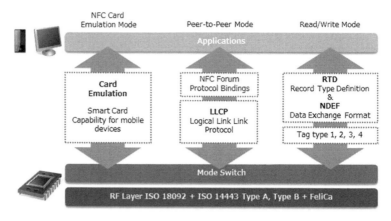

Fig. 1 NFC service's 3 types

3 Suggested Mobile RFID/NFC Linkage Method

3.1 Mobile RFID/NFC Linkage Environment and Service Type Selection

A specific setting for the service type and linking environment needs to be established to link the mobile RFID and NFC technologies. This paper deduces the linking methods based on the item information provision service that supports both mobile RFID and NFC technologies. In the case of linking environment, the device with mobile RFID and NFC chipsets is assumed to be distributed and the mobile RFID information delivery system is also assumed to be established. In the case of NFC, the NFC tag holds the URI information and the additional information delivery system is not assumed to be used and the tag attached on each item is 900 MHz mobile RFID interoperable tag and 13.56 MHz NFC interoperable tag. Lastly, the dual tag that support both frequencies based on necessity as well as the devices with dual band antennae are also assumed to be distributed. Such linking environment can be shown in the Fig. 2.

As shown in the Fig. 2, the mobile RFID and NFC environments that provide the item information provision services can be divided into tag, device, and information delivery system and depending on the information delivery route, various combinations are possible. The second clause examines the various scenarios to provide practical services in the actual operation environment among the information delivery route combination of the components. In addition, the linkage environment where the mobile RFID and NFC technologies are appropriately connected to create a synergy among the scenarios will be organized. Lastly, the third clause will deduce the requirements for the linkage based on the linking environment from second clause to satisfy the liking requirements.

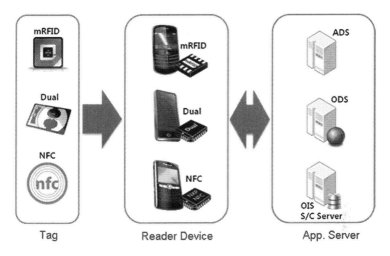

Fig. 2 Linking environment components of mobile RFID and NFC

3.2 Mobile RFID/NFC Linkage Service Methods

The linkage of the mobile RFID and NFC technologies occurs through the linking of the NFC type 3 tag memory structure of dual tag to the code system of mobile RFID. In detail, the mCode system, a mobile RFID code system, is encoded with a standard ISO/IEC 15961, ISO/IEC 15962, ISO/IEC 18000-6C and transitioning the results into the NDEF type message that can be stored in the NFC type 3 tag memory. The linking procedures using the dual tag for mobile RFID/NDC are as follows:

- Code system encoding: Follow the encoding method of mbile RFID standard to encode the data with mCode code system.
- NDEF encoding: re-encode into NDEF type so that the NFC device can read the mRFID encoding values.
- Storage and Linking: NDEF type data can be stored in the dual tag with tag memory structure in the type 3 shape from NFC forum to link.

(1) Mobile RFID code system encoding step
    The mCode among the mobile RFID code systems is used to find the linking method. Among the mCode, if the D class code is used, then the total length of the code system is 96 bt and 16 hexadecimal with TLC of E12, Class of 4, CC of 1234 1234 and SC of 5678 9012 are assumed.
(2) NDEF encoding step
    In the second step, the encoding data calculated from the first step are changed to NDEF type for the NFC device to recognize the data. When the data are changed, the standard record type from NFC forum of "T (Text Record Type)" is applied to allow the production of data in UTF-8 type strings and the completed data can be stored in the NFC forum type 3 tag memory structure.

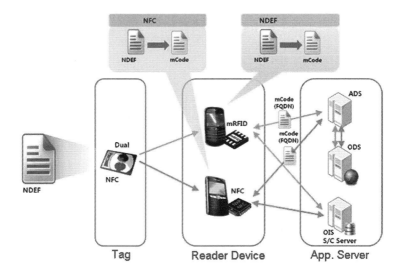

Fig. 3 Application of linking technology in associated environment using dual tag

(3) Data storage and linking step

The NDEF type record from step 2 is saved to dual tag comprised of NFC forum type 3 tag to complete the preparation for linking. The dual tag from this process stores the information on mobile RFID code system in the NDEF type. Therefore, both mobile RFID devices and NFC devices can approach these data. If NFC device received the NDEF encoding data, then the NFC chipset can immediately decode the NDEF data. On the other hand, if the mobile RFID device received the data, then the NDEF decoder that can decode the NDEF data should be used additionally to change the data to mCode type, which will subsequently be changed to FQDN shape according to the mobile RFID code management realization guideline to be sent to ADS or ODS. The figure below simplifies such a process.

3.3 Effect of Linking and Linking Requirement Satisfaction

As shown in the Fig. 3, the linking using the dual tag can be used to receive the services regardless of the type of the devices and the existing standards can be applied without corrections many times. In addition, the government and the companies that need to establish the information delivery system infrastructure can benefit economically because the service can be provided without additional information delivery system or expansion of the server.

4 Conclusion

Recently the mobile market is rapidly advancing based on the introduction of the various mobile devices and next-generation portable communication network with the Smartphones. Therefore, the mobile internet era has fully emerged to allow the users to connect to the internet anytime and anywhere and many services to make the life more convenient are emerging. Mobile RFID and NFC technologies are also working hard to standardize and widespread their uses in this era. Especially, the NFC technology with international standardization is rapidly expanding its market with the policies of mobile communication companies and device producers who are actively promoting its introduction.

This paper analyzed the standard statuses of NFC that is promoted internationally. Also, the mobile RFID related technologies that offer similar services as NFC were also analyzed to understand the potential linkage to NFC and the requirements for the linkage. In addition, the introduction of dual tag and the code system linkage suggested a linking method to maximally use the existing infrastructure and the requirement satisfaction was analyzed. Lastly, based on the analysis, the future direction for the new standard design was suggested.

To apply the linking method of two technologies suggested in this paper, the dual tag development with marketability and NDEF decoder development as well as additional technology development, standardization, and patents of related technologies are required. Based on these efforts, if the mobile RFID and NFC technologies are linked, the companies can use the existing infrastructures to reduce the initial technological costs. In addition, the services are provided regardless of the device types, which allow the consumers to use the services more actively. Such changes will be a catalyst to elicit a positive cycle for the industry and will aid the RFID technology development. In addition, it will increase the utilization of mobile RFID technology standardized in Korea and is expected to create a new high value market as well as initiating the international standards.

Acknowledgments This work was supported by the Industrial Strategic Technology Development Program funded by the Ministry of Knowledge Economy(MKE, Korea). (10035262, Development of Horticultural Crops' Quality Enhancement System with USN Technology). This paper is extended from a conference paper presented at The 9th FTRA International Conference on Secure and Trust Computing, data management, and Applications, Korea. The author is deeply grateful to the anonymous reviewers for their valuable suggestions and comments on the first version of this paper. The Corresponding author is Namje Park (namjepark@jejunu.ac.kr).

References

1. Choi, W., et al.: An RFID tag using a planar inverted-f antenna capable of being stuck to metallic objects. ETRI J. **28**(2), 216–218 (2006)
2. Needham, R.M., Schroeder, M.D.: Authentication revisited. Oper. Syst. Rev. **21**(1), 7 (1987)

3. Park, N., Kwak, J., Kim, S., Won, D., Kim, H.: WIPI mobile platform with secure service for mobile RFID network environment. In: Shen, H.T., Li, J., Li, M., Ni, J., Wang, W. (eds.) APWeb Workshops 2006. LNCS, vol. 3842, pp. 741–748. Springer, Heidelberg (2006)

4. Park, N.: Security scheme for managing a large quantity of individual information in RFID environment. In: Zhu, R., Zhang, Y., Liu, B., Liu, C. (eds.) ICICA 2010. CCIS, vol. 106, pp. 72–79. Springer, Heidelberg (2010)

5. Park, N.: Secure UHF/HF dual-band RFID: strategic framework approaches and application solutions. In: ICCCI 2011. LNCS, Springer, Heidelberg (2011)

6. Park, N.: Implementation of terminal middleware platform for mobile RFID computing. Int. J. Ad Hoc Ubiquitous Comput. **8**(4), 205–219 (2011)

7. Park, N., Kim, Y.: Harmful adult multimedia contents filtering method in mobile RFID service environment. In: Pan, J.-S., Chen, S.-M., Nguyen, N.T. (eds.) ICCCI 2010. LNCS(LNAI), vol. 6422, pp. 193–202. Springer, Heidelberg (2010)

8. Park, N., Song, Y.: AONT encryption based application data management in mobile RFID environment. In: Pan, J.-S., Chen, S.-M., Nguyen, N.T. (eds.) ICCCI 2010. LNCS(LNAI), vol. 6422, pp. 142–152. Springer, Heidelberg (2010)

9. Park, N.: Customized healthcare infrastructure using privacy weight level based on smart device. In: Communications in Computer and Information Science, vol. 206, pp. 467–474. Springer, Heidelberg (2011)

10. Park, N.: Secure data access control scheme using type-based re-encryption in cloud environment. In: Studies in Computational Intelligence, vol. 381, pp. 319–327. Springer, Heidelberg (2011)

11. Park, N., Song, Y.: Secure RFID Application Data Management Using All-Or-Nothing Transform Encryption. In: Pandurangan, G., Anil Kumar, V.S., Ming, G., Liu, Y., Li, Y. (eds.) WASA 2010. LNCS, vol. 6221, pp. 245–252. Springer, Heidelberg (2010)

12. Park, N.: The implementation of open embedded S/W platform for secure mobile RFID reader. J. Korea Inf. Commun. Soc. **35**(5), 785–793 (2010)

13. Park, N., Song, Y., Won, D., Kim, H.: Multilateral approaches to the mobile RFID security problem using web service. In: Zhang, Y., Yu, G., Bertino, E., Xu, G. (eds.) APWeb 2008. LNCS, vol. 4976, pp. 331–341. Springer, Heidelberg (2008)

14. Park, N., Kim, H., Kim, S., Won, D.: Open location-based service using secure middleware infrastructure in web services. In: Gervasi, O., Gavrilova, M.L., Kumar, V., Laganá, A., Lee, H.P., Mun, Y., Taniar, D., Tan, C.J.K. (eds.) ICCSA 2005. LNCS, vol. 3481, pp. 1146–1155. Springer, Heidelberg (2005)

15. Park, N., Kim, S., Won, D.: Privacy preserving enhanced service mechanism in mobile RFID network. In: ASC, Advances in Soft Computing, vol. 43, pp. 151–156. Springer, Heidelberg (2007)

16. Park, N., Kim, S., Won, D., Kim, H.: Security analysis and implementation leveraging globally networked mobile RFIDs. In: PWC 2006. LNCS, vol. 4217, pp. 494–505. Springer, Heidelberg (2006)

17. Kim, Y.: Harmful-word dictionary DB based text classification for improving performances of precesion and recall. Sungkyunkwan University, Ph.D. Thesis (2009)

18. Kim, Y, Park, N., Hong, D.: Enterprise data loss prevention system having a function of coping with civil suits. In: Studies in Computational Intelligence, vol. 365, pp. 201–208. Springer, Heidelberg (2011)

19. Kim, Y., Park, N., Won, D.: Privacy-enhanced adult certification method for multimedia contents on mobile RFID environments. In: Proceedings of IEEE International Symposium on Consumer Electronics, pp. 1–4. IEEE, Los Alamitos (2007)

20. Kim, Y., Park, N., Hong, D., Won, D.: Adult certification system on mobile RFID service environments. J. Korea Contents Assoc. **9**(1), 131–138 (2009)

21. MRF Forum: Application data format for mobile RFID services. MRFS-3-02 (2005)

Experimentation and Validation of Web Application's Vulnerability Using Security Testing Method

Taeseung Lee, Giyoun Won, Seongje Cho, Namje Park
and Dongho Won

Abstract The paper proposes a security testing technique to detect known vulnerabilities of web applications using both static and dynamic analysis. We also present a process to improve the security of web applications by mitigating many of the vulnerabilities revealed in the testing phase, and address a new method for detecting unknown vulnerabilities by applying dynamic black-box testing based on a fuzzing technique. The fuzzing technique includes a structured fuzzing strategy that considers the input data format as well as misuse case generation to enhance the detection rate compared to general fuzzing techniques. To verify the proposed approaches, we conducted an experiment using an open source web application (BugTrack) and web application server (JEUS 6). The experiment results show that our testing technique found 142 vulnerabilities of which we were able to remove or mitigate 138 by employing the principles of secure coding. These

T. Lee · D. Won
College of Information and Communication Engineering, Sungkyunkwan University,
300 Cheoncheon-dong, Jangan-gu, Suwon-si, Gyeonggi-do 440-746, Korea
e-mail: tslee@kisa.or.kr

D. Won
e-mail: dhwon@security.re.kr

G. Won · S. Cho
Department of Computer Science & Engineering, Dankook University, Suji-gu, Yongin-si,
Gyeonggi-do, Korea
e-mail: kgyoun4@gmail.com

S. Cho
e-mail: sjcho@dankook.ac.kr

N. Park (✉)
Department of Computer Education, Teachers College, Jeju National University,
61 Iljudong-ro, Jeju-si, Jeju Special Self-Governing Province 690-781, Korea
e-mail: namjepark@jejunu.ac.kr

S.-S. Yeo et al. (eds.), *Computer Science and its Applications*,
Lecture Notes in Electrical Engineering 203, DOI: 10.1007/978-94-007-5699-1_74,
© Springer Science+Business Media Dordrecht 2012

results imply that our proposed approaches are effective at detecting and mitigating vulnerabilities of web applications.

Keywords Web application · Security testing · Vulnerability · Security

1 Introduction

As the Internet is now widely used around the globe, web applications have become commonplace and users can engage in a variety of web-based activities simply by having a web browser on their PC or mobile device. E-mail, e-commerce, online auctions, wikis, social networks and blogs are all easily accessible and their use is only increasing.

As the number of web applications increase so do the number of vulnerabilities such as SQL injection, Cross Site Script (XSS), cracked authentication and session management intrusion [1–9]. According to the IBM X-Force 2010 Trend and Risk Report, 49 % of all vulnerabilities disclosed in 2010 were web application vulnerabilities. If vulnerabilities not disclosed for the sake of confidentiality are counted, the number of web application vulnerabilities is expected to far exceed the number of other types of security vulnerabilities [10]. As web application vulnerabilities are increasing continuously, the number of attacks abusing them and damages resulting thereof are also increasing. For example, in 2010 daily web-based attacks nearly doubled (up by 93 %) as compared to 2009 [11].

Most security incidents take place as the vulnerabilities inherent in the system or software are attacked. What is noteworthy here is that 90 % of software-related security incidents are said to take place on account of the vulnerabilities generated in the design or coding stage of software development [12]. Because software vulnerabilities are the sources of security incidents or attacks, it is necessary to detect them as early as possible according to the Software Security Testing and Secure Coding Guideline, and prevent security incidents by mitigating or removing the detected vulnerabilities.

Software security testing analyzes the security of applications from the viewpoint of attackers and is not really concerned with the functionality of the software. Such testing consists of static testing and dynamic testing [9, 13–16]. The advantage of code-based static testing is its ability to efficiently analyze software in its entirety, but this testing often has a high false detection ratio, and can hardly be applied to commercial software whose source code is not given. Execution-based dynamic testing can be applied to commercial software and has a low false detection ratio, but analysis time is long and code coverage is limited [17].

This paper proposes a vulnerability detection and mitigation technique to increase the security of web applications. The vulnerability detection technique used in this paper is a software security test that combines both static and dynamic analysis from the viewpoint of the attacker, not the developer; and the vulnerability mitigation technique used here is the secure coding method. To this end, automated static analysis tools and dynamic analysis tools are applied to the web application to detect known vulnerabilities [18–20]. Then a "fuzzing" technique for detecting new vulnerabilities is proposed. "Fuzzing" is a technique that generates/mutates input data either randomly or structurally and injects it into the application then monitors the results of application execution to detect vulnerabilities [21]. This paper proposes an abuse case generation and testing strategy for efficient fuzzing as well. Third, to mitigate or remove detected web application vulnerabilities, a process for applying the secure coding technique is shown.

To verify the proposed method, an experiment was conducted on web applications and a commercial Web Application Server (WAS) that are distributed as open sources. The experiment detected 139 known and 3 unknown vulnerabilities, and removed 95 % of detected vulnerabilities (135 vulnerabilities), thereby proving that the method proposed in this paper is effective at detecting and mitigating vulnerabilities.

2 Related Studies

2.1 Vulnerability Detection Methods

Methods for detecting vulnerabilities can be largely divided into two. The first is static analysis that makes a judgment based on source codes without executing the target program. The second is dynamic analysis that executes the program while subjecting it to several inputs for the purpose of detecting any vulnerability [17]. Each method can be subdivided into a black-box test and a white-box test. A black-box test does not look inside the program, i.e. source codes, while the source codes are included in a white-box test.

Static analysis detects vulnerabilities by analyzing given inspection rules, language conventions, inconsistencies between codes, flow control and program values. This method includes Code Auditing, Type Checking, Model Checking and Formal Verification, and as it can analyze the entire program, it offers high coverage. Also, this method helps discover problems likely to occur due to negligence when writing the program [22, 23]. However, static analysis has a high false detection ratio and is difficult to apply if source codes are not provided.

Dynamic analysis executes the program while substituting input values, and detects errors or defects. Software fault injection is an example of dynamic analysis, and can be used to discover new vulnerabilities. Dynamic analysis methods for web applications include scanning technology that automatically inspects for known security vulnerabilities, and the "fuzzing" technique used for discovering new vulnerabilities [21].

Vulnerability scanning uses an automated tool called a vulnerability scanner. There are two types, the 'pattern search scanner' and the 'structure and parameter analysis scanner.' The pattern search scanner substitutes vulnerability patterns for each server version one by one based on known vulnerability patterns and uses the responses to list vulnerabilities. The structure and parameter analysis scanner searches for vulnerabilities of applications based on not only pattern search, but also structure and parameter analysis of applications.

The fuzzing technique injects abnormal errors in the program so as to discover unexpected defects and vulnerabilities [21]. As a large number of test cases are used as input values, automation is essential. As the coverage and vulnerability detection ratio are low, it must be possible to generate test cases highly likely to inflict defects. Fuzzing can be classified into "dumb" fuzzing and "smart" fuzzing depending on whether the format structure is understood or not, and also into "generation" fuzzing and "mutation" depending on data processing methods.

2.2 Vulnerability Mitigation

As vulnerabilities of web applications can occur at any stage of the Software Development Life Cycle (SDLC), security at each stage is important. It is certainly better to consider errors or security issues at the early stages of development. According to a National Institute of Standards and Technology (USA) report, the cost of correcting errors in the distribution stage may be up to 30 times more than that of correcting errors in the design stage.

Accordingly, measures to consider mitigating vulnerabilities in the early stages of software development, e.g. 'secure coding' that considers security during the development stage, or at each stage of development, are becoming pertinent issues. Secure coding refers to a series of processes for developing safe software capable of responding to security threats such as hacking while minimizing security vulnerabilities inherent in the software due to developer mistakes and logic errors in the software development process. In a broad sense, secure coding includes all security activities required in each stage of the SDLC, and in a narrow sense, it means the activities for mitigating vulnerabilities in the implementation stage for writing codes during software development.

Secure SDLC is a methodology and guideline for development and operation of safe software, and provides a vulnerability mitigation technique for each stage from the start of software writing to the operating stage by considering security issues from the earliest stage. Conventional safe SDLC includes MS SDL, Cigital's Touchpoints and OWASP CLASP. MS SDL, developed by Microsoft, is a methodology for minimizing the number of security vulnerabilities due to implementation and documentation, and detecting and removing these vulnerabilities as early in the development lifecycle as possible. Cigital proposes Seven Touchpoints, which consists of seven security reinforcement activities, directly and indirectly related to security functions among the activities of each stage of the SDLC. OWASP CLASP, developed by Secure Software, provides a guideline for applying security activities to existing application development processes.

3 Experimentation and Validation

3.1 Experimental Environment and Tools

To prove the proposed method, an experiment was conducted for a Bug Tracking System, which is an open source. The experiment was conducted in Microsoft Windows 7, and the open source of the Bug Tracking System was downloaded from Gotocode (a site providing open sources of web applications), and JEUS 6, and implemented in a web application server. Vulnerability detection was conducted through Sparrow and Acunetix. Sparrow is a program that receives the source files of programs as input, and uses static analysis to automatically correct errors in the source codes. Acunetix is a web vulnerability scanner that can automatically check the vulnerabilities of the SQL web.

3.2 Experimentation and Validation Scenario

The experiment can be validated through the following procedure:

(1) Conduct code review
    Use Sparrow to detect the vulnerabilities of the source codes of the Bug Tracking System.
(2) Conduct vulnerability scanning
    Implement the source codes of the Bug Tracking System in JEUS 6, the web application server, and use Acunetix to detect vulnerabilities.
(3) Conduct fuzzing
    Use the abuse cases identified through the fuzzing process to detect vulnerabilities.
(4) Mitigate vulnerability
    Search the list of vulnerabilities that can be mitigated that is provided by the secure coding for the vulnerabilities detected through tests (1), (2) and (3), and apply the secure coding technique for mitigating the vulnerabilities.
(5) Check mitigated vulnerabilities
    Conduct tests (1), (2) and (3) again for the Bug Tracking System with the detected vulnerabilities mitigated, and check if vulnerabilities are mitigated.

3.3 Results of the Experiment

(1) Vulnerability detection

- Code Review (Static Analysis)

Sparrow was used to analyze the source codes of the Bug Tracking System. Sparrow is installed as a plug-in of Eclipse, a programming integration development tool, and used to analyze codes. As a result of the analysis using Sparrow, a total of 157 vulnerabilities in 4 types were discovered. Although 157 vulnerabilities were discovered in total, many identical vulnerabilities occurred as a result of there being redundant code. In other words, as inherited codes are inspected when the codes of a single file are analyzed, it was revealed that frequently used codes were analyzed redundantly.

- Vulnerability Scanning (Dynamic Analysis)

A total of 98 vulnerabilities in 5 types were detected when the Bug Tracking System was executed, and the threat level determined using this Bug Tracking System was high. If the detection result is analyzed, the vulnerability risk of XSS and SQL injection was high, and the vulnerabilities of these two accounted for 87 % of all vulnerabilities.

- Fuzzing

When fuzzing was done for the Bug Tracking System using JBroFuzz (a web application fuzzing tool provided by OWASP) no exception occurred, and the error codes were 400 and 414. 4XX error codes are client errors, meaning that the server cannot understand the request of the client or refused to respond. The request of the client was handled as an error because fuzzing was done with random data without the basic input format and web applications could not perceive it.

(2) Mitigating vulnerabilities through secure coding

- Removing NULL_DEREFERENCE

NULL DEREFERENCE occurs when the assumption 'In general, the object cannot be NULL' is violated. If the attacker executes NULL DEREFERENCE intentionally, the resulting exception can be abused for subsequent attacks.

- Removing USING_DYNAMIC_CLASS_SOADING

If the class is loaded dynamically, the class is likely to be a malicious code; so the class should not be loaded dynamically.

- Responding to EMPTY_CATCH_BLOCK

If no measure is taken for the errors, the program will continue to be executed in that state; so failing to do so may lead to an unintended consequence by the developer. It is very important that exceptions or errors be handled properly.

- MISSING_DEFAULT_IN_SWITCH_BLOCK

All switch statements must include a default case. If a default case is included, errors due to fall-through can be prevented. "Fall-through" means that, if no break is used in the switch statement, the following case statement will be executed. If there is no default case, the program will continue to be executed even if all conditions of the case are not met.

- Cross Site Scripting, SQL Injection, and Application error message

The causes of these three vulnerabilities were analyzed. If a result page is generated by an untested external input value, the script or query statement will be executed by special characters. To prevent this, methods like replaceAll() can be used to prevent dangerous character strings.

- Password type input with autocomplete enabled

The autocomplete attribute is a sentence that indicates whether to autocomplete. If autocomplete is in the password text field, the implied value will be inferred and the password can be exposed. To prevent this problem, the autocomplete function is not used in the password text field.

(3) Checking and verifying mitigated vulnerabilities

After applying the Secure Coding technique to the detected vulnerabilities, code analysis, vulnerability scanning and fuzzing were then conducted. Sparrow was used to analyze the Bug Tracking System with vulnerabilities mitigated, and no vulnerability was detected. Also, Acunetix was used to conduct vulnerability scanning, and GHDB vulnerability was not mitigated, but there was no risk of vulnerability either.

Fuzzing was conducted again for the Bug Tracking System, and the vulnerabilities detected in Sect. 3.3 were detected again. The vulnerabilities detected through fuzzing were vulnerabilities of the JEUS 6 server itself, and vulnerabilities could not be mitigated in the source codes of the web applications. 138 (95 %) of 142 detected vulnerabilities were removed. In other words, all vulnerabilities discovered in the source codes could be defended.

4 Conclusion

In this paper, static testing and dynamic testing were applied to web applications to detect security vulnerabilities which were then removed. In other words, code review, vulnerability scanning and fuzzing were conducted to detect vulnerabilities, and a process of using secure coding to mitigate vulnerabilities was proposed. Also, this paper proposed a fuzzing technique for discovering new vulnerabilities as well as an abuse case generation technique, which is the key to efficient fuzzing.

To verify the effectiveness of the proposed process, an experiment was conducted on open source web applications using commercial web application servers. The result of the experiment showed that 95 % of the vulnerabilities detected by security testing were mitigated by secure coding. The remaining unmitigated 5 %

were vulnerabilities related to the web application server, which were detected using the fuzzing technique proposed by the authors. As they were discovered in commercial software, the authors could not mitigate them.

Acknowledgments This work (Grants No.C0033984) was supported by Business for Cooperative R&D between Industry, Academy, and Research Institute funded Korea Small and Medium Business Administration in 2012. The corresponding author is Namje Park (namjepark@ jejunu.ac.kr).

References

1. Park, N., Song, Y.: AONT encryption based application data management in mobile RFID environment. In: Pan, J.-S., Chen, S.-M., Nguyen, N.T. (eds.) ICCCI 2010. LNCS(LNAI), vol. 6422, pp. 142–152. Springer, Heidelberg (2010)
2. Park, N.: Customized healthcare infrastructure using privacy weight level based on smart device. In: Communications in Computer and Information Science, vol. 206, pp. 467–474. Springer, Heidelberg (2011)
3. Park, N.: Secure data access control scheme using type-based re-encryption in cloud environment. In: Studies in Computational Intelligence, vol. 381, pp. 319–327. Springer (2011)
4. Park, N., Song, Y.: Secure RFID application data management using all-or-nothing transform encryption. In: Pandurangan, G., Anil Kumar, V.S., Ming, G., Liu, Y., Li, Y. (eds.) WASA 2010. LNCS, vol. 6221, pp. 245–252. Springer, Heidelberg (2010)
5. Park, N.: The implementation of open embedded S/W platform for secure mobile RFID reader. J. Korea Inf. Commun. Soc. **35**(5), 785–793 (2010)
6. Park, N., Song, Y., Won, D., Kim, H.: Multilateral approaches to the mobile RFID security problem using web service. In: Zhang, Y., Yu, G., Bertino, E., Xu, G. (eds.) APWeb 2008. LNCS, vol. 4976, pp. 331–341. Springer, Heidelberg (2008)
7. Park, N., Kim, H., Kim, S., Won, D.: Open Location-based Service using Secure Middleware Infrastructure in Web Services. In: Gervasi, O., Gavrilova, M.L., Kumar, V., Laganá, A., Lee, H.P., Mun, Y., Taniar, D., Tan, C.J.K. (eds.) ICCSA 2005. LNCS, vol. 3481, pp. 1146–1155. Springer, Heidelberg (2005)
8. Park, N., Kim, S., Won, D.: Privacy preserving enhanced service mechanism in mobile RFID network. In: ASC, Advances in Soft Computing, vol. 43, pp. 151–156. Springer, Heidelberg (2007)
9. Park, N., Kim, S., Won, D., Kim, H.: Security analysis and implementation leveraging globally networked mobile RFIDs. In: PWC 2006. LNCS, vol. 4217, pp. 494–505. Springer, Heidelberg (2006)
10. Ernst, M.D.: Static and dynamic analysis: synergy and duality. In: Proceedings of WODA 2003 (ICSE Workshop on Dynamic Analysis). (2003)
11. Godefroid, P., Levin, M.Y., Molnar, D.: Automated whitebox fuzz testing. NDSS (2008)
12. Kim, D.J., Cho, S.J.: Fuzzing-based vulnerability analysis for multimedia players. J. KIISE: Comput. Pract. Lett. **17**(2) (2011)
13. Kim, Y.: Harmful-word Dictionary DB based text classification for improving performances of precesion and recall. Sungkyunkwan University, Ph.D. Thesis (2009)
14. Kim, Y, Park, N., Hong, D.: Enterprise data loss prevention system having a function of coping with civil suits. studies in computational intelligence, vol. 365, pp. 201–208. Springer, Heidelberg (2011)

15. Kim, Y., Park, N., Won, D.: Privacy-enhanced adult certification method for multimedia contents on mobile RFID environments. In: Proceedings of IEEE International Symposium on Consumer Electronics, pp. 1–4. IEEE, Los Alamitos (2007)
16. Kim, Y., Park, N., Hong, D., Won, D.: Adult certification system on mobile RFID service environments. J. Korea Contents Assoc. 9(1), 131–138 (2009)
17. Kim, G., Cho, S.: Fuzzing of web application server using known vulnerability information and its verification. In: Proceedings of the KIISE Korea Computer Congress 2011, 38(1-B), 181–184 (2011)
18. Park, N., Kwak, J., Kim, S., Won, D., Kim, H.: WIPI mobile platform with secure service for mobile RFID network environment. In: Shen, H.T., Li, J., Li, M., Ni, J., Wang, W. (eds.) APWeb Workshops 2006. LNCS, vol. 3842, pp. 741–748. Springer, Heidelberg (2006)
19. Park, N.: Security scheme for managing a large quantity of individual information in RFID environment. In: Zhu, R., Zhang, Y., Liu, B., Liu, C. (eds.) ICICA 2010. CCIS, vol. 106, pp. 72–79. Springer, Heidelberg (2010)
20. Park, N.: Secure UHF/HF dual-band RFID: strategic framework approaches and application solutions. In: ICCCI 2011. LNCS, Springer, Heidelberg (2011)
21. SecurityFocus Vulnerability Database: Vulnerability Summary for BID : 32804, SecurityFocus. (2008)
22. Park, N.: Implementation of terminal middleware platform for mobile RFID computing. Int. J. Ad Hoc Ubiquitous Comput. 8(4), 205–219 (2011)
23. Park, N., Kim, Y.: Harmful adult multimedia contents filtering method in mobile RFID service environment. In: Pan, J.-S., Chen, S.-M., Nguyen, N.T. (eds.) ICCCI 2010. LNCS(LNAI), vol. 6422, pp. 193–202. Springer, Heidelberg (2010)

Teaching–Learning Methodology of STS Based on Computer and CAI in Information Science Education

Juyeon Hong and Namje Park

Abstract Application of STS approach at school can have quite a few benefits. STS(Science Technology in Society) emphasizes interaction between science and technology, impact of scientific and technological advancement on a society, and vice versa. Social issues usually have interdisciplinary elements, and form a complex system. Further, STS can reveal real-life aspect of science, as it addresses scientific facts and technologies that have a direct bearing upon the society.

Keywords STS · Science Technology in Society · Computer education · Science

1 Introduction

The term STS (Science Technology in Society) was first coined to show a close relationship among the three areas, indicating that scientific activities are part of social phenomena and should be understood in context of the social environment [1–3]. Since 1970s, modern philosophy of science embraced this idea as a new approach to science that is founded upon relativism, rationalism and idealism. The ideas of STS show that a scientific method has a social nature and is a process of democratic negotiation, and imply potential and practicality of scientific

J. Hong · N. Park
Science Technology in Society Research Center (STSRC), Jeju National University,
61 Iljudong-ro, Jeju-si, Jeju Special Self-Governing Province 690-781, Korea
e-mail: hongjuyeon@jejunu.ac.kr

N. Park (✉)
Department of Computer Education, Teachers College, Jeju National University,
61 Iljudong-ro, Jeju-si, Jeju Special Self-Governing Province 690-781, Korea
e-mail: namjepark@jejunu.ac.kr

S.-S. Yeo et al. (eds.), *Computer Science and its Applications*,
Lecture Notes in Electrical Engineering 203, DOI: 10.1007/978-94-007-5699-1_75,
© Springer Science+Business Media Dordrecht 2012

knowledge. STS emphasizes interaction between science and technology, impact of scientific and technological advancement on a society, and vice versa [3–13].

STS has a theoretical background in constructivism, in particular, in relativism. Scientific perspectives of STS raises an importance of collaboration and ethical training in science education, and recommends interactive learning between teachers and students, as well as among the students. The basic idea of STS is that participatory learning methods such as debate, role-playing, practice of decision making, research and problem-solving can be more effective than a traditional method of lecture and passive reading of textbooks [13–20].

Application of STS approach at school can have quite a few benefits. For example, by introducing social issues in science education, teachers can give a more comprehensive and larger picture of the scientific system for the students. Social issues usually have interdisciplinary elements, and form a complex system. Further, STS can reveal real-life aspect of science, as it addresses scientific facts and technologies that have a direct bearing upon the society.

2 Importance of STS in Information Science Education

First, there has been a growing agreement among science teachers that STS helps students acquire practical skills to cope with an industrialized society of the future, and nurtures their ability to make right decisions. Second, advancement of scientific technology, dramatic growth of the amount of knowledge and data, and scientific innovation combined to create environmental issues. This alone justifies the need to renovate goal, substance and methodology of traditional science education. As scientific advancement and subsequent mass production and consumption led to environmental pollution, people recognized the need for STS to take environmental issues seriously and strengthen an ability to make right decisions. Lastly, STS emerged amid a growing body of pessimistic outlooks for the survival of mankind. Further, it is necessary to adopt STS at school to better understand a relationship among science, technology and society (Fig. 1).

Combining science education with social issues would stimulate students' interest and curiosity in learning. Teachers should help students understand that science can be discovered in our daily lives, that it is not a distant and mysterious subject, but part of our daily living and activities. In this regard, social issues provide effective learning materials and methods for science education.

3 Teaching–Learning Methodology of STS

Methodology and procedures vary, and they can be largely categorized into a method that focuses on a relationship, career-oriented approach, transdisciplinary approach, historical research, philosophical analysis, a social investigation, and issue-based approach.

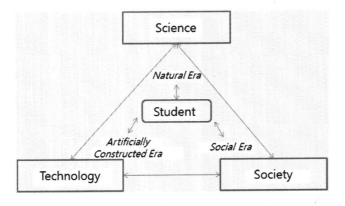

Fig. 1 The essence of STS education

3.1 Use of Internet materials for Science Education

As more schools use Internet and computer network for education, the educational environment is undergoing a dramatic change. It does not necessarily mean changes in learning materials or methods. However, it is evident that fundamental innovation of education requires changes in learning environment. Unlike conventional learning media that simply delivered knowledge and information, Internet enables a wide range of computer-related activities, expanding a possibility of altering teaching–learning methods.

Despite its powerful potentials and government policies to encourage use of Internet for education, such attempt is still in a nascent stage for a number of reasons: lack of teachers' skills, excessive work load, textbook-oriented teaching, inefficient school organization, cost burden, to name a few. Also, school education is largely biased toward preparation for college admission.

To promote use of Internet for science education, it is important to eliminate these barriers. However, some of them would take a long time, and it will be desirable to start with easier barriers to dismantle. The following lists some of the options to be considered.

First, teachers should change their perceptions toward Internet. To adjust to and cope with information society, students should be able to look up necessary information, and apply them to solve problems. To do so, they need research skills, and an ability to process them. As educational environment changes, teachers cannot meet students' needs and demands with traditional strategies and materials. They should be ready to utilize new learning methods, information and knowledge, and share them with students to meet fulfil expectations.

Second, teachers need substantial training and opportunities for advanced learning. To utilize Internet resources for science education, teachers need to acquire basic skills and knowledge first. This can be done at individual levels, but, a systematic and well-organized training program on the national level would be more effective. Many studies have shown lack of teachers' skill for Internet-based education, suggesting that the current training programs fail to offer customized, diverse training that teachers require. Also, it is necessary to expand the program to offer training to a greater number of teachers.

Third, a transition should be made from teacher-oriented to student-oriented learning. Exploration, research, constructivism, open education: all these educational trends pursue student-oriented education. And it implies that the conventional education has failed to do so. Modern society needs people who can search necessary information on their own and apply them. Student-oriented education, which can be facilitated by information media such as Internet, would help to promote students' independent thinking, creativity and cooperation.

Fourth, teachers need more discretion in designing and organizing curricula. Teachers are allowed to select and organize educational materials as long as they are in accordance with the national curricula. Thus, even within the same school, teachers could provide widely different learning materials and methodology. However, this is mostly in theory, and at school, the curricula hardly vary by grades or schools. Under this circumstance, use of Internet for science education is even more limited. To improve the condition, it is important to give discretion to science teachers to select and organize their own teaching materials.

Fifth, Internet-friendly learning environment should be established. To utilize Internet resources in science class, students should have easy access to Internet in all classrooms including computer room and science lab. The facility and space should be available to students and teachers as much as possible. In addition, an Internet-based science teaching–learning model and an evaluation tool should be developed.

3.2 Educational Benefit of Computer and CAI

Computer can be used for education in various manners, for example, CAI (Computer Assisted Instruction), CBT (Computer Based Training), and CMI (Computer Managed Instruction System). CAI indicates an education system that uses computer to deliver learning materials. CBT is widely used for corporate training. CMI uses a computer system to manage students' academic progress and information on learning resources, to enable customized education.

CAI has several benefits. First, it helps students make a progress at his or her own pace and level. Students receive instant feedback from a computer, monitor their progress, and make decisions accordingly. Second, it can provide interesting and diverse learning experiences that mostly lack in conventional education. Third, unlike a traditional one-to-many relationship between a teacher and students, CAI provides sufficient opportunities for interaction between students and the learning program. Fourth, it encourages students to make new attempts without fear of making a mistake. Fifth, latest information and technology can be available through Internet and communication network.

Meanwhile, there are a few limitations to consider in designing computer-assisted learning materials. First, it should be noted that computer lacks a human touch and response. Even the most excellent CAI cannot fully replace a good teacher. A good teacher can encourage shy students to engage in a group discussion, which might be hard for CAI to do. In fact, it might even turn away shy learners.

Also, as computer technology advances, multimedia CAI tends to grow heavy on graphics. Accordingly, students might not think deeply and thoroughly, with less interest in refining verbal skills, and might even lose power of concentration to some degree. All these should be carefully considered and the CAI materials should be well organized and subject to critical review.

4 Conclusion

For effective and successful science education, the national curricula, its goal and targets should be considered in designing science curricula, and assessment should be made for review and revision. The goal is to train students to be equipped with scientific knowledge, attitude and research skills. To offer effective science education, fully qualified teachers, facility and equipment are essential, along with rational science policy of the government that reflects the conditions and environment of a contemporary society.

Acknowledgments This paper is extended from a conference paper presented at the second international conference on Computers, Networks, Systems, and Industrial applications (CNSI 2012), Jeju Island, Korea. The author is deeply grateful to the anonymous reviewers for their valuable suggestions and comments on the first version of this paper. The corresponding author is Namje Park (namjepark@jejunu.ac.kr).

References

1. Mansour, N.: Impact of the knowledge and beliefs of Egyptian science teachers in integrating a STS based curriculum: a sociocultural perspective. J. Sci. Teacher Educ. **21**(5), 513–534 (2010)
2. Fensham, P.J.: Knowledge to deal with challenges to science education from without and within. The Professional Knowledge Base of Science Teaching. pp. 295–317 (2011)
3. Park, N., Kwak, J., Kim, S., Won, D., Kim, H.: WIPI mobile platform with secure service for mobile RFID network environment. In: Shen, H.T., Li, J., Li, M., Ni, J., Wang, W. (eds.) APWeb Workshops 2006. LNCS, vol. 3842, pp. 741–748. Springer, Heidelberg (2006)
4. Park, N.: Security scheme for managing a large quantity of individual information in RFID environment. In: Zhu, R., Zhang, Y., Liu, B., Liu, C. (eds.) ICICA 2010. CCIS, vol. 106, pp. 72–79. Springer, Heidelberg (2010)
5. Park, N.: Secure UHF/HF dual-band RFID: strategic framework approaches and application solutions. In: ICCCI 2011. LNCS, Springer, Heidelberg (2011)

6. Park, N.: Implementation of terminal middleware platform for mobile RFID computing. Int. J. Ad Hoc Ubiquitous Comput. **8**(4), 205–219 (2011)

7. Park, N., Kim, Y.: Harmful adult multimedia contents filtering method in mobile RFID service environment. In: Pan, J.-S., Chen, S.-M., Nguyen, N.T. (eds.) ICCCI 2010. LNCS(LNAI), vol. 6422, pp. 193–202. Springer, Heidelberg (2010)

8. Park, N., Song, Y.: AONT Encryption based application data management in mobile RFID environment. In: Pan, J.-S., Chen, S.-M., Nguyen, N.T. (eds.) ICCCI 2010. LNCS(LNAI), vol. 6422, pp. 142–152. Springer, Heidelberg (2010)

9. Park, N.: Customized healthcare infrastructure using privacy weight level based on smart device. Communications in Computer and Information Science, vol. 206, pp. 467–474. Springer (2011)

10. Park, N.: Secure data access control scheme using type-based re-encryption in cloud environment. Studies in Computational Intelligence, vol. 381, pp. 319–327. Springer, Heidelberg (2011)

11. Park, N., Song, Y.: Secure RFID application data management using all-or-nothing transform encryption. In: Pandurangan, G., Anil Kumar, V.S., Ming, G., Liu, Y., Li, Y. (eds.) WASA 2010. LNCS, vol. 6221, pp. 245–252. Springer, Heidelberg (2010)

12. Park, N.: The implementation of open embedded S/W platform for secure mobile RFID reader. J Korea Inf. Commun. Soc. **35**(5), 785–793 (2010)

13. Park, N., Song, Y., Won, D., Kim, H.: Multilateral approaches to the mobile RFID security problem using web service. In: Zhang, Y., Yu, G., Bertino, E., Xu, G. (eds.) APWeb 2008. LNCS, vol. 4976, pp. 331–341. Springer, Heidelberg (2008)

14. Park, N., Kim, H., Kim, S., Won, D.: Open location-based service using secure middleware infrastructure in web services. In: Gervasi, O., Gavrilova, M.L., Kumar, V., Laganá, A., Lee, H.P., Mun, Y., Taniar, D., Tan, C.J.K. (eds.) ICCSA 2005. LNCS, vol. 3481, pp. 1146–1155. Springer, Heidelberg (2005)

15. Park, N., Kim, S., Won, D.: Privacy preserving enhanced service mechanism in mobile RFID network. In: ASC, Advances in Soft Computing, vol. 43, pp. 151–156. Springer, Heidelberg (2007)

16. Park, N., Kim, S., Won, D., Kim, H.: Security analysis and implementation leveraging globally networked mobile RFIDs. In: PWC 2006. LNCS, vol. 4217, pp. 494–505. Springer, Heidelberg (2006)

17. Kim, Y.: Harmful-word dictionary DB based text classification for improving performances of precesion and recall. Sungkyunkwan University, Ph.D. Thesis (2009)

18. Kim, Y, Park, N., Hong, D.: Enterprise data loss prevention system having a function of coping with civil suits. Studies in Computational Intelligence, vol. 365, pp. 201–208. Springer, Heidelberg (2011)

19. Kim, Y., Park, N., Won, D.: Privacy-enhanced adult certification method for multimedia contents on mobile RFID environments. In: Proceedings of IEEE International Symposium on Consumer Electronics, pp. 1–4. IEEE, Los Alamitos (2007)

20. Kim, Y., Park, N., Hong, D., Won, D.: Adult certification system on mobile RFID service environments. J. Korea Contents Assoc. **9**(1), 131–138 (2009)

Encryption Scheme Supporting Range Queries on Encrypted Privacy Databases in Big Data Service Era

Jun Wook Lee and Namje Park

Abstract Security and privacy issues are magnified by velocity, volume, and variety of big data, such as large-scale cloud infrastructures, diversity of data sources and formats, streaming nature of data acquisition and high volume inter-cloud migration. Therefore, traditional security mechanisms, which are tailored to securing small scale static (as opposed to streaming) data, are inadequate. In this paper, we proposed Bucket ID Transformation that is a new encryption mechanism and the scheme can range search without order-preserving. Bucket ID Transformation is performed by recursive HMAC as many as a value of Bucket ID. As a future desk, we plan to carry out simulated experiments for performance evaluation and compare the results, and design and verify a provably secure encryption mechanism.

Keywords Big data · Encryption · Range query · Privacy data · Security

1 Introduction

When traditional encryption algorithm apply to the database, efficiency decline problem was occurred because order of encoded data are not equal to order of plaintext. To overcome this limit, Haciquimus proposed bucket based index [1]

J. W. Lee
Electronics and Telecommunications Research Institute (ETRI),
218 Gajeong-ro, Yuseong-gu, Daejeon 305-700, Korea
e-mail: junux@etri.re.kr

N. Park (✉)
Department of Computer Education, Teachers College, Jeju National University,
61 Iljudong-ro, Jeju-si, Jeju Special Self-Governing Province 690-781, Korea
e-mail: namjepark@jejunu.ac.kr

S.-S. Yeo et al. (eds.), *Computer Science and its Applications*,
Lecture Notes in Electrical Engineering 203, DOI: 10.1007/978-94-007-5699-1_76,
© Springer Science+Business Media Dordrecht 2012

that can bring performance improvement for queries over encrypted data. Besides, Order-Preserving Encryption scheme that is possible range queries over encrypted data without decryption was proposed by Sun [2], Agrawal [3], Ets. But, Encrypted data by Order-Preserving Encryption Scheme was exposed order of plaintext, As a result, the scheme cannot secure against inference attack. Especially, the scheme cannot used for rank scale [4–13]. Use of order-preserving function is desirable for efficiency. On the other side, obviously, order-preserving function cannot prevent the inference attack. Therefore, it needed stabilize trade-off to solve the problem.

In this paper, we proposed Bucket ID Transformation that is a new encryption mechanism and the scheme can range search without order-preserving. Bucket ID Transformation is performed by recursive HMAC as many as a value of Bucket ID.The proposed method, whose order is not exposed, has a more enhanced security than Sun and Agrawal and is also more efficient compared to Damiani's method as it can recover the original value by transmitting queries $\frac{d-n}{q}$ times (q: bucket size, d: number of transmitted queries of damiani, n: number of nodes) to the database.

2 Related Work

2.1 Bucket Based Index

Hacigumus et al [1] proposed the technique that queries encrypted data. This is based on the definition of the number of buckets in the attribute area. Let's assume that ri is the plaintext relation with schema Ri (Ai1, Ai2, . . ., Ain) and rki is the corresponding encrypted relation in Rki (Counter, E tuple). When a plaintext attribute Aij exists in RI where a domain is Dij, the bucket-based indexing technique can divide Dij without overlapping it. This is called "bucket". A bucket has a continual value. This procedure called "bucketing". The buckets are always created with same size. Each bucket is connected with a unique value and this value is a domain for connection between Ij and Aij. If a plaintext tuple t is given in ri, the value of attribute Aij for t should belong to a bucket. This is very important in keeping data confidentiality [14–20].

2.2 B+Tree

An untrusted DMBS can find encrypted data only and any B+-Tree defined on the index doesn't reflect the order of plaint text. This, in effect, makes the range search impossible. To overcome this problem, we can entrust a trusted front-end with the decision on B+-Tree information. This paper proposes encrypting the whole B+-Tree node. The original B+-Tree is represented as two attributes (Node ID and encrypted value) in an untrusted DBMS [21, 22].

The advantage of this method is that the content of B+-Tree node is not seen in an untrusted DBMS. The disadvantage is that B+-Tree traversal can be executed only by a trusted front-end. By intuition, To execute an interval query, the front-end must execute quires needed to go down the tree node. Once reached a leaf, the node ID within the leaf can be used to compose tuple [23–27].

For example, to use B+-Tree to organize all customers whose names begins with DF, the front-end creates several queries to access the sequence node 0,1,4 and 10 and then other queries can be used to 10, 11 and 12 nodes (needed to search for other leaf). However, this method requires reproductivity of B+-Tree for data insertion, modification and deletion and has a security problem as it needs decryption of a whole encrypted text. In addition, since, unlikely tables, the index is not included in the portion that users can modify, delete or enter, it cannot be used with universal database and a separate DBMS that supports those functions must be built.

To restore K pieces of data, SQL queries as many as K+N are required (N: number of nodes). This process is one of factors that lowers efficiency of database.

2.3 Order-Preserving Encryption for Numeric Data

For numeric data, it may be a serious problem, if the attacker can get a value close to plaintext p corresponding to encrypted text c, though he doesn't know p exactly. In other words, in ordinary Order-Preserving Mechanism, if the distribution is known, the plaintext can be inferred [3].

Since this paper considers the ciphertext only attack only, the proposed mechanism is secure from estimation exposure. However, when using this method, the order is exposed due to the nature of Order-Preserving. The p value can be inferred by designating a certain location. The fact that order is kept means consequently that the order is known and this means that some of information is exposed. Therefore this paper proposes an encryption mechanism that allows range query without keeping the order.

3 Suggested Mechanism

3.1 Notations

Notations that are used in this section is as follows:

m	Plaintext
r	Residue of plaintext
I_B	Bucket ID

S_B	Size of bucket
$T(x)$	Transform execution result of x
$IT(x)$	Inverse transform execution result of x
$T(I_B)^K$	Encrypted Bucket ID
$T(r)^{IB\|K}$	Encrypted residue (Ciphertext)
K	Key of keyed-HMAC

3.2 Overview of Our Scheme

In our mechanism, encryption and decryption procedure are constructed each two stage.

(1) Encryption process

- Pre-Processing: Pre-process stage for plaintext m, extract m in two integer and calculate bucket threshold in this stage.
- Transformation: Integer IB, r is transform into hashed value that can not recognize by attacker.

ms = m * There are 4 scales; nominal scale, ordinal scale, interval scale and ratio scale. The nominal scale is a scale that we cannot define the order and the scope of this paper is confined to ordinal scale, interval scale and ratio scale. This process executes integarization as follows. If it is ordinal scale and the ordinal value is rank, ms equals rank. If m is integer, then ms is equal to the m and if m has a decimal point, ms is equal to m * 10 1 (1:length of below decimal point).

(2) Decryption process

- Inverse Transformation: Calculate set of IB, r from each set of T(IB)K and T(r)IB∥K.
- Post-Processing: Calculate plaintext m based on IB, r.

According to these each two stage, authorized database manager can encrypt plaintext and decrypt ciphertext securely. Fig. 1

4 Analysis of Mechanism

4.1 Notations

(1) Comparing with the traditional mechanism

Anti-Tamper Open-Form [2] The encryption procedure of Anti-Tamper was constructed by summation of a set of pseudo-random number. but, security prove

Fig. 1 Illustrating our scheme

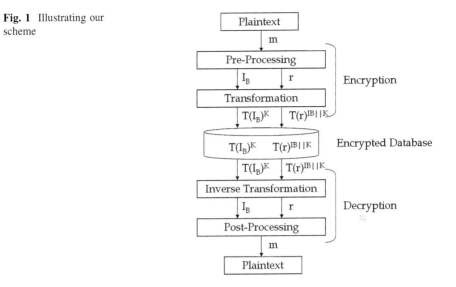

of pseudo-random number is very hard. On the other hand, according to incresement of plaintext, success probability of inference attack may increse too. For encryption of plaintext p, random sequence number Zi is added. That is, accoding to incresement of plaintext, The encrypted value is increase exponentially. For example, If the plaintext is 1,000,000 and average of random sequence is 10,000, then encrypted value E(p) may reach to 1,000,000 * 1000 + 1,010,000,000 = 50,000,050,000.

In this case, the attacker can be easily perform the inference attack because pseudo-random number is less excessively than plaintext value. In proposed mechanism, appearance extent is limited as many as SB. For example, If suppose the plaintext is 123456, Encryption procedure of Anti-Tamper Mechanism needs generation of pseudo-random number of 123456 times. but, our scheme only needed hash function execution of 123+456 = 579 times.

Order-Preserving Encryption [3] According to Characteristic of order-Preserving Function, order of ciphertext is perfectly equal to order of plaintext. This characteristic exposed some problem. If plaintext is organized in rank scale such as grade of student, the encrypted data is powerless about inference attack. On the other hand, Numeric ciphertext can be occur precision problem. For example, if an encrypted data is 1/3 = 0.333..., it is possible that the plaintext cannot be decrypt exactly. Proposed scheme is suitable for encryption of rank scale because the scheme is not expose order information. Besides, precision error is not exist because processing of plaintext only depends on Bucket ID transformation.

Bucketization [1] Bucketization is needed for special management for Bucket ID that unique value. If Bucket ID is exposed to an attacker, Information of order can be expose. Moreover, Aggregation Queries (MIN,MAX,COUNT) is not possible without plaintext decrypting.

Our mechanism calculate the arithmetic $r \equiv ms \bmod SB$ in Pre-processing stage, and r is transformed into T(r) by transformation. The T(r) is a ciphertext, and can MIN,MAX,COUNT queries from encrypted database by using of T(r) value. besides, our scheme can prevent inference attack because the key of each bucket for transforming r are each different.

B+-Tree [21, 22] B+-Tree is requred for many times of SQL for one range query. That is, the number of data that wish to find is 100, actually, it needs SQL quering more over than 100 times. For that reason, this method causes very much overhead. Moreover, it needs to decrypt entire data and reconstruct B+-Tree if user or data manager perform many times of Insert, delete and update queries. As a result, this problem brings very high overhead and can't perform update queries without decryption.

Hash Based Index [21, 22] Hash based Indexing method can prevent inference attack, but, the scheme has very difficult to range search and exist burdensomeness of re-filtering. our mechanism is possible range search by HMAC chained Bucket-ID.

Our Scheme Proposed mechanism can prevent frequence and order based inference attack because the mechanism is not preserve order. In addition, our mechanism can perform range query and aggregation queries (MIN,MAX,COUNT) over encrypted data. On the other hand, even if insert or update transactions are much repeat, plaintext decrypting is not needed. In case of range search, number of queries are greatly reduced better than Damiani's method.

5 Conclusion

Security and privacy issues are magnified by velocity, volume, and variety of big data, such as large-scale cloud infrastructures, diversity of data sources and formats, streaming nature of data acquisition and high volume inter-cloud migration. Therefore, traditional security mechanisms, which are tailored to securing small scale static (as opposed to streaming) data, are inadequate.

The security for database needs a separate consideration in addition to traditional cytological security. Especially, since various attacks such as inference attack, query execution attack or known-plaintext attack are possible according to the nature of database, an encryption mechanism suitable for database environment is required. This paper proposes a new encryption mechanism that can carry out range search without exposing the order. This method is more powerful than order-keeping methods of Sun and Agarwal and is expected to secure data more efficiently than Damiani method.

As a future desk, we plan to carry out simulated experiments for performance evaluation and compare the results, and design and verify a provably secure encryption mechanism.

Acknowledgments This paper is extended from a conference paper presented at 2007 International Conference on Convergence Information Technology, Korea. The author is deeply grateful to the anonymous reviewers for their valuable suggestions and comments on the first version of this paper. This work was supported by the Industrial Strategic Technology Development Program funded by the Ministry of Knowledge Economy (MKE, Korea). And, the National Research Foundation of Korea Grant funded by the Korean Government (NRF-2012S1A5A8024965). The Corresponding author is Namje Park (namjepark@jejunu.ac.kr).

References

1. Hacigumus, H., Iyer, B.R., Li, C., Mehrotra, S.: Executing SQL over encrypted data in the database-service-provider model. In: Proceedings of the ACM SIGMOD Conference on Management of Data, Madison, Wisconsin (2002)
2. Chung, S.S., Ozsoyoglu, G.: Anti-tamper databases: processing aggregate queries over encrypted databases, EECS Department, Case Western Reserve University, Cleveland Ohio, U.S.A., ICDEW'06, IEEE (2006)
3. Agrawal, R. et al.: Order preserving encryption for numeric data. In: Weikum, G., Konig, A., Deßloch, S. (eds.) In: Proceedings of the ACM SIGMOD 2004, Paris, France. ACM (2004)
4. Domingo, J., Ferror, i.: A new privacy homomorphism and applications. Inf. Process. Lett. **60**(5):277–282
5. Song, D.X., Wagner, D., Perrig, A.: Practical techniques for searches on encrypted data. In: The IEEE Symposium on Security and Privacy, Oakland, California (2000)
6. Iyer, B. et al.: A framework for efficient storage security in RDBMS. In Bertino, E. et al. (eds.) Proceedings of the International Conference on Extending Database Technology (EDBT 2004), volume 2992 of Lecture Notes in Computer Science, Crete, Greece. Springer, Heidelberg (2004)
7. Aggarwal, G. et al.: Two can keep a secret: a distributed architecture for secure database services. In Proceedings of the Second Biennal Conference on Innovative Data Systems Research (CIDR 2005), Asilomar, CA (2005)
8. Park, N., Kwak, J., Kim, S., Won, D., Kim, H.: WIPI mobile platform with secure service for mobile RFID network environment. In: Shen, H.T., Li, J., Li, M., Ni, J., Wang, W. (eds.) APWeb Workshops 2006. LNCS, vol. 3842, pp. 741–748. Springer, Heidelberg (2006)
9. Park, N.: Security scheme for managing a large quantity of individual information in RFID environment. In: Zhu, R., Zhang, Y., Liu, B., Liu, C. (eds.) ICICA 2010. CCIS, vol. 106, pp. 72–79. Springer, Heidelberg (2010)
10. Park, N.: Secure UHF/HF dual-band RFID: strategic framework approaches and application solutions. In: ICCCI 2011. LNCS, Springer, Heidelberg (2011)
11. Park, N.: Implementation of terminal middleware platform for mobile RFID computing. Int. J. Ad Hoc Ubiquitous Comput. **8**(4), 205–219 (2011)
12. Park, N., Kim, Y.: Harmful adult multimedia contents filtering method in mobile RFID service environment. In: Pan, J.-S., Chen, S.-M., Nguyen, N.T. (eds.) ICCCI 2010. LNCS(LNAI), vol. 6422, pp. 193–202. Springer, Heidelberg (2010)
13. Park, N., Song, Y.: AONT encryption based application data management in mobile RFID environment. In: Pan, J.-S., Chen, S.-M., Nguyen, N.T. (eds.) ICCCI 2010. LNCS(LNAI), vol. 6422, pp. 142–152. Springer, Heidelberg (2010)
14. Park, N.: Customized healthcare infrastructure using privacy weight level based on smart device. Commun. Comput. Inf. Sci. **206**, 467–474 (2011). Springer
15. Park, N.: Secure data access control scheme using type-based re-encryption in cloud environment. Stud. Comput. Intell. **381**, 319–327 (2011). Springer

16. Park, N., Song, Y.: Secure RFID Application data management using all-or-nothing transform encryption. In: Pandurangan, G., Anil Kumar, V.S., Ming, G., Liu, Y., Li, Y. (eds.) WASA 2010. LNCS, vol. 6221, pp. 245–252. Springer, Heidelberg (2010)

17. Park, N.: The implementation of open embedded S/W platform for secure mobile RFID reader. J. Korea Inf. Commun. Soc. **35**(5), 785–793 (2010)

18. Park, N., Song, Y., Won, D., Kim, H.: Multilateral approaches to the mobile RFID security problem using web service. In: Zhang, Y., Yu, G., Bertino, E., Xu, G. (eds.) APWeb 2008. LNCS, vol. 4976, pp. 331–341. Springer, Heidelberg (2008)

19. Park, N., Kim, H., Kim, S., Won, D.: Open location-based service using secure middleware infrastructure in web services. In: Gervasi, O., Gavrilova, M.L., Kumar, V., Laganá, A., Lee, H.P., Mun, Y., Taniar, D., Tan, C.J.K. (eds.) ICCSA 2005. LNCS, vol. 3481, pp. 1146–1155. Springer, Heidelberg (2005)

20. Park, N., Kim, S., Won, D.: Privacy preserving enhanced service mechanism in mobile RFID network. In: ASC, Advances in Soft Computing, vol. 43, pp. 151–156. Springer, Heidelberg (2007)

21. Damiani, E., di Vimercati, S.D.C., Jajodia, S., Paraboschi, S., Samarati, P.: Balancing confidentiality and efficiency in untrusted relational dbmss. In: Proceedings of the 10th ACM Conf. on Computer and Communications Security (CCS) (2003)

22. Damiani, E. et al.: Implementation of a storage mechanism for untrusted DBMSs. In: Proceedings of the Second International IEEE Security in Storage Workshop, Washington DC, USA. IEEE Computer Society (2003)

23. Park, N., Kim, S., Won, D., Kim, H.: Security analysis and implementation leveraging globally networked mobile RFIDs. In: PWC 2006. LNCS, vol. 4217, pp. 494–505. Springer, Heidelberg (2006)

24. Kim, Y.: Harmful-word dictionary DB based text classification for improving performances of precesion and recall. Sungkyunkwan University, Ph.D. Thesis (2009)

25. Kim, Y, Park, N., Hong, D.: Enterprise data loss prevention system having a function of coping with civil suits. studies in computational intelligence, vol. 365, pp. 201–208. Springer, Heidelberg (2011)

26. Kim, Y., Park, N., Won, D.: Privacy-enhanced adult certification method for multimedia contents on mobile RFID environments. In: Proceedings of IEEE International Symposium on Consumer Electronics, pp. 1–4. IEEE, Los Alamitos (2007)

27. Kim, Y., Park, N., Hong, D., Won, D.: Adult certification system on mobile RFID service environments. J. Korea Contents Assoc. **9**(1), 131–138 (2009)

Part V
New Technology Convergence, Cloud, Culture and Art

A Congested Route Discrimination Scheme Through the Analysis of Moving Object Trajectories

He Li, Hyuk Park, Yonghun Park, Kyoungsoo Bok and Jaesoo Yoo

Abstract In this paper, we propose a congested route discrimination scheme through the analysis of moving object trajectories in road networks. The proposed scheme divides the road into segments with different lanes and length. And then, it extracts congested road segments based on the moving speeds of moving objects and a saturation degree of each road segment. By doing so, we perform clustering method to find congested routes of the road network. Our experimental results show that our proposed scheme derives the directional congested routes through the clustering of the congested segments.

Keywords Location based service · Road network · Moving object · Clustering scheme

H. Li · H. Park · Y. Park · K. Bok
Department of Information and Communication Engineering, Chungbuk National
University, Cheongju 361-763, South Korea
e-mail: lihe@chungbuk.ac.kr

H. Park
e-mail: agodsun@naver.com

Y. Park
e-mail: yhpark1119@chungbuk.ac.kr

K. Bok
e-mail: ksbok@chungbuk.ac.kr

J. Yoo (✉)
Department of Information and Communication Engineering and CBITRC, Chungbuk
National University, Cheongju 361-763, South Korea
e-mail: yjs@chungbuk.ac.kr

S.-S. Yeo et al. (eds.), *Computer Science and its Applications*,
Lecture Notes in Electrical Engineering 203, DOI: 10.1007/978-94-007-5699-1_77,
© Springer Science+Business Media Dordrecht 2012

1 Introduction

With the increase of the use of the mobile devices, the location-based services are becoming increasingly popular. The rapidly increased satellites and tracking facilities have made it possible to collect a large amount of trajectory data of moving objects, such as the vehicle position data, hurricane track data, and animal movement data. The analysis over these trajectory data is becoming important for some applications, such as meteorological observation and forecast, animal habits observation, road traffic situation analysis, and navigation in transportations. In the road network based applications, the location information of each moving object can be recorded by the GPS equipped devices and the mobility of the moving object is road network constrained. According to these recorded trajectory data, the moving pattern, traffic situation and road recommendation services can be supported.

Recently, the route recommendation services are very important services for moving objects in the road network environments [1, 2]. Most of the existing methods try to monitor and forecast the traffic by using the recorded history trajectory data of vehicles equipped with GPS device. The index based schemes [3, 4] construct an index by adopting the trajectory data of the moving objects. And then the routes are recommended according to the history trajectory data of the related moving objects. Since the enormous amount of the trajectory data generated according to the time, the index structure can improve the retrieval speed. The clustering based schemes [5–8] generate the density regions of the road networks by analyzing the trajectory data of moving objects. The density regions of the road networks are evaluated by considering both the location and time of the moving objects. According to the trajectory data, the number of the moving objects within a specific road segment and timestamp is be used to identify the density regions of a road [9–12]. After that, the small density regions of each road are clustered and the final density regions in the road networks are generated. However, there are two problems of the existing methods: (1) the directions of the roads in the road networks are not considered; (2) the lanes and length of the roads in the road networks are not considered. In the real road network environments, each road is divided into two directions: positive direction and negative direction. The moving objects in the road toward to different directions do not affect each other. Furthermore, the lanes and lengths of each road segment is different in the road networks, which will affect the computation of the density regions of a road. Therefore, the existing schemes are not suitable for real road network environments.

To overcome such problems, we propose a congested route discrimination scheme in real road network environments. The proposed scheme divides the road into segments with different lanes and lengths. And then, the congested road segments are extracted by considering the moving speeds and directions of the moving objects in the road networks. By doing this, we perform clustering methods to find the final congested regions in the whole road networks.

Fig. 1 The definition of
neighbor segment

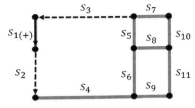

The remainder of the paper is organized as follows. Section 2 presents the details of the proposed method. Section 3 contains experimental evaluation that demonstrates the superiority of our proposed method. Finally, Sect. 4 concludes this paper.

2 The Proposed Method

2.1 Data Model of Road Network

We assume that the road network is represented by $G(N, E)$, where N denotes the node which is the intersection between different road segments and E denotes the edge connected between two adjacent nodes in the road network. The trajectory of a moving object (e.g. a car) is represented by Tr. Each node N_i in the road network is represented by a point $\{x_i, y_i\}$. E_i denotes a segment of the road network. '+' and '−' are used to represent the different directions of moving objects in the real road networks. Moreover, since the length and lane of each road are different, the length and lane of each road segment are stored. Therefore, each road segment is represented by $S_i(\pm) = \{N_i, N_j, \text{length}, \text{lane}\}$. As shown in Fig. 1, S_2 and S_3 are the neighbor segments of S_1 in a road network G.

In a road network G, each road segment S_i stores the information of directly connected road segments. This information can be used for the following clustering evaluations. Since the moving objects may move continuously or stay in a position, it is necessary to have the location knowledge of each moving object according to the timestamp. Suppose that the trajectory Tr of each moving object is as follows:

$$Tr_n = \; <(S_1(\pm), T_1), (S_2(\pm), T_2), \ldots, (S_k(\pm), T_j)>$$

where S_i denotes the segment ID and T is the timestamp. According to T, the location of each moving object can be retrieved easily.

2.2 The Computation of Congested Regions

The location and directions of each moving object can be retrieved according to the recorded trajectory data. After that, the complexity value of each road segment is computed. The complexity value of each road segment is evaluated by considering the moving speeds of the moving objects and the number of the moving objects in the road segment. The fast moving speed indicates that the complexity of the road segment is low. In the contrast, the low moving speed can indicate that the complexity is high. Moreover, the length and lanes of a road segment will also affects the computation of the complexity of the road segment. We use saturation to represent the complexity of a road segment. The saturation is the ratio of the number of the moving objects and the size of a road segment. We define that the congested regions of the road networks are the roads with high complexity values.

The average moving speed (Av) of the moving objects in a road segment according to different directions is computed by the following Formula 1, where $V(Ob_i)$ denotes the moving speed of object ob_i. The saturation (Sat) according to the lane (S_{lane}) and length (S_{length}) of a road segment is computed in Formula 2, where Ob_n denotes the number of the moving objects in a road segment. As a result, the complexity value of a road segment is computed by Formula 3 which combines Formulas 1 and 2, α denotes the weight value between the average moving speed and saturation of a road segment.

$$Av_i(\pm) = \frac{\sum_{Objecti=0}^{n} V(Ob_i)}{Ob_n(\pm) \times V_{max}} \tag{1}$$

$$Sat_i(\pm) = 1 - \frac{[ob_n](\pm)}{S_{lane} \times S_{length}} \tag{2}$$

$$Ri_i(\pm) = (\alpha \times Av_i(\pm)) + \{(1 - \alpha) \times Sat_i(\pm)\} \tag{3}$$

Since the complexity values of the road segments are change according to different timestamp, it has to be computed periodically. The congested regions of the road networks are indicated by the complexity values of each road segment. Figure 2 shows the congested regions (the dotted areas) of a road network in different timestamp T. According to the recorded trajectory data of moving objects at time $T = 0$, the congested regions of the roads of different directions are generated, such as $S_2(+)$ and $S_5(-)$ in Fig. 2a. As shown in Fig. 2b, when time $T = 1$ $S_5(-)$ and $S_6(-)$ are evaluated as congested regions. Since $S_5(-)$ and $S_6(-)$ are neighbor road segments and have the same direction in the road network, they are clustered together.

3 Performance Evaluation

In this section, we introduce the performance evaluation by comparing the proposed scheme with the existing scheme NETSCAN [7]. The moving objects are generated by the network-based generator [13]. The complexity values and clusters are generated according to the number of moving objects in each time interval. All

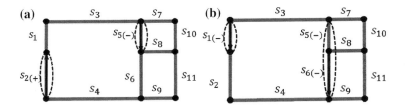

Fig. 2 The clusters according to different timestamp. **a** $T = 0$, **b** $T = 1$

Table 1 The values of parameters

Parameters	Values
Road network	Oldenburg city
The number of segments	7,035
The number of node	6,105
The number of object	10,000–100,000
Velocity	0–120

of the experiments are coded in Java and the experiments are performed in Intel i3 3.0 GHz CPU and 4G memory. Table 1 summarizes the parameters for this performance evaluation.

In the first experiment, we show the complexity regions of the road networks of Oldenburg city by using our proposed scheme. In this experiment, the total number of the moving objects in the road network is set to 50,000 and the saturation of the road is set to 30 %. As shown in Fig. 3, the results indicated that the congested regions of the road network are different according to the different directions of the road segments. The gray segment in Fig. 3a, b represent the congested regions of the roads of positive direction and negative direction respectively.

In Fig. 4, we compare the NETSCAN method with our proposed method. The number of the complexity regions of NETSCAN and the proposed method are evaluated according to the number of the moving objects. For the proposed method, the complexity regions are evaluated in different directions (positive direction and negative direction) and same direction respectively. The results show that the number of the complexity regions is increased when the number of the moving objects increases. The number of the complexity regions of the proposed method is similar when the number of the moving objects between 20,000 and 30,000. This is because the saturation of each road segment is considered in the proposed method, when the lane and length of a road segment is large the 20,000 and 30,000 moving objects is not large for the road. Therefore, most of the road segments are not identified as congested roads at first. For NETSCAN method, the number of complexity areas is increased proportionally with the increase of the number of moving objects.

Figure 5 shows the number of complexity regions according to the different timestamps, the time interval is set as 1 h. The complexity regions are evaluated in different directions (positive direction and negative direction) and same direction respectively. From Fig. 5, we can see that the number of the complexity regions of

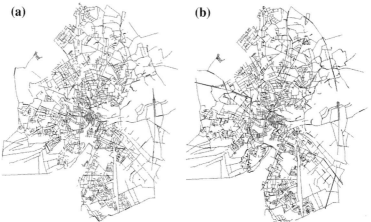

Fig. 3 The whole complexity areas of the Oldenburg city. **a** Positive direction, **b** negative direction

Fig. 4 The number of complexity areas according the number of moving objects

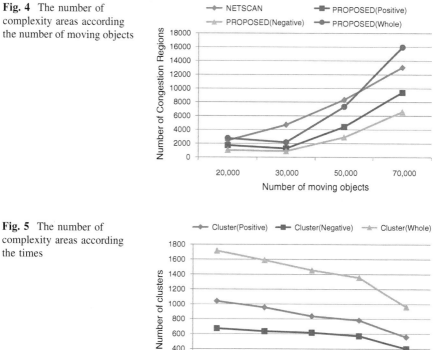

Fig. 5 The number of complexity areas according the times

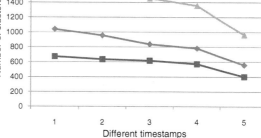

the positive direction is larger than that of the negative direction. And the number of the complexity regions that without considering the direction of the road segment is larger than the direction considered method.

4 Conclusions

In this paper, we propose a congested regions discrimination scheme in road networks. The proposed scheme divides the road into segments with different lanes and length. It extracts congested road segments based on the average velocity of the moving objects and a saturation degree within a road segment. Then, the congested regions are found by performing clustering method. The experimental results have shown that the proposed scheme derives the directional congested routes through the clustering of the congested segments. In the future, we will conduct more performance evaluation of our approach by using real car trajectory data.

Acknowledgments This work was supported by Basic Science Research Program through the National Research Foundation of Korea (NRF) grant funded by the Korea government (MEST) (No. 2009-0089128) and the Leaders in INdustry-university Cooperation (LINC) Program through the National Research Foundation of Korea (NRF) funded by the Ministry of Education, Science and Technology (2012-B-0013-010112).

References

1. Won, J.I., Kim, S.W., Baek, J.H., Lee, J.H.: Trajectory clustering in road network environment. In: IEEE Computational Intelligence and Data Mining, pp. 299–305 (2009)
2. Roh, G.P., Roh, J.W., Hwang, S.W., Yi, B.K.: Supporting pattern matching queries over trajectories on road networks. IEEE Trans. Knowl. Data Eng. **23**(11), 1753–1758 (2011)
3. Chang, J.W., Song, M.S., Um, J.H.: TMN-tree: new trajectory index structure for moving objects in spatial networks. In: IEEE Computer and Information Technology, pp. 1633–1638 (2010)
4. Huang, M., Hu. P., Xia, L.: A grid based trajectory indexing method for moving objects on fixed network. In: Geoinformatics, pp. 1–4 (2010)
5. Li, X., Han, J., Lee, J., Gonzalez, H.: Traffic density-based discovery of hot routes in road networks. In: The Seventh SIAM International Conference on Data Mining. LNCS, vol. 4605, pp. 441–459 (2007)
6. Pelekis, N., Kopanakis, I., Kotsifakos, E.E., Frentzos, E.F, Theodoridis, Y.: Clustering trajectories of moving objects in an uncertain world. In: IEEE Data Mining, pp. 417–427 (2009)
7. Kharrat, A., Zeitouni, K., Sandu-Popa, I.: Characterizing traffic density and its evolution through moving object trajectories. J. Signal Image Technol. Internet Based Syst. 257–263 (2009)
8. Mokhtar, H.M.O., Ossama, O., Sharkawi, M.E.: A time parameterized technique for clustering moving object trajectories. J. Data Min. Knowl. Manag. Process **1**(1), 14–30 (2011)

9. Chen, Z., Shen, H.S., Zhou, X.: Discovering popular routes from trajectories. In: IEEE Data Engineering, pp. 900–911 (2011)
10. Mauroux, P.C., Wu, E., Madden, S.: TrajStore: an adaptive storage system for very large trajectory data sets. In: IEEE Data Engineering, pp. 109–120 (2010)
11. Lee, S., Kim, T., Ko, H., Bok, K.: The evaluation of existing congestion indices applicability for development of traffic condition index. J. Korean Soc. Road Eng. **10**(3), 119–128 (2008)
12. Ester, M., Kriegel, H.P., Sander, J., Xu, X.: A density-based algorithm for discovering clusters in large spatial databases with noise. In: Knowledge Discovery and Data Mining, pp. 226–231 (1996)
13. Brinkhoff, T.: A framework for generating network-based moving objects. GeoInformatica **6**(2), 153–180 (2002)

An Efficient Hierarchical Multi-Hop Clustering Scheme in Non-uniform Large Wireless Sensor Networks

Chunghui Lee, Eunju Kim, Junho Park, Dongook Seong
and Jaesoo Yoo

Abstract In wireless sensor networks, an energy efficient data gathering scheme is one of core technologies to process a query. The cluster-based data gathering methods minimize the energy consumption of sensor nodes by maximizing the efficiency of data aggregation. However, since the existing clustering methods consider only uniform network environments, they are not suitable for the real world applications that sensor nodes can be distributed unevenly. Recently, a balanced multi-hop clustering method in non-uniform wireless sensor networks was proposed. The proposed scheme constructs a cluster based on the logical distance to the cluster head using a min-distance hop count. But this scheme also increases the occurrence of orphan nodes in case of expanding a cluster in large sensor network environments. To solve such a problem, we propose a hierarchical multi-hop clustering scheme that considers the scalability of a cluster

C. Lee · J. Park
School of Information and Communication Engineering, Chungbuk National University,
52 Naesudong-ro, Heungdeok-gu, Cheongju, South Korea
e-mail: reikan@naver.com

J. Park
e-mail: junhopark@chungbuk.ac.kr

E. Kim
Suresoft Technologies, Inc., #701-5, Banpo-1 dong, Seocho-gu, Seoul, South Korea
e-mail: ejkim0422@gmail.com

D. Seong
BOAS Electronics Inc., Industrial Technology Research Park, 52 Naesudong-ro,
Heungdeok-gu, Cheongju, South Korea
e-mail: seong.do@gmail.com

J. Yoo (✉)
School of Information and Communication Engineering, Chungbuk National University and
CBITRC, 52 Naesudong-ro, Heungdeok-gu, Cheongju, South Korea
e-mail: yjs@chungbuk.ac.kr

S.-S. Yeo et al. (eds.), *Computer Science and its Applications*,
Lecture Notes in Electrical Engineering 203, DOI: 10.1007/978-94-007-5699-1_78,
© Springer Science+Business Media Dordrecht 2012

in non-uniform large wireless sensor network environments. Our proposed scheme can decrease the number of orphan nodes and has sufficient scalability because our scheme selects member nodes by dividing an angle range in the regular hop count based on min-distance hop count. It is shown through performance evaluation that the orphan nodes of the proposed scheme decrease about 41 % on average over those of the existing methods.

Keywords Non-uniform large wireless sensor networks · Multi-hop clustering · Network lifetime · Energy efficiency

1 Introduction

In recent, with the development of the computing technology and the semiconductor MEMS technology, the minimization, low cost, and low power of a sensor node used in wireless sensor networks have become possible. These sensor nodes have a RF communication module and sensor modules which can collect environmental information such as temperature, humidity and illumination.

Sensor networks collect various information about interesting areas [1, 2]. The sensor networks have extended the range of various applications such as observations of land and aquatic animal habitats, monitoring of military purposes, and mobile healthcare [3]. However, the sensor nodes in the sensor network have several hardware restrictions such as the limited wireless communication bandwidth, low computing power and limited energy due to a small built-in battery.

The existing cluster-based data aggregation schemes have been proposed to efficiently aggregate similar collected values. The existing cluster-based schemes select a cluster head based on particular conditions. The cluster with the cluster head consists of the geographically close sensor nodes [4].

The existing clustering schemes form unbalanced clusters, since they do not consider the non-uniform dissemination of sensor nodes. Therefore, they cause the unbalanced energy consumption in the sensor network with the limited energy. The multi-hop clustering scheme based on a min-distance hop count was proposed to solve such problems [7]. However, in [7], some nodes are not contained as the member nodes of clusters when the clusters are constructed in large sensor network environments. The nodes are called orphan nodes. Since the orphan nodes cause the unbalanced energy consumption in the sensor network, the lifetime of the sensor network is shortened.

In this paper, we propose a novel multi-hop clustering scheme based on the min-distance hop count to solve the problems of the existing schemes in non-uniform large sensor network environments. Our proposed scheme increases energy efficiency because our scheme minimizes orphan nodes by dividing a angle range in regular hop count based on the min-distance hop count. It is shown through performance evaluation that the proposed scheme outperforms existing scheme.

The remainder of this paper is organized as follows. Section 2 presents the problems of the existing clustering schemes. Section 3 presents the characteristics and processes of the proposed scheme. Section 4 shows the superiority of the proposed scheme through performance evaluation. Finally, conclusions and future works are described in Sect. 5.

2 Related Work

Various works have been proposed to form efficient clusters in sensor network environments. Low Energy Adaptive Clustering Hierarchy (LEACH) [5] is a typical single-hop clustering scheme.

LEACH arbitrarily elects cluster heads every period to disperse the energy consumption of cluster heads in the cluster interior. It consists of clusters with member nodes that are geographically close nodes. LEACH determines the probability that each node becomes a cluster head periodically according to the number of nodes and fields known in advance. LEACH elects cluster-head nodes at the regular time to prolong the network lifetime. However, LEACH can form unbalanced clusters because it constructs clusters using the stochastic methods. It also does not consider energy when clusters are formed.

Multi-hop Cluster based Backbone Tree (MCBT) [6] which considers the residual energy of a node and the degree through the Max–Min D(hop-count) cluster algorithm forms the multi-hop clusters. Each node calculates a flooding value using its own residual energy and degree. It performs the Max–Min process D(hop-count) times when clusters are formed. After each node compares the saved flooding values in the Max–Min phase, nodes are chosen as the cluster head. However, MCBT forms unbalanced clusters depending on the density of sensor nodes in non-uniform wireless sensor networks.

Min-Distance Hop Count clustering [7] forms balanced clusters in non-uniform wireless sensor network environments. Min-Distance Hop Count clustering selects the closest node from a randomly selected point as a cluster head and then divides six direction areas of the hexagonal shape by the initial cluster head as the center. The initial cluster head selects member nodes using min-distance hop count information based on logical hop-counts in angle range and forms balanced clusters. When the scalability of clusters is demanded in large sensor network environments the angle range increasingly widens so orphan nodes which are not included as the member nodes of clusters occur. As a result, the lifetime of a whole sensor network is shortened because of the unbalanced energy consumption of the sensor nodes.

Fig. 1 Message structure of network initialization

3 The Proposed Clustering Scheme

The proposed scheme performs the phase of a network initialization process to form clusters. As shown in Fig. 1, the sensor nodes prepare initialized messages which have their own information in the network initialization process. Each sensor node broadcasts initialized messages to all of the neighbor nodes within a 1-hop communication range. All of the nodes which receive initialized messages can get the location information of neighbor nodes within a 1-hop range. This information is used as the basic information to form clusters.

When constructing clusters, we choose the closest node to the arbitrary coordinate as an initial cluster head. The selected node as the initial cluster head sets up the cluster head (CH) and other nodes are member nodes (CM) of the cluster. The initial cluster head divides the range into six direction areas of hexagonal shape.

In the existing clustering schemes, circular clusters have been commonly used. However, they have the overlapped areas and cause the uncovered areas in the communication radius. Therefore, the proposed scheme forms hexagonal clusters [8] because the hexagon is an ideal shape for clustering a large terrain into adjacent, non-overlapping areas.

The proposed scheme is based on [7] and elects a cluster head and then divides angle ranges with the initial cluster head as a center. If the ranges are determined, the cluster head sends a cluster join message to the closest 1-hop node.

Figure 2 shows how logical 5-hop member nodes are selected in the particular hop-count using a min-distance hop count within an angle range between 120° and 180°.

CH_1, the initial cluster head, sends the message of cluster participation request to the nearest 1-hop node N_2 such as {CH_1, N_1, 1, 120–180} between 120° and 180°. The N_2 that receives the message sends the reply message to the initial cluster head and then after increasing the hop-count information by 1, it sends {CH_1, N_2, 2, 120–180} to its nearest neighbor node N_3. The N_3 that receives the message of cluster participation request sends the reply message to the N_2.

In order to select member nodes, the process to send and receive the message of cluster participation request is repeated until 5-hop member nodes whose hop-count information is 5 are selected. The proposed scheme selects the divided nodes by the criterion <2^1, 2^2, 2^3, ..., 2^n>, divides the angle range by these nodes and searches their nearest nodes in the divided ranges. After the angle range is divided by 2-hop node N_3 between 120° and 180°, the N_3 sends {CH_1, N_3, 3, 120–150} that the hop-count is added by 1 to the nearest node N_4 between 120° and 150°. It also sends the message of cluster participation request such as {CH_1, N_3, 3,

Fig. 2 Member node selection of cluster

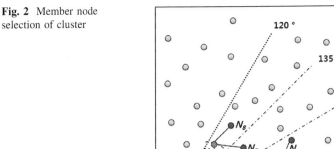

● cluster head   ○ member node   ◉ member node (divided)   ⊘ member node (terminal)

Fig. 3 Formation of initial cluster

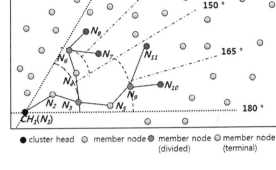
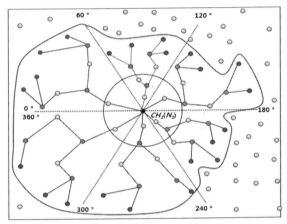

150–180} to the nearest node N_5 between 150° and 180°. The N_4 and N_5 as 3-hop nodes send the messages of cluster participation request to the nearest nodes within their own ranges. The 4-hop nodes N_6 and N_9 as the divided nodes send the messages of cluster participation request to the nearest nodes within the divided ranges. The N_7, N_8, N_{10} and N_{11} as last 5-hop nodes are selected within total four angle ranges. The 5-hop nodes which are recognized as terminal nodes through the hop-count information in the cluster participation request message send the information of member nodes within the angle range to cluster head using the path of messages. The cluster head which receives the reply message stores the node's ID which sends the reply.

When all member nodes are selected through these processes within six areas angle ranges, initial clusters are constructed as shown in Fig. 3.

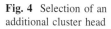 **Fig. 4** Selection of an
additional cluster head

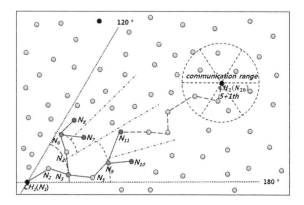

After constructing initial clusters, the N_{11} that the distance from the cluster head is the farthest among terminal nodes near the central axis of an angle range sends the cluster head request message to the (5+1)th node as shown in Fig. 4. The node N_{20} that receives the message becomes a new cluster head. The cluster heads receive messages from the member nodes of the clusters. The cluster heads assign time slots which can send the data based on the number of nodes in the clusters and make TDMA scheduling tables. And then the cluster heads broadcast the scheduling tables.

If the residual energy of a particular cluster head is decreased under the threshold in the proposed scheme, of the clusters is reconstructed. Cluster heads that have energy values under threshold compared to initial energy values at the time of constructing clusters forwards messages for reconstruction to all nodes in the sensor network.

After all of the sensor nodes receive messages for cluster reconstruction, cluster heads determine the delay times of all nodes in initial clusters according to residual energy and make schedules that send cluster heads beacon messages in sequence according to the residual energies of terminal nodes. That is, first of all, the sensor node which has the highest residual energy sends a beacon message to the cluster head. Cluster heads receiving the beacon messages inform terminal nodes that new cluster heads were elected for reconstruction and new cluster heads perform cluster reconstruction based on the elected cluster heads.

4 Performance Evaluation

We have developed a simulator to evaluate our proposed scheme and the existing proposed MCBT, Min-Distance Hop Count schemes. The simulator based on simulator J-Sim 0.6.0 version is performed using performance evaluation parameters shown in Table 1 [9].

Table 1 Performance evaluation parameters and values

Parameter	Value
Size of sensor network (m × m)	1000 × 1000
Radius of communication (m)	20
Number of distributed sensor nodes (EA)	700–1500
Initial energy per nodes (J)	2.5
Size of data packet (Bytes)	16

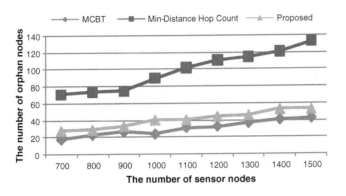

Fig. 5 The occurrence of orphan nodes according to the number of sensor nodes

In [5], when 5 % nodes of entire sensor nodes are elected as cluster heads, the energy efficiency of the sensor network is the best. In the proposed scheme, clusters are constructed based on 5-hop count by considering large sensor network environments. The optimal hop count of clusters depends on various network environments.

Figure 5 shows the occurrence of orphan nodes of the proposed scheme and the existing scheme according to the number of sensor nodes. In the case of MCBT scheme, when the neighbor nodes do not exist within communication range, orphan nodes occur. In because Min-Distance Hop Count clustering, the size of angle range is increased by increase of hop-count from initial cluster heads because Min-Distance Hop Count clustering selects member nodes using angle range so selecting member nodes of min-distance causes the increase of occurrence of orphan nodes. In contrast, our proposed scheme can decrease the number of orphan nodes because our scheme selects member nodes by dividing angle range in the regular hop count. Performance evaluation shows that the orphan nodes of the proposed scheme decrease about 41 % on average over those of the existing methods.

Figure 6 shows network lifetime of the proposed scheme and the existing scheme according to the number of sensor nodes. In the case of MCBT, unbalanced clusters are formed according to density in non-uniform sensor network. In the case of Min-Distance Hop Count clustering, scalability of clusters causes the number of orphan nodes because Min-Distance Hop Count clustering selects

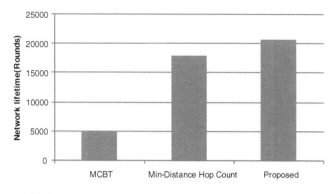

Fig. 6 Network lifetime

clusters by angle range in non-uniform wireless sensor network. In contrast, our proposed scheme can form balanced clusters in non-uniform wireless sensor network because our scheme selects member nodes by dividing angle range in the regular hop count. It is shown through performance evaluation that network lifetime increases by about 7 % compared with Min-Distance Hop Count scheme and by about 59 % compared with MCBT on average.

5 Conclusion

In this paper, we proposed a hierarchical multi-hop clustering scheme for energy efficient data collection to minimize the number of orphan nodes in non-uniform large wireless sensor networks. The proposed scheme divides angle range which use existing scheme based on nodes of constant hop-count to solve occurrence of orphan nodes and then minimizes the probability that occur orphan nodes through selecting each of member nodes in divided angle range. It is shown through performance evaluation that the orphan nodes of the proposed scheme decrease about 41 % on average over those of the existing methods and increase the whole sensor network lifetime about maximum 7 %. In future work, we will perform the experiment about optimal cluster size according to various environments and propose uniform clustering scheme in frequently changeable environments of network topology.

Acknowledgments This work was supported by Basic Science Research Program through the National Research Foundation of Korea(NRF) grant funded by the Korea government(MEST)(No. 2009-0089128) and the Leaders in INdustry-university Cooperation(LINC) Program through the National Research Foundation of Korea(NRF) funded by the Ministry of Education, Science and Technology (No.2012-B-0013-010112).

References

1. Akyildiz, F., Su, W., Sankarasubramaniam, Y., Cayirci, E.: A survey on sensor networks. IEEE Commun. Mag. **40**(8), 102–114 (2002)
2. Park, Y., Yoon, J., Seo, D., Kim, J., Yoo, J.: A time-parameterized data-centric storage method for storage utilization and energy efficiency in sensor networks. J. KIISE Databases **36**(2), 99–111 (2009)
3. Cerpa, A., Elson, J., Estrin, D., Girod, L., Hamilton, M., Zhao, J.: Habitat monitoring: application driver for wireless communications technology. In: Proceedings of the ACM Workshop on Data Communications in Latin America and the Caribbean, pp. 20–41 (2001)
4. Younis, O., Fahmy, S.: Distributed clustering in adhoc sensor networks: a hybrid, energy-efficient approach. In: Proceedings of the Annual Joint Conference of the IEEE Computer and Communications Societies (IEEE INFOCOM 2004), pp. 366–379 (2004)
5. Heinzelman, W.R., Chandrakasan, A.P., Balakrishnan, H.: Energy-efficient communication protocol for wireless microsensor networks. In: Proceedings of the Hawaii International Conference on System Sciences, pp. 3005–3014 (2000)
6. Shin, I., Kim, M., Mutka, M.W., Choo, H., Lee, T.: MCBT: multi-hop cluster based stable backbone trees for data collection and dissemination in WSNs. Sensors **9**, 6028–6045 (2009)
7. Kim, E., Kim, D., Seong, D., Yoo, J.: Min-distance hop count based balanced multi-hop clustering in non-uniform wireless sensor networks. In: Proceedings of the KIISE Korea Computer Congress 2011, vol. 38, no. A, pp. 416–419 (2011)
8. Wang, D., Xu, L., Peng, J., Robila, S.: Subdividing hexagon-clustered wireless sensor networks for power-efficiency. In: International Conference on Communications and Mobile Computing, pp. 454–458 (2009)
9. J-Sim 0.6.0., http://www.j-sim.zcu.cz/

An Energy-Efficient Data Compression and Transmission Scheme in Wireless Multimedia Sensor Networks

Junho Park, Dong-ook Seong, Byung-yup Lee and Jaesoo Yoo

Abstract In recent years, the demand of multimedia data in wireless sensor networks has been significantly increased for the environment monitoring applications that utilize sensor nodes to collect multimedia data such as sound and video. However, the amount of multimedia data is very large. Therefore, if the data transmission schemes in traditional wireless sensor networks are applied in wireless multimedia sensor networks, the network lifetime is significantly reduced due to excessive energy consumption on particular nodes. In this paper, we propose a novel energy-efficient data compression scheme for multimedia data transmission in wireless sensor networks. The proposed scheme compresses and splits the multimedia data using the Chinese Remainder Theorem (CRT) and transmits the bit-pattern packets of the remainder to the base station. As a result, it can reduce the amount of the transmitted multimedia data. To show the superiority

J. Park
School of Information and Communiacation Engineering,
Chungbuk National University, 52 Naesudong-ro,
Heungdeok-gu, Cheongju, South Korea
e-mail: junhopark@chungbuk.ac.kr

D. Seong
BOAS Electronics Inc., Industrial Technology Research Park, 52 Naesudong-ro,
Heungdeok-gu, Cheongju, South Korea
e-mail: seong.do@gmail.com

B. Lee
Department of E-Business, Paichai University,
155-40 Baejae-ro (Doma-Dong), Seo-Gu, Daejeon, South Korea
e-mail: bylee@pcu.ac.kr

J. Yoo (✉)
School of Information and Communiacation Engineering, Chungbuk National University
and CBITRC, 52 Naesudong-ro, Heungdeok-gu, Cheongju, South Korea
e-mail: yjs@chungbuk.ac.kr

S.-S. Yeo et al. (eds.), *Computer Science and its Applications*,
Lecture Notes in Electrical Engineering 203, DOI: 10.1007/978-94-007-5699-1_79,
© Springer Science+Business Media Dordrecht 2012

of our proposed scheme, we compare it with the existing scheme. Our experimental results show that our proposed scheme reduces about 71 % in the amount of transmitted data and increases about 188 % in the ratio of surviving nodes over the existing scheme on average.

Keywords Wireless multimedia sensor networks · Chinese remainder theorem · Energy efficiency

1 Introduction

In recent, with the development of hardware technologies and monitoring schemes, the applications for gathering multimedia data such as sound and image using multimedia sensors have been increased [1]. As the multimedia data are very large over simple data in traditional sensor networks, the network lifetime of the sensor network is significantly reduced due to excessive energy consumption in particular nodes for transmitting the data. In addition, the multimedia data increase the data transmission time and decline the data reception ratio. Consequently, the existing schemes based on the traditional sensor networks are not suitable for the environments to collect the multimedia data. Therefore, the multi-path transmission schemes have been used in the wireless multimedia sensor networks [1, 2].

The existing multi-path transmission schemes split the data into several segments and send them via several upper level nodes, in order to solve the problem of excessive energy consumption on particular nodes. Moreover, they distribute the load over the entire network during data transmission process [2, 3]. However, as the existing schemes split data into several segments, there are still large energy consumption and load on particular nodes due to the characteristics of the multimedia data. It is necessary to use compression schemes to alleviate such problems [4]. However, most of the existing compression schemes for sensor data are based on signal compression such as wavelet and variable quantization and code compression. These studies are not suitable for the environments based on wireless sensor networks. Also, the compression schemes specialized wireless multimedia sensor networks is in its infancy. Therefore, it is necessary to study an energy-efficient multimedia data compression and transmission scheme considering the characteristics of the wireless multimedia sensor networks.

In this paper, we propose an energy-efficient data compression and transmission scheme in wireless multimedia sensor networks. The proposed scheme splits and compresses the sensing multimedia data based on the Chinese Remainder Theorem (CRT) [5] algorithm considering their characteristics. Moreover, the proposed scheme decides the number of segments to be sent via each node based on the remaining energy of the nodes of the upper level in the path of the sending data. By doing so, it is possible to consume the balanced energy among the sensor nodes. As a result, the lifetime of the sensor network is increased.

The remainder of this paper is organized as follows. Section 2, we present our energy-efficient data compression and transmission scheme in wireless multimedia sensor networks. Section 3 shows the simulated experiments and compares the existing scheme with the proposed scheme. Finally, we present concluding remarks in Sect. 4.

2 The Proposed Multimedia Data Compression and Transmission Scheme

In this paper, we propose an energy-efficient data compression and transmission scheme to extend the network lifetime by reducing the energy consumption and load on particular nodes. The proposed scheme consumes the energy of the entire network in balance by transmitting the split packets via the multi-paths considering remaining energy of the upper level nodes. At first, the source node recognizes the number of nodes on the path to transmit the sensor readings and splits massive multimedia data into the segments based on the Chinese Remainder Theorem (CRT) algorithm. Then, it transmits the packets by the remaining energy of the upper nodes.

In order to carry out the proposed scheme, it is necessary to identify the transfer nodes during data transmission from the source node to the base station. Thus, all nodes identify information of the transfer nodes to send the multimedia data through the *network initialization stage*. The proposed scheme identifies routing levels on the basis of the number of hops for the nodes participating in data transmission. It also finds the level that the maximum number of nodes exist, called *Maximum Transfer Level (MTL)*, during routing process. The source nodes utilize *MTL* for data partition (Fig. 1).

Figure 2 shows the data partition of the proposed scheme. At the first, the source nodes collect the remaining energy of the transfer nodes at *level_N* corresponding to *MTL* that belongs to on the path from the source node to the base station. The number of split segments can be gotten by calculating the greatest common measure (GCM) after carrying out the simplification of the collected remaining energy information. It defines the number that is obtained by diving the remaining energy by the least common measure (LCM) as the final number of split segments and carries out the data partition based on it.

Figure 3 shows how the multimedia data using the Chinese Reminder Theorem (CRT) algorithm are partitioned. If the size of the original data is 40 Bits and the number of split segments is 7, the set of minimum prime numbers becomes {*43, 47, 53, 59, 61, 67, 71*}. For data of each packet, the bit-pattern of remainder by splitting the original data by the minimum prime number set is sent. As the remainder is always smaller than the divisor from the characteristics of modular operation, it is possible to get larger energy gain than transmitting the original data.

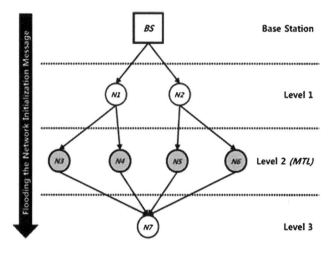

Fig. 1 Network initialization

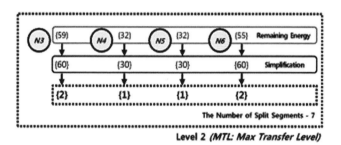

Fig. 2 Establishment of the number of data separation

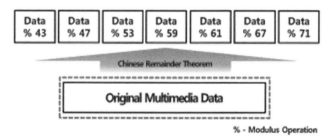

Fig. 3 Multimedia data separation

Table 1 Simulation parameters

Parameters	Values
Size of sensor network fields (m × m)	200 × 200
Number of distributed sensor nodes (EA)	200
Location of base station (X_{coord}, Y_{coord})	(0, 0)
Radius of communication (m)	18
Size of multimedia data (Bytes)	21

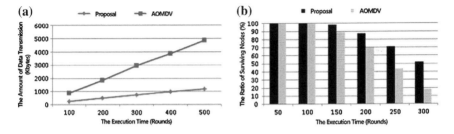

Fig. 4 Performance comparison. **a** The amount of data transmission, **b** the ratio of surviving nodes

3 Performance Evaluation

We have developed a simulator based on JAVA to evaluate our proposed scheme and the existing multimedia data transmission scheme without the compression algorithm [3]. We assume that 200 sensors are deployed uniformly in 200 × 200 (m) network field. The energy consumption for sending a message is determined by a constant function $S \cdot (C_t + C_a \cdot D^2)$, where S is the message size, C_t is the transmission cost, C_a is the amplification cost, and D is the distance of message transmission. We set $C_t = 50$ nJ/b and $C_a = 100$ pJ/b/m^2 in the simulation. The energy consumption for receiving a message is determined by a cost function ($S \cdot C_r$), where S is the message size and C_r is the transmission cost. We set $C_r = 50$ nJ/b in the simulation (Table 1).

Figure 4 shows the amount of the transmitted data and the ratio of the surviving nodes of the proposed scheme and the existing multimedia transmission scheme, called AOMDV over the execution times. Our proposed scheme compresses the multimedia data using the Chinese Remainder Theorem (CRT) algorithm and transmits the bit-pattern payload of the remainder to the base station. Therefore, the proposed scheme can reduce the amount of the transmitted data and the size of a packet using the data partition. In the result, our scheme reduces by about 71 % data transmission and increases the ratio of surviving nodes by about 188 % over the existing scheme on average.

4 Conclusion

In this paper, we have proposed a novel energy-efficient data compression scheme for multimedia data transmission in wireless multimedia sensor networks. Our proposed scheme splits and compresses the multimedia data using the Chinese Remainder Theorem (CRT) on the basis of the number of nodes in maximum transfer level and the remaining energy. And then it transmits the bit-pattern of the remainder to the base station. In addition, the entire energy could be balanced by differentiating the number of transmitted packets considering the remaining energy of the transfer nodes belonging to the upper level. Therefore, the proposed scheme can reduce the amount of data transmission and improve the network lifetime. As the results of performance evaluation, the proposed scheme showed that the amount of data transmission was reduced by about 71 % and the ratio of surviving nodes was increased by about 188 % on average over the existing scheme without the compression scheme. In the future work, we plan to extend our work to prolong the lifetime of a sensor network by making a detour route in consideration of the remaining energy of entire nodes.

Acknowledgments This work was supported by Basic Science Research Program through the National Research Foundation of Korea (NRF) grant funded by the Korea government (MEST) (No. 2009-0089128) and the Leaders in Industry-university Cooperation (LINC) Program through the National Research Foundation of Korea (NRF) funded by the Ministry of Education, Science and Technology (No.2012-B-0013-010112).

References

1. Akyildiz, I.F., Melodia, T., Chowdhury, K.R.: A survey on wireless multimedia sensor networks. Comput. Netw. **51**(4), 921–960 (2007)
2. Ehsan, S., Hamdaoui, B.: A survey on energy-efficient routing techniques with QoS assurances for wireless multimedia sensor networks. IEEE Commun. Surv. Tutor. **PP**(99), 1–14 (2011)
3. Yousef, C., Naoka, W., Masayuki, M.: Network-adaptive image and video transmission in camera-based wireless sensor networks. In: Proceedings of the ACM/IEEE Conference on Distributed Smart Cameras, pp. 336–343 (2007)
4. Chew, L.W., Ang, L.-M., Seng, K.P.: Survey of image compression algorithms in wireless sensor networks. In: Proceedings of the International Symposium on Information Technology (ITSim '08), pp. 1–9 (2008)
5. Chen, Y.-S., Lin, Y.-W.: C-MAC: an energy-efficient MAC scheme using Chinese-Remainder-Theorem for wireless sensor networks. In: Proceedings of IEEE International Conference on Communications, pp. 3576–3581 (2007)

A Sensor Positioning Scheme in Irregular Wireless Sensor Networks

Hyuk Park, Donggyo Hwang, Junho Park, Dong-ook Seong and Jaesoo Yoo

Abstract In wireless sensor networks, the positions of sensor nodes are very important for many applications. If each sensor node provides information with less positioning error, the positioning information will be reliable. One of the most representative positioning schemes, called DV-HOP is low for positioning accuracy in irregular network environments. Moreover, because it requires many anchor nodes for high accuracy, it is expensive to construct the network. To overcome such problems, we propose a novel sensor positioning scheme in irregular wireless sensor networks. By doing so, the proposed scheme ensures the high accuracy of sensor positioning in non-uniform networks. To show the superiority of our proposed scheme, we compare it with the existing scheme. Our experimental results show that our proposed scheme improves about 36 % sensor positioning accuracy over the existing scheme on average even in irregular sensor networks.

H. Park · D. Hwang · J. Park
School of Information and Communication Engineering, Chungbuk National University,
410 Seongbong-ro, Heungdeok-gu, Cheongju, Chungbuk, South Korea
e-mail: agodsun@naver.com

D. Hwang
e-mail: corea1985@gmail.com

J. Park
e-mail: junhopark@chungbuk.ac.kr

D. Seong
BOAS Electronics Inc., Industrial Technology Research Park, 52 Naesudong-ro,
Heungdeok-gu, Cheongju, South Korea
e-mail: seong.do@gmail.com

J. Yoo (✉)
School of Information and Communication Engineering, Chungbuk National University and
CBITRC, 52 Naesudong-ro, Heungdeok-gu, Cheongju, South Korea
e-mail: yjs@chungbuk.ac.kr

S.-S. Yeo et al. (eds.), *Computer Science and its Applications*,
Lecture Notes in Electrical Engineering 203, DOI: 10.1007/978-94-007-5699-1_80,
© Springer Science+Business Media Dordrecht 2012

Keywords Wireless sensor network · Positioning · Central limit theorem

1 Introduction

The wireless sensor network that is one of basic technologies to detect the event and to control the external human environment in the ubiquitous environment has been vigorously studied. The ad-hoc wireless sensor network is constructed autonomously and collects the diverse environment information through the communication between sensor nodes. The sink node receives the sensing values from the sensor nodes in the sensing area and transmits them to the user. The collected information is used for diverse purposes such as observation of wildlife's habitat, military affair, fire detection, environmental monitoring, medical service, and U-city [1].

In the sensor network, the positioning technology is one of the most required and basic technologies. In the positioning scheme using the wireless devices, every equipment is generally carrying the Global Positioning System (GPS) in order to collect the positioning information [2]. But it causes the problems such as excessive energy consumption due to GPS modules and high costs for their construction in the large-scale of the sensor network environment. Therefore, the positioning schemes to reduce the energy consumption in the sensor network with limited energy have been actively studying.

The recently studied typical positioning schemes are range-free schemes [3, 4]. Range-free schemes measure the distance and estimate the position through the connection information between nodes and the position information of an anchor node. In addition, it is efficient in the energy consumption and the cost to construct the network because only anchor node equipped GPS module. Therefore, the positioning schemes through the anchor nodes have been actively studying. The existing schemes estimated the distance between nodes and decided the position in the uniform sensor network environments without considering density. However, in real applications, since the sensors are distributed on the sensing fields randomly through aircrafts, missile, and so on the irregular sensor network environments are constructed in specific areas. Therefore, the positioning schemes for the uniform sensor network environments are not suitable for the actual situations since their error rates of density probability are very high in the irregular sensor network environments.

To solve the problems of the existing range-free schemes, we propose a novel positioning scheme by using the density probability model in the irregular network environment. In the proposed scheme, the minimum anchor nodes are used and the distance is estimated according to the density in the irregular sensor network environments. By doing so, the cost to construct the sensor network can be minimized and the positioning precision can be improved.

The remainder of this paper is organized as follows. Section 2, we present our sensor positioning scheme using density probability models in irregular wireless

Table 1 Shortest path and neighbor node list

Shortest path ID	Shortest path hop	Neighbor list
n_i	*Cumulated hop*	$n_i, n_{i+1}, \ldots, n_{i+n}$

sensor networks. Section 3 shows the simulated experiments and compares the existing scheme with the proposed scheme. Finally, we present concluding remarks in Sect. 4.

2 Neighbor Density Probability Positioning Scheme

In this paper, we propose a novel positioning scheme to reduce the positioning error and to decrease the construction costs in the irregular distributed sensor networks. The proposed scheme uses at least 4 anchor nodes placed at the boundary of the sensing fields. Thereby, the proposed scheme minimizes the cost of construction of the sensor network.

2.1 Network Model

The anchor nodes A1, A2, A3 and A4 are deployed in each corner of the sensing area. In the initial step, anchor nodes broadcast their positioning information messages (node ID, hop, coordinates) to all the nodes. The normal sensor nodes save the information of the anchor nodes and neighbor nodes like Table 1.

2.2 Distance Estimation Considering Neighbor Density Probability

Each node estimates 1-hop distance by using Central Limit Theorem [5] based on a normal probability distribution. The normal distribution or Gaussian distribution is a continuous probability distribution that has a bell-shaped probability density function. If the number of trials or samples objects increases, it shows the normal distribution curve. The theory that sensor network environment is consistent with the normal distribution model is the central limit theorem. The normal distribution model becomes approximate to average μ as the number of samples increases. In the sensor network environment that the thousands of sensors are deployed, samples are located in the center of the normal distribution curve. Therefore, on the basis of the central limit theorem and the normal distribution model, each sensor node estimates the distances to the neighboring nodes. For 1-dimension,

Fig. 1 The estimated
position of a sensor node

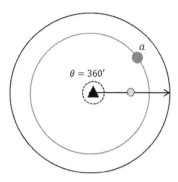

there is a point of the specific node that is an average of zero point. In other words, the point of 1/2 of the communication radius is the probability that node exists.

As shown in Fig. 1, if the node draws a circle for the communication range and angle ($\theta = 360°$), it is farther away than the estimated position of 1-dimension. Therefore, the estimated position for 2-dimension unlike the position of 1-dimension is a point in the circle that the area of its inside circle is equal to the area of its outside circle. As a result, it is possible to estimate the distance between nodes through the values of the normal distribution table.

Equation (1) is the distance calculation equation between neighbor nodes based on the values of the normal distribution table. R is a communication range of a sensor node and n is the number of its neighbor nodes. And then, the entire sensor nodes estimate the distance between them and their neighbor nodes through Eq. (1). It make possible to estimate more real distance in the irregular sensor network that the areas have different densities each other.

$$
\begin{aligned}
d_{Est(k)} &= \sqrt{\frac{\pi r^2/3}{\pi}} = \sqrt{\frac{r^2}{3}}, \quad n = 2; \\
d_{Est(k)} &= \sqrt{\frac{r^2}{n+1}}, \quad n \geq 3;
\end{aligned}
\tag{1}
$$

3 Performance Evaluation

We have developed a simulator based on JAVA to evaluate our proposed scheme and the existing scheme, DV-Hop [4]. The sensor network is based on random and irregular (Gaussian) models by considering the real distribution characteristics of sensors. The communication ranges for sensor nodes and anchor nodes are 10 m, 15 m, 20 m, 25 m and 30 m. The performance evaluation is performed based on J-Sim v.0.6.0. [6].

A positioning error rate is difference between a real coordinate and an estimated coordinate. Therefore, we evaluate the accuracy as the distance error rate between

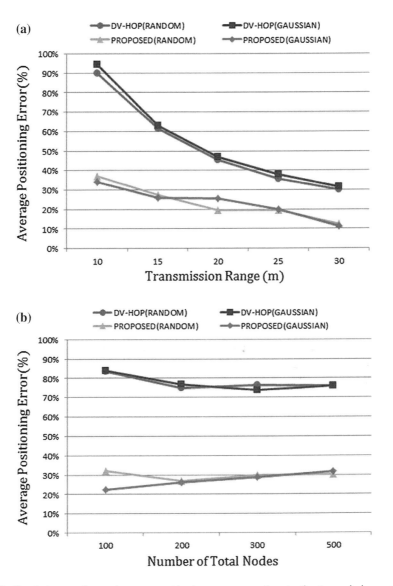

Fig. 2 Simulation results. **a** Average positioning error according to the transmission range, **b** average positioning error according to the number of total nodes

the real coordinate of a node and the coordinate of its estimated position. Equation (2) for a real distance and a positioning coordinate is used.

$$\text{Position Error}(\% \; R) = \frac{\sqrt{(x' - x)^2 + (y' - y)^2}}{r_{max}} \times 100 \qquad (2)$$

Figure 2a shows the average positioning error rate according to the communication range. With the same communication range, the position error rate of our proposed scheme is smaller than that of DV-HOP. In the case of DV-HOP, since the hop distance is proportional to the communication range, the accuracy of its positioning is low. In the random mode, the proposed scheme achieves about 30 % performance improvements over DV-HOP in terms of the positioning accuracy. In the irregular model that has very large deviation of density, the proposed scheme achieves about 36 % higher accuracy than DV-HOP. The reason is that the proposed scheme measures the positions of the nodes by considering their densities. As a result, our scheme improves the accuracy of positioning over the existing scheme, DV-Hop as a whole. As shown in Fig. 2b, our scheme improves the accuracy of positioning by about 49 % over DV-Hop on average. Unlike DV-Hop, the proposed scheme shows the high position accuracy as each node has a different 1-hop distance. As a result, our scheme has an advantage that it can be applied to various environments because it has high accuracy in the large scale network as well as the small scale network.

4 Conclusion

In this paper, we proposed a sensor positioning scheme using density probability models in irregular network environments. Our proposed scheme minimizes the construction cost of the sensor network by using only 4 anchor nodes. The proposed scheme estimates the distances from a node to neighboring nodes using the characteristics of density in irregular sensor network environments. The proposed scheme performs error correction between the estimated distance and the real distance. It was shown through performance evaluation that the positioning accuracy of the proposed scheme was significantly improved over that of the existing scheme, DV-Hop. In the future work, we plan to extend our work to estimate the positions of sensor nodes in the case of network-hole occurrence.

Acknowledgments This work was supported by Basic Science Research Program through the National Research Foundation of Korea (NRF) grant funded by the Korea government (MEST) (No. 2009-0089128) and the Leaders in INdustry-university Cooperation (LINC) Program through the National Research Foundation of Korea (NRF) funded by the Ministry of Education, Science and Technology (No.2012-B-0013-010112).

References

1. Yick, J., Mukherjee, B., Ghosal, D.: Wireless sensor network survey. Comput. Netw. (Elsevier) **52**(12), 2292–2330 (2008)
2. Schlecht, E., Hulsebusch, C., Mahler, F., Becker, K.: The use of differentially corrected global positioning system to monitor activities of cattle at pasture. Appl. Anim. Behav. Sci. (Elsevier) **85**(3–4), 185–202 (2004)

3. He, T., Huang, T., Blum, B.M., Stankovic, J.A.: Range-free localization schemes for large scale sensor networks. In: International Conference on Mobile Computing and Networking, pp. 81–95 (2003)
4. Niculescu, D., Nath, B.: DV Based positioning in ad hoc networks. Telecommun. Syst. (Springer) 22(1–4), 267–280 (2003)
5. William, A.: Central limit theorem, pp. 467–487. International Encyclopedia of the Social Sciences Publisher, London (2008)
6. J-Sim, http://www.j-sim.zcu.cz/

Secure Multipath Routing for WMSN

Sangkyu Lee, Junho Park, Dongook Seong and Jaesoo Yoo

Abstract In recent years, the requirements on the high quality environment monitoring by using the sensor nodes which can handle the multimedia data in WSN have been increased. However, because the volume of multimedia data is tremendous, the limited bandwidth of a wireless channel may incur the bottleneck of a system. To solve such a problem, most of the existing distributed multi-path routing protocols based on multimedia data just focused on overcoming the limited bandwidth in order to enhance the energy efficiency and the transmission rate. However, because the existing methods can not apply a key-based technique to the encryption for the multimedia data, they are very weak for aspects of the security. In this paper, we propose a secure disjoint multipath routing scheme for multimedia data transmission. Since our proposal scheme divides multimedia data (e.g. image) into pixels and send them through disjoint multipath routing, it can provide security to the whole network without using the key-based method. Our experimental results show that our proposed scheme reduces about 10 % the

S. Lee · J. Park
School of Information and Communication Engineering, Chungbuk National University,
410 Seongbong-ro, Heungdeok-gu, Cheongju, Chungbuk, South Korea
e-mail: jakysn4496@gmail.com

J. Park
e-mail: junhopark@chungbuk.ac.kr

D. Seong
BOAS Electronics Inc., Industrial Technology Research Park, 52 Naesudong-ro,
Heungdeok-gu, Cheongju, South Korea
e-mail: seong.do@gmail.com

J. Yoo (✉)
School of Information and Communication Engineering, Chungbuk National University and
CBITRC, 52 Naesudong-ro, Heungdeok-gu, Cheongju, South Korea
e-mail: yjs@chungbuk.ac.kr

S.-S. Yeo et al. (eds.), *Computer Science and its Applications*,
Lecture Notes in Electrical Engineering 203, DOI: 10.1007/978-94-007-5699-1_81,
© Springer Science+Business Media Dordrecht 2012

amount of the energy consumption and about 65 % the amount of the missed data packets caused by malicious nodes over the existing scheme on average.

Keywords Wireless sensor networks · Multimedia data · Distributed multipath routing · Security

1 Introduction

In recent years, with the development of wireless communication technologies based on wireless sensor nodes, various applications of sensor networks have been studied. Sensor nodes in a sensor network have various sensor modules which can gather information such as temperature, humidity, sound, and motion. Sensor networks perform status monitoring, information transmission and cooperative work between hundreds and thousands of nodes in order to collect data that is otherwise difficult for us to directly approach in various environments. They are used in a variety of fields such as environmental monitoring, object tracking, and intrusion detection [1]. Normally sensor nodes operate by the battery with a low capacity and it is almost impossible to change once the sensors are distributed. Because of that, excessive energy consumption in a specific node has a bad effect to the whole network. Therefore, many energy-efficient schemes for lifetime of the sensor network have been studied.

As the applications of sensor networks increase, the importance of security is also on the rise. Wireless sensor networks are very vulnerable to attacks such as wireless signal snooping, and message-copying. These security problems in sensor networks may lead to serious problems in the military and industry. Therefore, security in sensor networks should be considered. However, because each of the sensor nodes has limited energy and computing performance, many studies are being steadily conducted considering these characteristics. The representative scheme is an algorithm based on a key. As it can provide privacy and integrity for data through various applications in sensor networks, there are many studies of security schemes based on keys.

With the remarkable development of devices, sensor nodes basically include sensor modules for temperature and humidity on a main board, various modules can be combined using expansion slots. Especially if it mounts multimedia sensor modules such as cameras or microphones, it enables high-quality environmental monitoring. Therefore, the demand of multimedia data in wireless sensor networks has been significantly increased for environment monitoring. When multimedia data is handled by the existing transmission scheme based on Zigbee which uses low bandwidth, it can lead to a bottleneck, and then energy consumption may intensively increase in the specific nodes. Because the existing transmission scheme based on Zigbee uses low bandwidth, it is very difficult to transmit mul-timedia data which needs Mbps of bandwidth as in the existing scheme. To overcome the limitations of bandwidth, multipath transmission schemes have been proposed. The fact that these load-balancing multipath transmission schemes have reduced the load on the specific node and improved performance of the networks has been proven to be true. However, these transmission schemes actually just consider a distributed routing of data rather than security. As the amount of multimedia data is very large over simple data in traditional sensor networks, the lifetime of the sensor network is significantly reduced due to excessive energy consumption in particular nodes to secure the multimedia data using a key-based security scheme. Therefore, it is necessary to study an energy-efficient multimedia data encryption and transmission scheme considering the characteristics of wireless multimedia sensor networks.

In this paper, we propose a novel secure disjointed multipath routing scheme for multimedia data transmission in wireless sensor networks to overcome the limitations of communication bandwidth on existing distributed multi-path routing networks, and to keep data secure. The proposed scheme divides the image data in pixels and transmits it using different multipath. Therefore, even though a few packets are exposed or lost by malicious nodes, it is impossible to know all of the data. In the results, it is possible to provide security for the entire network without using an existing key-based scheme to use the characteristics of multimedia data and routing.

The remainder of this paper is organized as follows. Section 2 overviews the existing routing schemes for wireless multimedia sensor networks and analyzes the problems. In Sect. 3, we present our novel secure disjointed multipath routing for multimedia data transmission in wireless sensor networks. Section 4 shows the simulated experiments and compares the existing scheme with the proposed scheme. Finally, we present concluding remarks in Sect. 5.

2 Related Works

2.1 The Existing Multipath Routing Schemes

Multimedia data has a large size which is different from traditional sensor networks that handle simple numerical values. If multimedia data was sent through wireless sensor networks, because the frequent packet crashes according to the long time occupation of channel, the network does not perform well.

To solve the problems, many protocols and algorithms have been proposed. All the existing protocols take the minimum energy path, whereas the multi-path routing schemes distribute traffic among multiple paths instead of routing all the traffic along a single path. In multi-path routing, it is necessary to know the number of paths that are needed and choose the appropriate paths in the total number of available paths. Clearly, the number and the quality of the paths selected dictate the performance of a multipath routing scheme. The proposed work is intended to provide a reliable transmission of data for data synchronization at the destination on an environment with low energy consumption. This is done by efficiently utilizing the energy availability and the received signal strength of the nodes to identify multiple routes to the destination.

The proposed protocol spreads the traffic over the nodes lying on different possible paths between the source and the sink in proportion to their residual energy and received signal strength. The rationale behind traffic spreading is by considering the energy so that the overall lifetime of the network can be increased. The sequence number is assigned to each packet of data for data synchronization at the destination. The objective is to assign more loads to under-utilized paths and less load to over-committed paths so that uniform resource utilization of all available paths can be ensured.

Some of the related works in multipath routing are as follows. Most of the multipath routing protocols are extended versions of DSR [2] and AODV. A node-disjoint parallel multipath routing algorithm (DPMR) [3] has two key problems. It relies on the proposed algorithm and has the restrictions of using either clockwise regions or anticlockwise regions, which actually limits the number of routing paths.

Wang et al. [4] maintains additional paths to serve as backup on primary path failure. There is also Network Lifetime Maximization with Node Admission in Wireless Multimedia Sensor Networks [5], but due to stringent QoS requirements, it is not possible to admit all the potential sensor nodes into the network. Therefore, this work addresses the node admission into the network, in order to maximize the network lifetime.

In [6], to prevent the packet loss or delay caused by the limited bandwidth of source nodes and destination nodes, distributed multimedia data are sent to the neighbor nodes using Bluetooth, which has wider bandwidth than Zigbee. Then each neighbor node that receives the distributed data sends the data to the destination using Zigbee. If a multipath overlaps in the particular node, data loss arises because of the limited bandwidth of Zigbee. To solve the overlapping problem, they proposed the disjointed multipath scheme based on the competitive. Through this process, the information of neighboring nodes is listed on the routing table of each sensor node. If the set of all paths is done, the multimedia data are transmitted based on the routing information.

2.2 The Problem of the Existing Scheme

The existing schemes for multimedia data have just been studied from the standpoint of data processing or energy efficiency. The security of the data was not given significant consideration. Although these distributed routing schemes are energy-efficient approaches in data processing or transmission, the overhead is due to the joining of many paths. Also, it is almost impossible to apply a key-based method to secure each sensor node which has limited performance and energy. So

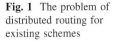

Fig. 1 The problem of distributed routing for existing schemes

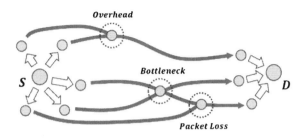

in terms of security, the network is very vulnerable to malicious nodes. For example, [6] provides the video data into the image frame and transfers the frames through the multipath. As a result, a load transferred to each sensor node is reduced so that the lifetime of the network is increased. Although a frame is exposed by a malicious node, the overall image is impossible to recover. However, a malicious node can infer the whole out of part. Even if [6] has the advantage in energy efficiency or limited bandwidth, it has a security problem specifically because security is not even considered. We propose a novel disjointed multipath routing scheme for multimedia data transmission to overcome bandwidth limitations and to solve the security problem that has not been considered in existing routing schemes.

3 The Proposed Secure Disjointed Multipath Routing Scheme

In this section, we describe a secure disjointed multi-path routing scheme for multimedia data transmission. First, we describe the characteristic of the proposed scheme and show how multimedia data is divided. Second, the formation process of the proposed scheme will be described in more detail.

In this paper, we proposed the method which is to provide the security for multimedia data in wireless sensor networks. To utilize the limited bandwidth of Zigbee efficiently, we divide large amounts of multimedia data and each split data is transmitted through a multipath. Figure 1 shows the problem of distributed routing for the existing schemes. Although these distributed routing schemes are energy-efficient approaches in data processing or transmission, the overhead is due to the joining of many paths. Also, it is almost impossible to apply the key-based method to secure each sensor node, because they have limited performance and energy. Therefore, in terms of security, the network is very vulnerable to malicious nodes. In order to solve these problems in transmitting multimedia data, we propose a secure disjointed multi-path routing scheme. The image data is composed by multiple pixels which are basic units, and the number of image frames such as video data is composed in the form of lists.

As shown in Fig. 2, the proposed scheme splits the image in pixels to generate packets for transmission. Each packet includes the source node's ID where data has occurred, and it specifies the ordering number so that base station can conduct

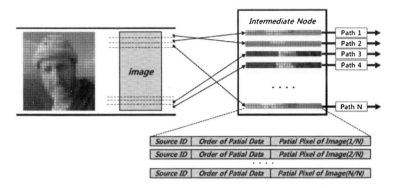

Fig. 2 Partition of the multimedia data

the re-alignment for an each packet. The reason for transmitting the image data in pixels, even though several packets are lost during transmission, is that if the rest of the packets are normally received, the image can be mostly recomposed. Conversely, if a small number of lost packets have been dropped by malicious nodes and some of the pixels of the image are not available, it is very difficult to recover the original data. In addition, even if it has succeeded in recovering some part of the data, it is almost impossible to identify the overall image. These packets generated by splitting the pixels are sent to the intermediate nodes that are located outside a specific distance.

Figure 3 shows the data transfer processing of the disjointed multipath routing scheme considering security. In the proposed scheme, when the first divided data is transmitted, each packet is sent to the random intermediate node a specific distance away from the source node. And at the intermediate nodes, data is sent to the destination node by a greedy-forwarding method. In the network initialization process, each node generates the routing table which has its information of neighboring nodes one hop away. Using the routing table, when multimedia data occurs, data is forwarded to their preferred direction as closely as possible. Because each piece of data is sent to the specific area around the destination in the horizontal direction as far as possible, the overlap between each path is minimized. The higher the value of minimum distance is specified by the user, the farther away each path is formed. As a result, intermediate nodes which have a certain angle and are certain distance away from the source node are selected, so that the maximum distance of each path interval can be found to transfer the data.

As each divided piece of data is transmitted through the multiple disjointed paths which are physically far away from each other, even if any malicious nodes in the middle of the route are invading or tapping some packets, these malicious nodes will find it difficult to get enough data for a meaningful result.

For example, when malicious nodes are placed around the destination node as shown in Fig. 4a, even if the divided data is transmitted through the disjointed multipath, malicious nodes can intercept almost all packets and recover approximate image data. Since the more image data is exposed to the invader the easier it

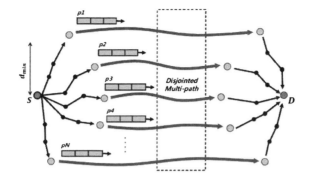

Fig. 3 The disjointed multipath routing scheme considering security

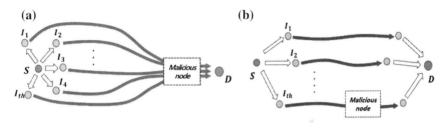

Fig. 4 Data packet loss due to the malicious node. **a** Case 1, **b** Case 2

is to guess the data, randomly-selected intermediate nodes transmit the data in a horizontal direction, as shown in Fig. 4b. When packet data reaches a certain area around the destination node, it is sent to the destination so that the possibility of lost data by malicious nodes is minimized.

4 Performance Evaluation

We have developed a simulator based on JAVA to evaluate our proposed scheme and the existing multimedia data transmission scheme [6]. The performance evaluation was carried out through the simulation parameters shown in Table 1. We assume that 200 sensors are deployed uniformly in a 300×300 (m) network field. The energy consumption for sending a message is determined by a constant function $S \cdot (C_t + C_a \cdot D^2)$, where S is the message size, C_t is the transmission cost, C_a is the amplification cost, and D is the distance of message transmission. We set $C_t = 50$ nJ/b and $C_a = 100$ pJ/b/m^2 in the simulation. The energy consumption for receiving a message is determined by a cost function $(S \cdot C_r)$, where S is the message size and C_r is the transmission cost. We set $C_r = 50$nJ/b in the simulation.

Figure 5 shows the amount of energy consumption of the proposed scheme and the existing multimedia transmission scheme over execution times. The proposed scheme provides a secure transmission unlike the existing schemes, by considering

Table 1 Simulation parameters

Criteria	Value
Size of sensor network fields (m × m)	300 × 300
Number of distributed sensor nodes (EA)	200
Radius of communication (m)	25
Size of multimedia data packet (Kbytes)	4–250

Fig. 5 Energy consumption over the execution time

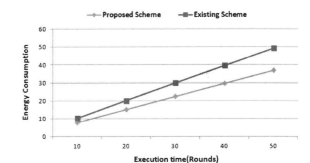

Fig. 6 Hijacked packets according to the malicious nodes

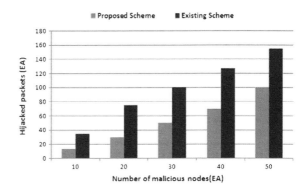

only the energy-efficient transmission of multimedia data. Nevertheless, the amount of energy consumption of the proposed scheme is similar to the existing scheme. In addition, because the proposed scheme does not need the process of path set-up, it can reduce energy consumption compared to the existing schemes. As a result, our scheme reduces energy consumption on average by about 10 % compared to the existing scheme.

Figure 6 shows the amount of hijacked packets by the malicious nodes according to the number of malicious nodes. As the existing scheme transmits multimedia data using multipath routing, it is energy-efficient and remarkable for data transmission rates. However, when a malicious node is placed in a central area, the scheme becomes very vulnerable, because the distance between the multiple paths is shorter than the center of the network. The proposed scheme forms transmission paths at a distance to each other. Therefore, the amount of hijacked packets is lower than the existing scheme. As a result, our scheme reduces the amount of packets hijacked by malicious nodes by about 65 % compared to the existing schemes.

5 Conclusion

In this paper, we proposed a secure disjointed multi-path routing scheme for multimedia data transmission in wireless multimedia sensor networks. The existing scheme enhanced the energy efficiency through the distributed method, but it needed an extra cost to prevent the overlap of paths. In addition, because of the characteristics of multimedia data, it is difficult to apply a security scheme. The entire network is vulnerable to malicious nodes. To solve this problem, the proposed scheme generates disjointed multipath to transmit large multimedia data and distributes the multipath over regular intervals. As a result, data exposure to malicious nodes is limited to only few data packets, so that the recognition of images is almost impossible. Through performance evaluation, the proposed scheme showed that the amount of energy consumption was reduced by about 10 %, and the amount of packets hijacked by malicious nodes was reduced by about 65 % on average over the existing scheme. In future work, we plan to extend our work to provide enhanced security through the dynamic partitioning of the pixels using a hash function.

Acknowledgments This work was supported by Basic Science Research Program through the National Research Foundation of Korea (NRF) grant funded by the Korea government (MEST) (No. 2009-0089128) and the Leaders in INdustry-university Cooperation (LINC) Program through the National Research Foundation of Korea (NRF) funded by the Ministry of Education, Science and Technology (No.2012-B-0013-010112).

References

1. Szewczyk, R., Osterweil, E., Polastre, J., Hamilton, M., Mainwaring, A., Estrin, D.: Habitat monitoring with sensor networks. Comm. ACM **47**(6), 34–40 (2004)
2. Johnson, D., Maltz, D.: Dynamic source routing in ad hoc wireless networks mobile computing, pp. 153–181. Kluwer Academic Publishers, Dordrecht (1996)
3. Waharte, S., Boutaba, R.: Totally disjoint multipath routing in multi-hop wireless networks. In: IEEE International Conference on Communications, vol. 12, pp. 5576–5581 (2006)
4. Wang, Z., Bulut E., Szymanski, B.: Energy efficient collision aware multipath routing for wireless sensor networks. In: Proceedings of the International conference on Communication, pp. 91–95. Dresden, Germany, June 2009
5. Francisco, R., Antnio, G., Pereira, P.R.: DTSN distributed transport for sensor networks. In: Proceedings of the IEEE Symposium on Computers and Communications, pp. 165–172. Aveiro, Portugal (2007)
6. Seong, D., Jo, M., Park, J. Yoo, J.: Disjointed multipath routing for real-time multimedia data transmission in wireless sensor networks. In: Proceedings of the Conference of Korea Information Processing Society, vol. 17, no. 2, pp. 1053–1056, Korea (2010)

Dynamic TDMA Scheduling for Data Compression in Wireless Sensor Networks

Myungho Yeo and Jaesoo Yoo

Abstract In this paper, we propose a novel compression approach using a dynamic TDMA scheduling. While the existing approaches exploit spatio-temporal correlation to suppress sensor readings, our approach tries to hide sensor readings into the TDMA schedule. The TDMA schedule consists of multiple frames that have been assigned with prefix symbols. Sensor nodes select their corresponding frames and transmit encoded symbols instead of raw sensor readings to the base station. Our scheme achieves the data compression, and occurs the latency in the range.

Keywords Sensor network · TDMA protocol · Data compression · Latency

1 Introduction

Recently, wireless sensor networks have found their way in many applications such as environmental monitoring, smart building, medical applications, and precision agriculture [1]. Sensor nodes collect useful information such as temperature, humidity, seismic, and acoustic. Thus, the energy is the most precious resource in the wireless sensor network. It is feasible to periodically replace batteries of sensor nodes deployed in large-scale field. Energy efficient mechanisms

M. Yeo
Agency for Defense Development, Mt. 25 Geoyeo-Dong, Song-pa-Gu, Seoul, South Korea
e-mail: myungho.yeo@gmail.com

J. Yoo (✉)
School of Information and Communication Engineering, Chungbuk National University and CBITRC, 52 Naesudong-ro, Heungdeok-gu, Cheongju, Korea
e-mail: yjs@chungbuk.ac.kr

S.-S. Yeo et al. (eds.), *Computer Science and its Applications*,
Lecture Notes in Electrical Engineering 203, DOI: 10.1007/978-94-007-5699-1_82,
© Springer Science+Business Media Dordrecht 2012

for gathering sensor data are indispensable to prolong the network lifetime as long as possible.

Data compression techniques are traditional and effective methods to reduce the communication cost and to prolong the network lifetime. Sensor readings are strongly correlated by both space and time [2]. There are many approaches with these correlations such as spatial compression and temporal compression. Temporal compression schemes transmit sensor readings to the base station if they take a change as compared with the latest reported data. The base station regards unreported data as the latest reported data [3]. Meng et al. [4] proposed data compression to utilize the mean operator. All nodes receive different time slots and transmit their own readings according to the received time slot. Sensor nodes can overhear the readings of neighbors that are transmitted to the base station while waiting for their time slots. At this time, each sensor node calculates an average from the overheard data. If this average is equal to its own reading, it does not transmit the reading to the base station. Also, in-network processing methods such as the clustering or the tree structure were proposed [5, 6]. In [5], sensor nodes make up clusters and each cluster member transmits its own reading to the cluster head node. The cluster head node may remove duplicated readings and reduce the size of readings with some compression techniques. However, conventional algorithms exploit just local information like historical data or neighbors in the range of communication.

In this paper, we propose a novel compression approach using a dynamic TDMA scheduling. While existing approaches exploit spatio-temporal correlation to suppress sensor readings, our approach tries to hide sensor readings into the TDMA schedule. The TDMA schedule consists of multiple frames that have been assigned with prefix symbols. Sensor nodes select their corresponding frames and transmit encoded symbols instead of raw sensor readings to the base station. We find out that our scheme achieves the data compression but incurs a time delay during data gathering. Thus, we derive a mathematical formula to predict and reduce the latency. The simulation results show that the latency is occurred in the specific range.

The remainder of this paper is organized as follows. Section 2 states our motivation. Section 3 describes our dynamic TDMA scheduling scheme to compress sensor readings. In Sect. 4, we show the superiority of our proposed compression scheme through performance evaluation and analysis. Finally, we conclude this paper in Sect. 5.

2 The Proposed Multimedia Data Compression and Transmission Scheme

We find out an interesting approach to hide partial symbols of sensor readings into the TDMA schedule. Each sensor node is an independent small computing machine and performs sophisticated behavior over a TDMA protocol. There are N sensor nodes in the sensor field. Every sensor node transmits its reading as d-bits in a round. While a TDMA schedule generally consists of a single frame that contains N time-slots, we assume that multiple frames are in the TDMA schedule. Each frame contains enough time-slots as the number of sensor nodes to transmit sensor readings into the base station. Each frame is assigned with a prefix symbol, which is allowed to abbreviate sensor readings in a round. Before transmitting sensor data, each sensor node determines its corresponding frame that the prefix symbol is identical to the prefix symbol of its current reading. Next, it just removes the prefix symbol from its sensor reading and transmits its shorted symbol in the corresponding frame. If the number of frames is extended to n, the length of an encoded symbol is shortened as $(d - log_2(n))$ bits. However, it incurs the latency for data gathering in a round. It requires the additional time-duration for data gathering in a round as $T \cdot N (n - 1)$. That is because every frame contains repeatedly time-slots for each sensor node. T denotes the time-duration for a single time slot. The response time is also an important factor for data gathering in sensor network applications. Thus, we should not only investigate to compress sensor readings, but also to improve the response time for data gathering.

3 Dynamic TDMA Schedule for Data Compression

We propose a dynamic TDMA schedule to compress sensor readings and to enhance the response time for data gathering. Note that the number of frames and the prefix symbol can be predefined before the sensor nodes are deployed in the sensor field. The response time in the sensor network depends on its applications. For example, reconnaissance surveillance application and gas detection applications require very fast responses, while environmental monitoring applications may allow late responses from sensor networks. There are n independent time-slots for each sensor node in the TDMA schedule. However, each sensor node must pick one time-slot among them, transmits its readings at the time-slot and maintains the sleep mode until the end of the current round. It causes the latency and prolongs the time-duration for data gathering.

To overcome this problem, we exploit the reservation mechanism into the TDMA scheduling scheme. We define the time-slot as three types: (1) TRS (TRansmission Slot) that sensor nodes transmit (compressed) sensor data; (2) RES (REservation Slot) that sensor nodes transmit a signal (or a beacon) for allocating a transmission slot into the current frame; (3) TES (TErmination Slot) that the base station broadcasts a signal to terminate data gathering of the current round into the sensor network. The length of TES and RES is short relatively over TRS, because sensor nodes just send a signal to inform the reservation or the termination to the sensor network. In the result, our scheme performs data gathering in the range of the response time R as Eq. (1). T_r denotes the time-duration for a reservation slot. Te denotes the time-duration for a termination slot.

$$N \cdot (T_r + T) + T_e \leq R \leq (n - 1) \cdot T_r + N \cdot (T_r + T) + T_e \qquad (1)$$

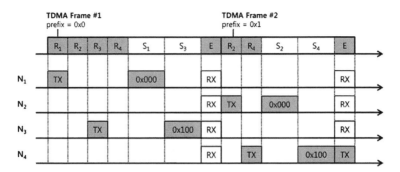

Fig. 1 Dynamic TDMA scheduling with reservation slots

Our dynamic TDMA scheduling scheme performs as follows:

(1) $k = N$, $F_c = T_0$.
(2) k reservation slots are allocated to the current frame F_c.
(3) m sensor nodes that should transmit their readings at the current frame broadcast a signal (or a beacon) and inform their reservation.
(4) m transmission slots are allocated adaptively into the frame corresponding to their reservation. m sensor nodes transmit sequentially their readings at TRS.
(5) $(k - m)$ reservation slots are just allocated to the next frame F_n because k reservation slots that have been exhausted in the current frame are unnecessary until the end of the current round.
(6) $k = k - m$, $F_c = F_n$.
(7) If $(k > 0)$, the termination slot is allocated to F_c. The base station broadcasts a termination signal into the sensor network and terminates data gathering in a round. Otherwise, repeat (2).

For example, suppose that a TDMA schedule consists of a single round with two frames as shown in Fig. 1. The *Frame #1* and the *Frame #2* are also assigned with the prefix symbol $\{0 \times 0, 0 \times 1\}$, respectively. Sensor nodes $\{N_1, N_3\}$ that have measured $\{0 \times 0000, 0 \times 0100\}$ make the reservation to transmit encoded symbols $\{0 \times 000, 0 \times 100\}$ in the *Frame #1*. N_1 and N_3 broadcast a reservation signal to the sensor network at the time slot S_1 and S_3 in the *Frame #1*, respectively. They transmit their encoded symbols at their own transmission slot in the *Frame #1*. Other sensor nodes recognize the data transmission of N_1 and N_3, because every sensor node maintains the active mode to overhear signals during the reservation slot. Next, sensor nodes N_2, N_4 that have measured $\{0 \times 1000, 0 \times 1100\}$ also make the reservation to transmit encoded symbols $\{0 \times 000, 0 \times 100\}$ in the *Frame #2*. N_2 and N_4 also broadcast a reservation signal to the sensor network at the time slot S_2 and S_4 in the *Frame #2*, respectively. They transmit their encoded symbols at their own transmission slot in the *Frame #2*. Finally, the final sensor node or the base station recognizes the termination of data gathering in a round and broadcasts the termination signal at the termination slot E.

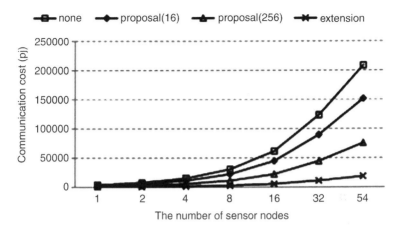

Fig. 2 The communication cost

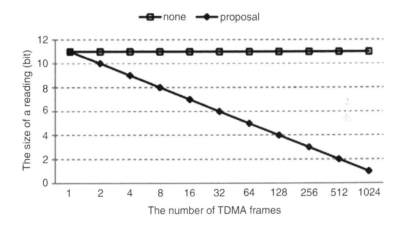

Fig. 3 The size of a sensor reading

4 Performance Evaluation

To show the superiority of our scheme, we have conducted intensive simulation experiments. In this simulation, we use real datasets collected from Intel Lab [7]. We consider a 100×100 network field, where the sink is placed in the center. We deploy 54 sensor nodes randomly in the field and sensors communicate in a multi-hop fashion. The communication range is set to 18 for good connectivity in a random topology. Figure 2 shows the communication cost as the number of sensor nodes increases from 1 to 54. We compare data gatherings with and without our dynamic TDMA scheduling scheme. Although the communication cost increases with the number of sensor nodes, our scheme reduces significantly the communication cost using dynamic TDMA scheduling. Figure 2 shows the size of each

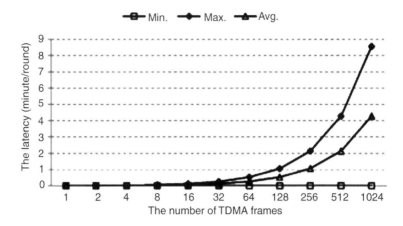

Fig. 4 The latency for data gathering

reading as the number of TDMA frames changes. As the number of TDMA frames increases, the size of a reading decreases linearly as $log_2(n)$. At the number of TDMA frames is set to 1024, the size of a reading is reduced from 12-bits to 1-bit; It also presents the latency in the range of the minimum 0.0144 (s) and the maximum 8.0897 (min). In the result, simulation results show the trade-off between the compression rate and the latency for data gathering (Figs. 3 and 4).

5 Conclusions

In this paper, we have proposed a novel compression approach over a TDMA protocol. While the existing approach have exploited spatio-temporal correlation to suppress sensor readings, our approach has hided the partial or entire symbols of sensor data into the TDMA schedule. It reduces the size of sensor data linearly as the number of TDMA frames is increased. We have also investigated an extended scheme to enhance the response time for data gathering. We have derived a analytic formula to predict the latency and have proposed an extended scheme to reduce it for data gathering. The experimental results show that the latency is occurred in the specific range and a communication protocol with consideration of both the compression and the response time for data gathering can be designed.

Acknowledgments This work was supported by Basic Science Research Program through the National Research Foundation of Korea (NRF) grant funded by the Korea government (MEST) (No. 2009-0089128) and the Leaders in INdustry-university Cooperation (LINC) Program through the National Research Foundation of Korea (NRF) funded by the Ministry of Education, Science and Technology (No. 2012-B-0013-010112).

References

1. Szewczyk, R., Osterweil, E., Polastre, J., Hamilton, M., Mainwaring, A., Estrin, D.: Habitat monitoring with sensor networks. Commun. ACM **47**(6), 34–40 (2004)
2. Yeo, M., Seo, D., Yoo, J.: Data correlation-based clustering algorithm in wireless sensor networks. KSII Trans. Internet Inf. Syst. **3**(3), 331–343 (2009)
3. Sharaf, M., Beaver, J., Labrinidis, A., Chryanthis, P.: Tina: A scheme for temporal coherency-aware in-network aggregation. In: Proceedings of the 2003 ACM Workshop on Data Engineering for Wireless and mobile Access, pp. 69–76 (2003)
4. Meng, X., Li, L., Nandagopal, T., Lu, S.: Event contour: an efficient and robust mechanism for tasks in sensor networks. In: Proceedings of Technical report, pp. 1–13 (2004)
5. Pattem, S., Krishnamachari, B., Govindan, R.: The impact of spatial correlation on routing with compression in wireless sensor networks. ACM Trans. Sens. Netw. **4**(4), 1–33 (2008)
6. Petrovic, D., Shah, R., Ramchandran, K., Rabaey, J.: Data funneling: routing with aggregation and compression for wireless sensor networks. In: Proceedings of the 2003 IEEE Sensor Network Protocols and Applications, pp. 156–162 (2003)
7. Intel Lab Data. http://berkeley.intel-research.net/labdata/

A Study of Forest Fire Correlation to Based on Meteorological Factors

Young-Suk Chung, Jin-Mook Kim and Koo-Rock Park

Abstract The forest fires that occur each year many lives, property, and ecosystems are destroyed. Approximately 64 % of the land, especially the Korea is forest. If you cause damage to large forest fires, it takes longer to repair the damage. Various studies to understand these forest fires process, but research related to forest fires and meteorological factors is lacking. In this paper, we use the correlation coefficient with forest fires and meteorological factors were investigated for the association. The result may be associated with forest fires and meteorological factors were confirmed.

Keywords Correlation coefficient · Forest fires · Simulation

1 Introduction

Korea's land area is 10,003,308. Forest occupies an area of 6,368,843 ha by 2010 to account for 63.67 % of land area [1]. In the area of land occupied by forest fires because of the high incidence of damage can be large. The forest fires that occur

Y.-S. Chung · K.-R. Park (✉)
Division of Computer Science and Engineering, Kongju National University 275,
8 Building 275, Budae-dong Seobuk-gu, Cheonan-si, ChungNam, Korea
e-mail: ecgrpark@kongju.ac.kr

Y.-S. Chung
e-mail: merope@kongju.ac.kr

J.-M. Kim
Division of Information Technology Education, Sunmoon University, #100, Galsan-ri
Tangjeong-myeon, Asan-si, ChungNam, Korea
e-mail: calf0425@sunmoon.ac.kr

S.-S. Yeo et al. (eds.), *Computer Science and its Applications*,
Lecture Notes in Electrical Engineering 203, DOI: 10.1007/978-94-007-5699-1_83,
© Springer Science+Business Media Dordrecht 2012

every year of life, property, and the enormous ecological damage are inflicted. And, it is also increasing the number of cases. A variety of forest fires in order to study the phenomenon has been studied. Forest fires occurred on affected areas using images taken with a digital camera to survey the damage, allowing many different processing methods were compared [2]. As the Forest fires have burned before and after examining satellite images by using statistical analysis to classify the degree of Forest fires damage had [3]. In addition, the point of Forest fires weather conditions at the time of the terrain and building databases identified a relationship between forest fires and the study was conducted [4]. To detect the forest fires, the smoke from the CCD camera video input on the Random Forest algorithm to apply the study was conducted [5]. But the fires of the factors that the relationship between weather conditions and Forest fires Study is lacking. In this paper, we apply a correlation factor of the weather conditions were analyzed and the relationship between Forest fires. This paper is organized as follows. Section 2 related research annual correlation analysis and the correlation coefficient are discussed. Section 3, using correlation analysis to model the relevance of Forest fires and meteorological factors and actual wildfire data is applied. In the final chapter on conclusions and future research are discussed.

2 Related Research

2.1 Correlation Coefficient

Referring briefly to the Pearson correlation coefficient is the correlation coefficient. Interval or ratio measured by applying a measure of proximity between two variables is a coefficient representing. Therefore seem to be related to each other, the relationship between two variables is applied in general to obtain more clearly. The correlation coefficient is expressed as r. And the correlation coefficient has a value between -1 and 1.

① A high correlation between two variables, the absolute value of r is close to 1.
② Low correlation between two variables the absolute value of r is close to 0.

The correlation coefficient comparing the two variables x and y to calculate the value, the formula is as follows:

$$S(xx) = \sum_{i=1}^{n} (x_i - \bar{x})^2 = \sum_{i=1}^{n} x_i^2 - \left(\sum_{i=1}^{n} x_i \right)^2 \bigg/ n \qquad (1)$$

$$S(yy) = \sum_{i=1}^{n} (y_i - \bar{y})^2 = \sum_{i=1}^{n} ny_i^2 - \left(\sum_{i=1}^{n} y_i \right)^2 \bigg/ n \qquad (2)$$

Table 1 Conditions list of the relationship between correlation coefficient

Condition	Coefficient
$0.8 \leq |r|$	A strong correlation
$0.6 \leq |r| \leq 0.8$	In correlation
$0.4 \leq |r| \leq 0.6$	In a weak correlation
$|r| \leq 0.4$	Little no correlation

$$S(xy) = \sum_{i=1}^{n} (x_i - \bar{x})(y_i - \bar{y}) = \sum_{i=1}^{n} x_i y_i - \left(\sum_{i=1}^{n} x_i \right) \left(\sum_{i=1}^{n} y_i \right) \Big/ n \qquad (3)$$

$$r = \frac{S(xy)}{\sqrt{S(xx)S(yy)}} \qquad (4)$$

where,

$S(xx)$ The sum of squared deviations of x
$S(yy)$ The sum of squared deviations of y
$S(xy)$ Multiplying the sum of the deviation of x and y
r Correlation coefficient

From the correlation coefficient to determine proximity to the general conditions are shown in Table 1 [6].

3 Association of Forest Fires Modeling and Meteorological Factors

Using Correlation Relationship Modeling Forest fires and meteorological factors are shown in Fig. 1.

Korea Forest Service annually publishes statistics relating to Forest fires. Forest Service to the public, based on data from weather conditions are analyzed based on number of forest fires. The first step, based on meteorological factors will collect the monthly incidence of Forest fires. Second step, the collected data is analyzed. The next step is applied to the analyzed data Correlation coefficient. Finally, by analyzing Correlation coefficient, and weather analysis, each factor is the relevance of forest fires.

3.1 Data Based on Applicable Forest Fires Meteorological Factors, and Modeling Results

The relationship between Forest fires and meteorological factors in 2008 and 2009 wildfires to see the data the number of days with dry weather factors were applied to select the number of days of precipitation. Table 2 in 2008 and 2009 is the

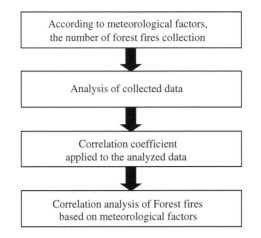

Fig. 1 Correlation relationship modeling forest fires and meteorological factors

Table 2 Number of forest fires

	2008 (year)	2009 (year)
January	24	64
February	30	32
March	22	119
April	51	206
May	45	52
June	21	17
July	0	0
August	0	0
September	0	13
October	5	23
November	44	23
December	40	21

number of Forest fires. Table 3 is the dry days. Table 4 is the number of days of precipitation.

Meteorological factors and should indicate the relevance of Forest fires. Each year's forest fires—dry days, and wildfires—modeling of precipitation days were treated with the association. Table 4 Relationship between the forest fires modeling results and meteorological factors is the correlation coefficient value (Table 5).

When analyzing the correlation coefficient Forest fires-drying days in 2008, 0.63, 0.72 in 2009, has emerged as a correlation. Forest fires—days of precipitation in 2008, −1.25, −0.42 in 2009 has emerged as a negative correlation. Days of dry weather of the factors that are relevant to the Forest fires were the result.

Table 3 Dry days

	2008 (year)	2009 (year)
January	23	28
February	10	13
March	5	18
April	12	21
May	11	11
June	0	0
July	0	0
August	0	0
September	0	0
October	5	4
November	24	9
December	22	20

Table 4 Precipitation days

	2008 (year)	2009 (year)
January	7.2	6.6
February	3.2	6.9
March	8.6	9
April	8.8	6.3
May	8.4	8.9
June	12.8	10.3
July	14.1	18.1
August	12.8	11
September	9	7.2
October	5.1	5.8
November	6.7	8.6
December	6.7	8.9

Table 5 The correlation coefficient

Type (year)		Correlation coefficient
2008	Forest fires-dry day	0.63
	Forest fires-Precipitation day	−0.39
2009	Forest fires-dry day	0.72
	Forest fires-Precipitation day	−0.42

4 Conclusion

In this paper, Forest fires and weather factors, correlation analysis model is proposed and applied to real data. Forest fires and dry days were concluded that there is an association. And forest fires—days of precipitation were a negative correlation coefficient. The results were associated with Forest fires and meteorological factors. Each year at Korea Meteorological Administration is presented for the prediction of weather conditions. Association published meteorological data modeling, when applied to meteorological factors, Forest fires prevention policy is expected to help.

In the future, more types of association between meteorological factors and the forest fires are going to be made for the study.

References

1. Korea Forest Service 2010 Forestry Statistics published data base, http://www.forest.go.kr/kfsweb/kfs/idx/SubIndex.do?mn=KFS_05
2. Soo, J.: A plan for estimation of damaged area from forest fire using digital photographs. J. Korean Soc. Geosp. Inf. Syst. **18**(4), 41–50 (2010)
3. Pil, C.S., Hee, k.D., Kun, L.S.: The abstraction of forest fire damage area using factor analysis from the satellite image data. J. Korean Soc. Geosp. Inf. Syst. **14**(1), 13–19 (2006)
4. Kawk, H.-B., Lee, W.-K., Lee, S.-Y., Won, M.-S., Lee, M.-B., Koo, K.-S.: The analysis of relationship between forest fire distribution and topographic, geographic, and climatic factors. In: CSIS Annual Autumn Conference 2008, pp. 465–470 (2008)
5. Kwak, J.Y., Kim, D.Y., Ko, B.C., Nam, J.Y.: Forest smoke detection using random forest. In: Proceedings of the KIISE Korea Computer Congress 2011, vol. 38, no. 1(C), pp 351–353 (2011)
6. Kim, J.-H., Kim, Y.-H.: Statistics and applied, Yunhaksa, pp. 238–246 (2007)

A Study on the Improvement of Interoperability in Rok C4I System for Future Warfare

Hyun-Jeong Cha, Jin-Mook Kim, Hwang-Bin Ryou
and Hwa-Young Jeong

Abstract The basic idea of Network Centric Warfare (NCW) is that data processing capabilities of computers and communication technologies of a network can be leveraged to guarantee information sharing, which will increase the efficiency of running a military. Amid advancements being made in IT, the Korea military is one of the militaries in the world doing research on ways of conducting more effective warfare, which it aims to achieve by connecting each of the entities together and sharing battlefield resources. However, maintaining interoperability between different systems is an issue, caused by the fact that different militaries have different tactics and weapon systems. In this paper, the current state of interoperability is examined, for when the Korean military builds a C4I system in preparation for future warfare. Furthermore, interoperability cases in the US are analyzed, and matters of consideration and various technologies for interoperability are proposed.

Keywords Future warfare · C4I · Interoperability

H.-J. Cha
Defense Acquisition Program, Kwangwoon University, Seoul, Korea
e-mail: Chj826@kw.ac.kr

J.-M. Kim (✉)
Divsion of Information Technology Education, Sunmoon University, Asan, Korea
e-mail: calf0425@sunmoon.ac.kr

H.-B. Ryou
Department of Computer Science, Kwangwoon University, Seoul, Korea
e-mail: ryou@kw.ac.kr

H.-Y. Jeong
Humanitas College of Kyunghee University, Hoegi-dong, Seoul 130-701, Korea
e-mail: hyjeong@khu.ac.kr

S.-S. Yeo et al. (eds.), *Computer Science and its Applications*,
Lecture Notes in Electrical Engineering 203, DOI: 10.1007/978-94-007-5699-1_84,
© Springer Science+Business Media Dordrecht 2012

1 Introduction

As information and communication technologies advance, the warfare is changing to one characterized by non-contact, non-linear, remote, network-based, parallel, simultaneous, effectiveness-oriented, mobile warfare, and integrated operations. The traditional warfare is weapon-systems based, but the future warfare is network-based, which has different weapons systems and other related systems connected together, a type of warfare called Network Centric Warfare (NCW). With this kind of innovative advancement, the environment in which the military conducts war is changing quickly and in diverse ways. The military had been working on an independent and dedicated command network, because of the unique requirements of military missions. Advancement in IT is essential for conducting the changed warfare of the twenty first century, in particular for military communications. Efforts are being made to maximize military might and to conduct a more efficient warfare through systematic gathering and sharing of battlefield information in real-time, as well as through strategical monitoring. Looking at how the war was conducted in Afghanistan or Iraq, it has been verified that the management of information like this has a great effect in securing victory. In the case of the US, it realized the importance of interoperability early on, thanks to its accumulated experience and knowhow when it comes to warfare. It has put in place procedures and guidelines for interoperability, and shares information among different systems and military forces. The Korea military is also working on a C4I system for commanding and controlling the army, the navy, and the air force. Currently it is expanding on the system and improving its performance. However, when building the system the Korea military did not take into account interoperability between the different systems, placing limitations on the usefulness of the joint command and control system.

In this paper, the current state of interoperability is examined, for when the Korean military builds a C4I system in preparation for future warfare. Furthermore, interoperability cases in the US are analyzed, and matters of consideration and various technologies for interoperability are proposed.

2 Related Work

2.1 Military Operation Environment

Network-Centric Warfare (NCW). As we ushered in the information age, the US made efforts for applying the latest technologies to the military in a military reform, one outcome of which is NCW. Innovations were made in the command and control system by introducing the modern computer technologies and networks to the military. Now, all units and individuals can be connected together via a network. From a weapons-system perspective as well, improved precision strikes allow immediate suppression of the target as long as its location and form can be identified, and with improved mobility of units, physical, spatial and temporal limitations are decreasing. Accordingly, NCW was proposed as one of the concepts that put together all these modern advancements for the military [1].

C4I System. C4I stands for Command, Control, Communication, Computer and Intelligence. The term C4I is used when discussing systems. Command and control, or the performance of the commander's commands according to the authorities, is key for successfully achieving a given mission. Communication and Computer is an element of supporting means for the command and control, a system that makes command and control possible. Intelligence is a set of results from information gathering, processing, integration, analysis and assessment.

Therefore, the C4I system refers to the overall command/control/communication/computer/intelligence system that support the privileges and commands that are given to one's own units in order to achieve a mission given to the commander. It involves command/control procedures for achieving the mission and is a means for carrying out these procedures. In a broader sense, it also includes personnel, facilities, supporting equipment and other procedures. That is, it refers to information about units and the strategic situation so that the current state of command and control and strength of forces can be assessed and adjusted as necessary [2].

Tactical Information and Communication Network (TICN). TICN is a system that aims to provide information communication paths that are high-speed and high-capacity to the sensor system, command and control system, and the combat system in NCW for smooth communication of information. The subsystems of TICN are the backbone transmission system, backbone switching system, network control system, wireless combat system, and mobile subscriber access point, the last of which aims to support means of mobile communications to subscribers of multi-function tactical mobile devices for command posts and nearby areas through voice, data, and multimedia services [3].

2.2 Interoperability

Generally speaking, interoperability is about providing integrated system capabilities by flexibly sharing services of heterogeneous systems. For DoD, it considers interoperability as that system which allows for sharing and use of services between heterogeneous systems (different units and military forces), so that they can take part in operations more effectively; also, the capabilities of the units and military forces for doing so is referred to as interoperability. While the US sees interoperability from both an operation perspective and a technical perspective, the Korea military sees interoperability as the capabilities for sharing, exchanging and operating specific services, information and data without any hindrance among heterogeneous systems or units, military forces and systems, putting the emphasis on the technical aspect.

Fig. 1 Concept of jointness, interoperability and linkage

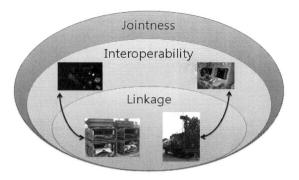

When it comes to national defense, the concept is defined as that system which allows for units or military forces to effectively take part in operations, or their capabilities for doing so, through sharing/exchanging/using information, data, and services involving heterogeneous systems, units or military forces. In addition, along with the technical exchange of information, one of the aims must be to maximize effectiveness of operations through information exchange needed for achieving a given mission. Also, it has to be more than just exchange of information, but systems, processing stages, processing procedures, organizations and tasks must be taken into account over the entire life cycle to achieve balance with information assurance. The purpose of interoperability for national defense is to guarantee a network-centric information sharing environment among the monitoring system, battlefield information management system, and the combat system [4, 5].

Concepts similar to interoperability are jointness and linkage. The relationship among them is shown in Fig. 1. Linkage is a concept that is smaller in scope than interoperability. It is the capabilities for data exchange by direct physical linkage between two entities. Jointness advances the concept of joint operations that coincidence with the image of future warfare. By effectively integrating the army, navy, and the air force, combat capabilities can be maximized, guaranteeing victory.

3 Current State of Interoperability

3.1 Interoperability of the US Military

The US military has put in place policies and guidelines for interoperability that are systematic and integrated in order to develop joint capability, including the DoD directive, DoD guidelines, chairman of the Joint Chiefs of Staff guidelines, and guidelines for different countries. The top-level directive is described in the document "DoDD 4630.5: Interoperability and Supportability of Information

Fig. 2 Policies and guidelines for interoperability of US military

Technology." DoDi 4630.8 describes the pertinent guidelines. According to these two policies and guidelines, CJCSI 6212.01C, which is the chairman of the Joint Chiefs of Staff guidelines, describes detailed standards and test/authentication procedures for interoperability, which need to be included in documents created in the process of DoDI 5000.2. CJCSI 3170.01 includes integration of interoperability, as well as procedures, policies, and assignment of responsibilities for the systems developed. DoDI 8100 consists of DoDD 8100.1, which is a policy document for the implementation of Global Information Grid (GIG); DoDD 8100.2, which describes commercial wireless devices, services, and technologies used in GIG; and DoDD 8100.3, which deals with a national defense audio network [6–9] (Fig. 2).

Interoperable organizations are characterized as follows: dedicated organizations are run for different branches of the military and according to the function in order to maintain professionalism, and segmentation are done into policy organizations, joint test/authentication organizations, and independent test/assessment organizations. Capability gaps are found using JCID analysis (Milestone A), and interoperability requirements are found as part of the process of finding requirements for weapon systems for eliminating the capability gaps. Interoperability requirements are authenticated in J-6 (Milestone B). There are different main organizations for standards, requirement/structure, network, dispersion technologies, combat services, resource management, and operational testing. Pertinent collaborative organizations dealing with the detailed areas are used to carry out interoperability tasks.

3.2 Interoperability of the Korea Military

The Korean military has the defense acquisition management rules as the top-level document of policies and guidelines related to interoperability. The information planning office of the national defense is responsible for reviewing/adjusting technologies and standards related to this, while the Joint Chiefs of Staff is mainly responsible for interoperability between the US and South Korea, as well as for adjusting and controlling work related to frequencies for the C4I system. Although items related to interoperability are included in the detailed acquisition and management guidelines for the automated information system, traceability still remains poor between different documents. Furthermore, although when it comes

Table 1 Matters of consideration when designing interoperability

Section	Consideration
Term	Urgent requirements and mid-to-long-term requirements
Range	Deliberated requirements and requirements for all military forces
Operation	Expected difficulties in all usage scenarios
Group	Development of individual component systems by other organizations
	Synchronization of independent programs
	Different organizations or units with different tendencies and technologies
System	Existing systems
	Difficulties of expanding the system with the existing equipment
Technology	Management of rapid technological changes
	Proprietary technology
Standard	Consideration of situations unsuitable to existing standards
Management	Strict management and control for changes in form

to interoperability and standardization management guidelines for the national information system, there is an interoperability and standardization (and technical review) committee run by the National Defense Computer Center, there is almost no substantial discussions taking place [10, 11].

4 Matters of Consideration When Implementing Interoperability

4.1 Matters of Consideration When Designing Interoperability

When implementing interoperability, the below Table 1 should be considered and the optimal conditions found and applied.

When interoperability is designed by taking into account the above items, in preparation for the battlefield environment of the future, the factors that hinder interoperability are as follows. First, usage can't be predicted. Second, joint-warfare. Third, flexibility that promotes change due to the command structure. Fourth, flexibility that meets the changing tactical situation. Fifth, short formation time. Sixth, building a horizontal system amid vertical organizations and programs.

4.2 Interoperability Technology

When it comes to designing interoperability, the larger the organization, the greater the difficulty. Large organizations already have a system in place that they use, so they can't start from a clean slate. Therefore, it's ideal for them to analyze the existing systems and tasks, looking at the work flow and detailed particulars, and then come up with a plan on how to implement interoperability.

A suitable architecture should be developed and implemented. Analysis and design should be done between systems involving interfaces, hierarchies, and middleware. Important environments and factors should be standardized, and system data and signals should be made interoperable.

5 Conclusions

With advancements in IT and a paradigm shift in warfare, the importance of NCW is being stressed. Accordingly, the development of a C4I system is also being stressed. This paper examined the current state of interoperability for the US military and the Korea military as a measure for obtaining interoperability in the area of national defense from the development of a C4I system.

While the US military has advanced functions and organizations when it comes to interoperability, thanks to their longtime experience in war and development of systematic systems, the Korea military has a problem with interoperability. Therefore, this paper proposed matters of consideration and pertinent technologies for the implementation of interoperability.

If work procedures are established with pertinent laws, regulations and guidelines related to interoperability for national defense and the private sector (industry, academia and research institutions) and the US and South Korea work together to build a collaborative system and continue to improve upon it by analyzing pending issues of interoperability, then a highly advanced battlefield management system would be able to be developed and efficiency would be maximized.

References

1. Bae, D.H., Cho, Y.G.: Operational attribute and principle of computer network operations (CNO). J. Nat. Def. Stud. **52**(2), 39–70 (2009) (Research Institute for National Security Affairs of Korea National Defense University)
2. Kim Y.G.: The theory and actuality of C4I, Korean institute for defense analysis Press, Korea (2006)
3. Ko S.J.: Smallness unit a commander telecommunication system embodiment plan research, PaiChai University, Korea (2005)
4. Ministry of National Defense: defense force development operation regulation 2006, Korea (2006)
5. Defense Acquisition Program Administration: defense capacity improvement project management regulation 2006, Korea (2006)
6. DoD: Department of Defense Directive 4630.5, Interoperability and supportability of information technology and national security systems (2004)

7. DoD: DoD Instruction 4630.8, procedures for interoperability and supportability of information technology and national security systems (2004)
8. DoD: Chairman of the Joint Chiefs of Staff Instruction 6212.01.C, interoperability and supportability of information technology and national security systems (2003)
9. DoD: DoD Instruction 5000.2, operation of the defense acquisition system (2003)
10. Ministry of National Defense: defense acquisition management regulation, Korea (2003)
11. Ministry of National Defense: instruction on the management of defense information system interoperability and standardization, Korea (2003)

Development Direction for Information Security in Network-Centric Warfare

Ho-Kyung Yang, Jin-Mook Kim, Hwang-Bin Ryou and Jong-Hyuk Park

Abstract As IT advances, warfare undergoing a paradigm shifts. It is going from a type that is weapon-based—making heavy use of conventional weapons—to one that is network-based, a type of warfare called network-centric warfare (NCW). In NCW, sharing of information can be guaranteed using communication technologies of the network. As the members that take part in missions can exchange and collaborate with other, they can be completed more successfully and efficiently, increasing the military might. Amid advancements being made in IT, the Korea military is also doing research for conducting more effective warfare, which it aims to do by connecting each of the entities together and sharing battlefield resources. This paper examines threats to information security in NCW, as well as requirements for information security. Furthermore, a development direction for information security is proposed.

Keywords Information security · NCW · TICN · Network security

H.-K. Yang
Defense Acquisition Program, Kwangwoon University, Seoul, South Korea
e-mail: porori2000@kw.ac.kr

J.-M. Kim (✉)
Divsion of Information Technology Education, Sunmoon University, Asan, South Korea
e-mail: calf0425@sunmoon.ac.kr

H.-B. Ryou
Department of Computer Science, Kwangwoon University, Seoul, South Korea
e-mail: ryou@kw.ac.kr

J.-H. Park
Department of Computer Science and Engineering, Seoul National University of Science and Technology, Seoul, South Korea
e-mail: parkjonghyuk1@hotmail.com

S.-S. Yeo et al. (eds.), *Computer Science and its Applications*,
Lecture Notes in Electrical Engineering 203, DOI: 10.1007/978-94-007-5699-1_85,
© Springer Science+Business Media Dordrecht 2012

1 Introduction

As information and communication technologies advance, the warfare is changing to one characterized by non-contact, non-linear, remote, network-based, parallel, simultaneous, effectiveness-oriented, mobile warfare, and integrated operations. The traditional warfare is weapon-systems based, but the future warfare is network-based, which has different weapons systems and other related systems connected together, a type of warfare called network-centric warfare (NCW). The military had been working on an independent and dedicated command network, because military missions have unique requirements. Advancement in IT is essential for conducting the changed warfare of the twenty first century, in particular for military communications. Efforts are being made to maximize military might and to conduct a more efficient warfare through systematic gathering and sharing of battlefield information in real-time, as well as through strategical monitoring. With the advancements in the military communications network, however, security threats are also increasing. As the network expands, the number of routes in which to attack increases, and as the amount of data that goes over the network increases, greater amounts of sensitive data are at risk of leakage. These days, attacks are not just for stealing data but there is an increasingly greater likelihood of cyber wars taking place over the network. For an organization such as the military for which keeping information and data secure is especially important, as even a small amount of their leakage may pose a huge threat, security is a number-one priority for the military network, more so than other types of networks.

This paper looks at the basic concepts of NCW, threats to information security in NCW, and requirements for keeping information secure in NCW. It also proposes a development direction for information security in NCW.

2 Related Work

2.1 Network-Centric Warfare (NCW)

The premise of NCW is that by connecting various computer systems and networks together that are involved in battlefield operations and systematically creating, storing, distributing, and managing battlefield information for all entities involved, missions can be carried out much more efficiently. As we ushered in the information age, the US made efforts for applying the latest technologies to the military in a military reform, one outcome of which is NCW. Innovations were made in the command and control system by introducing the modern computer technologies and networks to the military. Now, all units and individuals can be connected together via a network. From a weapons-system perspective as well, improved precision strikes allow immediate suppression of the target as long as its

Fig. 1 Concept of TICN

location and form can be identified, and with improved mobility of units, physical, spatial and temporal limitations are decreasing. Accordingly, NCW was proposed as one of the concepts that put together all these modern advancements for the military [1].

2.2 TICN

Tactical Information and Communication Network (TICN) is a system that aims to provide information communication paths that are high-speed and high-capacity to the sensor system, command and control system, and the combat system in NCW for smooth communication of information. The subsystems of TICN are the backbone transmission system, backbone switching system, network control system, wireless combat system, and mobile subscriber access point, the last of which aims to support means of mobile communications to subscribers of multi-function tactical mobile devices for command posts and nearby areas through voice, data, and multimedia services [2] (Fig. 1).

2.3 Aim of Security and Attack Techniques

To keep information secure, non-authorized access and non-authorized changes must be blocked, only allowing access and changes for privileged users. In passive attacks, attackers simply aim to obtain data, while in active attacks, they try to

Fig. 2 Security and attack techniques

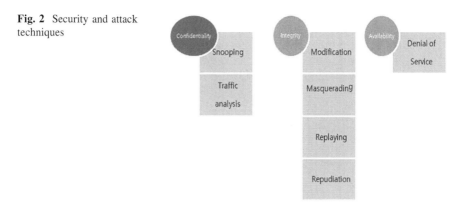

change data or cause harm to the system. Figure 2 above shows the aim of security and attack techniques that pose a threat to security [3].

3 Development Directions for Information Security

3.1 Threats to Information Security in NCW

The computer system at the Ministry of National Defense is implemented on top of commercial technology from the private sector, suffering from the same security vulnerabilities as systems from the private sector, which include unauthorized access to data, changes to data, and denial of service. The increase in the capabilities of the network in NCW will rapidly increase the amount of data distributed. To accommodate such an amount of data, the network must be expanded and as a result it becomes increasingly difficult to keep it secure. Also, as fast data processing is considered essential, if only low-level information security measures are used in order to obtain data in a timely manner, important data will get leaked. If linkage and connection are expanded in expanding the system, the number of attack or invasion routes will also increase, and there is also greater likelihood that unauthorized users will gain access. In addition, from a data sharing and distribution perspective, there is a greater likelihood that integrity and confidentiality will be damaged.

3.2 Information Security Requirements in NCW

Computer system security is considered mainly in two aspects. The first concerns physical security, protecting the communication and control system from physical destruction or jamming. The second concerns cyber security, protecting the system connected to the network. Many techniques become necessary for user authentication and access privilege control. In a network environment, data must be systematically managed and distributed according to different levels of security. Information security information should be monitored in real-time for each of the network systems. To maintain confidentiality, technologies and equipment must be applied to the network environment in a suitable way, and encryption keys, which serve as the basis for information security, must be systematically managed also, to guarantee organic networking that NCW seeks after, classification level must be expanded to integration or linkage of different computer systems or networks. A device like a guard must go with this, for guaranteeing this, and a information security solution must be established so that the information system is accurately linked and run [4, 5].

3.3 Development Direction for Information Security in NCW

In establishing a network system, things must go in parallel with the information security area in an organic way and in preparation for cyber terrorism, detection, block, and response capabilities should be strengthened against cyber attacks. Furthermore, encryption technologies and equipment should be developed and their use increased, and key management structure for encryption keys should be built and an identity management system should be built as well. After the expected changeover in the wartime operational control in 2015, so that joint operations can take place with the US army, the joint information system has to be distributed and linked for renewal. To implement NCW, efforts must be made to strengthen information security capabilities and measures against cyber warfare. Furthermore, information security systems such as an integrated security management system, authentication system, and a anti-virus system must be built and run [6–8].

4 Conclusions

With the growth of IT and changes in how warfare is conducted, the importance of NCW is being underscored. Amid this backdrop, the importance of information security is also being underscored, which deals with how to protect information and systems in this kind of a network environment. This paper examined threats to information security in NCW, as well as requirements for information security in NCW. It also proposed a development direction for information security, which would be able to improve efficiency in the area of information security if used as reference material when developing a network environment in the future.

References

1. Bae, D.H., Cho, Y.G.: Operational attribute and principle of computer network operations (CNO). J. Nat. Def. Stud. **52**(2), 39–70 (2009) (Research Institute for National Security Affairs of Korea National Defense University)
2. Ko, S.J.: Smallness unit a commander telecommunication system embodiment plan research, PaiChai University, Korea (2005)
3. Forouzan, B.A.: Cryptography and network security, McGraw-Hill, Seoul (2008)
4. Defense acquisition program administration, defense capacity improvement project management regulation (2006)
5. Ministry of National Defense: instruction on the management of defense information system interoperability and standardization, Korea (2003)
6. DoD: Department of Defense Directive 4630.5, interoperability and supportability of information technology and national security systems (2004)
7. DoD: DoD Instruction 4630.8, procedures for interoperability and supportability of information technology and national security systems (2004)
8. DoD: Chairman of the Joint Chiefs of Staff Instruction 6212.01.C, interoperability and supportability of information technology and national security systems (2003)

Part VI
Smart Grid Security and Communications

An IHD Authentication Protocol in Smart Grid

Ming-Yu Hsu, Yao-Hsin Chen, Shiang-Shong Chen, Wenshiang Tang, Hung-Min Sun and Bo-Chao Cheng

Abstract Recently, the security of Smart Grid has become an important issue. There are many security issues which should be addressed. With that, many previous studies have discussed these issues. However, most of them focus on the overall Smart Grid system rather than the Home Network Area (HAN) in Advanced Metering Infrastructure (AMI) network. The HAN has become a critical issue of Smart Grid due to save energy and to reduce energy peak loading. In-Home Display (IHD) is a useful device in the HAN. Therefore, in this paper, we design a secure authentication protocol between IHD and Head End. The protocol is compatible with IEC 62056-53 which is a Smart Grid standard. Furthermore, the proposed authentication is the first protocol which takes the relationship

M.-Y. Hsu (✉) · Y.-H. Chen · H.-M. Sun
Department of Computer Science, National Tsing Hua University, Hsinchu, Taiwan, R.O.C
e-mail: selhish@is.cs.nthu.edu.tw

Y.-H. Chen
e-mail: yaohsin.chen@is.cs.nthu.edu.tw

H.-M. Sun
e-mail: hmsun@cs.nthu.edu.tw

S.-S. Chen · W. Tang
Industrial Technology Research Institute, Chutung, Hsinchu, Taiwan, R.O.C
e-mail: ker.sschen@itri.org.tw

W. Tang
e-mail: wenshiang.t@itri.org.tw

B.-C. Cheng
Department of Communications Engineering, National Chung-Cheng University, Chiayi, Minhsiung, Taiwan, R.O.C
e-mail: bcheng@ccu.edu.tw

S.-S. Yeo et al. (eds.), *Computer Science and its Applications*,
Lecture Notes in Electrical Engineering 203, DOI: 10.1007/978-94-007-5699-1_86,
© Springer Science+Business Media Dordrecht 2012

between landlord and tenant into account. We leverage a cellphone to accomplish this purpose. In this paper, the security analysis of the proposed protocol is also given.

Keywords Authentication · In-home display · AMI · Smart grid

1 Introduction

A Smart Grid is proposed to enable energy savings and carbon emissions reductions. It is a digitally enabled electrical grid with communication ability. Also, it could gathers, distributes, and acts on information about the behavior of suppliers and consumers to improve the efficiency, reliability, economics, and sustainability of electricity services. In general, Smart Grid infrastructure includes power station, power substation, transmission line and Advanced Metering Infrastructure (AMI). AMI is viewed as the foundational building block and it enables a two-way communication between a utility and its customers and interconnections to distributed energy resources. In this infrastructure, the electrical meters are replaced with smart meters to achieve two-way communication function. Through this equipment, utility can rapidly receive consumers energy usage information. Rely on this information, utility could estimate or predict the total energy consumption to make energy generation efficient. Further, smart meters could communicate with in-home displays (IHDs) to make consumers more aware of their energy usage. Choi et al. though that the smart metering system is as an effective method for improving the pattern of power consumption of energy consumers [1]. Smart metering system can be categorized into two major types Advanced Metering Infrastructure (AMI)-related system and Energy Management System (EMS)-type system. The former usually is consisted of the AMI system and a IHD, which used to show information of power consumption and price rate change. The latter is constructed by the AMI and the EMS system which includes an IHD, where this IHD has more functions and typically has high resolution color display to present various kinds of information [1]. From the above discerption, we know that IHD plays an important role in power saving. In addition to an authentication between a IHD and a head end (HE), the user relationship between a landlord and a tenant is also considered. This user relationship is very common in society, but few previous studies discuss it. The tenant can obtain the power usage information through his own IHD if he is authorized by the landlord. Also, the landlord can query each tenant individual usage in order to charge the rent. The authentication protocol is secure and efficient.

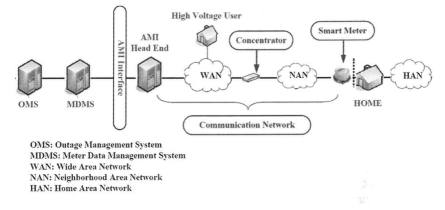

OMS: Outage Management System
MDMS: Meter Data Management System
WAN: Wide Area Network
NAN: Neighborhood Area Network
HAN: Home Area Network

Fig. 1 The Taiwan's AMI system topology [2]

2 Background and Related Works

Across the world, utilities are focusing on deploying AMI systems. Very often, however, they believe that AMI could pave path towards smart grid. In reality, that is not the case. AMI is important, although it is not synonymous with smart grid but it can be more accurately thought of as one of the early steps on the road to smart grid. Here, we will introduce an AMI systems in Taiwan [2]. This AMI system topology includes three kinds of the communication network, as shown in Fig. 1, and they are identified as follows.

- A wide area network (WAN) is a computer network in charge of a wide area. The WAN is used to connect an AMI head end and a data concentrator or a smart meter. The candidates of the communication protocols for WAN could be few wireless protocols like 3G and general packet radio service (GPRS) or few wired standards like optical fiber communication.
- A neighborhood area network (NAN) plays a role in the home-to-home or home-to-grid communication. Here, it serves a communication link between a data concentrator and smart meters.

A home area network (HAN) is a local area network (LAN) inside the house. It consists of some smart appliances, some intelligent digital devices, IHDs, and smart meter.

The IHD could be the center of the home energy management system to provide service for user like inquire the electric bill or show a chart about daily electric trend. Since the IHD would query the smart meter to respond the meter reading information, a data exchange protocol should be the same as that is used between smart meter and data concentrator or head end. It is due to the reason that smart meter could not be expensive such that it has limit processing ability and memory to support few protocols.

IEC 62056 is a set of international standards for electricity metering, including data exchange for meter reading, tariff and load control, and it is proposed by International Electrotechnical Commission (IEC). These standards are the international standard versions of the device language message specification (DLMS)/ companion specification for energy metering (COSEM) specification [3]. The DLMS defines the protocols into a set of four specification documents namely green book, yellow book, blue book and white book, where green book describes the architecture and protocols including the security mechanisms.

According to the Electric Power Research Institute (EPRI) [4], one of the biggest challenges facing the smart grid development is related to cyber-security of systems. Also, data security and privacy remain top concerns for utilities and consumers in Smart Grid [5–7]. It is due to the fact that the increasing potential of cyber attacks and incidents resist this critical sector as it becomes more and more interconnected [8]. An solution, PKI-based communication protocol, is proposed to verify each component identity [8]. However, the computing power to processing the asymmetric encryption and decryption algorithms requires large. Southern Company, one of the largest utilities in North America, used combination of commercial hardware and software, and additional software to build a foundation for the future of its substation data management system [9]. The system uses two-factor user authentication, via strong passwords and RSA tokens, to provide the data access security.

3 Problem Statement, Attack Model and Assumptions

Traditionally, the power company hires some people to check whether the electrical meters work, and record the energy usage from the electrical meters. That not only wastes much source but easily makes some mistakes as recording the energy usage value. As shown in Fig. 1, we can see the architecture of each smart grid component. Without human checking, smart meter will transmit the meter reading information to data concentrator (DC) and DC will automatically forward to HE. If a user wants to know the current energy fee, he obtains this information from the IHD, via smart phone or pad, after authenticating his access right and identification. Therefore, how to design a suitable and safe security protocol for authenticating user identity and authorizing access right is an important problem now. Unfortunately, there exists no proper security protocol, which is designed focus on the IHD and HE. And this is what we need to do.

3.1 Attack Model

The attacker's goal is to obtain the shared secret (i.e., SK) between the user's IHD and meter. With the shared secret, the attacker can read power consumption from the victim's meter via his own IHD. In this paper, we assume that the attackers could intercept the communication messages between each component. And attackers can manipulate or modify any message between the communication. A meter can be replaced with a fake one by the attackers. We assumed that the attackers cannot break basic cryptographic primitives. In another word, attackers cannot decrypt the AES algorithm without the secret key. Attackers also cannot break the hash function, in the other word, it is irreversible.

3.2 Assumptions

The assumptions in our system are described as follows. The service provider (i.e., Taipower) develops a website for authority. Through this website, a landlord can authorize his tenant to install an IHD for acquiring energy consumption information from the smart meter which is own by the landlord. The authority setup procedure in the website is assumed secure. A tenant has a cellphone with Short Message Service (SMS) ability. SMS messages are encrypted while transmitting. A landlord knows his tenant's phone number. IHDs can directly communicate with the HE via Internet.

4 The Proposed Scheme

In this section we describe the proposed system in detail. The system consists of setup procedure and IHD authentication protocol.

4.1 Setup Procedure

The aim of setup procedure is to allow a landlord to authorize his tenant for installing an IHD. In other words, before a tenant performs the IHD authentication procedure, his landlord has to authorize him. To achieve this, the landlord has to apply an account in the Tai-power website first. Please note that only the landlord has right to apply an account because he is the house and the meter owner in business perspective. In our system, we leverage phone numbers to achieve authorization. The landlord logs into the website and setup a phone number of his tenant. After that, the HE can authenticate the tenant's IHD via the phone number.

Fig. 2 Notations

Name	Description
$nonce_i$	A nouce sent from IHD to HE
$nonce_h$	A nouce sent from HE to IHD
ID_i	IHD's ID
ID_m	Meter's ID
SK	A shared secret between IHD and meter
PN	A tenant's phone number
$Challenge_i$	a random string generated by IHD
$Challenge_m$	a random string generated by meter
$\mathcal{H}(\circ)$	Hash function $\mathcal{H}$ with input $\circ$
$Enc\{\ \}_k$	symmetric encryption with key k

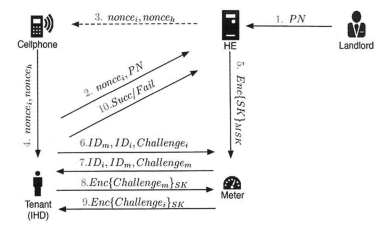

Fig. 3 The IHD authentication protocol

4.2 IHD Authentication Protocol

Figure 2 shows the notations used in our authentication protocol. Figure 3 depicts the IHD authentication protocol. At beginning, the landlord setups the tenant's phone number (*PN*) to HE via the Tai-power website. This step can be done at any time before the tenant starts to authenticate his IHD. The IHD authentication protocol can be roughly divided into two parts. The first part (step 2–5) is to generate a shared secret which is computed by the IHD and meter respectively. The second part (step 6–9) is to authenticate the IHD.

In step 2 of Fig. 3, the IHD sends a message, including a nonce (*nonce_i*) and tenant's phone number (*PN*), to the HE. After receiving the message, the HE checks the PN whether it is same as the phone number which is provided by the landlord or not. If the PN is different, the HE rejects the authentication request of IHD. Otherwise, the HE transmits the *nonce_i* and the *nonce_h* to the tenant via SMS

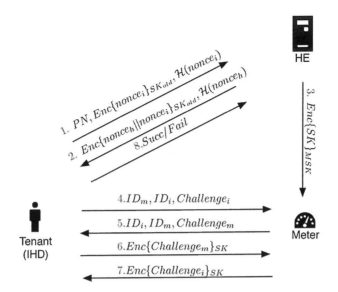

Fig. 4 The rekey protocol

channel according to the tenant's phone number. Once the tenant receives the SMS message on his cellphone, he correctly inputs the $nonce_i$ and $nonce_h$ into his IHD. At this point, the IHD has enough information to compute a shared secret by the following operation:

$$SK = \mathcal{H}(PN \| nonce_i \| nonce_h). \qquad (1)$$

In step 5, the HE also computes the same shared secret SK by the same operation. According to IEC 62056-53 standard, both HE and meter have a shared key called MSK. The HE encrypts the computed shared secret SK with the key MSK using AES-128. Then, the HE sends the encrypted SK to the meter. The meter can decrypt the encrypted message to obtain the important SK. The first part of IHD authentication is finished and it goes into second part.

The steps of second part are same as a high level security (HLS), introduced in [3]. The IHD and the meter can authenticate each other if they possess same SK. The IHD sends a request message, including ID_m, ID_i, and $Challenge_i$, to the meter. The meter responses a message, including ID_i, ID_m, and $Challenge_m$, to the IHD. Then, the IHD encrypts the $Challenge_m$ with the key SK using AES-128 and sends it out. After receiving the cypher text, the meter decrypts and verifies the authenticity of $Challenge_m$. If the received $Challenge_m$ equals the previously generated Challenges, the IHD is authenticated; otherwise, the meter will reject the IHD. Upon successful authentication, the meter sends back a encrypted $Challenge_i$ to the IHD. The IHD also can decrypt and verify the received $Challenge_i$. If pass, the IHD will send a success message to notify the HE; otherwise, the IHD rejects the whole authentication protocol. In this case, the meter might be replaced by a fake one which is controlled by attackers.

In addition to IHD authentication, we also design a rekey protocol. It is due to the fact that the shared secret should be changed periodically. In practice, the tenant maybe changes. Figure 4 is our rekey protocol which is different from the IHD authentication protocol. The tenant's cellphone is not involved in this rekey protocol. In our system, once IHD and meter communicate after a period of time, the IHD should send a rekey message to the HE. The rekey message includes *PN*, encrypted $nonce_i$, generated by the IHD for a current rekey process, with the old shared secret *SK*, and a hash value of $nonce_i$. After receiving the message, the HE decrypts the ciphed text and obtains the new $nonce_i$. The HE generates a new $nonce_h$ and securely sends back to the IHD. Once the IHD correctly gets the $nonce_h$, it can compute a new shared secret *SK* and sends *SK* to the meter in the secure way. At this point, both the IHD and the meter have same *SK*. They start to perform the HLS process which is same as the IHD authentication protocol. If there is no error, the IHD will send a successful message to notify the HE that the whole rekey process is finished correctly.

5 Security Analysis

In this section, we discuss that how our proposed IHD authentication and rekey protocols prevent the Man-in-the-Middle attack and replay attack in our system.

- **Man-in-the-Middle Attack**. The communication channel between IHD and HE is the public Internet; it is same between HE and meter. Therefore, an attacker has ability to intercept messages transmitted in this two channel. In fact, the attacker can capture the message 2 and message 5 in Fig. 3. The attacker's goal is to compute the shared secret *SK*. However, he cannot obtain the other information $nonce_h$ which is sent by secure *SMS* channel. Besides, the attacker also cannot decrypt the ciphered SK without *MSK*. Therefore, the attack cannot break the IHD authentication protocol. Furthermore, the attacker cannot obtain enough information to generate the new shared secret in the rekey protocol due to the fact that the information is all encrypted with the previous shared secret.

- **Replay Attack**. In this case, an attacker try to cheat the meter by replaying messages which are intercepted in other communications. This kind of attack still doesn't work, because the parameters, $nonce_i$, $nonce_h$, $Challenge_i$, and $Challenge_m$, are randomly generated in each authentication and rekey procedure. The captured messages are useless in another round. Hence, the both protocols prevent from replay attack.

6 Conclusion

In this paper, we proposed an authentication protocol between IHD and HE in smart grid system. In our best knowledge, the IHD authentication is the first protocol which takes the relationship between landlord and tenant into account in smart grid. The landlord can authorize his or her tenant to install an IHD for reading energy consumption from this smart meter. In our design, the landlord can arbitrarily revoke the authorization once the renting relationship is finished. The rekey protocol is also designed. Furthermore, our protocol is compatible with the IEC 62056-53 standard. For security, these two protocols resist to Man-in-the-Middle attack and replay attack.

Acknowledgments The financial support provided by Bureau of Energy is gratefully acknowledged.

References

1. Choi, T., Ko, K., Park, S., Jang, Y., Yoon, Y., Im, S.: Analysis of energy savings using smart metering system and in-home display (ihd). In: Transmission and distribution conference and exposition: Asia and Pacific 2009 IEEE, pp. 1–4 (2009)
2. Hsu, P.H., Tang, W., Tsai, C., Cheng, B.C.: Two-layer security scheme for ami system in Taiwan. In: Ninth IEEE international symposium on parallel and distributed processing with applications workshops (ISPAW), pp. 105–110, May 2011
3. DLMS user association, technical report: companion specification for energy metering green book 7, DLMS UA 1000-2 Ed. 7.0, 22 December 2009
4. Von Dollen, D.: Report to nist on the smart grid interoperability standards roadmap. EPRI, Contract No. SB1341-09-CN-0031–Deliverable 7 (2009)
5. Polonetsky, J., Wolf, C.: How privacy (or lack of it) could sabotage the grid. Smart grid news (2009)
6. Simmhan, Y., Kumbhare, A., Cao, B., Prasanna, V.: An analysis of security and privacy issues in smart grid software architectures on clouds. In: IEEE international conference on cloud computing (CLOUD), pp. 582–589 (2011)
7. McDaniel, P., McLaughlin, S.: Security and privacy challenges in the smart grid. Security Privacy, IEEE **7**(3), 75–77 May–June 2009
8. Metke, A., Ekl, R.: Security technology for smart grid networks. In: IEEE Transact. Smart Grid. **1**(1), 99–107, June 2010
9. Sologar, A., Moll, J.: Developing a comprehensive substation cyber security and data management solution. In: Transmission and distribution conference and exposition, 2008. T&# x00026; D. IEEE/PES, IEEE, pp. 1–7 (2008)

Flexible Network Design for Wide Area Measurement Protection and Control

Di Cao, Adam Dysko, Craig Michie and Ivan Andonovic

Abstract A Wide Area Measurement, Protection and Control (WAMPAC) strategy yields great potential for the upgrade of the supervision, operation, protection and control of modern power systems. A flexible, scalable network infrastructure is the key to enabling these enhanced functionalities. Information exchange among distributed synchrophasors in a national area requires a wide ranging capability for the acquisition and exchange of data. Although those phasor measurements are synchronized by the Global Positioning System (GPS), any time delay in monitoring and control services are damaging especially under fault conditions. The paper summarizes the communication needs of a range of advanced applications within the WAMPAC methodology. Network architectures capable of supporting the requirements of advanced power system control strategies are proposed. The performance of a number of candidate communication systems are analyzed through simulation using the deployed power system topology in United Kingdom as the basis.

Keywords WAMPAC · Phasor measurement unit · WAMPAC architecture · Centralized control and decentralized control strategies · Quality of service

D. Cao (✉) · A. Dysko · C. Michie · I. Andonovic
Department of EEE, University of Strathclyde, Glasgow, UK
e-mail: d.cao@strath.ac.uk

A. Dysko
e-mail: a.dysko@strath.ac.uk

C. Michie
e-mail: c.michie@eee.strath.ac.uk

I. Andonovic
e-mail: i.andonovic@eee.strath.ac.uk

S.-S. Yeo et al. (eds.), *Computer Science and its Applications*,
Lecture Notes in Electrical Engineering 203, DOI: 10.1007/978-94-007-5699-1_87,
© Springer Science+Business Media Dordrecht 2012

1 Introduction

Flexible communications is one of the core elements in the implementation of smart grid [1]. With the resultant migration to more complexity in power flows, established monitor and control strategies are becoming less effective and cannot fulfill increasing demands from generation and of consumers. The absence of flexible, easily configured, scalable communication platform results in the segmentation of the power system into islands lacking any situational awareness.

As a result of CO_2 reduction legislation, National Grid in the UK is targeting the deployment of up to 30 GW of wind generation, predominantly offshore, to the existing transmission system by 2020 [2]. It is understood that this level of wind will drastically change the dynamics of the grid and result in radical changes to the transmission power flows. This brings concomitant challenges in the manner the transmission system is monitored and controlled, the overall requirement being increasing levels of flexibility in inter-connection in order to maintain security of supply.

Therefore, Wide Area Measurement, Protection and Control (WAMPAC) is becoming increasingly significant for power systems and is widely accepted as a new competitive candidate to enable the smart transmission for smart grids [1, 3, 4]. Global Positioning Systems (GPS) enabled Phasor measurement units (PMUs) form the foundation of approach, monitoring the dynamic changes in power flow. As a consequence, a wide area communication network is required to support the changing dynamics in data exchanges among information resources, control centers and controllable devices. Any network design will have to take into consideration factors such as data volume, allowable delay and reliability according to different applications enabled by WAMPAC principles.

Control strategies are governed by the locations of multiple data sources, the decision making and the elements actuating control. Presently two strategies have been utilized widely viz. centralized and decentralized control and their advantages/disadvantages have been investigated in detail predominately within the context of traditional power systems [5]. Both will be considered within the paper.

This paper is organized as follows. Section 2 presents the communication requirements for WAMPAC based implementations. In Sect. 3, three potential control strategy driven network architectures are proposed. The performance of candidate communication networks are analyzed through simulated for the deployed UK power system in Sect. 4 together with the results of the characterization. Finally, conclusions are drawn in Sect. 5.

2 Communications for WAMPAC

2.1 Overview for WAMPAC

WAMPAC monitors the status of power systems via PMUs located at critical points along the power transmission path. The traditional usage of PMUs has been in post events processing. The logged phasor data in the server helps to aid understanding of the system operation and records the cause of any power fluctuations and blackouts. After enjoying years of development, currently a GPS based high-precision PMU can support an enhanced wide area monitoring, protection and control applications.

2.2 Quality of Service Requirements in WAMPAC

In addition to the coverage, the performance of the applications can be characterized by throughput, latency and reliability. Throughput represents the sum of the data rates to be delivered to all terminals in a network. The data volume of various applications range from a few hundred bytes to a few mega-bytes every second e.g. visualization requires a large volume of data from all PMUs whereas adaptive protection information requires less than 100 bytes directed to a specific controllable device.

Latency indicates the maximum end-to-end (ETE) delay that can be tolerated. In the case of monitoring applications, the ETE delay from the measurement device to the control center can be represented as:

$$T_{monitoring} = T_{process\_sens} + T_{trans} + T_{prop} + T_{queue} + T_{process\_cc} \qquad (1)$$

where $T_{process\_sens}$ and $T_{process\_cc}$ is the computation time at the sensor node and control center respectively; T_{trans} is the transmission time governed by the data rate of the link; T_{prop} is the time consumed on the link distance from transmitter to the receiver; and T_{queue} is the time for each packet is stored in a queue until it can be transmitted. In addition the delay owing to the control of relays can be presented as:

$$T_{control} = T_{monitoring} + T_{decision} + T_{trans'} + T_{prop'} + T_{queue'} + T_{operation} \qquad (2)$$

where the $T_{decision}$ and $T_{operation}$ is the execution time at the control center and end control device. Furthermore, the delay associated with the transmission of a control signal direct to controllable devices must also be taken into consideration.

Reliability is recognized to be the level of assurance in the delivery of data packets. The application demanding a high degree of reliability might result in the sacrifice of bandwidth in order to exchange acknowledgements.

2.3 Communication Requirements for WAMPAC Applications

The consistent development of WAMPAC has enabled the feasibility of several applications to be proposed and proved through theoretical analysis. The communication needs for each application have been discussed in [6] and [7]. Table 1 lists all the advanced applications supported by WAMPAC, categorized into wide area monitoring, wide area protection and wide area control.

The monitoring applications enable dynamic estimation of the system with synchronized phasor data and act as the first step in protection and control. Most of the power protection functions enhanced by WAMPAC have a slow response requirement [8]. Information for adaptive protection schemes needs less than 100 bytes to achieve the balance between the dependability and security in the system. Instead, intelligent load shedding and islanding are also important functions within smart grid self-healing environments. The effective management of fault conditions, normally reported through a series of alarms, requires a rapid decision and remedial action to be made in order to stop the propagation of the fault and lessen its impact on overall system operation. Wide area control is the most attractive application and the real time control functions significantly decrease the probability of large area blackouts.

3 Control Strategy Based Network Design

An extensive PMU-based WAMPAC solution acquires measurements from multiple, appropriately located and distributed PMUs across a national area. The processing of the data can be executed either locally or managed by a control center tens to hundreds of kilometers away.

With a centralized control strategy, all information resources send data to a Central Control Center (CCC). After processing the received data, appropriate decisions are made and related commands are transmitted to controllable devices. For example, the Supervisory Control and Data Acquisition (SCADA) system widely used today is designed as a tree topology where all data sources are scanned by the control center every 2–5 s. The advantage of such a system is no system synchronization is required since the entire network is managed through the same supervisory control system. However, major drawbacks of such an approach include traffic congestion, delay of the measurements and single-node failure.

With a decentralized control strategy, the entire system is segmented into small clusters, each comprising its own distributed control center (DCC) (see Fig. 1). The distributed control center processes the acquired data but unlike the CCC implementation, DCCs share data to enable situation awareness throughout the entire system. Thus a decentralized control scheme reduces the bottleneck at high

Table 1 WAMPAC applications and their associated communication requirements

Task	Throughput (Bytes/s)	Latency (ms)	Reliability	ACK requirement	Service priority
PMU estimator/Institute transformer calibration	3,060 × No. PMUs	<100	Medium	None	Level 4
Seams state estimates	102 × No. Boundary PMUs	Not critical	Medium	None	Level 5
Adaptive dependability and security	<100	Not critical	High	ACK required	Level 6
Monitoring approach of apparent impedance	<100	Not critical	Low	None	Level 7
Adaptive out-of-set	3,060 × No. Generators	50	Medium	None	Level 4
Supervision of back-up zones	<1,000	50	High	ACK required	Level 3
Adaptive loss-of-field	<100	50	High	ACK required	Level 3
Intelligent load shedding	3,000 × No. Tie lines	50	Medium	None	Level 2
Intelligent Islanding	100 × No. Circuit breakers	50	High	ACK required	Level 2
Control of inter-area oscillations	91,800	200	Medium	None	Level 1
Control of transient events	153,000	50	Medium	None	Level 1

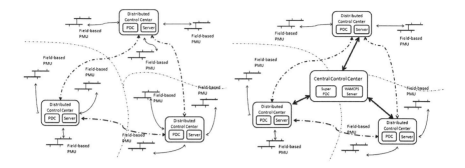

Fig. 1 Representation for decentralized control and hybrid control strategies

data volumes and provides robustness to single node failure; however network complexity increases due to inter-networking between DCCs.

A hybrid control strategy combines elements drawn from both centralized and decentralized algorithms to achieve a balance between the traffic bottleneck and network complexity. The data flow is decided by the WAMPAC application, predominately categorized into distributed monitoring service and central control service. In the former, state estimation is invoked locally to avoid large amount of data flow in the backbone network. However, some adaptive protection schemes require a more comprehensive understanding of the state of the entire network and that decision making can only be executed within the CCC. Moreover, it is essential for the entire system to operate holistically under fault conditions; thus protection applications, such as intelligent load shedding and islanding communicate directly with the CCC (Table 2).

4 PMU Deployments and Network Performance

4.1 OPNET

The simulator environment used to evaluate the performance of the candidate network solutions is the OPNET Modeler [9], a powerful and comprehensive modeling and simulation software dedicated to communication network research and development. OPNET adopts object-oriented analyses, provides graphic edit interfaces, large amount of models and network technologies and has location awareness functionality.

Table 2 Traffic flows for WAMPAC applications as a function of different control strategies

Applications	Centralized network	De-Centralized network	Hybrid network
PMU_UPLOAD	Distributed PMU → Centralized Control Center	Distributed PMU → Distributed Control Center	Distributed PMU → Distributed Control Center
Seams state estimate	None	Zone Boundary PMU → Distributed Control Center	Zone Boundary PMU → Distributed Control Center
ADS	Centralized Control Center → Distributed relay	Distributed Control Center → Distributed relay	Centralized Control Center → Distributed relay
AOS	Distributed Generator → Centralized Control Center	Distributed Generator → Distributed Control Center	Distributed Generator → Distributed Control Center
SBUZ	PMU → Backup Relay	PMU → Backup Relay	PMU → Backup Relay
ALF	Centralized Control Center → Loss-of-Field relay	Distributed Control Center → Loss-of-Field relay	Centralized Control Center → Loss-of-Field relay
ILS	ISO Boundary PMU → Centralized Control Center	Zone Boundary PMU → Distributed Control Center	Zone Boundary PMU → Centralized Control Center
II	ISO Boundary PMU → Centralized Control Center	ISO Boundary PMU → Distributed Control Center	ISO Boundary PMU → Centralized Control Center
Inter-area oscillation control	25 remote PMU → Centralized Control Center	25 remote PMU → Distributed Control Center +Information exchange between all Distributed Control Center	25 remote PMU → Distributed Control Center +Distributed Control Center → Centralized Control Center

Fig. 2 PMU deployments in
United Kingdom according to
an optimization based on ILP

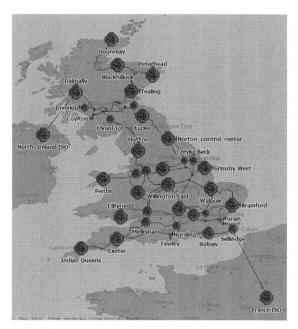

4.2 PMU Placements and Connection in the United Kingdom

Due to the high cost of a single PMU and supporting communication infrastructure, many studies on how to maximize the observability of the network with the minimum number of PMUs have been undertaken. PMUs are only required in 1/4 to 1/3 of network buses to ensure a full observability. Many algorithms that define optimum placement have been proposed such as Genetic algorithm [8], Integer Programming [10], binary searching [11]; the research in [12] proves that the Integer Linear Programming (ILP) algorithm is the most effective solution to determine the optimal position of the PMUs with least computation time. Thus in this paper, the ILP is used to determine PMU placements of for the UK system.

According to ILP, for a n-bus system, PMU placement can be formulated as:

$$min \sum_{i}^{n} f^{T}(i) \cdot x_i \quad \text{Subject to} Ax > 0 \tag{3}$$

where x is the binary decision variable vector, whose entries are defined as: x_i equals to 1 when a PMU is installed at bus i; otherwise, x_i is equal to zero. f^{T} indicates the cost of an installed PMU at the bus all set to be one. Figure 2 shows the results of PMUs placements derived from optimization. In total, 50 PMUs are required to achieve a full-observability of the network in UK.

Communication links interconnect all substations where PMUs are located and establish a backbone network. All backbone links use E1 links at 2.048 Mbit/s. Links internal to each substation interconnect the PMU, PDC and other

Table 3 ETE delay of each WAMPAC applications in different scenarios

	Centralized control-64 Kbps	Decentralized control-64 Kbps	Hybrid control-64 Kbps	Centralized Control-1 Mbps	Decentralized control-1 Mbps	Hybrid control-1 Mbps	Criteria
PMU_UPLOAD	546.9	249	254.6	543	44.5	39.3	100
SEAMS critical		67.4	52.8		18.2	10.2	Not
ADS critical	202.2	5.4	21.4	146.8	5	38.8	Not
ALF	211.7	2.3	23.1	79.3	1.8	35.3	50
AOS	461.3	232.9	232.7	456.8	42.7	42.9	50
SBUZ	9.6	96.9	56.8	19.1	25	17.3	50
II	400	8	29.5	156	29	60	50
ILS	453.1	230.8	231	372.9	39.5	67.4	50
Inter-area control	1256.4	237.9	239.2	710.4	270.1	294	200
Information exchange		75.2	26.1		94.4	32.4	

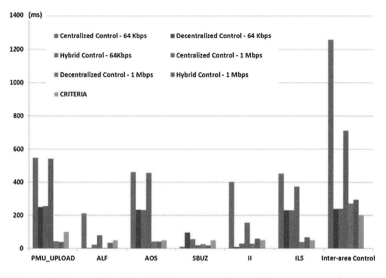

Fig. 3 Network delay as a function of different control strategies

controllable devices. Two scenarios at different channel capacity are simulated; 64 kbit/s is chosen to verify previous studies [13], where 64 Kb was estimated as the minimum capacity required; 1 Mbit/s is simulated to verify the existing link capacity for SCADA.

Based on the PMU topology and link topology described above, three control scenarios are simulated using OPNET. The 'Norton_control_center' is selected as the CCC by using the minimum spanning tree algorithm. In the cases of decentralized and hybrid control strategies schemes, the power system is segmented into 7 zones according to the National Grid SYS study divisions [2], with 2 zones in SHETL and SPT, 5 zones in the NGET system. A representative DCC is selected in each zone performing the same functionalities as the CCC.

4.3 Latency Comparison and Analysis

The End-to-End (ETE) delay criteria for different WAMPAC applications are listed in Table 3, used to verify the performance of candidate networks. Seams state estimation and Information Exchange are the two applications analyzed for decentralized and hybrid control strategies.

Figure 3 shows the comparison for ETE delay as a function of different time-critical applications. As shown, inter-area control application experiences the longest delay compared to the others attributed to the relatively long transmission distances between adjoining ISOs. In contrast, since the traffic flow in SBUZ is directed to controllable devices, less queuing delay will be added up to the total ETE delay time. The significant ETE delay for all centralized control based schemes does not meet the requirements of the system. However, both the decentralized and hybrid control based systems have acceptable performance for most applications, the hybrid scheme having the added feature of lower network complexity.

Also evident is that a 64 kbit/s channel capacity cannot fulfill the delay requirements for the UK power system, especially for those applications requiring high volume of data; traffic congestion at the router increases packet losses. A decentralized or hybrid control based network at 1 Mbit/s channel capacity can support the operation of all WAMPAC applications across the UK grid.

5 Conclusions

WAMPAC functionalities represent aspect core element facilitating efficient implementation of smart grid operation. The deployment of WAMPAC applications are enabled by a flexible, scalable networking platform inter-connecting widely spread distributed PMUs, the core component of WAMPAC. Large volume of data exchanges between measurement devices, control centers and controllable devices make the design of the network infrastructure challenging. The paper presents potential network solutions that fulfill the requirements of WAMPAC applications. Data sources, here governed by the locations of the PMUs, and data flow, influenced by the control strategy, all impact on the design of the network.

The proposed network solutions are evaluated in respect of delay through simulation for the case of the UK power transmission system. PMU-based WAMPAC applications are implemented and network performance results confirm the successful operation in terms of delay for all WAMPAC advanced monitoring, control and protection applications. The successful use of both de-centralized and hybrid control strategies are also verified, meeting the latency and data volume requirements. The ETE delay in monitoring and control schemes are within the 100 and 50 ms threshold respectively. The inter-area control signal is the most challenge application, which also has an acceptable 270 and 294 ms latency. Future work will consider Quality of Service (QoS) with the aim of enhanced control of a rich mix of data flows and services.

References

1. Ree, J.D.L., Centeno, V., Thorp, J.S., Phadke, A.G.: Synchronized phasor measurements and applications in power systems. IEEE Trans. Smart Grid 1(1), 20–27 (2010)
2. National Grid, 2011 NETS Seven Year Statement, 2011
3. Phadke, A.G., Thorp, J.S.: Synchronized Phasor Measurements and Their Applications. Springer, Heidelberg (2008)
4. Phadke, A.G.: The wide world of wide-area measurement. IEEE Power Energy Mag. 6, 52–65 (2008)

5. Shahraeini, M., Javidi, M.H., Ghazizadeh, M.S.: Comparison between communication infrastructures of centralized and decentralized wide area measurement systems. IEEE Trans. Smart Grid **2**(1), 206–211 (2011)
6. Phadke, A.G., Thorp, J.S.: Communication needs for wide area measurement application, Critical infrastructure (CRIS), 5th International Conference, September 2010
7. Deng, Y., Lin, H., Phadke, A.G., Shukla, S., Thorp, J.S., Mili, L.: Communication network modeling and simulation for wide area measurement applications. In: Proceedings of the 2nd IEEE International Conference on Smart Grid Communication, October 2011
8. Marin, F.J., Garcia-Lagos, F., Joya, G., Sandoval, F.: Genetic algorithms for optimal placement of phasor measurement units in electric networks. Electron **39**, 1403–1405 (2003)
9. OPNET Technologies, Inc. http://www.opnet.com
10. Xu, B., Abur, A.: Observability analysis and measurement placement for system with PMUs. In: Proceedings of the IEEE Power System Conference and Exposition, October 2004, vol. 2, pp. 943–946
11. Chakrabarti, S., Kyriakides, E.: Optimal placement of phasor measurement units for power system observability. IEEE Trans. Power Syst. **23**(3), 1433–1440 (2008)
12. Yuill, W., Edwards, A., Chowdhury, S., Chowdhury, S.P.: Optimal PMU placement: a comprehensive literature review. In: Power and Energy Society General Meeting, 2011
13. Chenine, M., Karam, E., Nordstrom, L.: Modeling and simulation of wide area monitoring and control system in IP-based networks. In: IEEE P&E Society Meeting, 2009

Green Communication and Corporate Sustainability of Computer Aided Audit Techniques and Fraud Detection

Ezendu Ariwa, Omoneye O. Olasanmi and Jaime Lloret Mauri

Abstract Fraud is prevalent in society and is increasing considerably over in recent years. Auditors have been criticized severally because they tend to reach conclusion based upon inadequate audit processes, and hence, the validity of the audit conclusion has often been questioned. This study thus sought to identify the various types of fraud encountered during financial transactions, evaluate the adoption of Computer Aided Audit Tools (CAATs) in fraud detection in an organization, appraise the impact of CAATs on performance of an organization, and ascertain the problems of CAATs' application within an organization. Data for the study was collected using a well-structured questionnaire which was distributed among 250 employees of an auditing company. The data were analysed using descriptive statistics and SPSS package, and the hypothesis of the study was analysed using regression analysis. The findings of the study showed that 72.8 % of the respondents agreed that CAATs have played a major role in fraud detection, and hence can be used to curb fraud to a minimal level in organisations. It was also discovered that CAAT helped to improve the auditors' performance as corroborated by 58.1 % of the respondents. However, quiet a handful of respondents (23.0 %) were undecided in this respect as they claimed that the improved performance has been associated with much stress and pressure. Furthermore, the cost of implementing CAAT, the required skills to be acquired for its usage, and

E. Ariwa (✉)
London Metropolitan University, London, UK
e-mail: e.ariwa@londonmet.ac.uk

O. O. Olasanmi
Department of Management and Accounting, Obafemi Awolowo University, Ile-Ife, Osun State, Nigeria

J. L. Mauri
Universidad Politecnica de Valencia, Valencia, Spain

S.-S. Yeo et al. (eds.), *Computer Science and its Applications*,
Lecture Notes in Electrical Engineering 203, DOI: 10.1007/978-94-007-5699-1_88,
© Springer Science+Business Media Dordrecht 2012

systems database requirements, were some of the problems found to be associated with the application of CAAT in organisations.

Keywords Fraud · CAAT · Audit software · Accounting information · Green communication · Corporate sustainability

1 Introduction

Fraud exists in all cycles of life and it is evident enough that if not checked, it could be on the increase day in day out. Fraud itself comprises a large variety of activities and includes bribery, political corruption, business and employee fraud, consumer theft, network hacking, bankruptcy and divorce fraud, and identity theft [5]. Fraud is an intentional act by one or more individuals among management, employees or third parties, which results in a misrepresentation of financial statements [6, 7]. Fraud can also be seen as the intentional misrepresentation, concealment, or omission of the truth for the purpose of deception or manipulation to the financial detriment of an individual or an organization which also includes embezzlement, theft or any attempt to steal or unlawfully obtain, misuse or harm the asset of an organization [1].

According to [9], fraud is an ever present threat to the effective utilization of resources and it will always be an important concern of management. Today, fraud it is a broad legal concept and auditors do not make legal determinations of whether fraud has occurred, rather the auditor's interest specifically relates to acts that result in a material misstatement of financial statements. The primary factor that distinguishes fraud from error is whether the underlying action that results in the misstatement of the financial statements is intentional or unintentional. There are different types of fraud but corporate fraud continues to be a pervasive problem and is seen to be costly [21, 23]. Most frauds occur due to the lack of adequate internal controls and the failures of auditors to easily detect fraud. The increased awareness of both fraud and the importance of transparent financial reporting have spurred the concern of regulatory bodies, as well as the accounting profession.

Over the years identifying and solving financial fraud, has continued to increase in different forms, both domestically and internationally within entities of all sizes and types. The forensic and Investigative Service Division of the independent accounting firm of KPMG conducted an extensive survey of fraud detection from a wide array of businesses and industries. According to various respondents it was discovered that 62 % of these entities had experienced at least one incident of fraud that year [20]. Among companies who responded to the KPMG survey in 2004, close to half had experienced a fraud costing them a total of $457 million [20]. The vast majority of the fraud reported in the KPMG survey related to misappropriation of assets. According to the Association of Certified Fraud Examinations (ACFE), an estimated $2.9 trillion worldwide were lost due to fraudulent financial statements, asset misappropriation, and corruption in 2009. Employee theft costs $40 billion a year with an average of $1,350 per employee, of which 75 % of them are undetected. The total losses of shoplifting amount to $13 billion a year with 800,000 incidents each day costing $142 loss per incident. However, only 5,000 incidents were caught. Identity theft is also becoming very serious. In the past five years there were 27.5 million victims of identity theft. These losses are ever increasing today. The above losses prove a point that detecting fraud is a constant challenge for any business [26].

Computer aided audit techniques (CAATs) is a growing field within the financial audit profession in Nigeria. CAATs are computer tools and techniques that an auditor (external or internal) uses as part of their audit procedure to process data of audit significance contained in an entity's information systems [25]. CAAT is the process of using computers to automate or simplify the audit process. Computer technology gives auditors a new set of techniques for examining the automated business environment. As early as 1982, CAATs was a powerful audit tool for detecting financial errors. It was in recent years, that analytical techniques became not only more powerful but also more widely used by auditors. Audit software allows its user to obtain a quick overview of operations of a business and to drill down into details of specific area of interest; it highlights individual transactions that contain features associated with fraudulent activities [24]. With audit software, millions of files can be examined, previous years' data can be used to identify anomalies, and it can also be used for comparison. Also computer aided audit technique is used in addressing suspected fraud situations. Audit software uses include comparing employee addresses with vendor addresses to identify employees that are also vendors; searching for duplicate check numbers to find photocopies of company checks; searching for vendors with post office boxes for addresses; analyzing the sequence of all transactions to identify missing checks or invoices; identifying vendors with more than one vendor code or more than one mailing address; finding several vendors with the same mailing address; and sorting payments by amount to identify transactions that fall just under financial control on contract limits [11]. It helps auditors focus their effort on areas of greater risk and transaction with high probability of fraud. Audit software also helps auditors extract information from several files with data base management system. Audit software also provides auditors with the ability to extract information from several files, with different database management systems, in order to search for underlying patterns or relationships among data. Computerized techniques assist the auditor in identifying symptoms early in the life of a fraud. This will serve to reduce the negative impact of many frauds before millions of money is lost or goodwill is destroyed [11].

CAAT is the practice of analyzing large volumes of data looking for anomalies. Using CAATs, the auditor will extract every transaction a business unit performed during the period reviewed. The auditor will then test that data to determine if there are any problems in the data. With audit software, millions of files can be examined, data's collected can be used to identify, analyze and comparisons can be made between different locations.

Computerized techniques and interactive software can help auditors focus their efforts on areas of greatest risk. Auditors can choose to exclude low risk transactions from their review and focus on those transactions that contain a higher probability of fraud. Audit software also provides auditors with the ability to extract information from several data's files with different database management systems, in order to search for underlying patterns or relationships among data [11]. Examples of CAATs include generalized audit software, custom audit software, test data, parallel simulation, and integrated test facility.

2 Statement of the Research Problem

Traditionally, auditors have been criticized because they reach conclusion based upon limited sample management of companies upon which research has been made to question the validity of the audit conclusion. Management realises that they conduct thousands of transactions a year and the auditor only sampled a handful. The auditor is then forced to defend their methodology. Auditors are concerned about assessing fraud detection accurately because of the high cost associated with too much or too little investigation. Doing too much investigation when fraud is not present adds little economic value to an auditor's job and failure to investigate sufficiently when fraud is present usually results in significant costs. This statement proves that fraud investigation, if not properly conducted, might lead to great negative consequences for an organisation. Thus, in order to prevent the criticisms and mistakes accompanying traditional methods of audit procedures, auditors are encouraged to incorporate computerized audit techniques in order to substantiate the results of the audit process and to be able to incorporate fraud detection into routine audit procedures without adding significant costs. Moreover, there is little evidence on the manner in which auditors employ CAATs in the pursuit of their audit objectives. This study thus makes a step to provide evidence on these issues.

3 Objectives of the Study

The objectives of this study are to:

1. identify the various types of fraud encountered during financial transaction;
2. evaluate the adoption of CAATs in organizations;
3. appraise the impact of CAATs on performance of organization; and
4. ascertain the problem of CAATs application in organizations.

4 Hypothesis of Study

H_o: CAATs have no role to play in the detection of fraud in an organization.

5 Literature Review

Fraud detection, unlike a financial statement audit, requires a unique skill set and forensic techniques developed for the sole purpose of detecting the evidence of fraud [14]. Specifically, it requires individuals who are skilled in the application of investigative and analytical tools related to the areas of accounting records, gathering and evaluating financial statement evidence, interviewing all parties related to an alleged fraud situation, and serving as an expert witness in a fraud case [25].

According to the Association of Certified Fraud Examiners (ACFE), a survey carried out on fraud detection as applied to the United State economy, it was shown that lack of adequate internal controls was most commonly cited as the factor that allowed fraud to occur [19]. The conclusion was that the implementation of anti-fraud controls appears to have a measureable impact on the organization's exposure to loss.

Additionally, two studies have explored the determinants of CAATs usage by generalist auditors [13]. These studies find that auditor willingness to employ CAATs in the financial statement audit is impacted by auditor perceptions of the usefulness of those procedures and concerns regarding the budget impact of CAATs usage. Firms play a significant role in these perceptions through training and other resources they provide, as well as through their communication of support for CAATs usage [12].

In Ireland, the Institute of Internal Auditors [17] conducted a survey touching on CAAT's usage. This survey revealed hesitation amongst those overseeing internal audit departments as relating to automation. The key reason cited for this was that they perceived there was a lack of software available in the market that met the needs of internal auditors. More than 40 % of respondents suggested that they would be willing to adopt an amended audit approach if they could find a package(s), that otherwise met their needs; this actually led to an increased usage of CAATs.

Debreceny, Lee, Neo, and Shuling [15] carried out a research on the usage of CAAT and Generalized Audit Software (GAS) in the banking industry in Singapore. GAS is a class of CAATs that allows auditors to undertake data extraction, querying, manipulation, summarization and analytical tasks [8]. It was discovered that the common uses of GAS in bank audits includes extraction of samples, identification of reactivated dormant accounts and verification of the completeness and accuracy of data. This study also found that GAS was frequently being used in special investigation audits and exceptional instances. From their findings, it was concluded that bank auditors do use GAS, but only to a limited extent because GAS was perceived as an interrogation tool to perform fraud investigations rather than as a general audit tool.

A study was carried out by Mahzan and Lymer [22] on the adoption of computer assisted audit techniques (CAATs) by internal auditors in the United Kingdom by Birmingham Business School. It was discovered that three top factors that influenced the respondents' decision whether to continue or not to continue using CAATs are the ability to train employees on the software usage, compatibility of the software with other departments' systems, and the ability of software to meet the data manipulation needs of the audit department.

Fraud attempts have seen a drastic increase in recent years, making fraud detection more important than ever. Fraud prevention and detection are related but not the same concepts. Prevention encompasses policies, procedures, training, and communication that stop fraud from occurring, whereas, detection focuses on activities and techniques that promptly recognize timely whether fraud has occurred or is occurring. While prevention techniques do not ensure that fraud will not be committed, they are the first line of defence in minimising fraud risk. Meanwhile, one of the strongest fraud deterrents is the awareness that effective detective controls are in place. Combined with preventive controls, detective controls enhance the effectiveness of a fraud risk management program by demonstrating that preventive controls are working as intended and by identifying fraud if it does occur. Although detective controls may provide evidence that fraud has occurred or is occurring, they are not intended to prevent fraud. Every organization is susceptible to fraud, but not all fraud can be prevented, nor is it cost-effective to try. It is thus important that organizations consider both fraud prevention and fraud detection [4].

6 Computer Assisted Audit Techniques

During the course of an audit, the auditor is to obtain sufficient, relevant, and useful evidence to evidence. In many instances, the computer can be used as an aid in obtaining audit evidence.

CAATs are 'techniques that use the computer as an audit tool' which are utilized in application of auditing procedures [10, 16, 18]. CAATs include tools that range from basic word processing to expert systems. Computerized audit techniques range from procedures as simple as listing the data in a given file to the use of Artificial Intelligence tools to predict financial failure or financial statement structures [15].

Arguably the most widely deployed class of CAATs is Generalized Audit Software (GAS). These packages are computer programs that contain general modules to read existing computer files and perform sophisticated manipulations of data contained in the files to accomplish audit tasks. They have a user-friendly interface that captures users' audit requirements and translates those user instructions or queries into program code. This is undertaken by interrogating the client's file systems or database and performing the necessary program steps [15].

CAATs are important tools for the auditor in gathering information from the organizational environment. They are referred to all techniques that make use of computer and computer programs. It consists primarily of test data packs and computer programs. CAATs are computer tools that extract and analyze data from computer applications [10]. They enable auditors to test 100 % of the population rather than a sample, thereby increasing the reliability of conclusions based on that test [2]. Recent audit standards encourage auditors to adopt CAATs to improve audit efficiency and effectiveness [3].

7 Fraud Detection and Prevention

Fraud detection is a topic applicable to many industries including banking and financial sectors, insurance, government agencies and law enforcement, and more. Fraud attempts have seen a drastic increase in recent years, making fraud detection more important than ever. Despite efforts on the part of the affected institutions, hundreds of millions of money are lost to fraud every year in different countries. In banking, fraud can involve using stolen credit cards, forging checks, misleading accounting practices, etc. In insurance, 25 % of claims contain some form of fraud, resulting in approximately 10 % of insurance payout dollars. Fraud can range from exaggerated losses to deliberately causing an accident for the payout. With all the different methods of fraud, finding it becomes a problem.

While the primary responsibility for fraud prevention and detection remains with the board and management of any organisation, the auditors can be a significant part of the organization's anti-fraud team. In a 2003 Fraud Survey published by KPMG, an international accounting and consulting firm, it was found that "almost two thirds of organizations surveyed reported discovery of fraud". The results of this and similar studies suggest that while an internal audit process does not prevent misappropriation of assets or misrepresentation of financial statements from happening, it does increase the probability of detecting fraud, thus resulting in smaller losses [20].

Fraud prevention and detection are related but not the same concepts. Prevention encompasses policies, procedures, training, and communication that stop fraud from occurring, whereas, detection focuses on activities and techniques that promptly recognize timely whether fraud has occurred or is occurring. While prevention techniques do not ensure that fraud will not be committed, they are the first line of defence in minimising fraud risks. Meanwhile, one of the strongest fraud deterrents is the awareness that effective detective controls are in place. Combined with preventive controls, detective controls enhance the effectiveness of a fraud risk management program by demonstrating that preventive controls are working as intended and by identifying fraud if it does occur. Although detective controls may provide evidence that fraud has occurred or is occurring, they are not intended to prevent fraud. Every organization is susceptible to fraud, thus it is important that organizations consider both fraud prevention and fraud detection [4].

8 Research Methodology

This research work is on the role of computer aided audit technique in the detection of fraud in an organization using KPMG, Lagos, Nigeria, which is an independent legal entity and is a member of KPMG International Co-operative, a Swiss entity. KPMG Nigeria is an industry which provides services which include include Audit, Tax, and Advisory services. KPMG was established in Nigeria in the year 1978. It has 23 partners, over 700 professionals and 5 office locations. It provides multidisciplinary professional services to both local and international organizations within the Nigeria business community. KPMG professionals are committed to delivering reliable, independent audit reports and in fulfilling this there has to be vigilance in making sure audit complies with changing regulations and the professional standards on the side of the auditor.

9 Data Collection Method

Data collection will be accomplished using primary sources through the administering of questionnaire. The questionnaire will be distributed to 250 randomly selected staff of KPMG Nigeria ranging from both junior and senior staff in the audit department.

10 Data Analysis Techniques

Data for this study will be analyzed using descriptive statistics of frequency tables and simple percentages of the SPSS statistical package. Regression analysis will be used to measure the relationship between the dependent and independent variables, and also analyzed the hypothesis of the study. In order to test the significance of the relationship between the dependent and independent variables of this study, the study was subjected to the Student t-test, Standard error test, and regression analysis, in which the Coefficient of Determinant (R^2) and P-value were determined.

11 Data Presentation and Interpretation

Out of the 250 questionnaires that were distributed for this research, 200 were retrieved back, giving a percentage of 66.7 % return. Hence, the analysis will be based on the 200 retrieved questionnaires.

Table 1 Distribution of respondents by age

Age distribution (years)	Frequency	Percentage (%)
21–30	85	42.5
31–40	82	41.0
41–50	32	16
51–60	1	0.5
Above 60	0	0
Total	200	100

A. Ages of Respondents

Table 1 shows that 42.5 % of the respondents are between 21 and 30 years, 41.0 % are between 31 and 40 years, 16 % are between 41 and 50 years, while 0.5 % of the respondents are over 50 years. This shows that majority of the respondents are youths and a dynamic work force.

B. Respondents' Qualification

Table 2 shows that 15 % of the respondents hold NCE/OND, 70.5 % hold B.Sc./HND degrees, 27 % hold Ph.D./M.Sc. degrees and 5 % others. This indicates that respondents are well learned and are fit to put opinion on the role of computer aided audit technique in fraud detection.

C. Respondents' Experience on the Job

Table 3 shows that 22.5 % have less than a year experience, 49.0 have about 1–5 year experience, 19 % have 6–10 years experience, and 9 % have 11–15 years experience and 0.5 % have 16 and above years of experience. This indicates that the respondents are well grounded in the process of auditing and hence, a reasonable result can be established from the responses.

D. Fraud as a Major Concern

Table 4 shows that 52.5 % strongly agreed that fraud is a major concern for businesses, 36.5 agreed, 6 % were uncertain, 5 % disagreed, while none strongly disagreed. This implies that most of the respondents believe that fraud is a major concern for business.

E. Management's Attitude to CAATs Adoption

Table 5 shows that 38.5 % of the respondent strongly agreed that management attitude towards CAATs is positive, 46 % agreed, 1.5 % were uncertain, 11 % disagreed, while 3 % strongly disagreed. Due to the responses gathered, it implies that management's reaction to the adoption of CAATs is positive. Thus, there will be a great advancement in record keeping and recording of various transactions and this will in a way reduce malicious practices of fraudster due to record keeping which will aid tracing of various transactions and this is a large contrast to tradition or manual audit.

Table 2 Distribution of respondents based on professional qualification

Qualification	Frequency	Percentage (%)
NCE/OND	30	15.0
B.Sc./HND	111	55.5
Ph.D./M.Sc.	54	27.0
Others	5	2.5
Total	200	100

Table 3 Distribution of respondents based on years of experience

Years of experience	Frequency	Percentage (%)
Less than a year	45	22.5
1–5 years	98	49.0
6–10 years	38	19.0
11–15 years	18	9.0
16 and above	1	0.5
Total	200	100

Table 4 Respondents' response as to whether fraud is a major concern for businesses

Response	Frequency	Percent	Valid percent	Cumulative percent
SA	105	52.5	52.5	52.5
A	73	36.5	36.5	89.0
U	12	6.0	6.0	95.0
D	10	5.0	5.0	100.0
SD	0	0	0	
Total	200	100.0	100.0	

Table 5 Respondents' response as to management's behavior to CAATs adoption

Response	Frequency	Percent	Valid percent	Cumulative percent
SA	77	38.5	38.5	38.5
A	92	46.0	46.0	88.5
U	3	1.5	1.5	94.0
D	22	11.0	11.0	96.5
SD	6	3.0	3.0	100.0
Total	200	100.0	100.0	

F. Using CAATs to Check Fraud Occurrence in Organizations

The Table 6 shows that 23 % of the respondents believe that CAATs can be used to check fraud occurrence in organization, 59 % agreed, 5.5 % were uncertain, 7.5 % disagreed, and 5 % strongly disagreed. This implies that a large proportion of the respondents believe that CAATs can be used to check and manage fraud thus reducing its occurrence in organizations.

Table 6 Respondents' reaction to whether CAATs can be meaningfully used to check fraud occurrence

Response	Frequency	Percent	Valid percent	Cumulative percent
SA	46	23.0	23.0	23.0
A	118	59.0	59.0	82.0
U	11	5.5	5.5	87.5
D	15	7.5	7.5	95.0
SD	10	5.0	5.0	100.00
Total	200	100.0	100.0	

G. Computer Audit Credibility

Table 7 shows that a total of 81.5 % respondents agreed that computer audit is credible. This implies that computer aided audit technique is a step worth taking in fighting fraud. This is shows advancement from the crude and traditional way of auditing, thus the change in audit is better than what it used to be. Due to the responses gathered, it can be implied that the adoption of CAATs is positive thus there is a great advancement in record keeping and this has reduced malicious practices of fraudsters due to record keeping in various transactions. It is evident that majority of the respondents claimed that there is large economies of scale in the adoption of CAATs. It was also affirmed that the validity of the truth and fairness of financial statement can be established through the use of computer audit.

H. Prevalence of Fraud Before CAATs

From the Table 8, can be seen that 50 % of the respondent strongly agreed that fraud has been prevalent in organization before the introduction of CAATs, 40.5 % agreed, 0.5 % were uncertain, 5.5 % disagreed, and 3.5 % strongly disagreed. This implies that fraud has been in existence and on the high side before the introduction of CAATs.

I. Effect of CAATs on Fraud Detection

Table 9 shows that 28.5 % strongly agreed that CAATs have played a major role in fraud detection, 44 % agreed, 6.5 % were uncertain, 11 % disagreed, 9.5 % strongly disagreed, while 0.5 % were missing. It is seen that majority of the response were in support of the fact that CAATs has effect on fraud detection and has been used to curb the occurrence of fraud in organizations. Respondents opined that auditors can use CAATs to help detect material misstatements in the financial statements, particularly in substantive tests of details of transactions and balances as part of meeting the general audit objectives of validity, completeness, ownership, valuation, accuracy, classification and disclosure of the data produced by the accounting system to support financial assertions.

Table 7 Computer audit credibility

Response	Frequency	Percent	Valid percent	Cumulative percent
SA	71	35.5	36.4	36.4
A	88	44.0	45.1	81.5
U	34	17.0	17.4	98.9
D	2	1.0	1.0	100
Total	195	97.5	100	
Missing value	5	2.5		
Total	200	100		

Table 8 The prevalence of fraud before the introduction of CAAT

Response	Frequency	Percent	Valid percent	Cumulative percent
SA	100	50.0	50.0	50.0
A	81	40.5	40.5	90.5
U	1	0.5	0.5	91.0
D	11	5.5	5.5	96.5
SD	7	3.5	3.5	100.0
Total	200	100.0	100.0	

Table 9 Role of CAATs in fraud detection

Response	Frequency	Percent	Valid percent	Cumulative percent
SA	57	28.5	28.6	28.6
A	88	44.0	44.2	72.8
U	13	6.5	6.5	79.3
D	22	11	11.1	90.4
SD	19	9.5	9.6	100.0
Missing value	1	0.5		
Total	200	100.0		

J. Impact of CAATs on Employees' Performance

Table 10 shows that 20.9 % of the respondents strongly agreed that computer aided audit techniques have positive effects on organization in terms of employee performance, 37.2 % agreed, 23.0 % were uncertain, 11.5 % disagreed, 7.3 % strongly disagreed. This shows that more than half of the respondents believed that CAATs has effect on employees performance in organization. The 23.0 % of the undecided respondents claimed that the improved performance had been associated with much stress and pressure. The respondents opined that CAATs reduced their audit risks as compared to traditional audit and the effects of litigation from different organizations were drastically reduced. It was further affirmed that CAATs brought a promising future to organizations in terms of the validity and fairness of record keeping. The respondents also believe that it can be used to place reliance on the internal control system as compared to others method so auditing.

Table 10 Respondents opinion on how employees performance have been affected due to CAATs

Response	Frequency	Percent	Valid percent	Cumulative percent
SA	40	20	20.9	20.9
A	71	35.5	37.2	58.1
U	44	22.0	23.0	81.1
D	22	11.0	11.5	92.6
SD	14	7.0	7.3	100.0
Missing value	9	4.5	100.0	
Total	200	100		

Therefore, with CAATs, auditors can perform substantive tests within a shorter time frame resulting in overall efficiency yet not compromising on the quality of audit effort.

K. Types of Fraud Encountered During Financial Transactions

From Table 11, different respondents selected more than one option. However, top on the list of fraud in financial transactions are falsification of records, diversification of securities, duplicating invoice numbers/dates, and terminated employees continuing to be paid.

L. Problems Encountered in the Adoption of CAATs

From Table 12, it is seen that majority of the respondent see the cost, competency in using software, and technical expertise of the system database as problems that will reduce the adoption of CAATs in organizations. Respondents claim that potentially expensive maintenance contracts are needed to make CAATs work efficiently. This is also supported by Coderre [11]. Moreover, CAATs will be limited depending on how well the computer system is integrate; the more integrated the system, the better the use of CAATs. Another problem is lack of technical expertise. It is believed that CAATs are too technical and complex for non-IT auditors, even if training is provided. This is also in accordance with the findings of Coderre [11]. CAATs are only useful methods of auditing only if the system can be relied on, so the auditor would have to assess the reliability first before adopting CAATs. Furthermore, respondents claim that clients are afraid that their systems and data will be compromised with the use of CAATs (100.0 %). This is strictly as a result of lack of trust on the part of the clients with their external auditors. In other to overcome the problem of cost in the adoption of CAATs, the respondents opined that a cost-benefit analysis from the audit point of view should be carried out prior to deciding to use the audit software. Majority of the respondents (92.0 %) however do not support the fact that lack of suitable computer facilities is a major problem to the adoption of CAATs.

Table 11 Types of fraud in financial transactions

Variables	Percentage (%)
Misuse of cash funds	45.8
Falsification of records	96.7
Diversification of securities	87.1
Falsification of employees size	55.2
Pay date precedes employment date	67.3
Terminated employees continuing to be paid	82.1
Money laundering	57.5
Internet fraud	32.7
Duplicating invoice numbers/dates	84.3
Disbursement to false vendor	12.4
Producing reports of debit balances	74.5
Unrecorded or understated sales or receivables	34.3
Theft of cash receipts	55.7
Theft of inventory	11.6

Table 12 Respondents' view on the problems encountered in the adoption of CAATs

Variables	SA	A	U	D	SD	Percentages (%)				
Cost	102	61	5	29	Nil	51.0	30.5	0.25	14.5	Nil
Software competency	34	112	19	20	13	17.0	56.0	09.5	10.0	6.5
Technical expertise	45	98	38	18	1	22.5	49.0	19.0	9.0	0.5
Lack of suitable computer facilities	4	12	Nil	45	139	2.0	6.0	Nil	22.5	69.5
Clients are afraid that their systems and data will be compromised with the use of CAATs	151	49	Nil	Nil	Nil	75.5	24.5	Nil	Nil	Nil

12 Testing of Hypothesis

In order to test the hypothesis, an econometric model showing the relationship between the fraud prevalence (FRAUD) and Computer Aided Audit Technique (CAAT) will be adopted and it is stated thus;

FRAUD = f (CAAT) [Functional Relationship]

FRAUD = $b_0 + b_1$ CAAT + U

Where FRAUD = Fraud prevalence and CAAT = Computer Aided Audit Technique

The expectation concerning the sign of the coefficients is; $b_0 > 0$ and $b_1 > 0$.

12.1 Regression Analysis

ANOVA[b]

Model		Sum of squares	df	Mean square	F	Sig.
1	Regression	112.073	1	112.073	375.406	0.000[a]
	Residual	58.513	196	0.299		
	Total	170.586	197			

[a] Predictors: (constant), does computer aided audit techniques have any effect on fraud detection in organizations?
[b] Dependent variable: has fraud been prevalent in organizations before the introduction of CAATs?

Coefficients[a]

Model	Unstandardized coefficients		Standardized coefficients	t	Sig.
	B	Std. error	Beta		
(Constant)	0.106	0.090		1.176	0.241
Does computer aided audit technique have any effect on fraud detection in organizations?	0.670	0.035	0.811	19.375	0.000

[a] Predictors: (constant), does computer aided audit techniques have any effect on fraud detection in organizations?

13 Evaluation of Estimates

In this section, the estimates of the parameters are evaluated in order to ascertain whether they are theoretically meaningful and statistically satisfactory.

$FRAUD = b_0 + b_1 \, CAAT + U$
$FRAUD = 0.106 + 0.670 \, CAAT$
$S.E = (0.090), (0.035)$
$T\text{-test} = (1.176), (19.375)$
$R^2 = 0.657$
$F\text{-test} = 375.406$
$N = 200$

The intercept (b_0) of 0.106, indicates that when the independent variable indicated in this study is zero (CAAT), the rate of fraud (FRAUD) will be about 0.106 %.

From the model summary, the **R**-square shows the extent to which the independent variable explains the variation in the dependent variable, while the **R** value shows the correlation value. The above the R value is 0.811. Obviously, this implies that there is a strong positive relationship between fraud detection and CAATs. More so, the R-square value explains that computer aided audit technique explains exactly 65.7 % of the variation of fraud.

13.1 Coefficient of Determinant (R^2)

This shows the percentage of the total variation of the dependent variable being explained by the changes of the explanatory variables. From the result, the R^2 is 0.657. This means that 65.7 % of the changes in FRAUD are explained by the changes in CAAT. Thus, Computer aided audit techniques have a significant role to play in fraud detection in organizations.

13.2 Student t-Test

The Student t-test which is equivalent to the standard error was used to test the significance of the parameter estimates. The student t-test is at 5 % level of significance. To obtain $T_{tabulated}$, the degree of freedom has to be ascertained.

DF = N–K
Where N = Number of respondents
K = Number of parameters
Therefore we have
DF = 200 − 2 = 198
According to the regression result, $T_{calculated}$: $b_1 = 19.375$

From the T distribution table, we have $T_{tabulated} = 1.96$

Decision Rule
If $T_{calculated} < T_{tabulated}$ Accept H_0
If $T_{calculated} > T_{tabulated}$ Accept H_1
According to the regression results
b_1: 19.375 > 1.96

From the result, we can therefore conclude that $T_{calculated} > T_{tabulated}$ and which leads to the acceptance of our Alternative hypothesis (H_1) i.e. computer aided audit techniques have a role to play in the detection of fraud in organizations.

13.3 P-value Test

The p-value is 0.000 and the standard p-value is 0.05 which indicates 95 % assurance of the research leaving 5 % for uncertainty. Since the p-value is less than 0.05, we reject the Null hypothesis and accept the Alternative hypothesis and conclude that there is a significant relationship between CAATs and Fraud.

13.4 Standard Error Test

This test states that for an estimate to be statistically meaningful, the standard error of its coefficient must be less than half the value of the coefficient, (i.e. S.E $b < b/2$). For the regression result we have;

FRAUD $= 0.106 + 0.670$CAAT
S.E $= (0.090), (0.035)$
Decision Rule
SE (b) $< b/2$ Accept H_1
SE (b) $> b/2$ Accept H_0
SE for $b_2 = 0.035$
$B_2/2 = 0.670/2 = 0.335$
$0.035 < 0.335$

From the result, we therefore conclude that SE $(b_2) < b_2/2$ and which leads to the acceptance of our Alternative hypothesis (H_1) and the rejection of the Null Hypothesis (H_0).

Based on the findings of the analysis, the coefficient of determinant (R^2) is 65.7 %, which implies that 65.7 % of the changes in FRAUD are explained by the changes in CAAT, the t-statistics and standard error tests were also statistically significant. Thus, Computer Aided Audit Techniques play a significant role in fraud detection in manufacturing organizations.

To further buttress on the primary data obtained, it is seen that information technology is now a way of life and for organizations to operate well in this computer age and reduce the existence of fraud, computer aided audit techniques enables auditors or investigators on the spot overview of business operations, gain in-depth understanding of the relationship among various data elements and easily drill down to specific areas of interest; therefore computer aided audit techniques do have significant roles to play in fraud detection.

14 Summary of Findings

In the analysis above we can extract that in auditing, an auditor presenting reliable audited financial reports to shareholders and customers must have an important use of computers to carry out such audit work in other to achieve speed, accuracy, credibility and reliability of every information which aids management and auditors in quick decision making relating to financial policies.

From the result of the data collected and analyzed, the standard error was S.E = (0.090), (0.035) of which the decision rule was if SE (b) < b/2 Accept H_1 and if SE (b) > b/2 Accept H_0. The SE for $b_2 = 0.035$ and $B_2/2 = 0.670/2 = 0.335$ which gave the value of $0.035 < 0.335$, so H_1 was accepted. The t-test was (1.176), (19.375); $T_{calculated}$: $b_1 = 19.375$ and $T_{tabulated} = 1.96$, so H_1 was accepted. The p-value = 0.000 and since the standard one is 0.05, the null hypothesis was rejected. The use of computer aided audit techniques can thus be seen as very effective in the detection of fraud in organizations from the result obtained through the tests. The coefficient of determinant R^2 showed a positive significant relationship of 65.7 %. This implies that the changes in fraud are explained by the introduction of the use of computer aided audit techniques (CAAT) and the remaining variation of 34.3 % would be explained by other factors which have effect on fraud detection.

15 Conclusion

The findings of the study showed that 72.8 % of the respondents agreed that CAATs have played a major role in fraud detection, and hence can be used to curb fraud to a minimal level in organisations. It was also discovered that CAAT helped to improve the auditors' performance as corroborated by 58.1 % of the respondents. However, quiet a handful of respondents (23.0 %) were undecided in this respect as they claimed that the improved performance has been associated with much stress and pressure. Furthermore, the cost of implementing CAAT, the required skills to be acquired for its usage, and systems database requirements, are some of the problems found to be associated with the introduction and application of CAAT to organisations. Moreover, top on the list of fraud in financial transactions are falsification of records, diversification of securities, duplicating invoice numbers/dates, and terminated employees continuing to be paid.

It is therefore concluded that the adoption of computer aided audit techniques in the preparation and presentation of financial transaction and reports has enhanced the quality of financial reports. Thus, the right information and computer technique solutions, matched with intelligence, untiring crime detection and prevention, will undoubtedly make any organization a pride of all. In a bid for organizations to operate in a fraud free environment, it is thus recommended that the use of computer aided audit techniques be encouraged in the preparation of financial reports.

References

1. Adeniji, A.: Auditing and Investigation. Value Analysis Publishers, Lagos (2004)
2. AICPA: Web TrustTM Programme-Security Principle and Criteria-Version 3.0, American Institute of Certified Public Accountant/Canadian Institute of Chartered Accountants. http://www.aicpa.org, http://www.cica.org (2001)
3. AICPA: Consideration of Fraud in a Financial Statement Audit 169, AICPA, 2006
4. Association of Certified Fraud Examiners, ACFE: Report to the Nation on Occupational Fraud and Abuse, http://www.acfe.com/resources/publications.asp?copy=rttn (2006). Accessed 20 Aug 2011
5. Albrecht, C.C., Albrecht, W.S., Dunn, J.G.: Can auditors detect fraud: a review of the research evidence. J. Forensic Account. 2(1), 1–12 (2001)
6. Albrecht, W.S., Albrecht, C.C., Albrecht, C.O.: Fraud and corporate executives: agency, stewardship and broken trust. J. Forensic Account. 5(1), 109–124 (2004)
7. Albrecht, C.C.: Fraud and Forensic Accounting in a Digital Environment, Brigham Young University, 2008
8. Boritz, J.E.: Computer Control and Audit Guide, 12th edn. University of Waterloo, Waterloo (2003)
9. Brink, V.Z., Witt, H.: Internal Auditing. Wiley, New York (1982)
10. Braun, R.L., Davis, H.E.: Computer-assisted audit tools and techniques: analysis and perspectives. Manag. Aud. J. 18(9), 725–731 (2003)
11. Coderre, D.: Computer assisted techniques for fraud detection. CPA J. 8(69), 57–59 (1999)
12. Curtis, M.B., Jenkins, J.G., Bedard, J.C., Deis, D.R.: Auditors' training and proficiency in information systems: a research synthesis. J. Inf. Syst. 23(1), 79–96 (2009)
13. Curtis, M.B., Payne, E.A.: An examination of contextual factors and individual characteristics affecting technology implementation decisions in auditing. Int. J. Account. Inf. Syst. 9(2), 104–121 (2008)
14. Davia, H.R.: Fraud 101: Techniques and Strategies for Detection. Wiley, New York (2001)
15. Debreceny, R., Lee S.L., Neo, W., Shuling, J.T.: Employing generalized audit software in the financial services sector: challenges and opportunities. http://raw.rutgers.edu/docs/wcars/8wcars/contassurronaldjdaigle_files/DebrecenyCAATSBanks_paper.pdf (2004)
16. IAASB: International auditing practice statement 1006 audits of the financial statements of banks. In: IAASB (ed.) Handbook of International Auditing, Assurance, and Ethics Pronouncements, pp. 663–751. International Auditing and Assurance Standards Board, International Federation of Accountants, New York (2003)
17. IIA: Internal Audit Automation Survey. The Institute of Internal Auditor UK and Ireland, London (2003)
18. ISACA: Use of Computer Assisted Audit Techniques (CAATs). Information Systems Audit and Control Association, Rolling Meadows (1998)
19. Keller & Owens: LLC, Certified Public Accountants. Preventing and Detecting Fraud in Not-Profit-Organizations (2008)
20. KPMG: Fraud risk management: developing a strategy for prevention, detection, and response. KPMG Professional Service, Capability Statement. www.ngkpmg.com (2004). Accessed 18 Aug 2011
21. Lanza, R.B.: Using technology to mitigate fraud mal practice claims. Trust. Prof. 10(1), 2–3 (2007)
22. Mahzan, N., Lymer, A.: Adoption of Computer Assisted Audit Tools and Techniques (CAATTs) by Internal Auditors: Current Issues in the UK. Birmingham Business School, Birmingham (2006)
23. Mansoury, A., Salehi, M.: Firm Size, Audit Regulation and Fraud Detection: Empirical Evidence from Iran. J. Manag. 29(1), 53–65 (2009)
24. Olowokere, K.: Fundamentals of Auditing. Silicon Publishing Company, Ibadan (2007). Revised Edition 2007, pp. 272–280, 283–284

25. Singleton, T.W., Singleton, A.J., Bologna, J., Lindquist, R.J.: Fraud Auditing and Forensic Accounting, 3rd edn. Wiley, New Jersey (2006)
26. Yang, J.G.S.: Data mining techniques for auditing attest function and fraud detection. J. Forensic Invest. Account. **1**(1), 4–10 (2006)

Green Communication and Consumer Electronics Sustainability in Delivering Cost Benefit Business Federation in Professional Service Firms

Ezendu Ariwa, Carsten Martin Syvertsen and Jaime Lloret Mauri

Abstract We present the business federation as a new organizational form using insights from business literature as means to tailor make services to clearly defined market segments. We illustrate how professional service firms can achieve economic growth through operating locally within an international network. Within the business federation local offices gain access to resources through an extreme form of delegation as it is not top management that delegates to local offices but rather local units that give top management the permission to handle certain tasks because it is most efficient this way. With these new realities existing strategies do not seem to provide the necessary cure. Firms must constantly innovate using insights from new organizational models.

Keywords Professional service firms · New organizational forms · The business federation · Knowledge creation · Innovation

1 Problem Definition

In this article we illustrate how to combine knowledge creation with innovation in order to achieve a competitive advantage within professional service firms.

E. Ariwa (✉)
London Metropolitan Business School, London Metropolitan University, London, UK
e-mail: e.ariwa@londonmet.ac.uk

C. M. Syvertsen
Østfold University College, Halden, Norway

J. L. Mauri
Universidad Politecnica de Valencia, Valencia, Spain

S.-S. Yeo et al. (eds.), *Computer Science and its Applications*,
Lecture Notes in Electrical Engineering 203, DOI: 10.1007/978-94-007-5699-1_89,
© Springer Science+Business Media Dordrecht 2012

2 Introduction to Professional Service Firms

It was not till the 1980s and 1990s that professional service firms received a significant interest from the academic world. This limited research interest is striking since the diversity and number of these firms has increased substantially in industrialized countries in the last decades. The professional service industry accounts for approximately 17 % of employment in Europe and USA. In total it had revenues of about 7000 billion British Pounds worldwide in 1997 with a growth around 15 % per year [81].

One reason for this limited research interest on professional service firms can be the secretive nature of these firms, as many of them remain under private ownership. This secrecy contributes to an "aura of professionalism" that prevents the public from gaining deeper insights. Traditional management techniques also seem to play a less important role in these firms as well-established management tools are no longer as relevant as they once were. In addition, professional service firms can be regarded as firms with too specific set of features to have received much attention from scholars interested in general management problems [8, 11–13].

Professional service firms rely on a high degree of autonomy, frequently linked to scientific knowledge. Authority is in the hands of professionals in the sense that they ultimately make the key decisions. Professionals attain, develop and retain authority through the use of knowledge rather than through managerial titles. Although management is vital for the success of professional service firms' success, the paradoxical fact is that professionals do not like others have authority over them, supporting the idea that value creation best can take place not supervising and interfering with their decision making tasks. The reason for this is found in the reversed power structure, as control over the most critical resources for value creation is found in the minds of the professionals, and not with the owners of the firm's equity [57].

3 Towards the Business Federation in Professional Service Firms

The business landscape has changed dramatically since the 1980s and 1990s when the business idea was to match capital, ideas and talent. Large corporations were like elephants unable to dance to the tones of the jazzing entrepreneurial economy. Professional service firms, operating as advisors, used slogans such as "*small is beautiful*", "*slim is in*" and "*less is more*". Executives became interested in improving operational efficiency in order to maximize shareholder value. The heroes were the Leverage Buy Out (LBO) artists who broke up client firms into smaller units [17, 20, 24, 31].

Today other medicines are necessary to cure the patient, with an increased focus on untraditional models of meting complex needs of clients. The economic shock of 2008, and the Great Recession that followed, created uncertainty of the direction of the global economy. Leading economics entered into a stage of flux, which resulted in a reduced confidence in the financial industry and doubts if professional service firms were able to help clients in the harder economic times that we are facing.

Managers of professional service firms can feel uncomfortable when trying to apply classical organizational models within complex and turbulent business environments. In the past economic regimes rewarded firms for maintaining strong organizational hierarchies and strict work rules. Newer organizational forms, on the other hand, put a strong emphasis on extracting the potentials of creativity, internal and external collaboration, resource sharing, and responsiveness to clients. The philosophy and rules of newer organizational forms fit nicely with the requirements of professional service firms [50, 79, 88, 91, 93].

At the same time we have witnessed a shift in the management literature from the industrial organizational perspective [75, 76] towards the resource based perspective [15, 100] and, finally, towards the dynamic capability approach [95]. Teece et al. [95] argue that the dynamic capability approach can be "…. especially relevant in a Schumpeterian world of innovation-based competition, price/performance rivalry, increasing returns and the creative destruction of existing competencies".

Researchers have often chosen to view organizations as autonomous entities striving for competitive advantage from either external industry sources (e.g. [75]) or from internal capabilities (i.e. [15]). However, the image of atomistic actors competing against each other in an impersonal market place is getting increasingly inadequate in a world in which organizations are embedded in networks of professional relationships.

Alliances with multiple partners are used extensively in professional services to provide highly complex and customized services. Examples of professional services using alliances include investment banking [30, 71, 72], management consulting, design engineering firms [4, 78], global media services [69] and architecture services [1].

Authors have gone far to promote clearly defined organizational models for creating knowledge, as in the case of the hyper dynamic structure provided by Nonaka and Takeuchi [66] and the N-form structure proposed by Hedlund [43]. Our research is related to the this line of thinking as professional service firms operating at the international stage will try to achieve local adaptation and scale effects as a result of being a part of an larger internal network. This way of thinking should suit well into the times we live in. We want to be a part of a strong entity at the same at the same time as a great degree of autonomy is appreciated in our professional lives.

We regard our model as a strategic internal network consisting of local units. Local units co-operate due to the availability of complementary resources (e.g. [9, 16, 27]). Local units can help each other overcome resource limitations by pooling resources as well as realizing synergies (i.e. [56]). Mowery et al. [62] reported that sharing complementary resources positively impact learning.

4 Introduction of The Business Federation

We call our model the business federation as the head office is given power on the basis wishes from the local levels. The business federation can in this way regarded as a sort of an extreme form of delegation as local units are the main changes agents through an entrepreneurial orientation. Local offices will try to meet complex and demanding needs of clients though tailor making of services at local levels [40, 41].

Within our model the role of the local manager is similar to that of an entrepreneur. The entrepreneur searches for new services in order to serve local markets. An entrepreneur can be regarded as motor combing ideas and resources in new ways [85]. The entrepreneurial character is important since local manager need to be quick in making and implementing decision in response to demands set by complex and dynamic client wishes. Our model explicitly mentions the need for tailoring services to local needs.

Being entrepreneurial also means that professionals are allowed to use autonomy as far as circumstances permit [65]. Through such steps they can determine the availability of assignments and clients [19, 89]. Rather than relying on carefully planned strategies that might support long-term success, these professionals can rely on more emerging strategies. Such a way of thinking suits nicely into an evolutionary perspective as it opens for local units to create, maintain and sustain strategic advantages [23, 61].

The business federation consists of a variety of highly autonomous units, operating under the influence of cohesive forces found at a central office. The units can be regarded as loosely coupled parts working together, because there are some things they do better jointly rather than individually [28, 58, 87].

Within the business federation local offices gain access to resources through an extreme form of delegation as it is not top management that delegates to local offices but rather local units that give top management the permission to handle certain tasks because it is most efficient this way.

Corporate control is infrequent and exceptional at the corporate level. Traditional means of corporate control is infrequent and exceptional since it makes little sense to interfere in entrepreneurial processes at local levels.

Social means of control become more important to support entrepreneurial process at local levels, through the use of culture, ideology and sanctions. Culture refers to shared cognitive maps that guide actions, decisions and inferences in a professional community [53, 80]. Ideology can be regarded as an integrated sets of belies that bind professionals to a firm. Ideology can help to reduce mental uncertainty as it identifies and crystallizes a firm's agenda [92]. In addition, there may be collective sanctions, which involve punishing group members to violate common group goals, norms and values (i.e. [38]). Culture, ideology and sanctions can be regarded as congruent mechanisms. They can reinforce one another to promote coordination within a given firm.

However, a certain degree of standardization might take place at the corporate level through the use of supply services. Such services are used in order to develop products and services, and for strengthen the competitive position of local units. Their role is to stimulate the constant transformations of activities within the local units, in many ways similar to the machine adhocracy in "the university" introduced by Bowman and Carter [21].

Each supply unit is a profit centre, not a cost centre. The local units can buy internally or externally. The pricing aspects are governed by market rules. However, corporate level can make some of the goods that have been developed by supply services compulsory for the all the local units. Normally this applies to just essential features, such as when a logo describes a corporate image.

5 How Knowledge Can Make Professional Service Firms More Innovative

Professional service firms are in a position where the ability to innovate is regarded as a source to develop competitive advantage. Scholars explain that the development of competitive advantage requires the mastery of two divergent tasks [59, 64]. A professional service firm can center its attention on sets of techniques to cultivate valuable and commercially viable products and services, often referred to as the exploitation of knowledge [54, 59]. In addition, professional service firms can continually acquire sets of knowledge that can serve as seeds for future competitive advantage, in the business literature often referred to as the exploration of knowledge [59, 63]. We focus on the exploration of knowledge as it can be a source of competitive advantage, using local offices in internationally oriented professional service firms as the empirical setting.

The creation and sharing of knowledge is by nature dynamic in professional service firms. By definition the use of dynamic capabilities involve adaptation and change, building on foundations provided by Schumpeter [82–85], Penrose [70], Nelson and Winter [63], Barney [14], Teece et al. [95], Zollo and Winter [102], and Helfat and Peteraf [44].

Capabilities imply a link between knowledge, skills and tasks [49]. Knowledge is in our opinion the most important element in the capability concept as it indicates how professional service firms can apply, develop and integrate capabilities in a strategic manner [37, 42, 55].

There are a number of conceptual avenues that can be followed when studying knowledge as a dynamic force in professional service firms. We use an approach derived from evolutionary economics, illustrating that the use of dynamic capabilities is dependent upon organizational routines (i.e. [6, 63]).

We find routines interesting as they can create new knowledge. By focusing on routines, professional service firms can use information filters that constrain the range of knowledge being explored [45]. If a professional service firm decides to search outside a given core knowledge, it can be difficult to comprehend and apply [25, 63, 96].

When studying how knowledge can be created and shared we make a distinction between explicit and tacit knowledge [73, 74]. Explicit knowledge is relatively easy to imitate [66]. Explicit knowledge seems to be more accessible for a growing number of professional service firms, for example when using the internet and public registers. This underlies the importance of focusing on tacit knowledge as it to a greater extent is based on intuition, and it is difficult to communicate to others as information [73, 74].

In the research literature, the definition of innovation includes concepts of novelty, commercialization and/or implementation. In other words, if an idea has not been developed or transformed into a product, process or service, or it has not been commercialized, it cannot be regarded as an innovation.

Definitions of innovation is conducted by Schumpeter [85, 101], Kirzner [51, 52], Utterback [98], Damanpour [26], Afuah [3], Fischer [33], Garcia and Colantone [36], McDermott and O' Conner [60], and Benner and Tushman [18].

Based on writings on innovation we distinguish between market based innovations and technological innovations. By combining market-based innovations with technological innovations it is possible to illustrate how professional service firms can move from a situation with unclear priorities to a situation with a more clean-cut strategic focus. However, this is a statement with modifications. Even if we can explain, ex post, how and why a professional service firm moved from archetype X to archetype Y, or from position A to position B, it will not be fine-tuned enough to show how, de facto, change takes place [32, 48, 67].

Market innovation refers to how new knowledge is embodied in distribution channels, product applications, as well as customer expectations, preferences, needs and wants [3]. We are of the opinion that market innovation mostly takes place at local levels within the business federation framework. It is at this organizational level that clients needs are tried to be meet through entrepreneurial processes.

By technical innovation we mean the extent to which a new product fulfills key customer needs better than the existing one [2, 22]. Technological knowledge can be destroyed as existing knowledge has a tendency to become outdated, particularly in complex business environments. We believe that technological innovation mostly takes place at the local level as the need for crafting of business processes is high. Corporate level can also play a role as there might be a need for a certain degree of standardization of products and services.

We are of the opinion that only through radical change in marketing and technological practices it is possible for professional service firms to focus deeply on how innovation can be used as a weapon in a more complex and turbulent market place (i.e. [7, 46, 86]).

6 The Relationship Between Innovation and Knowledge Creation

The relationship between innovation and knowledge creation is complex, and little research is conducted in this line of research. We regard innovation as an enabler to create value in professional service firms (i.e. [99]). We suggest that innovation depends on knowledge creation. We believe that innovation is more related to market-based activities, often involving collective efforts.

Knowledge creation can be seen as a spiral moving through epistemological and ontological dimensions [66]. Professionals work in dynamic networks, in many ways in conflict in many ways in conflict with traditional organizational principles, having a narrower view on the dynamics of knowledge.

We are of the opinion that explicit knowledge is related to knowledge exploitation, while there is a closer link between tacit knowledge and the exploration of knowledge. Tacit knowledge can help local units to be engaged in new business processes, while head office to a greater extent will be engaged in existing processes through the use of explicit knowledge. Exploration involves discovery and experimentation, which can lead to increased productivity through repeated practices.

We believe that the exploration of tacit knowledge within new processes is the main driver of creation of knowledge within professional service firms using innovation as a tool. Such a framework can be used as basis for offering tailor made services to clearly defined market segments, which again can lead to path-breaking innovations [34, 35, 63, 64, 85]. In this way professional service firms can become more entrepreneurial using new processes as tools.

How explicit knowledge can be transferred into tacit knowledge within local offices is illustrated in Fig. 1, built on inspiration from Popaduik and Choo [68] and Syvertsen [94].

7 Discussion and Conclusion

The business federation can be of relevance for professional service firms. How abstract knowledge best can be organized at local levels is a great challenge in many professional service firms. New markets have challenged professional firms for example in accounting, consultancy and law. Professional service firms have to follow clients into international markets, and in doing so, need to rethink the adequacy of their organization [5, 10, 39].

The business landscape is reshaped by forces such as globalization, deregulation, technological advances and social unrest. Managers are increasingly realizing that the basis of their competitive advantage in many situations is related to the

Knowledge creation

Innovation	Exploitation of knowledge	Exploration of knowledge
(technology and marketing)	**Traditional ecotourism (nature)** Use of technology/ (the resource based view)	**Newer forms of ecotourism (spiritual dimensions)** Use of marketing/ (dynamic competencies)
New processes	Regular innovations (A) Incremental innovations (B)	Revolutionary innovations (A) Major process innovations (B) Technological innovations (D)
Existing processes	Niche innovations (A) Modular innovations (B)	Radical innovations (B) Process/product and service innovations (C) Market breakthroughs (D)

Fig. 1 Innovation and knowledge creation in professional service firms: A [2], B [45], C [97], D [22]

performance of professionals [29]. Better use of existing knowledge and more effective acquisition and assimilation of new knowledge becomes a business imperative [77]. New organizational forms are of academic interest as manager's experiment with new organizational arrangements [21, 47, 90].

Research into the business federation, using professional service firms as the empirical setting, is a new and unexplored field. The findings must therefore at this research stage be regarded as highly exploratory.

The business federation strongly support that entrepreneurship is allowed to grow at local levels, allowing professionals to operate with a great degree of autonomy. This also illustrates that it might be necessary to change the strategic focus of the firm gradually, as local initiatives are given sufficient space.

A route for managers wishing to implement federalist principles is to try to let professionals work with a high degree of autonomy at local levels. Thereafter, firms can analyze if it is possible to organize local units as own profit-centers, with a high degree of interaction between local units. The internal market for local units cannot be regarded as a laissez-faire market but as an alliance between entrepreneurs. Key to its effectiveness can be a collaborative culture in which knowledge can move freely among and between units, and be organized around challenges in the client market and the talent market.

References

1. Abbott, A.: The system of professionals: an essay on the division of expert labor. University of Chicago Press, Chicago (1988)
2. Abernathy, W., Clark, K.B.: Mapping the winds of creative destruction. Res. Policy **14**, 3–22 (1985)
3. Afuah, A.: Innovation Management: Strategies, Implementation and Profits. Oxford University Press, New York (1998)
4. Aharoni, Y.: Coalitions and Competition: The globalization of professional services. Routledge, London (1993)
5. Aharoni, Y.: A note on the horizontal movement of knowledge within organizations. Paper presented at the conference on the management of change in knowledge-based organizations, University of Alberta, Edmonton, Canada (1995)
6. Argyris, C., Schon, D.: Organizational Learning. Addition-Wetly, Reading (1978)
7. Argyris, C.: Learning and Action: Individual and Organizational. Jossey-Bass, San Francisco (1982)
8. Armbrecht, F.M.R. Jr., Chapas, R.B., Chappelow, C.C., Farris, C.F., Friga, N, Hartz, C.A, McLlvaine, M.E., Postle, S.R., Whitwell, G.E.: Knowledge management in research and development. Res. Technol. Manag. **44**(4), 28–48 (2001)
9. Badaracco, J.L.: The Knowledge Link: How Firms Compete through Strategic Alliances. Harvard Business School Press, Boston (1991)
10. Baden-Fuller, C.: The globalization of professional service firms: Evidence from four case studies. In: Aharoni, Y. (ed.) Coalitions and Competition, pp. 102–120. Routledge, London (1993)
11. Baden-Fuller, C., Volberda, H.W.: Strategic renewal in large complex organizations: a competence-based view. In: Heene, A., Sanchez, R. (eds.) Competence-Based Strategic Management. Wiley, Chicester (1997)
12. Baker, W.E.: The social structure of a national securities market. Am. J. Sociol. **89**(4), 775–811 (1984)
13. Barley, R.S., Kunda, G.: Bringing work back in. Organ. Sci. **12**, 76–95 (2001)
14. Barney, J.: Strategic factor markets: expectations, luck, and business strategy. Manage. Sci. **21**, 489–506 (1986)
15. Barney, J.: Firm resources and sustained competitive advantage. J. Manag **17**(1), 99–120 (1991)
16. Barringer, B.R., Harrison, J.S.: Walking a tightrope: creating value through interorganizational relationships. J Manag **26**(3), 367–403 (2000)
17. Bartlett, C.A., Ghoshal, S.: Managing Across Borders: The Transnational Solution. Harvard Business School Press, Boston (1988)
18. Benner, M.J., Tushman, M.L.: Exploitation, exploration, and process management: the productivity dilemma revisited. Acad. Manag. Rev. **28**(2), 238–256 (2003)
19. Bettencourt, L.A., Ostrom, A.L., Brown, S.W., Roundtree, R.I.: Client Co-Production in Knowledge-Intensive Business Services. Calif. Manag. Rev. **44**(4), 100–128 (2002)
20. Block, Z., MacMillan, I.C.: Corporate Venturing-Creating New Businesses within the Firm. Harvard Business School Press, Boston (1993)
21. Bowman, C., Carter, S.: Organizing for competitive advantage. the machine adhocracy. In: O'Neil, T.D., Chetman, M. (eds.) Strategy, Structure and Style. Wiley, Chichester (1997)
22. Chandy, R.K., Tellis, G.J.: Organizing for radical product innovation: the overlooked role of willingness to cannibalize. JMR J. Market. Res. **35**(4), 119–135 (1998)
23. Christensen, C.M.: The Innovator's Dilemma: When New Technologies Cause Great Firms to Fail. Harvard Business School Press, Boston (1997)
24. Choo, C.: The Knowing Organization: How Organizations Use Information to Create Meaning, Create Knowledge, and Make Decisions. Oxford University Press, Oxford (1998)

25. Cohen, W.H., Levinthal, D.A.: Absorptive capacity: a new perspective on learning and innovation. Adm. Sci. Q. **35**, 128–152 (1990)
26. Damanpour, F.: Organizational complexity and innovation: developing and testing multiple contingency models. Manage. Sci. **42**(5), 693–716 (1996)
27. Das, T.K., Teng, B.S.: A resource-based view of strategic alliances. J. Manag **26**(1), 31–61 (2000)
28. Drucker, P.: The Practice Management. Harper & Row, New York (1954)
29. Drucker, P.: The Age of Discontinuity. Harper & Row, New York (1969)
30. Eccles, R.G., Crane, D.B.: Doing Deals: Investment Banks at Work. Harvard Business School Press, Boston (1987)
31. Etzioni, A.: A comparative analysis of complex organizations. Free Press, New York (1961)
32. Feldman, M.: Organizational routines as a source of continuous change. Organ. Sci. **11**(6), 611–629 (2000)
33. Fisher, M.: Innovation, knowledge creation and systems of innovation. Ann. Reg. Sci. **35**, 199–216 (2001)
34. Fleming, L.: Recombinant uncertainty in technical search. Manage. Sci. **47**, 117–132 (2001)
35. Galinic, D.C., Rodan, S.: Resource recombination's of the firm: knowledge structures and the potential of Schumpeterian innovation. Strateg. Manag. J. **19**, 1193–1201 (1998)
36. Garcia, R., Calantone, R.: A critical look at technological innovation typology and innovativeness terminology: a literature review. J. Prod. Innov. Manag. **19**(2), 110–132 (2002)
37. Grant, R.M.: Towards a knowledge-based theory of the firm". Strateg. Manag. J. **17**, 109–122 (1996)
38. Greenwood, R., Hinings, C.R.: Understanding radical organizational change: bringing together the old and the new institutionalism. Acad. Manag. Rev. **21**(4), 1054–1922 (1996)
39. Greenwood. R., Hinings, C.R, Cooper, D., Brown, J.: The global management of professional services: the example of accounting. Paper presented at the 6th APROS international colloquium, Cuernavaca, Mexico (1995)
40. Handy, C.: Balancing corporate power: a new federalist paper. Harv. Bus. Rev. **70**, 59–72 (1992)
41. Handy, C.: The Age of Paradox. Harvard Business School Press, Cambridge (1994)
42. Hamel, G., Prahalad, C.K.: Strategy as a field of study: why search for a new paradigm. Strateg. Manag. J. **15**, 5–16 (1994)
43. Hedlund, G.: A model of knowledge management and the N-form Corporation. Strateg. Manag. J. **15**, 73–90 (1994)
44. Helfat, C.E., Peteraf, M.A.: The dynamic resource-based view. Capability lifecycles. Strateg. Manag. J. **24**, 997–1010 (2003)
45. Henderson, R.M., Clark, K.B.: Architectural innovation: the reconfiguration of existing product technologies and the failure of established firms. Adm. Sci. Q. **35**(1), 9–22 (1990)
46. Huber, C.P.: Organizational learning: the contributing process and the literature. Organ. Sci. **2**(1), 88–115 (1991)
47. Ilinitich, A.Y., D'Aveni, R.A., Lewin, A.Y.: New organizational forms and strategies for managing in hypercompetitive environments. Organ. Sci. **7**(3), 211–220 (1996)
48. James, W.: A Pluralistic Universe. University of Nebraska Press, Lincoln (1996)
49. Johannessen, J.-A., Olsen, B., Olaisen, J.: Intellectual capital as a holistic management philosophy: a theoretical perspective. Int. J. Inf. Manag. **25**, 151–171 (2005)
50. Jones, P.G., Hesterly, W.S., Borgatti, S.P.: A general theory of network governance: exchange conditions and social mechanisms. Acad. Manag. Rev. **22**, 911–945 (1998)
51. Kirzner, I.M.: On the method of Austrian economics. In: Dolan, E.G. (ed.) The Foundation of Modern Austrian Economics, pp. 40–51. KS Shed and Ward, Kansas City (1976)
52. Kirzner, I.M.: Discovery and the Capitalist Process. University of Chicago Press, Chicago (1985)
53. Kotter, J.K., Heskett, J.: Corporate Culture and Performance. Free Press, New York (1992)

54. Leonard-Barton, D.: Core capabilities and core rigidities: a paradox in managing new product development. Strateg. Manag. J. **13**, 111–125 (1992)
55. Leonard-Barton, D.: Wellsprings of Knowledge: Building and Sustaining the Sources of Innovation. Harvard Business School Press, Boston (1995)
56. Lunnan, R., Haugland, S.A.: Predicting and measuring alliance performance: a multidimensional analysis. Strateg. Manag. J. **29**, 545–556 (2008)
57. Løwendahl, B.R.: Strategic Management in Professional Service Firms. Copenhagen Business School Press, Copenhagen (1997)
58. March, J.G., Simon, H.A.: Organizations. Wiley, New York (1958)
59. March, J.G.: Exploration and exploitation in organizational learning. Organ. Sci. **2**, 71–87 (1991)
60. McDermott, C.T., O'Conner, G.C.: Managing radical innovation: an overview of emergent strategy issues. J. Innov. Manag. **19**(6), (2002)
61. Mintzberg, H., McHugh, A.: Strategy formulation in an adhocracy. Adm. Sci. Q. **30**, 160–197 (1985)
62. Mowery, D.C., Ixley, J.E., Silverman, B.S.: Strategic alliances and interfirm knowledge transfer. Strateg. Manag. J. **17**, 77–91 (1996)
63. Nelson, R.R., Winter, S.G.: An Evolutionary Theory of Economic Change. Belknap Press, Cambridge (1982)
64. Nerkar, A., Roberts, P.W.: Technological and product-market experience and the success of new product introductions in the pharmaceutical industry. Strateg. Manag. J. **25**, 779–799 (2004)
65. Nonaka, I.: A dynamic theory of organizational knowledge creation. Organ. Sci. **5**(1), 14–37 (1994)
66. Nonaka, I., Takeuchi, H.: The Knowledge-Creating Company: How Japanese Companies Create the Dynamics of Innovation. Oxford University Press, Oxford (1995)
67. Orlikowski, W.J.: Improvising organizational transformation over time: a situated change perspective. Inf. Syst. Res. **7**, 63–92 (1996)
68. Papadiuk, S., Choo, C.W.: Innovation and knowledge creation: how are these concepts related? Int. J. Inf. Manag. **26**, 302–312 (2006)
69. Parisotti, A.: Transnational corporations and the emerging global media markets. In: Aharoni, Y. (ed.) Changing Roles of State Intervention in Services, pp. 199–254. State University of New York Press, Albany (1997)
70. Penrose, E.T.: The Theory of the Growth of the Firm. Wiley, New York (1959)
71. Podolny, J.: A status-based model of market competition. Am. J. Sociol. **98**, 829–872 (1993)
72. Podolny, J.: Market uncertainty and the social character of economic exchange. Adm. Sci. Q. **39**, 458–483 (1994)
73. Polanyi, M.: Personal Knowledge: Towards a post-critical philosophy. University of Chicago Press, Chicago (1962)
74. Polanyi, M.: The Tacit Dimension. Anchor Books, Garden City (1967)
75. Porter, M.: Competitive strategy. Free Press, New York (1980)
76. Porter, M.: Competitive Advantage: Creating and Sustaining Superior Performance. Free Press, New York (1985)
77. Quinn, J.B.: Intelligent Enterprise. The Free Press, New York (1992)
78. Sabbagh, K.: Twentieth-First Century Jet: The Making and Marketing of the Boeing 777. Scriner, New York (1996)
79. Scarborough, H.: Knowledge management in HRM and the innovative process. Int. J. Manpow. **24**(5), 501–516 (2003)
80. Schein, E.H.: Organizational Culture and Leadership. Jossey Bass, San Francisco (1985)
81. Scott, M.C.: The Intellect Industry-Profiting and Learning from Professional Service Firms. Wiley, Chichester (1998)
82. Schumpeter, J.A.: The Theory of Economic Development. Harvard University Press, Cambridge (1934)

83. Schumpeter, J.A.: Business Cycles: A Theoretical, Historical and Statistical Analysis of the Capitalist Process. McGraw-Hill, New York (1939)
84. Schumpeter, J.A.: Capitalism, Socialism, and Democracy. Harper & Brothers, New York (1942)
85. Schumpeter, J.A.: The creative response in economic history. J. Econ. Hist. **7**, 149–159 (1947)
86. Senge, P.: The Fifth Discipline. MIT Press, Cambridge (1991)
87. Simon, H.A., Smith, A.J., Thompson, C.B.: Modern organization theories. Adv. Manag. **15**, 2–14 (1950)
88. Silvestri, G.T.: Occupational employment projections in 2006. Mon. Labor Rev. **12**, 58–83 (1997)
89. Skjølsvik, T., Løwendahl, B.R., Kvålshaugen, R., Fosstenløkken, S.M.: Choosing to learn and learning to choose-strategies for client co-production and knowledge development. Calif. Manag. Rev. **49**(3), 110–128 (2007)
90. Snow, C.C., Miles, R.E., Coleman, H.J.: Managing 21st Century Network Organizations, pp. 5–20. Organ. Dyn., Winter (1992)
91. Soliman, F., Spooner, K.: Strategies for implementing knowledge management: role of human resources management. J. Knowl. Manag. **4**(4), 337–345 (2000)
92. Starbuck, W.H.: Congealing oil: investing ideologies to justifying ideologies. J. Manag. Stud. **19**(1), 3–27 (1982)
93. Storey, J., Quntas, P.: Knowledge management and HMR. In: Storey, J. (ed.) Human Resource Management: A Critical Text, 2nd edn, pp. 3–20. Thompson Learning, London (2001)
94. Syvertsen, C.: What is the future of business schools? Eur. Bus. Rev. **20**(2), 142–151 (2008)
95. Teece, D.J., Pisano, G., Shuen, A.: Dynamic capabilities and strategic management. Strateg. Manag. J. **18**(7), 509–533 (1997)
96. Tushman, M.L., Anderson, P.C.: Technological discontinuities and organizational environments. Adm. Sci. Q. **31**(3), 439–465 (1986)
97. Tushman, M.L., Anderson, P.C., O'Reily, T.: Technological cycles, innovation streams and ambidextrous organizations: organizational renewal through innovation and strategic change. In: Tushman, M.L, Anderson, P.C. (eds.) Managing Strategic Innovation and Change: A Collection of Readings. Oxford University Press, New York (1997)
98. Utterback, J.M.: Mastering the Dynamics of Innovation. How Companies can Seize Opportunities in the Face of Technological Change. Harvard Business School Press, Boston (1994)
99. Von Krogh, G., Ichijo, G., Nonaka, I.: Enabling Knowledge Creation. Oxford University Press, Oxford (2000)
100. Wernerfeldt, B.: A resource-based view of the firm. Strateg. Manag. J. **5**(2), 171–180 (1984)
101. Zaltman, G., Ducan, R.R., Holbeck, J.: Innovations and Organizations. Wiley, New York (1973)
102. Zollo, M., Winter, S.G.: Deliberate learning and the evolution of dynamic capabilities. Organ. Sci. **13**(3), 339–351 (2002)

SMATT: Smart Meter ATTestation Using Multiple Target Selection and Copy-Proof Memory

Haemin Park, Dongwon Seo, Heejo Lee and Adrian Perrig

Abstract A smart grid is verging on a promising technology for reforming global electrical grids. Currently, attackers compromise security and privacy by maliciously modifying the memory of smart grid devices. To thwart such attacks, software-based attestation protocols ensure the absence of malicious changes. A verifier and a target device locally generate their own checksums by memory traversal, and the verifier attests the target device by comparing the checksums. For smart grids, however, two challenges are arise in practically deploying the attestation protocol: verification overhead for large-scale networks and evasion of attestation by memory replication. To address these challenges, we propose a novel software-based attestation technique, termed SMATT (Smart Meter ATTestation), to address the aforementioned two challenges by leveraging multiple target selection and copy-proof memory. A verifier randomly selects multiple smart meters, and receives checksums. The verifier only compares the checksums instead of performing memory traversal, thereby remarkably reducing the computational overhead. To prevent memory replication, we design a customized copy-proof memory mechanism. The smart meter outputs garbage values when copy-proof memory sections are being accessed, and thus, attackers cannot replicate the memory. Furthermore, we define an SI epidemic model considering two

H. Park · D. Seo · H. Lee (✉)
Division of Computer and Communication Engineering, Korea University, Seoul, Korea
e-mail: heejo@korea.ac.kr

H. Park
e-mail: gaiger@korea.ac.kr

D. Seo
e-mail: aerosmiz@korea.ac.kr

A. Perrig
CyLab/ECE, Carnegie Mellon University, Pittsburgh, PA, USA
e-mail: adrian@ece.cmu.edu

S.-S. Yeo et al. (eds.), *Computer Science and its Applications*,
Lecture Notes in Electrical Engineering 203, DOI: 10.1007/978-94-007-5699-1_90,
© Springer Science+Business Media Dordrecht 2012

attestation parameters, the number of infectious smart meters and the number of selected smart meters by a verifier, to enhance the malware detection accuracy of SMATT. In our experimental environments, SMATT takes only 20 s for a verifier to attest millions of smart meters. In addition, SMATT detects malware with over 90 % probability, when the malware tampers with 5 % of the memory.

1 Introduction

A smart grid, an ongoing modernization of a power grid, is a system that integrates existing electricity infrastructure with IT networks. This provides further benefits to customers and power utilities: reducing electricity costs for customer and controlling power generation for power utilities [1, 2]. Especially, a smart meter plays a key role to facilitate such beneficial services, and thus it becomes an attractive target for an adversary to gain an advantage and cause social chaos [1, 3–5]. For example, malware infiltration on easily accessible smart meters is a perplexing problem on a faraway utility, because it can manipulate electricity costs or fabricate metering information for a targeted home. This vulnerability can be monetized by attackers.

A technology for malware detection in an embedded device is to check the integrity of the device's firmware. In particular, software-based attestation techniques remotely verify whether the memory of the embedded device is maliciously changed [6, 7]. These techniques are based on a challenge-response protocol. A verifier sends a challenge to a target device, and then it reads entire memory in a pseudo-random fashion. The checksum as a result of pseudo-random memory traversal is sent from the target device to the verifier. Finally, the verifier computes the checksum in the same fashion, and compares it with the checksum from the target device.

However, the existing attestation techniques cannot be easily applied to smart meters. The complexity and scale of power grids cause new challenges: management of millions of devices and routine maintenance [2]. Especially, from attestation perspective, it is difficult for the verifier to attest millions of smart meters due to the heavy processing burden caused by checksum computation. Furthermore, recent research indicates that malware can subvert memory [8], and contribute to evasion of attestation [9].

In this paper, we propose a Smart Meter ATTestation mechanism, termed SMATT, that reduces computational overhead of a verifier and prevents memory replication to evade attestation. It leverages multiple target selection and copy-proof memory sections. Multiple meters manufactured with the same configuration, namely *identical smart meters,* are randomly selected by the verifier. Then, the verifier only compares the checksums without pseudo-random memory traversal. As a result, SMATT reduces the verification overhead. To prevent memory replication attacks, the smart meter sets the copy-proof memory sections and monitors memory-related APIs. If the copy-proof memory sections are being accessed through the APIs, then the smart meter outputs garbage values as if they are real ones. Attackers obtain memory values ruined by the garbage, and the replicated smart meter fails to attest. Moreover, we define attestation parameters to enhance the malware detection accuracy of SMATT by using the popular SI epidemic model [10]. Our experiments show that SMATT reduces verification overhead, and detects memory falsification by malware with a high probability.

The main contribution of our work is twofold. First, SMATT reduces the verification overhead on the verifier; it is efficient for large-scale smart grids that consist of a large number of smart meters with a small number of verifiers. Second, SMATT enhances the robustness of attestation by preventing memory replication attacks.

The remainder of this paper is organized as follows. In Sect. 2, we introduce several problems, and describe the detail attestation procedure in the next section. Security analysis and evaluation are followed in Sects. 4 and 5. Finally, Sect. 6 presents our conclusion.

2 Problem Definition

In this section, we describe the weaknesses in software-based attestation protocols and state the problem that we aim to solve throughout this paper.

2.1 Weaknesses in Software-Based Attestation Protocols

Software-based attestation protocols have been proposed for embedded devices. Due to the constrained resources of embedded devices, the existing protocols pursue a lightweight design, and it can be suitable for the smart meter [2]. Nevertheless, there are several remaining issues yet to be solved in terms of verification overhead and malware infiltration [2].

Immense verification overhead by large-scale networks. Considering the complexity and scale of smart grids, a large number of smart meters will create a great computational overhead for a verifier. That is, a verifier linked to many smart meters should compute the checksums of each smart meter. For example, the verifier can become exhausted when many attackers send an overwhelming amount of miscalculated checksums. Therefore, it is a challenge to design an efficient software-based attestation protocol for the verifier that manages a large number of smart meters.

Evasion of attestation by memory replication attacks. Malware utilizes a linear scan of memory to create trial data from all possible memory positions [9]. An attacker can obtain whole memory values including critical information, and can produce replicated smart meters. The attacker modifies the firmware of one's smart meter in order to make illegal profits such as modifying electricity usage. Meanwhile, the replicated smart meter transmits correct responses to verifiers and success to attest. For example, Song et al. [11] designed a one-way attestation protocol that leverages built-in values as challenges to compute a checksum. However, the challenges can be predicted by a memory replication attack, and thus, the attacker can compute the correct checksum.

2.2 Problem Statement

A viable attestation mechanism needs to address the following requirements:

- To reduce verification overhead on the verifier that handles a large number of smart meters.
- To prevent evasion of attestation by the memory replication attack.

  Our goal is to design the attestation protocol that satisfies the two requirements.

3 SMATT: Smart Meter ATTestation

We introduce three assumptions that comply with the NIST guidelines for smart grids [3]. Then, we describe SMATT in detail. Finally, we determine two attestation parameters to enhance the malware detection accuracy of SMATT.

3.1 Assumptions

SMATT should incorporate the following three technical assumptions.

1. Time synchronization: Accurate and reliable time synchronization is essential to ensure that equipments operate correctly in the smart grid. Especially, a metering service based on time-of-use pricing is the reason that time synchronization is necessary.
2. Encrypted communication: Existing key management solutions such as Public Key Infrastructure (PKI) are considered as the basis of further innovation for secure communication between smart meters and verifiers [3].
3. Identical smart meters: A manufacturer produces identical smart meters installed an exact same firmware and configuration. The identical smart meters generate same checksums as the result of memory traversal. A verifier has information that indicates the groups of identical smart meters.

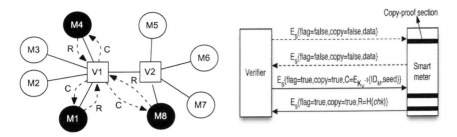

Fig. 1 Randomly selected smart meters (M1, M4, M8) exchange challenges (C) and responses (R) with a verifier (V1) (*left*). During the attestation period (*solid arrows*), V1 inserts a challenge (C), and receives a response (R) from the smart meter (*right*)

3.2 SMATT Operations

SMATT operations are based on a challenge-response protocol, which consists of four steps: multiple target selection, copy-proof memory generation, checksum generation and verification.

Multiple target selection. In existing attestation mechanisms, a verifier should perform checksum computing operations to attest a single target (smart meter), and it causes very high computational overhead for the verifier in large-scale networks like smart grids. Multiple target selection is designed to reduce the overhead problem by comparing checksums instead of computing them.

A verifier selects multiple identical smart meters to be attested, and receive checksums. Then, the verifier compares the checksums and is able to identify which smart meter sends a different checksum. Hence, the verifier can detect compromised meters without checksum computing operations. A detailed description is as follows:

1. A verifier has a private key (K_V^{-1}). Then, it shares a public key (K_V) and a symmetric key (S) with smart meters. All packets are encrypted by the symmetric key $(E_S\{\cdot\})$.
2. During usual period, the verifier occasionally transmits the encrypted packet, $E_S\{flag = false, copy = false, data\}$, to its managed meters. The $flag = false$ means that the verifier is in usual period.
3. During attestation period, the verifier (V1) determines a group to be attested. In Fig. 1 (left), there are two identical groups, G1 = {M1, M4, M8} and G2 = {M2, M3, M5, M6, M7, M8}, and V1 selects G1. Among the G1, V1 randomly selects multiple smart meters M_N to be attested.
4. The verifier sets the $flag = true$ and inserts a challenge, $C = E_{K_V^{-1}}\{ID_M, seed\}$. After decryption using K_V, ID_M is utilized to identify each selected meter. The seed becomes the input for the memory traversal algorithm.
5. The selected meter returns a response $(R = H(chk))$ to the verifier, where $H(\cdot)$ denotes a cryptographic hash function and chk denotes a checksum.

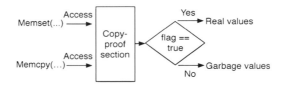

Fig. 2 Memory-related APIs are monitored by SMATT. When copy-proof selections are being accessed through those APIs, SMATT determines whether the outputs become real values according to the flag

Copy-proof memory section. An attacker can evade attestation using the memory replication attack (viz., Sect. 2). The effective scheme to prevent the attack is to utilize memory randomization. The values of special memory sections are periodically changed, so that the attacker cannot obtain the real values of whole memory. However, since the smart meter uses a flash memory as storage, the frequent changes of specific memory section can cause the "wear-out" problem that shortens flash memory life.

Therefore, we define copy-proof memory sections in the smart meter that addresses the wear-out problem. The copy-proof sections are scattered throughout the memory, and they are filled with garbage values. The smart meter monitors memory-related APIs, such as *memset* and *memcpy*. During attestation period, the memory-related APIs output real values of the copy-proof sections. During usual period, however, the APIs output random garbage values instead of real ones. Consequently, the attacker cannot replicate the copy-proof sections, and thus, the attacker's smart meter containing garbage copy-proof sections fails to attest. Figure 2 shows conceptual operations that output different memory values depending on the flag.

Moreover, the smart meter sets *copy* = *true* if the memory-related APIs access to the copy-proof sections during attestation period. Thus, the verifier can recognize the illegal access to the copy-proof sections if the smart meter sends packets including *copy* = *true* during usual period.

Checksum generation. SMATT adopts a block-based pseudo-random memory traversal technique that generates a checksum over the memory cells of the smart meter [7]. A seed (S) substitutes for the input value of $RC4(\cdot)$, which is a pseudo-random number generator. The memory traversal addresses (A) meets *size_of_memory* by modulo operation. *memory_traversal(A, size_of_block)* reads memory from address A to $A+size\_of\_block$, and outputs its result to MEM_i, where MEM_i denotes the memory values of the ith traversal. The *chk* is derived from the concatenation of MEM_i and CP_MEM_i, where CP_MEM_i denotes the memory values of copy-proof sections of the ith traversal. After generating the *chk*, A becomes the new seed for the next step. These four sequences are repeated until the number of iterations reaches a predefined maximum iteration number (T). Finally, the *chk* is hashed by a cryptographic hash function such as SHA-1 to avoid exposing the original *chk*. Equation (1) represents the checksum generation.

$$H(chk) = H\left(\sum_{i=1}^{T} (MEM_i \| CP\_MEM_i)\right) \tag{1}$$

Checksum verification. Selected meters send a $H(chk)$ to a verifier. In the verifier, the $H(chk)$ is compared with that of each selected meter. If these $H(chk)$'s are identical, the verifier determines that the smart meters are valid. Thus, SMATT reduces verification overhead because the verifier does not need to perform memory traversal.

3.3 Determining Attestation Parameters

SMATT leverages randomization and multiple target selection to reduce verification overhead and prevent evasion of attestation. However, if there are a sufficient number of compromised meters, only compromised meters can be selected by a verifier.

In other words, the verifier cannot distinguish between legitimate meters and compromised meters, since the compromised meters can provide the same incorrect checksum. To avoid the situation, a verifier should select one legitimate meter at least; Eq. (2) shows the condition

$$k > I(t), \tag{2}$$

where $I(t)$ denote the malware spreading ratio at a time t, and k denote the number of meters to be selected by a verifier. Here, we define two attestation parameters, $I(t)$ and k, in order to enhance the malware detection accuracy of SMATT.

Determining $I(t)$. Supposing a smart meter becomes either susceptible (S) or infected (I). We utilize a deterministic epidemic model, the SI model to analyze malware propagation in the smart grid. The SI epidemic model [10] for smart meters is described in Eq. (3).

$$\frac{dI}{dt} = \alpha SI, \tag{3}$$

where α is the infection rate of smart meters. It depends on the number of susceptibles (S) and infectives (I). Also, Eq. (4) explains α:

$$\alpha = \frac{\beta x}{N}, \tag{4}$$

where x means contacts per unit of time. β is the probability of infection when an infective contacts a susceptible. N means the total number of smart meters. Since we assume that a smart meter can be either S or I, S becomes $N - I$. Now, we can derive Eq. (5):

$$\frac{dI}{dt} = \alpha \cdot (N - I)I = \alpha NI - \alpha I. \tag{5}$$

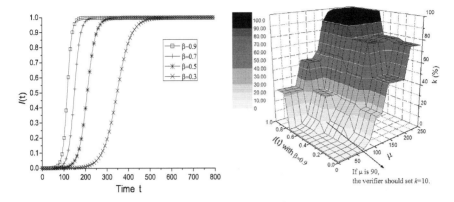

Fig. 3 According to research on malware propagation [17], malware has the characteristics of slow start and exponential propagation (*left*). The malware spreading ratio at a time t ($I(t)$) determines the number of meters to be selected (k). That is, k should be greater than the number of infected smart meters at a time μ, where μ means the attestation cycle (*right*)

Then, Eq. (6) is also derived by the general solution.

$$I(t) = \frac{N}{1 + cNe^{-\alpha Nt}}, \tag{6}$$

where c is a constant that depends on the initial conditions.

We use the same environment considered in the reports [12, 13]. The number of smart meters per house is 3 (gas, electricity, and water meters), and there are 105,000 meter/km^2 in an urban area such as London. The maximum contacts per unit of time is set to 0.1/s. Hence, we set $I_0 = 3$, $N = 105,000$ and $x = 0.1$, and obtain Eq. (7). The result of Eq. (7) is shown in Fig. 3 (left).

$$I(t) = \frac{N}{1 + \frac{1}{3}(N-3)e^{-\alpha Nt}}. \tag{7}$$

Determining k. Next, we determine the k. As we mentioned, a verifier should select at least one legitimate meter. That is, k is greater than the number of infected smart meters during attestation. By using the parameter, $I(t)$, the verifier sets $k > I(\mu)$, where μ means the attestation cycle. The k related to $I(\mu)$ is shown in Fig. 3 (right). Depending on the availability of network resources, an administrator can consider two possible cases to determine the k:

- In the case of a low-bandwidth smart grid network, a verifier should set k high. For example, if $\beta = 0.9$ and $\mu = 90$, then $k = N \times 0.1$.
- In the case of a high-bandwidth smart grid network, a verifier should set k low. For example, if $\beta = 0.9$ and $\mu = 60$, then $k = N \times 0.01$.

4 Security Analysis

In this section, we analyze the efficiency of SMATT in terms of time complexity, and we describe how SMATT prevents evasion of attestation by malware.

4.1 Verification Overhead

In practice, estimated numbers of meters per sector range from approximately 12,000 to 35,000 in London [12]. It can be seen that there is a large variation in the number of meters per sector. However, SMATT does not need memory traversal of the verifier. The time complexity of checksum verification in SMATT takes $O(N)$, while that of the existing attestation protocols takes $O\left(\frac{Nm \ln m}{b}\right)$, where N is the number of smart meters, b means the size of a block, and m denotes the memory size [7, 14].

4.2 Malware Infiltration

We define feasible attack scenarios, and then we discuss how SMATT thwarts them.

Attack scenarios and countermeasures. An attacker with the objective of monetization aims to inject malware into easily accessible smart meters. The malware contributes to evasion of attestation by using a proxy attack and a memory attack.

A proxy attack scenario. The attacker installs a proxy meter that generates a correct checksum instead of a compromised meter, and evades attestation. The attacker abuses 2 m: compromised and proxy meters. The compromised meter simply forwards the attestation request from the verifier to the proxy meter. Then, the proxy meter returns the correct checksum to the compromised meter. This enables the compromised one to evade attestation by forwarding the checksum to the verifier.

Unfortunately, the ID of each meter prevents against the proxy meter. The compromised meter should encrypt the challenge that includes the proxy ID, and sends it to the proxy meter. Because malware cannot infer the private key for the encryption, it has no chance to generate the new challenge, and evasion of attestation by the proxy meter is defeated.

A memory replication attack scenario. The attacker abuses the compromised meter that sends the correct checksum to the verifier, and evades attestation. By the memory replication attack, the attacker obtains knowledge of the entire program memory and the correct checksum. The compromised meter returns the checksum to the verifier when attestation begins.

In SMATT, however, copy-proof sections prevent the memory replication attack. Since SMATT differentiates the output values of the copy-proof sections when the attacker attempts to access, the attacker cannot replicate the smart meter memory and fails to attest. A sophisticated attacker attempts to access the copy-proof sections only during attestation period, because it is the only time that SMATT gives real values. Nevertheless, the attacker cannot anticipate when attestation period begins (*false = true*) because of the packet encryption; therefore, the attacker cannot help guessing the timing of attestation period, and it leads *flag = false* and *copy = true*, which means illegal access to the copy-proof sections during usual period.

5 Experiment and Results

We first establish an experimental environment. Then, we examine the computational overhead of the verifier in the environment, and measure the malware detection rate.

5.1 Experimental Environment

SMATT is implemented on two Android devices (as smart meters) and a laptop (a verifier). Verifier's hardware specification is Intel(R) Atom 1.60 GHz CPU and 1 GB RAM, while the two Android devices are the Google development smartphone (Nexus-dev1), Scorpion 1 GHz CPU and 512 MB RAM. The Android smartphone provides similar hardware specifications to the smart meter [15]; moreover, Google announced that the Android OS can be applied at home in the smart grid [16].

5.2 Verification Overhead

We examine verification overhead by measuring the time overhead to attest lots of smart meters. The Android device simulates lots of smart meters by sending *k* checksums at once. And, we assume several variables related to smart grid requirements and the attestation parameters that we mentioned in Sect. 3. Based on

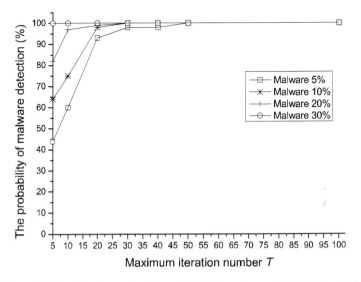

Fig. 4 The probability of malware detection: SMATT thwarts over 90 % of the attacks, if the malware tampers with 5 % of the memory of the smart meter

the requirements [12, 13], the network latency between the verifier and a smart meter is 5,000 ms and the data size is 35 bytes. In addition, we set $k = N \times 0.2$ and $N = 105,000$. In our environment, SMATT takes 20,047 ms, which is more efficient than the time of the typical software-based attestation as the verifier respectively sends a challenge to each smart meter. The time is approximately 6 days in the worst case. Therefore, the experimental result shows that SMATT has low verification overhead.

5.3 Malware Detection

We vary the fraction of malware in smart meter memory to measure the probability of malware detection. Figure 4 shows the probability of malware detection. SMATT on an Android device has a different fraction of malware ranging from 5 to 30 %, since the attacker can exploit the available space, approximately 11.6 %, on average [8] to inject malware. If malware takes 10 % of program memory, a smart meter should traverse memory contents approximately 30 times. Although malware changes a small amount of memory contents, such as 5 % of program memory, it can be detected by traversing memory 50 times. The experimental result shows that SMATT detects malware with a high probability.

6 Conclusion

In this paper, we propose a novel software-based attestation protocol for smart meters, SMATT, which provides a technique to reduces the processing burden of a verifier and prevents evasion of attestation by memory replication attacks. Multiple target selection contributes to simplify the verification operation of a verifier, and copy-proof memory protects against memory replication attacks. We analyze the attestation parameters to enhance the malware detection accuracy of SMATT. In our experiments, SMATT is remarkably faster than typical attestation protocols due to its reduction of verification overhead, and it has a high detection probability with low memory falsification rates.

Acknowledgments This research was supported by the National IT Industry Promotion Agency (NIPA) under the program of Software Engineering Technologies Development and Experts Education. Additionally, this research was supported by Seoul R & BD Program (WR080951).

References

1. McDaniel, P.D., McLaughlin, S.E.: Security and privacy challenges in the smart grid. IEEE Secur. Priv. **7**(3), 75–77 (2009)
2. Khurana, H., Hadley, M., Lu, N., Frincke, D.A.: Smart-grid security issues. IEEE Secur. Priv. **8**(1), 81–85 (2010)
3. National Institute of Standards and Technology (NIST): NIST IR 7628: Guidelines for Smart Grid Cyber Security (2010) http://csrc.nist.gov/publications/PubsNISTIRs.html
4. McLaughlin, S.E., Podkuiko, D., McDaniel, P.: Energy theft in the advanced metering infrastructure. In: CRITIS, pp. 176–187 (2009)
5. Mo, Y., Kim, T.H., Brancik, K., Dickinson, D., Lee, H., Perrig, A., Sinopoli, B.: Cyber-physical security of a smart grid infrastructure. In Proceedings of the IEEE (2011)
6. Seshadri, A., Perrig, A., van Doorn, L., Khosla, P.K.: SWATT: Software-based attestation for embedded devices. In: IEEE Symposium on Security and Privacy. (2004)
7. AbuHmed, T., Nyamaa, N., Nyang, D.: Software-based remote code attestation in wireless sensor network. In: GLOBECOM, pp. 1–8 (2009)
8. Castelluccia, C., Francillon, A., Perito, D., Soriente, C.: On the difficulty of software-based attestation of embedded devices. In: ACM Conference on CCS, pp. 400–409 (2009)
9. Hargreaves, C., Chivers, H.: Recovery of encryption keys from memory using a linear scan. In: ARES (2008)
10. Khelil, A., Becker, C., Tian, J., Rothermel, K.: An epidemic model for information diffusion in MANETs. In: Proceedings of the 5th ACM international workshop, MSWiM '02, pp. 54–60 (2002)
11. Song, K., Seo, D., Park, H., Lee, H., Perrig, A.: OMAP: One-way memory attestation protocol for smart meters. In: EEE ISPA Workshop SGSC, pp. 111–118 (2011)
12. Himayat, N., Johnsson, K., Talwar, S., Wang, X.: Functional requirements for network entry and random access by large number of devices. Technical Report IEEE802.16ppc-10/0049r1, IEEE 802.16 Broadband Wireless Access Working Group (2010)
13. Himayat, N., Talwar, S., Johnsson, K.: Smart grid requirements for IEEE 802.16 M2MNetwork. Technical Report IEEE C802.16ppc-10/0042r2, IEEE (2010)
14. Mitzenmacher, M., Upfal, E.: Probability and Computing—Randomized Algorithms and Probabilistic Analysis. Cambridge University Press, Cambridge (2005)

15. SmartSynch: Functions and Features i-210+c SmartMeter. http://smartsynch.com/pdf/i-210+c_smartmeter_e.pdf (2012)
16. Berst, J.: Will Google destroy Zigbee? http://www.smartgridnews.com/artman/publish/Business_Strategy/Will-Google-destroy-ZigBee-3681.html. (2011)
17. Zhu, Z., Cao, G., Zhu, S., Ranjan, S., Nucci, A.: A social network based patching scheme for worm containment in cellular networks. In: IEEE INFOCOM (2009)

Part VII
Computer Convergence

Design of a Structured Plug-in Smart Education System

Jaechoon Jo, Youngwook Yang and Heuiseok Lim

Abstract With a recent emergence and growing interest in smart education and rapid growth of the related market, we established Structured Plug-in Smart Education System for effective smart education, based on the concept of smart education, research, and technology. This system consists of a Smart Contents Service System that links learning contents to producing, managing, and learning, as well as a School and Home Learning System that supports cooperating, intellectual, and life-long learning by creating learning spaces in school and home, and effective learning correlation. In order to realize this system for smart education, we plan to verify the effectiveness of the Structured Plug-in Smart Education System by applying this system to formal education and analyzing its effectiveness.

Keywords Smart education · Smart learning · Contents service · School and home learning · School of the future

J. Jo · Y. Yang · H. Lim (✉)
Department of Computer Science Education, Korea University, Seoul, Korea
e-mail: limhseok@korea.ac.kr

J. Jo
e-mail: Jaechoon@korea.ac.kr

Y. Yang
e-mail: brilliant_yyw@korea.ac.kr

S.-S. Yeo et al. (eds.), *Computer Science and its Applications*,
Lecture Notes in Electrical Engineering 203, DOI: 10.1007/978-94-007-5699-1_91,
© Springer Science+Business Media Dordrecht 2012

1 Introduction

With the launch of smart phones in Korea in November 2009, smart phone fever started to sweep the country and it affects various sectors, such as culture and education, along with other cutting edge information and communication technologies. In particular, focus has been put on education using smart phones, and discussion regarding this is under way in many countries. There are new approaches to make use of smart phones, which is a new form of mobile computer unlike existing cell phones, along with other smart devices. Several teachinglearning models using smart devices have been suggested as an academic approach. However, there are few concepts or educational environments for the actual application of smart education [13]. Under the circumstances, there are increasing cases of approaches based only on instrumental environments (smart device, network environment).

Although the size of the market for smart education is expected to amount to 3.5 trillion won by 2015 [4], some problems in the e-Learning market are emerging, such as lack of content resources and rising production costs [5]. The content production costs should decrease, because smart education content can be reproduced into a variety of formats through collaboration, sharing, and participating instruments. In addition to this, the educational content industry is expected to grow further, as the educational ecosystem itself is digitalized. Responding to this trend, the Ministry of Education, Science, and Technology in Korea announced Way to Talent Rich CountryStrategy on the Implementation of Smart Education. This is to nurture creative global talent for future societies. The strategy will be implemented through smart education that can support customized teachinglearning services instead of delivering knowledge, based on the standardized educational curriculum of industrialized society. The Ministry of Knowledge Economy invited the public to a contest for Development of Plug-in type Software for Smart Education, as a WBS (World Best Software) project, highlighting many issues and necessity of smart education in Korea [6, 7].

With the development of IT technology, distribution bases such as books, textbooks, film, newspaper, and broadcasting have been integrated into the Internet network, which led to the crisis of the traditional media industry. Furthermore, tablet PCs and smart phones that successfully entered the market are leading the smart media market by integrating contents. The educational environment is also changing as the need for IT convergence that can replace the existing knowledge delivery form (based on books) grows, classes increasingly focus on cooperation, sharing, and participation, and support for self-directed learning expands. These changes garner attention from developing countries seeking to improve their economy through education reform, as well as advanced countries. There are an increasing number of contacts to introduce Koreas successes combining education and IT technology.

This study suggests a Plug-in Smart Education System that can maximize the teaching and learning effect for teachers and students by introducing research and development trends of smart education that are emerging in Korea, and reflecting on the learning characteristics of digital natives who are familiar with sharing, openness, cooperation, and participation culture. This system is a general system that can be applied in not only Korea, but also other countries that has infrastructure for smart education, such as wireless internet and smart devices.

2 Backgrounds

2.1 Concept of Smart Education

Smart education (smart learning) is generally known as education and learning that makes use of smart phones. In Wikipedia (Korea), smart education is defined as learning that makes use of smart phones, and furthermore, as an overall approach that enhances the effectiveness of education based on new learning methods which include characteristics such as participation, sharing, and customization, by introducing smart teaching methods, smart contents, and smart devices [8]. Smart education or smart learning is a commonly used term in Korea, while in other countries, Mobile Learning, Ubiquitous Learning, and Future School are used.

While a definite concept of smart education has yet to be established, various research is under way, including research on teaching methods, plans to establish smart learning systems, development of curriculum using smart devices, and analysis of the effectiveness of smart education [9–11].

2.2 Market Trends of Smart Education in Korea

The e-Learning sector is rapidly growing at a 36.7 % growth rate in the Korean mobile content market [12]. Applications dealing with books and education register as the first and forth most downloaded genre among iPad applications, respectively. As smart phone users surpass 15.60 million (Sept. 2011), smart education-related content market is soaring [13]. While content service is an important factor in smart education systems, securing and managing contents is currently emerging as a big challenge.

As markets for smart education content and electronic publications expand, many publication and e-Learning companies enter the smart education market, and companies specializing in Smart Education started to emerge. The smart education market is gaining energy as Korea Telecom (KT Co.) and other conglomerates enter the market, and government support expands. Although an enormous amount of money is invested in the smart education market by conglomerates, they enter the market by joining hands mainly with content specialists. In addition, the conglomerates use mostly smart devices or existing content, not new teaching methods or a system specialized for smart education. Thus, effective smart education has not yet been established. Thats why the Structured Plug-in Smart Education System for smart education is needed. Through this system, not only content management and services, but also customized, intellectual, mutual, and cooperative smart education can be accomplished.

2.3 Related Research and Technology

Establishing a system for smart education does not necessarily mean developing a new one, but creating a system by integrating and sharing existing smart technologies. Many educators and developers in various sectors have been expanding and developing technologies for effective education. The technologies for smart education systems include technology for content management and services, wireless communication, searching, and learning management.

NeoLMS of DaulSoft is a LMS (Learning Management System)/LCMS (Learning Contents Management System) solution complying with international standards suggested by SCORM (Sharable Content Object Reference), and can be used to produce and run educational content and construct courseware. The operator of the solution manages content and supports the learners study and evaluation of teachers [14]. Learning Wave of Seoul Cyber University (SCU) is a teaching and learning system, developed based on IMS common cartridge, which is an international standard of e-Learning. This system solved the problems of the e-learning system concerning learning instruments, limits on number, as well as closed and passive learning instruments [15]. Lee Jong-Gi explains the direction and characteristics of support by introducing the successful cases of learning management systems that makes use of Moodle (module object oriented developmental learning environment), an open source system [16]. Aspera developed FASP v2.0 transport technology supporting an enterprise server (AES). This technology for sending files in bulk form eliminates bottlenecks such as FTP (File Transfer Protocol), HTTP (Hypertext Transfer Protocol), and CIFS (Common Internet File System), and transports in a safer and faster way under the poor networking environment, such as remote transmission, wireless transmission, and satellite transmission [17]. In order to achieve smart education, these technologies should be expanded, developed, integrated, and shared, while the integrated smart education system should be customized for learners.

As interest in smart phones and smart education market expands in Korea, the government and conglomerates launched research and related activities to lead smart education. SK Telecom Co. made T-Smart Learning, the first domestic education platform based on tablet PC, in conjunction with 12 education companies. The platform uses a mobile communication network to support two-way learning and provide individuals with an optimized education service. It recommends a customized learning plan by suggesting curriculum and methods based on the learners level and learning style [18]. E-Edulib of Seoul Metropolitan Office of

Education provides people with electronic books, audio books, and other music-related services. The program also provides services such as customized reading diagnosis based on the readers level that chooses a book for readers, reading education programs, multicultural education, and translation [19].

Interest in smart education and its market continues to grow overseas as well. Smart Classroom Suite of Smarttech provides an interactive electronic black board, so that teachers and students run mutual classes. The mutual class running system, a four-step system including SmartNote, Smart Sync, Smart Notebook, and Smart Response, covers various areas from learning to evaluation [20]. SlideRocket is a useful instrument being utilized in establishing web-based cooperative presentation. It can manage the content on one screen and invite several people to cooperate and conduct group work [21]. Tutoring Club in the U.S. provides individuals, from kindergarteners to adults, with a personal education service. It developed software named Tutor Aid, which updates customized lesson plans every day by using the information gathered from all students. The lesson plan is evaluated by experts [22]. Many other governments, companies, and universities are developing smart education technologies to apply to e-Learning or u-Learning [23–26].

3 A Structured Plug-in Smart Education System

Identifying the concepts of smart education, developing related technology, and conducting research are attempts to create efficient and effective learning. Now, not only smart environment and smart teachinglearning methods, but also smart education systems that actually implement smart education are needed. This chapter designs a Structured Plug-in Smart Education System that can reflect the details of smart education by using smart technology. The system is based on the concept of smart education and the learning model.

As shown in Fig. 1, Structured Plug-in Smart Education System consists of the Smart Content Service System and School and Home Learning System. Smart Content Service System is an integrated education content service that manages the life cycle of learning content based on international standards, and connects the content to publication and even education. The system can also connect the content to service, and supports content transformation of Inter Screen that automatically changes the content format based on the characteristics of the equipment. School & Home Learning System forms a learning space involving both school and home, to normalize and strengthen public education.

In additional, The Structured Plug-in Smart Education System was designed for each system to organically plug-in, so that the system aims to be a developmental system, not a closed system. Under the plug-in enabling system, it can provide customized smart education system, by assembling system modules for each institution, company, or any group that uses it. Figure 1 shows that the service platform consists of Smart Content Service System, School & Home Learning

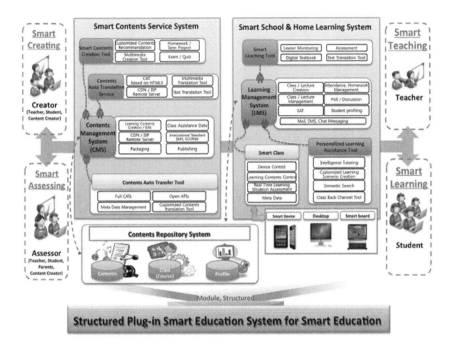

Fig. 1 Structured plug-in smart education system for smart education

System and Contents Repository System, while all other systems (such as module and web interface) are structured to enable plug-in and assemble them according to each requirements, so that a customized smart education system can be provided.

In smart education, through the Structured Plug-in Smart Education System, a variety of peoplefrom instructors, learners and parents to content producersachieve in the formation of education, teaching, learning, content production, etc. through smart creation, smart teaching, smart learning and smart assessment. The four-phased activities interact with one another and, through the various services of the Structured Plug-in Smart Education System, smart education can be achieved.

3.1 Smart Contents Service System

Smart Contents Service System consists of international standards, management system for non-standard content, smart textbook production system, content package/test, and interface management system.

As the content unit for education/learning is changing from an existing course ware type to a segmented type that is appropriate to Just-in-time and Workflow Learning, support capacity for Asset-type content and the border between publication and e-Learning is disappearing because smart phones and tablets are spreading. Under the circumstances, a management system for international standard & non-standard content is needed, so that various forms of content such as e-Learning course ware content (standard/non-standard), e-Book, Asset type contents (Image, Video, Animation, 3D, VR), Document, Web-Source can be managed as a complex of knowledge. In addition, support should be provided not only to SCORM 2004 of e-Learning and IMS Common Cartridge v1.0 of IMS GLC standards, but also IDPF standards.

The content publication instrument of smart content service system is a web-based instrument, unlike existing types of desktop application instruments. The system enables the content created by the publication instrument to run on multi-devices, and realizes a smart textbook system that can meet the International Digital Publishing Forum (IDFP) and International Standards. Smart Content Service System provides an environment in which you can produce content on the web, not based on programs. The system also provides additional functions in the plug-into facilitate attaching additional functionalities, such as animation, images, and events.

The content package/test and interface management system covers import/export, packaging, beta-testing, and the management of Data I/F linking outside system module. It also controls every process that unit content and Plug-in Module connect. The system supports connection of contents auto-conversion and auto-transmission system (CATs), Common Cartridge API, and LMS. It realizes a complete Smart Content Service System after running tests in multi-devices.

3.2 Smart School and Home Learning System

The Smart School and Home Learning System consists of a Smart Device Learning Support System, a Cooperative System to enhance the learning effect, a Learner Profiling System, a Real-time Learning Evaluation System, a Real-time Learning Monitoring System, a Learning Scenario System, an Intellectual Tutoring System, a Customized e-Portpolio System to support learning, a Parent Support System, and an Inter-Screen Sharing Video System.

The Smart Device Support System enables smart devices to include various learning content, so that learners can use the same learning content in various device environments. To this end, the system realizes functionality running smart device applications, as well as conversion module transmitting to HTML5 web content and viewers. Smart learning can be achieved by making use of various functionalities, such as the information from users, curriculum, or groups, management of user rights, management of assignment, forums and bulletin boards, quizzes, file-uploading, web links, questionnaires, blogs, chatting, and wiki.

The cooperative system to enhance learning effects uses an HTML 5-based drawing instrument that can draw a basic figure and attach or edit images, along with a real-time white board synchronized system, in which users can work on the same online white board in real time. The system realizes a 1:1 learning method that shows how teachers or a certain learner work on the learners screen, and a 1:1 learning method for individual teaching and learning. In addition, a user can let other users participate in the work. All these works can be stored with various image formats such as JPEG, PNG, Power Point, MS Word, and PDF.

The Learner Profiling System connects with several systems in the Structured Plug-in Smart Education System suggested by this paper, and stores the capacity and learners status shown in each system. Based on the stored data, the system evaluates learning status, monitors learning in real-time, offers customized content and intellectual tutoring. The Learner Profile includes basic information, learner capacity (attention, understanding, performance, and participation), and learner status (progress, score, and course status). The capacity and status of the learner are evaluated based on the data that Smart Education System sends.

The Real-Time Learning Evaluation System and Monitoring System are there to improve the effect of the class by identifying information on learners and classes in real-time, and applying it to the class. The learners attention, understanding, and participation can be evaluated by analyzing the problem-solving time of the learner, the rate of correct/wrong answers, interaction, and Q&A. The results are sent to a teacher via learning status notice module in real-time, and therefore the teacher can control the learners in real-time and make the class more effective.

The Learning Scenario Establishment System is a system that provides a customized learning scenario in an environment without classes or teaching. The system automatically generates a customized learning scenario based on the learners attention, understanding, performance, and participation offered by Real Time Learning Evaluation System and Real Time Monitoring System, as well as basic information and evaluation information in the learners e-portfolio. Using this system, learners can study the content not only at school, but pick up from home from where they left off, or they can prepare for or review their lessons.

The Intellectual Tutoring System consists of the learner/status module, learning module, and smart intellectual tutor module. It is a more expanded module than the leaner module, expert module, and tutor module of the general tutoring system, and provides customized education and feedback. The Smart Intellectual Tutor module automatically evolves by identifying the weak points of the learner and evaluates mistakes and problems based on diagnosis algorithms using the Bayesian Network. It applies these techniques assuming the current learning status of the learner and sends back content to make up for the identified weak points.

The Customized Learning Support e-Portfolio System is a personalized e-Portfolio that includes additional information, such as learner profile data, course information, and evaluation guidelines. The profiles of the portfolio consist of dates, content, scores, and courses, as well as basic information on students. The date is a period that includes the start date and end date. The content includes the learning content that a teacher and learner work on until a specified date. Score and course are the scores given during the learning process and the courses that the learner enrolled in. The Evaluation Guideline is used as a meter data for the learners life-long portfolio, customized learning, customized content, and intellectual tutoring.

The Parent Support System is used to resolve the problems that parents face because of limits in time and space when they manage their childs learning, by providing them monitoring of the childs learning. The system also enhances the quality of education by offering communication between teachers and parents.

The Inter Screen Sharing Video Education System uses open source to create video data streams, such as a digital stream or PC smart device, and establish a media server that can share the stream in real-time between users. The system also realizes smart education that enables schools and students to use smart devices anytime and anywhere.

4 Discussion

In order to verify the effectiveness of the Structured Plug-in Smart Education System, this paper develops evaluation indicators for the effects of the smart education system, as well as analyzes and verifies the quantitative effects based on system evaluation using a questionnaire and other data. The evaluation indicator considers four factors, such as hardware (infra), software, human-ware, and system-ware. Hardware (infra) evaluates the appropriateness of the device, the smoothness of the wireless network, the appropriateness of server, and after-service. Software evaluates learning content, teachinglearning support, teachinglearning activities, and learning operation. System-ware evaluates security or system support. Human-ware evaluates convenience, access, and efficiency in terms of learners, teachers, and parents.

In order to conduct quantitative effectiveness analysis, this paper suggests effect evaluation indicator of achievement of education target and learning performance, and applies the indicators to public schools to analyze the effect of the Structured Plug-in Smart Education System. It also produces additional factors that are necessary for the indicators. Based on the developed indicators, the smart education system will be verified and completed, and finally applied to domestic smart education and exported to the overseas market. In addition, we analyzed the cognitive capacities such as brain waves and learning activities with psychology researchers to conduct quantitative effect analysis and verify the effectiveness in terms of cognitive science. The effectiveness of the Structured Plug-in Smart Education System is analyzed through this process.

5 Conclusion

With the recent emergence of smart devices and media, demands for an improved educational environment have expanded, and there have been several attempts to realize this. There are various researches under way, such as studies on the concept of smart education, the teachinglearning model, and analysis of the learning effect of smart education. As the number of smart phone users and the electronic publication market grows, in addition to publication companies and e-Learning companies, new smart content companies enter the smart education market. Furthermore, interest in smart education and its market are also growing as the Korean government and conglomerates enter the market. However, there is a current shortage of related systems and approaches to smart education, which are limited to instrumental.

Based on previous studies and conceptual theories, this paper suggests a Structured Plug-in Smart Education System that can be established and implemented in a smart education environment. Smart education in Korea emerged with the appearance and development of smart devices. There has been a lot of related research and activities performed in not only Korea, but also overseas. We established a partnership with Samsung SDS, Intel, and other oversea companies to realize the Structured Plug-in Smart Education System and analyze the effect of the system. We applied smart education to two elementary schools, one middle school, and one college and developed effectiveness evaluation indicators to verify the systems efficiency and effectiveness. We will attempt to enter the overseas market and export this system.

Acknowledgments This work was supported by the Technology Innovation Program funded by the Ministry of Knowledge Economy (MKE, Korea) [1004850, Smart Education을 위한 Plug-In 구조형 SW 개발].

References

1. Lim, K: Research on developing instructional Design models for enhancing smart learning. Korea Association of Computer Education, (2011)
2. Leem, J.H.: A study on the design strategies of teaching and learning model for mobile learning. J. Korean Edu. Forum **8**(1), 101124 (2008)
3. Koo, J.-K.: Model for self-directed learning in a mobile environment: With the main focus on the SmartPhone. Chung Ang University of Graduate School of Education (2010)
4. 20092010 e-Learning White Paper. Ministry of Knowlegde Economy & National IT Industry Promotion Agency & Korea Association of Consilience Education, Republic Korea (2010)
5. Gartner Higher Education E-lerarning Survey. Gartner (2009)
6. Lee, J.: The Way to Matching TalentA Smart Strategy to Promote Education. Ministry of Education Science and Technology, Republic Korea (2011)
7. Developing Constructed Plug-in Software for Smart Education. Ministry of Knowlegde Economy, Republic Korea, Project No. 2011-507, 10 Oct 2011
8. Wikipedia: Retrieved from http://ko.wikipedia.org/wiki/스마트러닝
9. Lee, O.-H.: A Study on Development of the Smart-Learning System for Korean Language. Chungbuk University (2011)
10. Jo, J.-C., Lim, H.-S.: Comparative analysis of learning effect on lexical recognition in the e-learning and s-learning. Future Inf. Technol. Commun. Comp. Inf. Sci. **185**(5), 304308 (2011)
11. Jo, J.-C., Lee, S.-B., Lim, H.-S.: Analysis based on brain wave of learning effect on vocabulary using smartphone. Human & cognitive language technology conference (2010)

12. Lee, Y.-M.: The strong of mobile contents is e-learning. Retrieved from http://biz.heraldm. com/common/Detail.jsp?newsMLId=20101009000035, 9 Oct 2010
13. E-newspaper Media Tech Trend team.: Retrieved from http://www.etnews.com/news/detail. html?id=201109210041&mc=m_014_00001, 21 Sept 2011
14. Daul Soft.: Retrieved from http://www.daulsoft.com
15. Seoul cyber university.: Retrieved from http://www.iscu.ac.kr/
16. Lee, J.: A case Study of learning management system (LMS) implementation using open source program: For collaborative learning Promotion through ease of use
17. Aspera.: Retrieved from http://www.asperasoft.com
18. SK Telecom, T Smart Learning.: Retrieved from http://www.tsmartlearning.com/cls-main-web/index.html
19. Seoul metropolitan office of Education, e-edulive.: Retrieved from http://e-lib.sen.go.kr/index.php/
20. Smarttech.: Retrieved from http://smarttech.com/classroomsuite/
21. SlideRocket.: Retrieved from http://www.sliderocket.com/
22. Tutoring club.: Retrieved from http://www.tutoringclub.com/
23. Saba, T.: Implications of E-learning systems and self-efficiency on students outcomes: A model approach. HCIS **2**, 6 (2012)
24. Hsiao, K.-F., Rashvand, H.F.: Integrating body language movements in augmented reality learning environment. HCIS **1**, 1 (2011)
25. Pan, R., Xu, G., Fu, B., Dolog, P., Wang, Z., Leginus, M.: Improving recommendations by the clustering of tag neighbours. JoC **3**(1), 1320 (2012)
26. Chuan, D., Wang, L., Ma, L., Cao, Y.: Towards a practical and scalable trusted software dissemination system. JoC **2**(1), 5360 (2011)

Impact of Background Utilization and Background Traffic on the Foreground Applications in a Wide Area Network

Jia Uddin and Jong Myon Kim

Abstract In the real scenarios, network utilization widely depends on the background traffic and the number of active users. This paper presents the effect of background traffic on the QoS of foreground traffic by varying the utilization of LAN devices, server, and bandwidth. We designed a WAN (Wide Area Network) using the OPNET simulator, which consists of different Local Area Networks (LANs) located in different cities. We considered a foreground GSM quality voice traffic, and FTP heavy and video conference background traffic. The experimental results demonstrated that the channel bandwidth utilization has much effect on the QoS of foreground traffic. In addition, heavy background traffic highly affects the QoS of foreground traffic due to the network congestion.

Keywords Background traffic · Background utilization · Foreground traffic · OPNET · WAN (Wide area network)

1 Introduction

A device load is usually defined as a percentage of total capacity on a server, LAN, or other node. Background utilization specifies the load as a percentage of the total utilization on that node. For example, the LAN background utilization specifies the baseline load on the modeled LAN. Usually a baseline load is a static throughput

J. Uddin · J. M. Kim (✉)
Department of Computer Engineering and Information Technology, School of Electrical Engineering, University of Ulsan, 93 Daehak –ro, Nam-gu, Ulsan 680-749, Korea
e-mail: jmkim07@ulsan.ac.kr

J. Uddin
e-mail: jia@mail.ulsan.ac.kr

S.-S. Yeo et al. (eds.), *Computer Science and its Applications*,
Lecture Notes in Electrical Engineering 203, DOI: 10.1007/978-94-007-5699-1_92,
© Springer Science+Business Media Dordrecht 2012

on a link, node, or connection. In any scenario management, one of the key components is to generate the realistic background traffic to simulate the load in network caused by various traffic sources [1, 2]. The consideration of background traffic gives more accurate features of the actual load in the network [3]. It has been worth mentioning interest in understanding the characteristics of network services under the sensible deployment conditions. There are two high-level options for the application developers to assess their network prototypes such as live deployment, and emulation or simulation [4]. Example of a live deployment on a testbed is PlanetLab, which permits the developers to focus their systems to realistic network conditions and failures. The emulation and simulation simplify the experimental management and formulate it relatively straightforward to con-quer the reproducible outcomes. However, the recent emulation environments allow unmodified applications and simulate in some target network conditions including topology, failure characteristics, routing, and background traffic [5–7].

Several recent works have reported aspects of background traffic in different network scenarios [2, 3, 8–10]. Peter et al. measured the impact of fragmentation threshold tuning on speech quality and background traffic throughput in mixed of voice and data transmission in an environment of WLAN [3]. Matthew et al. pro-posed an efficient (M, P, S) background traffic model for generating large scale background load [2]. Zheng et al. examined the relationship between jitter and the characteristics of the background traffic; the jitter increases when the background traffic becomes more burst [8]. Park et al. measured the performance of ATM networks considering the background traffic [9]. He et al. proposed a model to observe the impact of background traffic on the performance of a mobile station [10].

It has been observed that a number of research works which have been done in different environments with statistically defined constant execution time of each individual service, CPU utilization of LAN devices, and channel BW considering only the foreground traffic [3, 10–22].

This paper investigates the impact of background traffic on the QoS of GSM quality voice foreground traffic in a WAN [23] using an OPNET simulator considering various utilizations. Our contributions of the present works are as follows:

1. Design and deploy a large WAN in a simulator, which consists of several LANs placed in different distant cities.
2. Observe the impact of background utilization of channel BW, Server and LAN devices on the QoS parameters of foreground traffic without considering the effect of background traffic.
3. Evaluate the impact of background traffic on the QoS parameters of GSM quality voice foreground traffic in the large WANs without considering the variation of utilization.
4. Investigate the impact of background traffic on the QoS parameters of fore-ground traffic considering the variation of background utilization. In this paper, FTP heavy and video conference traffic considered as background traffic; and different utilizations such as 50, 75, and 90 % of LAN devices, Server and channel BW are considered for the low, medium and high load scenarios.

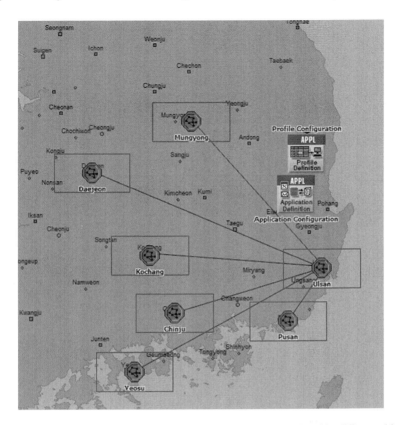

Fig. 1 A scenario of WAN model, which consists of seven LANs placed in different cities in South Korea

The rest of the paper is organized as follows. Section 2 describes the experimental environment and Sect. 3 presents experimental results and analysis. Section 4 concludes the paper.

2 Experimental Environment

In this section, we designed a WAN as illustrated in Fig. 1. A WAN scenario consists of seven individual LANs. Each LAN is placed in different cities in South Korea using the world map of OPNET modeler [24]. Each LAN is configured using a subnet. All subnets are connected with each other via a router using PPP DS0 duplex link. In our experiment, 25 workstations are connected with each other via 10baseT physical links to form an individual LAN. A voice server is placed in ULSAN's LAN and it is connected with the local router via a switch as presented in Fig. 2. This server generates the GSM quality foreground voice traffic for all workstations of different LANs within the WAN. Two different servers- video and

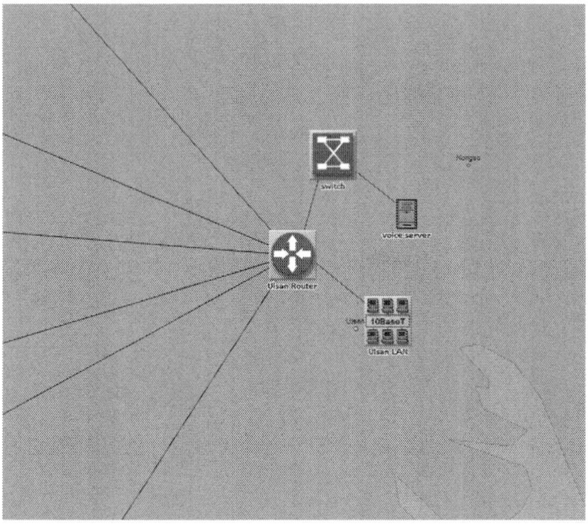

Fig. 2 A scenario of LAN
model in ULSAN without
background traffic

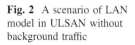

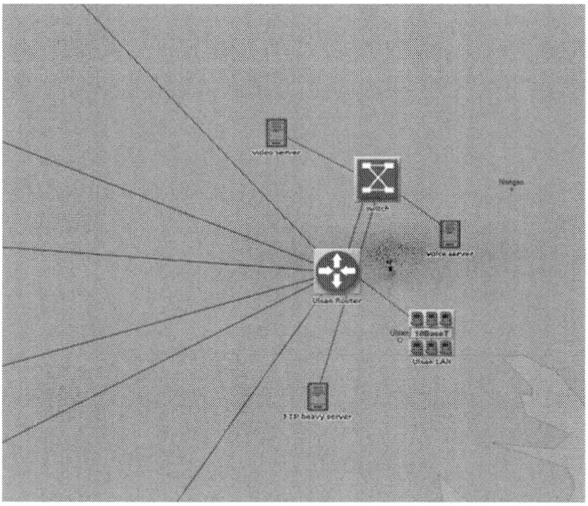

Fig. 3 A scenario of LAN
model in ULSAN with
background traffic

FTP are also placed in ULSAN's LAN for generating FTP heavy and video conference background traffic as shown in Fig. 3.

Various utilizations of channel BW, server, and LAN devices are considered for designing four different types of networks such as simple, low, medium, and busy. For example, no background traffic is considered for the simple network scenarios. On the other hand, 50, 75, and 90 % of different utilizations are considered for simple, medium, and busy network scenario, respectively. Table 1 presents the lists of parameters and their corresponding values used in the experimental evaluation. All scenarios were compiled for 600 s to collect the results.

The step-by-step procedure to configure a WAN in OPNET simulator is as follows:

Table 1 Simulation parameters and values

Network parameters	Values
Network scale	World
Selected area	Different cities in Korea
Foreground traffic	GSM quality voice
Background traffic	FTP heavy and video conference
No. of LAN devices within a LAN	25
No. of LANs	7
Physical links among subnets	PPP-DS0 (duplex, 64 kbps)
Physical links among workstations	10baseT (10 Mbps)
Simulation time	10 min

(a) Drag an application configuration object in the project workspace, and then set the name and different parameters by configuring the edit attribute. In our experiment, we set the application definition attribute to default. The default mode of application definition usually supports eight standard applications including database access, e-mail, file transfer, file-print, telnet session, VoIP call, video conference, and web browsing.

(b) Drag a profile configuration object in the project workspace. Similar to the application configuration we can configure the different parameters of profile by configuring the edit attribute. In our experiment, we create a new profile LAN client which supports the GSM quality voice foreground traffic, video conference and FTP background traffic. All traffics set with start time offset uniform 0, 300 s.

(c) Using the global map of OPNET modeler placed a subnet in the Ulsan city. We set x span and y span by 0.036 and 0.0311, respectively, which indicates the geographic area covered by the subnet. We placed one server of foreground traffic and two servers of background traffic within the ULSAN subnet via the local router using a switch.

(d) Configure the different LANs placed in different cities using the different devices such as workstations, LAN switches, physical links, etc. The workstations are interconnected via the 10baseT physical link.

(e) In an individual city, each LAN is configured within a subnet. These subnets are connected interconnected via the PPP DS0 duplex physical link.

3 Experimental Results and Analysis

3.1 Scenario 1: Investigate the Effects of LAN Devices Utilization on the Performance of Network and QoS Parameters of Foreground Traffic

Figures 4, 5, 6, and 7 present the effect of LAN devices utilization on the performance of network and QoS parameters of GSM quality voice foreground traffic.

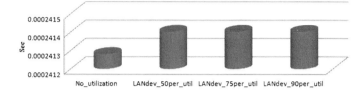

Fig. 4 Ethernet delay of various utilizations of LAN devices

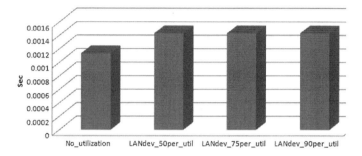

Fig. 5 Jitter of various utilizations of LAN devices

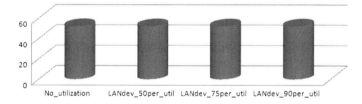

Fig. 6 Point to point utilization of Chinju to Ulsan link for various utilizations of LAN devices

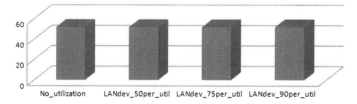

Fig. 7 Point to point utilization of throughput of Mungyong to Ulsan for various utilizations of LAN devices

In this scenario, we have collected some global statistics of WAN including Ethernet delay, jitter along with some object statistics such as point-to-point link utilization, and throughput. The experimental results demonstrate that the LAN devices utilization does not considerable influence on the QoS parameters of foreground traffic. Different QoS parameters of GSM quality voice traffic maintain almost steady characteristic curve for various LAN devices utilizations. For example, we experienced approximately average 0.000241 ms delay, 0.001431 s jitter, and 3.796926 s packet end-to-end delay for different LAN devices utilizations (50, 75, and 90 %). We also observed that the average point-to-point link utilization from the Chinju to Ulsan was 52.03756. Similar to the individual link utilization, almost equal throughput from Mungyong to Ulsan had experienced for various LAN devices utilizations as depicted in Fig. 7.

3.2 Scenario 2: Investigate the Effects of Server Utilization on the Performance of Network and QoS Parameters of Foreground Traffic

In scenario 2, the effect of server utilization on the QoS parameters of foreground traffic is illustrated in Figs. 8, 9, 10 and 11. Four different utilizations such as no utilization, 50, 75, and 90 % of server utilization are considered. Similar to the LAN devices utilization, the variation of server utilization also does not vastly affect the QoS parameters of foreground traffic.

For example, the average Ethernet delay was observed 0.000233 s for various server utilizations except the no utilization scenario; however, it was approximately 0.000224 s for no utilization scenario as depicted in Fig. 8. The average Jitter also shows characteristics curse similar to the Ethernet delay as illustrated in Fig. 9. The average jitter was experienced 0.001431 s, approximately for various server utilizations. On the other hand, almost equal amount of point to point link utilization and throughput among the different LANs were observed for various server utilizations as illustrated in Figs. 10 and 11.

3.3 Scenario 3: Investigate the Effects of Channel Bandwidth Utilization on the Performance of Network and QoS Parameters of Foreground Traffic

The effect of channel BW utilization has observed in scenario 3. The simulation results of different QoS network parameters for various channel BW utilizations are presented in Figs. 12, 13, 14, 15, 16 and 17. Compare to the LAN devices and server utilization, the variation of channel BW utilization highly affects the performance of overall network and the QoS parameters of foreground traffic. The average Ethernet delay and jitter were automatically increased with the increment of BW utilization as demonstrated in Figs. 12 and 13. On the other hand, the average Mean Opinion Score (MOS) value of foreground voice traffic was gradually decreased with the increment of channel BW utilization as shown in Fig. 14.

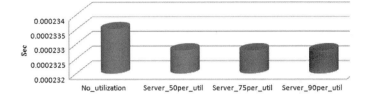

Fig. 8 Ethernet delay of various server utilizations

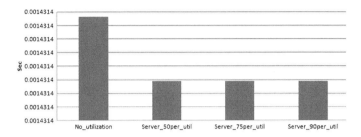

Fig. 9 Jitter of various server utilizations

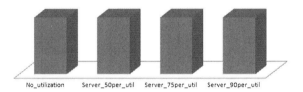

Fig. 10 Point to point utilization of Daejeon to Ulsan link for various server utilizations

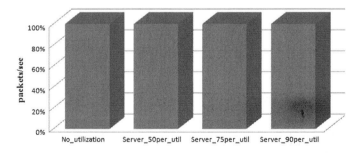

Fig. 11 Point to point utilization of throughput of Ulsan to Yeosu for various server utilizations

Similar to the QoS of foreground traffic, the variation of BW utilization also affects the different QoS parameters of overall network. The throughput of different LANs was followed decreasing trend with the increment of BW utilization as depicted in Figs. 15 and 16. On the other hand, the queuing delay

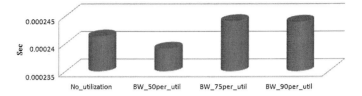

Fig. 12 Ethernet delay of various bandwidth utilizations

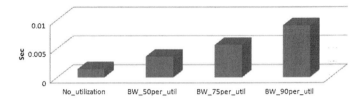

Fig. 13 Jitter of various bandwidth utilizations

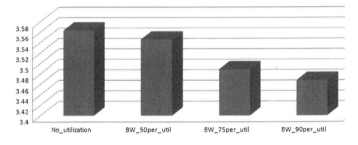

Fig. 14 MOS of various bandwidth utilizations

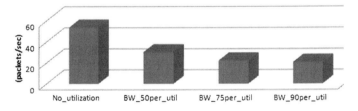

Fig. 15 Point to point throughput of Ulsan to Yeosu for various bandwidth utilizations

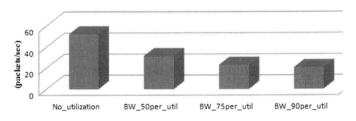

Fig. 16 Point to point throughput of Chinju to Ulsan for various bandwidth utilizations

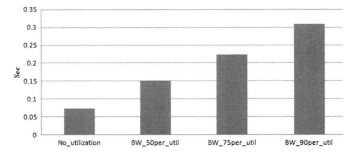

Fig. 17 Point to point Queuing delay of Kochang to Ulsan for various bandwidth utilizations

between different LANs was automatically increased with the increment of channel BW utilization as illustrated in Fig. 17.

3.4 Scenario 4: Investigate the Effects of Background Traffic on the Performance of Network and QoS Parameters of Foreground Traffic

The Sect. 4 presents the effect of background traffic on the performance of network and QoS parameters of foreground traffic. No utilization, 50, 75, and 90 % of utilization of LAN devices, server, and BW were considered for four different network scenarios such as simple, low, medium and high load. To observe the effect of background traffic, FTP heavy and video conference traffics were considered as background traffic for the GSM quality voice foreground traffic in all scenarios. The simulation results of different QoS parameters of different load scenarios with and without background traffic are presented in Figs. 18, 19, 20, 21, 22 and 23.

The simulation results showed that the heavy background traffic influenced the performance of overall network and the QoS parameters of foreground traffic. For example, the average Ethernet delay, jitter, and packet end to end delay were increased with the presence of background traffic in different load scenarios as demonstrated in Figs. 18, 19 and 20. However, the average MOS value of GSM quality voice was slightly decreased at the presence of background traffic as depicted in Fig. 21.

Similar to the QoS parameters of foreground traffic, the presence of background traffic also affects the QoS parameters of overall network including point to point queuing delay and throughput. The average queuing delays between the LANs were increased at the presence of background traffic. Figure 22 shows the average queuing delay of Ulsan to Chinju in different load scenarios. On the other hand, the average throughput of different LANs was decreased at the presence of background traffic in different load scenarios. The Fig. 23 shows the average throughput of Daejeon to Ulsan link Fig. 19.

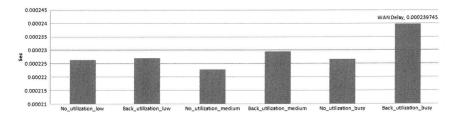

Fig. 18 Ethernet delay of different scenarios with and without background traffic

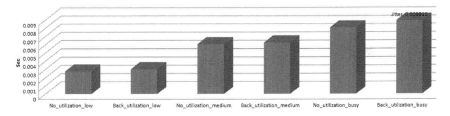

Fig. 19 Jitter of different scenarios with and without background traffic

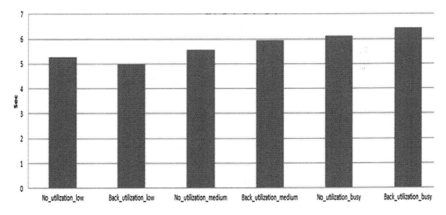

Fig. 20 Packet end-to-end delay of different scenarios with and without background traffic

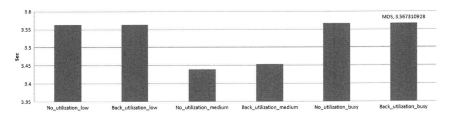

Fig. 21 MOS of different scenarios with and without background traffic

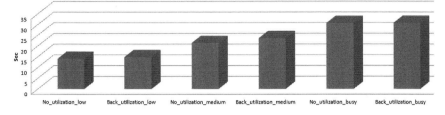

Fig. 22 Point to point Queuing delay of different scenarios between Ulsan to Chinju with and without background traffic

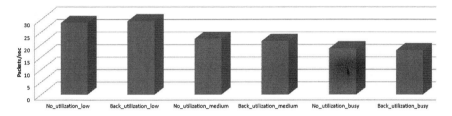

Fig. 23 Point to point throughput of different scenarios between Daejeon to Ulsan with and without background traffic

4 Conclusion

In this paper, we evaluated the impact of two important network parameters-background utilization and background traffic on the QoS parameters of foreground traffic. The GSM quality voice traffic was considered as foreground traffic, and FTP heavy and video conference traffic were considered as background traffic. For experimental validation, we designed a WAN, which consists of different LANs placed in different distant cities of South Korea using the global map of OPNET simulator. The effect of background utilizations on the QoS parameters of foreground traffic was observed in scenarios 1-3 considering three different utilizations including LAN devices, server and channel BW. The experimental results demonstrated that the channel BW utilization highly affected the QoS parameters of foreground traffic.

Scenario 4 presents the impact of background traffic on the QoS parameters of foreground traffic. The simulation results illustrated that the heavy background traffic exceedingly affected the performance of overall network and the QoS parameters of foreground traffic.

Therefore, before deploying the real network to reduce the hardware and installation cost, designers should consider the effect of variation of channel utilization and background traffic for designing an efficient new routing algorithm, routing protocol, network topology, MAC design, etc. using a simulator.

Acknowledgments This work was supported by the National Research Foundation of Korea (NRF) Grant Funded by the Korean Government (MEST) (No. 2012-0004962).

References

1. Nicol, D.M., Yan, G.: Discrete event fluid modeling of background TCP traffic. J. ACM Trans. Model. Comput. Simul. **14**(3), 211–250 (2004)
2. Lucas, M.T., Dempsey, B.J., Wrege, D.E., Weaver, A.C.: (M, P, S)-an efficient background traffic model for wide-area network simulation. In: The Proceeding of the International Conference on Global Telecommunications, GLOBECOM 1997, pp. 1572–1576 (1997)
3. Pocta, P., Bilsak, M., Rousekova, J.: Impact of fragmentation threshold tuning on performance of voice service and background traffic in IEEE 802.11b WLANs. In: The Proceeding of the 2010 20th International Conference on Radioelektronika (RADIOELEKTRONIKA), April 2010, pp. 1–4 (2010)
4. Vishwanath, K.V., Vahdat, A.: Evaluating distributed systems: Does background traffic matter? USENIX 2008 Annual Technical Conference, ATC 2008, June 2008, pp. 1–14 (2008)
5. Emulab.Net: The Utah Network Testbed. [Online]: http://www.emulab.net
6. Vahdar, A., Yocum, K., Walsh, K., Mahadevan, P., Kostic, D., Chase, J., Becker, D.: Scalability and accuracy in a large-scale network emulator. In: The Proceeding of the 5th Symposium on Operating Systems Design and Implementation, OSDI 2002, 36(SI), pp. 271–284 (2002)
7. Malakuti, S., Aksit, M., Bockisch, C.: Runtime verification in distributed computing. J. Converg. **2**(1), 1–10 (2002)
8. Zheng, L., Zhang, L., Xu, D.: Characteristics of network delay and delay jitter and its effect on voice over IP (VoIP). In: The Proceeding of the IEEE International Conference on Communications, ICC 2001, June 2001, pp. 122–126 (2001)
9. Park, C.G., Han, D.H., Jung, J.I.: Interdeparture time analysis of CBR traffic in AAL multiplexer with bursty background traffic. Proc. IEE Proc. Commun. **148**(5), 310–315 (2001)
10. Yong, H., Yuan, R., Ma, X., Li, J.: Analysis of the impact of background traffic on the performance of 802.11 power saving mechanism. IEEE Commun. Lett. **13**(3), 164–166 (2009)
11. Kenesi, Z., Szabo, Z. Belicza, Z. Molnar, S.: On the effect of the background traffic on TCP's throughput. In: The Proceeding of the 10th IEEE Symposium on Computers and Communications, ISCC 2005, June 2005, pp. 631–636 (2005)
12. Moldovansky, A.: A utilization of modern switching technology in ethernet/IP^{TM} networks. CIP Technical paper, pp. 61–68 (2012)
13. Nie, W., Zhang, J., Lin, K.J.: Estimating real-time service process response time using server utilizations. In: The Proceeding of IEEE International Conference on Service-Oriented Computing and Applications, SOCA 2010, pp. 1–8 (2010)
14. Yabusaki, H., Matsubara, D.: Scheduling Algorithms on optimizing spatial and time distribution of bandwidth utilization. In: World Telecommunications Congress, WTC 2012, pp. 1–6 (2012)
15. Fu, Q., White, L.: The impact of background traffic on TCP performance over indirect and direct routing. In: The Proceeding of the 8th International Conference on Communication Systems, ICCS 2002, pp. 594–598 (2002)
16. Nychis, G., Licata, D.R.: The impact of background Network traffic on foreground network traffic. In: The Proceeding of the IEEE Global Telecommunications Conference, GLOBECOM 2001, pp. 1–16 (2001)

17. Wang, S.C., Ahmed, H.: BEWARE: Background traffic-aware rate adaptation for IEEE 802.11. In: The Proceeding of the 2008 International Symposium on a World of Wireless, Mobile and Multimedia Networks, WoWMoM 2008, pp. 1–12 (2008)

18. Lagkas, T.D., Papadimitriou, G.I., Nicopolitidis, P., Pomportsis, A.S.: A novel method of serving multimedia and background traffic in wireless LANs. IEEE Trans. Veh. Technol. **57**(5), 3263–3267 (2008)

19. Zunino, C.: Performance measurements of 802.11 WLANs with burst background traffic. In: The Proceeding of the IEEE Conference on Emerging Technologies and Factory Automation, ETFA 2007, pp. 1392–1395 (2007)

20. Hopkinson, K., Roberts, G., Wang, X., Thorp, J.: Quality of service considerations in utility communication networks. IEEE Trans. Power Deliv. **24**(3), 1465–1474 (2009)

21. Luo, H., Shyu, M.L., , M.: Quality of service provision in mobile multimedia- a survey. Journal of Human- centric Computing and. Inf. Sci. **1**(5), 1–15 (2011)

22. Bhattacharya, A., Wu, W., Yang, Z.: Quality of experience evaluation of voice communication: an affect based approach. J. Hum. Centric Comput. Inf. Sci. **2**(7), 1–18 (2012)

23. Javed, K., Saleem, U., Hussain, K., Sher, M.: An enhanced technique for vertical handover of multimedia traffic between WLAN and EVDO. J. Converg. **1**(1), 107–112 (2010)

24. Salah, K., Alkhoraidly, A.: An OPNET-based simulation approach for deploying VoIP. Int. J. Netw. Manage. **16**(3), 1–23 (2006)

Load Balancing in Grid Computing Using AI Techniques

Nadra Tabassam Inam, M. Daud Awan and Syed Shahid Afzal

Abstract Work load and resource management are two important factors that have to manage across the grid environment. To increase the overall efficiency of grid based infrastructure the work load across the grid environment has to manage. Hence the work load must be evenly scheduled across the grid nodes so that grid resources can be properly exploited. The technique that we have investigated in this paper is based upon the combination of genetic algorithms which is an evolutionary algorithm and artificial neural networks. Both of these techniques are applied for local grid load balancing. Genetic algorithm selects the optimal set of jobs for assigning to the grid nodes which overall minimizes the total execution time. Afterwards when optimal set of jobs is selected they are assigned to artificial neural network which selects the minimum loaded grid processor for further processing of this optimal set of jobs. We compare our proposed technique with the already existing strategies for load balancing like random algorithm, round robin algorithm, decreasing time algorithm and least connection algorithm. Results shows that our strategy gives optimal results in terms of overall time efficiency. So we can overall conclude that GA's and ANN's increase overall efficiency of job scheduling especially in case where the tasks coming for scheduling and processing nodes are continuously increasing.

Keywords Genetic algorithm · Artificial neural network · Load balancing

N. T. Inam (✉) · M. Daud Awan · S. S. Afzal
Department of Computer Science, Preston University, Islamabad, Pakistan
e-mail: nadrainam@hotmail.com

S.-S. Yeo et al. (eds.), *Computer Science and its Applications*,
Lecture Notes in Electrical Engineering 203, DOI: 10.1007/978-94-007-5699-1_93,
© Springer Science+Business Media Dordrecht 2012

1 Introduction

With the advent of internet we encounter an explosion of information. To convert this huge plethora of information into meaningful data, we need increasingly reciprocally faster processing. Hence this faster processing can be limit to single processing unit and can be expanded to variety of processing types like multiple processing units, distributed processing units and parallel processing. Distributed computing gives one of the modern form of computing known as grid computing. It is a network in which resource of a particular computer can be used by any other node which is united to that particular network. If we look then we can find that sharing resources has been an important fact over the internet or networks [1, 2] There are number of resources like computing powers, storage capacities, processing powers that are present in grid environment, that can be use by all those people who are member of grid environment. A grid computing design can have several computers running on a network having same hardware or software or it can have multiple diverse forms of hardware or software even one can imagine. In grid computing all systems are incorporated in such a way that someone is using one computer for the admittance of the entire set of resources which are physically scattered. So on the whole it appears that last part users feels that their computer has powers analogous to super computers.

This diverse nature of gird has many challenges like resource management, Heterogeneity of machines as well as software and operating systems, security issues [3] when connecting to the grid. Quality of service in grid environment and load balancing in distributed environment. So on the whole the multifaceted nature of grid has generated some serious concerns needed to be concentrate on. Load balancing is the set of algorithms that deals with equalization of loads across the network, so that total execution time can be minimized by increasing overall efficiency of the processors. One of the major apprehensions is load balancing across the nodes during parallel processing. Whenever server requires to schedule the jobs across the grid, node with the least load must be selected or load across multiple nodes must remain balanced.

Abstractly, load balancing algorithms can be distributed into two classes: *static* or *dynamic*. Centralized or decentralized.

- In static load balancing client that wants to request for some resource, always propel its request to the one same server.
- Dynamic load balancing is reliant on the current condition of the system. Every time the client requests for a resource, its request is managed by new handler depending upon the state of system.
- In case of centralized load balancing the responsibility for making overall decisions about load balancing present at one centralized location.
- In distributed load balancing the information of the overall system and load balancing statistics are present at multiple nodes across the network.

2 Related Work

There is lot of examples in literature which deals with parallel processing in distributed environment.

In order to make well managed collaborative applications writer has given the infrastructure and call it middleware. This paper gives idea on a middleware concept and points the problems of service discovery, organization of communication between devices, harmonization, data-security and minimization of communication between the devices within distributed and collaborative environments [4].

In this paper authors have discuss the distributed systems and taken in account the concept of P2P systems. Taken in consideration the concept of mutual trust and understanding between different peers, they have discuss that how the group of reliable peers can be created by using different protocols that are based on agreement among the peers [5].

The work presented by Writer in [6] is a decay load balancing algorithm based on priority for a heterogeneous computing environment. This technique resolves the task priority graph of the jobs that are running in parallel with dynamism at run-time and allocates suitable priorities to the processes in order to resolve the dependencies. A heterogeneous program model has been design to test the heuristic algorithm

In [7] writer considers work load and resource management the two important issues to be managed at the service level infrastructure of grid. Tree based technique is presented in order to manage the problem of load balancing across the grid nodes. This technique which makes use of neighborhood property.

Many writers have explored the concept of genetic algorithm in load balancing specially in grid environment. Much diverse work has done in this regard because genetic algorithm gives an optimized solution especially in cases where search space is quite exhaustive.

In [8] writer proposes the genetic algorithms to solve the problem of load balancing resourcefully. This algorithm taken in account multi purposes in its solution assessment and solves the scheduling problem in a way that concurrently decreases maxspan and overall inters communication cost, and maximizes processor utilization and efficiency.

In [9] authors has proposed genetic algorithm that utilizes dynamic load-balancing method. Genetic algorithm proposes here gives good results when size of tasks start getting bigger. Concept of central scheduler in this scenario overall decreases the cost by taken in account the concept of minimum communication for load balancing decisions.

In [10] authors of the paper presents a back propagation neural network in order to predict the overall runtime of tasks. This design has many benefits as it is uncomplicated and have speedy learning abilities. This neural network has also taken in account scheduling of tasks in terms of time given by user. This method is verified to be resourceful and perfect when results are investigated.

There are many short comings in overall techniques described above. In above given load balancing methodologies message passing and communication overhead amplifies when on demand load balancing is focused. Writers of different papers have not taken in account the execution time of a task up to its completion and waiting time of a task. The direct effects of execution time and waiting time on task scheduling is also never considered.

Above discussed methods have different draw backs. Authors of papers waits Genetic algorithm to converge, this has raises on the whole execution time in search of optimized solution. Majority of writers concentrates on the selection of optimized set of tasks for nodes; they do not concentrate on the selection of minimum loaded node. Majority of solutions are based upon the selection of minimum loaded node on the basis of some threshold values.

In the above given research techniques in case of artificial neural networks, the output layer's activation functions for neurons are not visibly mentioned which do not provides us an information about the consequential output values for artificial neural network. Less number of training examples are being input ANN's which do not properly trained the neural networks.

Our main motivation behind this research is to focus on not only the selection of optimized set of tasks but also to implement an infrastructure which also focuses on the selection of minimum loaded processing node. No one has implemented the neural networks for the selection of appropriate processor for assigning tasks. The infrastructure gives collaboration between genetic algorithm and neural networks in an effective manner which overall increases the efficiency in terms of time. Genetic algorithm is good option where number of solutions can be unlimited and neural network is good option for non linear problems and they are easy to implement.

3 Design

3.1 Genetic Algorithm

Whenever the different set of jobs come to the scheduler, the scheduler contains a queue of tasks which contains the tasks that have to be scheduled to the processor From the queue of unscheduled tasks the certain numbers of tasks are selected for assigning to the genetic algorithm. Out of these tasks the optimal set of tasks are being selected by genetic algorithm for assigning to the processor. In case of genetic algorithm input strings the earlier population could be the set of tasks that has been arrived at the scheduler for assignment to the processor. Each task is taken on the basis of its size and task number in a particular queue. This set of tasks in the form of strings turn into new population for genetic algorithm. These strings pass through the fitness function that we have defined. So that we can find that which input set of string (population) is fit for survival into the next generation.

3.2 Fitness Function

Fitness function based on our algorithm comprise of important factors. The following equation gives our fitness function

$$FF = \sum_{i=1}^{LS} \{1/TWTQ_i * 1/TT_i\} * LP \qquad (1)$$

TWTQ   Time taken by tasks in which it waits for its turn in queue
TT   Time of a particular task to complete
LP   Load of a particular processor
LS   Length of string.

Load on the processor is the tasks that are currently being executed by processor as well as size of all tasks that are waiting in processors queue. This load on overall processor can find by using following equation.

$$LP = \sum_{i=1}^{k} (CETRT_i + STPQ_i) \qquad (2)$$

LP   Load of a particular processor
TCET   Currently executing task's remaining time
STRQ   Size of tasks in terms of time present in processor's queue
K   Number of processors.

Input string after passes through the fitness function the resultant values are divided into two categories either they can approach towards zero or towards one. Hence fitness function defines it in the following manner.

$$FF \rightarrow 0$$
$$FF \rightarrow 1$$

Each string has to pass through this fitness function in order to select for the future population. The following table shows the selection of strings (Table 1).

3.3 Crossover and Mutation

After fitness function of application on input strings the next step is crossover of the fittest parents. Hence cross over produces the off springs from these fittest parents and these off springs survive in the next pool of generation there are many types of cross over but we will take the cycle cross over for parent fittest strings. Cycle cross over has the property that it selects each member of the string once. So

Table 1 Fitness values of selected strings

No of string	Fitness value	Probability
1	0.00157	0.047
2	0.00231	0.070
3	0.00430	0.013
4	0.00819	0.248
5	0.0156	0.473
Total	0.03297	1.000

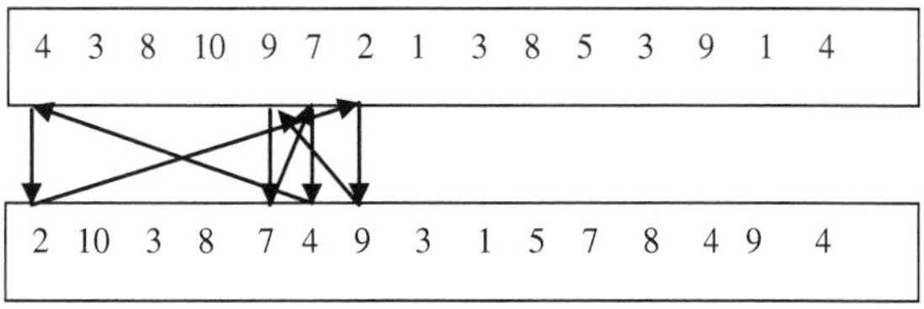

Fig. 1 Cross over

in our case in order to select each task once we use cycle cross over. The figure below gives the cycle cross for the two parents that have been selected to be fittest (Figs. 1 and 2).

After crossover the final step in algorithm is mutation which produces the consequential new strings. This new population is then appended into the pool of strings and hence those set of strings which do not satisfy the required fitness values do not survive in the next pool of generation.

Replacement method for mutation is used in order to complete the last step of genetic algorithm (Fig. 3).

This new population replaces the old strings of population within the next generation. Then again the fitness value of each string within the new population is calculated. Each individual in the generation of population will have a fresh fitness value and a fresh probability of survival into upcoming generation of set of strings. Genetic algorithm do converges for the better solution but we d not wait for genetic algorithm to converge we will run the five cycles of genetic algorithm so after five cycles we will get the optimized strings which would be assigned to the certain processor. These are the set of optimized tasks which can be assigned to the processor for further processing. These strings will be the input to the artificial neural network. Hence our ANN finds the minimum loaded processor for task assignment.

3.4 Artificial Neural Network

Our neural network is based on an input layer, two hidden layer and one output layer. Our inputs to the hidden layers are based upon the three basic factors (Fig. 4).

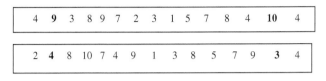

4	10	3	8	9	7	2	3	1	5	7	8	4	9	4
2	3	8	10	7	4	9	1	3	8	5	3	9	1	4

Fig. 2 Resultant children after cross over

4	9	3	8	9	7	2	3	1	5	7	8	4	10	4
2	4	8	10	7	4	9	1	3	8	5	7	9	3	4

Fig. 3 Mutation

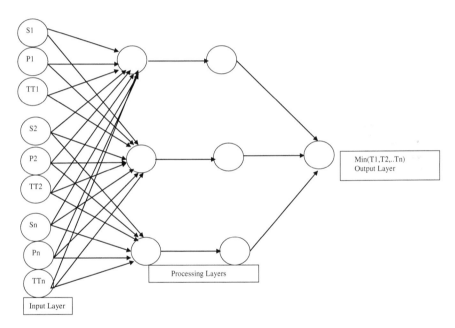

Fig. 4 Artificial Neural Networks

- Size of tasks
- Processor number for assigning these tasks
- Overall Time taken by certain task to reach the neural network.

Training set for genetic algorithm is based on one thousand set of tasks that we have generated by using genetic algorithm. These tasks are use to trained the neural network for the selection of processor. Our neural network takes the set of optimal jobs and selects the processor for assignment of jobs on the basis of fact that all the processors within the network must have balance loads. This is the feed

Table 2 Set of tasks as an input to the Neural networks for processor 1

Size of tasks	Waiting time for tasks	Processor No
2	9	1
3	10.9	1
8	11.9	1
4	8.7	1
SUM	40.5	

forward neural network with two hidden layers. The inputs to neural networks comprised of following factors

- S = Size of each task
- P # = Processor no
- TT = Overall Time taken by certain task to reach the neural network.

The optimal sets of jobs are assigned to neural network in order to find the processor with minimum load to assign these jobs. For every job we select the processor randomly. This processor number is also an input to the neural network. WTT is the overall waiting time taken by a particular job to reach the neural networks.

After these inputs the neural networks multiply the weights with all the inputs. Hidden layer of neural net takes the load information in terms of time from all the processors and add this load to the already existing waiting time of tasks. The same process is done with all the processors. At the end, processor with minimum work load is selected to assign the jobs.

Let us take an example

Suppose we have an optimal set of jobs which contains the following jobs (Table 2).

Suppose we consider that we are assigning this set of jobs to processor #1. So our input to the neural network is shown in above table.

Now after passing through hidden layers of neural networks, our NN taken in account the load of each processor and add this load to already existing waiting time of tasks. The same is done for each processor.

Suppose we have the following time remaining for each processor

- Remaining Time left for Processor # 1 22
- Remaining Time left for Processor # 2 11
- Remaining Time left for Processor # 3 8.

Then we add the sum of remaining time with the remaining time left for processor #1

T1 = 40.5 + 22 = 62.5

If we done the same procedure with the input tasks and assign these tasks to Processor # 2 (Table 3).

Then we add the sum of remaining time with the remaining time left for processor #2

Table 3 Set of tasks as an input to the Neural networks for processor 3

Size of tasks	Waiting time for tasks	Processor No
2	9	2
3	10.9	2
8	11.9	2
4	8.7	2
SUM	40.5	

Table 4 Set of tasks as an input to the Neural networks for processor 3

Size of tasks	Waiting time for tasks	Processor No
2	9	3
3	10.9	3
8	11.9	3
4	8.7	3
SUM	40.5	

T2 = 40.5 + 11 = 51.5

If we done the same procedure with the input tasks and assign these tasks to Processor # 3 (Table 4).

Then we add the sum of remaining time with the remaining time left for processor #3

T3 = 40.5 + 8 = 48.5

So out of these three time intervals neural network selects the smallest one with the given output function as

Min (T1, T2, T3)

Hence the result will be the T3 which is the processor # 3.

Hence the following diagram shows overall proposed strategy (Fig. 5)

4 Results

Hence our strategy gives two solutions in terms of better load balancing. One is, it gives set of optimal tasks and then secondly it helps in selecting the processor with the minimum loads. We will compare our strategy with already three existing scheduling algorithms.

The first strategy we consider is random load balancing (RA) in which load is distributed randomly across the network picking one via random number generation and sending the tasks to that processor. The second strategy we consider is round robin (RR) that is mainly fundamental of load balancing strategy used where tasks are assigned with a certain slice of time. Task will process in this given time. The third strategy we consider is the Least Connections (LC) load balancing. With this method, the system passes tasks to the processor that has the least number of current load third scheduling strategy we consider is decreasing time algorithm. The DTA is based on a simple strategy. Do the longer jobs first and save the

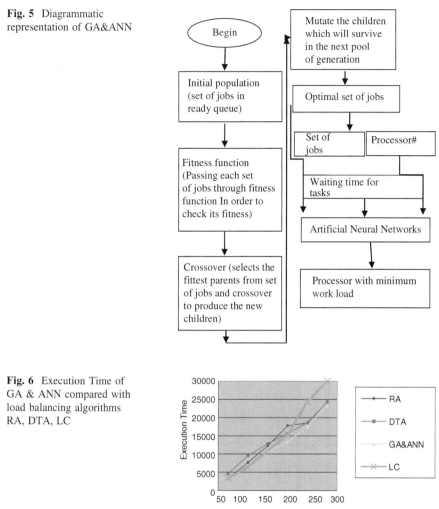

Fig. 5 Diagrammatic representation of GA&ANN

Fig. 6 Execution Time of GA & ANN compared with load balancing algorithms RA, DTA, LC

shorter jobs last. This technique simply lists the tasks in decreasing order of processing times. Longest tasks first smallest tasks last. Tasks with equal processing time can be listed in any order.

Hence we compare our GA and ANN's based strategies with the above describe load balancing algorithms we have consider some important factors for experimental results. These factors are completion time of task, waiting time and total execution times for particular tasks. We see the effects of these factors as the number of tasks increases.

The first factor that we taken in account is execution time for tasks. Graph shows that GA and ANN based strategy gives the optimal results when we compare other algorithms with them (Fig. 6).

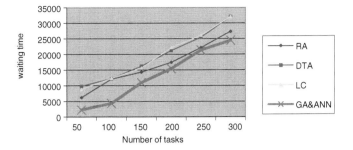

Fig. 7 Waiting Time of GA & ANN compared with load balancing algorithms RA, DTA, LC

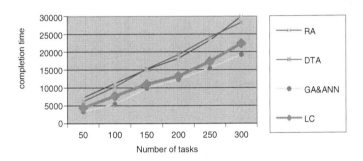

Fig. 8 Completion Time of ANNs GA & ANN comparing with load balancing algorithms RA, DTA, LC

The second factor that we consider is the waiting time for tasks. Graph shows that GA and ANN based strategy gives the optimal results when we compare other algorithms with them (Fig. 7).

The above given graph shows that the GA and ANN based strategies give optimized results in case number of tasks kept on increasing. Overall waiting time for tasks is minimized in our proposed strategy as compare to other load balancing algorithm (Fig. 8).

Now we consider a certain scenario and effects on overall efficiency in terms of completion time. As the number of processors increases and then number of tasks also increases we can see that GA & ANN based strategy gives optimal results (Table 5).

Hence we can see that as the number of processors and number of tasks increasing the overall efficiency in terms of time keep on increasing as the overall time kept on decreasing.

No of processors	No of tasks	Efficiency in terms of Time
3	10	125 ms
5	20	122 ms
10	30	115 ms
20	50	108 ms
40	100	95 ms
60	200	70 ms

Table 5 Overall efficiency with respect to no of processors and No of tasks

5 Conclusion

Hence our algorithm is efficient in manner that it decreases the overall execution time for tasks in case where the numbers of tasks are continuously increasing. The selection of minimum loaded node in grid computing overall increases the efficiency of proposed algorithm. So in case of grid computing where the resources have to exploit and nodes are quite heterogeneous this strategy gives the optimal results and of great use. Both priorities and sizes of tasks are of uneven length. This algorithm is implemented on the local grid created during this research work.java based software gridsim is use to implement on the grid infrastructure for checking the load balancing issues and results in terms of time. As a future work some new techniques for load balancing, like Ant colony optimization and particle swarm optimization will be taken into consideration.

References

1. Yen, N.Y., Kuo, S.Y.F.: An intergrated approach for internet resources mining and searching. J. Converg. **3**, 37–44 (2012)
2. Pyshkin, E., Kuznetsov, A., Approaches for web search user interfaces: How to improve the search quality for various types of information. J. Converg. **1**, 1–8 (2010)
3. Ai Ling, A.P., Masao, M.: Selection of model in developing information security criteria for smart grid security system. J. Converg. **20**, 39–46 (2011)
4. Schmid, O., Hirsbrunner, B.: PerComw middleware for distributed collaborative ad-hoc environments. In: IEEE International Conference on Pervasive Computing and Communications Workshops, pp. 435–438, (2012)
5. Aikebaier,A., Enokido, T., Takizawa, M.: Trustworthy group making algorithm in distributed systems. Hum. centric Comput. Inf. Sci. **1**, 6. doi:10.1186/2192-1962-1-6,:22 (2011)
6. Maheshwari, P.: A dynamic load balancing algorithm for a heterogeneous computing environment. In: HICSS 29th Hawaii International Conference on System Sciences (HICSS'96) Volume 1: Software Technology and Architecture, p. 338 (1996)
7. Yagoubi, B., Slimani, Y.: Task load balancing strategy for grid computing. J. Comput. Sci. Science Publications, USA, vol. 3 (2007)
8. Zomaya, A.Y.: Observations on using genetic algorithms for dynamic load-balancing. Parallel Distrib. Syst. IEEE Trans. **12**(9), 899–911 (2001)
9. Nikravan, M., Kashani, M.H.: A genetic algorithm for process scheduling in distributed operating systems considering load balancing. J. Parallel Distrib. Comput.**70**(1), 1–6 Netherlands
10. Yuan, J., Ding, S., Wang, C.: Tasks scheduling based on neural networks. Grid Conf. Nat. Comput. **3**, 372–376

LiQR: A QR Code-Based Smart Phone Application Supporting Digital Marketing

Jong-Eun Park, Jongmoon Park and Myung-Joon Lee

Abstract A Facebook Page is considered as an essential place for businesses to build connections with people. The page is often used for building up relations with Facebook users through a function called 'Like'. In this paper, we propose a method supporting digital marketing based on Facebook pages and QR code, presenting a smart phone application developed using the proposed method. The developed application named LiQR provides a function transforming the information of a Facebook page into a corresponding QR code. When a user takes a picture of the printed QR codes with the smartphone camera, the application enables the user to support the Facebook page with the same effect as clicking the 'Like' button, providing functions useful for managing the information on the Facebook page.

Keywords Digital marketing · QR code · Facebook page · Like · LiQR

1 Introduction

Social Network Service (SNS) has been widely used as smart phones become populated. Also, it has contributed to create and develop markets in various areas such as social games and social commerce based on the accumulated social information [1–3].

J.-E. Park · J. Park · M.-J. Lee (✉)
School of Electrical Engineering, University of Ulsan, Ulsan, Korea
e-mail: mjlee@ulsan.ac.kr

J.-E. Park
e-mail: cjswowhddms@nate.com

J. Park
e-mail: monster28g@gmail.com

S.-S. Yeo et al. (eds.), *Computer Science and its Applications*,
Lecture Notes in Electrical Engineering 203, DOI: 10.1007/978-94-007-5699-1_94,
© Springer Science+Business Media Dordrecht 2012

Facebook [4, 5] is the world's biggest SNS, maintaining the most subscribers and active members. Through its social platform F8, it provides environments for the development and execution of various applications. Real-world objects like stores, companies, and products are expressed as Facebook pages, which make social relations between users [6]. A page can be created by any Facebook account holders, and the created page is activated as a structural element of social relations. Also, many social relations can be made because the relations with supporters are immediately formed when users click the Facebook 'Like' button [7] associated with the page. Owing to the power of communication, it has been widely used for managing a celebrity's fans or product marketing. In addition, the marketing effects would increase if the relations with supporters could be made more easily during offline activities.

In this paper, to easily establish such relations, we propose a method for associating Facebook pages and Quick Response Codes (QR) [8]. Based on the method, we present a smartphone application named LiQR, which provides a function transforming the information of a Facebook page into a QR code. Also, with LiQR, a user can support the Facebook page with the same effect as clicking the 'Like' button by taking a picture of the printed QR codes with his smartphone camera, and can post his message on the page into his Facebook space. LiQR is implemented using HTML5 [9] technology, working on various smart operating systems.

This paper is organized as follows. In Sect. 2 we give brief description of QR code, Facebook page and the Graph API. In Sect. 3, we describe a method to associate a Facebook page with a corresponding QR code. In Sect. 4, we explain the functions of LiQR in relation to Facebook. Section 5 shows the user interface of LiQR. Finally, we summarize and conclude in Sect. 6.

2 Background

2.1 The Graph API and Facebook Page

The Graph API [10] provides all information of relations among the objects in the Facebook. It is requested via HTTP GET or POST methods in the form of URL, returning a message in the form of JSON. The Graph API supports an easy development of an application utilizing social information in Facebook. Every social object in the social graph has a unique ID. Developer can access the properties of a social object requesting 'https://graph.facebook.com/ID'. For example, the official page for the Facebook Platform has id '19292868552', so developer cat fetch the object at 'https://graph.facebook.com/19292868552'. Alternatively, people and pages with names can be accessed using their as an ID. Table 1 shows using the Graph API.

Table 1 The example of the request using graph API

Users: Users: https://graph.facebook.com/btaylor (Bret Taylor)
Pages: https://graph.facebook.com/cocacola (Coca-Cola page)
Events: https://graph.facebook.com/251906384206 (Facebook developer Garage Austin)
Groups: https://graph.facebook.com/195466193802264 (Facebook developers group)
Friends: https://graph.facebook.com/me/friend
News feed (this is an outdated view, does not reflect the news feed on facebook.com): https://graph.facebook.com/me/home
Profilefeed (Wall): https://graph.facebook.com/me/feed
Likes: https://graph.facebook.com/me/likes

Using Facebook pages, institutions, companies, and public people can communicate with each other. The relations between pages and users are made via the Facebook 'Like' function. These pages contain their own ID and URL similar to a user account.

2.2 QR Code

A QR code is a 2-D image that can contain up to 2,953 bytes and has a faster recognition speed, and higher success ratio than a 1-D barcode. QR codes are widely used in marketing, and advertisement associated with online and offline communication environment through QR code applications in smartphones.

ZXing [11] is an open library for processing 1-D barcodes and QR codes, and supports the development of applications working on various platforms such as Java, Android, IOS, and so forth.

3 Generation and Recognition of QR Code for Facebook Pages

Since mass marketing is the one-way communication from companies or institutions to users, it shows essential difficulty in communicating with users. In this section, to build up social relations between companies and users easily, we propose a method for associating Facebook pages with QR codes.

3.1 Design of QR Codes Associated with Facebook Pages

Facebook pages help businesses, organizations and brands share their stories and connect with people. If users are the official representative of an organization, business, public people, or band, can create a page by putting the information for advertising as shown in Table 2. Table 2 describes the request and response message for an advertising Facebook page.

Table 2 Information on an Advertising Facebook Page

	Message format
Request	"https://graph.facebook.com/YOUR PAGE ID"
Response	{"id": "Page's id(String)",
	"name": "Page's name(String)",
	"picture": "Link to the Page's profile photo(String)",
	"link": "Link of the page on Facebook(String)",
	"likes": The number of users who like the Page(Number),
	"category": "Page's Category(String)",
	"location": {The page's street address(String)},
	"phone": "The phone number for the page(String)",
	"description": "the advertising paragraphs edited by the marketer"
	...
	}

The information from the response message in Table 2 is used for creating the associated QR code whose message format is defined in Table 3. When a user take picture of the QR code through LiQR, the stored information in the QR code is displayed to the user. Then the user can make an immediate relationship with the advertising Facebook page by clicking a confirmation button.

3.2 Generation and Recognition of QR Codes

The QR codes for advertising pages are generated and recognized with the help of the ZXing Library. The ZXing Library provides a QRCodeWriter class in order to generate QR codes. The procedure to generate QR codes containing information on Facebook pages is listed below:

1. Requesting the target page information through the Graph API
2. Generating a QR code protocol message based on the target page information and the advertised
3. Transforming the generated QR code protocol message into a bitmap image with the
4. Printing out the bitmap image and Distributing the printed image.

The recognition of QR codes is performed with the QRCodeWriter Class in the Zxing Library. The QRCodeReader Class activates the smart phone camera and returns the results of the recognition in a string format. The procedure to recognize QR codes is listed as follows Fig. 1:

1. Reading a QR code with the QRCodeReader class in the Zxing Library.
2. Confirming the correctness of the format of the recognized string in accordance with the defined QR code protocol.
3. Converting the recognized string into a JSON object.
4. Extracting the property values from the JSON object.
5. Displaying the extracted values to the use.

Table 3 Message format consisting QR codes

{ "id": "Page original ID",
"name": "Page name",
"picture": "Page picture",
"link": "Page link",
"likes": a number of total likes,
"talking_about": the number of people talking the page
"category": "Page Category",
"address": "Address",
"zip": "Post code",
"phone": "Phone number",
"description": "the advertising praragraphs edited by the marketer" }

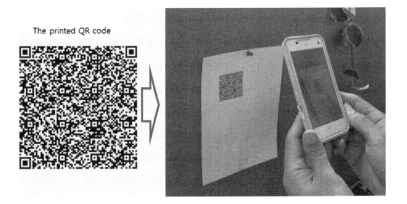

Fig. 1 Scanning the printed QR code

4 Association with Facebook

In this section, we discuss how to utilize the Facebook 'Like' function and the message posting facility, presenting the structure of LiQR.

4.1 Like Button and Posting Paragraphs

LiQR is associated with the Facebook 'Like' function. Facebook provides a source code for embedding the 'Like' function [12] in any kind of HTML-based applications, as shown in Fig. 2.

The messages posted into Facebook by LiQR users who find the advertising page interesting are open to the friends who have relations with the users. Since these friends might circulate the advertising page, marketing could be more effective from the viewpoint of the page owner. In addition, the users in

```
<iframe "src=https://www.facebook.com/plugins/like.php?href='YOUR
FACEBOOK PAGE URL'" ></iframe>
```

Fig. 2 Source codes for embedding the 'Like' function

Table 4 Format of a message posted on facebook

URL of the advertising page
Relation between the advertising page and the posting user
The advertising message edited by the page distributor
The message edited by the posting user

association with the advertising page could have closer relationship through forming a supporting community around the page. So it could provide a new opportunity for the additional marketing. The function for posting messages is provided through the Graph API. Table 4 shows the format of a message that is posted on Facebook.

4.2 LiQR Structure

LiQR supports digital marketing based on an advertising page on Facebook and the associated QR code. LiQR uses Graph API for providing the Facebook-related functions and the ZXing Library for providing the QR code-related functions. Also, to work independently of particular smartphone operating systems, its implementation is based on HTML5. Figure 3 depicts the structure of LiQR.

The GUI module in Fig. 3 consists of HTML and CSS, displays the information on the advertising page and the result of the QR code recognition, and performs interaction between the users. The Business Logic module reconfigures the user interfaces through communication with the GUI module, and hands over an action to the control module. The control module calls a method in the library at the smartphone, returning the resulting value to the Business Logic module. PhoneGap [13] is a library for accessing phone resources such as phone number and camera in HTML-based smartphone applications. The generated QR code image files and the information related to the authentication are saved in the file storage of the smartphone.

5 User Interface and Useful Function

In this section, we explain the user interface of LiQR and some useful functions provided by LiQR.

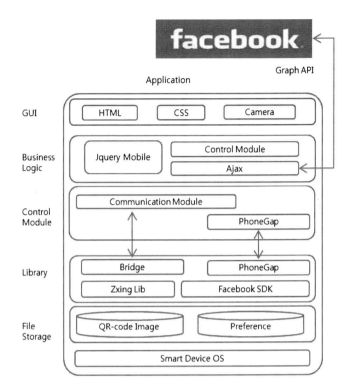

Fig. 3 Structure of LiQR

5.1 User Interface for Reading QR Codes and Building Social Relations

The request for the information on the advertising pages is performed in an asynchronous way using AJAX [14]. From the message responded in a JSON format, LiQR extracts necessary values, generates HTML source codes, and displays them to the user as shown in Fig. 4.

Figure 4a shows the screenshot for requesting the list of concerning Facebook pages. Figure 4b shows the screenshot when an advertising page in the screenshot in (Fig. 4a) is selected. When the user clicks the QR code generation button, the LiQR generates the QR code associated with the page.

After printing the QR code, the marketer can put the printed codes to places such as the front doors or customer tables where customers can easily take pictures. Figure 5 shows the screenshot of the result from the recognition of a QR code.

As shown in Fig. 5a, the 'Like' button is provided through the HTML Source Code distributed by Facebook. LiQR also provides the button for posting messages

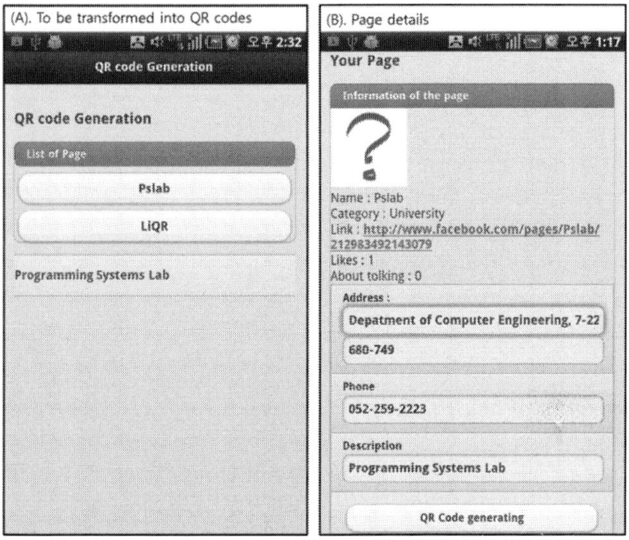

Fig. 4 User interface for handling advertising facebook pages

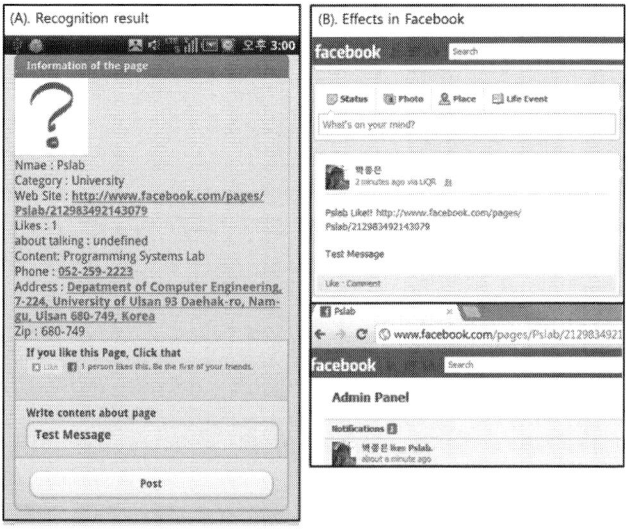

Fig. 5 Recognition result of a QR code

about the advertising page on Facebook. Through the 'Like' and 'Post' buttons, the user can make social relations with the advertising page. Figure 5b shows the screenshot of the result from clicking the 'Like' button and the 'Post' button.

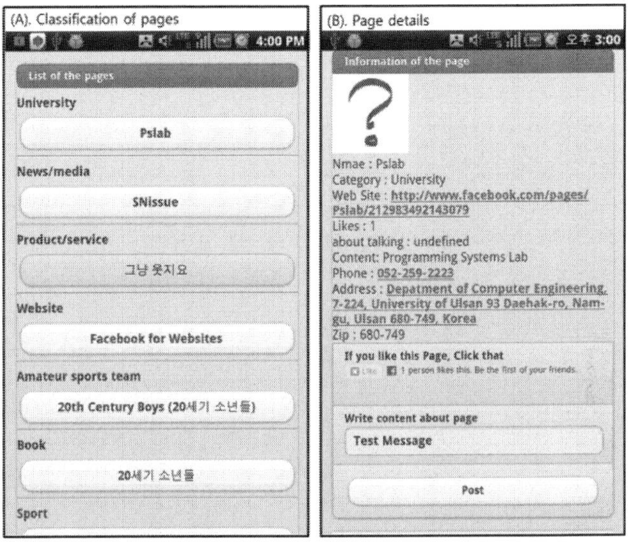

Fig. 6 List of facebook pages with social relations

5.2 Useful Services Utilizing PhoneGap and Google Map

Through the Graph API, LiQR provides the information on the Facebook pages that have social relations with a user. The category property in the provided information is used to classify the characteristics of each page. LiQR classifies the pages according to the category property as in Fig. 6a.

Figure 6b shows the screenshot of the results from the selection of a page on the screen in Fig. 6a, where HTML links on a web site, a phone number, and an address. The functions related to these links are shown in Table 5.

The function for the contact details link is implemented via the PhoneGap library. The function for making a phone call is provided by putting 'tel:' and the phone number of the advertising page in 'a href' HTML tag [15]. Also, the function for saving the phone number is provided by the Contacts object in the library containing the phone number information. The 'find' method in the Contacts object supports a search for a phone number. If the phone number is not saved in the list, the phone is newly stored in the smart phone. Figure 7 shows the overall process.

The function for the address link is provided through the Google Map API. Figure 8a shows the map from selecting the address link. Figure 8b shows the stored phone information.

Table 5 Functions related to three links

Link	Function
Web site	Moves on to the Facebook page URL
Contact details	Makes an immediate phone call Saves the phone number into the smartphone
Address	Displays the map around the specified address

```
//The method for searching the contacts
 function callContacts() {
    var options = new ContactFindOptions();
    options.filter = PHONENUMBER;
    contactFields = ["displayName","name];
navigator.contacts.find(contactFields, onSuccess, onError, op-
tions);
}
function onSuccess(contacts) {
if(contacts.length > 0){
        alert("contacts already exists");
    }else{
        createNewContact();
    }
}
//The method for creating new contact
 function createNewContact() {
    contact = navigator.contacts.create();
    contact.name = PAGENAME;
    var phoneNumbers = PHONENUMBER;
    phoneNumbers[0] = new ContactField('home', PHONENUMBER,
false);
    contact.phoneNumbers = phoneNumber;
    contact.save(onSaveSuccess,onSaveError);
 }
```

Fig. 7 Implementation of Storing a Phone Number

5.3 Related Works and Evaluation

Since SNS is widely used in marketing, SNS has become one of the ways supporting high efficacy with low cost. Recently, as a service for establishing the 'Like' relations between Facebook pages and users, the likify [16] service is provided.

Fig. 8 Map and stored phone information

Table 6 Comparison of LiQR and likify

	LiQR application	Likify service
QR code generation	LiQR application	Likify service
Information stored in QR code	Page information (id, picher, address, phone..etc.)	URL generated by likify
QR code recognition	LiQR application	Any QR code scanning applications
Creating Like relation	LiQR application	URL generated by likify
Extra features	Saving contacts, showing map	N/A

The likify service is very similar to the LiQR because the likify service establishes the Like relations between Facebook pages and users through QR codes. But, to use the likify service, users should put information on the associated Facebook pages into the likify homepage. A QR code generated by the service includes the URL specific to the associated Facebook page. When a user takes picture of the QR code, the QR code scanning application make a request to the URL that enables the user to use the 'Like' function. On the contrary, LiQR supports the 'Like' function in the same application without requesting URL, which makes LiQR to run faster than the likify service. In addition, LiQR automatically saves the phone number of the associated Facebook pages into the smartphone and can show the map designating the location of the store in association with the Facebook page. These features are considered very useful for real life. Table 6 summarizes the comparison of the LiQR application and the likify service.

6 Conclusions

In this paper, we proposed a method for a digital marketing strategy using QR codes and the Facebook 'Like' function, presenting a smartphone application LiQR.

LiQR provides an environment where digital marketing strategy is supported through Facebook pages for advertisement. A user concerning the advertisement of a store or a product only distributes the QR code created from the advertising page. When the QR code is photographed by a user who finds the product interesting, a new social relation is built up naturally in Facebook. In addition, the user automatically acquires the information related to the store such as phone number and location via the functions provided by LiQR. Furthermore, the message on the store written by the user can be posted into Facebook, which is usually available to the friends of the user. We believe that the proposed method would be very successful in the field of digital marketing.

Acknowledgments This research was supported by Basic Science Research Program through the National Research Foundation of Korea (NRF) funded by the Ministry of Education, Science and Technology (No. 2012-0004747).

References

1. Buter, B., Dijkshoorn, N., Modolo, D., Nguyen, Q., van Noort, S., van de Poel, B., Salah, A.A., Akdag Salah, A.A.: Explorative visualization and analysis of a social network for arts: the case of deviantART. J. Convergence **2**(1), 87–94 (2011)
2. Shtykh, R.Y., Jin, Q.: A human-centric integrated approach to web information search and sharing. Hum. Centric Comput. Inf. Sci. **1**, 2 (2011)
3. Yen, N.Y., Kuo, S.Y.F.: An intergrated approach for internet resources mining and searching. J. Convergence **3**(2), 37–44 (2012)
4. Luo, H., Shyu, M.-L.: Quality of service provision in mobile multimedia—a survey. Hum. Centric Comput. Inf. Sci. **1**, 5 (2011)
5. Ellison, N.B., Steinfield, C., Lampe, C.: the benefits of facebook friends: social capital and college students' use of online social network sites. J. Comput. Mediat. Commun. **12**(4), 1143–1168 (2007)
6. Chan, C.: Using online advertising to increase the impact of a library Facebook page. Libr. Manag. **32**(4/5), 361–370 (2011)
7. Kathyn, K.: JEB on facebook. J. Exp. Biol. **213**(15), 2549 (2010)
8. Sun, Y., Liu, C.: The QR-code reorganization in illegible snapshots taken by mobile phones. Computational Science and its Application, ICCSA 2007, pp. 532–538 (2007)
9. Kymin, J.: Sams teach yourself HTML5 mobile application development in 24 hours. books.google.com (2011)
10. Graph API http://developers.facebook.com/docs/reference/api/
11. Zxing—a rough guide to standard encoding of information in barcodes. http://code.google.com/p/zxing/wiki/BarcodeContents(2009). Accessed Feb 2009
12. Like Button http://developers.facebook.com/docs/reference/plugins/like/

13. Allen, S., Graupera, V., Lundrigan, L.: PhoneGap. Pro smartphone cross-platform development part2, pp. 131–152 (2010)
14. Paulson, L.D.: Building rich web applications with Ajax. IEEE Comput. Soc. **38**(10), 14–17 (2005)
15. Schulzrinne, H.: The tel URI for telephone numbers. Network Working Group. http://mail.tools.ietf.org/html/rfc396 (2004)
16. likify http://likify.net/

Ensuring Minimal Communication Overhead in Low Bandwidth Network File

Muhammad Ahsan Habib, Waqas Nasar, Shehzad Ashraf Ch and Ahsan Jamal Khan

Abstract All wireless networks are resource constrained, the battery power, memory and bandwidth are main constraints along with security of information in wireless networks. Running of ordinary file systems on such wireless networks can result in short battery life and bandwidth dilemma for large file sharing, which results in high latency and unbearable delays which user do not consider while working on wireless networks, also the security is second main issue in wireless networks as any false node can inject his packet or private information can be divulged to someone who is not intended, so there is paramount need of an efficient and secure file system for wireless networks. The paper describes the need of secure and bandwidth efficient file system for wireless networks and also proposes a file system which is secure and bandwidth efficient and can become a part and parcel of wireless networks. Simulation results are evident of the performance achieved for our proposed scheme. Proposed methodology can save maximum bandwidth by taking advantage of cross file commonalities, in best case analysis it will just transmit a packet containing just hash tree.

Keywords NFS · Hash tree · Cross file commonalities

M. A. Habib (✉)
Comsats Institute of Information Technology, Islamabad, Pakistan
e-mail: ahsan@comsats.edu.pk

W. Nasar · S. Ashraf Ch
Department of Computer Scienece, Software Engineering International Islamic University,
Islamabad, Pakistan
e-mail: vaqasnasar@yahoo.com

S. Ashraf Ch
e-mail: shahzad@iiu.edu.pk

A. J. Khan
Qassim University, Buraidah, Kingdom of Saudi Arabia
e-mail: ahsanjamalk@yahoo.com

S.-S. Yeo et al. (eds.), *Computer Science and its Applications*,
Lecture Notes in Electrical Engineering 203, DOI: 10.1007/978-94-007-5699-1_95,
© Springer Science+Business Media Dordrecht 2012

1 Introduction

The consideration of high bandwidth consumption over slow network is an important feature for performance and efficiency, the issue got less attention as for as user perspective is considered [1–3].

It became more important in interactive session on slow networks, where performance is key issue, over slower networks interactive programs freeze, like during file I/O, batch commands the execution time increases many fold. In case of File I/O either users have to keep local copies of each or they can use remote login and edit the same file where it is placed. Remote login becomes frustrating for long latency networks [2, 3].

The paper [3] proposed a new network file system Low bandwidth network file system (LBFS), the main idea behind LBFS is to exploit the similarities between the versions of a file placed on different systems over LAN. LBFS avoids the sending of data that is already present in client's cache. The method is close-to open consistency. After a file has been written and closed by a client the edited file will be available to all the subsequent users. Moreover, once a file is successfully written and closed, the data resides safely at the server [4, 5]. LBFS assumes that the entire working of a client will remain on client's cache and the communication is solely for maintaining consistency. When client modifies a file, it must be transmitted to server, similarly when a client opens a file the server sends him the latest version of file.

Low bandwidth network file system creates chunks of a file and keep the indexes of these chunks in a chunk data base which will reside on both server and client, the indexes are created using SHA-1, which is collision resistant, the chunk boundaries are created on the basis of file contents if any of client or server modifies the file, the modification only affects the surrounding chunks. Chunk boundaries called region are created using Robin fingerprints, expected chunk size will be $2^{13} = 8192 = 8$ KBytes $+$ 48-byte breakpoint window size.

Mostly applications in the network require 10 Mbps or more bandwidth over LAN or campus area network therefore there is the need to develop a network file system which requires comparatively less bandwidth [1, 2, 6]. In the wireless network there is a big issue of the bandwidth due to many factors which does not effect in wired communication medium [7, 8]. These factors may include fading, lightning and attenuation. Delays in the propagation are one of the many factors due to wireless nature. Wireless network now becomes the need of the moment. There are many application of wireless network in simple it can be said that the future is of wireless network. As the wireless networks are the desires of the time but there are two main issues to be resolved for wireless networks [7, 8].

1. Bandwidth
2. Security

Due to wireless nature propagated signals faces reduction in power. Because signals are transmitted into the space so there are much chances of the intrusion.

This intrusion must be avoided, especially when you need to transmit some security sort of information over the wireless network.

2 Related Work

The consideration of high bandwidth consumption over slow network is an important feature for performance and efficiency, the issue got less attention as for as user perspective is considered [2]. In [1] the author proposed WAN acceleration model for developing world, the idea is based on reducing network traffic by avoiding data which is already present in receiver cache. The data is broken into variable size chunks; each chunk will have a corresponding integer number, these integer numbers will be used for decision of sending the chunk onto WAN. Author proposed Multi resolution chunking (MRC) which use robin finger prints for chunk boundaries, MRC creates a hierarchal aligned chunk boundaries, all of the smallest boundaries are detected, and then larger chunks are generated by matching different numbers of bits of the same boundary detection constraint, here the largest chunk is the root, and all smaller chunks share boundaries with some of their leaf chunks. Performing this process using one fingerprinting pass not only produces a cleaner chunk alignment, but also requires less CPU. Bhagwat and Eshghi [6] described a scalable de-duplication technique for non-traditional backup workloads that are made up of individual files with no locality among consecutive files in a given window of time. Bhagwat and Eshghi [6] also claimed that due to lack of locality, existing techniques perform poorly on these workloads. Extreme binning exploits file similarity instead of locality, and makes only one disk access for chunk lookup per file, which gives reasonable throughput. In scheme of [6] each file is allocated using a stateless routing algorithm to only one node, allowing for maximum parallelization, and each backup node is autonomous with no dependency across nodes,

Article [9] is about to make Grid Security Infrastructure (GSI), provides new capabilities to be used on top of different grid middle wares. The solution is specifically implemented on gLite and accomplishes the access to data storage Grid resources in a uniform and transparent way, the proposed approach of [9] is based on XML data encapsulation to store encrypted data and metadata as well as data owners, X509v3 certificates, AES keys, data size, and file format.

Ledlie [2] Proposed a file system support for low-bandwidth channels typically the wireless and mobile ad-hoc networks where devices are mobile, typically the problem arises for wireless networks where browsing, downloading and exchanging files can halt the network traffic and a situation of deadlock occurs, the problem is increased twofold when mobile device users wants to share large files or want to collaborate each other. Two approaches were adopted before the proposed solution (1) Fetch the file metadata only and (2) Fetch both metadata and file contents Metadata just gives information about file size, type etc. which in some cases is not appropriate (e.g. video files etc.), option 2 is impractical because of resource constraints.

In proposed remote file system [2] the metadata of the file is modified and thumbnail information is added, when a user want to access a file remotely, first of all only metadata of the file will be sent to him, as metadata also contains thumbnail so after seeing thumbnail user can decide whether to open the file or not, so author claimed to reduce significantly bandwidth consumption.

In [10] a new technique for protecting sensitive contents in view only file system is proposed. The authors intended to meet the challenges such as spyware, removable media, and mobile devices which make the sensitive contents disclosed. VOFS relays on trusted computing primitives and virtual machine technology to provide a much greater level of security in the current system. VOFS clients disable non essential device output before allowing the user to view the data to his or her non-secure VM. When the user has finished his work, VOFS reset the machine to previous state and resume normal device activity. It is difficult to prevent data linkage from authorize user, but VOFS do it to some extent using VMM (virtual machine monitor) technology. The VMM has complete control over the computer. In the client architecture above the VMM are the guest virtual machine VM, and is referred to as primary guest VM. This machine has not sufficient access to the resources. Another guest virtual machine i-e domain 0, has complete administrative access to all the resources. A third guest say SVFS VM is responsible for downloading sensitive contents, storing it, and telling the domain 0, when to save or restore primary guest state and when to enable or disable device output. The author claimed that the VOFS clients also get benefit of trusted platform module (TMP) technology. The clients also take the advantage by using an integrity measurement architecture which uses trusted boot to enable remote verification of trusted components by content providers. If a user wishes to view a view only file in VOFS, then the primary VM will make a request to SVFS VM. It will contact the content provider to start an authorization session.

In [11] author claimed that the conventional method (reference monitor) used for authentication was purely centralized, now the world is moving towards distributed environment for this requirement to be fulfilled centralized approach should not be change. In conventional reference monitor method there was the problems of scalability and resource protection. In traditional reference monitor method ACL (access control list) was maintained, every user must be entered in the ACL to have access to the resources. ACL was not scalable well due to ACL, so there should be a way to modify the ACL approach to give access to the new user. Another drawback of ACL was the need of the full time monitoring the requests for resources. Protection and scalability were also the issues to be discussed in conventional approach. The author also proposed a method of cryptographic Access Control. This method eliminates the centralized approach and handles the resource protection as well as the scalability issues. In proposed method encryption is applied at the client and the encrypted data is stored at the server without decrypting the file. The creator of any file is the generator of the keys for encryption and decryption and he is the only who can modify the file. The server saves the file along with keys and the server also maintains the binary tree at server side so fast access can be done in case of searching of the keys. The benefit of saving the key along the file is if any trouble with the binary tree happened then the key is not compromised. Due to the distributed in the nature the proposed method provides scalability, controlled access to the resources, gets rid from constant monitoring because of the encryption the only user can decrypt the file who owned the decryption key (only the trusted user). The author claimed that authentication, integrity are achieved through digital signature and hash function.

3 Proposed Solution

By keeping the two issues of the wireless networks there should be some sort of network file system which is efficient and bandwidth affected as well [1, 3, 5, 6]. In the perspective of the end user security in some cases is not negligible but the bandwidth can be compromised to some extent. It is more considerable in the view of the network file system designers. Especially when there are interactive sessions the less bandwidth results into less efficiency and can halt too, like batch commands the execution time increases manifold cryptographic access control. File copying form remote site becomes a frustrating for long latency networks.

Our proposed solution work as, First the hash code of the file is generated. Then the file is broken into two chunk of equal size and their hash code is generated. These chunks is further divided be into four chunks and hash code of each chunk is calculated and so on till an appropriate level k where $k = \log 2n$ (Fig. 1).

Hash code calculated for each chunk forms a hierarchal hash tree of chunks. The file is encrypted each time before write back. Encrypted Hash code of the whole file is sent to client. Client decrypts it and match with the Hash code of file already present in his memory. If result is same, this implies that server and client having same version of the file, if result is different than client asks next level hash code of hash tree present in server memory, at each level hash codes are compared and the chunk having different data are identified, after identification of chunk whose data has been changed, client asks server to send the specific chunk. Encrypted data is transmitted on communication channel by server and client, If client and server data chunks produce same code, they do not transmit data.

The file is encrypted each time before write back then encrypted hash code of the whole file is sent to client. Client decrypts it and match with the Hash code of file already present in his memory. If result is same, this implies that Server and Client having same version of the file, if result is different than client asks next level hash code of hash tree present in server memory, at each level hash codes are compared and the chunk having different data are be identified, after identification of chunk whose data has been changed, client asks server to send the specific chunk, encrypted data transmitted on communication channel by server and client, If client and server data chunks produces same code, they do not transmit data.

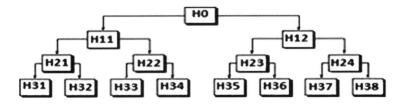

Fig. 1 Hash tree generation

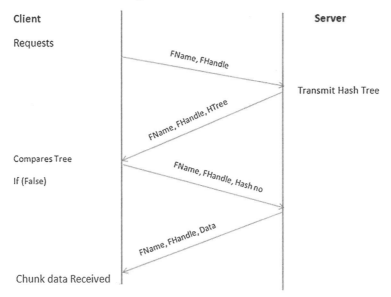

Fig. 2 Time diagram of proposed solution

If a client brought a file from the server in its local cache and did not do any work on it and started some other work. After some time it need to do some work on that file, but it is a possibility that may be during that time some other client had taken that file and updated that file or may not. To ensure consistency and freshness, it needs to consult the server to check whether the file is it in the same state when it took or it has been updated. Then it consults the server and send it file name and file handle (Fig. 2).

4 Comparative Analysis

The proposed architecture is content-based, that is it takes advantages of common contents between a versions of a file residing on different machines one on server others on clients. Proposed solution can only reduces bandwidth usage if and only

Table 1 Comparison of new data and data transmitted in proposed solution

Network type	File	Extension	Data (kb)	New data (kb)	Overlap (^%)	No of comparison	Data transmitted	
							KB	%
NFS	MS word	.Doc	238	23	90	1	238	100
LBFS						9	68	28.6
Proposed						7	60	26
NFS	Notepad	.txt	71	3	79	1	71	100
LBFS						13	6.45	9.1
Proposed						11	4.44	6.25
NFS	Acrobat	.pdf	43	7	84	1	43	100
LBFS						15	12.3	28.6
Proposed						7	10.7	25
NFS	MS access	.mdb	129	52	60	1	129	100
LBFS						7	77.4	60
Proposed						5	65	50

Fig. 3 Comparison between NFS, LBFS and proposed System

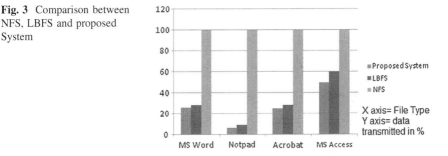

if there is some overlapping in contents of version of a file, the amount of network traffic reduced is the only the common contents.

The proposed architecture is content-based, that is it takes advantages of common contents between a versions of a file residing on different machines one on server others on clients. Proposed solution can only reduces bandwidth usage if and only if there is some overlapping in contents of version of a file, the amount of network traffic reduced is the only the common contents. Table 1 summarizes the application type, file extension, original file size, and amount of new data, overlapping between two versions of a file, number of chunks/comparisons and percentage of transmitted data.

A comparison of network file system, low bandwidth network file system and proposed solution for four different applications is provided in Fig. 3.

Comparisons from results for NFS, LBFS and proposed system and shows that proposed system transfer less data from LBFS and NFS.Comparisons from results for NFS, LBFS and proposed system and shows that proposed system does fewer comparisons as compare to LBFS, as in NFS no comparison is done but complete file is sent every time.

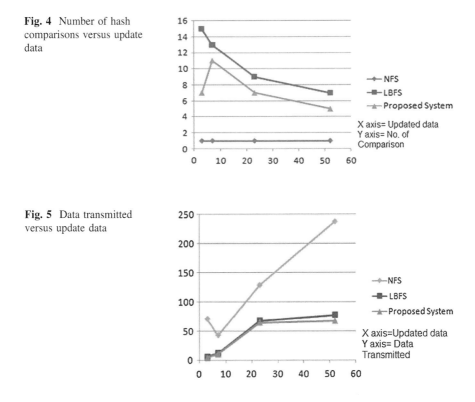

Fig. 4 Number of hash comparisons versus update data

Fig. 5 Data transmitted versus update data

Graph in Fig. 4 shows that proposed system does fewer comparisons as compare to LBFS, as in NFS no comparison is done but complete file is sent every time (Fig. 5).

Comparisons from results for NFS, LBFS and proposed system and shows that proposed system transfer less data from LBFS and NFS

5 Conclusions

The proposed architecture can save bandwidth by taking advantage of commonality between files, ensures integrity and freshness of a remote file, Operations such as editing documents and compiling software, proposed architecture can consume over an order of magnitude less bandwidth than traditional file systems and performs efficient use of communication link. It also makes transparent remote file access a viable and less frustrating alternative to running interactive programs on remote machines. To ensure minimum running time for Hash matching we used hierarchal hash tree.

Simulation results are evident of the performance achieved for our proposed scheme. Proposed methodology can save maximum bandwidth by taking advantage of cross file commonalities, in best case analysis it will just transmit a packet containing just hash tree.

References

1. Ihm, S., Park, K., Pai, V.S.: Wide area Network acceleration for the developing world, USENIXATC'10 Proceedings of the 2010 USENIX conference on USENIX annual technical conference, pp. 18 (2010)
2. Ledlie, J.: File System for Low-Bandwidth Thumbnails. Nokia research center, Cambridge, US, 6 May 2008
3. Muthitacharoen, A., Chen, B., Mazi 'eres, D.: A low-bandwidth network file system. In: Proceedings of the 3rd International Symposium on Information Processing in Sensor Networks (IPSN'04), pp. 81–88, 26–27 April 2004
4. Nizamuddin, S.A.C., Nasar, W., Javaid, Q.Efficient signcryption schemes based on hyperelliptic curve cryptosystem. ICET 2011, Sept. 2011. doi:10.1109/ICET.2011.6048467
5. Ashraf Ch, S., Nizamuddin, M.S.: Public verifiable signcryption schemes with forward secrecy based on hyperelliptic curve cryptosystem. ICISTM 2012, CCIS 285, pp. 135–142, 2012. Springer (2012)
6. Bhagwat, D., Eshghi, K.: Extreme binning: Scalable parallel deduplication for chunk based file backup, Modeling, Analysis & Simulation of Computer and Telecommunication Systems, 2009. MASCOTS '09. IEEE International Symposium on, pp. 1–9 (2009)
7. Nizamuddin, S.A.C., Amin, N.: Signcryption schemes with forward secrecy based on hyperelliptic curve cryptosystem. HONET 2011, Dec. 2011. doi:10.1109/HONET.2011.6149826
8. Amin, R., Ashraf Ch, S., Akhter, M.B., Khan, A.A.: Analyzing performance of ad hoc network mobility models in a peer-to-peer network application over mobile ad hoc network. In: International Conference on Electronics and Information Engineering (ICEIE 2010) Kyoto, Japan, 1–3 Aug. 2010. doi:10.1109/ICEIE.2010.5559795
9. Tusa, F., Villari, M., Puliafito A.: Design and Implementation of an XML-based grid file storage system with security features, Enabling Technologies: Infrastructures for Collaborative Enterprises, 2009. WETICE '09. 18th IEEE International Workshops on, pp. 183–188 (2009)
10. Borders, K., Zhao, X., Prakash, A.: Securing Sensitive Content in a View-Only File System. University of Michigan 30 Oct 2006 ACM
11. Harrington, A., Jensen, C.: Cryptographic Access Control in a Distributed File System, SACMAT '03 Proceedings of the eighth ACM symposium on Access control models and technologies, pp. 158–165 (2003)

Preventing Blackhole Attack in DSR-Based Wireless Ad Hoc Networks

Fei Shi, Weijie Liu and Dongxu Jin

Abstract Wireless ad hoc networks become more and more popular and significant in many fields. However, the deployment scenarios, the functionality requirements, and the limited capabilities of these types of networks make them vulnerable to a large group of attacks, such as blackhole attacks. In this paper, we proposed a scheme to prevent blackhole attacks in wireless ad hoc networks. Our scheme consists of three mechanisms which are trust management mechanism, detection mechanism and location mechanism. The performance simulation analysis confirms the availability and efficiency of our scheme.

Keywords Wireless ad hoc networks · Blackhole attack · Detection · Location · Trust

1 Introduction

Wireless Ad hoc networks [1] are attracting great interest. Compared to other types of wireless networks, wireless ad hoc networks are more vulnerable to attacks due to their unique characteristics. This paper focuses on preventing blackhole attacks

F. Shi · D. Jin
Department of Computer Science, Yonsei University, Seoul, Korea
e-mail: shifei@emerald.yonsei.ac.kr

D. Jin
e-mail: jjjddx@emerald.yonsei.ac.kr

W. Liu (✉)
Department of Information and Industrial Engineering, Yonsei University, Seoul, Korea
e-mail: weijie@yonsei.ac.kr

S.-S. Yeo et al. (eds.), *Computer Science and its Applications*,
Lecture Notes in Electrical Engineering 203, DOI: 10.1007/978-94-007-5699-1_96,
© Springer Science+Business Media Dordrecht 2012

Fig. 1 Blackhole attack
scenario

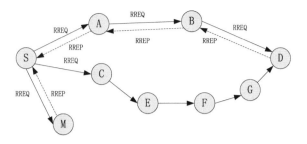

[2]. The malicious node sends fake Route Reply message (RREP) including fake routing path to the source node and claims a short route to the destination. Consequently, the source node is attracted to select the route path with the blackhole nodes. Afterwards, the malicious node can misuse the packets. As shown in Fig. 1, source S intends to perform transmission to destination D. According to Dynamic Source Routing (DSR) protocol, node S broadcasts Route Request message (RREQ), the destination node or the intermediate node, which has the route path to destination, replies RREP to the source node piggybacking the route path. As receiving the RREQ, the malicious node M sends RREP immediately piggybacking a fake routing path (e.g., S-M-D) which is shorter than the normal path (e.g., S-A-B-D), so data traffic will be attracted through the malicious node M, i.e., all the packets sent by node S are drawn into the malicious node M. Then node M can misuse or simply drop those packets. In this paper, we assume that DSR protocol is used in the network. However, our scheme can be extended to other routing protocols, like Ad hoc On-Demand Distance Vector (AODV), Destination-Sequenced Distance Vector (DSDV), etc.

The organization of this paper is as follows. The next section summarizes and discusses the related works. In Sect. 3, we present the details of our scheme. First, we present a trust management mechanism to introduce trust value to each node. Through this trust management mechanism, the LMT node can be elected. Then we present a detection mechanism to detect blackhole attacks. Next, we present a location mechanism to identify the blackhole nodes. A performance simulation is presented in Sect. 4. Finally, we conclude the paper and present the future work in Sect. 5.

2 Related Work

In [3], Kurosawa et al. believe that the destination sequence number in the RREP generated by the malicious node must be larger than the number in other RREPs, so as to overwhelm other RREPs and make the source node to use the route path with the malicious node. Thus, the authors present an anomaly detection approach based on statistical analysis to detect the blackhole attacks by compared the received RREPs' destination sequence numbers. This scheme does not incur any extra routing traffic, and it does not modify the existing routing protocols. However, the main weakness of this scheme is that its value of false positives is relatively high which is inevitable due to the utilization of anomaly detection.

In [4], Mistry et al. propose a scheme similar with that of [3]. Source node is required to verify RREPs' destination sequence numbers which are collected during the predefined waiting period using the heuristic method. If the sequence number of a RREP is found to be extraordinary high, the corresponding node sending the RREP will be identified as the malicious node. One weakness of this approach is the delay incurred in the route discovery process. Since the source node is required to defer for the waiting time period for the collection of RREPs' destination sequence numbers, even if there is no malicious node, it still suffers from delay.

In [5], Tsou et al. propose a DSR based secure routing protocol, named Baited-Black-hole DSR (BDSR) which can detect and avoid the black hole attack. BDSR requires source node to send bait (RREQ') which piggybacks a non-existent target address, and attract blackhole node to reply the fake RREP. However, the source node fabricating a non-existent target address must base on the premise that the source node has the knowledge about the number of nodes in the network and all their addresses. This premise is difficult to achieve considering the distribution and mobility characteristic of MANETs. Hence, the fabricated target address by the source node may not be non-existent, occasionally. In this case, the malicious node has the warrant to claim it has the knowledge to reach the target address.

The same authors proposed another scheme named as CBDS in [6]. It is similar with the scheme in [5], namely, it uses the baiting RREQ' packet to attract blackhole node to reply the fake RREP. The difference is that the scheme in [6] requires that the source node randomly cooperates with a stochastic adjacent node. By using the address of the adjacent node as the bait destination address, it baits malicious nodes to reply RREP. However, when the number of malicious nodes increases, namely, when the percentage of the network comprised of malicious node increases, the randomly selected adjacent node, which cooperates with the source node, may be a malicious node. In this case, the source node may suffer from deceiving by the randomly selected malicious node.

Both the schemes in [5] and [6] make the assumption that the destination node sends the alarm message when it is conscious of the severe packet loss. However, since the scenario is MANETs, all the nodes have the coordinative priority level in terms of security. Even as for the destination node, it may perform misbehaviors, e.g., the destination node may disregard the packet loss in the network or report the fabricated alarm even in the normal network. Thus, the destination node should not be assumed as the well-behaved node.

In this paper, our scheme not only detects blackhole attacks but also locates the blackhole nodes. And our scheme can prevent collusive blackhole attack. Our scheme does not use any encryption/decryption mechanism which causes huge amounts of resource consumption. In both the detection and location phases of our scheme, behaviors of any suspects (potential blackhole nodes) are monitored, so the fabricated information forged by malicious nodes can be prevented.

3 Proposed Scheme

First of all, the overview of the proposed scheme is described as follows. (1) For each node, there exists one node whose trust value is the largest among the one-hop neighbors, and we name this node as local most trustworthy (LMT) node. (2) We present a detection mechanism to detect the existence of blackhole attack. We consider the main cause of blackhole attack is the fake RREP message sent by intermediate malicious node. We require that the intermediate node that generated the RREP must send a TRACE REQUEST (TREQ) message to its next-hop node towards the destination. Meanwhile, the corresponding LMT node of the inter-mediate node sets a timer. Once the next-hop node receives the TREQ message, it must send a TRACE REPLY (TREP) back and forward the TREQ message to its next-hop node towards the destination. Once the TREP message arrives at the previous intermediate node, timer in LMT node stops. Otherwise, if the previous intermediate node that generated the RREP does not receive the TREP, timeout occurs. It means that the RREP is faked and there is blackhole attack. During the detection phase, the corresponding LMT nodes monitor those intermediate nodes' behavior. (3) We present a location mechanism to identify the blackhole nodes, i.e., the nodes that send the fake RREP can be located by our location mechanism. When intermediate node does not receive the TREP from its next-hop node towards the destination, the intermediate node must send a Route Error message (RERR) to its former-hop towards the source. This process is monitored by the corresponding LMT node. Thus, if any intermediate node refuses to transfer this RERR message, it can be considered to be the blackhole node detected by the corresponding LMT node.

In both the proposed detection and location mechanisms, LMT nodes act as the monitors. Hence, first of all, in the following subsection, we present the procedure for electing LMT node of each intermediate node.

3.1 Trust Management Mechanism

We propose a trust management mechanism that introduces a parameter, i.e., trust value, to evaluate each node's trustworthiness. The trust value T ($T = f(C, S)$) is a function of credit value (C) and stability value (S). The first component, C, is evaluated based on node's transmission behaviors. The second component, S, indicates if the node moves fast or slowly, or remains stable. In short, the node that performs normal behavior and moves slowly owns the larger trust value, T. Conversely, the node that performs malicious behavior (i.e., dropping packets) or moves fast owns the smaller T.

Fig. 2 A scenario of
neighbors' connectivity

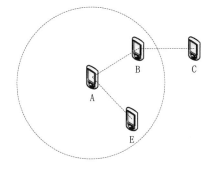

3.1.1 Evaluation of Credit Value (C)

There are two types of credit value, i.e., direct credit value (C_d) and recommended credit value (C_r). If a node can directly observe another node's behaviors, direct credit can be established. $C_d(i,j)$ denotes that there are direct interactions between node i and j, i.e., node i can monitor the behaviors of node j and evaluate node j's credit value. As shown in Fig. 2, node A can directly observe the behaviors of node B and E and obtain their direct credit values. On the other hand, if a node receives recommendations from other nodes about another (third-party) node, recommended credit can be established. There are two types of recommended credits. One is that there is no direct interaction between node i and node j, while there may be indirect interactions between them: (1) there is an intermediate node k between nodes i and node j; and (2) there are interactions between node i and node k and between node j and node k. Thus, node i can obtain the credit value of node j through node k. In Fig. 2, node A and node C are not immediate neighbors, but node A can obtain the credit value of node C through node B's recommendation. Another type of recommended credit value is that there is direct interaction between node i and node j, and node i can obtain both the direct value and recommended credit value of node j. In Fig. 2, node A can obtain the direct credit value of E, while node B can recommend the recommended credit value of node E to A; in this case, node A will obtain the recommended value of node E through node B. The recommended credit value is represented by $C_r(i,j)$.

The total credit value $C(i,j)$ can be obtained by integrating $C_d(i,j)$ and $C_r(i,j)$ using the following formula:

$$C(i,j) = w_1 \cdot C_d(i,j) + w_2 \cdot C_r(i,j) \tag{1}$$

where w_1 and w_2 denote the weight factors for $C_d(i,j)$ and $C_r(i,j)$, respectively. We adopt $w_1 > w_2$, and $w_1 + w_2 = 1$. The reason for adopting factor w_1 larger than w_2 is that we consider that the direct trustworthiness of one node is more trustful than the recommended trustworthiness from other nodes. Since malicious nodes may provide dishonest recommendation, the recommended credit value should be treated separately from regular direct credit value. Thus, we set the factor w_2 a relative smaller value to make this part of value less important. Even if

Table 1 Parameters for evaluating direct credit value (C_d)

Number of packets	Explanation
N_j^{act}	Number of packets actually forwarded by node j
N_j	Number of packets to be forwarded by node j
N_j^{out}	Number of packets that come out from node j
N_j^{src}	Number of packets with node j as the source
N_j^{in}	Number of packets that go into node j
N_j^{dest}	Number of packets with node j as the destination

there was dishonest recommendation, the damage caused by recommended credit value will be small.

Frequently exchanging recommended credit value between nodes will definitely cause more traffic and the opportunity of transmission collision will be increased. Recommended credit value could be recommended only if it is larger than a threshold to decrease the traffic in network and avoid congestion. If not, the recommendation is useless and ignored. In this way, the traffic caused by the recommendation will be decreased, and congestion will be avoided.

To make the issue simple and clear, we make the assumption that the misbehaving nodes only drop packets and they do not modify the content of the packets. It's considered that the nodes in the network should not only share the medium fairly but also carry out their obligations actively. Thus, we consider that the behavior of dropping packets is misbehavior. For example, some nodes dropping packets to cut off the network are malicious nodes and some nodes dropping packets for the purpose of saving their energy are also malicious nodes. The corresponding credit value of the node that drops packets is smaller. The direct credit value is established upon observations of whether or not the previous interactions between node i and node j are successful. In other words, $C_d(i,j)$ is node i's evaluation value of node j by directly monitoring the packet communication of node j. $C_d(i,j)$ can be calculated by node i, using the following formula:

$$C_d(i,j) = \frac{N_j^{act}}{N_j} = \frac{N_j^{out} - N_j^{src}}{N_j^{in} - N_j^{dest}} \tag{2}$$

where the corresponding parameters are interpreted in Table 1.

Equation (2) measures node j's ability to forward packets. Based on packet transmission direction, there are two types of packet related to each node. One type of packet is the packet that "goes into" the node (the number of this type of packet is represented by N_j^{in}); another type of packet is the packet that "comes out" the node (the number of this type of packet is represented by N_j^{out}). Further, the former type of packet ("go into" packet) is divided into two subtypes. One type of "go into" packet is the packet with node j as the destination (the number of this type of packet is represented by N_j^{dest}). Because the destination is node j, this type of packet should not be forwarded. Another type of "go into" packet is the packet that should be forwarded by node j (the number of this type of packet is represented by N_j). So by subtracting the number of packet with node j as the destination (N_j^{dest}) from the number of packet that "goes into" the node (N_j^{in}), the number of packet to be forwarded by node j is obtained (N_j). Furthermore, the type "come out" packet is divided into two subtypes too. One type of "come out" packet is the packet with node j as the source (the number of this type of packet is represented by N_j^{src}). This type of packet is not forwarded but generated by node j. Another type of "come out" packet is the packet that actually forwarded by node j (the number of this type of packet is represented by N_j^{act}). By subtracting the number of packet with node j as the source (N_j^{src}) from the number of packet that "comes out" the node (N_j^{out}), the number of packet that to be forwarded by node j is obtained (N_j^{act}).

3.1.2 Evaluation of Stability Value (S)

An explanation of the need for the stability value is first called for. Consider the following scenario. One node is moving, and it's so fast that other nodes cannot connect to and communicate with it. In this case, this node is useless and cannot be trusted. In our scheme, we intend to choose the relatively stable LMT node, which may stay in the transmission range of the originator node for a longer time compared to the node that has the high mobility rate. Considering this, the stability value is introduced to weigh up the stabilization of the nodes.

To reasonably describe node stability, our paper uses a similarity computation method [7] and the graph theory [8] to calculate the stability of each node. The network formed by nodes and links can be represented by a directed graph, $G(t) = (V, E(t))$, called neighbor relation graph, wherein $V = \{1, 2, \ldots, N\}$ denotes the number of nodes, and $E(t) = \{e_1, e_2, \ldots, e_m\}$ denotes the number of wireless links. If node i can receive information which is sent from j, there is a directed edge $e(i, j)$ between node i and node j, i.e., node j is the neighbor of node i. $E_i(t_j)$ and $E_i(t_{j+1})$ denote the wireless links situation of node i at two adjacent time points, t_j and t_{j+1}, as shown in Fig. 3. According to the similarity theory, stability value can be represented by the mean similarity value between $E_i(t_j)$ and $E_i(t_{j+1})$ as shown in the following Eq. 3.

$$S_i = \frac{1}{n-1} \sum_{j=1}^{n-1} \cos \theta_j = \frac{1}{n-1} \sum_{j=1}^{n-1} \left[\frac{E_i(t_j) \cdot E_i(t_{j+1})}{|E_i(t_j)||E_i(t_{j+1})|} \right] \tag{3}$$

where θ_j is the included angle between the vector $E_i(t_j)$ and $E_i(t_{j+1})$, as shown in Fig. 4. S_i denotes the similarity between node i's vicinity situations status at different time points, e.g., t_j and t_{j+1}. If S_i is larger, the angle between $E_i(t_j)$ and $E_i(t_{j+1})$, namely θ_j will be smaller which implies more similarity between $E_i(t_j)$ and $E_i(t_{j+1})$, i.e., the neighbors of node i do not change dynamically at time t_j and

Fig. 3 A scenario showing the change of node i's neighborhood due to i's mobility

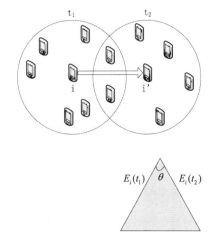

Fig. 4 The angle θ representing the similarity between two vectors: $E_i(t_1)$ and $E_i(t_2)$

t_{j+1}, namely the node i is relatively stable. On the contrary, if the S_i is smaller, θ will be larger, which expresses less similarity between $E_i(t_j)$ and $E_i(t_{j+1})$, i.e., the neighbors of node i change dynamically at time t_j and t_{j+1}, namely the node i moves fast. Hence, in our scheme, similarity S_i is used to represent the stability of node i.

3.1.3 Evaluating the Trust value (T)

The reward and penalty mechanism is used to evaluate the trust value of each node which depends on the node's credit and stability values. We utilize three mathematic functions to evaluate the trust value of each node. These three functions are the exponential function, the logarithmic function and the logarithmic function.

If $C \geq C_{Max\_threshold}$, the logarithmic function is used as shown in Equation 4.

$$T = \log_{2-C}(1 + S) \tag{4}$$

where $C_{Max\_threshold}$ is set to 0.7 for the simulation in Sect. 4.

If $C \leq C_{Min\_threshold}$, the exponential function is used as shown in Eq. 5.

$$T = (1 + C)^S \tag{5}$$

where $C_{Min\_threshold}$ is set to 0.3.

If $C_{Min\_threshold} \leq C \leq C_{Max\_threshold}$, the linear function is used as shown in Eq. 6.

$$T = C \cdot S \tag{6}$$

First, when the credit value of the node is larger than the threshold, e.g. 0.7, the logarithmic function ($y = \log_a x$) is used to measure its trust value. When value a decreases, y increases with the same x. Logarithmic functions increase quickly with an increase in x, when x is a small number, as exactly the case in our scheme. Therefore, logarithmic functions have fast increasing shapes. In our scheme, logarithmic functions are used to measure the nodes with larger credit values as shown in Eq. 4.

Second, when the credit value of the node is smaller than the threshold, e.g. 0.3, the exponential function is used to measure its trust value. Exponential functions $(y = a^x)$ are characterized by their growth rate that is proportional to their value. We only use $a > 1$ to describe a node's trust increase in our scheme. Thus, the exponential function $(a > 1)$ has a slow increase shape when x is not a large number, and y will increase with an increase in a. Such functions are suitable for measuring nodes with smaller credit value as shown in Eq. 5.

Finally, when the credit value of the node is in a medium period, the linear function $(y = a \cdot x)$ is used to measure its trust value. With the same value x, function value y increases along with a's increase. Since linear functions have stable increasing shapes, they are used to measure nodes with a stable change in trust or with constantly cooperative behavior. Thus, linear functions have modestly increasing shapes unlike logarithmic functions and exponential functions. Such functions are suitable for measuring nodes with medium credit value as shown in Eq. 6.

In short, if the node has a small credit value, or if it moves faster (stability value is small), its trust value is small. If the node has a large credit value, or if it moves slowly (stability value is large), its trust value is large.

In some case, a malicious node may hide its maliciousness by not moving fast. However, this malicious node will not be assigned as the LMT node due to our trust value evaluation mechanism. Since this malicious node does not move fast, its stability value (S) is relatively large. However, its credit value (C) is relatively smaller than the normal node because of its misbehaviors (i.e., dropping packets). Thus, the trust value (T) of the malicious node is smaller than that of the node which moves slowly and shows good behaviors. It means that this type of malicious node could not be the LMT node in our scheme.

3.2 Detection Mechanism of Blackhole Attack

We make the two assumptions. (1) All the links are symmetric. (2) The network has a proper density of node, i.e., any node has the corresponding LMT node among its neighbors.

First of all, an explanation of how to choose the LMT node follows. Using the aforementioned trust management mechanism, the neighbor nodes' trust value in each neighborhood can be obtained. In our scheme, each node maintains a table containing its neighbor nodes' trust values. By comparing with the neighbor nodes' trust values through the trust table, each node can judge itself to be the LMT node or not. If it has the biggest trust value compared with other neighbor nodes in the trust table, it will act as the LMT node.

Fig. 5 A scenario of
originators and their
corresponding LMT node

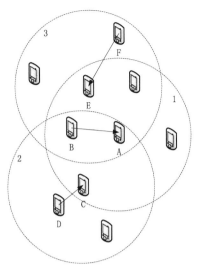

There could be one potential issue that some other misbehaving nodes could pretend to be the LMT node of the intermediate node and give the faked alarm of blackhole attack to either treat the normal intermediate node as the blackhole node or harbor the real blackhole attack. However, by using trust value table and watchdog mechanism, this potential issue can be prevented. If a misbehaving node pretends to be the LMT node and sends some control messages to the intermediate node, according to the watchdog mechanism [9], other nodes in the neighborhood of the misbehaving node will overhear those messages sent by the misbehaving node and know who the source of those messages is. After comparing the trust value of that misbehaving node in the trust value table, it will be revealed that the misbehaving node's trust value is not big enough to be the LMT node of the corresponding originator, i.e., there is other node in the trust table whose trust value is bigger than the misbehaving node's. Thus, this potential LMT node impersonation issue can be prevented.

There could be another potential issue that LMT node may probably consume much more battery power in the trust management mechanism. Since the LMT node performs the supererogatory blackhole detection and location task, its battery may be consumed faster than other nodes. Fortunately, as to different originators, their corresponding LMT nodes will probably change as shown in Fig. 5. In local neighborhood 1, node A's corresponding LMT node is node B. But as to node C and E, since their neighborhoods change (neighborhood 2 and 3), their corresponding LMT node will probably change to other nodes (i.e., node D and F are the corresponding LMT nodes to node C and E respectively in Fig. 5).

Now let's see how the detection mechanism works. The key idea of detection mechanism is to use intermediate nodes' LMT node to detect the existence of blackhole attacks and locate blackhole nodes. We introduce TRACE message (including TREQ message and TREP message) traveling along the routing path, which is between the intermediate node that announced it has the path to destination, and the destination. LMT nodes are required to monitor the process of detection. As requirement in DSR routing protocol, intermediate node can reply the RREP message in response to the RREQ if the intermediate node has the path to the destination. In the detection process, the node replying the RREP is required to send TREQ message piggybacking the routing path, which is the same with the one piggybacked in RREP, to its next-hop neighbor (e.g., node e is the next-hop of node d) towards the destination. The TREQ is required to be re-forwarded by other intermediate nodes along the routing path until it reaches the destination. The intermediate node receiving the TREQ message is also required to send TREP message to its former-hop node (e.g., node d is the former-hop of node e) towards the source, if its address is in the piggybacked list of routing path. When the intermediate node sends the TREQ to its next-hop node, a timer in the corresponding LMT node is required to begin counting. The timer stops as soon as the LMT node hears the TREP replied by the next-hop node reaching the previous intermediate node. The implement of this monitor process can be supported by watchdog mechanism [9]. If the next-hop neighbor node does not reply TREP in a time threshold, timer's timeout occurs. In this paper, we make the assumption that packet loss is only caused by packet dropping misbehavior performed by malicious node. Issues related to access channel, like Media Access Control (MAC), packet collision, etc., are out the scope of this paper.

If the LMT node can hear the TREP in the time threshold, the link between the intermediate node and its next-hop node exists. The link checking process is required to be done by all LMT nodes along the routing path.

If timeout occurs in the corresponding LMT node of one link through the routing path, it gives the fact that the link does not exist and the RREP sent by the intermediate node is faked, which means there is blackhole attack in the routing path. If blackhole attack is detected, LMT node sends ALARM message via the inverse routing path to the source node piggybacking the routing path where the blackhole attack exists.

To describe our scheme more clearly and intuitively, we use the scenario in Fig. 6 as the example for explanation. Source node a intends to transfer data to destination node h, so node a sends out RREQ message to start the routing discovery process. Node b, c, d, e, f and g are intermediate nodes. Node 1, 2, 3, 4, and 5 are corresponding LMT node to node d, e, f, g, and h, which can be elected by our trust management mechanism mentioned in Sect. 3.1. We assume that intermediate node d claims that it has the path to the destination node h by replying RREP to the source node a. If the RREP is a normal reply in response to the corresponding RREQ, we can judge that the routing path is secure and there is no blackhole attack in the routing path. Conversely, if the RREP is faked, we can say that there is blackhole attack in the routing path d-e-f-g-h. The proposed detection mechanism is to determine whether the RREP is faked or not.

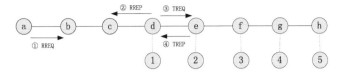

Fig. 6 The process of detection mechanism

The process of our detection mechanism is as follows.

(1) Node a broadcasts RREQ to start the routing discovery process.
(2) Node d replies RREP to claim that it has the routing path to the destination node h. The routing path (a-b-c-d-e-f-g-h) is piggybacked in the RREP.
(3) Node d is required to send TREQ to its next-hop neighbor node e along the routing path towards the destination node h. This process is monitored by d's LMT node, i.e. node 1.
(4) Node 1 is required to initialize a timer as soon as it hears the TREQ sent by the node d. If node e receives the TREQ, it replies the TREP back to its former-hop neighbor node d. Depending on whether LMT node 1 can hear this TREP reaching the node d or not, LMT node can judge whether there is blackhole attack in the path or not.

    (4.1) If LMT node 1 hears the reception of the TREP by node d, it stops the timer. It means the link between node d and e exists, and the process of link checking as mentioned above is continued by other intermediate nodes and their corresponding LMT nodes until the TREQ reaches the destination node h.

    (4.2) Otherwise, if LMT node 1 cannot hear the TREP reaching node d, timeout occurs, which means the link between d and e does not exist. It also means that the RREP sent by node d is faked. So we can say that there is blackhole attack in the path from d to h.

(5) If blackhole attack is detected, LMT node sends ALARM message via the inverse routing path to source node piggybacking the routing path where the blackhole attack exists. In Fig. 6, LMT node 1 finds there is blackhole attack in the path (d-e-f-g-h), then node 1 sends ALARM message via the inverse routing path (d-c-b-a) piggybacking the routing path to inform source node a the existence of the blackhole attack. There is a potential issue that if d is the attacker in Fig. 6, it may drop the ALARM message sent by d's LMT node 1. In this case, the source node a cannot receive the blackhole attack alarm from LMT node 1. To solve this potential issue, we specify that the inverse routing path is one of the candidate routes for LMT node (e.g., LMT node 1) to report the existence of blackhole attack. In other word, the LMT node can eventually report the ALARM information to the source. It is an assumption which can be easily achieved when the proportion of malicious nodes is relatively low. In our future work, we will try to weaken this assumption, i.e., consider the solution when the network is occupied by numerous blackhole nodes. We

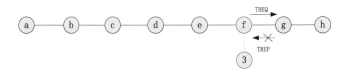

Fig. 7 A scenario of discovery of blackhole attack

specify that after receiving the ALARM message, the source node should reply ALARM REPLY to the corresponding LMT node that sent the ALARM message. After a period since sending the ALARM message, if the LMT node does not receive the ALARM REPLY from the source node, timeout for the ALARM message occurs. Then, the LMT node 1 initiates a routing request process to find another routing path. In one word, if there is blackhole attack in the routing path, the source node can be informed the existence of blackhole attack.

In our scheme, LMT nodes are required to monitor intermediate nodes' behaviors of including both packet receiving behaviors and packet sending behaviors. As shown in Fig. 7, if timeout occurs in LMT node 3, it means that after sending the TREQ to node g, node f does not receive the expected TREP from g. Consequently, LMT node 3 identifies that the link between f and g does not exist, which means there is blackhole attack in this routing path (d-e-f-g-h). Till now, the blackhole attack in the routing path can be detected by LMT node. In next Sect. 3.3, we will introduce how to identify the blackhole nodes, namely location of the blackhole nodes.

3.3 Location Mechanism of Blackhole Attack

As mentioned in the above scenario in Fig. 7, we can figure that the routing path piggybacked in RREP sent by node d is fabricated. However, we cannot simply treat the node d as the blackhole node, as specified in DSR routing protocol, each node's routing cache can be impacted by other nodes [10]. In other word, node d may get the fabricated routing path from other intermediate node. Thus, we present a location mechanism to identify the blackhole nodes.

Now let's see how the proposed location mechanism works. The location mechanism is triggered if the blackhole attack is detected by the aforementioned detection mechanism. By the aforementioned detection mechanism, we can find the break link, i.e., the link does exist between two intermediate nodes, e.g., the link between f and g in the scenario as shown in Fig. 7. The process of location mechanism is monitored by the corresponding LMT nodes. The key idea of the proposed location mechanism is that LMT node inquires the intermediate node that whether it has the path to the destination or not. If the link between the intermediate node and its next-hop node does not exist, but the intermediate node

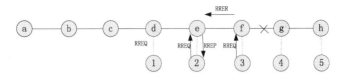

Fig. 8 The process of location mechanism

answers to the LMT node that it has the path to the destination, then we can judge that this intermediate node is the blackhole node.

By the above detection mechanism, the path with blackhole nodes can be found out. We use the same scenario as the example to describe the location mechanism shown in Fig. 8. Through the aforementioned detection mechanism, it is found that the link between node f and g does not exist. The detailed process of the proposed location mechanism is described as follows.

In Fig. 8, LMT node 3 sends RREQ to the node f piggybacking the routing path. The purpose of this RREQ is to inquire node f about whether it can reach destination node f via the piggybacked routing path, which is d-e-f-g-h in the scenario in Fig. 8. Node f may have two possible answers to reply: one is to reply RREP back to LMT node 3; another is to broadcast RERR. The RREQ message sent by LMT node represents the meaning "Can you connect with the destination using the routing path?" The RREP message sent by intermediate node represents the meaning "Yes, I can reach the destination via the given routing path." On the other hand, the RERR broadcasted by intermediate node represents the meaning that "The path is fabricated". In the scenario in Fig. 8, if node f replies RREP, it indicates that node f is the blackhole node. If node f broadcasts RERR, it represents that node f is a normal node.

In Fig. 8, we assume that node f sends RERR along the inversed routing path reporting the fake path, so node f is labeled as a normal node according to our mechanism. When the LMT node of the previous former-hop intermediate node observes the reception of the RERR, the LMT node will perform the same aforementioned check process. As the case in Fig. 8, after hearing the arrival of RERR at node e, LMT node 2 sends the RREQ to node e. We assume that node e replies RREP to LMT node 2 instead of the RERR. Consequently, LMT node 2 can identify that node e is the blackhole node in the scenario.

4 Simulation

We compare the performance of the proposed scheme with that of DSR, BDSR and CBDS using Network Simulator version 2 (NS-2) [11]. The 802.11 MAC layer implemented in NS-2 is used for simulation. The parameters for simulation are presented in Table 2. The malicious nodes are randomly selected to perform blackhole attack, whose proportion ranges over (0, 40 %).

Table 2 Parameters for simulation

Parameters	Value
Simulation time	1200 s
Simulation area	2000 by 2000 m
Number of nodes	100
Speed	Random (0–20 m/sec)
Packet size	256 bytes
Traffic type	CBR
Mobility model	RWM
Radio range	150 m

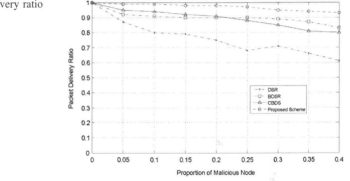

Fig. 9 Packet delivery ratio versus traffic load

Figure 9 shows that the performance of our scheme is much better than that of other schemes. Under the situation with blackhole attacks, using DSR protocol, the delivery ratio decreases significantly to around 60 % as the percentage of blackhole nodes increases. This is because the blackhole attacks cause hotspots at blackhole end points due to consumption of local channel bandwidth for replays, i.e., blackhole attacker attracts around 40 % of the data traffic and discards those data packets. Since DSR was not designed as a robust routing protocol against blackhole attack, its performance is worst among all the protocols. As to BDSR and CBDS, when the proportion of malicious nodes is less than 25 %, the performance of CBDS is better than that of BDSR. The reason is that in CBDS one assistant node is employed to cooperate with the source node to detect the blackhole nodes, which can mitigate the deceiving from the malicious node. On the other hand, when the proportion of malicious nodes is more than 25 %, BDSR works better than CBDS. Since the assistant node is randomly selected, the chance for employing a malicious node as the assistant node is relatively high when the proportion of malicious nodes is large. In this case, the source node may be deceived by the malicious node and cannot identify the blackhole node. The common weakness of BDSR and CBDS is that they cannot prevent the blackhole attack with collusive nodes that work in collusion. Conversely, in our scheme, due to the cooperation among LMT nodes, the collusive blackhole nodes can still be detected and identified. All in all, our scheme works much better than DSR, BDSR and CBDS.

5 Conclusion and Future Work

In this paper, we proposed a trust-based scheme to prevent blackhole attacks in wireless ad hoc networks. The performance simulation analysis confirms the availability and efficiency of our scheme.

We plan to extend our scheme in the near future as follows.

(1) We plan to consider more parameters into the calculation of each node's trust value, like cumulative time during which the node acts as a LMT node. Cumulative time implies node stability that increases the stability LMT node. The purpose of including more parameters is not only to reduce the frequency of re-election for LMT node, but to balance the power consumption of each node, especial LMT node. We plan to measure more parameters and pick more important and rightful ones in order to make the balance between the rightness of LMT node and efficiency of trust value's calculation and exchange.

(2) As mentioned before, the proposed scheme is based on the DSR protocol. In the expected extended version of our work, we plan to develop the scheme to make it suitable for other well-known routing protocols, such as DSDV, AODV.

(3) As we assumed that the packet loss is caused by the packet dropping behavior of malicious node. In the near future, we plan to extend the scheme by considering other causes that can incur the packet loss. The improved version of the proposed scheme is expected to involve MAC protocols for the competition for channel access and prevention against transmission collision.

(4) We only simulated our scheme regarding to proportion of malicious node. In our future work, we plan to simulate the performance of the proposed scheme in the scenario where nodes are moving around with certain velocity and direction. We also plan to simulate the performance considering traffic load and other parameter.

(5) In this paper, we only compare the proposed scheme with the DSR and other two DSR-based protocols, BDSR and CBDS. In the near future, we will to extend the simulation by comparing the proposed scheme with some other latest related researches [12–15].

References

1. Jurdak, R., Lopes, C.V., Baldi, P.: A survey, classification and comparative analysis of medium access control protocols for ad hoc networks. Commun. Surv. Tutor. **6**(1), 2–16 (2009)
2. Bala, A., Bansal, M., Singh, J.: Performance analysis of MANET under blackhole attack. Netw. Commun. 141–145 (2009)
3. Kurosawa, S., et al.: Detecting blackhole attack on AODV-based mobile ad hoc networks by dynamic learning method. Int. J. Netw. Secur. **5**(3), 338–346 (2007)
4. Mistry, N.H., Jinwala, D.C., Zaveri, M.A.: MOSAODV: solution to secure AODV against blackhole attack. Int. J. Comput. Netw. Secur. **1**(3), 42–45 (2009)
5. Tsou, P.-C., Chang, J.-M., Lin, Y.-H., Chao, H.-C., Chen, J.-L.: Developing a BDSR scheme to avoid black hole attack based on proactive and reactive architecture in MANETs. In: 13th International Conference on Advanced Communication Technology, Gangwon-Do, pp. 755–760 (2011)
6. Chang, J.-M., Tsou, P.-C., Chao, H.-C., Chen, J.-L.: CBDS: a cooperative bait detection scheme to prevent malicious node for MANET based on hybrid defense architecture. In: 2nd International Conference on Wireless Communication, Vehicular Technology, Information Theory and Aerospace and Electronics Systems Technology, Chennai, pp. 1–5 (2011)
7. Faragó, A.: Scalable analysis and design of ad hoc networks via random graph theory. In: Workshop on Discrete Algorithms and Methods for MOBILE Computing and Communications, New York, pp. 43–50 (2002)
8. Tsumoto, S., Hirano, S: Visualization of rule's similarity using multidimensional scaling. In: 3rd IEEE International Conference on Data Mining, Florida, pp. 339–346 (2003)
9. Rodriguez-Mayol, A., Gozalvez, J.: On the implementation feasibility of reputation techniques for cooperative mobile ad-hoc networks. In: European Wireless Conference, Tuscany, pp. 616–623 (2010)
10. Johnson, D.B., Maltz, D.A., Broch, J.: DSR: the dynamic source routing protocol for multi-hop wireless ad hoc networks. In: Perkins C.E. (ed.) Ad Hoc Networking, Chapter 5. Addison-Wesley, Boston, pp. 139–172 (2001)
11. Fall, K., Varadhan, K.: The ns manual. http://www.isi.edu/nsnam/ns/doc/index.html
12. Zhou, X., Ge, Y., Chen, X., Jing, Y., Sun, W.: A distributed cache based reliable service execution and recovery approach in MANETs. JoC **3**(1), 5–12 (2011)
13. Shbat, M.S., Tuzlukov, V.: Dynamic frequency reuse factor choosing method for self organizing LTE networks. JoC **2**(2), 13–18 (2011)
14. Tseng, F.-H., Chou, L.-D., Chao, H.-C.: A survey of black hole attacks in wireless mobile ad hoc networks. HCIS **1**(4), 1–4 (2011). doi:10.1186/2192-1962-1-4
15. Hiyama, M., Kulla, E., Oda, T., Ikeda, M., Barolli, L.: Application of a MANET testbed for horizontal and vertical scenarios: performance evaluation using delay and jitter metrics. HCIS **1**(3), 1–3 (2011). doi:10.1186/2192-1962-1-3

Fundamental Tradeoffs for Ubiquitous Wireless Service: A QoE, Energy and Spectral Perspective

Yueying Zhang, Fei Liu, Yuexing Peng, Hang Long and Wenbo Wang

Abstract Recent advances in computing and communication technologies enable the popularity of versatile mobile devices, which accounts for a substantial amount of energy use. The energy consumption has become of unprecedented importance, particularly in a cloud computing systems which provide ubiquitous services. Conventional designs of wireless networks mainly focus on spectral efficiency (SE) enhancing. Given the variety of media services in the cloud environment, a green computing network, which meets the quality of experience (QoE) requirements for users and also improves energy efficiency (EE), is the most appropriate solution. In this paper, we first propose the unit QoE per Watt, which is termed QoE efficiency (QEE), as a user-oriented metric to evaluate EE. Then, we investigate the fundamental tradeoffs between QEE and SE for typical ubiquitous services. Our analytical results are helpful for network design and optimization to access a flexible and desirable tradeoff between the QoE and energy conservation.

Keywords Energy efficiency (EE) · Spectral efficiency (SE) · Quality of experience (QoE) · Ubiquitous services, tradeoff

1 Introduction

The escalation of power consumption in ubiquitous computing networks results in great increase of greenhouse gas emission, which is becoming a major threat for sustainable development. European Union has acted as a leading flagship in energy

Y. Zhang (✉) · F. Liu · Y. Peng · H. Long · W. Wang
Key Laboratory of Universal Wireless Communications, Ministry of Education, Beijing University of Posts and Telecommunications, Beijing, China
e-mail: zhangyueying@bupt.edu.cn

S.-S. Yeo et al. (eds.), *Computer Science and its Applications*,
Lecture Notes in Electrical Engineering 203, DOI: 10.1007/978-94-007-5699-1_97,
© Springer Science+Business Media Dordrecht 2012

saving over the world and targeted to have a 20 % greenhouse gas reduction [1]. With increasingly high datarates supported by cloud computing networks, the interactions between ubiquitous services and users take an important position [2].

For these reasons, the concept of quality of experience (QoE) has the potential to become the guiding paradigm for managing quality in the Cloud [3]. QoE combines user perception, experience and expectations with non-technical and technical parameters, such as network-level quality of service (QoS). It is likely to be the main factor of multimedia service quality metrics for customers [3] and has attracted many interest from academia and industrial areas [4, 5]. How to maintain excellent QoE perceived by users and reduce energy consumption from the perspective of green computing network design makes more sense.

Energy consumption issues draw attentions in green computing either to limit battery energy [6], or to reduce energy utilization in data centers [7]. Quality has become critical to provide personalized services in ubiquitous computing [8]. User-centered interactivity evaluation metrics of ubiquitous service attributes are developed in [2]. Energy efficiency (EE) and spectral efficiency (SE) tradeoff in downlink OFDMA networks are investigated in [9]. More fundamental tradeoffs are analyzed and extended to green wireless networks in [1]. However, the relations among QoE, power and SE for ubiquitous services in ubiquitous computing networks is not clear yet.

In this paper, we first introduce QoE per Watt, termed QEE as a user-oriented metric to evaluate EE in QoE perspective, which is simply defined as the ratio of the QoE perceived by users and the signal power expenditure. The difference between the two metrics lies in that QEE characterizes real acceptability of users with unit power for versatile multimedia services, combining user satisfaction and operator satisfaction (energy consumption) together. SE is a network-level metric emphasizing the network performance to the contrary. We then address the QEE-SE tradeoff issue for ubiquitous services, such as datarate-sensitive services and delay-sensitive services.

The remainder of this paper is organized as follows. In Sect. 2, we describe the system model and analyze the QEE-SE tradeoff. In Sect. 3, we explore QoE mapping functions and investigate fundamental tradeoffs for several typical multimedia services. Finally, we conclude the paper and point out future works in Sect. 4.

2 Qee

SE is defined as the system datarate for unit bandwidth, is a widely accepted criterion for wireless network optimization [10]. However, EE is previously ignored by most of communication standards and research efforts until very recently green computing has become a vital trend.

From Shannon's formula, the up-bound datarate of an AWGN channel is given by

$$R = W \log_2(1 + \frac{P_t h}{WN_0}) \tag{1}$$

where W is the system bandwidth, P_t is the given transmit power, N_0 denotes the power spectral density of AWGN, and h is the channel gain. Hence, transmission power can be formulated as $P_t = (2^{R/W} - 1)WN_0/h$. SE and EE can be expressed by $\eta_{SE} = R/W$, and $\eta_{EE} = R/P_t$, respectively. Determined by (1), SE is given by

$$\eta_{SE} = \log_2(1 + \frac{P_t h}{WN_0}) \tag{2}$$

From [11], circuit energy consumption increases with the bandwidth. The circuit power includes all electronic power consumption except transmit power, expressed by

$$P_c = WP_{cir} + P_{sb} \tag{3}$$

where P_{cir} is the part of dynamic circuit power consumption which is related to bandwidth, and P_{sb} is the average static power in the transmit mode. Obviously, the overall power turns out to be $P = P_t + P_c$. EE can be defined as

$$\eta_{EE} = \frac{W \log_2(1 + P_t h/WN_0)}{P_t + WP_{cir} + P_{sb}} \tag{4}$$

QEE is simply defined as the ratio of the perceptual QoE to the signal power expenditure. Contract to EE, as a user-centric metric, QEE differentiates specific types of services. Moreover, QEE is not only a quality metric, but also a measure of energy efficiency since it decreases with energy consumption increasing. The two contributors indicate the reason why QEE is suitable to be utilized in analyzing ubiquitous services. Furthermore, there are many kinds of resources can be traded to attain higher QoE efficiency, such as bandwidth, delay, power, etc. QEE denoted by η_{QEE} is

$$\eta_{QEE} = \frac{\Delta Q}{P_t + WP_{cir} + P_{sb}} \tag{5}$$

where $\Delta Q = Q - Q_{Min}$, Q denotes current perpetual QoE, and Q_{Min} is the minimum Q, which is relevant to the assessment system.

3 QoE and QEE-SE Tradeoff

To characterize the QEE-EE tradeoff for point-to-point communication in ubiquitous computing systems, QoE performance for typical multimedia services should be addressed. In this section, we will first overview QoE functions for

Fig. 1 QoE functions of data-sensitive services

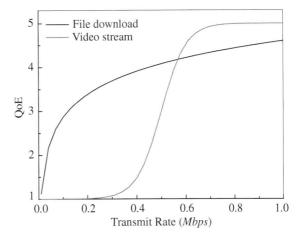

services, such as file download and video streaming. Then the QEE-SE tradeoffs are investigated.

File Download Service.

For file download service, user's satisfaction is most related to the effective transmit rate. Moreover, file download service is considered as one kind of elastic applications, whose QoE curve is concave and monotonically increasing of effective datarate R. As Fig. 1 shows, QoE improves quickly at the start and approaches 5 as the transmit rate approaches infinite. The logarithmic function introduced in [12] is adopted as follows

$$Q_{ftp}(R) = a_1 + b_1 \log(R) \tag{6}$$

Streaming Video Service.

From QoE perspective, streaming video service is sensitive to delay and rate, and is less elastic than data services such as file transfer. When the transmission rate decreases below a certain threshold, a significant drop in the QoE will be resulted in. However, when the transmission rate increases above the certain threshold, the QoE will be greatly improved. As a result, the MOS of streaming video service can be modeled as a sigmoidal-like function of the average transmission rate [13] given by

$$Q_{video}(R) = \frac{a_2}{1 + e^{b_2(R - c_2)}} \tag{7}$$

We observes from Fig. 1, a portion of the QoE curve is convex, representing the fact that, once the effective transmit rate is below a certain value, the user satisfaction drops sharply. As datarate increases, QoE approaches the maximum value.

3.1 QEE-SE Tradeoff

The goal of optimization in computing network is to deliver the applications to the end user at high quality, at best while minimizing the costs of the energy consumptions. Next, we discuss QEE-SE tradeoff to balance QoE and energy consumption.

File Download Service.

We start with file download services whose QoE is the function of R. Plugging (6) into (5), we obtain

$$\eta_{QEE} = \frac{a_1 + b_1 \log(R + c_1) - Q_{\min}}{P_t + WP_{cir} + P_{sb}} \tag{8}$$

where $Q_{\text{Min}} = 1$. According (1), the transmit power can be given by

$$P_t = \frac{(2^{R/W} - 1)WN_0}{h} \tag{9}$$

From (9), we observe that transmit power monotonically decreases with W. However, the operating power cannot be ignored in ubiquitous networks. In this case, we derive the overall power consumption as follows:

$$P = P_t + P_c = \frac{(2^{R/W} - 1)WN_0}{h} + WP_{cir} + P_{sb} \tag{10}$$

For a fixed transmit distance $d = 1000\,m$, Fig. 2 shows the relationship by trading bandwidth and datarate for power. The minimum power consumption exists when both datarate and allocated bandwidth is extremely small. In the scenario with a narrow bandwidth, adequate power provision is necessary to enhance the daterate and improve QoE at the same time. That is why power scales with the datarate when given a fixed bandwidth. However, when quite a wide bandwidth is available, the circuit power dominates the budget of energy costs, and consequently leads to overall power increase. Therefore, when a certain datarate required, i.e., a QoE level for datarate-sensitive service, there is an optimal bandwidth for the minimal transmit power.

According to (10, 8) can be simplified by

$$\eta_{QEE} = \frac{a_1 + b_1 \log(R) - Q_{\text{Min}}}{(2^{R/W} - 1)WN_0/h + WP_{cir} + P_{sb}} \tag{11}$$

According to (1), transmit power can be expressed by

$$P_t = \frac{(2^{\eta_{SE}} - 1)WN_0}{h} \tag{12}$$

As a result, we derive QEE-SE relation using (8) as follows:

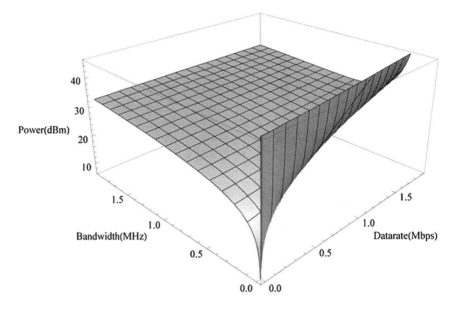

Fig. 2 Power consumption versus bandwidth and datarate

$$\eta_{QEE} = f_1(\eta_{SE}, W) = \frac{a_1 + b_1 \log(W\eta_{SE}) - Q_{\text{Min}}}{(2^{\eta_{SE}} - 1)WN_0/h + WP_{cir} + P_{sb}} \tag{13}$$

It is noted that QEE is not only related to SE, but also related to allocated bandwidth. It is possible to find a way to achieve satisfying QEE by jointly adjusting bandwidth and SE. Figure 3 illustrates a visual example of the 3-dimension relation among SE, bandwidth and QEE. From the figure, we have the following observations.

- Full utilization of bandwidth resource may not be the most energy efficient way under fixed SE in the QoE perspective, since circuit power scales with the transmission bandwidth. When SE is low, bandwidth can be trade with certain gains in terms of QEE. However, when SE is above a certain value, QEE achieves a satisfying level given a quite small bandwidth and decrease afterward with more bandwidth used.
- Given a target QEE, the bandwidth-SE relation is non-monotonic. Because QoE does not monotonically rises by increasing SE for fixed bandwidth.
- In a bandwidth limited system, a proper SE obtains highest QEE. Thus, there are at most two corresponding SE to achieve the same QEE. We will choose the larger one with higher datarate in limited bandwidth. Because a higher perceptual QoE can be obtained despite of more energy costs.

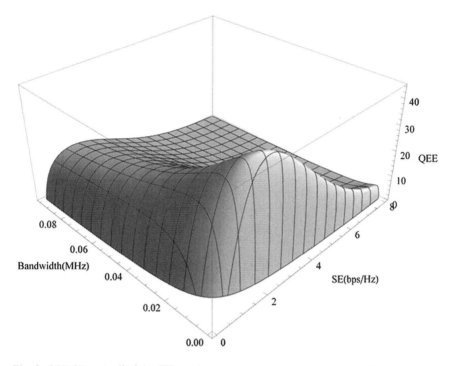

Fig. 3 QEE-SE tradeoff of the FTP service

Streaming Video Service.

In the similar way, the relationship by trading bandwidth and datarate for QEE of the streaming video service is expressed by

$$\eta_{QEE} = \frac{a_2(1 + e^{b_2(R-c_2)})^{(-1)} - Q_{\text{Min}}}{(2^{R/W} - 1)WN_0/h + WP_{cir} + P_{sb}} \qquad (14)$$

The QEE-SE relation for streaming video service is denoted as

$$\eta_{QEE} = f_2(\eta_{SE}, W) = \frac{a_2(1 + e^{b_2(W\eta_{SE}-c_2)})^{(-1)} - Q_{\text{Min}}}{(2^{\eta_{SE}} - 1)WN_0/h + WP_{cir} + P_{sb}} \qquad (15)$$

Figure 4 illustrates the graphical results of (15). It is a belt-like function of bandwidth and SE and has the optimal value with allocated proper resources. In addition, with the same bandwidth, it requires more power consumption to providing higher SE. But it is beneficial to achieve a more satisfying QoE. Thus there is a tradeoff with QEE.

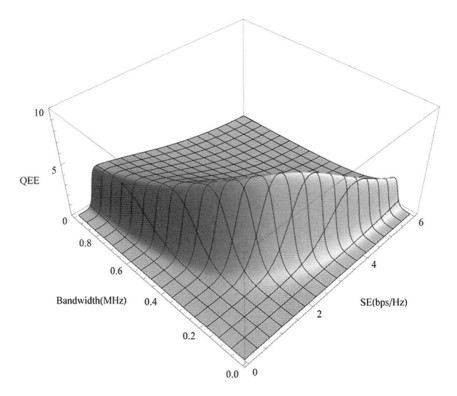

Fig. 4 QEE-SE tradeoff of the video service

4 Conclusion

We proposed QEE, as an energy efficient metric focusing on user satisfaction efficiency in computing networks. As higher QEE can be achieved, the system becomes more energy efficient in that it gains more QoE per unit of consumed power. For green computing networks, we propose that the data rate and bandwidth resources can be traded to save energy consumption under certain perceptual quality demands. In addition, the relationships of QEE and SE are thoroughly studied for some typical ubiquitous service. The analytical results indicate QEE and SE do not always coincide and may even conflict sometimes. However, the optimum QEE can always be achieved for diverse multimedia services as long as resources such as power and bandwidth are appropriately allocated.

Overall, this paper gives a deep insight into fundamental tradeoffs for computing networks in a comprehensive perspective. Correspondingly, many optimization schemes could be designed for achieving a flexible and desirable tradeoff between the performance and energy saving in the next generation networks. However, the analysis of this work addresses the scenario of point to point communications. There are more open issues to be investigated in the view of computing networks, including the tradeoff between the unified QEE and SE, and even the tradeoff between deployment cost and QEE.

Acknowledgments This research was supported by the National Basic Research Program of China (973 Program) under Grant 2012CB316005, National Key Technology R&D Program of China under Grant 2012ZX03004001, Graduate Innovation Fund of HUAXING CHUANGYE & SICE, BUPT, 2011, and the Program for New Century Excellent Talents in University (Grant No. NCET-11-0600).

References

1. Chen, Y., Zhang, S., Xu, S., et al.: Fundamental trade-offs on green wireless networks. IEEE Commun. Mag. **49**(6), 30–37 (2011)
2. Lee, J., Song, J., Kim, H., et al.: A user-centered approach for ubiquitous service evaluation: an evaluation metrics focused on human-system interaction capability. In: 8th Asia-Pacific Conference on Computer-Human Interaction, pp. 21–29. IEEE Press, Seoul (2008)
3. Brooks, P., Hestnes, B.: User measures of quality of experience: why being objective and quantitative is important. IEEE Netw. Mag. **24**(2), 8–13 (2010)
4. Bhattacharya, A., Wu, W., Yang, Z.: Quality of experience evaluation of voice communication: an affect-based approach. Human-centric Comput. Inf. Sci. **2**(7), 1–18 (2012)
5. Viswanathanand, V., Krishnamurthi, I.: Finding relevant semantic association paths through user-specific intermediate entities. Human-centric Comput. Inf. Sci. **2**(9), 2–11 (2012)
6. Liang, W.Y., Lai, P.T., Chiou, C.W.: An energy conservation DVFS algorithm for the android operating system. J. Convergence **1**(1), 93–100 (2010)
7. Beloglazov, A., Abawajy, J., Buyya, R.: An energy-aware resource allocation heuristics for efficient management of data centers for cloud computing. Future Generation Comp. Syst. **32**(28), 755–768 (2012)
8. Vakili, A., Grégoire, J.C.: Modelling the impact of the position of frame loss on transmitted video quality. J. Convergence **2**(2), 43–48 (2011)
9. Xiong, C., Li, G.Y., Zhang, S.: Energy- and spectral-efficiency tradeoff in downlink OFDMA networks. IEEE Trans. Wireless Commun. **10**(11), 3874–3886 (2011)
10. Wu, J., Mehta, N.B., Molisch, A.F., et al.: Unified spectral efficiency analysis of cellular systems with channel-aware schedulers. IEEE Trans. Wireless Commun. **59**(12), 3463–3474 (2011)
11. Cheng, W., Zhang, X., Zhang, H.: On-demand based wireless resources trading for green communications. In: INFOCOM WKSHPS, pp. 283–288. IEEE Press, Shanghai (2011)
12. Thakolsri, S., Khan, S., Steinbach, E., et al.: QOE-driven cross-layer optimization for high speed downlink packet access. J. Commun. **4**(9), 669–680 (2009)
13. Lee, J.W., Mazumdar, R.R. et al.: Non-convex optimization and rate control for multi-class services in the internet. IEEE Trans. Netw. **13**(4), 827–840 (2005)

Multi-Policy Collaborative Access Control Model for Composite Services

Bo Yu, Lin Yang, Yongjun Wang, Bofeng Zhang, Linru Ma and Yuan Cao

Abstract Service composition has become the main style of cross-domain business collaboration environments, and security issues prohibit the widespread use of composite services. Based on attribute, this paper presents a multiple policies collaborative access control model which combines the attribute policies of composite service, component services and user domain. This model can provide fine-grained access control for service composition and support collaborative authorization based on business attributes in business collaboration environments while keeping the standalone of component service access control. The analysis result shows that this model not only satisfies the access control requirements of business process in composite service, but also provides fine-grained access control for component services.

Keywords Composite services · Access control · Multi-policy · Collaborative authorization

B. Yu (✉) · Y. Wang · B. Zhang · Y. Cao
School of Computer, National University of Defense Technology, Changsha, China
e-mail: yubo0615@sina.com

Y. Wang
e-mail: wwyyjj1971@163.com

B. Zhang
e-mail: bfzhang@nudt.edu.cn

Y. Cao
e-mail: dolf.cao@gmail.com

B. Yu · L. Yang · L. Ma
Center for Security, Institute of EESEC, Beijing, China
e-mail: yanglin61s@yahoo.com.cn

L. Ma
e-mail: malinru@126.com

S.-S. Yeo et al. (eds.), *Computer Science and its Applications*,
Lecture Notes in Electrical Engineering 203, DOI: 10.1007/978-94-007-5699-1_98,
© Springer Science+Business Media Dordrecht 2012

1 Introduction

With the popularity of Service Oriented Architecture (SOA) and Software as a Service (SaaS), software applications are developed by composing now existing services in open environment. Service composition, which is one of the keystones of SOA, considers function of information systems as composition of loosely coupled, modularized services [1]. However, service composition design not only needs to describe the control flow and data flow of composite services, but also needs to provide the security, reliability and scalability of composite services in order to satisfy confidentiality, integrity, privacy, controllability, and availability requirements of service applications [2]. How to design security mechanisms to ensure the security and reliability of composite services becomes the key of wide use of composite services. Recent years, development and design of new security mechanisms for composite services to support the security of cross-domain business collaboration is widely concerned, among which, access control is one important dimension of securing composite services.

Access control system of composite services is constructed by linking different security components in different domains regarding security functions [3]. It is a challenging task to design an access control system which considers the comprehensive requirements of security components in cross-domain collaboration environments. The access control of composite services is different from the traditional access control model such as RBAC [4]; the former is not longer a simple Subject-Action-Resource style [5]. New access control requirements in collaborative systems are identified by Tolone et al. [6]. The new features, such as distributed collaborative access control, collaborative authorization, scalability of access control, and dynamic access control, must be taken into consideration when designing new access control model in collaborative environments. Access control model of composite services considers not only the requirements of service consumer and service provider, the composite service which acts as an intermediary of component services, is needed to be considered into the access control. Access control model becomes more complex in collaboration environment. Currently there is no comprehensive access control model for composite service business application [7].

Menzel et al. [8] discusses the classification of security frameworks regarding secure multi-domain service composition, and presents a two-layer security architecture to identify security requirements of access control in service domain. The model ACM4WSC proposed by Zhu et al. [9] discusses whether a user is allowed to access a composite service by collecting and analysis the security policies of component services. Huang [10] proposes a three-level access control model for business process. This model offers a policy layer to facilitate managing and enforcing security policies of component services, also the policy inference and negotiation are achieved in this layer. However, the previous approaches don't consider the access control requirements between user and composite services

Moreover, the centralized policy management [9, 10] may result in information leakage of component services.

Because of the wide use and strong ability of RBAC, quite a few approaches focus on the extension of RBAC in composite services [11–14]. By extending traditional RBAC with role-mapping, and role collaboration [15], RBAC can provide cross-domain access control for composite services. Paci et al. [16] have defined the related concepts of RBAC in composite services, such as RBAC-WS-BPEL, and so on. The methods of RBAC extension include context-based and attribute-based extension [17]. A XACML-based collaboration access control framework is proposed by Kim et al. [18]. In this framework, Role Enablement Authority (REA) is introduced to assign role attributes to users or enable role attributes delegation during a user's session, also collaborative separation of duty is support in this framework. A RBAC-based workflow authorization model (WAM) is proposed by Atluri et al. [19], and an authorization flow is associated with each task in WAM model which is used to keep pace with workflow. Liu et al. [20] construct a composite service access control model named WS-RBAC by extending classic RBAC. However, the assumption of RBAC adopted for all component services domain in open environment is too far-fetched; also the role-mapping among service domains is a difficult task.

Quite a few approaches focus on security enhancements of BPEL. Business Process Execution Language (BPEL) combines the advantages of XLANG and WSFL, and can provide complex application by composing services into a powerful business process. However, BPEL does not involve access control mechanism, cannot provide rights management to protect critical resources, which largely affects the reliability and security of BPEL-based business processes [16]. Many studies, including AO4BPEL [21, 22], Semantic-based BPEL [17], try to provide enhanced security mechanisms for the BPEL, and associate security mechanisms into the activities in BPEL for business process access control. Bertino et al. [23] extend the BPEL with authorization constraint and propose the RBAC-based business process access control RBAC-WS-BPEL. Nevertheless, the new features of access control in composite services are not highlighted in these approaches.

In addition, combination of BPEL and RBAC for access control is an optional solution. Hummer et al. [14] propose a RBAC-based BPEL access control mechanism by adding security annotation for BPEL activates to specify the access control policies for BPEL. The different between this mechanism and our approach is that our approach support the access control of business process and provide attribute-based multiple policies authorization decision.

In summary, centralized access control model deprives the access control of autonomous domain and cannot provide fine-grained access control for component services. Traditional decentralized access control model cannot meet the access control requirements of composite services and cannot support the collaborative authorization of composite services, and are not suitable for composite service access control in cross-domain business collaborative environment. Therefore, a comprehensive access control model which considers all security requirements of all entities in composite service is need. The goal of our work is to address such limitations by proposing distributed access control model for composite services in collaborative environments. We propose an attribute-based multi-policy authorization model for composite services and present a two-layer access control architecture based on the idea of security services.

The paper is structured as follows: In Sect. 2 we introduce multi-policy collaborative authorization model, MPCAM for short. Section 3 presents an access control architecture based on MPCAM, and experimental results and graphs are provided in this section. Some conclusions and future works are reported in Sect. 4.

2 Multi-Policy Collaborative Authorization Model

2.1 Constructing the MPCAM

Traditional access control model considers only the subject attribute and object attribute, which is unable to support for business collaborative access control of composite service, is not suitable for expressing the policy constraint and collaboration between composite services and component services. Toward the heterogeneous access control environment of component service, we present a multi-policy collaborative authorization model (MPCAM) in this Section. By attribute assignment and attribute-based authorization policies, a unified cross-domain access control mechanism for component services is proposed. As shown in Fig. 1, business execution unit (BEU) and business authorization unit (BAU) are the core components of business layer access control, and subtask execution unit (SEU) and subtask authorization unit (SAU) are the core components of component service layer access control.

The main idea of MPCAM is that all security aspects of component services and composite services are expressed by well-designed attributes, and access control policies are defined based on these attributes. The business layer and the component service layer are two layers in MPCAM, the business layer is responsible for authorization of composite services, and the component service layer is responsible for authorization of component services. When a user requests a composite service in a virtual domain, the business layer will authorized the user based on subject attributes. If the authorization result is positive, the business layer will assign the business attribute to the composite service requested by this user. The subject attribute, together with the business attribute, becomes the basic of component service authorization for cross-domain users. When a user requests a component services, inter-domain trust attribute, together with subject attribute and business attribute, are being considered for access control of component services. The collaboration between composite service and component services is achieved by transmission of business attribute and the collaboration among

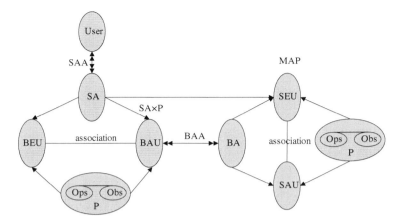

Fig. 1 The structure of MPCAM

component services is achieved by transmission of trust attribute. The authorization policies based on these attributes forms the multi-policy collaboration of overall access control of composite services.

The formal definition of MPCAM is show as follows. The Definition 1 describes the basic components of MPCAM, including component services, composite services and so on.

Definition 1 The basic components of MPCAM

- Service Domain SD is a set of domains $\{d_1, d_2, \ldots\}$, where each domain d_i includes a set of services $\{s_1, s_2, \ldots\}$ and user set U_i.
- Virtual Domain VD is a set of domains $\{d_1, d_2, \ldots, d_n\}$, where each domain d_i stands for a component service domain, and n equals to the total number of domain in VD.
- The user of VD is defined as $U = \cup_{i=1}^{n} U_i$ where U_i is the user of domain d_i. The user of composite service is known as business user who is composed by multiple users or owns multiple credentials of different domains. And the composite service is invoked by the business users.
- The component service set is defined as $CNS = \{cns_i\}$ where cns_i is a component service in domain d_i, and provides shared resources or functions.
- The composite service set is defined as $CSS = \{css_j\}$, where css_j is a composite service and for each css_j, $css_j = \{cns_i\}$, where cns_i is a component service of domain d_i.
- For each user $u \in U$, there exists at least one subject attribute certificate (SAC) issued by the AA, and each SAC includes an assertion of some attributes of user u.
- Trust Attribute (TA) describes trust relationships of domains. For each component service domain in VD, trust relationships between domains are needed to be defined. In our model, the trust attribute is adopted to describe trust relationships and the trust value can be acquired from a public trust authority.

- Business Attribute (BA) describes basic attributes of business service in the form of Business Attribute Certificate (BAC). And each BAC is issued by the AA in VD.
- Resource Attribute (RA) describes basic attributes of service or resource in the form of Resource Attribute Certificate (RAC). And each RAC is issued by the attribute authority of a service domain.
- Object set is defined as $Obs = \{ob_1, ob_2, ...\}$, where each ob_i is a composite service css or component service cns, and each ob_i is either data resource or function.
- Operation set is defined as $Ops = \{read, write, access\}$, where $read$ and $write$ are defined for resource access, and $access$ is defined for service access.
- Permission set is defined as $P = \{p_1, p_2, ..., p_k\} = 2^{Ops \times Obs}$.
- Subject Attribute Assignment (SAA) defines the attribute assignment relationship of users, SAA is a many to many relationship. User requires a SAC by SAA authority.
- Business Attribute Assignment (BAA) defines the attribute assignment relationship of business, BAA is a many to many relationship.
- Trust Attribute Assignment (TAA) defines the attribute assignment relationship of users or component services, TAA is a many to many relationship. Each user or component service requires a TAC by TAA authority.
- Multi-Attribute Permission MAP is defined as $MAP = 2^{UA} \times 2^{BA} \times 2^{TA} \times 2^{RA} \times P$. MAP describes multiple attributes assignment relationship of permission. For user who has subject attribute sa access the component service cns_i with resource attribute ra under the composite service css with business attribute ba, a MAP defines the authorization policies of component service access.

According to the attributes and the attribute permission assignment defined in Definition 1, a multiple attributes permission mapping relationship is provided for MPCAM.

2.2 Authorization Process of MPCAM

According to MPCAM, we design the timing diagram of authorization process in Fig. 2. There are four important phases in authorization process of composite services. We discuss these phases in detail as follows.

- Subject Attribute Assignment.
  The first process is the subject attribute assignment of system administrator, subject attribute is expressed as triple $<U, C, SA>$, where U is the business user of this service execution process, and C is credential set of U, and SA is the attribute set of U. $A(U) = \{a(u)|u \in U, a(u) = d_i.a(u)\}$, where d_i is the domain of each u, and $d_i.a(u)$ is the subject attribute assigned by domain d_i.

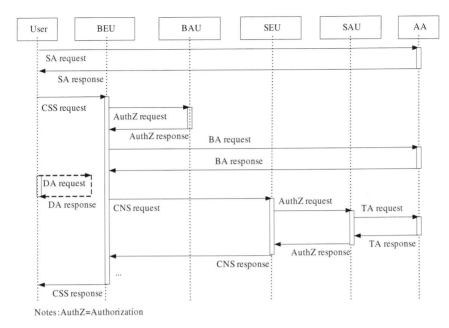

Notes: AuthZ=Authorization

Fig. 2 Timing diagram of authorization process of MPCAM

- Authorization Decision of BAU.
  In this process, the BAU will check the subject attribute, including user credentials. For business user U, composite service css, a subject attribute decision (SAD) process is defined to check the subject attributes sa of user U.
- Business Attribute Assignment.
  The business attribute is assigned after the authorization process is finished, and the business attribute is expresses as tuple $<U, CSS, BA>$, where U is authorized user set of composite service CSS, and BA stands for business attributes. The tuple $<U, CSS, BA>$ defines the relationship between the business user and the attributes. It is noteworthy that the strength of BA will impact the authorization decision of SAU.
- Authorization Decision of SAU.
  Several attribute decisions are needed in the authorization decision process of SAU, including subject attribute decision (SAD), trust attribute decision (TAD) and business attribute decision (BAD). And each attribute decision process includes attributes decision based on authorization policies. The comprehensive consideration of three attributes is the key of component service access control.

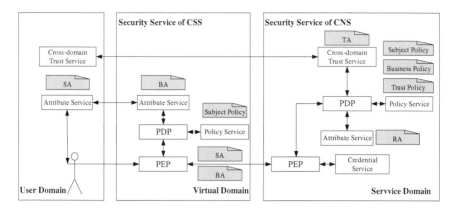

Fig. 3 A two-layer access control architecture

3 System Implementation and Evaluation

In this Section we describe the prototype implementation of MPCAM of composite services. This section is divided into two parts: firstly, we outline the architecture of two-layer access control architecture of MPCAM system. We design this architecture based on idea of security services which are widely used in SOA. In the second we compare the features of our approaches with existing classic approaches.

3.1 Access Control Architecture

According to the security requirements of users, composite services and component services, we present a two-layer access control architecture based on MPCAM. This architecture is based on the mainstream view of security service and can be divided into two layers: business layer and component service layer. The business layer is responsible for authorization of composite services, while the component service layer is responsible for cross-domain access control of component services. There are three kinds of security policies: subject policy, trust policy and business policy. The subject policy defines the policies on subject attributes, the trust policy defines the policies on trust attributes of users and the business policy defines the policies based on business attributes of composite services. The collaboration of these policies becomes the integrated access control policies of component services.

The two-layer access control architecture based on MPCAM is shown in Fig. 3, the security services involved in each domain include attribute service, policy service, inter-domain trust service and credential management service. The attribute service provides the basic subject attributes, business attributes, and resource

Table 1 Comparison of access control approaches

Approaches	Centralized or Distributed	Authorization granularity	Collaborative authorization	Scalability	Dynamic nature
MPCAM	Distributed model	Two level fine-grained authorization policy based on attributes	Supported by business attribute	High, security service enabled	High
Menzel's [8]	Distributed model	Fine-grained authorization policy based on attributes	No	Medium	High
Zhu's [9]	Centralized model	Coarse-grained	No	Low, depended on centralized controller	Low
Srivatsa's [11]	Centralized model	Fine-grained security policy based on LTL	No	Low, depended on centralized controller	Low
Hummer's [14]	Distributed model	Coarse-grained, only RBAC supported	No	High	High
Paci's [16]	Centralized model	Coarse-grained, only RBAC supported	No	Low, depended on centralized controller	Low
Han's [17]	Distributed model	Fine-grained authorization policy based on attributes	Supported by solid contract or complex negotiation	Medium	High

Note: *LTL* Linear Temporal Logic

attributes and so on. The policy service is responsible for policy administration and policy retrieval. The policy service and the PDP serve as an authorization authority of service domain. The inter-domain trust service is responsible for trust negotiation and trust provision. And the credential service is responsible for credential management and delegation management. These four security services provide the basic security functions for service providers in each domain. The PEP of the virtual domain needs to make an authorization decision request to the PDP. The PDP of the virtual domain request subject attributes and perform the policy decision according to the policies retrieved from the policy service. However in the component service domain, the PDP needs to retrieve more attributes according the defined policies in its policy service. For the authorization decision in component service domain, the subject attribute, business attribute and trust attribute are needed to be retrieved, and the PDP will check these attributes according to the defined policies.

3.2 Comparison of Comprehensive Access Control Approaches

This Subsection mainly discusses the comparison between our model and now existing approaches. Table 1 summarizes the features of these access control approaches. Based on their authorization model, these approaches can be further categorized into two models: a centralized model and a distributed model [5]. And the goal of an access control model in collaborative environment is to provide a fine-grained, scalable, dynamic authorization mechanism. Therefore, we compare these approaches based on the definitions: authorization granularity, collaborative authorization, scalability, and dynamic nature. We can see that the centralized models [9, 11, 16] shown in Table 3 have disadvantages in scalability and dynamic nature. Comparing with approaches [8, 14, 17], our model provides fine-grained based on subject attributes and service attributes, also enables flexible collaborative authorization via business attribute. Another advantage of our model is security service-based access control to provide the scalable and custom security function.

4 Conclusion

In this paper we have proposed a novel attribute-based multi-policy authorization model for composite service applications. And a two-layer access control architecture is presented to secure composite service application based on the mainstream view of security services, also collaborative authorization is supported by business attributes. And the system architecture and implementation details of MPCAM are discussed, the comparison results show that our model is flexible and can provide fine-grained collaborative access control for composite services in collaborative environment. Further, we will consider more security functions, such as privilege delegation, to be supported in our model.

Acknowledgments This work is sponsored by the National High Technology Research and Development Program of China (863 programs) under the Grant No. 2009aa01z426, partially supported by The National Natural Science Foundation of China, under the Grant No. 60873215, the Program for Changjiang Scholars and Innovative Research Team in University (NO.IRT1012), Aid Program for Science and Technology Innovative Research Team in Higher Educational Institutions for Human Province "network technology", and Human Provincial Natural Science Foundation of China, under the Grant No. 11jj7003 and No. S2010J5050.

References

1. Fethi, R., Yu, H., Feras, D., Wu, S.: A service-oriented architecture for financial business processes: a case study in trading strategy simulation. Inf. Syst. E-Busi. **5**, 185–200 (2007)
2. Carminati, B., Ferrari, E., Huang P.C.: Security conscious web service composition. In: IEEE International Conference on Web Service 2006, pp. 489–496. IEEE press, Chicago (2006)
3. Milanovic, N., Malek, M.: Current solutions for web service composition. IEEE Internet Comput. **8**, 51–59 (2004)
4. Wonohoesodo, R., Tari Z.: A role based access control for web services. In: IEEE International Conference on Services Computing 2004, pp. 49–56. IEEE press, Shanghai (2004)
5. Jie, W., Arshad, J., Sinnott, R., Townend, P., Lei, Z.: A review of grid authentication and authorization technologies and support for federated access control. ACM Comput. Surv. **43**, 12–37 (2011)
6. Tolone, W., Ahn, G.J., Pai, T.: Access control in collaboration systems. ACM Comput. Surv. **32**, 29–42 (2005)
7. Tiwari, S., Singh, P.: Survey of potential attacks on web services and web service compositions. In: 3rd International Conference on Electronics Computer Technology, pp. 47–51. IEEE Press, Kanyakumari (2011)
8. Menzel, M., Wolter, C., Meinel, C.: Access control for cross-organizational web service composition. J. Inf. Assur. Secur. **2**, 155–160 (2007)
9. Zhu, J., Zhou, Y., Tong W.: Access control on the composition of web services. In: International Conference on Next Generation Web Services Practices 2006, pp. 89–93. IEEE press, Souel (2006)
10. Huang, P.C.: Semantic policy-based security framework for business processes. In: Semantic Web and Policy Workshop 2005 , pp. 142–1147. IEEE Press, Galway (2005)
11. Srivatsa, M., Iyengar, A., Mikalsen, T., Rouvellou, I., Yin J.: An access control system for web service compositions. In: International Conference on Web Services 2007, pp. 1–8. IEEE Press, Salt Lake City (2007)
12. Nasirifard, P.: Context-aware access control for collaborative working environments based on semantic social networks. In: 6th International and Interdisciplinary Conference on Modeling and Using Context, pp. 260–275. Springer Press, Denmark (2007)
13. Shravani, D., Varma, P.S., Rani, B.P., Kumar, D.S., Kumar, M.U.: Web services security architectures composition and contract design using RBAC. Int. J. Comput. Eng. Sci. **2**, 2609–2615 (2010)
14. Hummer, W., Gaubatz, P., Strembeck, M., Zdun, U., Dustdar, S.: An integrated approach for identity and access management in a SOA context. In: 16th ACM Symposium on Access Control Models and Technologies, pp. 21–30. ACM Press, New York (2011)

15. Du, S., Joshi B.D.: Supporting authorization query and inter-domain role mapping in presence of hybrid role hierarchy. In: 11th ACM Symposium on Access Control Models and Technologies, pp. 228–236. ACM Press, Monterey (2006)
16. Paci, F., Bertino, E., Crampton, J.: An access control framework for WS-BPEL. Int. J. Web Serv. Res. **5**, 20–43 (2008)
17. Han, R.H., Wang, H.X., Xiao, Q., Jing, X.P., Li, H.: A united access control model for systems in collaborative commerce. J. Netw. **4**, 279–289 (2009)
18. Kim, K.I., Ko, H.J., Choi, W.G., Lee, E.J., Kim U.M.: A collaborative access control based on XACML in pervasive environments. In: 3rd International Conference on Convergence and Hybrid Information Technology, pp. 7–13. IEEE Press, Daejeon (2008)
19. Atluri, A., Shin, H., Vaidya, J.: Efficient security policy enforcement for the mobile environment. J. Comput. Secur. **16**, 439–475 (2008)
20. Liu, P., Chen, Z.: An access control model for web services in business process. In: Web Information Systems Engineering 2004, pp. 292–298. Springer Press, Brisbane (2004)
21. Charfi, A., Mezini, M.: Using aspects for security engineering of web service compositions. In: International Conference on Web Services 2005, pp. 59–66. IEEE Press, San Diego (2005)
22. Wang, X., Zhang, Y., Shi, H., Yang, J.: BPEL4RBAC, An authorization specification for WS-BPEL. In: Web Information Systems Engineering 2008, pp. 381–395. Springer Press, Auckland (2008)
23. Bertino, E., Crampton, J.: Access control and authorization constraints for WS-BPEL. In: International Conference on Web Services 2006, pp. 275–284. IEEE Press, Chicago (2006)

Author Index

S.-S. Yeo et al. (eds.), *Computer Science and its Applications*,
Lecture Notes in Electrical Engineering 203, DOI: 10.1007/978-94-007-5699-1,
© Springer Science+Business Media Dordrecht 2012

Printed by Publishers' Graphics LLC